The Essence of Geotechnical Engineering: 60 Years of Géotechnique

HONORARY EDITOR

J. H. Atkinson

ice

Institution of Civil Engineers

Published by Thomas Telford Publishing, Thomas Telford Ltd, 1 Heron Quay, London E14 4JD.
URL: http://www.thomastelford.com

Distributors for Thomas Telford books are
USA: ASCE Press, 1801 Alexander Bell Drive, Reston, VA 20191-4400
Japan: Maruzen Co. Ltd, Book Department, 310 Nihonbashi 2-chome, Chuo-ku, Tokyo 103
Australia: DA Books and Journals, 648 Whitehorse Road, Mitcham 3132, Victoria

First published 2008

A catalogue record for this book is available from the British Library

ISBN: 978-0-7277-3536-2

Typeset by Keytec Typesetting Ltd, Bridport, Dorset
Printed and bound in Great Britain by MPG Books, Bodmin, Cornwall

Preface

J. H. ATKINSON

To celebrate the first issue of *Géotechnique* in 1948, the June 2008 issue contained a set of papers which traced the history of *Géotechnique* and described the personalities of the engineers who were its founders. It also summarised the important contributions that *Géotechnique* papers have made to the theory and practice of ground engineering and gave a glimpse into the future. These papers, and the editorial that accompanied them, are reproduced in this book.

In addition, a collection of seminal papers from previous issues of *Géotechnique* has been reproduced. They are here because of our belief that everyone – students, professors and engineers – should, from time to time, rediscover the origins of geotechnical engineering. These are some of the papers which set the direction of modern geotechnics and to which current theory and practice can be retraced. There was interesting and important discussion published on several of these papers and we have included this as an essential part of the contribution. The papers are reproduced as they originally appeared in *Géotechnique*.

Choosing these papers was an interesting and rewarding experience and we debated their merits long and hard compared with other equally important early *Géotechnique* papers. We excluded papers included in other anthologies such as *A century of soil mechanics* (Skempton *et al*., 1969) and *Selected papers on soil mechanics* (Skempton, 1984). Readers are strongly recommended to browse these anthologies where there are many equally significant seminal papers. We have not included Rankine lecture papers because these are rarely seminal and those up to 1990 have been published separately as collections (Institution of Civil Engineers, 1975, 1983, 1992). We also limited the number of appearances by any one author.

Our selection contains papers from 1950 to 1990, the years in which modern ground engineering developed from the early work of Terzaghi and others. These papers and those in the June 2008 issue really are the essence of geotechnical engineering. They cover a range of topics including theoretical soil mechanics, numerical analysis and practical design. Others would probably have chosen differently, such is the wealth of seminal contributions to geotechnical engineering that have appeared in *Géotechnique*, but this is our list and we commend it to you.

Diamond Jubilee issue Sub-Committee Chairman
Prof. John Atkinson, City University

Sub-Committee Members
Dr Beatrice Baudet, University College, London
Prof. John Burland, Imperial College London
Prof. Stephen Brown, University of Nottingham
Mr Mike Chrimes, Institution of Civil Engineers
Dr Bill Craig, University of Manchester
Dr Chris Martin, University of Oxford
Dr Paul Morrison, Ove Arup & Partners

REFERENCES

Institution of Civil Engineers. (1975). *Milestones in soil mechanics: the first ten Rankine lectures*. London: ICE.

Institution of Civil Engineers. (1983). *Developments in soil mechanics: the second ten Rankine Lectures*. London: Thomas Telford.

Institution of Civil Engineers. (1992). *Landmarks in soil mechanics: the Rankine lectures 1981–1990*. London: Thomas Telford.

Skempton, A. W. (1984). *Selected papers on soil mechanics*. London: Thomas Telford.

Skempton, A. W., Cooling, L. F. & Little, A. L. (eds) (1969). *A century of soil mechanics*. London: Institution of Civil Engineers.

Contents

Part 3. Seminal papers and discussion: 1950–1990

Atkinson, J. (2008). *Géotechnique* **58**, No. 5, 313 [**doi:** 10.1680/geot.2008.58.5.313]

Editorial

This special issue of *Géotechnique* marks the 60th anniversary of the first issue in June 1948. It has been edited by a sub-committee of the *Géotechnique* Advisory Panel consisting of Beatrice Baudet, Stephen Brown, John Burland, Mike Chrimes, Bill Craig, Chris Martin, Paul Morrison and myself with assistance from Ben Ramster, Craig Schaper and Kathleen Hollow from the publishers, Thomas Telford. Although the papers were written by authors invited by the sub-committee, they were reviewed by members of the Panel following normal *Géotechnique* practice.

The first four papers trace the history of *Géotechnique* and of ground engineering, describe the often colourful personalities of its founders, and give a glimpse of the future. As the generations move on and memories fade, these papers, especially those by Stephen Brown and John Burland, record both the factual history and personal memories. As well as being important historical documents, they give fascinating insights into the origins of *Géotechnique* and, together with the paper by Mike Chrimes, the early days of British geotechnical engineering.

This issue also contains a collection of shorter articles, each of which reviews the contributions that papers published in *Géotechnique* have made to the development of theory and practice over the past 60 years. To some extent the variations in the lengths of these reflects the numbers of papers that have been published. There is a popular but inaccurate perception that *Géotechnique* papers are too theoretical and of little importance to practice. Almost without exception, those published here show this perception to be false and the authors conclude that *Géotechnique* has made major contributions to ground engineering practice and continues to do so.

I should like to thank all those who have contributed to the production of this issue: authors, sub-committee, Advisory Panel, secretariat and all those who have published papers in *Géotechnique* over the past 60 years. I should also like to thank the publisher, Leon Heward-Mills, for his support and encouragement for this special issue. Following its publication there will be a public meeting on 9 July 2008 at ICE headquarters, One Great George Street, for long the home of *Géotechnique*, followed by a dinner to celebrate its diamond jubilee.

Finally I wish *Géotechnique* 'many happy returns'.

John Atkinson

Brown, S. F. (2008). *Géotechnique* **58**, No. 5, 315–326 [**doi:** 10.1680/geot.2008.58.5.315]

Géotechnique 1948–2008: more than just a journal

S. F. BROWN*

The establishment of *Géotechnique* in 1948 as the first specialist soil mechanics journal and its development over 60 years is described, building on an earlier paper that dealt with the initial 30 volumes. The work and inspiration of the pioneers whose efforts created and nursed the journal through its early years are illustrated by reference to both published and private recollections. The role of the Geotechnical Society, both in the foundation of *Géotechnique* and in linking the members of its Advisory Panel, is described with reference to some of its traditions. The expansion of *Géotechnique* and its evolution to become and remain the leading international journal in its field are discussed, based on a study of the Advisory Panel's minutes and recollections of past chairmen. Interactions with the British Geotechnical Association (formerly Society), the Institution of Civil Engineers and the publishers, Thomas Telford Ltd, are recorded. Details are given of the significant number of awards and prizes that have been attracted by *Géotechnique* papers over the years.

KEYWORDS: historical review

La fondation de *Géotechnique*, en 1948, en tant que premier journal spécialisé dans la mécanique des sols, et son évolution au cours des 60 dernières années, sont décrites ; ceci fait suite à une communication précédente qui se penchait sur les 30 premiers volumes. L'oeuvre et l'inspiration des pionniers, dont le travail a permis de créer et de guider soigneusement le journal pendant ses premières années, sont illustrées par des témoignages à la fois publiés et privés. On y découvre également une description de la mission de la Geotechnical Society, aussi bien dans la fondation de Géotechnique, que pour son rôle de trait d'union entre les membres du Groupe Consultatif, en évoquant certaines de ses traditions. On y discute également de l'expansion et de l'évolution de *Géotechnique*, qui a su devenir et rester le principal journal international dans son domaine, et cette discussion s'appuie sur une étude des procès-verbaux de réunions du Groupe Consultatif, et sur le témoignage d'anciens présidents. On relève également les interactions avec la British Geotechnical Association (l'ancienne Society), l'Institution of Civil Engineers et l'éditeur, Thomas Telford Ltd. Enfin, on dresse un bilan du nombre important de prix qui ont été décernés à des communications de *Géotechnique*, au fil des années.

INTRODUCTION

In June 1948, following a great deal of hard work and consultation in soil mechanics circles around Europe by a group of friends, who were also British pioneers in this subject, the first issue of a new journal, entitled *Géotechnique*, was produced by a private organisation known as the Geotechnical Society. It was priced at ten shillings, but the initial costs were underwritten on behalf of the Society by one of its members, Rudolph (Silas) Glossop, via an overdraft at his bank. By the time of the second issue, in December 1948, 500 subscriptions had been sold and Glossop's financial risk consequently reduced. Today, 60 years later, *Géotechnique* is firmly established as the leading soil mechanics journal in the world, with a circulation of about 1500 and an annual subscription of £369 or about £37 per issue (£1 10s in 1948 money; journal inflation!).

The birth and evolution of the journal and its progress over the first 30 volumes have been described by Brown (1982). A study of this paper will reveal that the story of its early development was more than just the introduction of a much-needed publication to fulfil the needs of a rapidly expanding civil engineering discipline in the post-war years. The personalities involved, and the social interactions they established in the form of their friendships, love of dinner parties and innovative hard work, have become a feature of the soil mechanics fraternity, not only in the UK but throughout the world. The five founders of *Géotechnique* and the original members of the Geotechnical Society were Silas Glossop and Hugh Golder, then directors of Soil Mechanics Ltd, Leonard Cooling and Bill Ward of the Building Research Station, and Alec Skempton of Imperial College. All are now household names in the subject, all made very significant contributions to the development of geotechnical engineering, and all are now, sadly, dead. They formed the Geotechnical Society specifically to launch the journal, which was adopted by the Institution of Civil Engineers (ICE) in 1950 following an inaugural meeting of the Advisory Panel on 30 November 1949 under the chairmanship of W. K. Wallace, Chief Civil Engineer of the London Midland and Scottish Railway. Details about the lives and achievements of all these men and others, who were instrumental in the birth and development of *Géotechnique*, are presented in a companion paper by Burland (2008). The earlier history of publications in ground engineering is recounted by Chrimes (2008), who has clearly demonstrated how great the need for an international specialist journal in this subject was by the late 1940s.

The interest in researching and recording the history of *Géotechnique* originated in 1979, when the *Géotechnique* Advisory Panel (GAP), as it became known, was under the chairmanship of Peter Wroth, who suggested that a celebration to mark the first 30 volumes would be in order. As a member of the panel at that time, the author started looking into the journal's past and became fascinated by what he discovered. This led to publication of the paper noted above, following a celebration dinner at ICE, held on 21 October 1981, which was attended by the four surviving originators of *Géotechnique*. Sadly Leonard Cooling had passed away four years earlier.

This paper summarises the early development of *Géotechnique* and traces its continuing development and interactions with the British Geotechnical Society (BGS), now Association (BGA), and ICE. The important and continuing role of the Geotechnical Society is also described.

Discussion on this paper closes on 1 December 2008, for further details see p. ii.
* School of Civil Engineering, University of Nottingham, UK.

THE FIRST HALF: 1948 TO 1980

This section presents, for completeness, key points from the information published by the author in 1982 (Brown, 1982), which drew on earlier brief contributions by Golder (1969) and by Cooling *et al.* (1975) at the time of the 25th anniversary of BGS. In addition, the surviving founders of *Géotechnique* (Glossop, Golder, Skempton and Ward) were extremely helpful in providing information from their files, diaries and memories, and in the speeches by Glossop and Golder to the October 1981 dinner (Ward came prepared to speak, complete with his diaries, but was not called upon, owing to lack of time following an excellent *ad hoc* interjection by Hugh Sutherland, another British soil mechanics pioneer). Some additional points have been added, based on recent research.

The 'conception and infancy' of *Géotechnique* was fostered by the fertile, pioneering soil mechanics environment of the Building Research Station from the 1930s, and was brought into focus by a European tour of soil mechanics laboratories in 1946 undertaken by Glossop, Golder, Ward and D. J. Maclean from the Road Research Laboratory. Conversations with their various hosts identified the need for a new journal, and led to the idea of 'doing it themselves' really taking shape during an evening at a nightclub in either The Hague or Rotterdam attended by Glossop, Golder and E. C. W. A. Geuze. Records are, understandably, somewhat confused on this point (cf. Golder, 1969, and Glossop in Cooling *et al.*, 1975) as a consequence of the Dutch *genever* that was consumed. However, as Glossop reported that Geuze lived in Rotterdam, it seems likely that this was the probable venue of the conception, which resulted from what Golder (1969) described, in his own inimitable style, as '*no mean orgasm'*. The title for the journal derives from ideas generated by the team on this tour, notably in Lausanne at the home of J. P. Daxelhofer, and during the subsequent planning period. The final choice was probably brought into focus by Glossop following Daxelhofer's proposal to use 'La *Géotechnique*'. Glossop (1982), in private correspondence with the author, described his launching of the term 'geotechnology' in 1945, which was as follows.

> When I took over the management of the Mowlem-Siemens Schukert partnership from Harold Harding in 1937, I became interested in the wider application of special processes in Europe and soon found that the French were ahead of us and referred to them as 'Procédés Géotechniques'... I eventually discovered that the word had first been used in Sweden, in the 1920s and was probably coined by Fellenius. I first introduced it, in print, to English engineering literature in a lecture to the ICE in 1945; one of four lectures by Cooling, Skempton, Marwick and myself which became known as 'The Four Apostles', the result of a quiet dinner with Skempton at the old Queen's Restaurant in Sloane Square.

The familiar design of the front cover owes its origins to the artistic work of Alec Skempton's wife, Nancy, and incorporates the familiar 'Coulomb colophon', which was the printer's ornament from his 1776 paper on earth pressures. It was Nancy Skempton who decided that the definite article should be dropped from the selected title of 'La *Géotechnique*' for aesthetic reasons, much to the chagrin of the original French collaborators. However, the French flavour to the chosen title reflected the aims of the founders to publish papers in both French and English, a policy that has remained in place to this day, though not apparently one that is strongly marketed. The first four issues, constituting Volume 1, contained four papers in French but, over the years, use of the language has been represented mainly by the synopses. In 2002 there was concern expressed by the GAP's continental Europe representative that the accuracy of the translations into French left something to be desired. Though the colour, size and detail of the cover have changed over the years, not without periodic complaints from older subscribers, the essential features of Nancy Skempton's design remain. The evolution of the cover design was shown in the earlier paper (Brown, 1982).

A significant boost for the launch of *Géotechnique* was the very strong support given by Karl Terzaghi, who wrote the first foreword, which is still a piece worth studying (Terzaghi, 1948). The first editorial, presumably written by Glossop & Golder (1948) in their role as 'Temporary Editors', stated the aims of the journal.

> The Journal is intended to fulfil four purposes: to promote international collaboration between workers in Soil Mechanics and related sciences; to publish papers on specialised aspects of these subjects, such as might not be acceptable to the Proceedings of Institutions and Societies devoted to Civil Engineering; to encourage the pursuit of Engineering Geology; and to make the results of research available to the practising Civil Engineer. In this first number the emphasis is, perhaps, on the theoretical Soil Mechanics, but it is intended in the second number to devote more space to Foundation Engineering, and thus to maintain the balance of theory and practice.

This statement still captures the objectives of the journal and the ongoing concerns over the balance between theory and practice, which successive Editors and their Panels have sought to address. Introduction of the 'Symposia in Print' in 1975 was intended to provide detailed practical information on topical issues for design and construction engineers. The first, on 'Ground treatment by deep compaction' was followed a year later by 'Piles in weak rocks'. The reference to publication of papers that might not be attractive to Institution *Proceedings* could be misinterpreted as the pursuit of lower standards, but this has been far from the case. The review process for *Géotechnique* has remained of the highest calibre, a tradition that has, in no small measure, helped to create its world-leading status. The idea, presumably, was that the 'establishment' might not yet be ready for papers in this newly emerging subject.

The GAP, under the chairmanship of an Honorary Editor and supported by secretarial services from ICE and publication by Thomas Telford Ltd (TTL), once this company was created, is what drove the journal and still does today. Following the *ad hoc* arrangement in 1948–1949, when Glossop and Golder acted as joint Temporary Editors, W. K. Wallace assumed the role of Honorary Editor and chairman of the GAP when ICE assumed responsibility for the journal. He had been a driving force in persuading ICE Council to take the journal under its wing. The terms of reference that were then put in place have remained essentially the same to the present day. They were endorsed by ICE Council on 12 October 1949, and read as follows.

> To advise the Secretary in obtaining and selecting articles and in other editorial matters relating to 'Géotechnique' and to make recommendations, through the Publications and Library Committee, to the Council if any major change is proposed in the nature of the publication.

The Publications and Library Committee minutes of 16 September 1949 recorded that the costs to ICE of assuming responsibility for *Géotechnique* were estimated to be £250 per year initially. Given the large profits that it attracted, once established, this represents one of ICE's better investments.

The GAP was quite small in those early days, and during Wallace's term only Glossop, Golder, Skempton, Cooling,

Guthlac Wilson and Kevin Nash were involved. Nash succeeded Wallace as chairman in 1960 and served a five-year term, after which two- or three-year periods of office were established, and have remained the norm. Nash was followed by Leonard Cooling and then Alec Skempton before new blood emerged in the form of Bob Gibson, followed by David Henkel and then Peter Wroth. Bob Gibson uniquely served two terms (1973–1975 and 1987–1990), and his many contributions to the health of *Géotechnique* and his enthusiasm for the Geotechnical Society deserve proper recognition. A full list of GAP members and their terms of office can be found on the *Géotechnique* website (located via www.thomastelford.com/journals). Interestingly, during the early 1960s, when new members were recruited to deal with the increasing workload and range of topics, Ken Roscoe and Peter Rowe joined the GAP. They served in parallel for seven years, which must have been an interesting experience for all concerned, given the rivalries between their respective research teams at Cambridge and Manchester, an issue that was (according to the folklore) eventually settled during an alcoholic evening in the North towards the end of this period, when the respective merits of critical state soil mechanics and stress dilatancy theory became mutually appreciated in a smoke-filled room.

Up to 1980, 59 people served for various periods on the GAP, and the methods of working became firmly established. The dual review process for each paper, involving a GAP member as an 'Assessor' and an external referee has largely remained as practice up to the present day, and has stood the test of time and increased academic scrutiny. There have been periodic attempts to solicit papers from particular engineers, particularly on practical subjects and case histories, but all have been properly reviewed. The theory/practice balance of the papers has always been of concern to the GAP, but the continued success of the journal indicates that the market likes the product which has, for many years, been one of the most profitable parts of the TTL portfolio.

Prior to the introduction of the Symposia in Print, two conference proceedings were published: that held in London on the shear strength of soils (1950) and the slope stability event in Sweden (1955). Fig. 1 shows a signed menu given to the author by Silas Glossop, from a dinner held at the former event in June 1950 at Verreys Restaurant in Regent Street. It is populated with famous soil mechanics names, most of which are decipherable.

Following feedback from overseas readers in 1961, conveyed in a forthright report by Alan Bishop after a visit to the USA, the Panel appointed 20 overseas advisers and also published their own names for the first time. These advisers served for various periods up to 1976, although it is not at all clear from the deliberations of the GAP precisely what role they fulfilled, other than to assure the American cousins that the rest of the world, particularly that beyond Imperial College, was being taken seriously.

During the 1970s the Panel size increased, and stood at 13 by 1980, close to the present population of 15 members. During this period, too, the size of each issue increased and the number of GAP meetings per year was raised to eight in order to cope with the expanding workload.

Periodic reviews were undertaken to assess standards and coverage of the papers, and in 1982 the author (Brown, 1982) concluded, after a period of service on the GAP and a study of the minutes since its inception, that *Géotechnique* had been born with a silver spoon in its mouth and had been nourished by the best of international geotechnical engineering. Its state of health remained much as Cooling had concluded at the time of its 21st birthday: a bit too much theory, but standards are high, and efforts continue to ensure a balanced diet (Cooling, 1970).

Fig. 1. Signed menu from the 1950 London Shear Strength Conference dinner held at Verreys Restaurant (courtesy of the late R. Glossop)

The dinner held at ICE on 21 October 1981 to celebrate the first 30 volumes of *Géotechnique* attracted about 50 past and present members of the GAP, together with the ICE staff who were then involved with its work. Fig. 2 shows the assembled group, and provides images of most of the leading figures in British soil mechanics over the period being celebrated, and of several whose time had still to come.

THE SECOND HALF: 1981 TO 2008

Over this period, *Géotechnique* expanded its size and the number of papers that it published, continued to be a highly successful part of the TTL portfolio, eventually embraced the electronic age, and went through some periodic difficulties in the GAP's relationships with both ICE and TTL. The creative tension between these three elements of the *Géotechnique* production process—GAP, ICE and TTL—has always existed, because each has slightly different priorities from time to time. It is apparent that the GAP has always maintained its principles and observed its terms of reference, with its primary responsibility being to the excellence of the papers that it approves for publication and in keeping a close eye on the time taken from submission to publication.

The workload of the GAP has increased significantly since 1980, and therefore, to maintain the delivery schedule, more work has been required, not only from the Panel and its

Fig. 2. *Géotechnique* Advisory Panel Dinner to celebrate the first 30 Volumes; October 1981
***From left to right, front row*: Dickie Bassett, John Burland, Michael Tomlinson, Bob Gibson, Bill Ward, Hugh Golder, Silas Glossop, Alec Skempton, Nancy Skempton, Peter Wroth and Hugh Sutherland. *Second row*: Ken Cole, Ken Fleming, Sam Thorburn, Gilliane Sills, Andrew Schofield, Alan Meigh, David Henkel, Tom Hannah, Stewart Littlejohn, Arthur Marsland, Tom Whitaker, Maurice Kirkpatrick, Arthur Penman, Dick Parry, David Hight, Alan Bishop, Michael Kennard, Pam Skanthan. *Back rows*: Martyn Stroud, John Rigden, Noel Simons, David Croney, Ian Smith, Ian Christie, John Hooper, David McKinnlay, Robin Arthur, Ian Wood, Jimmy James, Bill Dearman, David Naylor, Phil Green, Paul Thompson, John McKennar, Alan Little, Noel Hobbs, Peter Vaughan, Stephen Brown, Anne Thompson, Joyce Davis**

reviewers, but also from the ICE secretariat and the TTL production staff. Under David Muir Wood's chairmanship (1991–1993) the level of secretarial support from ICE started to become an issue, which came to a head under the leadership of Roy Butterfield (1994–1996) at a GAP meeting in December 1993, when there were some rather frank exchanges of views with the ICE representatives in attendance. The GAP has, on many occasions, had to remind both ICE and TTL of *Géotechnique's* importance. It was always considered, by the GAP, a false economy, not to provide it with the level of support that the workload demanded. As a Past President of ICE, Professor Rex Coates (1977–1978) always remarked in relation to the performance of his staff at the University of Nottingham: one should support and reward 'he that bringeth home the corn'.

At a more mundane level, during ICE's cost-cutting phase of the 1990s, the threat of removing the GAP's complementary lunch (not pursued) and the provision of a free glass of wine (stopped) were matters that grated with some more than others. Both these privileges were regarded as minor ways in which the hard voluntary work of the Panel could be rewarded.

Attempts to reduce ICE/TTL secretarial support have been resisted by the GAP in recent years. There was even a suggestion from a very high level in ICE that perhaps reviewing of papers was not really necessary. The minutes reported a firm but probably well-toned-down version of their response. Over the years a feature of the GAP minutes has been the rather bland use of language to report controversial issues, usually in response to disaffected authors, difficult senior former members and both ICE and TTL senior officials.

Despite periodic and clearly enunciated differences of opinion between the Panel and TTL or ICE, overall relationships have worked well and, at the operational level, with great cordiality and cooperation, although tempers did fray a little at the time of the '1996 Butterfield confrontation' noted above. The various ICE and TTL staff who have worked on *Géotechnique* over the years are shown in Table 1.

The paper review form was improved in 1981, following a study of best practice elsewhere, and was changed again in 1986. The advice to authors was improved in 1985, and an agreed set of 100 keywords introduced by David Muir Wood (1991–1993). This process of revisiting key elements of the GAP's work has been a feature of its activities over the years. In 1982 steps were taken to review circulation figures in Europe, concluding that there was then very little serious competition. The US market was taken seriously under John Burland (1982–1985) by offering a reduction in subscriptions to retain and expand this market. In 1985, and again in 2005, the practice of only using one external reviewer in addition to the Panel Assessor was considered, but no change was thought to be necessary, although back-up reviewers became identified as reserves. It was not until recently, under Guy Houlsby (2003–2005), that two formal reviewers were appointed for each paper. The idea of reviewing each issue of *Géotechnique* was introduced in 1986, as a means of quality control, and this is a practice that is still followed, the task being undertaken by different members of the Panel. The first of these reviews, by a distinguished member of the GAP, who should perhaps be left unnamed, revealed that the subtitle 'The International Journal of Soil Mechanics' had been dropped from the cover. This matter was again raised in 1990, but it was not until the following year that the Panel were reminded that this practice actually ceased in 1964, after being originally introduced in 1950 when ICE assumed responsibility. So much for attention to detail!

The workload increase did not lead to a much larger membership for the GAP, or to any change in the number of meetings per year beyond eight, which had become the norm by 1980. Difficulties with attendance on some occasions prompted Brian Simpson to hold interim meetings for members who were close to ICE in order to deal with non-controversial business. Guy Houlsby (2003–2005) replaced these with 'email meetings' in 2003 to ensure wider participation. This was one small manifestation of the gradual move to embrace digital technology, which had started to be discussed during Gilliane Sills' chairmanship (1997–1999) and was taken seriously under Brian Simpson (2000–2002). The minutes were not produced electronically until 1999 but, since the paper versions for the previous seven years

Table 1. ICE and TTL staff who have supported the GAP

ICE/TTL staff	Dates
Alex Bannister	1949–1959
R. F. Farmer	1959–1962
Paul Thompson	1965–1984
Pam Skanthan	1978–1996
Margaret Summers	1982–?
G. Owen	1984–1987
Alan Dawson	1985–1990
Bill Pope	1990–2000
Graham Jones	1991–1993
Sharon Iliffe	1997–1998
Lesley Wilson	1998–1999
Mary Henderson	1999–2006
Zoe Holland	2000–2003
Richard Sands	2004–
Ben Ramster	2005–
Kathleen Hollow	2006–
Craig Schaper	2006–

seem to have mysteriously disappeared, perhaps someone assumed the change had taken place somewhat earlier. There was a struggle to obtain good electronic data management for papers, to ease the task of the Panel, and this is an ideal still being pursued. From 1998 the Panel received a stream of reports on progress with development and improvements to the website, then being set up by TTL. Teething problems were reported in late 1999, the site was then to go live in April 2000, and it was reportedly operational in September but not working by December. However, by March 2001 everything was functioning. Use of the website has increased, but its use to promote discussion of papers and for the submission of additional technical data from authors has not been successful. In 2000 suggestions from TTL to introduce a sister electronic journal and then a bimonthly companion journal were not taken up. There was a feeling that the currency should not be devalued. In 2001 a set of CDs of past issues was made available, and led to very good initial sales.

Changes have been made to *Géotechnique* over the past 25 years to accommodate the increased number of papers and to maintain an attractive appearance. Quarterly issues, which had become the norm since 1952 but had grown in size, continued until 1997, when five issues appeared and then six in 1998, finally increasing to ten in 2001. A recommendation that the cover should be printed with a glossy finish and that A4 format should be introduced was first suggested in 1986, although there had been discussions and consultations five years earlier. The glossy cover idea was implemented immediately, but the A4 format was not adopted until 2000 and led to some complaints from the USA.

A feature of the past 10 years has been an increase in the number of Technical Notes that have been submitted. They have traditionally been reviewed by the Honorary Editor, and this practice continued until 2003, when the pressure of work persuaded Guy Houlsby to seek assistance from external reviewers in order to speed things up. The popularity of Technical Notes came, at least partly, from the need by academics to get their work published more quickly than the 22 months or so that was the delay that had developed between submission and publication for full papers. This desire had in turn been fuelled by the introduction, in the early 1990s, of the Higher Education Funding Councils' periodic Research Assessment Exercises in the UK and similar moves in other countries to measure the quality of academic research performance linked to funding. *Géotechnique*, as the leading soil mechanics journal with a very high reputation for quality, was clearly the publication outlet of choice by the academics in order to demonstrate their excellence. A review of impact factors for journals in 2003 revealed that *Géotechnique* was leading its field.

The continuous concern by the Panel to increase the number of papers on practical subjects resulted in periodic calls for such work, particularly case studies. Under both Peter Wroth (1979–1981) and John Burland (1982–1985) serious attempts were made to solicit papers from key people. Indeed, the idea of publishing the Bjerrum Lectures on behalf of the Norwegian Geotechnical Institute was the subject of much correspondence in the early 1980s, but to no avail. In 2005 it was reported that, over the previous decade, there had indeed been more papers on theoretical concepts than on other topics. However, a recent review under John Atkinson, the current GAP Chairman, revealed that the balance was good, and that the time to publication was better than some and no worse than most other comparable journals.

Over the years, the GAP has always kept a healthy eye on the competition, so it was with much concern, in 1991, that it learned about ICE's proposal to change the format of its *Proceedings* and to introduce a geotechnical engineering journal, bearing in mind that the magazine *Ground Engineering* already existed to cover practical topics. Joint discussions were held with the ICE Ground Engineering Board against the background of one of the Panel's periodic attempts to solicit more papers on case histories, which, they thought, the new *Proceedings* was likely to attract. In the end the ICE plans went ahead, and *Géotechnique* has coexisted with *Proceedings of the Institution of Civil Engineers, Geotechnical Engineering* without either suffering significantly. This mutual success is another manifestation of the increased demand for publication space by both UK and overseas authors.

Since 1980 much time and effort has been expended by the Panel on the organisation of Symposia in Print (SIP). Following the launch of this initiative in 1975 by the GAP under Michael Tomlinson (1974–1976), a total of 14 have been published. While many papers have been offered following a general call, a major effort was always needed by the GAP to solicit papers from key people working in the particular field. Reviewing of all papers was, however, always regarded as *de rigueur*. A full list of the SIP titles can be found on the *Géotechnique* website.

The circulation figures for *Géotechnique* rose steadily in the early years, peaking at about 4000 in 1969 and stabilising just below this level for 10 years. By 2000 they had fallen below 2000 as electronic publication had become more prevalent. However, it was reported that this reduction was no more marked than that of competitor journals. By 2006 the circulation was just over 1500. These figures need to be interpreted in the context of the availability on the website of paper abstracts and download facilities for full papers (for a fee). In September 2007 the site attracted 12 319 visits, and 5934 abstracts were downloaded—pleasingly high numbers. The subscription rates are summarised in Fig. 3 in terms of 2007 prices, from which it can be seen that there was a gentle increase over the years until the late 1990s and then something of an upward surge. This coincided with the increase from four to ten issues per year between 1997 and 2001.

Other major events over the past 28 years included the retirement, after 19 years' outstanding service and wise advice to the Panel, of Paul Thompson, who started as the ICE staff member servicing the GAP in 1965, was the formal Editor and became the TTL man. His interesting background was described in Brown (1982). Summaries of UK PhD theses have been published since 1989 to inform readers of the latest university research. In reviewing the GAP membership over the first 30 volumes, Brown (1982) noted that few people had served more than one term, unless appointed as Chairman. This tradition was questioned under John Burland's watch (1982–1985), and the practice changed, a necessity brought on by the needs of the Panel, while recognising that new blood was always required. David Muir

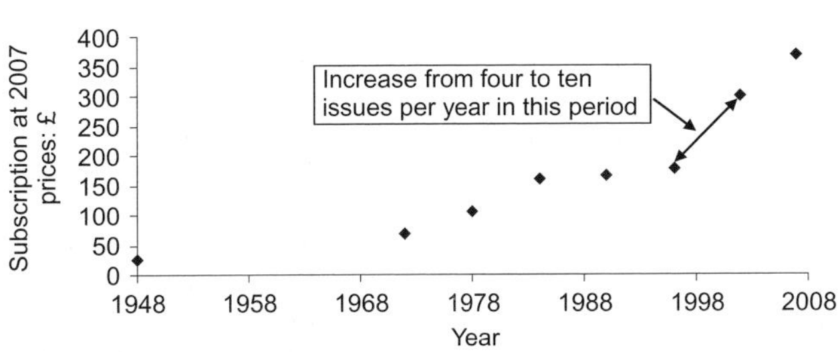

Fig. 3. Subscription rates

Wood introduced an 'information pack' for new GAP members in 1991, with some help from the author based on his research for the earlier paper (Brown, 1982). It set out a summary of the history of *Géotechnique* so that new recruits could appreciate the real nature of the organisation that they had joined. This information and much more is now available on the website following an initiative by Guy Houlsby during his period as Chairman (2003–2005).

RELATIONSHIPS WITH THE BRITISH GEOTECHNICAL SOCIETY/ASSOCIATION

The British Geotechnical Association (BGA) is a specialist society within ICE that was formed in 2000 from the former British Geotechnical Society (BGS) and the ICE Ground Board. BGS in turn grew from the British National Committee of the International Society for Soil Mechanics and Foundation Engineering (as it was then known), which was founded in 1947. The scope of this original group was expanded in 1964 and it was renamed as BGS. The relationship between BGS/BGA and the GAP has always been close and cordial, with representation on each other's committees and a certain amount of regular common activity, notably the annual prestigious Rankine Lecture, the biannual *Géotechnique* Lecture, and the Symposia in Print. This collaboration between a world-class journal and a very active learned society with a full programme of well-attended meetings has been to their mutual advantage. The chemistry has been good, not least because many of the same people have been involved in the leadership of both organisations and in serving on their committees.

Until the late 1990s the Rankine Lecture was generally published in the September issue of *Géotechnique*, but in recent years, the Panel has been frustrated by the late or non-appearance of the written versions from busy lecturers. Although occasional presentation of *Géotechnique* papers had been facilitated by the British National Society in the early days, the idea of a '*Géotechnique* Lecture' arose from a major review of their activities by the BGS committee under the author's chairmanship in 1987. The concept was to allow the authors of a suitable paper published in *Géotechnique* to present their work at a BGS meeting, and for this to take place on alternate years. After concerns about it detracting from the importance of the Rankine Lecture had been dispelled, on the understanding that it would take place in those years when the Rankine Lecturer was from overseas, the idea was accepted. The lecturers are nominated by the Panel, and BGS organise the meeting. It has been very successful, and to date eight lectures have been delivered. Details can be found on the *Géotechnique* website. The BGS review of activities that spawned the lecture also lead to the creation of the Skempton Gold Medal, referred to below, and the BGS Touring Lecture series, both significant developments at the time.

THE GEOTECHNICAL SOCIETY

The Geotechnical Society was founded by the creators of *Géotechnique* (Glossop, Golder, Skempton, Cooling and Ward) as a vehicle to launch publication of the journal in 1948. Its formation was stimulated, at least in part, by the need for the new journal to be posted to subscribers at the newspaper rate, which was available only to journals of societies and, presumably, to newspapers (Penman, 2007). Glossop and Golder used the office of Soil Mechanics Ltd, 123 Victoria Street, as the Society's address. Following adoption of the journal by ICE in 1949, the Society had no immediate purpose to serve, but was not formally wound up. The financial arrangements that were established, and their sequel, have become part of *Géotechnique* folklore. They are described in detail elsewhere by Glossop (Cooling *et al.*, 1975) and Golder (1969), and were summarised by Brown (1982).

Initial financing was arranged by Silas Glossop, who persuaded his bank manger, a Mr Esmond Barclay (or was it Berkeley?) of Williams Deacon's Bank in South Kensington, to grant a £300 overdraft facility for the Society under his guarantee. Together with the £150 raised in subscriptions, the costs of the first issue were covered, and it was printed by the Marshall Press. A French agent was also used to collect subscriptions, with a note that read

> Pour la France et les Colonies françaises, les souscriptions doivent être adressées à notre Agent de vente, exclusif: LIBRAIRIE REYNAUD, 52, rue d'Assas, Paris (6)'.

According to Bill Ward's diaries (Ward, 1981), the Society met on only two occasions to deal formally with publication matters for Volume 1: on 27 May 1948 and 6 May 1949 at the Victoria Street office. He reports that all the work was done by Glossop and Golder, with the help of their secretaries and encouragement from the other members. Since there were no financial problems, the profits were used to host dinners for important friends, such as an occasion at Kettner's restaurant in Soho for Terzaghi and W. K. Wallace and the well-documented 70th birthday party for Terzaghi at the Good Intent restaurant in Chelsea (Cooling *et al.*, 1975). However, not all was spent, as Glossop and Golder were contacted by the bank in 1969 with the news that a long-abandoned account existed in the name of the Geotechnical Society, and did they know anything about it? The residual funds were used, with the agreement of the other founders, to pay for a pair of silver Armada bowls for ICE at the time of its 150th anniversary. In making the presentation, Glossop stated that it was 'in recognition of the encouragement and help which it had given to soil mechanics and geotechnology since the early days of these subjects' (Cooling *et al.*, 1975)

The revival of the Geotechnical Society in 1983 owes its origins to the dinner at ICE in October 1981 held to celebrate the 30th volume. Following speeches by Glossop and Golder, who spoke of the early days and the Society's activities, a conversation between the author, Glossop, Golder and Skempton, as they walked down Great George Street, took place about the idea of an annual dinner along the lines of the one that had just been enjoyed. The concept was taken further by the author, stimulated by the thought-inducing job of lawn-mowing during the following spring. A case was prepared for the GAP and presented to a meeting in October 1982, and a sub-committee of Robert Mair, Roy Wakeling and the author was set up to take the matter further. A questionnaire was sent to all past and present members of the Panel to ascertain their views, which proved to be generally positive and were reported to the Panel at its January 1983 meeting. Of the 47 responses received, only David Croney voted against the idea, stating that, since we had BGS, 'there was no logic in attempting to resuscitate it (the Geotechnical Society) as a dining club, particularly if a lecture from one of our number is to provide the penance before the indulgence.' All the Society founders responded, and only Golder (1982) was somewhat negative, apparently wanting to reinvent something along the lines of the British Tunnelling Society, having, apparently, overlooked the existence of BGS. He also wanted it to be open to all, and wrote: 'Who do you think you are? You are the Advisory Panel! You were elected to do a job of work, NOT to become a snotty-nosed ELITE!' Glossop (1982), having seen the letter, responded 'Need I say that I find his manner deplorable and in the worst possible taste.' Matters progressed, and Golder's support was obtained.

In the light of this feedback, proposals were drawn up and, after minor revision, were approved by the GAP in February 1983. The details are shown in Fig. 4. Robert Mair was appointed as the first secretary, with Alec Skempton as Chairman for the first dinner, which was held at ICE on 26 October 1983. A record of the various venues, chairmen and secretaries is shown in Table 2.

In addition to the terms of reference shown in Fig. 4, an important part of this first dinner, and a feature that was repeated for many years, was the drinking of a toast to 'The immortal memory of Karl Terzaghi', later interpreted as 'to The Old Master'. At this first dinner there were 44 members and 18 guests present. Bill Ward entertained the gathering with his recollections of working on the Isle of Skye shortly after the Second World War to investigate the use of peat to produce lightweight building blocks. In order to deal with the high water content a washing mangle fitted with 2 inch nails was adopted to shred the peat. The idea was taken no further.

The toast to 'The Old Master' was put to verse when Gilliane Sills was asked to propose it in the early 1990s. She used a superb piece written by her husband Richard. It ran as follows.

A young engineer called Karl studied mud, gravel, sand, clay and marl.
To allay builders' panic he wrote 'Erdbaumechanik', this ingenious Austrian, Karl.
Examining things at their base he travelled all over the place.
From Vienna to Turkey then to Boston to work, he lived life at a furious pace.
Now settled with Uncle Sam—and far from the Isis or Cam
His output was huge—without centrifuge!—and he worked on the Aswan High Dam.
Soil mechanics, before its foundation, was foreign to every nation.
Nobody could guess how effective the stress, nor estimate consolidation.
Terzaghi then opened the doors by examining pressure in pores.
With great perspicacity he defined bearing capacity, and such geotechnical laws.
This great engineer, Terzaghi, famous son of the city of Prague, he fully earned all his fees and his umpteen degrees. The toast is 'The Master—Terzaghi!'

Proposals for the 'Geotechnical Society'

1. The Society should exist for the benefit of *Géotechnique* and the members of the Advisory Panel and its secretariat.
2. Membership should be restricted to present and former members of the *Géotechnique* Advisory Panel, together with the Panel Secretary.
3. A chairman should be selected for one year and a secretary/treasurer for two years. The secretary/treasurer should be a member of the current GAP. The chairman should not also be the current chairman of GAP. Secretarial assistance will be provided by the *Géotechnique* secretariat.
4. The Society should meet annually, the first meeting to be in mid-October 1983, at the Institution of Civil Engineers. Subsequent meetings to be held in different parts of the country at universities which are active in the geotechnical field.
5. The annual meeting should take the form of a dinner followed by a brief (15 minutes) talk with a geotechnical flavour. It should be anecdotal and 'off the record' but have a useful message or provide historical or other information. An informal discussion should then follow on matters related to *Géotechnique*.
6. Appropriate guests (one per person) may be invited to the annual meeting.

Fig. 4. Rules for the revival of the Geotechnical Society, agreed by the GAP in 1983

Table 2. 'Revived' Geotechnical Society dinner venues and office holders

Date of meeting	Venue	Chairman	Secretary/Treasurer
26 Oct 1983	ICE	A. W. Skempton	R. J. Mair
12 Sept 1984	Magdalen College, Oxford	R. Glossop	R. J. Mair
18 Sept 1985	ICE	W. H. Ward	A. E. Skinner
12 Sept 1986	Churchill College, Cambridge	A. L. Little	A. E. Skinner
23 Oct 1987	The Athenaeum	P. R. Vaughan	R. E. Gibson
28 Oct 1988	King's College	S. F. Brown	C. R. I. Clayton
19 Oct 1989	Imperial College	M. J. Tomlinson	D. M. Potts
26 Oct 1990	The Athenaeum	C. P. Wroth	R. E. Gibson
25 Oct 1991	City University	J. B. Burland	R. N. Taylor
30 Oct 1992	Imperial College	N. Simons	R. J. Chandler
9 Oct 1993	Imperial College	A. N. Schofield	R. J. Chandler
28 Oct 1994	Imperial College	R. Butterfield	R. J. Jardine
3 Nov 1995	Imperial College	R. Butterfield	R. J. Jardine
1 Nov 1996	ICE	S. F. Brown	GAP Secretariat
1997	ICE	G. C. Sills	GAP Secretariat
1998	ICE	G. C. Sills	GAP Secretariat
1999	ICE	G. C. Sills	GAP Secretariat
2000	ICE	B. Simpson	GAP Secretariat
2001	ICE	B. Simpson	GAP Secretariat
2002	ICE	B. Simpson	GAP Secretariat
2003	Brasenose College, Oxford	G. T. Houlsby	GAP Secretariat
2004	Imperial College	G. T. Houlsby	GAP Secretariat
2005	Rhodes House, Oxford	G. T. Houlsby	GAP Secretariat
2006	Imperial College	J. H. Atkinson	GAP Secretariat
2007	Imperial College	J. H. Atkinson	GAP Secretariat

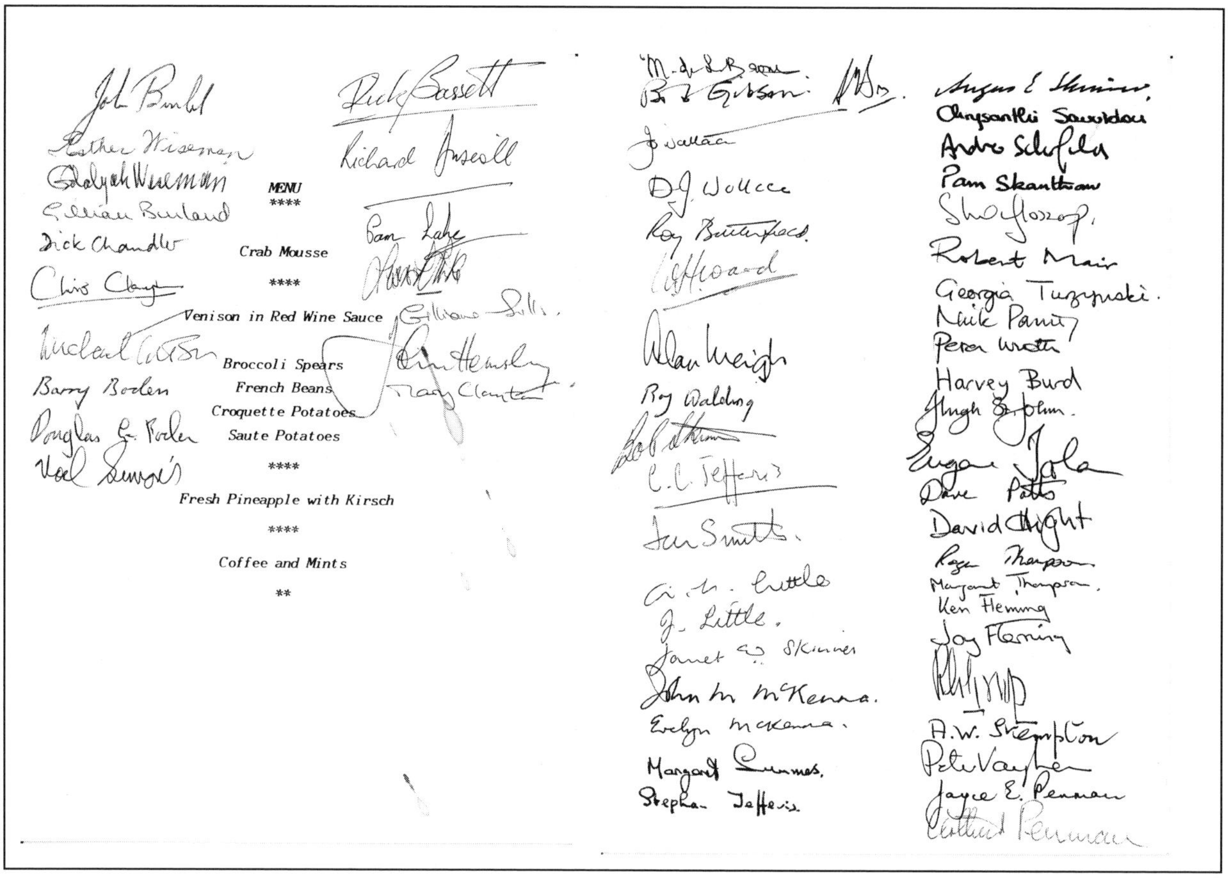

Fig. 5. Menu from the 1988 Geotechnical Society Dinner

It was so well received that it was reprised at later dinners. In recent years the toast has been fittingly expanded to embrace 'The Founders', whose achievements and characters are eloquently described by Burland (2008).

David Muir Wood also took to verse, or doggerel as he describes it, in paying a tribute to Pam Skanthan following her long period of service as the GAP secretary at the 1997 dinner. It accompanied the presentation of a cheque, which was to be used to buy a rocking chair and a camera.

The 1988 dinner at King's College, London, was notable for the launch of the BGS Skempton Gold Medal, witnessed by 66 members and guests. The author, then the BGS Chairman, had the privilege of also serving as Chairman for the Society on that occasion. Of the 'Founders', Silas Glossop and Bill Ward were in attendance in addition to 'Skem'. In making the presentation of the first medal to Alec Skempton, the author said

> The criteria for the award of this gold medal are that it be awarded to a BGS member who has made an outstanding contribution to the practice of geotechnical engineering over a sustained period of time and that the work should be closely associated with, though not necessarily all carried out in, the United Kingdom. It is the Society's intention that the award should be made no more than four times in a decade. In deciding on a precise name for this award, the BGS Committee very quickly concluded that it should be known as the BGS Skempton Medal in recognition of all that Skem has done and is continuing to do for British Geotechnical Engineering, notably leading by example.

A signed menu from this occasion is reproduced, appropriately complete with wine stains, in Fig. 5.

The dinner at the Athenaeum in 1990 became, in retrospect, a sad occasion, as it marked one of the last public appearances, before his untimely death a few months later, of Peter Wroth, who that year served as the chairman. He had only recently taken up his appointment as Master of Emmanuel College, Cambridge following his outstanding career at Cambridge and Oxford, during which he made a huge number of friends and helped in the formation of many successful geotechnical specialists. An excellent tribute to him can be found in Burland (2005). His wise advice helped so many engineers with aspects of their careers, and a very small example is illustrated by the note from the back of a seating card at a dinner reproduced in Fig. 6, relating to oral presentations, at which he was masterful.

In recent years the detailed rules set out in Fig. 4 have not been strictly applied, as the occasion has been chaired

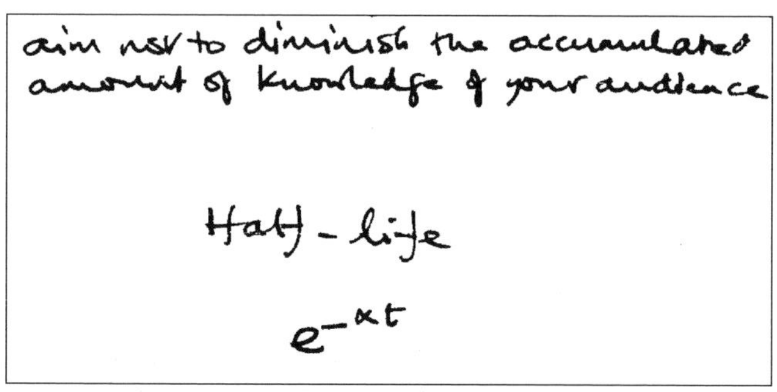

Fig. 6. Advice to a speaker: C P Wroth, 1980s

Table 3. British National Committee and British Geotechnical Society prizewinners (1958–1976)

Year	Authors	Title	Vol./No.
1958	Roscoe, Schofield and Wroth	On the yielding of soils	8/1
1959	Skempton	Cast in situ bored piles in London clay	9/4
1960	Bishop and Morgenstern	Stability coefficients for earth slopes	10/4
1962	Lumb	The properties of decomposed granite	12/3
1963	Gibson	An analysis of system flexibility and its effect on time-lag in pore water pressure measurements	13/1
1964	Cornforth	Some experiments on the influence of strain conditions on the strength of sand	14/2
1965	Rowe and Peaker	Passive earth pressure measurements	15/1
	Knill and Jones	The recording and interpretation of geological conditions in the foundations of the Roseires, Kariba and Latiyan dams	15/1
	Morgenstern and Price	The analysis of the stability of general slip surfaces	15/1
1966	Christie	The solution of general consolidation problems by general purpose analogue computer	16/2
1967	Ambrayses and Sarma	The response of earth dams to strong earthquakes	17/3
1968	Ward, Burland and Gallois	Geotechnical assessment of a site at Mundford, Norfolk, for a large proton accelerator	18/4
	Gibson and Shefford	The efficiency of horizontal drainage layers for accelerating consolidation of clay embankments	18/3
1970	James and Bransby	Experimental and theoretical investigations of a passive earth pressure problem	20/1
1972	Bjerrum, Nash, Kennard and Gibson	Hydraulic fracture in field permeability testing	22/2
	Palmer	Undrained plain strain expansion of a cylindrical cavity in clay: a simple interpretation of the permeability test	22/3
1973	Vaughan and Walbancke	Pore pressure changes and the delayed failure of cutting slopes in overconsolidated clay	23/4
1975	Arthur and Phillips	Homogeneous and layered sand in triaxial compression	25/4
1976	Hyde and Brown	The plastic deformation of a silty clay under creep and repeated loading	26/1
	Hobbs and Robins	Compression and tension tests on driven piles in chalk	26/1

Table 4. British Geotechnical Society and British Geotechnical Association prizewinners (1977–2005)

Year	Authors	Title	Vol./No.
1977	Marsland and Randolph	Comparison of the results from pressuremeter tests and large in situ plate tests in London clay	27/2
1978	Parry	Estimating foundation settlements in sand from plate bearing tests	28/1
1979	Cooke, Price and Tarr	Jacked piles in London Clay: a study of load transfer and settlement under working conditions	29/2
	Thornton	The conditions for failure of a face-centred cubic array of uniform rigid spheres	29/4
1980	Cooke Price and Tarr	Jacked piles in London Clay: interaction and group behaviour under working conditions	30/2
1981	Been and Sills	Self weight consolidation of soft soils: an experimental and theoretical study	31/4
	Randolph and Wroth	Application of the failure state in undrained simple shear to the shaft capacity of driven piles	31/1
	Randolph	The response of flexible piles to lateral loading	31/2
1983	Hooper	Non-linear analysis on a circular raft on clay	33/1
	Milligan	Soil deformations near anchored sheet-pile walls	33/1
1989	Coop and Wroth	Field studies of an instrumented model pile in clay	39/4
1990	Craig and Chua	Deep penetration of spud-can foundations on sand and clay	40/4
1992	Hight	Contributions to Bothkennar Soft Clay test site: characterisation and lessons learned (four papers)	42/2
1996	Vaughan, Tika and Lemmos	Fast shearing of pre-existing shear zones in soil	46/2
1997	Taylor and Stallebrass	The development and evaluation of a constitutive model for the prediction of ground movements in overconsolidated clay	47/2
1998	Cooper, Bromhead, Petley and Grant	The Selbourne Cutting stability experiment	48/1
1999	Cuccovillo and Coop	On the mechanics of structured sands	49/6
2001	Poulos	Piled raft foundations: design and applications	51/2
2004	White and Bolton	Displacement and strain paths during plane-strain model pile installation in sand	54/6
2005	Hossain, Hu, Randolph and White	Limiting cavity depth for spudcan foundations penetrating clay	55/9

by the current GAP Chairman. In fact, some minutes record items associated with the event as referring to the '*Géotechnique* Dinner'. However, the spirit of the occasion remains unchanged, and it is still very much focused on the health of the journal and how it might be enhanced, while members enjoy a convivial environment. Associated with several dinners have been debates, initiated by Brian Simpson, on the following subjects: 'Géotechnique – the way forward' (1999), 'This House prefers paper to pixels' (2000) and 'The house believes that modern soil mechanics is irrelevant to

Table 5. Institution of Civil Engineers' prizes and awards (1983–1992)

Award year	Award	Authors	Title	Vol./No.
1983	George Stephenson Medal	Potts and Martins	The shaft resistance of axially loaded piles in clay	32/4
	Baker Medal	Bolton and Pang	Collapse limit states of reinforced earth retaining walls	32/4
1984	George Stephenson Medal	Voight, Janda, Glicken and Douglass	Nature and mechanics of the Mount St Helens rockslide-avalanche of 18 May 1980	33/3
	Telford Prize	Ward, Tedd and Berry	The Kielder experimental tunnel: final results	33/3
	T. K. Hsieh Award	Mynett and Mei	Earthquake-induced stresses in a poro-elastic foundation supporting a rigid structure	33/3
1985	George Stephenson Medal	Tedd, Chard, Charles and Symons	Behaviour of a propped embedded retaining wall in stiff clay at Bell Common Tunnel	34/4
	Coopers Hill Memorial Prize	Hubbard, Potts, Miller and Burland	Design of the retaining walls for the M25 cut and cover tunnel at Bell Common	34/4
	Telford Prize	Jardine, Symes and Burland	The measurement of soil stiffness in the triaxial apparatus	34/3
1986	Telford Prize	Gibson, Goldsworthy and Chandler	Differential evapo-transpiration from a clay surface and its influence on equilibrium pore water pressure	35/2
	Telford Prize	Leroueil, Kabbaj, Tavenas and Bouchard	Stress–strain-strain rate relation for the compressibility of sensitive natural clays	35/2
1987	George Stephenson Medal	Cooke	Piled raft foundations on stiff clays—a contribution to design philosophy	36/2
	Frederick Palmer Prize	Baligh	Undrained deep penetration, 1: shear stresses	36/4
1988	Telford Prize	Sagesata	Analysis of undrained soil deformation due to ground loss	37/3
1989	Geotechnical Research Medal	Houlsby and Withers	Analysis of the cone pressuremeter test in clay	38/4
	Telford Prize	Leroueil, Kabbaj and Tavernas	In situ and laboratory stress–strain relationships	38/1
1990	Geotechnical Research Medal	Coop and Wroth	Field studies of an instrumented model pile in clay	39/4
	Trevithick Prize	De Buhan, Mangiavacchi, Nova, Pellegrini and Salencon	Yield design of reinforced earth walls by a homogenisation method	39/2
	T. Hsieh Award	Bolton and Wilson	An experimental and theoretical comparison between static and dynamic torsional soil tests	39/4
1991	Telford Medal	Potts, Dounias and Vaughan	Finite element analysis of progressive failure of Carsington embankment	40/1
1992	Geotechnical Research Medal	Nova and Montrasio	Settlement of shallow foundations on sand	41/1

geotechnical design' (2001), while a 'Question Time' was held in 2002. The House did indeed prefer paper to pixels.

AWARDS AND PRIZES

The very high standard of papers published in *Géotechnique* is reflected in the large number of prizes and awards that they have attracted over the years. The British National Committee, later BGS and then BGA, introduced its Annual Prize in 1958, and the inaugural winners were Ken Roscoe, Andrew Schofield and Peter Wroth for their seminal paper on critical state soil mechanics. For the period to 2005, 40 *Géotechnique* papers attracted the prize out of a total of 72 during the 48 years involved. The full list of winning papers is shown in Tables 3 and 4, together with the *Géotechnique* references. It should be recognised that there have been several prizewinners in some years, and that some prizes have been awarded for outstanding contributions to geotechnical engineering over a period, often involving the publication of several papers in *Géotechnique*. These have not been included.

In 1983 ICE decided to allow papers published in *Géotechnique* to become eligible for its awards and prizes for the first time. The GAP responded by arranging, each year, to send nominations, many of which were successful. They had occasion to complain several times that insufficient of their suggestions were taken up, given the large number of papers that they published relative to other ICE journals. The full list of 61 *Géotechnique* papers that have attracted a wide range of medals and prizes is listed, with the journal references, in Tables 5 to 7. Fifteen categories of award have been attracted of which only one, the Geotechnical Research Medal, is specifically for papers in geotechnical engineering. In all other categories *Géotechnique* papers have competed with other disciplines. A remarkable achievement has been the award of seven Telford Medals, the Institution's most prestigious prize for a paper, and the same number of George Stephenson Medals. In addition to these, there have been 15 Geotechnical Research Medals out of a total of 18, 13 Telford Prizes and six T. K. Hsieh awards. Over the 24 years since *Géotechnique* papers became eligible for ICE awards there have been about three attracted each year. Table 8 presents the leading award recipients, the list being dominated by Imperial College and Cambridge.

CONCLUDING REMARKS

With the perspective available in a 60-year review such as this, it is clear that the achievements of *Géotechnique* in becoming and remaining the leading journal in its field are indeed remarkable. This is due largely to the foresight and energy of its founders, and to the maintenance of their high standards by successive GAP members under their experienced and devoted chairmen. As this review has revealed, the road has not always been easy, but the results speak for themselves, and the success fully merits an appropriate

Table 6. Institution of Civil Engineers' prizes and awards (1993–2002)

Award year	Award	Authors	Title	Vol./No.
1993	Telford Prize	Fleming	A new method for single pile prediction and analysis	42/3
	Geotechnical Research Medal	den Haan	The formulation of virgin compression of soils	42/3
1994	Telford Medal	Alonso, Gens and Lloret	The landslides of Cortes de Pallas, Spain	43/4
	George Stephenson Medal	Griffiths and Fenton	Seepage beneath water retaining structures founded on spatially random soil	43/4
	T. K. Hsieh Award	Papastamatiou, Pappin, Richards and Sweeney	Earthquake response at Grangemouth	43/4
1995	Telford Prize	Skempton and Chrimes	Thames Tunnel: geology, site investigation and geotechnical problems	44/2
	Telford Prize	Lee, Jones, Sullivan and Trolinger	Failure and deformation of four reinforced soil walls in eastern Tennessee	44/3
	Telford Prize	Saada, Bianchini and Liang	Cracks, bifurcation and shear bands propagation in saturated clays	44/1
1997	Geotechnical Research Medal	Vardoulakis	Deformation of water-saturated sand: Part II. Effect of pore water flow and shear banding	46/3
	T. K. Hsieh Award	Makris, Gazetas and Delis	Dynamic soil–pile–foundation–structure interaction: records and predictions	46/1
1998	Telford Medal	Potts, Kovacevic and Vaughan	Delayed collapse of cut slopes in stiff clay	47/5
	Geotechnical Research Medal	Mandolini and Viggiani	Settlement of piled foundations	47/4
1999	Geotechnical Research Medal	Hicks and Boughrarou	Finite element analysis of the Nerlerk underwater berm failure	48/2
2000	Telford Medal	Merrifield, Sloan and Yu	Rigorous plasticity solutions for the bearing capacity of two-layered soils	49/4
	Geotechnical Research Medal	Skinner, Charles and Watts	Ground deformations and stress redistribution due to a reduction in volume of zones of soil at depth	49/1
	John Henry Garrood King Medal	Guo and Randolph	An efficient approach for settlement prediction of pile groups	49/2
2001	Geotechnical Research Medal	Lemos and Vaughan	Clay-interface shear resistance	50/1
	Telford Prize	Consoli and Sills	Soil formation from tailings: comparison of predictions and field measurements	50/1
	T. K. Hsieh Award	Juang, Chen, Rosowsky and Tang	CPT-based liquefaction analysis, Part 1: Determination of limit state function and Part 2: Reliability for design	50/5
2002	Geotechnical Research Medal	Poulos	Piled raft foundations: design and applications	51/2
	Telford Prize	Klotz and Coop	An investigation of soil state on the capacity of driven piles in sand	52/8
	Overseas Prize	Long and O'Riordan	Field behaviour of very soft clays on the Athlone embankments	51/4

celebration in July 2008, when no doubt the spirits of Glossop, Golder, Cooling, Skempton and Ward, together with the other early leaders, will hover benevolently over the port.

ACKNOWLEDGEMENTS

The author is grateful to many people for providing information to assist with the preparation of this article. Unfortunately, some of the notes and records collected for his 1982 paper have gone astray, as have the GAP minutes between 1992 and 1998, but several items of correspondence with the founders of *Géotechnique* have been scrutinised afresh and quoted in this paper. Mike Chrimes and Carol Morgan assisted in unearthing old GAP minutes and information about ICE prizes, Jade Donovan and Neil Smith helped with BGA information, and Kathleen Hollow and Ben Ramster with GAP/TTL input. The reflections of several former GAP Chairmen have been invaluable and rather revealing, so grateful thanks are due to David Muir Wood, Roy Butterfield, Gilliane Sills, Brian Simpson and Guy Houlsby, who, it is hoped, can share some of the blame for inaccuracies or misinterpretations! The author is particularly grateful to John Atkinson, the current GAP Chairman, for providing the opportunity to write this paper, and for his advice. John Burland and Robert Mair both made helpful suggestions, which were greatly appreciated.

Table 7. Institution of Civil Engineers' prizes and awards (2003–2007)

Award year	Award	Authors	Title	Vol./No.
2003	Telford Medal	Zdravkovic, Potts and Hight	The effect of strength anisotropy on the behaviour of embankments on soft ground	52/6
	James Watt Medal	Collins and Kelly	A thermodynamical analysis of a family of soil models	53/6
	Geotechnical Research Medal	Foti, Lai and Lancellotta	Porosity of fluid-saturated media from seismic wave velocities	52/5
	David Hislop Award	Houlsby and Cassidy	A plasticity model for the behaviour of footings on sand under combined loading	52/2
2004	Telford Medal	Wheeler, Sharma and Buisson	Coupling of hydraulic hysteresis and stress–strain behaviour in unsaturated soils	53/1
	Geotechnical Research Medal	Randolph and Puzrin	Upper bound limit analysis of circular foundations on clay under general loading	53/9
	Crampton Prize	Collins and Muhunthan	On the relationship between stress–dilatancy, anisotropy and plastic dissipation for granular materials	53/7
2005	T. K. Hsieh Award	Bhattacharya, Bolton and Madabhushi	An alternative mechanism of pile failure in liquefiable deposits during earthquakes	54/3
	Crampton Prize	Lings and Dietz	An improved direct shear apparatus for sand	54/4
	Geotechnical Research Medal	White and Lehane	Friction fatigue on displacement piles in sand	54/10
	George Stephenson Medal	Coop, Sorensen, Freitas and Georgoutsos	Particle breakage during shear of a carbonate sand	54/3
2006	Crampton Prize	Alonso, Olivella and Pinyol	A review of Beliche dam	55/4
	George Stephenson Medal	Ratnam, Soga and Whittle	A field permeability measurement technique using a conventional self-boring pressuremeter	55/7
	Geotechnical Research Medal	Collins	The concept of stored plastic work or frozen elastic energy in soil mechanics	55/5
	Halcrow Prize	Houlsby, Kelly, Huxtable and Byrne	Field trials of suction caissons in clay for offshore wind turbine foundations	55/4
2007	Telford Medal	Gens and Alonso	Aznalcollar dam failure: stability conditions and failure mechanisms	56/3
	Geotechnical Research Medal	Jardine, Standing and Chow	Some observations of the effects of time on the capacity of piles driven in sand	56/4
	Telford Prize	Merrifield, Lyamin and Sloan	Three-dimensional lower-bound solutions for the stability of plate anchors in sand	56/2

Table 8. Leading prizewinners

Name	Medals	Prizes
David Potts	4	1
Mark Randolph	2	4
Matthew Coop	2	4
Peter Vaughan	3	2
Malcolm Bolton	1	3
Peter Wroth	1	3

REFERENCES

Brown, S. F. (1982). The development of *Géotechnique* 1948–1980. The first 30 volumes. *Géotechnique* **32**, No. 2, 95–110.

Burland, J. B. (2005). Soil mechanics at Emma: elegant, rigorous and relevant. *Emmanuel College Magazine*, 2004–2005 **87**, 154–174.

Burland, J. B. (2008). The founders of *Géotechnique*. *Géotechnique* **58**, No. 5, 327–341.

Chrimes, M. (2008). Geotechnical publications before *Géotechnique*. *Géotechnique* **58**, No. 5, 343–355.

Cooling, L. F. (1970). Correspondence. *Géotechnique* **20**, No. 1, 107–109.

Cooling, L. F., Skempton, A. W., Glossop, R. & Golder, H. Q. (1975). British Geotechnical Society Twenty-fifth Anniversary Report. *Géotechnique* **25**, No. 4, 629–656.

Glossop, R. (1982). Private correspondence.

Glossop, R. & Golder, H. Q. (1948). Editorial. *Géotechnique* **1**, No. 1, p. 6.

Golder, H. Q. (1969). Correspondence. *Géotechnique* **19**, No. 2, 313–315.

Golder, H. Q. (1982). Private correspondence.

Penman, A. D. M. (2007). Private correspondence.

Terzaghi, K. (1948). Foreword. *Géotechnique* **1**, No. 1, 3–5.

Ward, W. H. (1981). Private correspondence.

Burland, J. B. (2008). *Géotechnique* **58**, No. 5, 327–341 [**doi:** 10.1680/geot.2008.58.5.327]

The founders of *Géotechnique*

J. B. BURLAND*

The first issue of *Géotechnique* was published in June 1948, and the story of its origins is well known. Publication followed the wide circulation of a letter in July 1947 from 'The Geotechnical Society' to every western European country in which the proposal for such a publication was made. Listed at the bottom of the letter were eleven 'Founders', four from the UK and the remainder from elsewhere in Europe. The function of the 'Founders' was to contribute to the early issues of the journal and to assist with identifying other contributors. This paper gives brief vignettes of all these 11 eminent geotechnical engineers, together with two others who contributed significantly to the early success of the journal. It is hoped that these vignettes will serve to inspire present and future generations of geotechnical engineers, not only in their professions, but also in their enjoyment of working with fellow enthusiasts.

KEYWORDS: historical review

Le premier numéro de *Géotechnique* a été publié en 1948 ; les circonstances de sa fondation sont bien connues. La publication fit suite à la diffusion, au mois de juillet 1947, d'une lettre de la Geotechnical Society à chacun des pays de l'Europe de l'Ouest où la proposition de la création de cette publication avait été formulée. Au bas de la lettre se trouvait le nom des onze « fondateurs », quatre desquels étaient britanniques, les sept autres de différents pays d'Europe. La fonction de ces « fondateurs » était de contribuer aux premières éditions du journal, et d'assister dans l'identification d'autres collaborateurs. Dans la présente communication, nous présentons une esquisse sur chacun des onze éminents ingénieurs géotechniciens qui, avec deux collègues, contribuèrent de façon considérable au succès rapide de cette publication. Nous espérons que ces portraits serviront à inspirer les générations actuelles et futures d'ingénieurs géotechniciens, non seulement dans leur profession, mais également dans le plaisir de travailler avec des collègues passionnés.

INTRODUCTION

The fascinating and colourful story of the origins of *Géotechnique* has entered the annals of modern soil mechanics, and is well known. Some of the personalities involved were larger than life, and many aspects of their professional and personal lives have been well documented. But there are others who contributed to the conception and early days of *Géotechnique* who are less well known. The purpose of this paper is to gather together some details of the 'Founders' of *Géotechnique* together with a few others who significantly contributed to its early success. The story of *Géotechnique* is brought up to date by Brown (2008) in a companion paper.

The birth of *Géotechnique* flowed from the outburst of interest in ground engineering and soil mechanics that immediately followed the Second World War, together with the desire to establish international contacts between experts and practitioners. The driving force behind the Second International Conference on Soil Mechanics and Foundation Engineering, held in Rotterdam in June 1948, stemmed from the same interests and desires. Therefore, to some extent, the birth of *Géotechnique* and the planning and running of the Rotterdam Conference are inextricably linked, as many of the key people were involved simultaneously in both.

At the First International Conference in Cambridge, USA, in 1936 it was agreed to hold the second one in Holland, but preparations for that were interrupted by the onset of the war. Soon after cessation of hostilities planning was resumed, contact was made with Karl Terzaghi (President of the International Conference on Soil Mechanics and Foundation Engineering), and an Organising Committee was formed. The Netherlands government and the Municipality of Rotterdam sent an invitation to the Permanent Committee of the ICSMFE to hold the second conference, and this was accepted. A. W. Skempton was appointed as Vice-President and T. K. Huizinga was Secretary to the Conference. In May 1947 the Organising Committee issued a bulletin announcing the Conference and the subjects to be treated. The Institution of Civil Engineers anticipated this bulletin, presumably because of 'inside information', by establishing a British National Committee in March 1947 'to concern itself with the preparation of papers for presentation at the Rotterdam Conference'. This committee held its first meeting in April 1947, and by September 1947 it had accepted 76 synopses – such was the vitality of soil mechanics within the UK.

In the meantime, in October 1946 Glossop, Golder, Maclean and Ward had made their memorable visit to Europe during which *Géotechnique* was conceived, as described by Golder (1969), Cooling *et al.* (1975) and Brown (1982). After wide consultation, including Karl Terzaghi, 'The Geotechnical Society' was formed with Golder as Secretary and Glossop as Treasurer. A letter dated 7 July 1947 was widely circulated in English and French to every western European country where contact had been made. This letter is reproduced in the paper by Brown (1982) and suggests the formation of a European Society with two objects: (a) to found a journal for circulation in Western Europe; and (b) to hold a conference once a year in rotation in various western European capitals. Listed at the bottom of this letter are three Patrons (Dr K. Terzaghi, Ir T. K. Huizinga and Sir G. M. Burt) and 11 'Founders' (L. F. Cooling, J. P. Daxelhofer, E. E. De Beer, J. Florentin, E. C. W. A. Geuze, R. Glossop, H. Q. Golder, R. Haefeli, A. W. Skempton, A. von Moos and W. H. Ward).

In the final paragraph of the July 1947 letter it is stressed that it was not intended that the Society should cut across or conflict with any world organisation that might be set up at the International Conference at Rotterdam in 1948, being more local in character. As it happens there was a strong possibility of conflict. In his capacity as Secretary to the Rotterdam Conference, Ir T. K. Huizinga circulated the draft

Discussion on this paper closes on 1 December 2008, for further details see p. ii.

* Imperial College London, UK.

statutes for the permanent organisation of the International Conference (Vol. 2, pp. 325–327 of the Proceedings of the Rotterdam Conference). This draft included a proposal that the Society should 'publish a periodical'. Terzaghi dealt firmly with this proposal as follows (Vol. 6, p. 167):

> . . . I wish to draw your attention to the efforts which are being made by Hugh Q. Golder and his associates in London to establish a pan-European unit, to be served by a periodical, 'La Géotechnique'. This appears to me as a worthwhile experiment which deserves wholehearted support; it requires a much less elaborate apparatus than an international organisation and execution of program is much easier, because the participants are neighbours, at least geographically.

At the ensuing meeting to discuss the draft statutes (Vol. 6, pp. 171–179) no mention was made of the article referring to 'the periodical'. How wise it was to bring Terzaghi into the project early on!

The function of the Patrons of the Geotechnical Society was clearly to give the initiative credibility. Huizinga was an engineer of great significance, being Director of the Delft Soil Mechanics Laboratory and Secretary of the Rotterdam Conference. Sir George Burt was a director of the construction firm John Mowlem & Co. Ltd, and on the boards of both the Building Research Station and the Road Research Laboratory. The wisdom of inviting Karl Terzaghi as a Patron has already been demonstrated. His Foreword to the first issue of *Géotechnique* is both penetrating and thought provoking: like his Presidential Address to the First International Conference it certainly warrants frequent close study.

But what about the 'Founders'? Glossop, in his account of the origins of *Géotechnique* (Cooling *et al.,* 1975), wrote:

> But, if our expedition had committed us to starting a journal, it had also introduced us to most of the leading workers in soil mechanics in Western Europe, and this was of immense help when the time came to solicit contributions to the first number. Indeed, the list of authors in volume one is remarkable.

Thus an important function of the 'Founders' was to assist with the contents of the early issues.

Brief vignettes will now be given of these eminent geotechnical engineers. I had the good fortune to know some, but by no means all, of the 'Founders'. Where possible I have tried to include my own personal memories, but in many cases I have had to rely on published material and the recollections of others. The vignettes are by no means biographies, but brief sketches, which it is hoped will give life and character to the technical contributions of these remarkable engineers.

THE BRITISH FOUNDERS

The 'Geotechnical Society' referred to previously was set up primarily to publish the journal. The membership consisted of Golder (Secretary), Glossop (Treasurer), Cooling, Skempton and Ward. We can refer to this group as the 'British Founders', whose vignettes will be presented first.

Leonard Frank Cooling (1903–1977)

While Professor Sir Alec Skempton was the father of British soil mechanics, Dr Leonard Cooling (Fig. 1) was its founder. Cooling's first degree was in physics, and his early years at the Building Research Station (BRS) were spent in the Physics and Chemistry Divisions, where he worked on the capillary properties of porous materials and on the weathering of building stone. Skempton (1977) describes how, in 1933, a soil physics section was established at BRS and Cooling was put in charge of it. He set up the first proper soil mechanics laboratory in Britain, equipped with the apparatus necessary to classify soils, measure their basic mechanical properties, and carry out sampling. By 1935 the first investigations of civil engineering problems had begun, and the group was moved to the Engineering Division of BRS, being renamed the Soil Mechanics Section. He was the sole representative of the United Kingdom at the First International Conference, at Harvard, in 1936, at which he published two papers and a report on the soil mechanics laboratory at BRS. It was in August 1937 that the well-known Chingford dam failure occurred that brought Terzaghi to England. This led to the acceptance of soil mechanics as a key discipline in civil engineering. That the young Soil Mechanics Section managed the investigation of the Chingford failure so well, with staff who had no experience of soil mechanics prior to joining the Section, was due largely to Cooling's abilities and leadership.

Fig. 1. Leonard Frank Cooling (1903–1977)

During the next 10 years leading up to the 1948 Rotterdam Conference the BRS team occupied a central and formative position. Engineers and academics came there to visit, to study, to work and above all to seek advice. A typical example of the influence that Cooling played is related in my description of the early days of soil mechanics at Cambridge (Burland, 2005). In 1928 John Baker (later Lord Baker) was sent to BRS to undertake research for the steel construction industry. At BRS Baker got to know Cooling, and was so impressed with the work of the soil mechanics group that, when he went to Bristol University, one of his first actions was to set up a small soils laboratory. When Baker then moved to Cambridge he found that the subject was virtually unknown, so he suggested that Kenneth Roscoe, who had recently joined the staff, should carry out his research work on soil mechanics. He advised Roscoe to visit BRS to discover 'what soil mechanics was about'. There he met Cooling and was introduced to the work of Hvorslev and Rendulic, whose seminal papers were published in German. Roscoe, who had spent his war years in a prisoner-of-war camp in Germany, set about translating the papers. Thus the seeds of critical state soil mechanics were unwittingly planted by Cooling, someone for whom Roscoe had the highest regard.

As emphasised by Skempton (1977), Cooling carried out an immense amount of work for the general good both of the subject and of the profession. He was on the drafting committees of the three basic soil mechanics codes: site investigations, foundations and earth-retaining structures. It is clear from a study of the minutes of these committees that Cooling took responsibility for drafting large sections of these codes. It was he who started the informal discussion meetings at ICE in 1940, where he also took part in the influential lectures on soil mechanics that were held in 1945. He was a leading member of the Soil Mechanics and Foundations Committee set up by ICE in 1947, and of the British National Committee of the ISSMFE established in the same year. He was on the committee of what was to become the British Geotechnical Society for many years, holding the position of Chairman of the Society from 1955 until 1959. In 1957 he was elected an Associate of the Institution of Civil Engineers in recognition of his outstanding contributions to civil engineering – at that time a singular honour for someone who had trained as a physicist. Very importantly for this article, he served on the *Géotechnique* Advisory Panel from its first meeting in 1949 for 20 years without a break, and was Chairman from 1966 to 1969. In a letter to *Géotechnique* on his retirement as chairman, responding to Golder's question 'Are we doing what we set out to do'? (Golder, 1969), Cooling concludes with the memorable statement (Cooling, 1970):

> Finally I think Dr Golder can rest assured that the infant he helped to nurture during the early stages of its existence, has now come of age as a strong virile member of society. During adolescence it acquired features of a pronounced individual character and we must trust a personality will continue to develop and improve.

I had the privilege of working under Cooling for a few years at BRS. He must have been the last of the generation that addressed colleagues by their surnames. There was something friendly and affectionate about his greeting 'Hello Burland, how are things going?' He was very approachable and good humoured, and his advice was steeped in common sense. Perhaps the best piece of advice he gave to me was to 'publish your original work in journals and use conferences to draw attention to it and for describing developments of the work'. He went on to explain that in this way the research would remain readily accessible, not getting buried in conference proceedings. It is particularly appropriate that the British Geotechnical Society chose to honour him by establishing, in 1970, the Cooling Prize to be given annually for the best paper presented at a competition by a young engineer. Cooling himself drew up the criteria for assessment and chaired the panel of judges for the first few years of what has become a most prestigious competition.

Rudolph Glossop (1902–1993)

According to Skempton (1993), Glossop (Fig. 2) was the driving force in establishing *Géotechnique*. He also contributed hugely to the cause of engineering geology, a subject that was always close to his heart, no doubt stimulated by his training as a mining engineer. Building on the work of Sir Harold Harding, he championed 'geotechnology' as a broad discipline, publishing widely on the subject, including his Rankine Lecture (Glossop, 1968) – the first to be given by a contractor! Williams & Norbury (2008) give a full and fascinating account of his career and his contributions to geotechnology. He never lost sight of the importance of bringing together the practical and academic aspects of both geology and soil mechanics.

Fig. 2. Rudolph Glossop (1902–1993)

Rudolph Glossop studied mining engineering at the Royal School of Mines (Imperial College), graduating in 1924. At college he gained the nickname 'Silas' after Silas Q. Porter, the absent-minded professor in the novel *Tarzan of the Apes*, and he retained the name throughout his life. After gaining experience as a mining engineer in Canada and Mexico, followed by a brief spell on the staff at Birmingham University, he joined the firm of John Mowlem & Co. Ltd, where he worked on the construction of Leicester Square Underground station and other civil engineering projects. From 1933 to 1937 he returned to mining, working as a mine manager on the Gold Coast.

Glossop's geotechnical career was finally launched in 1937 when he rejoined Mowlems to manage a department concerned with specialist processes such as groundwater lowering and chemical grouting. He freely admitted that he knew nothing about soil mechanics (Cooling *et al.*, 1975), but he avidly studied the French and German literature on geotechnical processes. It was, of course, in 1937 that the famous slip occurred during the construction of the earth dam being built by Mowlems at Chingford. When work on the dam was resumed, following the investigations carried out by BRS and Terzaghi's subsequent recommendations, Glossop was appointed sub-agent. He immediately set up a small laboratory on site to carry out control tests on the

London Clay foundations and the embankment fill. It was at this time that Glossop was introduced to Cooling, Skempton and Golder at the BRS, where he received his training in soil mechanics, the principles of which he soon mastered.

In fact Glossop and Skempton had met, coincidentally, just before Chingford in quite different circumstances. The story is delightfully related by Glossop (1984). It is reproduced in full here as it reveals much about his interests, determination and sense of humour.

> One day during a party conversation with an old friend, she, knowing something of my activities said, 'I met a remarkable young man the other day who should interest you. He is doing research on the properties of clay, and bought a picture from John Tunnard's exhibition at Peggy Guggenheim's gallery.' My reply was, 'If such a man exists, I must meet him at once, What is his name?' She had forgotten it, so as soon as possible I went to Guggenheim Jeune, in Cork Street. Mrs Guggenheim was away, and an assistant was in charge. 'Can you tell me the name of a young man who bought a Tunnard and is researching on clay?' I asked. This simple question led to an extraordinary conversation, based on total miscomprehension, which threatened to go on forever. At last, it dawned on me that while I was talking about clay she, not unnaturally, was talking about Klee, the Swiss painter. I tried again, and suggested that we should look through the firm's books. There I found a recent purchaser called Skempton, whose address was Watford, near the Building Research Station. 'That's my man,' I said, and so it proved to be.'

Silas and Skem soon became lifelong friends.

When in 1941 Glossop was sent as agent to the construction of an airfield at Leiston in Suffolk he hired the science laboratory in a local school and transferred the equipment from the Chingford site laboratory. Hugh Golder joined Mowlems in 1942 and was sent to join Glossop at Leiston. Together they devised a simplified method of airfield pavement design that overcame the serious problems caused by the heavy bombers being deployed by the American Air Force. This work formed the basis of a paper that was published at the 1948 Rotterdam International Conference (Glossop & Golder, 1948). The soil mechanics laboratory had proved its worth, and it was moved to a flat in Victoria Street owned by the chairman of Mowlems. There, in November 1943, the firm of Soil Mechanics Ltd was founded, with Glossop, Golder and Harding (later to be Sir Harold Harding) as directors. As related by Skempton (1993) in his obituary of Glossop, the firm prospered: geologists as well as engineers were added to the staff; enlarged premises were found at Chelsea, then at Bracknell, and ultimately in a new building, appropriately named Glossop House, at Wokingham.

Silas Glossop was someone of broad vision, who inspired others to join him in achieving it. He delegated wisely, thereby encouraging the career development of others.

> Generous and helpful, drawing on an extraordinary store of knowledge and quick to grasp the essence of a problem at his desk or on site, a man of wide interests enjoying discussions on all manner of subjects, enlivened by entirely fresh ideas, he enriched the lives of all who knew him. (Skempton, 1993)

Hugh Quentin Golder (1911–1990)

Hugh Golder (Fig. 3), with Silas Glossop, played a key role in the birth of *Géotechnique*. Gibson (1969), who worked with him for many years, describes him as a person of great intellectual capacity who was practical, inventive, and insistent on going back to first principles. He also had a most colourful turn of phrase, as displayed in his classic description of the conception of *Géotechnique* (Golder, 1969).

Fig. 3. Hugh Quentin Golder (1911–1990)

Golder studied engineering at Liverpool University, graduating with a BEng in 1932 and staying on to carry out research on the causes of deterioration of concrete, for which he was awarded a master's degree. He then joined the Forest Products Research Station, where he studied the mechanical properties of timber and the effect of various agencies on these properties. His geotechnical career began in May 1937 when he joined the Soil Mechanics Section of the BRS.

The Chingford embankment dam failure occurred in July 1937. Golder was involved in the investigation from the start, introducing a composite slip surface approach for analysing the failure and for checking the redesign proposed by Terzaghi (Cooling & Golder, 1942). Thereafter he undertook a number of geotechnical studies arising out of the requests for help and advice that arrived at BRS. These studies were of a thoroughly practical nature, involving both field and laboratory work. One of the most notable jobs was the Huntspill River cut in the Somerset Levels, where he made use of a portable compression apparatus that he had designed (Cooling & Golder, 1940).

Golder began to hanker after more scope for his activities than BRS could offer. The manner of his replacement by Bill Ward has entered the folklore of British soil mechanics (Penman, 1996). At a time of war there were restrictions on job movements, and Golder was told by the Establishment Office that, without a replacement, he was not permitted to leave. He therefore placed an advertisement for the post in *The Times* to which Ward, who was working for Freeman Fox & Partners, replied. Ward took up his post at BRS in September 1942, whereupon Golder was permitted to leave. Cooling (1975) confessed somewhat wryly that he was never really able to fathom the details of this arrangement! It is typical of the man that Golder circumvented the civil service bureaucracy by such a simple and direct expedient. He then approached John Mowlem & Co. Ltd, civil engineering contractors, who took him on, posting him to Leiston to work under Glossop (see vignette on Glossop). One cannot help but suspect that much of this had been prearranged behind the scenes!

As mentioned previously, the firm of Soil Mechanics Ltd was formed in November 1943 with Golder as a director together with Glossop and Harding. Golder remained with the firm until 1958, keeping his close links with the personnel at BRS during this time. For the 1948 Rotterdam Conference he was author of five papers, many of which described projects that he had undertaken at BRS. He also contributed a paper to each of the first two volumes of *Géotechnique*.

In 1958 Golder resigned from Soil Mechanics Ltd 'on a point of principle' in protest against what he understood was the arbitrary dismissal of his lifelong mentor Harold Harding by Sir George Burt following disputes between them concerning current road contracts in Iran (Milligan, 2007). He emigrated to North America, where Professor Arthur Casagrande offered him a teaching post in engineering geology at Harvard University for a year as a stand-in for Professor Terzaghi, who was going on sabbatical leave. He formed a deep and lasting friendship with Arthur Casagrande, and frequently used him as a consultant thereafter. At the same time he was invited to head a team formed by a consortium of engineering consultants to carry out a detailed feasibility study of a causeway and bridge from the Canadian mainland to Prince Edward Island.

In 1959 Golder moved to Toronto to live with his family, and began working as an independent consultant. Victor Milligan relates how the formation of the firm carrying Golder's name came about. While advising the Department of Highways of Ontario, Golder met Larry Soderman. It was Larry who, while at dinner in his home eating geese shot earlier in Hudson's Bay, and drinking copious wine, suggested to Hugh Golder and his wife Molly (his guests, along with Victor Milligan and his wife Mary Ann) that Hugh should form a partnership. In this way he would cease to be an individual practitioner who just advised, but would create a company of a size capable of project design and implementation. Thus it was, in late 1960, that H. Q. Golder & Associates (now Golder Associates) was born, with Golder and Milligan as shareholders. There is an interesting footnote to the Prince Edward Island project that first brought Golder to Canada. Although shelved for many years, the Confederation Bridge was opened in May 1997, with Golder Associates acting as geotechnical consultants throughout its period of design and construction.

Hugh Golder was a great spinner of yarns, and particularly enjoyed one-liners. Two that I use in my teaching are:

> Any design that relies for its success on a precise calculation is a bad design.

and, when designing to limiting settlement,

> Always ask who is limiting the settlement and why.

As mentioned previously, Golder had a great admiration for the famous tunnelling engineer Sir Harold Harding. He frequently related the story of Sir Harold's response when he was told of a young engineer whose high academic qualifications led him into such over-confidence that it was feared that he would produce a folly: 'Send him to me and I will put him down a great hole in the ground and teach him humility' growled Sir Harold. Hugh Golder, who wrote verse, composed the following tribute to Sir Harold to mark his election as President of ICE in November 1963, which makes reference to this story:

Sonnet to a Tunnel Engineer
To virtue, modesty, and stubborn worth,
I pay my tribute now to this my friend,
Who taught humility, who loved the earth.
Within his powers the laws of nature bend.
Foundations, tunnels, deep holes in the ground,
Tube railways, sewers, his monument shall be.
We know his worth, treasure his judgement sound,
Our debt is great to those who set him free.
Other sad people, architects and such
Vaunting achievements like a high-rise tower,
Or business men high-placed (not overmuch),
Covet his fame, as they in public glower,
Laying foundation stones with golden trowels,
Change London's face; but he has changed its bowels.

Alec Westley Skempton (1914–2001)

Professor Sir Alec Skempton (Fig. 4), or 'Skem' as he has always been affectionately called, was one of the most eminent engineers of the twentieth century and the father of British soil mechanics. How does one even begin to write a vignette of such a person? Fortunately his daughter, Judith Niechcial (2002), has written a biography of him in a delightful book *A Particle of Clay.* The title is taken from two events. The first was his inaugural lecture at Imperial College, when he ended the hour-long lecture quoting two lines from William Blake:

> To see a World in a grain of sand...
> And eternity in an hour.

He then concluded:

> The first, I trust, may be considered apt: the second, I hope is not true, at least so far as my audience is concerned.

He returned to this quotation years later when awarded the Gold Medal of the Institution of Structural Engineers, with an addition:

Fig. 4. Alec Westley Skempton (1914–2001)

To see the World in a grain of sand ... or in my case a Particle of Clay.

Judith Niechcial skilfully weaves into her narrative Skem's many great achievements and accolades, placed in the context of his family life and background, his friends and colleagues, his all-consuming passions and enthusiasms, his ways of working and his strong but tightly controlled emotions. We read about his student days at Imperial College, where he captained the rugby team and represented City and Guilds in boxing as a heavyweight. His passion for jazz was replaced by an even stronger passion for classical music. We learn of the meeting between the devastatingly handsome but serious and remote Skem and the beautiful young Royal College of Art student, Mary Nancy Wood, who knew immediately that she wanted him as her husband, and set out to win him.

The book is illuminated by many perceptive personal insights. We read of the turmoil and struggle Skem experienced when, having decided to leave BRS and accept a teaching fellowship at Trinity Hall, he came to realise that Cambridge was not remotely 'his scene'. His final decision to abandon Cambridge was appallingly difficult—but turned out to be Imperial College's gain.

There are many humorous illustrations of Skem's difficulties in dealing with emotionally delicate situations: for example the occasion when the horn player in a lunchtime concert at Imperial College kept blowing raspberries and finally went berserk. Skem was the senior academic present, but opted out and left the hall. Or another very different situation, when he received a furious letter from a distinguished professor of soil mechanics accusing him of undermining his life's work. Again, Skem asked a colleague (in this case me) to deal with it. Yet Skem could be very emotional himself. On one occasion I was describing to Skem an important meeting I had just had with a very senior official in the Scientific Civil Service over an exciting initiative. The official adopted the heavy hand of bureaucracy, and snuffed out the idea. Skem became extremely angry, and stormed out of the room, saying that the story had spoiled his day. Having read his biography it is now apparent to me that the reason for Skem's deep dislike for civil servants and bureaucracy lay in the refusal of the War Office to accept liability for his father's health and ultimate death from being gassed during the First World War. This same suspicion and impatience in dealing with civil servants may explain the delay in award of the knighthood that he so richly deserved and which ultimately he received in the Queen's Millennium New Year's Honours.

Professor Bob Gibson (1984), in his article on 'Working with Skempton', brilliantly encapsulates Skem's approach as follows:

> Anyone having the good fortune to work with Skem immediately became aware of his outstanding ability to reduce a problem to its essentials, whether this was a consulting job, at a research seminar or over a discussion at the lunch table. Not content merely to accept what he was told, he adopted the mantle of student asking penetrating questions that seemed always to illuminate the path towards the ultimate conclusions. He never hesitated to acknowledge his inability initially to follow an argument, but he persisted until its essence had been grasped and he had formed his own opinion. This unassuming and scholarly approach was entirely free of pomposity and made a profound impression on his students and colleagues. To attend a lecture by Skem was an unforgettable experience. It would be prepared with great care and given with the utmost formality but without notes. The delivery itself was reflective—on occasion almost hesitant—but imbued with a compelling logic and totally gripping.

In the field of the history of engineering, Skempton's work has almost single-handedly transformed the subject from that of the enthusiastic amateur historian, focused largely on steam engine technology, to a rigorous academic discipline, challenging orthodox assumptions. His major contribution was the study of the work of early civil engineers, more especially those of the eighteenth century, whose achievements had been ignored by previous historians.

Skempton retired in 1981, when he became Professor Emeritus and Senior Research Fellow at Imperial College. Up to a few weeks before his death he remained remarkably active, coming into the department every day to work on the latest biography of an early engineer, attending and contributing to research seminars, lecturing to postgraduate students, and meeting local and overseas visitors. Wherever one travelled internationally, engineers would ask after Skem, giving their own personal testimonies on the way he had influenced their careers. There can be few civil engineers who inspired such affection, had such a joy of life, and who made such a profound and lasting contribution in so many ways. An obituary of Skempton was published in *Géotechnique* (Vol. 51, No. 10, 829–834).

Let the last word rest with Ralph Peck (1977):

> Skempton is a physically big man; indeed everything about him is big—his smile, his laugh, his voice, his gestures. He even writes with a broad pen. Whatever he does, whatever he discusses, is pervaded with an air of intense excitement. He never describes large objects as merely large, they are <u>enormous</u>.

William Hallam Ward (1917–1996)

Dr Bill Ward (Fig. 5) was an inspirational figure, whose interests and knowledge ranged across the whole spectrum of practical soil and rock mechanics into glaciology. His memory of previous case histories was prodigious, and his ability to recall relevant details years later was much admired, proving invaluable in many of his consultancies.

Ward was an archetypal Londoner, born in Pimlico and educated at the Polytechnical Secondary School. He studied civil engineering at Imperial College, London, graduating in 1937. As a student he became a keen rock climber, and he also took part in the Imperial College expedition to Jan Mayen Island in the Greenland Sea, in the capacity of surveyor and assistant geologist. After graduating he stayed on to do research for the PhD on seepage flow through granular media, but the work was interrupted by the outbreak of the Second World War. He joined Freeman Fox & Partners as assistant resident engineer for the construction in South Wales of the world's largest cordite factory. The story of Ward's appointment at BRS in 1942 has entered the folklore of British soil mechanics, and is described in the accompanying vignette of Hugh Golder.

Dr Bill Ward had an intuitive sense of the physics of a problem, which he seemed always to be able to call upon instantly, even during arduous fieldwork (which he loved). An illustration of his ability to identify the fundamentals is given by an event that occurred when I was assisting him in his role as an advisor on the foundations of Hartlepool Nuclear Power Station. Constructing the foundations involved boring each shaft through about 35 m of water-bearing soft alluvium under mud down to Bunter sandstone. This was followed by inserting and grouting a steel casing, casting a concrete plug at the bottom, pumping out the mud,

Fig. 5. William Hallam Ward (1917–1996)

and then breaking through the concrete plug so as to hand-excavate an under-reamed socket into the sandstone. One of the first casings was not properly grouted, and as the mud was being pumped down, the 35 m long 3 m diameter steel tube suddenly launched itself upwards out of the ground. I was with Bill Ward when he received the telephone call from site. Suitably prefixed, his instantaneous response was:

. . ., haven't they ever heard of Archimedes?

Like Haefeli, another 'Founder' of *Géotechnique*, Ward was an experienced glaciologist. Professor John Nye FRS, in a letter of appreciation after his death, describes another example of this immediate grasp of the fundamental physics of a problem at an historic meeting in London 1947 (Nye, 1996). An international expert introduced, for the first time, the concepts of modelling the flow of ice by means of plasticity theory, and pointed out that, on this basis, a tall slender column of ice could not attain a height of more than about 10 m. In the discussion that followed, Bill Ward suggested that this was why crevasses in glaciers attained roughly that depth. Subsequently he was shown to be right.

I joined the BRS in 1966, and my first project was to work under Ward on the experimental investigation of the suitability of a site on chalk at Mundford, Norfolk, for the construction of a 300 GeV proton accelerator (Ward *et al.*, 1968). The conception of the large-scale loading test using a million-gallon steel water tank was due entirely to Ward. The nuclear physicists had been advised that, in order to assess the magnitudes of ground movements outside an area loaded by heavy radiation shielding, all that was required was to carry out laboratory tests on specimens of the rock to measure Young's modulus of elasticity. Ward pointed out that natural rock in situ did not behave like an ideal homogeneous isotropic elastic material. The response could be determined only by means of a large-scale test. Once again, Ward was proved to be absolutely right in his intuitive judgement. It is well known that the measurements at Mundford demonstrated that the movements were concentrated much closer around the loaded area than predicted by simple elastic theory. Moreover, the experiments proved to be seminal in a number of ways, even though, in the event, the proton accelerator was never built.

Bill Ward contributed three papers to the Rotterdam Conference. One of them discusses a slip on a flood defence bank constructed on a peat bog (Ward, 1948). This was Borth Bog (or Cors Fochna), at Cardigan Bay in South Wales. It is interesting that David Henkel describes his first clear memory of Ward as watching him put down hand auger holes at Borth Bog in 1945 (Henkel, 1978). Ward's first and last papers in *Géotechnique* typify his unique and distinctive contributions. In 1950 he published the first paper that dealt with the thermal effects of industrial plant (Ward and Sewell, 1950). This paper describes a number of valuable case histories, and reveals clearly Ward's fundamental understanding of heat flow problems, in relation both to frost and to heat penetration. I witnessed this deep understanding of heat flow when, at Mundford, I was attempting to minimise the effects of temperature change on a reference beam for the plate loading tests. Ward rightly insisted that the best way of doing this was to surround the beam in a double-walled box, painted white throughout, and with the outer cavity fully ventilated.

Ward's last paper in *Géotechnique* describes the final results of the Kielder experimental tunnel (Ward *et al.*, 1983). As with the Mundford experiment, he had impressed on the authorities the importance of driving an experimental tunnel prior to the construction of the main Kielder tunnel in order to investigate the effectiveness of various support systems in problematic mudstones. Like Mundford, the results of this field experiment have value way beyond the primary objective of the trial itself. The papers describing the work in the Kielder experimental tunnel are required reading for engineers concerned with tunnelling through mudstone strata.

Bill Ward was a larger-than-life character. His views were often expressed in a forthright and blunt manner, and for those who did not know him this could cause offence. His manner was really a reflection of the passion he felt for getting at the truth, and he never bore malice. Indeed, in my experience, when taken in the right spirit (and sometimes firmly stood up to!) he always responded warmly and with good humour. The tributes that are given in the obituaries by Christine Cooling (1996) and Penman (1996) give a much fuller picture of Bill Ward than could possibly be contained in a brief vignette.

EUROPEAN FOUNDERS

Jean-Pierre Daxelhofer (1907–1998)

Professor J. P. Daxelhofer (Fig. 6) played a prominent role in the naming of *Géotechnique*. Golder (1969) relates that one of his suggestions was the Greek word for soil ($\varepsilon\delta\alpha\phi o\sigma$). Brown (1982, 2008) reports that it was Daxelhofer who made the suggestion that the name should be *La Géotechnique*.

Born on 20 April 1907, Jean-Pierre Daxelhofer obtained the Diploma in Civil Engineering from l'École Polytechnique de l'Université de Lausanne (EPUL) in 1929. He joined the Italian contractor Rodio working in Paris, and then Milan and Rome, as an expert in geotechnical engineering, and gaining wide and varied practical experience. In 1945 he returned to Switzerland, where he was appointed director of research at the geotechnical laboratory of EPUL. He was made Professor Extraordinaire in 1949, directing his expertise towards rock mechanics and construction materials. In 1963 he was promoted to full professor, and taught construc-

Fig. 6. Jean-Pierre Daxelhofer (1907–1998)

tion materials, particularly in relation to water-retaining structures. He retired in 1972, and died on 8 June 1998.

Professor François Descoeudres remembers Daxelhofer as a remarkable engineer with a broad knowledge of civil engineering, materials science, chemistry and mechanics. Undoubtedly he made important contributions to the development of geotechnics, but he published very little. Known as having a sharp temper, he also seems to have been something of a perfectionist, always pushing on to the next development and unwilling to commit himself to print. His move from geotechnics to construction materials in 1949 took place fortuitously. The École d'Ingénieurs de Lausanne was looking for a concrete specialist to advise on the construction of concrete and arch dams in Switzerland, there being something of a construction boom of these at that time. He was nominated because of his wide knowledge and experience of geotechnical and construction materials.

A notable exception to his lack of publications was a major paper that he published in 1944 on the shearing resistance of soils, with examples of its importance (Daxelhofer, 1944). In this report he describes a variety of shearing apparatus, distinguishing clearly between undrained and drained conditions. He discusses a number of practical examples of stability, including Fort Peck dam. Many of his recommendations are applicable today: for example, laboratory results should indicate the apparatus used to obtain them; do not expect the laboratory to provide all the answers; minor geological features can dominate behaviour; the formation of plastic zones can be dangerous as they have a tendency to propagate; field measurements are essential, but are difficult to carry out in geotechnics because they must be made over a number of years and require collaboration between contractor, designer and laboratory. In 1995 Daxelhofer donated to the 'Archives de la Construction Moderne EPFL' 367 reports covering the period 1943 to 1989, which deal mainly with accounts of failures of foundations and damage to buildings.

Edward E. De Beer (1911–1994)

At the 1989 International Conference in Rio de Janeiro Edward E. De Beer (Fig. 7) was awarded the Kevin Nash Gold Medal of the International Society. The terms of the medal read as follows:

> The Medal should be awarded to a person who, through his distinction as an engineer, his international contribution to engineering practice and education, his contributions to international good will, and his service to the Society, has made a major contribution to fostering the ideals and goals of this International Society throughout the world.

This statement could have been drafted with Professor De Beer in mind, for it is a most apt synopsis of his personal achievements.

De Beer graduated in civil engineering from the State University of Ghent in 1934. He immediately sat for the entrance examination into the Corps of Engineers, for which he was ranked first, and was offered a post at the Ministry of Public Works. His future career in soil mechanics was firmly charted when he was sent to the Technical University of Delft to work in the soil mechanics laboratory under the guidance of Professors Buisman and Broekman. On returning to Ghent in 1938 he set up a soil mechanics laboratory at the Ministry of Public Works, which soon became the State Geotechnical Institute with De Beer as Director—a post that he held until 1978. In this capacity he was not only responsible for geotechnical research, but was also a consultant on numerous

Fig. 7. Edward E. De Beer (1911–1994)

civil engineering projects involving an immensely wide range of geotechnical problems. This extraordinary range of topics is reflected in his many publications.

At the Rotterdam Conference De Beer was the author of seven papers, and he was the General Reporter for Session 12—Classification of soils. In his capacity as a 'Founder', he published a paper in the first issue of *Géotechnique* on the interpretation of the shearing resistance of soils from deep sounding tests (De Beer, 1948)—a topic to which he returned several times. He was also a lively participant in the 1950 London Conference on the Measurement of Shear Strength of Soils in Relation to Practice, the proceedings of which were published in Volume 2 of *Géotechnique*. Indeed, he presented a paper on the application of the Dutch cell-test (De Beer, 1950) and, while chairing the session at which it was presented, also participated in a heated discussion on the interpretation of results from this apparatus.

De Beer made important contributions to education in Belgium. From 1952 till 1972 he taught at the Université Catholique de Louvain, and from 1958 to 1974 at the Katholieke Universiteit Leuven. He also taught soil and rock mechanics from 1962 to 1981 at the State University of Ghent, his *alma mater*. He was very active internationally, attending many international conferences and seminars—usually playing an official role in one capacity or another. It was at Professor De Beer's instigation that the Permanent Coordinating Secretariat was set up in 1973 to coordinate the activities of the three sister geotechnical societies ISRM, IAEG and ISSMGE, and he acted as Secretary for many years.

Not so well known in geotechnical circles are his many contributions to public life. These include his help in administering a technical school and a hospital in his home town of Aalst, and his work for the National Housing Agency. He was chief of staff to the Minister of Public Works and Reconstruction, and later to the Minister of Health. In his homage to De Beer, the Honorary Rector of the State University of Ghent described him as a person of 'matchless energy' with a 48 hour day (Vandepitte, 1982). I had the good fortune to work with him between 1983 and 1987 on a research project concerned with improving the efficiency of steel H-bearing piles. Throughout the project I witnessed for myself De Beer's enthusiasm, intense curiosity and unquenchable desire to understand the mechanisms governing the behaviour of the piles in relation to the results of in situ cone tests—passions that he exhibited all of his life. His obituary was published in Vol. 44, No. 4 of *Géotechnique*.

Jacques Florentin (1912–1975)

Jacques Florentin (Fig. 8) was a student at the École Polytechnique, after which he graduated as a civil engineer from the École Nationale des Ponts et Chaussées in 1937. His interest in soil mechanics dates from that time, when he was inspired by the teaching of Albert Caquot. After graduation he joined the staff of the Laboratoire du Bâtiment et des Travaux Publics (LBTP) where, under the genial guidance of Armand Mayer, he contributed to the development in France of laboratory-based research into the physical behaviour of soils.

After the Second World War he was appointed assistant director of the LBTP together with Guy L'Hériteau. As joint authors they published three papers at the Rotterdam Conference. Like De Beer, as a 'Founder', Florentin published a paper with L'Hériteau in the first issue of *Géotechnique* (Florentin & L'Hériteau, 1948). They were soon to leave LBTP to set up a consulting engineering office, which in 1949 became the company Mécasol, and which, within a few years, had established an international reputation.

Both in his time at LBTP and at Mécasol the name of

Fig. 8. Jacques Florentin (1912–1975)

Jacques Florentin is associated with some of the most prestigious French projects of the day. These include the grouting of the dam at Génissiat, Serre-Ponçon (the highest earth dam in France), the barrette foundations for the nuclear power station at Saint-Laurent-des-Eaux, and the foundations for the complex steelworks at Fos-sur-Mer, to name but a few examples in France. At the same time he was involved in many important projects in other parts of the world.

Florentin had an abiding passion for soil mechanics, and was able to combine it with his knowledge of physics, chemistry and geology in tackling the practical problems encountered in the design and construction of civil engineering works.

He was a founder member of the French National Society in 1948, and remained on the Council, becoming successively Vice-President and the President in 1973. At the same time he was a member of the Executive Council of the French National Society of the International Congress on Large Dams. He also gave generously of his time to ISSMGE.

I was to share with Jacques Florentin the task of preparing the State of the Art Report for Session 2 of the 1977 Tokyo International Conference together with Professors Bengt Broms and Victor de Mello. We had a delightful dinner together to begin to plan the report, but sadly he died shortly after that.

Emmericus Carel Willem Adriaan Geuze (1906–ca. 1987)

Known to his friends as Wim, Geuze (Fig. 9) was born on 27 March 1906 in Dordrecht, The Netherlands. After leaving high school in 1923 he studied civil engineering at the Technical High School Delft (THD), two of his teachers being Professor van Mourick Broekman (coastal and river engineering) and Professor Keverling Buisman (mechanics). After graduation he stayed on to work under Buisman (who

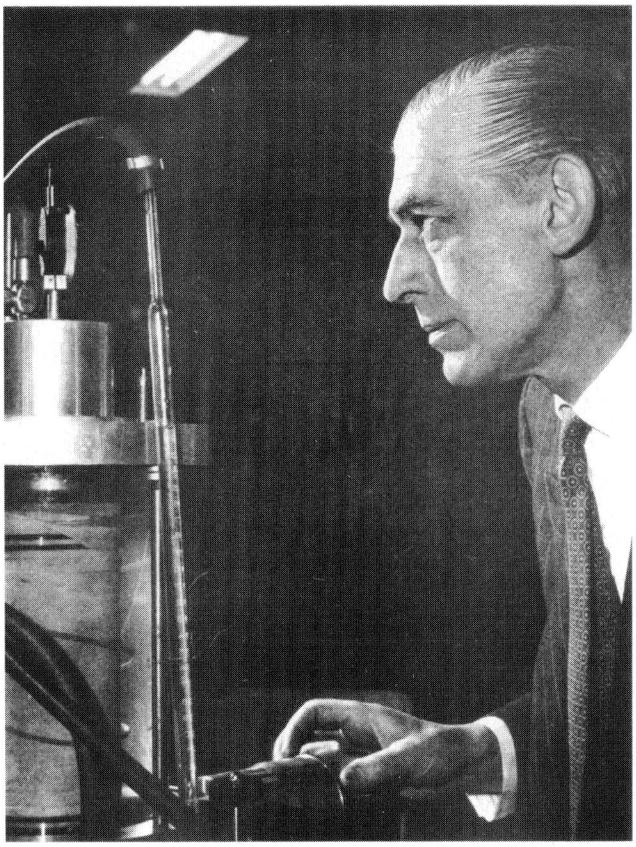

Fig. 9. Emmericus Carel Willem Adriaan Geuze (1906–ca. 1987)

sadly perished in a prisoner of war camp in Indonesia in 1942). When the Laboratorium voor Grondmechanica Delft (later called GeoDelft) was established in 1936, Geuze was a founder member of staff, becoming Head of the Research Department in 1949. He was appointed as a lecturer in soil mechanics at THD in 1946, after which he was promoted to the position of Professor in 1952—an exceptional appointment at the time in recognition of his strong links with geotechnical practice.

Professor de Josselin de Jong remembers Wim Geuze as a charming, friendly and persuasive man who was fluent in four languages—Dutch, English, French and German. An outstanding lecturer, popular with his students, he won a best teacher award. He was also an accomplished saxophonist, and was a founder member of the Dutch Swing College Band. During the Second World War he was arrested and deported to Germany to work in a labour camp. He escaped, returning to Holland hidden in a truck.

Both Golder (1969) and Glossop (Cooling *et al.*, 1975) refer to Geuze as a most entertaining person. They were with Geuze on that momentous night in late 1946, first at a restaurant and later in a night-club, when the suggestion was first mooted of starting a specialist journal: in Golder's words '*Géotechnique* was conceived in a night-club'. Golder (Cooling *et al.*, 1975) also relates another occasion at the 1948 Rotterdam Conference when Professor and Mrs Geuze entertained about 20 guests (including Terzaghi) to an evening *rijsttafel* at their home. Terzaghi not only kept everyone entertained with stories until about 2 a.m., but also finished the Geuzes' Dutch gin!

Geuze's early research was on the study of the flow of groundwater through dykes using a Hele-Shaw model, and on the Dutch cell test, originally designed by Buisman. He is best known for his studies of the cone penetration test (CPT) and the critical density of sands. He performed tests on various sizes of cone, ranging from 3 cm to 15 cm, in order to gain an understanding of the scaling rules. Professor de Josselin de Jong relates that providing heavy enough transportable reactions required a lot of creativity! On the subject of transport, Geuze owned a stylish car but did not have a driving licence. He therefore employed a chauffeur, who would drive him to various investigation sites, often loaded up with testing equipment.

Geuze took an active part in the organisation of the 1948 Rotterdam International Conference, in which he published four papers. He was, in fact, Secretary of the Organising Committee of the parallel conference on Underground Town Planning. He also contributed a paper to the 1950 London Shear Strength Conference that was published in Volume 2 of *Géotechnique* (Geuze & Tan Tjong Kie, 1950). This paper, together with the one by De Beer (1950), gave rise to the heated discussion mentioned earlier over the differences between the Dutch cell test and the triaxial test. It was Geuze who responded to this discussion (Geuze, 1950).

In 1960 Geuze was sent to Jordan by the Dutch Ministry of Foreign Affairs to set up the Arab Potash Company, which ultimately grew to be a large industry. He then emigrated to the USA, where he eventually joined the permanent staff of Rensselaer Polytechnic Institute, Troy, New York State. He is believed to have died in the late 1980s.

Robert Haefeli (1898–1978)

Robert Haefeli (Fig. 10) was born in Lucerne, Switzerland, in August 1898, and loved the Alps and the open-air life. He revelled in rock climbing and skiing. When they met during that memorable summer of 1946 he and Bill Ward (see the accompanying vignette) would have discovered that they had much in common because of their shared interests in rock climbing and glaciology as well as in soil mechanics.

Fig. 10. Robert Haefeli (1898–1978)

Haefeli graduated in civil engineering from the Swiss Federal Institute of Technology (also known as the Eidgenössische Technische Hochschule ETH in Zurich) and worked for a short period as a bridge engineer. Then followed some years abroad, particularly in Spain, where he was engaged in work on water power development schemes. His interest in soil mechanics and the mechanical properties of snow and ice were kindled as a result of working on the construction of dams in the Alps. In 1935 he established a soil mechanics laboratory at ETH Zurich, where he carried out fundamental studies on the shearing resistance of soil, snow and ice. Indeed his doctoral thesis, published in 1939, was entitled *Snow mechanics with reference to soil mechanics*. He was appointed a professor at ETH Zurich in 1947, and lectured on soil and snow mechanics together with avalanche mitigation and preventive measures.

Haefeli wrote a large number of papers: a full bibliography is given in the obituary published in *Géotechnique* (**29**, No. 1, 103–110). The Proceedings of the Rotterdam Conference contain no less than six of his papers. Reference is made in the *Géotechnique* obituary to his persistence, his untiring energy, and his ability to tackle problems in a fundamental, scientific manner. This approach is amply demonstrated in his seminal paper 'Investigation and measurements of the shear strengths of saturated cohesive soils', which he presented at the 1950 Shear Strength Conference in London (Haefeli, 1951), and which was published in *Géotechnique* along with the other papers to the conference. The paper summarises the developments that took place at ETH, under his direction, in the understanding of the fundamental strength of clays. Even today, a study of the paper is rewarding, in that it deals with a wide range of phenomena, including residual strength, progressive failure and creep.

The above paper acknowledges the contributions of a certain Mr L. Bjerrum, who studied at ETH between 1947 and 1951. Bjerrum had intended to work on snow mechanics. However, Haefeli, who tragically suffered lengthy periods of depressive ill health, was unable to supervise him in this work, and Bjerrum changed to geotechnical studies (Flaate *et al.*, 2003). Bjerrum also presented a seminal paper at the London Shear Strength Conference, describing the results of the research he had carried out at ETH in Haefeli's laboratory (Bjerrum, 1951).

In recognition of his interest in glaciology, Haefeli was invited to write a paper for *Géotechnique* on snow mechanics, but sadly this did not materialise—perhaps for the health reasons mentioned previously. However, one did appear in the *Journal of the Institution of Engineers India* (Haefeli, 1952). It was largely due to Haefeli that a major expedition was organised to study the Greenland ice sheet. The rheology of the Greenland ice sheet engaged his thinking right up to his death. An obituary describing his contributions to glaciology was published in the *Journal of Glaciology* (de Quervain, 1979). It is very evident from the *Géotechnique* and *Journal of Glaciology* obituaries that Haefeli was much loved and greatly respected by his colleagues. Dr John Nye FRS remembers him as a most endearing, enthusiastic and somewhat eccentric person, referred to by his Swiss colleagues as 'our Haefeli'.

Armin von Moos (1907–1981)

Armin von Moos (Fig. 11) was born on 20 February 1907 in Wülflingen, Switzerland. He began his studies at the University of Zurich in 1926, and later studied at the Swiss Federal Institute of Technology (ETH). Like many students of geology at that time, after graduation he had intended to travel to foreign countries to conduct field studies for an oil exploration company. However, in 1928 a severe accident on a geological field trip to the Swiss Lugnez Valley thwarted his plans. He slipped on a schistose slab and fell down a rock face. During the following two years he had to undergo several operations, but his hip joint was to remain stiff for the rest of his life.

Fig. 11. Armin von Moos (1907–1981)

In 1931 he spent three months at the Royal School of Mines, Imperial College, London, where, under the guidance of the professor of geology, Percy Boswell, he made himself familiar with the current techniques of sedimentary petrography. One year later, after graduating in geology from the ETH in Zurich, he became an assistant lecturer to Professor N. Niggli, who encouraged him to carry out sedimentary petrographical investigations of heavy minerals in the Molasse sandstone. His doctoral thesis on this subject was published in 1935. Having completed his PhD in geology, and still restricted by his poor health, he had to face additional adversities during the world economic crisis of the 1930s. He therefore gladly accepted Professor Niggli's proposal to contribute his expertise in petrographical methods in helping to set up a soil mechanics laboratory. The laboratory was established at the ETH in 1935, in close collaboration with the civil engineer Robert Haefeli (another 'Founder') under the guidance of Professors Meyer-Peter and Niggli. This marked the beginning of an enriching and fruitful collaboration between geology and civil engineering that was to last over 20 years. Although in the beginning the laboratory activities were mainly scientific, over the years they became more and more oriented towards engineering practice.

Von Moos made important contributions to the education of engineering geologists in Switzerland. In 1948 he was formally appointed as a university lecturer at the ETH, where, until 1968, he taught engineering geology, site investigation and hydrogeology. Like his friend Glossop in England, von Moos always promoted close collaboration between geologists and geotechnical engineers. Indeed, with Bjerrum as a co-author, von Moos published a paper at the 1948 Rotterdam Conference setting out clearly the benefits of close collaboration between geotechnical engineers and geologists (von Moos & Bjerrum, 1948). He had been provoked into writing the paper because the word 'geology' had not appeared in the invitation bulletin of the conference! His passion was recognised and appreciated by K. Terzaghi, who wrote in a letter to von Moos dated March 1949:

> I always enjoy hearing about the intimate cooperation between Soil Mechanics and Engineering Geology in Switzerland and I wish you all success in connection with your interesting assignments.

In 1953 the Third International Conference on Soil Mechanics and Foundation Engineering was held in Zurich with von Moos as General Secretary and also Editor of the Proceedings.

As a co-founder and the first President of the Swiss Society of Soil Mechanics and Foundation Engineering, which was established in 1956, Armin von Moos maintained close professional contacts with this organisation during the whole of his life. From 1966 to 1970 he acted as Vice President of the International Society for Rock Mechanics.

Armin von Moos published many papers (mainly in German), on both geological and geotechnical topics. In the first issue of *Géotechnique* he published a comprehensive review of the literature on engineering geology in Switzerland (von Moos, 1948). He also wrote several books, including a well-reviewed one on technical petrography with Professor de Quervain as co-author (von Moos & de Quervain, 1948).

It was no easy decision for him to make when, at the end of 1955, von Moos left the Swiss Federal Institute of Technology to take on the new challenge of setting up a geological consulting office. Thanks to his well-founded and extensive technical experience, the venture proved to be very successful. Not only was his company one of the first in Switzerland at which geologists and engineers worked side by side in tackling challenging geotechnical problems, but 50 years later it remains a leading international consultancy.

TWO FOUNDER MEMBERS OF THE *GÉOTECHNIQUE* ADVISORY PANEL

There are two engineers who joined the Advisory Panel of *Géotechnique* when it was formed in 1949 and who played a significant part in the early success of the journal. These are W. K. Wallace and Guthlac Wilson. No account of the 'Founders' would be complete without including them.

William Kelly Wallace (1883–1969)

In the various accounts of the beginnings of *Géotechnique* frequent mention is made of the name of Mr W. K. Wallace (Fig. 12). In particular, not only was he the first Chairman of the *Géotechnique* Advisory Panel after it was taken over by ICE in November 1949 but he was also the first Chairman of the British Section of the ISSMFE (see Cooling *et al.*, 1975). Yet he is not well known in soil mechanics circles.

Wallace was born in Belfast, and was educated privately. In 1903, after a three-year period of practical training, he started his long and varied career in railway engineering. Between 1904 and 1930 he worked for the Belfast and Northern Counties Railway, becoming Chief Engineer in 1924. In 1930 he was transferred to London, and in 1933 he became Chief Civil Engineer of the London, Midland and Scottish Railway, a position he retained until his retirement in 1948.

Fig. 12. William Kelly Wallace (1883–1969)

Wallace was remarkably active within the Institution of Civil Engineers, and became President in 1955, having served for 11 years on its Council, during which time he probably served on all its standing committees and many of its *ad hoc* committees. For his Presidential Address he gave a fascinating account of the geotechnical challenges and difficulties faced by Robert Stephenson (son of George Stephenson) in the construction of the London to Birmingham railway (Wallace, 1956). Graphic descriptions are given of slope instabilities, embankment failures and tunnelling problems, together with the solutions adopted by Stephenson. The paper is clearly the work of a civil engineer with considerable experience and knowledge.

A flavour of the man can be glimpsed from remarks made at the beginning and end of his Presidential year. In the vote of thanks to Wallace for his Presidential Address, Mr A. C. Hartley (Vice President) remarked that

> he had had the pleasure of serving with him on some of his many committees and had always enjoyed the way in which his remarks had enlivened and encouraged the proceedings...

He went on to say that under his chairmanship he

> had always admired very greatly the extremely efficient way in which he had conducted the affairs of the meeting and got the right answers in the minimum of time.

In his vote of thanks to Wallace at the end of his Presidency, Sir Herbert Manzoni remarked that

> he would remember Mr Wallace, and he thought that most of those who had sat at his feet in Council or Committee or on other occasions would remember him, by an almost paternal kindness and wisdom, usually put over with a delightful humour, and because he had instilled into his presidential year a quite unusual element of personality

which had commanded not only gratitude but also a considerable amount of affection.

This vote of thanks was seconded by Mr H. Shirley Smith, who said that

> [he too had] been present at meetings presided over by Mr Wallace for a number of years, and ... all had been impressed by the twinkle in his eye and by his amusing Irishisms and asides, which masked an extraordinary degree of determination and considerable shrewdness.

Before the Second World War, Mr Wallace served on the Sub-Committee on Earth Pressures of ICE's Research Committee: this Sub-Committee had recently taken over the work of the British Association Earth Pressures Commission (Cooling *et al.*, 1975). In the ICE *Proceedings* of 1938/39 (Vol. 11, p. 172) it is reported that the programme of experimental work of the Sub-Committee had been considerably extended by the financial support of four main-line railway companies and the London Transport Board, amounting to £1000 per annum for three years. There can be no doubt that Wallace, in his capacity as Chief Civil Engineer of the London, Midland and Scottish Railway, was very influential in this decision.

After the Second World War ICE set up a small *ad hoc* committee under the chairmanship of Wallace to discuss the resumption of the work of the Earth Pressures Sub-Committee. This *ad hoc* committee recommended to the Council of ICE the establishment of a Soil Mechanics and Foundations Committee. As described by Skempton (Cooling *et al.*, 1975) it was the establishment of this Committee that eventually led to the formation of the British Section of the ISSMFE in 1949 with Wallace as its first Chairman—the forerunner of the present British Geotechnical Association.

Brown (1982) stresses the close relationship that existed between Wallace and the Soil Mechanics Section at the BRS. He lived near the BRS, and involved himself in their work in various ways. From 1949 to 1954 he was Chairman of the Building Research Board of the Department of Scientific and Industrial Research, and in that capacity alone would have had very close contact with BRS at the highest levels. Skempton (1948), in his paper to the Rotterdam International Conference *The rate of softening of stiff fissured clays, with special reference to London clay*, gives an acknowledgement to the Chief Engineer of the London and Scottish Railway.

As originally suggested by Brown (1982), it is now clear from the minutes of ICE Council that Wallace played a key role in persuading ICE to take over the publication of *Géotechnique* in October 1949 (Brown, 2008). The first meeting of the Advisory Panel took place in November 1949 with Wallace as Chairman. The importance that Wallace attached to soil mechanics is reflected in the fact that he retained his chairmanship of the *Géotechnique* Advisory Panel during his year of office as President of ICE. However, it seems that to retain his chairmanship of the British section of the ISSMFE as well was too much to expect, and on 20 January 1955 a resolution was passed recording 'the appreciation of the Committee and the British National Society to Mr Wallace for his work in founding the Society and undertaking the chairmanship for six years'. It is clear from the foregoing that Wallace played a key role in guiding the evolution of the organisations that enabled post-war British soil mechanics to thrive.

In 1946 Wallace was appointed Commander of the Order of the British Empire (CBE), and in 1956 he was awarded a DSc in Applied Technology (Honoris Causa) from Queen's University, Belfast. He died on 23 May 1969, and his obituary is published in the *Proceedings of the ICE*, Vol. 43, pp. 525–526.

Guthlac Wilson (1902–1953)

Like Wallace, the name Guthlac Wilson (Fig. 13) keeps appearing in the early history of *Géotechnique*. Although not a 'Founder' in the formal sense, he played an important role, being on the Advisory Panel when ICE took on the running of the journal.

Guthlac Wilson was educated privately, and studied engineering at East London College (now Queen Mary College, London) where he graduated with first-class honours in 1921. After serving under articles with Sir Robert McAlpine and Sons, he worked for some years for Braithwaite and Co. in India, where he was responsible for the design and construction of several bridges. His submission to ICE for membership shows that he was much involved in the design and installation of screw piles at this time.

In 1938 he went Harvard University to study for a masters degree in soil mechanics under Casagrande and Terzaghi. A fellow student on the course with him was Henry Grace, who later became a partner in Wilson's firm. At Harvard, Guthlac Wilson was named as outstanding student of the year, and was selected by Terzaghi to act as his personal assistant on consultancy work. But the outbreak of the Second World War forced his return to Britain. Folklore has it that an august elder of ICE asked him: 'Where have you been, Wilson?' 'I have been in the USA studying soil mechanics under Professors Casagrande and Terzaghi at Harvard' came the reply. 'That will be of little use to you, Wilson! We have such good soils in England!'

On his return to Britain, after being engaged on the

Fig. 13. Guthlac Wilson (1902–1953)

design of a Royal Ordnance factory with Sir Alexander Gibb and Partners, Wilson was appointed Director of Constructional Design for the Ministry of Works and Buildings. In 1943 he went into private practice and established his own soils laboratory. Two years later he joined W. L. Scott to form the firm of Scott & Wilson (now Scott Wilson), becoming the senior partner after Scott's death in 1950.

The firm grew rapidly, both in Britain and overseas. Perhaps Guthlac Wilson's greatest achievement was the structural design, with his partners W. L. Scott and E. O. Measor, of the Royal Festival Hall in London, which was opened in 1951. With Henry Grace, Guthlac Wilson became deeply involved in soil mechanics work in the Far East and Africa. Projects included the runway for Kai Tak Airport in Hong Kong, the Blantyre-Zomba and Blantyre-Luchenza roads, Chileka airport, and the New Mudi dam. Guthlac Wilson was also responsible for the design of major foundation works in many parts of the world. On 29 March 1953, while returning to England with his wife from a trip to Hong Kong, and en route from Nairobi to Blantyre, the aircraft crashed, tragically killing all on board.

Guthlac Wilson was a dynamic and brilliant engineer, who threw himself into whatever project he was involved in with huge enthusiasm and intensity. While with the Ministry of Works and Buildings he was a member of all three principal British soil mechanics codes of practice drafting committees: site investigation, earth-retaining structures and foundations. In 1943 he published, with Henry Grace, the seminal paper on the settlement of London due to underdrainage (Wilson & Grace, 1942). Golder (Cooling *et al.*, 1975) tells how arguments about the paper at Skempton's house ended with Guthlac entering Skempton's bedroom at 3 a.m., shaking him awake, and exclaiming 'You are quite right, Skem!' In 1950 he published a major paper on the bearing capacity of screw piles, in which he drew on his early experiences in India (Wilson, 1950). I recall devouring this paper when attempting to understand the mechanisms of behaviour of under-reamed bored piles. At the Rotterdam Conference Guthlac Wilson published two papers and a report on his laboratory at Scott & Wilson. He was General Reporter for the sessions on 'Problems in road and runway construction' and 'Improvements of the mechanical properties of the soil'. Both reports were very thorough and critical, provoking discussion from the floor. He was elected to the Council of ICE in 1949, and was awarded the DSc degree by the University of London in 1951. The early minutes of the *Géotechnique* Advisory Panel refer to a paper to be prepared by Guthlac Wilson on the application of Mohr's Circles. We are left to speculate on its contents, but there can be no doubt that it would have been rigorous and practical. Obituaries of Guthlac Wilson have been published in *Géotechnique*, Vol. 3, No. 6, 265–266 and in the *Proceedings of the ICE*, General, Part 1, Vol. 2, 372–373.

CONCLUDING REMARKS

The conception and birth of *Géotechnique* resulted from a recognition of the practical importance of geotechnics, the challenge of developing the underpinning experimental and theoretical disciplines, and above all the strong desire to share experience. It would not have happened without the boundless enthusiasm, ability and passion of a number of outstanding individuals who were able to combine in a most remarkable way. It is hoped that these vignettes will serve to inspire present and future generations of geotechnical engineers, not only in their professions, but also in their enjoyment of working with fellow enthusiasts.

ACKNOWLEDGEMENTS

I am indebted to the many people who contributed reflections, memories and information. In particular I would like to thank the following: Frans Barends, Christine Cooling, Francois Descoeudres, Pierre Florentin, Roger Frank, Ann Grace, Evert Hoek, Noël Huybrechts, Bob McKittrick, Victor Milligan, John Nye, Beat Rick, Sarah Springman and Ron Williams. Mike Chrimes and ICE Library staff kindly assisted in finding published obituaries and minutes of various ICE committees. Stephen Brown and Jamie Standing offered very helpful suggestions.

REFERENCES

Bjerrum, L. (1951). Fundamental considerations on the shear strength of soil. *Géotechnique* **2**, No. 3, 209–218.

Brown, S. F. (1982). The development of *Géotechnique* 1948–1980. The first 30 volumes. *Géotechnique* **32**, No. 2, 95–110.

Brown, S. F. (2008). *Géotechnique* 1948 to 2008: more than just a journal. *Géotechnique* **58**, No. 5, 315–326.

Burland, J. B. (2005). Soil mechanics at Emma: elegant, rigorous and relevant. *Emmanuel College Magazine 2004–2005* **87**, 154–174.

Cooling, L. F. (1970). Correspondence. *Géotechnique* **20**, No. 1, 107–109.

Cooling, L. F. (1975). British Geotechnical Society Twenty-Fifth Anniversary Report. *Géotechnique* **25**, No. 4, 629–656.

Cooling, C. M. (1996). Obituary of W. H. Ward. *Proc. Instn Civ. Engrs. Geotech. Engng* **119**, No. 4, 246–247.

Cooling, L. F. & Golder, H. Q. (1940). A portable apparatus for compression tests on clay soils. *Engineering* **149**, 56–58.

Cooling, L. F. & Golder, H. Q. (1942). The analysis of the failure of an earth dam during construction. *J. Instn Civ. Engrs* **19**, No. 1, 38–55.

Cooling, L. F. , Skempton, A. W. , Glossop, R. & Golder, H. Q. (1975). British Geotechnical Society Twenty-fifth Anniversary Report. *Géotechnique* **25**, No. 4, 629–656.

Daxelhofer, J. P. (1944). Remarques sur la résistance au cisaillement des sols et son importance dans quelques cas particuliers. *Bulletin Technique de la Suisse Romande, Lausanne*, Nos. 25 et 26.

De Beer, E. E. (1950). The cell-test. *Géotechnique* **2**, No. 2, 162–172.

De Beer, E. E. (1948). Données concernant la résistance au cisaillement déduites des essais de pénétration en profondeur. *Géotechnique* **1**, No. 1, 22–39.

de Quervain, M. R. (1979). Obituary of Robert Haefeli, *J. Glaciol.* **22**, No. 87, 405–407.

Flaate, K., DiBiagio, E. & Senneset, K. (eds) (2003). *Laurits Bjerrum: more than engineer.* NGI, Oslo, pp. 37–39.

Florentin, J. & L'Hériteau, G. (1948). Remarques sur quelques marnes fortement préconsolidées. *Géotechnique* **1**, No. 1, 59–65.

Geuze, E. C. W. A. (1950). Reply to discussion on 'The shearing properties of soils'. *Géotechnique* **2**, No. 2, 181–182.

Geuze, E. C. W. A. & Tan Tjong Kie (1950). The shearing properties of soils. *Géotechnique* **2**, No. 2, 141–161.

Gibson, R. E. (1969). Obituary of H. Q. Golder. *Géotechnique* **41**, No. 3, 475–477.

Gibson, R. E. (1984). Working with Skempton. In *Selected Papers on Soil Mechanics by A. W. Skempton*, pp. 9–11. London: Thomas Telford.

Glossop, R. (1968). The rise of geotechnology and its influence on engineering practice. Eighth Rankine Lecture. *Géotechnique* **18**, No. 2, 107–150.

Glossop, R. (1984). A short biographical essay. In *Selected Papers on Soil Mechanics by A. W. Skempton*, pp. 5–8. London: Thomas Telford.

Glossop, R. & Golder, H. Q. (1948). The shear strength method of determination pavement thickness. *Proc. 2nd Int. Conf. Soil Mech. Found. Engng, Rotterdam* **4**, 164–167.

Golder, H. Q. (1969). Correspondence, 'In the beginning'. *Géotechnique* **19**, No. 2, 313–315.

Haefeli, R. (1951). Investigation and measurements of the shear strengths of saturated cohesive soils. *Géotechnique* **2**, No. 1, 186–208.

Haefeli, R. (1952). Snow mechanics. *J. Inst. Engrs India* **32**, No. 3, 138–141.

Henkel, D. J. (1978). Introduction to the Eighteenth Rankine Lecture. *Géotechnique* **28**, No. 2, 134–171.

Milligan, V. (2007). Private correspondence.

von Moos, A. (1948). Engineering geology in Switzerland. *Géotechnique* **1**, No. 1, 40–48.

von Moos, A. and Bjerrum, L. (1948). Soil mechanics and geology. *Proc. 2nd Int. Conf. Soil Mech. Found. Engng, Rotterdam* **3**, 1–3.

von Moos, A. & de Quervain, Fr. (1948). *Technische Gesteinkunde*. Basel: Birkhäuser.

Niechcial, J. (2002). *A particle of clay: The biography of Alec Skempton, civil engineer*. Caithness: Whittles Publishing.

Nye, J. (1996). Letters of appreciation: Obituary of William Hallam Ward. *Géotechnique* **46**, No. 3, 576–585.

Peck, R. B. (1977). Vignettes of four Presidents, 1936–1969. *Proc. 9th Int. Conf. Soil Mech. Found. Engng, Tokyo* **3**, 285–290.

Penman, A. D. M. (1996). Obituary of William Hallam Ward. *Géotechnique* **46**, No. 3, 576–585.

Skempton, A. W. (1948). The rate of softening of stiff fissured clays, with special reference to London clay. *Proc. 2nd Int. Conf. Soil Mech. Found. Engng, Rotterdam* **2**, 50–53.

Skempton, A. W. (1977). Obituary of L. F. Cooling. *Géotechnique* **27**, No. 2, 265–270.

Skempton, A. W. (1993). Obituary of Rudolph Glossop. *Géotechnique* **43**, No. 4, 623- 625.

Vandepitte, D. (1982). *Life and work of Em. Prof. Dr Eng. De Beer*, pp. 9–16. Brussels: Comité d'hommage au Professeur E. De Beer.

Wallace, W. K. (1956). Presidential Address. *Proc. ICE* **5**, Part 1, 8–22.

Ward, W. H. (1948). A slip on a flood defence bank constructed on a peat bog. *Proc. 2nd Int. Conf. Soil Mech. Found. Engng, Rotterdam* **2**, 19–23.

Ward, W. H. & Sewell, E. C. (1950). Protection of the ground from thermal effects of industrial plant. *Géotechnique* **2**, No. 1, 64–81.

Ward, W. H. , Burland, J. B. & Gallois, R. W. (1968). Geotechnical assessment of a site at Mundford, Norfolk, for a large proton accelerator. *Géotechnique* **18**, No. 4, 399–431.

Ward, W. H. , Tedd, P. & Berry, N. S. M. (1983). The Kielder experimental tunnel: final results. *Géotechnique* **33**, No. 3, 275–291.

Williams, R. E. & Norbury, D. (2008). Rudolph Glossop and the development of 'Geotechnology'. *Q. J. Engng Geol. Hydrogeol.*, in press.

Wilson, G. (1950). The bearing capacity of screw piles and screwcrete cylinders. *J. Instn Civ. Engrs* **34**, No. 5, 4–93.

Wilson, G. & Grace, H., (1942). The settlement of London due to underdrainage of the London Clay. *J. Instn Civ. Engrs* **19**, No. 2, 100–115.

Chrimes, M. (2008). *Géotechnique* **58**, No. 5, 343–355 [**doi:** 10.1680/geot.2008.58.5.343]

Geotechnical publications before *Géotechnique*

M. CHRIMES*

Geotechnical engineering emerged as a recognised engineering discipline in the late 1930s. This paper reviews its development over the previous 200 years by examining the nature and shortcomings of the literature of geotechnics, and analysing trends. It looks at the nature of exchange of information in the period, and barriers to effective advancement in the subject before 1948. The absence of an effective international journal in the field before *Géotechnique* is revealed.

KEYWORDS: historical review

Le génie géotechnique a vu le jour, en tant que discipline technique reconnue, vers la fin des années 1930. Dans la présente communication, nous passons en revue son évolution au cours des derniers 200 ans, en examinant la nature et les lacunes des ouvrages sur la géotechnique, et en analysant les tendances. Nous nous penchons également sur la nature des échanges d'information au cours de cette période, et des obstacles à la progression efficace de cette discipline avant 1948, et nous révélons l'absence d'un authentique journal international jusqu'à l'avènement de *Géotechnique*.

INTRODUCTION

Modern geotechnics developed in the period 1900–1940 in Scandinavia, the USA, and central Europe, with isolated but significant developments in India and Japan. However, in geotechnical engineering historically one is confronted with a paradox of the practical solution of many problems before they had been 'properly' analysed, and the successful execution of works that subsequent investigation might have forbidden. As Peck relates in his discussion of the gestation of *Soil mechanics in engineering practice* (Terzaghi and Peck, 1948):†

> After much painful assessment it rather suddenly came to our minds that much of the practice of foundation and earthwork engineering actually had a quite satisfactory empirical basis, sometimes stretching back for generations, and that this empirical basis contained a vast amount of knowledge that the engineer could and still should use. Soil mechanics could point out the circumstances under which the body of empirical knowledge might be unreliable and could furnish the means for coping with problems for which traditional practice would be unsuitable. (Peck, 1993)

Although there were a number of forums for engineering debate by the late nineteenth century, it was not until the 1920s that there was any real international exchange between experts in the geotechnical field. In fact, in the 1930s there was even a degree of hostility in debate that arguably hindered progress in the subject.

The year 1948 represents a milestone in geotechnical engineering. In that year the Second International Conference on Soil Mechanics was held in Rotterdam. This revealed that, despite the Second World War, enormous advances had been made in the subject since the 1936 Harvard Conference, and it had become a significant discipline within civil engineering. In June of that year *Géotechnique* was published by a group of British engineers as the first international journal in the field. That its home was in the UK bears great testimony to the way in which the British engineering community had embraced the discipline since May 1939, when Terzaghi had given his James Forrest Lecture and ICE President W. J. E. Binnie had remarked: 'In Great Britain rather slow progress had been made in soil research . . .' (Terzaghi, 1939).

Table 1 details early serial publications in the geotechnical field, and demonstrates that before *Géotechnique* only research reports associated with specialist bodies were available, with no refereed journals capable of acting as an international forum for debate and advancing the subject. This paper will demonstrate how the history of the subject reveals the need for such a journal, and thus the historical significance of *Géotechnique*'s foundation.

EARLY ORIGINS

It is evident that ancient civilisations had design methods and administrative systems for what would be now considered civil engineering work: for instance, the Romans and their road network (Chevallier, 1989; Davies, 2002). One struggles, however, to trace any continuity with modern practice. In that sense, in Europe, a strong case can be made for the medieval hydraulic engineers of the Low Countries being the first practical geotechnical engineers. There, in the period 800–1250, a large area was drained and enclosed by dikes, and administrative bodies responsible for their construction and maintenance emerged in the Middle Ages. 'Dutch' drainage and hydraulic engineers worked in other Lowland areas of Europe. A sixteenth-century manuscript (De Hullu & Verhoeven, 1920) by one of the leading Dutch hydraulic engineers, Andres Vierlingh, described practice standard at that time, and that remained so into the nineteenth century. Perhaps this 'Low Countries' practice can be regarded as the first geotechnical code of practice in the West.

Outside Europe are other examples of early standard practice with earth dam profiles established in India, with soil selection and dimensions found on an inscription of 1369 AD (Rao, 1960; Kulkarni, 1985).

THE LITERATURE OF SOIL MECHANICS 1900–1948

Despite this prehistory, there is eloquent testimony to the shortcomings of the literature on soil mechanics in the first part of the twentieth century. Corthell (1907) discovered a lack of consensus among engineers over the safe bearing pressure of soils, and published a paper revealing the range of bearing capacities used in over 70 examples (Sutherland, 2001). In 1917 Terzaghi wrote on 'The inadequacy of

Discussion on this paper closes on 1 December 2008, for further details see p. ii.

* Institution of Civil Engineers, London, UK.

† Most references are available in the ICE Library.

Table 1. Early geotechnical serials

1906–1924	*Geotekniska Meddelanden** (Helsingfors)	Finland
1917–	*Swedish Railway Geotechnical Committee Proceedings*	Sweden
1929–	*Geologie und Bauwesen* (now *Rock Mechanics*)	Germany
1931–1938	Japan Government Railways Geotechical Committee reports	Japan
1933–	*Degebo Mitteilungen*	Germany
1948–	*Géotechnique*	UK
1948–	Inst. Tech. Bat. Trav. Pub. Annales: *Sols et fondations*	France
1950–	Royal Swedish Geotechnical Institute Publications	Sweden
1954–	*Geotecnica*	Italy
1956–	*DGI Bulletin*	Denmark
1956–	*Journal of Soil Mechanics and Foundations Division ASCE* (*Proceedings*, Vol. 82, No. SM1)	USA
1959–	*Osnovaniya, fundamenty I mechanika gruntov*	USSR
1960–	*Soils and Foundations*	Japan

* The meaning seems to be 'technical properties of earth materials', and the reports are generally concerned with, for example, properties of clay for brickmaking, the lime industry.

published construction descriptions...' (Terzaghi, 1917), in which he railed against published literature for failing to provide data that would enable an experienced engineer to design a similar structure.

The situation to which Corthell and Terzaghi allude is indicated by Tables 2 and 3. At the start of the twentieth century subject coverage and interest were patchy. There was an enormous literature on earth pressure and retaining wall design, but in English there were hardly any textbooks, even on the practical aspects of foundation design. Nonetheless, one feels that Terzaghi's critique underestimates the valuable case studies and collective knowledge available. The literature on dams and tunnels (Table 2) is overwhelmingly project related. Technical journals had been published since the 1790s, and civil engineering journals since the 1820s (Chrimes, 1995). Geotechnical content is reflected in Table 4.

Few collections of nineteenth-century engineers' reports survive to enable a proper assessment of engineering knowledge to be made; further research in archives is still required. However, archives at ICE relating to eighteenth- and nineteenth-century engineers are indicative of informed practice, exemplified by the evolution of embankment dam design (Binnie, 1981, 1987; Skempton, 1990). Without this knowledge resource one doubts whether the rapid progress of the pioneers could have been made.

That progress was subsequently uneven is indicated in Glossop's (1975) remarks on the situation in 1930s Britain.

> There was no adequate textbook of soil mechanics in the English language, the only sources of information were the *Proceedings of the First International Conference*, held at Harvard in 1936; this was unobtainable and one was lucky to have access to a copy as I had at BRS. There were also the valuable papers in the *Proceedings of the Boston Society of Engineers*, by Terzaghi, Arthur Casagrande, and others ...'

It is certainly true that the situation then was very different from today, when specialist journals are produced in many countries and in most languages, and textbooks are available from competing publishers and authors. However, an enormous body of literature was produced in the first 40 years of the twentieth century, particularly in the US and in the German language (Tables 5 and 6). Much was available at ICE and the BRS.

Glossop's statement about the availability of the 1936 conference is misleading. Two copies were acquired by ICE in 1936, one of which was loaned out over 40 times in the period 1937 to 1942. On the other hand, a review of the British Library holdings of early twentieth-century authors such as Terzaghi suggests the subject had yet to be recognised by that great institution, and anybody without access to the ICE would have depended on personal knowledge and contacts.

Of the early twentieth-century contributors, Terzaghi is the leading author, but many contributions were made by others generally forgotten today, notably Hogentogler with 33 publications between the wars, Loos 25, Koegler 23, Press with 17, Housel 14, Gilboy 15, and Krynine 26, and a large number in Russia. There is an abundance of evidence of the ignorance of other work in the field, even among academics. When the ASCE Soil Mechanics and Foundation Division compiled a bibliography in the late 1930s, 945 publications were initially nominated by the 26-strong advisory board; only 320 had more than one sponsor (ASCE, 1940). Only two had been identified by 50% of the contributors, and only 43 by 25%, suggesting parlous ignorance even among experts, or the disparate nature of the subject. Today one might assume Terzaghi's *Erdbaumechanik* (Terzaghi, 1925a) to be the seminal work. Clearly it was not seen as such by his contemporaries.

French writers on earth pressure show virtually no awareness of work in the German language. Awareness among Germans of French work is better, but even Terzaghi was unaware of the work of the Scandinavian pioneers until 1924, and his *Erdbaumechanik* (Terzaghi, 1925a) makes reference only in an appendix. In his review of French work on uplift in the early twentieth century, Leliavsky (1958) provides another example.

> It might appear surprising that in 1928, Fillunger's ideas were still unknown to French designers, but this is by no means the only instance of the striking lack of interest in foreign literature on the part of even the most advanced technical scientists.

However, Wilson‡ (Institution of Structural Engineers, 1943) had a comprehensive knowledge of the literature, as indicated by the bibliography of 494 references he produced for the committee that drew up the UK's first code of practice for earth-retaining structures.

In the period 1870 to 1914 all ICE members received

‡ See Burland (2008) in this issue.

Table 2. Geotechnical publications in ICE library catalogues

	1700–1750	1750–1799	1800–1809	1810–1819	1820–1829	1830–1839	1840–1849	1850–1859	1860–1869	1870–1879	1880–1889	1890–1899	1900–1909	1910–1919	1920–1929	1930–1939	Totals
Underwater foundations		3	1	2	3	3	2	7	1	7	7	4	0	1	1	1	43
Bearing pressure													2	1	0	1	4
Dynamics											3	1	1	0	0	1	6
Earth pressure/retaining walls		1	2		2		2	2	2	4	12	3	4	6	6	4	50
Tunnels		3	4	1	27	6	4	5	19	41	47	26	13	15	20	20	251
Groundwater								1			1			2	2	5	11
Slope stability							3	1	0	0	0	1	1	0	1	1	8
Dams				1		1	1	1	12	21	13	14	30	27	48	73	242
Piles							1			2	2	1	1		4	7	18
Foundations						1			2	4	3	4	5	3	7	10	39
Specialist techniques						1		1	3	2	5	1	1	2	1	3	20
Plant										1	3	0	1	3	5	2	15
Earthwork tables						2	7	1	4	6	5	2	2	2	6	5	42
Soil mechanics														1	3	3	7
Soil properties															2	8	10
Embankments							2	3	3	4	6	5	5	2	4	8	42
Engineering geology											1			1	8	8	18
Soil tests															2	2	4
Total		7	7	4	32	14	22	22	46	92	108	62	66	66	120	162	

abstracts of relevant papers from the world's engineering literature. Key papers by Forchheimer (1882) and Kurdumov (1892) were abstracted in this way. In the US the Bureau of Reclamation and Corps of Engineers had the resources to translate papers in the 1930s (e.g. Fellenius, 1936). Similar work was done on a smaller scale at UK research laboratories. Technical visits such as those of British roads engineers to the US and Germany in the 1930s resulted in some information exchange. The British Intelligence Objectives Subcommittee reports after the war summarised key German work in areas such as soil stabilisation (BIOS, 1946).

The geotechnical contributions of early Russian and Japanese engineers are hard to trace. In Japan, modern earthquake-resistant design is associated with the work of Sano (1916, 1917). His observations of the 1906 San Francisco Earthquake led him to propose the use of seismic inertia force in design. This was developed in the field of geotechnics by Okabe (1924, 1926) and Mononobe & Matsuo (1929) in their theory of seismic active earth pressure. Dissemination of Okabe's work was hampered because it was in Japanese, but the World Engineering Congress in Tokyo in 1929 must have had some positive influence. The work of the Japanese Railways Soil Mechanics Investigation Committee led to perhaps the first regular soil mechanics publication as their annual report (Table 1).

In Russia in the late 1880s Kurdumov, well aware of the work of Boussinesq, investigated the area of influence of foundations, and with the aid of photography was able to demonstrate the failure curve and develop a formula. The work was published as a lecture in Russian in 1889 and then as a book (Kurdumov, 1891), and summarised in Germany in 1892 (Kurdumov, 1892). Although Terzaghi mentioned this work in 1925 (Terzaghi, 1925a), it had greater influence in Russia, where it was used by Sokolovskii (1960) and others. However, contributions of Russian engineers between the wars are recorded only at the international conference of 1936, and some specialist studies.

Language was of course not the only barrier; the proliferation of engineering journals and societies made it very difficult to keep track of everything published, and the absence of an obvious home for the new subject meant it cropped up everywhere: in established engineering society journals, specialist technical journals on water engineering,§ technical reports of government agencies and research institutes, and, in the case of specialist techniques such as ground treatment, trade literature. Not all of these were abstracted in the US-focused *Engineering Index*, while research interests drove the abstracting activity of research stations such as BRS. Then, as now, personal contact was probably the key. The bibliographic scatter of Terzaghi's publications (Table 7) reveals how difficult it would have been to keep abreast of his thinking, particularly in the early period.

It is of some interest to consider the US material recently selected by the ASCE (Marr, 2003) as still worth reading from the interwar period. Of 31 items, 18 appeared in *Engineering News Record* (*ENR*) (effectively three series of articles), three were ASCE papers and four from regional engineering societies, three were conference papers, and two were from the *Journal of applied physics*. Strangely, *Public Roads* and *HRB Proceedings* do not feature. This suggests that few were likely to have access to all the literature, and *ENR* had a pivotal role in the USA.

---

§ Thiem's work (Thiem, 1870, 1881, 1886) on groundwater flow was published in a German journal on gas lighting and water.

Table 3. Geotechnical publications in ENPC catalogue (Paris)

	1700–1749	1750–1799	1800–1850	1850–1870	Totals
General works*	3	10	31	20	64
Retaining walls/earth pressure			4	2	6
Specialist techniques			1		1
Tunnels			7	15	22
Earthwork tables			17	12	29
Dams				4	4
Foundations			4	2	6
Underwater foundations		2	8	8	18
Piles			1	1	2
Slope stability			1	2	3
Earthworks				3	3

*All held at ICE 1895.

Table 4. Subjects of papers in *ICE Proceedings* 1836–1948

	1830–1849	1850–1869	1870–1889	1890–1899	1900–1909	1910–1919	1920–1929	1930–1939	1940–1948	Total
Cuttings, embankment, earthworks	5	2	7	3	1	1	3	2	4	28
Bridge and subaqueous foundations	3	2	1			1	4	9	7	27
Foundations			2	2	1		1	2	6	12
Road foundations							1	1	8	10
Specialist processes	2	1		1					6	8
Earth pressure and retaining walls			5	3	1	2	6	7	3	27
Tunnels	2		7	2	2	3		1	3	20
Subsidence				1	1					2
Piling	2	1	2	1			3	4	1	14
Percolation and groundwater flow				2	1					3
Natural landslides	1		1					1		3
Coasts and coast works		3	1	2	2			2	3	13
Dams	2	2	1	1				3	3	12
Engineering geology				1	2			2		5
Soil properties							1	2	3	6
Soil mechanics								1	7	8
Total	12	11	27	19	11	7	19	37	54	

THE LITERATURE OF FOUNDATION PRACTICE BEFORE 1900

Our knowledge of the foundation practice of ancient civilisations relies largely on archaeological evidence, and this continues into the medieval period (see bibliography). There is some literary evidence: for example, Vitruvius and Caesar describe Roman practice, which in much of Europe would have represented best practice down to the eighteenth century. Detailed case studies using archives have been published of medieval and renaissance bridge foundations (Parsons, 1939, Boyer 1966, 1984, 1985).

From the sixteenth century onwards illustrated books describe piling and other techniques to complement archival and physical evidence (Jensen, 1969). One has to look in encyclopedias and scientific journals as well as books on building and bridges to get a full picture of what is going on.

Belidor produced arguably the first engineering textbooks in the early eighteenth century. Heavily illustrated, they contained what geotechnical science was readily available. Perronnet (1788) and Cessart (1806) described bridge construction, providing piling records and evidence of how loads were calculated on the foundations, all indicative of great thoughtfulness and experience. At that time some attempts were made to develop piling formulae. Thinking is confused, and it is not really until *c.* 1800, when Woltmann (1791–1799) and Eytelwein and Gilly (1802–1809) published their hydraulic encyclopedias, that dynamic formulae were developed (Woltmann, 1799a, 1804; Eytelwein, 1805) that were used by some engineers into the twentieth century (Chellis, 1961; Little, 1961). There is much common sense in their deliberations, as there is earlier in the writings of the Swedish technologist Polhem (1947). All recognised that different ground needed a varying approach, and both Eytelwein and Woltmann drew up tables indicative of the bearing capacity of different soils, and piled foundations (Fig. 1).

Both Eytelwein and Perronet were also involved in teaching, and in the second half of the nineteenth century the rapid proliferation of technical education across Europe saw growing numbers of textbooks by academics.

Table 5. Geotechnical publications by subject 1910–1946

	1910–1919	1920–1929	1930–1939	1940–1946	Total to 1946
Engineering geology	1	8	13		22
Soil mechanics		9	21	24	54
Soil classification		4	9	3	16
Cohesionless soils		9	31	28	68
Cohesive soils	3	14	52	34	103
Loess			1	4	5
Sedimentious rocks		1	7	3	11
Site investigation		17	105	80	202
Laboratory tests		28	50	57	135
Field tests	1	35	125	65	227
Soil properties	1	83	129	110	323
Strength	3	7	16	1	21
Compressive strength		2	123	4	9
Shear strength	2	0	24	11	37
Soil dynamics		7	51	12	70
Pressure beneath foundations	2	52	91	106	251
Uplift		14	35	6	55
Bearing capacity	2	35	78	39	154
Earth pressure	2	39	56	32	129
Retaining walls		30	61	41	132
Slope stability		38	72	85	195
Groundwater seepage		21	122	70	2·3
Pavements		32	60	190	282
Dams		23	99	40	162
Ground improvement		9	185	230	424
Foundations		10	62	49	121
Settlement analysis		24	60	23	107
Consolidation			10	14	24

Table 6. Geotechnical publications by country 1910–1946

	1700–1799	1800–1849	1850–1899	1900–1909	1910–1919	1920–1929	1930–1939	1940–1946
USA				2	6	197	481	537
Germany					1	85	274	65
France	1	1	2			19	83	28
USSR						10	38	13
UK		2	3	3	5	63	126	233
Scandinavia					5	6	31	17
Other Europe						24	34	25
Latin America						4	7	32
India						1	32	2
Japan						2	15	2
China							10	
Africa						2	2	10
Australasia						1	12	21
Canada							14	17

Table 7. Terzaghi publications 1900–1939

Decade	Conferences		Journals		Monographs	Languages
	No. of different conferences	No. of papers	No. of different journal titles	No. of papers		
1900–1909	0	0	2	4	1	1
1910–1914	0	0	7	8	2	1
1920–1929	5	5	20	46	4	3
1930–1939	4	25	31	62	9	6

Timber was the normal foundation material until the twentieth century, although iron and concrete were used from around 1800 (Glossop, 1976; Sutherland, 2001). The history of these developments can be traced in patent literature and discussions in the architectural and engineering press through the nineteenth century, with a few textbooks appearing in the second half of that century. Although iron was used for sheet piles from the early nineteenth century, it is

Last, welche der Pfahl tragen kann. (R – Q)	Gewicht des Rammklotzes (P)								
	6 Centner			8 Centner			10 Centner		
	Gewicht des Pfahls (Q)			Gewicht des Pfahls (Q)			Gewicht des Pfahls (Q)		
	4 Centn.	8 Centn.	12 Centn.	4 Centn.	8 Centn.	12 Centn.	4 Centn.	8 Centn.	12 Centn.
Centner.	Wie viel Zoll der Pfahl auf 20 Schläge eindringen muß, um die Last zu tragen (20e)								
50	24,5	17,5	13,6	38,1	28,5	22,8	53,6	41,7	34,1
100	11,5	8,2	6,3	17,4	13,0	10,4	23,8	18,5	15,2
150	7,5	5,3	4,2	11,2	8,4	6,7	15,3	11,9	9,7
200	5,6	3,9	3,1	8,3	6,2	5,0	11,3	8,7	7,2
250	4,4	3,1	2,4	6,6	4,9	3,9	8,9	6,9	5,7
300	3,7	2,6	2,0	5,5	4,1	3,3	7,4	5,7	4,7
350	3,1	2,2	1,7	4,9	3,5	2,8	6,3	4,9	4,0
400	2,8	1,9	1,5	4,1	3,1	2,4	5,5	4,2	3,5
450	2,4	1,7	1,3	3,6	2,7	2,2	4,9	3,8	3,1
500	2,2	1,6	1,2	3,2	2,4	2,0	4,4	3,4	2,8

(a)

Load, which the pile can bear (R – Q) (in cwt)	Weight of ram (P)								
	6 hundredweight			8 hundredweight			10 hundredweight		
	Weight of pile (Q)			Weight of pile (Q)			Weight of pile (Q)		
	4 cwt	8 cwt	12 cwt	4 cwt	8 cwt	12 cwt	4 cwt	8 cwt	12 cwt
	How many inches the pile after a volley of 20 blows must enter to carry the load (20e)								
50	24·5	17·5	13·6	38·1	28·5	22·8	53·6	41·7	34·1
100	11·5	8·2	6·3	17·4	13·0	10·4	23·8	18·5	15·2
150	7·5	5·3	4·2	11·2	8·4	6·7	15·3	11·9	9·7
200	5·6	3·9	3·1	8·3	6·2	5·0	11·3	8·7	7·2
250	4·4	3·1	2·4	6·0	4·9	3·9	8·9	6·9	5·7
300	3·7	2·6	2·0	5·5	4·1	3·3	7·1	5·7	4·7
350	3·1	2·2	1·7	4·9	3·5	2·8	6·3	4·9	4·0
400	2·8	1·9	1·5	4·1	3·1	2·4	5·5	4·2	3·5
450	2·4	1·7	1·3	3·6	2·7	2·2	4·9	3·8	3·1
500	2·2	1·6	1·2	3·2	2·4	2·0	4·4	3·4	2·8

(b)

Fig. 1. (a) Eytelwein's (1805) load tables relating size of pile hammer, pile size, depth of penetration and load; (b) translation of by M. Chrimes (NB: units are Prussian)

only as steel began to be rolled widely in the 1890s that the profiles familiar today were developed, particularly in the US and Germany.

From handbooks at ICE, soil mechanics advice did not appear until the 1940s. The advent of reinforced concrete saw a proliferation of trade literature and then specialist journals in most Western countries between 1890 and 1920. Emperger's *Handbuch fuer Eisenbetonbau* (Emperger, 1906) is a great source of information, continuing the tradition of Belidor and Eytelwein.

EARTH PRESSURE AND RETAINING WALL DESIGN 1700–1940

One classic engineering problem identified by Gautier (1717) was that of earth pressure and retaining wall design. Subsequent literature is extensive and illustrative of progress, or the lack thereof, in soil mechanics. The works of Coulomb (1773) and Rankine (1857) are well known, and their theories remain influential, and with some justification can be regarded as the origin of our subject. Their ideas have been reviewed by Heyman (1972), Skempton (1979) and Corradi (1995).

Vauban is frequently cited as the engineer who first formulated rules for the design of retaining walls, for fortifications, but he was building on the experience of Spanish and Italian engineers (Corradi, 1995). Their 'rule of thumb' approach to retaining wall design, published in military manuals on fortification in many languages, persisted through the nineteenth century (Curie, 1870; Dubosque, 1880).

There was, however, an extensive literature on retaining wall design that referred to Coulomb, and subsequently Rankine. Coulomb's influence was restricted. His initial publication in an obscure part of the Académie des Sciences will not have helped. For some the mathematics may have been challenging, while others had not seen Coulomb's work in the original, and rejected it without a proper understanding. In a number of treatises educationists attempted to develop Coulomb's ideas and make them more accessible, with mixed success. In Britain it was probably Moseley (1843) who introduced Coulomb to a broader audience. A useful contribution by Neville (1845) was probably known in Irish universities, and anticipated Rebhann (1871), but seems to have been generally ignored.

Aside from Coulomb, Woltmann's (1794) research is of particular interest. It agreed, where cohesion is zero, with Coulomb, but he introduced the angle of friction in place of the coefficient of friction. Unfortunately, lack of clarity in his explanations devalued his findings, and in a subsequent part (Woltmann, 1799b) of his hydraulic engineering treatise he produced another formula. His work was taken up by Hagen (1831) and later Curie (1870) and Skibinski (1885).

Rankine (1857) developed his general theory in a scientific journal; as with Coulomb, it was slow to attract attention in other countries. Most commentators were academic engineers or scientists, conducting a literary debate. Coulomb and Rankine were challenged or ignored because practising engineers could not match the mathematical conditions with their practical experience. This is most evident in the paper to the ICE by Baker in 1881 (Hamilton, 1958; Skempton, 1979). In Britain, in particular, engineering education may have left many practitioners without the knowledge of theoretical mechanics to engage with the literature.

An obvious response to these shortcomings was tests to measure earth pressure. Although Mayniel (1808) provided a useful summary of unpublished eighteenth-century work, subsequent work, published in a variety of journals and treatises, was not collated, many researchers proceeding *de novo*. Graphical methods became common in the second half of the nineteenth century. Poncelet (1840) in France, followed by Cuhlmann (1866) and Winkler (1867), developed them for retaining wall design. Mohr's name is still associated with this approach, which remained current into the twentieth century.

Koetter (1892) was responsible for a comprehensive review at the turn of the century. Soon after, he developed his novel equation for a circle of rupture to determine internal stresses in rupture lines (Koetter, 1902). Skempton (1979) has identified the work of Resal (1910) as the culmination of this period of 'classical soil mechanics', while in Germany the near-contemporary review and systematic tests of Mueller-Breslau (1906) are seen as the dawn of modern soil mechanics. Neither Resal nor Mueller-Breslau broadened their discussion of soil properties and behaviour in a way one would expect in a modern geotechnical textbook. A more modern approach can be seen around 1900, when Bell was designing the monoliths at Rosyth Dockyard in clay strata (Skempton, 1958, 1979). Bell (1915) was a pioneer in his use of shear test data on clay in his calculations, and correlating this to pressure cell observations.

Bell's paper, unlike many earlier contributions, was published in the ICE *Minutes of Proceedings*. He also introduced the topic of bearing capacity at the 1921 engineering conference. Despite his work being in the mainstream of engineering publishing in the UK, little was done to follow it up. When the BRS was established after the First World War it was slow to take an interest in geotechnics (Cooling, 1975).

Some problems with earlier earth pressure theories were addressed by Boussinesq (1885), who also provided elastic solutions for a number of pressure and settlement problems. However, when Darwin (1883) carried out his tests he found discrepancies. Boussinesq had assumed a constant angle of internal friction, which may have been reasonably satisfactory in a practical sense, with Darwin getting a better correlation than with Rankine, but in a scientific sense the need for a theory taking account of a variable internal angle of friction had not been addressed. It was Caquot (1934), in perhaps the most important French contribution to geotechnics between the two world wars, who produced a full solution to some of the issues raised in Boussinesq's work (Kerisel, 1956). This was largely unknown outside France, and Koetter's contribution was hardly known outside Germany (Ritter, 1936); it had to wait until after the Second World War, when Hansen (1953) made expert use of his ideas.

THE BIRTHPLACE OF GEOTECHNICAL ENGINEERING: SWEDEN 1900–1930

The tortuous story of earth pressure and retaining wall design in the nineteenth century gives no hint of the rapid progress to be made in Sweden in the early twentieth century, which can be regarded as the birthplace of geotechnical engineering. For the first time, rational solutions were found to a number of geotechnical problems, and great advances were made in the understanding of soil properties and testing of soils (Petterson, 1955; Bjerrum & Flodin, 1960; Flodin & Broms, 1981). Close links with Germany and the development of international conferences meant that the obscurity of Scandinavian languages proved no barrier to dissemination, in central Europe at least. Atterberg's work in soil science (1900–1916) led to his classification of soils according to grain size (Atterberg, 1913) being adopted as an international standard. His work on clays led to the development of liquid and plastic limits. His work was published in the leading international journal of soil science. Without this outlet it could not have been disseminated so rapidly.

Similarly, the solution to the engineering problems of the soft clays found across much of southern Sweden might have remained of local concern were it not for engagement with German colleagues. The March 1916 failure of the Stigberg quay at Gothenburg led to a commission comprising four Swedish engineers, including Pettersen and Fellenius, and overseas input from Max Moeller, who had been involved with the Kiel Canal and written on earth pressure (Moeller, 1902), and Ysselsteyn from the port of Rotterdam. The significance of circular sliding surfaces was demonstrated, and the commission report was published in 1916 (Handlingar, 1916). Papers in the leading Swedish engineering journal by Petterson (1916) and Hultin (1916) publicised their methods. Moeller (1922) publicised them in Germany, where Fellenius' work (Fellenius, 1927) was published. By the 1930s the method was generally adopted in Scandinavia.

The Gothenburg investigations were more or less contemporary with those of the Geotechnical Commission of the Swedish State Railways. The first achievement was the design of standard site investigation equipment and procedures (SJGK, 1917). The laboratory established at Stockholm in 1914 was almost certainly the first dedicated geotechnical laboratory in the world (Table 8). The final report of the Commission was published in 1922 (SJGKS, 1922). By then the Swedish State Railways had established a geotechnical department, and soil mechanics can be seen to have been firmly established in Sweden. Elsewhere in Scandinavia the work was followed up in Finland, where a Geotechnical Committee was set up in 1922. There the term Geotekniska appeared in 1906 (Table 1).

THE KIEL CANAL AND EARLY SOIL MECHANICS RESEARCH IN GERMANY

The Kiel Canal (Fuelscher, 1898a) was a fine example of earthwork design on the cusp of modern soil mechanics. The case was cited in Terzaghi & Peck (1948) as an example of the displacement method. The project was widely reported in the German technical press and abstracted in the ICE *Minutes of Proceedings*. It also featured in international conference papers (Fuelscher, 1893, 1898b).

Partly in response to problems on the construction of the

Table 8. Geotechnical laboratories

Year	Country	Laboratory	Number in country	
			1936	1948
1903	Germany	Mueller-Breslau		
1904	Germany	Preussiche Versuchsanstalt fur Wasserbau und Schiffbau	5	2
1914	Sweden	Geotechnical Commission for Railways	2+	14
1919	Turkey	Roberts University	unknown	1?
1920	USA	Bureau of Public Roads	15+	79
1923	Denmark	Danish Geological Survey (Geoteknisk Institut)	2	3
1926	Finland	Finnish Geotecnical Committee	1 (?)	unknown
By 1930	Italy	G Rodio spa, Milan	1	unknown
1930	Hungary	Technical University, Budapest	1	1
1930	Netherlands	Delft University	2	unknown
1931	Japan	Japan Government Railways	10	unknown
1931	USSR	GOSSTROY, Moscow	3?	unknown
c. 1932	India	Punjab Irrigation Research Institute	2	12?
1933	UK	BRS, Garston	1	15
1934	Egypt	University for Cairo	1	2?
1935	France	Laboratorie du Sol et Fondations, Paris	1	unknown
1936	Mexico	National Irrigation Commission	0	2
1937	Brazil	IPT, Sao Paulo	0	3
1938	Belgium	Ghent University	0	1
c. 1946	South Africa	CSIR	0	6
1944	Spain	Escuela de Ingenieros de Caminos	0	3
1944	Australia	Victoria State Rivers and Water Supply Commission	0	9
pre 1948	Algeria		0	1

Canal, the Versuchanstalt für Wasserbau und Schiffbau (VWS) in Berlin was, from 1910, charged with an increasing volume of geotechnical research. Under the leadership of Hans-Detlef Krey it developed equipment, and from 1927 had a soil mechanics facility. This was one of a number of research facilities that developed in Germany before 1936 (Table 8) and produced reports, generally summarised in journals such as *Bauingenieur* and *Bautechnik*. *Bauingenieur* was also the vehicle for reporting the activities of the soil mechanics section of the German Society of Construction Engineers in the late 1920s (see Table 9). Mysteriously this group disappeared in the 1930s.

SITE INVESTIGATION AND TESTING

Early civil engineers were well aware of the need for adequate site investigation, with frequent reference to this in the reports of engineers such as Smeaton. Such reports were the normal means of information exchange among engineers before the arrival of engineering journals in the 1820s. As the nineteenth century proceeded, site investigation equipment improved, but tests to provide reliable information on soil properties remained in their infancy. Soon after Terzaghi went to Constantinople, in 1917, he began to build equipment of his own to investigate earth pressure, and in 1919 he began to develop more fundamental equipment for investigating soil properties (Goodman, 1998). It was he who proposed the standard penetration test in 1926. Terzaghi was not the only investigator, and some of the important developments in this area were pioneered by others.

In the early twentieth century Frontard (1910) and Bell (1915) carried out shear tests in isolation. Major advances in site investigation and laboratory testing were first made in Sweden (Bjerrum & Flodin, 1960). However, the general lack of coordinated progress is indicated by the triaxial test, the full significance of which was not realised until the late 1930s, despite having been first attempted in the 1890s. Adams & Nicolson (1901) and Adams (1910) trialled it for rock testing. Von Karman developed a triaxial testing machine (Karman, 1911). Westerberg adapted the method for soft clays, which enabled him to prove experimentally that $\phi = 0$ for saturated clays under rapid loading (Westerberg, 1921; Skempton, 1960). Ehrenberg at the VWS introduced the test for soil mechanics generally in 1927 (Ehrenberg & Tiedemann, 1928). Arthur Casagrande saw the equipment in 1929 and realised its potential. From 1930 Terzaghi used it and, with Rendulic, he developed it to measure the shear strength of clays (Rendulic, 1935, 1936).

When the Franklin Falls Dam experienced seepage problems in 1936, direct shear tests were producing erratic results. Casagrande suggested the triaxial test. In 1938 the US Corps of Engineers, reflecting on the rapid and uncoordinated progress in soil mechanics, organised a conference at Boston, which determined the need for a research programme specifically on triaxial testing, underseepage, and pressure cell analysis (Fatheree, 2006). Rutledge reported on the triaxial test programme (US Waterways Experiment Station, 1944).

In the US much soils research was the result of the development of the country's highway network, and was published in the journal *Public Roads* and the *Proceedings of the Highway Research Board*. The establishment in 1936 of the ASTM Committee on Soils for Foundations is an indication of the maturity of the field. By the late 1930s there were soils laboratories in many parts of the world that could carry out direct shear, triaxial and consolidation tests (Table 8). From 1933 the Autobahn teams all had access to their own laboratories. This was associated with a growth in standard procedures and related codes of practice (Table 9).

THE EARLY WORK OF TERZAGHI

At the 1936 Harvard Conference Terzaghi (1936) welcomed efforts in Boston to create a modern building code, reflecting advances in soil mechanics, advances due in no small part to him. In the autumn of 1916 he had gone to Istanbul to take up a teaching post, and began to address issues that had frustrated him in his work as an engineer in the field of 'technical geology' (Terzaghi's autobiographical notes, quoted in de Boer, 2005). By June 1918 he had

Table 9. Development of standards and codes

Year(s)	Standard/code	Country
1917	Swedish Railways. Recommendations for boring, etc.	Sweden
1921	US Bureau of Public Roads. Tests for subgrade soils	USA
1928–1929	Deutscher Baugrundausschuss. DGB Richtlinien (site investigation). Revised 1935, 1937	Germany
1929–1936	Boston Building Code. Redrafted to embrace soil mechanics	USA
1931–1933	Development of Proctor tests/Los Angeles Water Department (*ENR*, Vol. 111, 1933)	USA
1934–1936	Deutscher Strassenwesen. Guidelines for earthworks	Germany
1934	DIN 1054: Permissible loading of subsoil	Germany
1934	US Bureau of Reclamation. *Laboratory Procedures Manual*	USA
1935	DIN 1179: Grain sizes for sand, pebbles etc.	Germany
1935	ASTM D420–D427: Tentative methods ... sampling soils for highway grades, etc.	USA
1935	DIN 4021: Subsoil and groundwater	Germany
1935	USWES. *Laboratory Procedures Manual*	USA
1937	DIN 4107: Guidelines for observation of movement of buildings	Germany
1938	DIN 4021: Ground investigation by excavation, boring etc.	Germany
1938–1944	US triaxial test program	USA
1938	Swedish Roads Board. *Code of practice for earthworks*	Sweden
1938	Reichsautobahnen. Technical guidance on earthworks	Germany
1938	DIN 4022: Standards for labels of soil types	Germany
1939	DIN 4150: Vibration from construction work	Germany
1939	ASTM D421: Standard method for mechanical analysis of subgrade soil (draft 1935)	Germany
1939	ASTM D422: Standard method for mechanical analysis of soils (draft 1935)	Germany
1939	ASTM D423: Standard method of test for liquid tunnels of soils (draft 1935)	USA
1944	ASTM. Procedures for testing soils	USA
1948	BS 1377: Methods of test for soil classification	UK

formulated a new earth pressure theory (Terzaghi, 1919, 1920). Despite his relative geographical isolation he was able to secure publication in Germany and the USA. In 1919 he moved on to consider the permeability of soils due to changes of water pressure, constructing equipment at Robert College. He realised time had a decisive role in the deformation of clays. He began studying sand under increasing upward hydraulic pressure, built apparatus for measuring permeability and saturation, and developed an oedometer. Over the next three years Terzaghi wrestled with his data, trying to formulate a satisfactory theoretical framework, but he could investigate only one-dimensional load states as he was ignorant of triaxial devices.

Terzaghi gave a paper on the partial differential equation of the consolidation process in 1923 (Terzaghi, 1923). In this he was concerned largely with calculation of the permeability of clay, rather than settlement problems, which were addressed with Froehlich (Terzaghi & Froehlich, 1936). At times his thinking was obscure, as it was in *Erdbaumechanik* (Terzaghi, 1925a), on which he was soon working. It was in fact his articles for *Engineering News Record* (Terzaghi, 1925b) that provided much-needed clarity for an engineering audience, perhaps befitting from the editorial hand of *ENR*. Although generally positively received, Terzaghi could be criticised from the viewpoint of theoretical mechanics, and he was not abreast of all relevant literature (de Boer, 2000, 2005).

Progress in getting his ideas disseminated was slow. Despite the number of his publications, there was no obvious single vehicle in which to promulgate them (Table 7). However, as a result of his articles in *ENR*, Terzaghi was invited to advise the US Bureau of Public Roads on their testing equipment and procedures. From then on his work was increasingly as an international consultant. That gave him numerous opportunities to test and refine his ideas (Goodman, 1998).

THE GROWTH OF THE INTERNATIONAL COMMUNITY OF GEOTECHNICAL ENGINEERS, AND CONFLICT

The organisation and success of the Harvard Conference on Soil Mechanics by Arthur Casagrande in 1936 can be seen as an indication of the maturity of soil mechanics as a discipline, and of the existence of a community in soil mechanics. However, coming after 30 years of rapid development in the discipline, unsurprisingly not all of these developments or those responsible were represented.

The sense of an international community of civil engineers developed in the late nineteenth century.

From the 1830s engineering societies proliferated. They corresponded, their meetings and journals providing arenas for debate, while international exhibitions drew engineers together. By the final quarter of the nineteenth century international conferences were being organised in specialist disciplines such as railways and materials testing. Inevitably the field of geotechnics featured at such gatherings. By the late 1930s significant discussions took place (Table 10). The early navigation congresses contain a number of practical reports. The International Engineering Congress of 1915 included a huge amount of material on Panama (Goethals, 1915). Surprisingly the interwar Permanent International Association of Road Congresses (PIARC) congresses do not refer to geotechnics until after the Harvard Conference. Of greater significance in terms of the geotechnical community was the First International Congress on Applied Mechanics, where papers were presented by Terzaghi (1924) and Reissner (1924). This brought Terzaghi into contact with a number of European engineering scientists, after his relative isolation in Turkey. Direct influence on practising engineers would have been small, and in this regard the World Power Conferences and International Congresses on Large Dams were of more significance, bringing together a number of engineering geologists and practising engineers (Tables 10 and 11).

In 1933, at the first World (ICOLD) Congress on Large Dams, Terzaghi was reporter on the session on materials for earth dams, drawing up a table that illustrated the range of tests in use, and calling for standardisation. The need for more work was noted by Pavlovsky (1933), who said of soil mechanics that 'the theory itself is still to a great extent unexploited'. In contrast the 1936 ICOLD reflected rapid advances in the subject (ICOLD, 1936). In a way the conference was as much a landmark in the development of geotechnics as that at Harvard, suggesting that in one specialist engineering discipline soil mechanics was now in the mainstream. The PIANC event of 1935 and the International Association of Bridge and Structural Engineering Congress in 1936 were others.

This was the background to the first international conference at Harvard. Unlike other such events it came not out of collaborative discussion with existing engineering societies, but rather through Casagrande's realisation that the network of correspondents and researchers around Terzaghi could now be considered a new discipline, and a conference would enhance the reputation of Harvard in the field. This rather sectarian gestation meant that, undoubted success though it was, the conference had a negative side, with conflicts among pioneers of the subject, many of whom had developed ideas independently of the Terzaghi–Harvard school (Peck, 1988; de Boer, 2000, 2005). Schofield (2001) argues that ideas in Taylor's work (Taylor, 1948) were ignored by Terzaghi and his followers, and that there was a narrowness of debate at the time and failure to move on through theory.

Unfortunate as these disputes were, none can compare with the tragic consequences of the public dispute between Terzaghi and Fillunger, originator of the modern theory of flow in porous media. This provides abundant evidence of the difficulties in effective dissemination of geotechnical ideas at the time.

Fillunger had been looking at ideas of interest to geotechnical engineers for as long as Terzaghi. From 1910 his work on dams and uplift led to development of the porous media theory of liquid saturated porous bodies (Fillunger, 1913) and the discovery of the concept of effective stress. He wrote:

> It is less the actual individual stress problems than the complete argument of the final results that lead straight to the conviction that a pressure carrying liquid penetrating the masonry construction creates a pressure in the material equal in all directions.

In the 1920s Terzaghi became interested in the uplift problem, and in *Erdbaumechanik* (Terzaghi, 1925a) gave the correct formula for the uplift force in water saturated sand. In the early 1930s he came to the problem afresh, and disputed Fillunger's findings. He discussed the findings with Fillunger, who was uncooperative. The following year they were published (Terzaghi & Rendulic, 1934). Fillunger (1934) disputed the theoretical proof of Terzaghi's formula, and attacked the basis of Terzaghi's work in a polemical manner (Fillunger, 1936). Terzaghi had little option but to challenge Fillunger's professional conduct through the academic body—the Technical University in Vienna—that employed them both. The result was, from Fillunger's position, disastrous: his conduct was censured, and his theories were subjected to a more substantial critique than Terzaghi's. The immediate consequence was

Table 10. International conferences with geotechnical content prior to 1940 (*significant content)

Year	Conference	Location
1892	PIANC (Permanent International Association of Navigation Congresses)	Paris, France
1892	Congress of Engineers and Architects	Palermo, Italy
1893	International Engineering Exhibition	Chicago, USA
1897	ICE Engineering Conference	London, UK
1898	PIANC*	Brussels, Belgium
1899	ICE Engineering Conference	London, UK
1904	International Engineering Exhibition	St Louis, USA
1907	ICE Engineering Conference	London, UK
1908	PIANC	St Petersburg, Russia
1908	PIARC (Permanent International Association of Roads Congresses)	Paris, France
1910	PIARC	Brussels, Belgium
1912	PIANC	Philadelphia, USA
1913	Engineering Conference*	Sind, India
1915	International Engineering Conference*	Stamford, USA
1921	ICE Engineering Conference	London, UK
1921	Highway Research Board (now TRB) Proceedings (annual)	USA
1923	PIANC*	London, UK
1924	Applied Mechanics*	Germany
1924	World Power Conference*	London, UK
1926	PIANC*	Cairo, Egypt
1929	World Engineering Conference*	Tokyo, Japan
1930	World Power Conference*	Berlin, Germany
1931	PIANC*	Venice, Italy
1932	IABSE (International Association for Bridge and Structural Engineering)	Paris, France
1933	ICOLD* (International Commission on Large Dams)	Stockholm, Sweden
1935	PIANC*	Brussels, Belgium
1936	ICOLD*	Washington DC, USA
1936	IABSE*	Berlin, Germany
1936	1st International Conference on Soil Mechanics and Foundation Engineering*	Harvard, USA
1937	Highway Research Board Proceedings*	USA
1938	PIARC*	Munich, Germany
1938	Applied Mechanics	Cambridge, USA

Table 11. International conferences without geotechnical content prior to 1940

Year	Conference	Location
1883	PIANC	Brussels, Belgium
1885	IRCA (UIC Railways)	Nothing traced before 1940
1888	PIANC	Frankfurt, Germany
1894	PIANC	The Hague, The Netherlands
1900	PIANC	Paris, France
1901	Glasgow Exhibition	Glasgow, UK
1902	PIANC	Dusseldorf, Germany
1903	ICE Engineering Conference	London, UK
1905	PIANC	Milan, Italy
1913	PIARC	London, UK
1923	PIARC	Seville, Spain
1926	IABSE	Zurich, Switzerland
1926	PIARC	Milan, Italy
1928	IABSE	Vienna, Austria
1928	ICE Engineering Conference	London, UK
1930	PIARC	Washington D C, USA
1930	Applied Mechanics	Stockholm, Sweden
1934	PIARC	Munich, Germany
1934	Applied Mechanics	Cambridge, UK
1938	Glasgow Exhibition	Glasgow, UK

the suicide of Fillunger and his wife. In a posthumous pamphlet Fillunger recognised the validity of some of Terzaghi's work that he had previously criticised (Fillunger, 1937).

As an addendum to this sad tale it appears that when Fillunger (1913) and Terzaghi (1923, 1924) first discussed effective stress, neither realised the full implications. When Terzaghi came to do so, as a result of work in the 1930s, he was in no mind to acknowledge Fillunger's contribution. The episode rankled until Terzaghi's death. When A. Casagrande saw Skempton's draft essay for *From theory to practice*, where Skempton (1960) acknowledged Fillunger's contribution to the theory of effective stress, he asked Skempton to rephrase it because he felt Fillunger had threatened to sabotage the birth of soil mechanics (Niechcial, 2002: 122; De Boer, 2005: 269–270).

One cannot help but reflect that the absence of an obvious vehicle for the debate, and the publication of material in a variety of sources—journals, conferences, pamphlets—contributed to these problems.

PAVEMENT DESIGN AND THE GENERAL ACCEPTANCE OF SOIL MECHANICS

During the Second World War, the need to design airfields for larger aircraft, particularly bombers, meant that all combatant nations had to have design methods readily learned and applied, and the work of soil mechanics experts, which had developed over the previous 20 years, suddenly acquired strategic importance (US Corps of Engineers, 1992; Fatheree, 2006). By the end of the war many engineers had been trained in the rudiments of soil mechanics and pavement design.

Practical road pavement design was long understood (Lay, 1992). Winkler introduced the idea of subgrade reaction in his textbook on elasticity, which can be regarded a key early text in soil mechanics (Winkler, 1867). As described by Terzaghi (1955) it had been applied to a number of geotechnical problems. Improved standards of highway design and maintenance were associated with expansion of road transport in the interwar years. In Germany and the US this led to great advances in soil mechanics. It is clear from the 1938 PIARC congress how important US work was. Notable pioneers were Westergard, Porter and Proctor.

In the US, the Waterways Experiment Station took the lead in carrying out research to provide the Army with reliable design methods. After discussions with Casagrande it was decided to concentrate on the CBR as the basis for design. Design curves for flexible pavements were developed. Research was associated with lecture programmes by Casagrande and others (Fatheree, 2006). Casagrande also made a major contribution to soil classification at this time. Previous work had depended on grain size scales, not necessarily helpful with regard to engineering attributes. He developed the Airfield Classification System—the basis of the Unified Soils Classification System.

Much of this research came too late for the British engineers. The Royal Engineers *Manual for Airfield Construction* did not appear until 1946, including reference to US design. In the same year the Road Research Laboratory (RRL, now TRL) published its series of wartime lectures (RRL, 1946) summarising progress. Some idea of the situation that British engineers found is given by Glossop & Golder (1944). Their method for clays made use of the Boussinesq theory of stress distribution. It was criticised by Marwick of the Road Research Laboratory and others in the ICE discussion for its limited applicability, but served well in wartime conditions.

CONCLUSIONS

Tremendous work had been done in difficult ground conditions before 1900, and from as early as 1700 theoretical models had been developed to address a number of geotechnical problems, notably earth pressure and retaining wall design, and pile design. However, there was little general acceptance of early soil mechanics theories among practising engineers. Much of the debate was in general scientific journals, although some theories were included in engineering textbooks that appeared in increasing numbers as the nineteenth century progressed.

In the early twentieth century a number of factors, such as the great challenges of the Panama and Kiel Canals, loss of life on Swedish railways, and the growth of road transport, encouraged engineers to look at geotechnical problems afresh and develop new solutions. Over 40 years great advances were made in geotechnical understanding, but language, distance and the absence of a recognised forum for discussion and dissemination hampered progress. Key ideas were scattered in a range of publications. By the late 1930s momentum was building to solve these problems.

Géotechnique provided a means, for the first time, of overcoming problems in advancing the subject that were the result of the absence of an internationally recognised publication for debate and peer review of ideas. It is ironical that it was launched in a country that, prior to 1936, had shown little interest in the subject, but from 1948 took a vigorous lead.

ACKNOWLEDGEMENTS

The author would like to thank his colleagues Annette Ruehlmann, for her help in tracking down German-language literature, and Paul Parkes, for wrestling with the text, and collaborators John Atkinson, F. Tatsuoka, Renato Lancellotta, Matti Loyander, Minna Karstunen, John Hutchinson, N. R. Morgernstern, Michael Brodhead, Chuck Carnillo and Stephen Brown for their suggestions. Particular thanks are due to Dr Karl Kurrer for supplying material on the Fillunger dispute, and Philippe Mestat and Jean Pierre Magnan for advice on developments in France.

REFERENCES

Adams, F. D. (1910). An experimental investigation into the action of differential pressure on certain minerals and rocks. *J. Geol.* **18**, 489–525.

Adams, F. D. & Nicolson, J. T. (1901). An experimental investigation into the flow of marble. *Phil. Trans. R. Soc. London Ser. A* **195**, 363–401.

ASCE (1940). *Bibliography on soil mechanics.* New York: American Society of Civil Engineers.

Atterberg, A. (1913). Die Plastizitaet und Bindigkeit liefernde Bestandteile der Tone. *Int. Mitt. Bodenkd.* **3**, 291–330.

Bell, A. L. (1915). Lateral pressure and resistance of clay and the supporting power of clay foundations. *Min. Proc. ICE* **199**, 233–272.

Binnie, G. M. (1981). *Early Victorian water engineers.* London: Thomas Telford.

Binnie, G. M. (1987). *Early dam builders in Britain.* London: Thomas Telford.

BIOS (1946). *German roads and soil stabilisation.* London: British Intelligence Objectives Subcommittee.

Bjerrum, L. & Flodin, N. (1960). The development of soil mechanics in Sweden 1900–1925. *Géotechnique* **10**, No. 1, 1–18.

Boussinesq, J. (1885). *Applications des potentials a l'étude e l'équilibre et du movement des solides élastiques.* Paris: Gauthier-Villars.

Boyer, M. N. (1966). Rebuilding the bridges at Albi, 1408–1410. *Technology and Culture* **7**, No. 1, 24–37.

Boyer, M. N. (1984). A fourteenth century pile driver: the engine at Orleans. *History of Technology* **9**, 38.

Boyer, M. N. (1985). Resistance to technological innovation in the history of the piles driven through the eighteenth century. *Technology and Culture* **26**, 1, 56–58.

Burland, J. (2008). The founders of *Géotechnique*. *Géotechnique* **58**, No. 5, 327–341.

Caquot, A. (1934). *Equilibre des massifs pulverents a frottement interne.* Paris: Gauthier-Villars.

Cessart, L. A. de (1806). *Description des travux publics.* Paris: Collin.

Chellis, R. D. (1961). *Pile foundations.* New York: McGraw-Hill.

Chevallier, R. (1989). *Roman roads.* London: Batsford

Chrimes, M. (1995). Introduction to *Catalogue of periodical publications, 1665–1994.* London: ICE Library.

Cooling, L. F. (1975). 'The early history of soil mechanics at the Building Research Station.' In British Geotechnical Society Twenty-fifth Anniversary Report. *Géotechnique* **25**, No. 4, 629–656 (629–634).

Corradi, M. (1995). De la statique des demi-fluides à la théorie de la poussée des terres. In *Between mechanics and architecture* (eds P. Radelet de Grave and E Benvenuto), pp. 220–254. Basel: Birkhauser.

Corthell, E. L. (1907). *Allowable pressures on deep foundations.* New York: Wiley (abstracted in *Min. Proc. ICE* **165**, 1905–1906).

Coulomb, C. A. (1773). Essai sur une application des règles de maximes et minimes à quelques problèmes de statique. *Academie Royale des Sciences Memoires Presentes par Diverses Savants* **7**, No. 1776, 343–382.

Cuhlmann, K. (1866). *Die graphische Statik.* Zurich, 547–633.

Curie, J. (1870). *Nouvelle théorie de la poussée des terres et de la stabilité des murs de revêtement.* Paris: Gauthier-Villars.

Darwin, G. H. (1883). On the horizontal thrust of a mass of sand. *Min. Proc. ICE* **71**, 350–378.

Davies, H. (2002). *Roads in Roman Britain.* Stroud: Tempus.

De Boer, R. (2000). *Theory of porous media.* Berlin: Springer.

De Boer, R. (2005). *The engineer and the scandal: A piece of science history.* Berlin: Springer.

De Hullu, J. & Verhoeven, A. G. (eds) (1920). *Andries Vierlingh: Tractaet van Dyckagie.* 'S-Gravenhage: Martinus Nijhoff.

Dubosque, J. (1880). Etudes théoriques et pratiques sur les murs de soutènement et les ponts et viaducs en maconniere. Paris: Beranger.

Ehrenberg, J. & Tiedemann, B. (1928). Baugrundforschung der Erdbauabteilung der Staatlichen Versuchsantalt fuer Wasserbau und Schiffbau, Berlin. *Zentralblatt der Bauwerwaltung*, 48, 287.

Emperger, F. v. (1906). *Handbuch fuer Eisenbetonbau.* Berlin: Ernst (four editions by 1940).

Eytelwein, J. A. & Gilly, D. (1802–1809). *Praktische Anweisung zur Wasserbaukunst.* Berlin: RealSchulbuchhandlung (Seven parts in four volumes. Vol. 1 covers piles; Vol. 3 retaining walls and piledriving; Vol. 4 locks and sluices).

Eytelwein, J. A. (1805). *Praktische Anweisung zur Wasserbaukunst*, Vol. 3, 137–146.

Fatheree, B. H. (2006). *The history of geotechnical engineering at the Waterways Experiment Station 1932–2000.* Vicksburg: USWES.

Fellenius, W. (1927). *Erdstatische Berechnung mit Reibung und Kohasion.* Berlin: Ernst.

Fellenius, W. (1936). On computation of earth pressure against walls with earth having cohesion. *Tekn. Tid.* **39** (*US Bureau of Reclamation Technical Memorandum* **559**, 1937).

Fillunger, P. (1913). Der Auftrieb in Talsperren. *Oest. Woch. oeffentlichen Baudiens*, **19**, 31–34, 532–556, 567–570.

Fillunger, P. (1934). Der Kapillardruck in Talsperren. *Wasserwirtschaft*, **27**, Nos 13/14, 129–131.

Fillunger, P. (1936). *Erdbaumechanik.* Privately published (thanks to Dr Karl Kurrer for this).

Fillunger, P. (1937). *Erdbaumechanik und Wissenschaft.* Privately published (not located in UK).

Flodin, N. & Broms, B. (1981). History of civil engineering in soft clay. In *Soft clay engineering* (eds E. W. Brand and R. P. Brenner), Developments in Geotechnical Engineering, Vol. 20, pp. 27–156. Elsevier.

Forchheimer, P. (1882). Ueber Sanddruck und Bewegungs-erscheuningen un Innern trockenen Sandes. *Z. Oesterr. Ing. Archit.-Ver.* **34**, 111–136 (abstracted in *Min. Proc. ICE*).

Frontard, J. (1910). Note sur l'accident de la digue de Charmes. *Ann. Ponts Chaussées* **22**, 173–280.

Fuelscher, J. (1893). The North and East Sea Canal. *ASCE Trans.* **30**, 420–444 (Chicago engineering conference).

Fuelscher, J. (1898a). *Bau der Kaiser Wilhelms Kanal.* Berlin: Ernst.

Fuelscher, J. (1898b). Moyens de consolidation des talus des canaux maritimes. *PIANC.* **VII**, **2**, Q. 23, paper 10 (English summary by Corthell).

Gautier, H. (1717). *Dissertation sur les cullées des ponts.* Paris.

Glossop, R. (1975). Geotechnology and *Géotechnique*. *Géotechnique* **25**, No. 4, 646–653.

Glossop, R. (1976) The invention and early use of compressed air to exclude water from shafts and tunnels. *Géotechnique* **26**, No. 2, 253–280.

Glossop, R. & Golder, H. Q. (1944). *The construction of pavements on a clay foundation soil.* ICE Road Paper 15; Engng Div. Paper II, 1943–4.

Goethals, G. W. *et al.* (1915). The Panama Canal (2 vols). In *International Engineering Congress Transactions*. San Francisco: Stanford University.

Goodman, R. E. (1998). *Karl Terzaghi: The engineer as artist.* NY: ASCE.

Hagen, G. (1831). Untersuchung uber der Druck und die Reibung des sandes. *Ann. Phys. Chem.* **28,** 17–48, 297–323.

Hamilton, S. B. (1958). Sir Benjamin Baker: with particular reference to his contribution to the study of earth pressures. *Géotechnique* **8**, No. 2, 105–112.

Handlingar, R. (1916). *Goteborgs Hamnstyrelse.* Gothenburg: Harbour Trust.

Hansen, J. B. (1953). *Earth pressure calculations.* Copenhagen: Technical Press.

Heyman, J. (1972). *Coulomb's memoir on statics.* Cambridge: Cambridge University Press.

Hultin, S. (1916). Grusfyllningar foer kajbyggnader. *Tek. Tidskr.* **46**, 292–294.

Institution of Structural Engineers (1943). *Minutes of Committee on Code of Practice ... Earth retaining structures*, CP/SE/EPS(P)1-P(7) (unfortunately the final bibliography does not seem to have survived).

ICOLD (1936). *Proc. 2nd Int. Cong. on Large Dams.* Paris: CIGB.

Jensen, M. (1969). *Civil engineering around 1700.* Copenhagen: Danish Technical Press.

Karman, T. von (1911). Festigkeitsversuche unter allseitigem Druck. *VDI-Z.* **55**, 1749–1758.

Kerisel, J. (1956). Histoire de la mecanique des sols en France jusqu'au 20e siècle. *Géotechnique* **6**, No. 4, 151–166.

Koetter, F. (1892) Bericht ueber die Entwicklung der lehre von Erddrunk. *Jahresbericht Deutsche Math.-Ver.* **2,** 76–154.

Koetter, F. (1902). Ueber das Problem der Erddruckbestimmung. *Verh. Dtsch. Phys. Ges.* **7**.

Kulkarni, R. P. (1985). Indian soil mechanics, ancient to medieval period. In *Indian contribution to geotechnical engineering*, pp. 5–10. Meerut: IGS.

Kurdumov, V. I. (1891). *On the question of the resistance of foundations. A photographic method for understanding the failure process.* St Petersburg.

Kurdumov, V. I. (1892). Zur frage des widerstandes des Grundungen auf naturlichem Boden. *Civilingenieur* **292** (abstracted in *Min. Proc. ICE*).

Lay, M. G. (1992). *Ways of the world.* New Brunswick: Rutgers UP.

Leliavsky, S. (1958). *Uplift in gravity dams.* London: Constable

Little, A. L. (1961). *Foundations.* London: Arnold.

Marr, W. A. (2003). *A history of progress: Selected US papers in geotechnical engineering.* Reston, VA: ASCE.

Mayniel, K. (1808) *Traite expérimental, analytique et pratique de la poussée des terres.* Paris: Colas.

Moeller, M. (1902) *Erddruck tabellen.* Leipzig: Hirzel.

Moeller, M. (1922) *Erddruck tabellen*, 2nd edn. Leipzig: Hirzel.

Mononobe, N. & Matsuo, H. (1929). On the determination of earth pressure during earthquake. *World Eng. Conf.* **9**, 177–185.

Moseley, H. (1843). *The mechanical principles of engineering and architecture*, pp. 440–458. London: Longman (pp. 443–461 in 1855 edn).

Mueller-Breslau, H. F. B. (1906). *Erddruck und Stuetzmauern.* Leipzig: Kroener.

Neville, J. (1845). Investigation of some formulae for finding the maximum amount of resistance required to sustain banks of earth. *Trans. ICE Ireland* **1**, 108–144.

Niechcial, J. (2002). *A particle of sand: the biography of Alec Skempton, civil engineer.* Caithness: Whittles.

Okabe, S. (1924). General theory of earth pressure and seismic stability of retaining wall and dam. *JSCE Proc.* **10**, Part 6, 1277–1330.

Okabe, S. (1926). General theory on earth pressure and laboratory testing on seismic stability of retaining walls. *JSCE Proc.* **12**, No. 1, 123–134.

Parsons, W. B. (1939). *Engineers and engineering in the Renaissance*, pp. 116–117, 150. Baltimore, MD: Williams & Wilkins.

Pavlovsky, N. (1933). Discussion on 'Hydraulic fill dams'. *Proc. ICOLD, 1st Cong.*, Q2b, **4**, 305.

Peck, R. B. (1988). Growing pains of a new profession. In J. Dunnicliff and N. P. Young (eds) (2006). *Ralph B Peck: Educator and engineer*, pp. 139–145. Vancouver: Bitech.

Peck, R. B. (1993). The coming of age in soil mechanics: 1920–1970 1st Spencer Buchanan Lecture, Texas, A & M. In J. Dunnicliff and N. P. Young (eds) (2006). *Ralph B Peck: educator and engineer.* Vancouver: Bitech, 146–154.

Perronnet, J. R. (1788). *Œuvres*, 2nd edn. Paris: Didot.

Petterson, K. E. (1916). Kajraset in Goteborg den 5 Mars 1916. *Tek. Tidskr.* **46**, 471–478.

Petterson, K. E. (1955). The early history of circular sliding surfaces. *Géotechnique* **5**, No. 4, 275–296.

Polhem, C. (1947). (Collected works) *Teknologiska skrifter.* Uppsala: University Press.

Poncelet, J. V. (1840). Mémoire sur la stabilité des revêtements. *Mem. l'Off. Génie* **XIII**.

Rankine, W. J. M. (1857). On the stability of loose earth. *Phil. Trans. R. Soc. London* **147**, 9–27.

Rao, K. L. (1960). Presidential address. *Proc. 1st Asian Reg. Conf. Soil Mech. Found. Engng, New Delhi*, 14–20.

Rebhann, G. (1871). *Theorie der Gewoelbe: Futtermauern und eisernen Bruecken.* Brunswick: Schulbuchandlung.

Reissner, H. (1924). Zur Erddruckproblem. *Proc. 1st Int. Conf. Appl. Mech., Berlin*, 295–311.
Rendulic, L. (1935). Der hydrodynamische Spannungsausgleich in zentral entwaesserten Tonzylindern. *Wasserwirtschaft* **2**, 250–253, 269–273.
Rendulic, L. (1936). Porenziffler und Porenwasserdruck in Tonen. *Bauingenieur* **17**, 557–564.
Resal, J. (1910). *Poussée des terres*. Paris: Beranger.
Ritter, M. (1936). Research concerning building ground: General report. *Proc. 2nd IABSE Congr., Berlin*, Final report session VIII, 821–822.
Road Research Laboratory (RRL) (1946). *Soils, concrete and bituminous materials*. London: HMSO.
Sano, T. (1916). Seismic resistant design of buildings: Part 1. *Research Institute for the Mitigation of Earthquake Disasters*, **83-1** (report).
Sano, T. (1917). Seismic resistant design of buildings, part 2, *Research Institute for the Mitigation of Earthquake Disasters*, **83-2** (report).
Schofield, A. C. (2001). Re-appraisal of Terzaghi's *Soil mechanics*. *Proc. 15th Int. Conf. Soil Mech. Geotech. Engng, Istanbul* **4**, 2473–2480.
SJGK (1917). Statens Jarnvaegars Geotekniska Kommision. Vaegledningar vid jordborringar... *Medd.* **1**.
SJGKS (1922). Statens Jarnvaegars Geotekniska Kommision Slutbetaenkande. *Medd* **2**.
Skempton, A. W. (1958). Arthur Langtry Bell and his contribution to soil mechanics. *Géotechnique* **8**, No. 4, 143–157.
Skempton, A. W. (1960). Significance of Terzaghi's concept of effective stress. In *From theory to practice in soil mechanics*, pp. 42–53. New York: Wiley.
Skempton, A. W. (1979). Landmarks in early soil mechanics. *Proc. Eur. Conf. Soil Mech. Found. Engng, Brighton* **5**, 1–26.
Skempton, A. W. (1990). Historical development of British embankment dams to 1960. In *Proceedings of the conference on clay barriers for embankment dams*, London, 15–52.
Skibinski, K. (1885). Theorie des Erdrucks. *Z. Oesterr. Ing. Archit.-Ver.* **37**, 65–76.
Sokolovskii, V. V. (1960). *Statics of soil media*. London: Butterworth.
Sutherland, J. (2001). *Historic concrete: Background to conservation*. London: Telford.
Taylor, D. W. (1948). *Fundamentals of soil mechanics*. New York: McGraw-Hill.
Terzaghi, K. (1917). Die Unzulanglichkeit veroeffentlicher Baubeschreibungen. *Z. Oesterr. Ing. Archit.-Ver.* **69**, 325–328.
Terzaghi, K. (1919). Die Erddruckerscheinungen in oertlich beauspruchten Schuettungen... *Oesterr. Woch. F. d.oeff. Baudienst.* 194–199, 206–210, 218–223.
Terzaghi, K. (1920). Old earth pressure theories and new test results. *Engng News Rec.* **25**, 632–634.
Terzaghi, K. (1923). Die Berechnung der Durchlassigkeitsziffer des Tones. *Akad Wiss. Wien. Sitz. Math.-Naturwiss.* **IIa**, No. 3–4, 125–138.
Terzaghi, K. (1924). Die Theorie der hydrodynamischen Spannungsercheinungen und ihr erdbautechnisches Anwendungsgebiet. *Proc. Int. Cong. Appl. Mech., Delft* **1**, 288–294.
Terzaghi, K. (1925a). *Erdbaumechanik*. Vienna: Deuticke.
Terzaghi, K. (1925b). Principles of soil mechanics. *Engng News Rec.* **95**, 742–746, 796–800, 832–836, 874–878, 912–915, 987–999, 1026–1029, 1064–1068 (published by McGraw-Hill as a book in 1926).
Terzaghi, K. (1936). Discussion on proposed new Boston Building Code. *Proc 1st Int. Conf. Soil Mech. Found. Engng, Harvard* **2**, 240–242.
Terzaghi, K. (1939). Soil mechanics: A new chapter in engineering science. The James Forrest Lecture. *ICE Journal* **12**, 106–142.
Terzaghi, K. (1955). Evaluation of coefficients of subgrade reaction. *Géotechnique* **5**, No. 4, 297–326.
Terzaghi, K. & Froehlich, O. (1936). *Theorie der Setzung von Toschichten*. Leipzig: Deuticke.
Terzaghi, K. & Peck, R. B. (1948). *Soil mechanics in engineering practice*. New York: J. Wiley & Sons.
Terzaghi, K. & Rendulic, L. (1934). Die wirksame Flaechenporositaet des Betons. *Z. Oesterr. Ing. Archit.-Ver.* **86**, 1–9.
Thiem, A. (1870). Die Ergiebigkeit arteisischer Bohrloecher. *J. Gasbeleuchtung verw. Beleuchtungsarten Wasserversorgung* **14**, 450–467.
Thiem, A. (1881). Beitrag zur Kenntniss der Grundwasserverhaltnisser. *J. Gasbeleuchtung verw. Beleuchtungsarten Wasserversorgung* **24**, 686–695.
Thiem, A. (1886) Neue Messungsart naturlicher Grundwassergeschwindigkeiten. Die Ergiebigkeit arteisischer Bohrloecher. *Journal fur Gasb. Wasserversorgung*, **31**, 18–28
US Corps of Engineers (1992). *Builders and fighters*. Washington, DC: US Corps of Engineers.
United States Waterways Experiment Station (1944). *Review of the co-operative triaxial research program*. Vicksburg: WES.
Westerberg, N. (1921). Jordtryck I kohaesionare jordarter. *Tek. Tidskr.* **51**, 25–29, 61–65.
Winkler, E. (1867). *Die Lehre von der Elasticitaet und Festigkeit*. Prag: H. Dominiens.
Woltmann, R. (1791–1799). *Beytrage zur hydraulischen Architektur*. Gottingen: J. C. Dietrich.
Woltmann, R. (1794). *Beytrage zur hydraulischen Architektur*, Vol. **3,** pp. 145–213. Gottingen: J. C. Dietrich.
Woltmann, R. (1799a). *Beytrage zur hydraulischen Architektur*, Vol. 4, pp. 371–389. Gottingen: J. C. Dietrich.
Woltmann, R. (1799b). *Beytrage zur hydraulischen Architektur*, Vol. 4, pp. 295–371. Gottingen: J. C. Dietrich.
Woltmann, R. (1804). *Recherches théoriques et expérimentales sur l'effet des machines et outils, dont on se sert pour produire des mouvements instantanes; principalement sur l'effet du mouton pour l'enfoncement des pieux*. Gottingen: H. Dieterich.

FURTHER READNG

The above paper on the early development of soil mechanics is a much abbreviated account of major developments before 1948. In the course of researching that paper a considerable number of papers, almost all held in the ICE library, were reviewed. The bibliography that follows reflects that research and includes many key papers and books published before 1948, as well as some more recent work that provides a useful insight into practice before that time. As is clear from the journal paper, thousands of articles had been published in this area before 1948. For those interested in the subject I strongly recommend they visit our library and see for themselves.

Algoma Steel Corporation. (1942). *Algoma steel sheet piling*. Ontario: Algoma Steel Corporation.
American Society for Testing and Materials (ASTM). (1944). *Procedures for testing soils*. Philadelphia: ASTM.
American Society of Civil Engineers (ASCE). (1915–1926). Special Committee to codify present practice on the bearing value of soils for foundations. Progress reports. *ASCE Proc.* **41-52,** *passim*.
American Society of Civil Engineers. (1938). Practical aspects of soil mechanics: symposium. *ASCE Trans.* **103**, 1377–.
American Society of Civil Engineers (1944). Military airfields: a symposium. *ASCE Trans.* **110**, 669–848.
Anon. (1845a). Terrace new pier; Gravesend. *Illustrated London News* **6**, 5 April 1845, 1–2.
Anon. (1845b). Lighthouse on the Goodwin Sands. *Mechs Mag.* **43**, 9 August.
Anon. (1847). Pneumatic pile driving. *Civ. Engr. Arch. J.* **10**, December, 385.
Anon. (1850). Shannon iron bridge. *Civ. Engr. Arch. J.* **13**, December, 392.
Anon. (1865). Jennings' mode of constructing caissons, cofferdams, etc. *The Engineer* **19,** 43.
Anon. (1895). *A short history of the acquisition and restoration of the Royal Terrace Pier by pilots in the year 1893*. London.
Anon. (1905a). American steel sheet piling. *The Engineer* **100**, 435–436.
Anon. (1905b). Steel sheet piling. *Engng Record*, 23 November, 545–546.
Anon. (1906). Palplanches metalliques, systeme Krupp. *Genie civil,* **49**, 109 (*Centralblatt der Bauwerwaltung,* 4 April 1906).

Anon. (1932). Steel-pile foundations in Nebraska. *Civ. Engng* **2**, 553–557.

Appleby Frodingham. (1948). *Steel sheet piling*. Scunthorpe: Appleby Frodingham.

Arnold, H. H. (1941). Air forces and military engineers. *Proc. ASCE* **70**, 27–89.

Associazione Geotecnica Italiana (AGI). (1985). *Geotechnical engineering in Italy: an overview, 1985*. Rome: AGI.

Atterberg, A. (1910). Die Plastizitaet und Cohaerenz der Tone. *Chemische Zeitschrift* **34,** 369–371; 379–380.

Atterberg, A. (1911). Lerornas foerhallande till vatten, deras plasticitetsgraenser och plasticitetsgrader. *Int Mitt fuer Bodenkunde* **1,** 10–43.

Atterberg, A. (1916). Konsisternslaeran, en ny fysikalisk laera. *Sv. Kemisk. Ts.* **28**, 29–37 und Bindigkeit liefernde Bestandteile der Tone. *Int Mitt fuer Bodenkunde* **3,** 291–330.

Baker, B. (1881). On the actual lateral pressure of earthwork. *Min. Proc. ICE* **64**, 140–186.

Barentsen, P. (1936). Short description of a field testing instrument with a cone shaped sounding apparatus. *First ICSMFE*, **1**, 7–10.

Barlow, C. (1896). Discussion on Greathead (1895). *Min. Proc. ICE* **123**, 75–76.

Barre de Saint Venant, A. J. C. (1870a). Sur une determination rationelle par approximation de la poussee des terres. *Comptes rendus Acad. Sci.* (Paris) **70**, 228–235; 281–286.

Barre de Saint Venant, A. J. C. (1870b). Recherche d'une deuxieme approximation dans le calcul rationnel de la poussee exercee contre un mur. *Comptes rendus Acad. Sci.* (Paris) **70**, 717–724.

Barre de Saint Venant, A. J. C. (1874). Examen d'un essai de la theorie de la poussee des terres. *Comptes rendus Acad. Sci.* (Paris) **77**, 778–781.

Bateman, J. F. L. (1884). *History and description of the Manchester waterworks*. Manchester: T. J. Day.

Bazalgette, J. W. (1877–78). The Victoria, Albert and Chelsea embankments. *Min. Proc. ICE* **54**, 1–60.

Belidor, B. F. (1729). *La science des ingénieurs dans la conduite des travaux de fortification et d'architecture civil*. Paris: Jombert.

Belidor, B. F (1747). *Architecture hydraulique*. Paris: Jombert.

Bell, A. L. (1921). The bearing power of soils. *Proc. ICE Engng Conf.* **2,** 74–76.

Bérigny, C. (1832). *Mémoire sur un procédé d'injection propre à prévenir ou arrêter les filtrations sous les fondations des ouvrages hydrauliques*. Paris.

Beste, H. J. (2003). Foundations and wall structures in the basement of the Colosseum in Rome. *1st Int. Congr. Const. History* **1,** 373–380.

Bidwell, H. S. (1906). Outer barrier, Hodbarrow iron mines, Millon, Cumberland. *Min. Proc. ICE* **165**, 165–173, 193.

Bordoni, A. (1829). *Degli argini di terra*. Pavia.

Borthwick, M. A. (1836). Memoire on the use of cast iron in piling, particularly at Brunswick Wharf, Blackwall. *ICE Trans.* **1**, 195–206.

Bosch, A. & van der Hamm, W. (1998). *Twee eeuwen Rijkswaterstaat*. Zalthommel: Europeese Bibliothetek.

Bossut, C. & Viallet, G. (1764). *Recherches sur la construction la plus advantageuese des digues*. Paris: Firmin Didot.

Boussinesq, J. (1876). Essai theorique sur l'equilibre des massifs pulvurents et sur la poussee des terres sans cohesion. *Mem. Couronn.* **40**, **4**, 12, 112, 150.

Boussinesq, J. (1882). Sur la determination de l'epaiseur minimum que doit avoir un mur vertical. *Annales des ponts et chaussees* **6**, No. 3, 625.

Boussinesq, J. (1883). Note sur la poussee horizontale d'une masse de sable. *Annales des ponts et chaussees* **6**, No. 6, 494, 510.

Boussinesq, J. (1884). Compliment a de precedentes notes sur la poussees des terres. *Annales des ponts et chaussees* **6**, No. 7, 443.

Boycott, G. W. M. (1909). *Compressed air work and diving*. London: Crosby Lockwood.

Boyd, J. R. (1925). Procedure for testing subgrade soils. *Public Roads* **6**, 34–39, 41.

Brees, S. C. (1838). *Railway practice,* 1st series. London: J. Williams.

Brereton, R. P. (1861). Description of the centre pier of the Saltash Bridge on the Cornwall Railway, and the means employed for its construction. *Min. Proc. ICE* **21**, 268–276.

British Steel Piling Company. (1920). *BSP pocketbook*. Ipswich: British Steel Piling Company.

British Steel Piling Company. (c.1929). *Nag Hammadi barrage*. Suffolk: British Steel Piling Company.

British Steel Piling Company. (c.1940). *Larsen sheet piling*. BSP publication 173; *The durability of Larsen steel piling*. BSP publication, 176; *BSP pocketbook*, 5th ed. Suffolk: British Steel Piling Company.

Brown, R. A. (1983). *History of the King's works: the middle ages*. London: HMSO, p. 434.

Brunlees, J. (1858). Description of the iron viaducts erected across the tidal estuaries of the rivers Leven and Kent, in Morecambe Bay, for the Ulverston and Lancaster Railway. *Min. Proc. ICE* **17**, 442–448.

Brunlees, J. & Eckersley, W. (1868). Discussion on supporting power of piles. *Min. Proc. ICE* **27**, 309–319.

Bulleid, A. (1894). The lake village near Glastonbury. *Somerset Archaeological and Natural History Society, Proc.* **40**, 141–181.

Bulleid, A. (1938). *The lake village of Somerset*, 3rd edn. Glastonbury: Glastonbury Antiq. Soc.

Burmister, D. M. (1943). Theory of stresses and displacements in layered systems and application to the design of airport runways. *Proc. HRB* **23**, 126–148.

Caesar, J. (1917). *The Gallic war*. London: Heinemann, (trans. H. J. Edwards) pp. 200–203.

Cain, W. (1911). Experiments of retaining walls and pressures on tunnels. *ASCE Trans.* **72,** 403–448.

Candrelier. (1913). Emploi des palplanches sur les chantiers de la Compagnie Parisienne de Distribution d'Electricité. *Annales des Ponts et Chaussées*, **17**, 445–453.

Carson, A. B. (1965). *Foundation construction*. New York: McGraw-Hill, pp. 154–157.

Casagrande, A. (1932). Research on the Atterberg limits of soils. *Public Roads* **13**, 121–130.

Casagrande, A. (1947). Classification and identification of soils. *ASCE Proc.* **73**, 783–809.

Chrimes, M. (2003). Building the London & Birmingham Railway. In *Robert Stephenson – the eminent engineer.* (M. R. Bailey (ed.)). London: Ashgate, pp. 241–261.

Christow, C. (2003). Die erste fotographische Untersuching der Grusebruchvorgange inter einen Fundament. *OIAZ* **148**, 177–180.

Collin, A. (1846). *Recherches expérimentales sur le glissement spontané des terrains argileux*. Paris: Delmont.

Considere, A. (1870). Note sur la poussee des terres *Annales des ponts et chaussees* **IV s. 19,** 547–594.

Cooling, L. F. & Smith, D. B. (1935). The shearing resistance of soils. *ICE J.* **3**, 333–343.

Cooling, L. F. & Smith, D. B. (1936). The shearing resistance of soils. *1st ICSMFE*, **1**, 37–41.

Copperthwaite, W. C. (1906). *Tunnel shields and the use of compressed air in subaqueous works*. London: Archibald Constable & Co. Ltd.

Cornfield, G. M. (1974–1999). *Steel bearing piles*, 4 edns. Croydon and Ascot: Constrado and SCI.

Cummings, E. M. (1960). Cellular cofferdams and docks. *ASCE Trans.* **125**, 13–45.

Dahlberg, C. (1913). Skred och saettningar eller s.k. ras vid jaernsvaegbyggnadr. *Tek. Tidskr.* **43**, 129–141.

Darcy, F. G. (1856). *Les fontaines publiques de la ville de Dijon*. Paris: Dalmont.

Darcy, H. (1856). *Les fontaines publiques de la ville de Dijon*. Paris: Victor Dalmont.

Davis, S. N. (1973). Teodoro Ardemanns pioneer water-supply engineer of Spain. *AWRA J.* **9**, 1028–1034.

Day, R. A. & Potts, D. M. (1989). *A comparison of design methods for propped sheet pile walls*. Ascot: SCI publication 77.

De Boer, R. & Ehlers, W. (1990). The development of the concept of effective stress. *Acta mechanica*, **83**, Nos 1–2, 77–92.

De Boer, R., Schiffman, R. L. & Gibson, R. E. (1996). The origins of the theory of consolidation: the Fillunger–Terzaghi dispute. *Géotechnique*, **46**, No. 2, 175–186.

Delprat, J. P. (1821). *Verhandeling over de zijdelingsche drukking der aarde*. Delft: Pile Groot.

Donath, A. D. (1891). Untersuchungen ueber den Erdruck auf Stuetzwande. *Zeit. F. Bauwesen*, **41**, 491–518.

Dorman Long (1938). *Handbook*. Middlesbrough, pp. 125–129.

Dunnicliff, J. & Young, N. P. (2006). *Ralph B Peck: educator and engineer*. Vancouver: Bitech.

Dupuit, A. J. E. J. (1863). *Etudes theoriques et pratiques sur le movement des eaux dans les canaux decouverts et a travers les terrains permeables*. Paris: Dunod.

Dupuit, A. J. E. J. (1865). *Traite theorique et pratique de la conduite et la distribution des Paris*. Paris: Dunod.

Durkee, E. L. and McIntosh, R. C. (1937). Structural steel bearing piles: their use and capacity. *Boston Soc. Civ. Engrs J.* **24**, 78–104.

EAU 1990 (1992). *Recommendations of the Committee for Waterfront Structures, Harbours and Waterways,* 6th English edition. Berlin: Ernst.

Ekwall, R. (1911). Beraekning av grundpalningar. *Tek. Tidskr.* **41**, 79–83.

Engesser, F. (1880). Geometrische Erddruck-Theorie. *Zeitschrift fur Bauwessen*. **31,** Col 190–210, Blatt E.

Esselborn, K. (1910). *Lehrbuch des Tiefbaues,* 4 Aufl, Band I., p. 149. Leipzig: Engelmann.

Ewart, P. (1822). Specification for a new method of making a cofferdam. *Repertory of arts and manufactures*, 2nd series, **43**, 193–202 and plate 9.

Eytelwein, J. A. (1800). *Praktische Anweisung zur Konstruktion der Faschinenwerke an Flüssen und den dazu gehörigen Anlagen an Flüssen und Strömen*. Berlin.

Eytelwein, J. A. (1822). *Handbuch der Mechanik fester Koerper und der Hydraulik etc.*, 3rd edn. (1842). Leipzig: Fleischer.

Fahlbusch, H. (2001). *Historical dams*. New Delhi: ICID.

Fang, H. Y. & Dismuke, T. D (1970). *Design and installation of pile foundations and cellular structures*. Lehigh: Envo.

Fannin, J. (2007). Karl Terzaghi: from theory to practice in geotechnical filter design. *ASCE J. Geotech. Geoenviron. Eng.* **134**, No. 3, 267–276.

Fellenius, W. (1910). Om beraekning af kajbyggnader. *Tek. Tidskr.* **40**, No. 3, 29–33.

Fellenius, W. (1918). Kaj och jordrasen I Goeteborg. *Tek. Tidskr.* **48**, 17–19.

Fellenius, W. (1919). Utredning betraefflande kajbyggnader foer Trondhjems hamm. *Tek. Tidskr.* **49**, 1–19.

Fellenius, W. (1926). Jordstatiska beraekningar med friction och kohaesion. *Kungl Vaeg och Vattenbyggarkarens 75 arsskrift*, pp. 79–127.

Fillunger, P. (1914). Neuere Grundlagen fur die statische Berechung von Talsperren. *OIAVZ*, 441–447.

Fillunger, P. (1915). Versuche ueber die Zugfestigkeit bei alltigem Wasserdrueck. *Oest.Woch.oeffentlichen Baudienst*, **29,** 443–448.

Fillunger, P. (1929). Auftrieb und Unterdruck inTalsperren. *Die Wasserwirtschaft* **22**, 334–336.

Fillunger, P. (1930). Auftrieb und Unterdruck in Staumauern. *World Power Congress* **9**, 323–329.

Findlay, A. G. (1847). On lighthouses and beacons. *Trans. Soc. Arts* **56,** 269–271.

Fleming, S. (1876). *The Intercolonial*. Montreal: Dawson, pp. 200–202.

Forchheimer, P.(1886). Ueber die Ergiebigkeit von Brunnen-Anlagen. *Zeitschrift AIV Hannover*, 539–586.

Forchheimer, P. (1898). Grundwasserspiegel bei Brunnenlagen. *OIAV-Z*, 629–635, 645–648.

Forchheimer, P. (1914). *Hydraulik*. Leipzig: Teubner.

Fowler, C. E. (1914). *A practical treatise on subaqueous foundations*. New York: John Wiley.

Fox, C. (1850). *Minutes of evidence, 9 July, Select Committee on the Westminster Temporary Bridge Bill,* pp. 23–24. London: HMSO.

Français, J. F. (1820). Recherches sur la poussée des terres, sur la forme et les dimensions des murs de revêtements et sur les talus d'excavation. *Mémorial de l'officier du génie* **4**, 157–206.

Francke, A. (18–). Die inneren Krafte eines durch Ebenen begrenzten Erdkorpers, nebst Anwendung auf die Ermittelung des Druckes gegen stutz – und Druckwande. *Zeitschrift f Bauwessen*, AIVH-Z, **34**, 707-726.

Franco, L. (1996). History of coastal engineering in Italy. In *ASCE. History and heritage of coastal engineering.* New York: ASCE.

Franzius, O. (1927). *Der Grundbau*. Berlin: Springer.

Freeman, J. R. (1903). *Report of Committee on Charles River dam.* Boston: Water Commission.

Frick, P. (1926). *Fouilles et fondations*. Paris: Dunod.

Fuss, N. (1797). Examen theorique des revetements a dos incline. *Nova Acta Academiae Scientarum Imp Petropolitanae* **13**, 80–100.

Gadroy, F. (1746). Mémoire sur la poussée des terres. *Archives du corps du Génie*, summarised in Mayniel (1806): see paper.

Galloway, R. L. (1882). *A history of coal mining in Great Britain*. London: Macmillan.

Gaudard, J. (1867–1877). On foundations. *Min. Proc. ICE*, **50**, 112–147.

Gauthey, E. M. (1784). Mémoire sur l'epaisseur que l'on doit donner aux murs de soutenement pour resister de la poussee des terres. *Nouv. Mem de Dijon* **II**, 238–266; (1785). **I**, 1–45.

Germany (1938). *Die Reichsautobahnen*. Berlin: Volk und Reich Verlag.

Glossop, R. (1960–61). The invention and development of injection processes. *Géotechnique* **10**, No. 3, 91–100; **11**, No. 4, 255–279.

Gobin, A. M. (1883). Determination precise de la stabilite des murs de soutenement et de la poussee des terres. *Annales ponts et chausses* **6**, No. 6, 98–231.

Goldbeck, A. T. (1925). Researches on the structural design of highways by the USBPR. *Trans. ASCE* **88,** 264–300.

Goldbeck, A. T. & Jackson, F. H. (1921). Tests for soils with relation to their use in the subgrade of highways. *ASTM Proc.* **21**, 1057–1072.

Golder, H. Q. (1946). Relationship of runway thickness and undercarriage design to the properties of subgrade soil. *ICE Airfield paper 4, Engng. Div.*, Vol. V.

Golder, H. Q. (1948). Coulomb and earth pressure. *Géotechnique* **1**, No. 1, 66–71.

Goodrich, E. P. (1904). Lateral earth pressure and related phenomena. *ASCE Trans.* **53**, 272–304.

Gray, H. S. G. (1911–17). *The Glastonbury lake village*, 2 vols. Glastonbury: Glastonbury Antiq Soc.

Greathead, J. H. (1896). The City and South London Railway. *Min. Proc. ICE* **123**, 39–123.

Greenwell, G. C. (1855). *A practical treatise on mine engineering*. Newcastle-upon-Tyne: M & M. W.Lambert.

Gullander, P. (1902). Bidrag till teorien foer grundpalning. *Tek. Tidskr.* **32**, 51–52; 58–62; 69–70.

Hagen, G. (1847–1865). *Handbuch der Wasserbaukunst,* (2nd edn, 1874). Berlin: Ernst.

Handbuch der Ingenieur Wissenschaften (1884). 2 Aufl, 1, 2, pp. 329–334. Leipzig: Endelmann.

Handbuch der Ingenieur Wissenschaften (1906). 4 Aufl, I, 3,45. Leipzig: Endelmann.

Harding, H. J. B. (1981). *Tunnelling history and my own involvement*. Toronto: Golder.

Hayter, H. (1863). The Charing Cross Bridge. *Min. Proc. ICE* 512–517.

Hazen, A. (1895). Water supply of various European cities. *J. Assoc. Engng. Socs.***14**, 112–130.

Heinrich, G. (1938). WissenschaftlicheGrundlagen der Theorie der Setzung von Tonschichten. *Wasserkraft und Wasserwirtschaft* **33,** 97–98.

Hersent, H. J. (1889). *Travaux publics. Ouvrages exécutés au moyen de l'air comprime. Dragage, dérochements, terrassements, outillage. Description des moyens d'exécution, machines, engins et installations diverses.* Paris.

Hersent, H. J. & G. (1899). *Entreprises de travaux publics et maritimes, fondations a l'air comprime, dragages, dérochements, bassins de radoub, etc.* Paris: Imprimerie Chaix.

Hersent, H. J. & G. (1906). *Note sur l'emploi de l'air comprime pour l'exécution des ouvrages hydrauliques et spécialement des fondations. Expériences faites a Bordeaux pour démontrer qu'il est possible de travailler a de plus grandes profondeurs que celles usitées.* Paris: Dunod.

Hill, C., *et al.* (1979). *The Roman riverside wall and triumphal arch: excavations in Baynard's Castle, 1974–76*. London: London & Middlesex Archaeological Society.

Holzmuller, G. (1882). *Einfuehrung in die theorie der Isogalen Verwandschaften und der Konformen Abbildungen.* Leipzig: Teubner.

Hool, G. A. & Kinne, W. S. (eds). (1923). *Foundations, abutments and footings*. New York: McGraw Hill (pp. 22–25 summarise allowable bearing capacity according to US building codes).

Hope, C. W. (1845). Experiments on the pressure of earth against

revetments and the best form of retaining walls. *Papers Corps of Royal Engineers* **7**, 69–86.
Horslev, M. J. (1949). *Subsurface sampling of soils for civil engineering*. Vicksberg: USWES.
Housel, W. S. (1929). A practical method for the selection of foundations based on fundamental research in soil mechanics. *Univ. Michigan Engng Res. Bull.*, No. 13.
Housel, W. S. (1935). A penetration of measuring soil resistance. *ASTM Proc*. **35**, 472–490.
Housel, W. S. (1936). Research in foundations and soil mechanics. *IABSE, Proc. 2nd Congress, Final report*. Zurich: IABSE, pp. 852–873.
Hughes, J. (1851). On the pneumatic method adopted in constructing the foundations of the new bridge across the Medway at Rochester. *Min. Proc. ICE* **10**, 356.
Hughes, J. D. (1840). *A practical inquiry into the laws of excavations and embankments upon railways*. London.
Hultin, T. (1911). Om berakning av grundpahlning. *Industrietidn. Noren* 370–372.
Hultin, T. (1937). Forsoek till bestaemning av Goeteborgslarens hallfasthet. *Tekn. Samfundets Handlingar* **2**, 87–105.
Humber, W. (1863). *Record of modern engineering*, pp. 42–44.
Humber, W. (1870). *A complete treatise on cast and wrought iron bridge construction*, 3rd edn, pp. 180–181, 247. London: Crosby Lockwood.
Hveem, F. N. & Crmany, R. M. (1948). The factors underlying the rational design of pavements. *Proc. HRB* **28**, 1–36.
ICOLD (1933). *Proc. 1st Int.l Congr. on Large Dams*. Paris:CIGB.
Institution of Civil Engineers (1949). *Conference on biology and civil engineering*. London: ICE.
Jacoby, H. S. & Davis, R. P. (1941). *Foundations of bridges and buildings*. New York & London: McGraw Hill, pp. 198–215.
Jankilowski, J. (1973). Mosty w polsce i mostowcy polsey. *Polska Akademia Nank Zalkad Histori,Nauki, Techniki. Monografia* **83**, 9–11, 301.
Jenkin, C. F. (1931). The pressure exerted by granular material: an application of the principles of dilatancy. *Proc. Royal Soc.* **131A**, 53–89.
Jenkin, C. F. (1933). The pressure on retaining walls. *Min. Proc. ICE* **234**, 103–154.
Jenkin, C. F. & Bevan, R. C. (1934). *Earth pressure tables*, BRS SR 24. Garston: BRS.
Josef, D. (1980). A comparison of rare, remarkable and historical bridges in Bohemia, Moravia and Poland. *Sbornik vysokeho Uceni Tech, v Brno, Rocnik* **97**.
Kempe, H. R. (1923). *Engineer's yearbook*. London: Kempe, pp. 505–509.
Khosla, A. (1936). *Design of weirs on permeable foundations*, CBIP publication 12. Pune: CBIP.
Kingsbury, W. J. (1858–1859). Description of the entrance lock, and jetty works of the Victoria (London). Docks, *Min. Proc. ICE* **18**, 447.
Kjellman, W. (1936). Report on an apparatus for consummate investigation of the mechanical properties of soils. *Proc. 1st[t] ICSMFE*, **2**, 16–20.
Kiersch, G. A. (1991). *The heritage of engineering geology: the first hundred years*. Boulder: Geological Society of America.
Klanica, Z. (1968). The results of the 14th period of investigations at Mikilcike. *Prehlad vyskumu CSAU, Brno Arch. Inst*, 61. Brno.
Koegler, F. & Scheidig, A (1927). Drueckverteilung im Baugrund. *Bautechnik* **4; 5**, 418-421, 445-447; 6, 205-208, 229-232; **7**, 268-272, 828-830.
Koegler, F. & Scheidig, A (1938). *Baugrund und Bauwerk*. Berlin: Ernst.
Koetter, F. (1903). Die Bestimmung des Druckes an gekruemsten Gleitfaechen. *Kon Preuss Ak Wiss Sitzungsbericht.*
Krey, H. D. (1912). *Erddruck*. Berlin: Ernst. (The 3rd–5th editions (1926, 1933, 1936) are those that contain reference to advances in soil mechanics.)
Lefranc, E. (1937). La théorie des poches absorbantes et son application à la détermination de la perméabilité. *Génie Civil.*
Lehuerou-Kérisel, J. (1936). *Contribution à l'étude du frottement des milieux pulvérulents*. Paris: Editions Gauthier-Villars.
Leliavsky, S. (1942). Uplift. *Nature*, 31 January, 135.
Leliavsky, S. (1942). *Uplift in dams*. Cairo: Egypt. Min Pub Works Projects Dept.
Levy, M. (1873). Essai sur une theorie nouvelle de l'equilibre des teres. *Journal des Mathematiques* **18**, 241–300.
Leygue, L. (1885).Nouvelle recherché sur la poussee des terres. *Ann Pont Chauss*, **6**, **10**, ii, 788–1003.
Lugeon, M. (1933). *Barrages et géologie*. Paris: Editions Dunod.
McAlpine, W. J. (1868). The supporting power of piles; and on the pneumatic process for sinking iron columns, as practised in America. *Min. Proc. ICE*, **27**, 275–293.
Mackley, F. R. (1971). A history of sheet piling. ICE Southern Association, Chairman's address. *ICE Tracts*.
Mehrtens, G. (1904). *Vorlesungen ueber Statik der Baukonstruktionen*, vol. II.
Mendonca de Oliveira, M. & Posada Presa, E. (2003). Reinforcing foundations with wood piles: origin and historical development. *Proc. 1st Int. Congr. Const. History, Madrid* **3**, 1537–1545.
Middlebrooks, T. A. & Bertram, G. E. (1950). Adaptation to the design of airfield pavements, *ASCE Trans*. **155**, 468–471.
Mohr, O. (1871–1872). Bitraege zur Theorie des Erdrucks. *ZAIV Hannover* **17**, 67–, 344–, 245–.
Mohr, O. (1882). Ueber die darstellung des spannungszustandes und des deformationszustandes eines koerperelements. *Civilingenieur* **28**, 113–156.
Mohr, O. (1914). *Abhandlungen aus dem Gebiete der technischen Mechanik*. Berlin: Ernst (3rd edn, 1928), pp. 220–240.
Mosher, R. L. (1992). *Three dimensional finite element analysis of sheet-pile cellular cofferdams*. Vicksburg, MI: US Army Corps of Engineers Waterways Experiment Station, USWES report TR ITL-92-1.
Mueller-Breslau, H. F. B. (1881). *Elemente der graphische Statik*. Berlin: Polytechnische Buchhandlung.
Muhs, H. (1969). *1928–1968 40 Jahre DEGEBO*. Berlin: Degebo.
Muhunthan, B. & Schofield, A. N. (2000). Liquefaction and dam failures. *ASCE Geo. Denver 2000*, CUED/D-SOIL.SITR310. Denver, CO: ASCE.
Navier, C. L. M. H. (1833). *Résumé des leçons données à l'Ecole des Ponts et Chaussées sur l'application de la mécanique*, vol. 1. Paris: Editions Carilian-Goeury.
Newman, J. (1893). *Notes on cylinder bridge piers*. London: Spon.
Nygard, B. (1944). *Sarpsfosen*. Oslo: Tell Forlag, pp. 25–32. (Thanks to NGI for a copy of this.)
Okabe, S. (1924). General theory on earth pressure and seismic stability of retaining wall and dam. *JSCE Proc.* No. 591, 283–297.
Olsson, J. (1923). Statens Jaernvaegers Geotekniska Kommission. *Tekn. Uke*. **41**, 310–311, 344.
Olsson, J. (1925). Kolvborr. *Tek. Tidskr.* **52**, 13–16.
Pasley, C. W. (1822). *A course of elementary fortification*, 2 vols. London: Murray (2nd edn, vol. 2 discusses retaining wall experiments and the use of reinforced earth).
Pavlovsky, N. N. (1922). *The theory of the movement of groundwater beneath hydraulic structures*.
Peck, R. B. (1940). Foundations of buildings. In *Chicago Illinois Univ. Exp. Sta. Bull.*
Pepper, S. (2000). Sword and spade: military construction in renaissance Italy. *Construction History* **16**, 13–32.
Perry, J. (1721). *An account of the stopping of Dagenham Breach*. London: T Tooke.
Petermann & Boedeker (1937). Schrifttum ueber Bodenmechanik. *Schriftenreihe der Forschungsgesellschaft fuer das Strassenwesen Heft* **12**.
Polubarinova-Kochina, P. Y. (1962). *Theory of groundwater flow*. Princeton: Princeton University Press.
Porter, O. J. (1938). The preparation of subgrades. *Proc. HRB* **18**, 1–36.
Porter, O. J. (1942). Foundations for flexible pavements. *Proc. HRB* **22**, 100–136.
Porter, O. J. (1950). Development of the original method for highway design. *ASCE Trans*. **115**, 461–467.
Poskitt, T. J. (1997). Discussion: the origins of the theory of consolidation: the Fillunger–Terzaghi dispute. *Géotechnique* **47**, No. 4, 893–895.
Potts, L. H. (1844). *Minutes of evidence, 10 June. Commission upon the subject of harbour of refuge*, pp. 119–122. HMSO: London.
Potts, L. H. (1847). On a pneumatic process for forming foundations for piers, breakwaters and similar structures. *Trans. Soc. Arts* **156**, 441–443.

Proctor, R. R. (1933). Fundamental principles of soil compaction. *Engng News Record*, **111**, 245–248; 28–289; 348–351; 372–375.

Prony, G. C. M. R. (1790). *Nouvelle architecture hydraulique*, Vol. 1. Paris: Firmin Didot.

Prony, G. R. De. (1802). *Recherches sur la poussee des terres*.

Prony, G. C. M. R. (1802). *Recherches sur la poussée des terres, et sur la forme et les dimensions à donner aux murs de revêtement*. Paris: Imprimerie de la epublique.

Prosser, R. D. (1896) & Chrimes, M. M. (1997). Laurence Holker Potts. In *Oxford dictionary of national biography*. Oxford: Oxford University Press.

Ramelli, A. (1588). *Le diverse et artificiose machine*. Paris: Ramelli, chapter 111–112, pp. 171–174.

Rankine, W. J. M. (1862). *Manual of civil engineering*. London: Griffin.

Redman, J. B. (1845). Account of the new cast iron pier, at Milton-on-Thames, near Gravesend. *Min. Proc. ICE* **4**, 222–250.

Rendulic, L. (1937). Ein Grundgesetz der Tonmechanik und sein experimentale Bewies. *Bauingenieur* **18**, 349–467.

Richardson, L. F. (1908a). A freehand graphic way of determining streamlines and equipontentials. *Phil Mag*. **15**, 237–269.

Richardson, L. F. (1908b). The lines of flow in saturated soils. *Royal Dublin Soc. Sci. Proc*. **11**, 295–316.

Road Research Laboratory. (1952). *Soil mechanics for road engineering*. London: HMSO.

Roberts, T. P. (1905). Construction of cofferdams. *Engng News* **54**, 138–140.

Rossow, M., Demsky, E. & Mosher, R. (1987). *Theoretical manual for design of cellular pile structures*. Vicksburg, MI: US Army Corps of Engineers Waterways Experiment Station, USWES technical report, ITL, 85-5.

Rowe, P. W. (1952). Anchored sheet pile walls. *Proc. ICE* **I**, No. 1, 27–70.

Royal Academy of Engineering (RAE). (2007). *Report on status of engineers*. London: RAE.

Salzman, L. F. (1952). *Building in England down to 1540*. Oxford: Clarendon Press, pp. 86, 328–329.

Scheffler, H. (1851). Ueber den Druck im Innern einer Erdmasse. *Crelle's J. Baukunst* **30**, 185–223.

Scheffler, H. (1857). *Theorie der Gewoelbe, Futtermauern und eisernen Bruecken*. Brunswick: Schulbuchhandlung.

Schlumberger, C. (1936). L'auscultation des terrains. *Proc. 2nd IABSE Congess*. Zurich: IABSE.

Schroeder, H. (1955). *Grundbau Taschenbuch*, 2 vols. Berlin: Ernst (Vol. 2 contains standards).

Schnitter, N. J. (1994). *A history of dams*. Rotterdam: Balkema.

Schultze, E. (1953). Hans-Detlef Krey and the experimental approach to soil mechanics. *Géotechnique* **4**, No. 3, 93–96.

Shirley Smith, H. (1976). Royal Albert Bridge, Saltash. In *The works of Isambard Kingdom Brunel* (ed. A Pugsley). Cambridge: Cambridge University Press, pp. 163–170.

Sibley, R. (1832). Motives which induced the adoption of cast iron piles and panels to face the wharf of the lead works at Limehouse. *ICE Original Communication*, **140**, ICE archives.

Sidelor (c. 1952). *Steel sheet piling*. Paris: Steel sheet piling central office.

Sidelor (c. 1957). *Steel sheet piling*, 2nd edn. Paris: Steel sheet piling central office.

Siegler (1885). Experiences nouvelles sur la poussee des terres, *Ass. Franc, Av. Sci. Comptes rendus* **13**, No. 2, 73.

Simms, F. W. (1844). *Practical tunnelling*. London: Troughton & Simms (4th edn, 1896).

Singer, C. (1956). *History of technology*, vol. 2. Oxford: Clarendon Press.

Skelton, R. A. (1944). *Handbook no.22*, 3rd edn. Sheffield: Skelton, pp. 165–169.

Skempton, A. W. (1946). Alexandre Collin pioneer in soil mechanics. *Newc. Soc. Trans*. **25**, 91–104 (see *Géotechnique* **1**, No. 4, 216–221, (1949)).

Skempton, A. W. (1995). Embankments and cuttings on the early railways. *Construction History* **11**, 33–49.

Skempton, A. W. & Chrimes, M. M. (1994). Thames tunnel: geology, site investigation nad geotechnical problems. *Géotechnique* **44**, No. 2, 191–216.

Slichter, C. S. (1899). Theoretical investigation of the motion of ground waters. *USGS 19th Annual Report*, **2**, 295–384. Washington, DC: USGS.

Slichter, C. S. (1902). The motion of underground waters. *USGS WSP 67*. Washington, DC: USGS.

Smith, N. A. F. (1971). *A history of dams*. London: P. Davies.

South Durham Steel and Iron Company (1956). *Larssen piling pocketbook for site engineers*. Middlesbrough.

South Durham Steel and Iron Company (c. 1960). *Larssen piling pocketbook for site engineers*. Middlesbrough.

Spangler, M. G. (1933). The supporting strength of rigid pipe culverts, *Iowa Eng. Exp. Sta. Bull*. No. 104.

Spangler, M. G. (1941). The structural design of flexible pipe culverts. *Iowa Eng. Exp. Sta.. Bull*. No. 157.

Statens Jarnvaegars Geotekniska Kommision. (1917). Vaegledningar vid jordborringar. *Medd*. **1.**

Statens Jarnvaegars Geotekniska Kommision (1922). Slutbetaenkande. *Medd* **2.**

Stefan, J. (1871). Ueber das Gleichgewicht und die Bewegung, insbesondere die Diffusion von Gasmengen. *Kais. Akad. Wiss. Wien, Math.-nat. Kl. Abt IIa*, **63**, 63–124.

Stephens, G. H. (1904). The barrage across the Nile at Asyut. *Min. Proc. ICE* **158**, 30–36; 71, plate 2.

Streiff, A. (1938). Application of soil mechanics. *Symposium on Practical Aspects of Soil Mechanics, ASCE Trans*. **103**, 1492.

Stroyer, J. P. R. N. (1927–1928). Earth pressure in flexible appraisal walls. *Min. Proc. ICE* **226**, 94– .

Suers, E. (1862). *Der Boden der Stadt Wien*.

Taylor, T. U. (1900). *The Austin Dam*. USGS WSP, 40.

Telford, T. (1814). Article on bridge practice. In *Edinburgh encyclopaedia*. Edinburgh: Brewster.

Tennessee Valley Authority (TVA) (1957). *Steel sheet piling cellular cofferdams on rock*. TVA technical monograph, p. 75.

Terzaghi, K. (1922). Der Grundbruch der Stauwerken an Stauwerken und seine Verhutung. *Die Wasserkraft* **17**, 445–449.

Terzaghi, K. (1930). Die Tagfaehigkeit von Pfahlgruendungen. *Bautechnik*, **8**, 475–478; 517–521.

Terzaghi, K. (1933). Austrieb und Kappillerdruck am betonierten Talsperren. *ICOLD*, **1**, Q2b, 5–13.

Terzaghi, K. (1933). Investigation of the characteristics of soils in respect to their suitability for the construction of earth dams. *ICOLD*, **1,** Q 2a, 317-324.

Terzaghi, K. (1935). The actual factor of safety in foundations. *Struct. Engr* **13**, 126–160.

Terzaghi, K. (1936). The shearing resistance of saturated soils. *Proc. 1st Int. Conf. Soil Mechanics and Foundation Engng* **1**, 54–56.

Terzaghi, K. (1938). *Report on the stability of the embankment for the reservoir no.2 Chingford*. Mowlem archive, ICE.

Terzaghi, K. (1943). *Theoretical soil mechanics*. New York: John Wiley.

Terzaghi, K. (1944). Stability and stiffness of cellular cofferdams. *ASCE Trans*. 1083–1202.

Terzaghi, K. (1954). Anchored bulkheads. *ASCE Trans*. **119**, 1243–1324.

Thiem, A. (1906). *Hydrologische metoden*. Leipzig.

Tiedemann, B. (1937). Ueber die Schubfestigkeil bondiger Boden. *Bautechnik* **15**, 400–433.

Tomlinson, M.J. (1995). *Foundation design and construction*, 6th edn, pp. 232–264. Harlow, UK: Longman.

Tschebotarioff, G. P. (1948).Large scale model earth pressure tests on flexible bulkheads, *ASCE Proc*. **74**, 9–48.

Tschebotarioff, G. P. (1949). *Large scale earth pressure tests with model flexible bulkheads*, final report. US Bureau of Yards and Docks. Princeton: Princeton University Press.

Tschebotarioff, G. P. (1951). *Soil mechanics, foundations and earth structures*. New York: McGraw Hill.

Triger (1841). Memoire sur un appareil a air comprime pour le percement des puits de mines et autres travaux sous les eaux et dans les sables submerges. *Comptes rendus Acad. Sci*. (Paris) **30,** July, December.

Tsinker, G. P. (1983). Anchored sheet pile bulkheads: design practice. *ASCE J.Geotech. Engng* **109**, No. GT8, 1021–1038.

Tuerke, H. (1999). *Statik im Erdbau*. Berlin: Ernst.

Turhollow, A. F. (1992). Airfields for heavy bombers. In *Builders and fighters*, pp. 207–217. Washington, DC: US Army Corps of Engineers.

United States Bureau of Reclamation. (1934, 1936). *Laboratory procedure in testing earth dam materials*, reports No. TM 410; TM 533 (1936). Denver: USBR.

United States Corps of Engineers. (1939). *Report of a slide of a portion of the upstream face of the Fort Peck Dam*. Washington, DC: USGPO.

United States Corps of Engineers. (1996). *Design of sheet pile walls*. New York: ASCE.

United States Public Roads Administration. (1943). *Principles of highway construction as applied to airports*. Washington, DC: USBPR.

United States Waterways Experiment Station. (1945). *The California Bearing Rotation test as applied to the design of flexible pavements*, WES TM 213-1. Vicksburg, USWES.

United States Waterways Experiment Station. (1945). *Certain requirement for flexible pavement design for B29 planes*. Vicksburg, USWES.

Van de Ven, C. P. (ed.). (2004). *Man-made lowlands,* 4th edn. Utrecht: Stichting Matrijs for ICID.

Van Ysselsteyn, H. A. (1908). *The port of Rotterdam*. Rotterdam: Nijgh & Van Ditmar.

Vauban, S. (1737). *Travaux de soutènement des fortifications. Traité de la défense des places*. The Hague: de Hodt.

Virgin, A. (1918). Marktans hoejdfoeraendrungar hos loess jordlager. *Tek. Tidskr.* **48**, 113–115.

Vitruvius, (1931). *On architecture* (ed. and transl. K Ganger), vol. 1, book 3, chapter iv, pp. 180–183; book 5, chapter xii, pp. 311–357; vol. 2, book x, p. 275. London: Heinemann.

Von Martony de Koszegh. (1828). *Versuche ueber den Sectendruck der Erde*. Vienna. (Copy in British Library.)

Wagner, C. J. (1881a). Absitzungen und Rutschungen an Schutthalden von Gesteinen aelterer Formationen. *Allgemeine Bauzeitung* **46**, 21–24. 3 Blaetter.

Wagner, C. J. (1881b). Uferschutzbauten der Saltzburg-Tiroler Bahn. *Allgemeine Bauzeitung* **46**, 85–88. 3 Blaetter.

Wagner, E. (1884). *Beziehungen der Geologie zu den Ingenieur Wissenschaften*. Vienna: Spielhage. (Copy in Bodleian Library.)

Walker, F. C. (1937). *Development of earth dam design in the Bureau of Reclamation*. Denver, CO: US Bureau of Reclamation.

Weingarten, J. (1869). Vortrag ueber Erddrueck. *Zeit. fuer Bauwesen*. **20**, 122–123.

Wendel, E. (1900). Om Profbelastning pa palar. *Tekn. Samfundets Handlingar* **7**, 3–62.

Westergaard, H. M. (1926). Stresses in concrete pavements. *Public Roads* **7**, 25–35.

Westergaard, H. M. (1939). Stresses in concrete runways of airports. *Highways Research Board Proc.*, 197–202.

Weyrauch, A. (1878). Zur Theorie des Erdruckes. *Zeitschrift f Baukunde* **X**, cols. 193–208.

Weyrauch, A. (1881). Theorie des Erdruckes, *Allgemeine Bauzeiting* **45**.

Wilhelmi, J. (1910). Die eiserne Spundwand von Larssen, *VDI-Zeitschrift* **54**, 2094–2098.

Williams, B. P. & Waite, D. (1993). *The design and construction of sheet-piled cofferdams,* CIRIA SP95. London: Construction Industry Research and Information Association.

Williams, R. & Norbury, D. (2008). Rudolph Glossop and the development of geotechnology. *QJEG*, anniversary issue, in press.

Winkler, E. (1871). Die neue Lehre Erdruck *OIAV-Z* **43**, 79–89, 117–125.

Winkler, E. (1872b). *Die neue Lehre des Erdruckes nebst einer Geschichte der Theorie*. Wien: Waldheim.

Winkler, E. (1880*). Vortraege ueber dir Theorie des Erddruckes.* Berlin: Technische Hochschule (Lith. notes).

Winkler, E. (1885). Ueber Erddruck auf Gebroschen und gekrumonte Wandflachen. *Centralblatt der Bauwerwaltung,* **5**, 73–76.

Wittman, W. (1877). *Beitrag zur Theorie des Erddrueckes*. Bay: AIV-Z.

Woodworth, R. B. (1909). Steel sheeting and steel piling, *ASCE Trans*. **64**, 476–519.

Simpson, B. & Tatsuoka, F. (2008). *Géotechnique* **58**, No. 5, 357–368 [**doi:** 10.1680/geot.2008.58.5.357]

Geotechnics: the next 60 years

B. SIMPSON* and F. TATSUOKA†

The authors were set the challenge of thinking about the next 60 years, not only of *Géotechnique* but of geotechnical engineering in general. They have attempted to do this by talking to colleagues, by dreaming a little, but mainly by trying to extrapolate the trends they see in society and in present-day geotechnical activity. Although the principal emphasis of *Géotechnique* tends to be research, this is driven by the requirements of design and construction. Hence, in looking to the future, even the future of research, likely trends in geotechnical construction activities are considered first, particularly noting the emerging importance of energy conservation and carbon dioxide reduction. This leads to discussion of the characterisation of material behaviour and investigation of the ground, and the development of design processes. The future of dissemination of geotechnical information is also discussed.

KEYWORDS: constitutive relations; environmental engineering; ground improvement; laboratory tests; numerical modelling and analysis; site investigation; soil stabilisation

Le défi, pour les auteurs, c'était de réfléchir sur les 60 années à venir. Non seulement sur le plan de *Géotechnique*, mais sur celui de la géotechnique en général. Ceci, ils se sont efforcés de le faire en s'entretenant avec des collègues, un peu dans un rêve, et surtout en essayant d'extrapoler les tendances qu'ils perçoivent dans la société et dans les activités actuelles dans le domaine de la géotechnique. Bien qu'en géotechnique, on mette généralement l'accent sur la recherche, la motivation est principalement les exigences de l'étude et de la construction. Par conséquent, lorsqu'on se penche sur l'avenir, même l'avenir de la recherche, on examine en premier lieu les tendances probables dans les activités de construction géotechnique, en relevant notamment l'importance croissante de la conservation de l'énergie et la réduction du gaz carbonique. Ceci porte à des discussions sur la caractérisation du comportement des matières et des études du sol, et sur le développement de méthodes d'étude. On discute également de l'avenir dans la divulgation d'informations de géotechnique.

INTRODUCTION

In this paper, the authors try to look into the future of geotechnical engineering, for the next 60 years. The term 'geotechnical engineering' is taken to include all forms of engineering activity in the ground, and in which the behaviour of the ground is relevant; investigation, design, construction, maintenance and related research are all included. Besides soil mechanics, other necessary skills might include hydrogeology, geology, biology, chemistry, rock mechanics, structural engineering, thermodynamics, and many others. As pointed out by Henkel (1982), an overlap of skills is essential to successful geotechnical engineering.

Although the principal emphasis of *Géotechnique* tends to be research, this is driven by the requirements of design and construction. Hence, in looking to the future, even the future of research, likely trends in geotechnical construction activities are considered first: these lead to consideration of the characterisation of material behaviour and investigation of the ground, and the necessary design approaches.

In preparation for the paper, the authors sent out a questionnaire to a large number of colleagues, asking questions about their views of the future of the industry and of research developments. The authors are very grateful for the answers returned, which helped them to choose topics for inclusion, and made useful suggestions about likely progress. However, the authors accept responsibility for the views expressed in the paper. Obviously, any view of the future is speculative and uncertain, but for brevity, with no intended presumption, statements about the future are made in a simple future tense, generally omitting wording such as 'the authors expect (or guess) that . . .'.

Discussion on this paper closes on 1 December 2008, for further details see p. ii.

* Arup Geotechnics, London, UK.
† Tokyo University of Science, Japan.

The potential scope of the discussion is vast, so the paper is restricted to a small selection of topics. References have been provided to current information sources, partly in the hope that they may be of historic interest at some time in the future. The authors' overall expectation is that the future will involve advances in geotechnology, many of them beyond present-day imagination. But it is acknowledged that it could also involve stagnation or even retrogression in both technology and society: all of these have been experienced in human history. To quote Niels Bohr: 'Prediction is very difficult, especially about the future.'

GEOTECHNICAL CONSTRUCTION

The aspirations and achievements of society have always been characterised by strong interaction with developments in civil engineering. Hence developed nations will continue to want improved infrastructure, commercial and residential buildings, and higher environmental standards. Advances can be expected in all forms of geotechnical construction, including tunnelling and foundations. Ground improvement and reinforcement are likely to be particularly important, together with more advanced techniques for utilisation of brownfield sites. Developments in computation, measurement and communication will provide new opportunities and enhance existing processes.

Noting the recent reports of the Intergovernmental Panel on Climate Change (2007), it is likely that the production and use of energy, limiting its effect on production of harmful gases, and mitigating the effects of climate change will be dominant themes in coming years. These will necessitate development of new geotechnical skills and wider dissemination of currently rare skills, and they will also require a considerable increase in conventional civil engineering activity. This topic will be considered extensively below.

GROUND IMPROVEMENT AND SOIL REINFORCEMENT

Ground improvement technology is still developing, and its applications in practice are expanding, involving soil reinforcement, nailing, mechanical and chemical stabilisation. Ground improvement will be critically important in future geotechnical practice, for the following purposes in particular.

(*a*) More cost-effective construction of infrastructure, with better balance of cut and fill. Reinforcement enables reduction in both cutting and filling quantities, with construction much more efficient than the use of conventional steel or concrete retaining walls.
(*b*) Reduction of the total construction energy used and CO_2 emissions.
(*c*) Cost-effective prevention and mitigation of natural disasters, including those by landslides, soil erosion, flooding, heavy rainfall, or earthquakes.
(*d*) Cost-effective solutions of environmental issues, including treatment and recycling of industrial and urban wastes, waste disposal in landfill, and prevention of groundwater pollution. Continued research and development of geosynthetics is anticipated.
(*e*) Effective use of industrial by-products and demolition materials, including the use of crushed concrete and ash from power plants as construction materials. Use of these as very high-quality fills, well compacted and possibly with some cement added, merits further research (e.g. Aqil *et al.*, 2005).
(*f*) Remediation techniques of polluted ground and soils.
(*g*) Maintenance and rehabilitation of constructions that are decaying or which were not constructed to currently accepted standards. Compared with concrete and steel structures, deterioration due to ageing is generally less significant for soil structures constructed by compacting unbound backfill. Soil reinforcement can be used effectively to provide protection for natural slopes and to augment embankments, and this may become more important in future if climate change leads to more severe weather events.

Ground improvement techniques are usually used either to reduce the stresses in the ground, particularly the shear stress, or to enhance the shear strength. Stresses in the soil can be reduced by the use of lightweight embankments, micropiling and tensile reinforcement. For compaction of existing unbound backfill, apart from the conventional compaction technologies discussed above, new cost-effective compaction methods, such as heavy tamping and underground and underwater compaction, will be developed.

Chemical methods of increasing shear strength

It is anticipated that more permanent soil structures, such as bridge abutments and fill dams, will be constructed by using cement-mixed backfill. Also, more permanent superstructures will be constructed on and in existing soil strata improved by cement-mixing in-place and embankments constructed using cement-mixed soil. The boundary between cement-mixing technology in geotechnical engineering and concrete engineering will become less important.

Supporting these developments in construction practice, research will provide better understanding of the stress–strain properties of chemically improved geomaterial by properly taking into account relevant influence factors that are not important for untreated geomaterials, such as the effects of ageing, heat and chemical reactions, and their combined effects and interactions with the elasto-viscoplastic properties. Only once the strength and deformation characteristics of chemically improved geomaterials are properly understood and evaluated will their use to construct important primary soil structures become standard geotechnical practice.

Tensile reinforcement

Tensile reinforcing technologies include: (a) those used to reinforce existing ground and slopes (e.g. nailing); and (b) those used to reinforce backfill, usually by placing metal strips or geosynthetic reinforcement layers during construction. Given confirmed long-term durability, both types of tensile reinforcing technology are becoming used routinely in many countries.

Ongoing development of type (b) is anticipated for use in permanent structures, such as bridge abutments, in place of conventional concrete structures supported by piles. For example, Tatsuoka *et al.* (2007) describe a reinforced concrete integral bridge with a backfilled abutment reinforced with geosynthetics. The design of this type of structure requires the integration of geotechnical engineering and structural engineering, as in the foundation design.

Combined chemical–mechanical methods

The technologies described above are often used in isolation at present, but their use in combination could be particularly powerful. Combinations of mechanically and chemically stabilised ground will be found useful, such as mechanically stabilised backfill structures constructed on chemically improved soft ground. Permanent soil structures, including foundations, made of cement-mixed backfill or existing ground, reinforced with steel or fibre-reinforced plastic rods or geosynthetic reinforcement, will become popular in some situations.

DEVELOPMENT OF BROWNFIELD SITES

Over the next 60 years there will be an ongoing need to develop existing brownfield sites: that is, previously developed land, which may or may not be contaminated. The increasing amounts of electrical, electronic, chemical, household and even nuclear waste will present major challenges.

Engineers and scientists have been enabling the development of brownfield sites for hundreds of years, by the simple processes of digging and dumping elsewhere, or by breaking the pathway of pollution, and covering up. Legislation will continue to get tighter, requiring innovative remediation solutions, possibly using transferable skills and techniques from other sciences, such as medicine, biochemistry, physics and biology. In addition to development in the use of geosynthetics, as noted above, such solutions could include

(*a*) more sophisticated, possible even ‘intelligent’, permeable reactive barriers
(*b*) heat treatments, perhaps harnessing solar power or geothermal power
(*c*) injected chemical treatments that would either ‘fix’ contaminants or create barriers
(*d*) the use of plants, possibly biomass fuel plants, to be grown on contaminated sites to ‘adsorb’ contaminants
(*e*) nanotechnology, such as the nano zero-valent iron technologies, in which highly reactive nano iron particles cause less reactive metals to precipitate onto them (If present uncertainties about safety can be tackled, this technique may be suitable for the remediation of very deep plumes of contaminated groundwater, injecting nano-sized iron particles deep into the ground through wells.)

(*f*) better risk assessment techniques, using a better understanding of the toxicology of isolated and interacting chemicals, possibly eliminating some current concerns.

It is possible that landfilling as currently practised will eventually disappear, with improved treatment technologies allowing the reuse and sharing of surplus soils, perhaps through regional treatment and storage centres. Alternatively, instead of remediation, it might become the norm to leave contamination in the ground, develop over it safely by isolating the end user and the environment from it, allow groundwater to become polluted, and remediate or treat it only when it is extracted for use.

Interpolating levels of pollution between a small number of scattered samples may be very misleading, so new methods of investigation, detection and measurement are needed, combined with statistical analysis. We anticipate the development of remote sensing techniques and of advanced soil probes that can carry out continuous sampling and testing in situ.

Hopefully, society will learn the lessons of the past and avoid creating new brownfield sites. Whatever happens, engineers will use a considerable variety of solutions, remembering that sometimes the simple ones are best.

UNDERGROUND SPACE AND TUNNELLING

Greater use of underground space is anticipated, both as accommodation and to improve the efficiency of transportation, requiring novel and more powerful construction techniques. Some recent developments are discussed by Simpson (2007a), noting that construction below the surface of the ground offers not just more space but the advantages of a whole new dimension. Admiraal (2006) has rightly pointed out, however, that 'underground space is not a virtually limitless reservoir waiting to be exploited'. Arguing that humans have an urgent need to exploit the space below the ground surface, he also adds 'but neither can it be preserved as a nature reserve'.

Major excavations in cities are disruptive and potentially damaging, so it is important to monitor stresses and displacements in structures and the ground. We can expect much greater application of recent developments in technologies, such as fibre optics, wireless communication, measurement from satellites and the use of internet databases, and further developments will make them more common in this and other aspects of geotechnical engineering.

Technical developments are anticipated both in sprayed concrete tunnelling and in shield tunnelling to reduce ground deformation and provide cost-effective construction, with automation of excavation and lining erection. Innovative methods of forming underground spaces such as stations will include multiple-tube tunnels, as described for the Shanghai Rail Transit Project by Chow (2006), or the multi-micro shield tunnelling described by Hirosue & Kagami (2006), a technique that allows the construction of large rectangular underground spaces by tunnelling, without the use of cut-and-cover.

ENERGY AND CARBON CONSIDERATIONS

Helping energy users to reduce carbon dioxide emissions

In the United Kingdom, heating and cooling of living space and water account for about 34% of human energy usage (Department of Trade and Industry, 2003). The geotechnical profession, along with many others, is seeking ways of making this process more efficient, and the possibilities of using the ground as a source, or at least as a store, of heat and cold show some promise. Admiraal's statements on the limitations of space below ground, noted above, are equally applicable to its use for thermal purposes, and a balance of heat input and output is usually necessary.

For relatively small buildings surrounded by a large amount of open land, it is possible to pump water in pipes through the ground at shallow depth, typically about 1·5 m below the ground surface, in order to exchange heat. Concerns about the effect of this process on vegetation, including, in the extreme, freezing the ground at shallow depth, require careful consideration. In principle it may be possible to use this type of system in an 'unbalanced' manner, that is, for unequal amounts of heating and cooling, relying on heat interchange with the atmosphere to make up the difference. This requires, in particular, greater understanding of the thermal and hydraulic properties of the layers of soil close to the ground surface, which are generally unsaturated.

Many authors (e.g. Brandl, 2006; Law & Nicholson, 2006; Hofinger & Adam, 2007) have discussed systems for interchange of heat with the ground at depth, using either wells or conduction through concrete elements such as piles and retaining walls. They have generally shown that, in most cases involving large buildings, the ground can be used only as a temporary store of heat or cooling, on either an annual or a daily basis. Only in rare cases where there is a significant flow of water through the ground, and no environmental restrictions preventing water at a modified temperature flowing from the site, can heat be taken from or transmitted into the ground without a need for long-term balancing.

Transfer of heat or cold into the ground is usually by flow of fluid from the affected structure, either to structural elements with large surface areas in contact with the ground, such as foundation slabs, piles or retaining walls, or by direct flow of the fluid, in this case water, into and out of the ground. In the latter case, flow may be through the matrix of a soil or through the joints of a rock. There is a clear overlap here between the disciplines of soil mechanics, rock mechanics and hydrogeology, and combinations of expertise in all three will be needed.

Where it is not possible for one structure or installation to use the heat it produces within the annual cycle, the heat could potentially be shared with other users. For example, in the United Kingdom, many large buildings currently produce excess heat over the whole year, but most domestic homes do not have air conditioning and use energy only for heating. One solution could be to re-plan much of the urban areas, and beyond, so as to bring producers and users of heat nearer each other (e.g. Ward, 2008). Reorganisation of the built environment on such a scale would inevitably lead to a massive amount of civil engineering construction, including the necessary geotechnical engineering. While much of this would be conventional in nature, economy in the use of energy and in carbon dioxide emissions would be very important, as discussed further below.

Some simple opportunities to take heat from producers to users may already exist, and could be geotechnical in nature. For example, the possibility has been discussed of cooling underground rail lines by conducting the heat into the ground, and later extracting it for domestic heating in homes close to the underground lines. It could be taken out to homes on the surface above by water circulating in piles or wells. Hofinger & Adams (2007) describe the heating of a school in this way, and mention slabs, piles, tunnel linings and also ground anchors as devices that can be used to transmit heat into the ground.

This brief review of ways in which geotechnical engineering might help society to reduce carbon dioxide emissions shows that geotechnical engineers will need to develop a better understanding of the thermal behaviour of the ground, in terms of both its capacity to transmit and retain heat, and the effects of temperature change on other properties such as strength, stiffness and solubility of solid components in groundwater. Unsaturated soil mechanics and biological issues may also be relevant, and hydrogeology will be very important. It is unlikely that single individuals will be fully competent in all these disciplines, but engineers might develop competence in various combinations of them, and education courses could also combine them in some way.

To date, most of the published data on heat interchange with the ground relate to trials rather than to full-scale constructions, and a full explanation of the science is rarely provided. It is therefore very timely that *Géotechnique* will hold a Symposium in Print in 2009 on the subject 'Applications affected by the thermal characteristics of the ground'.

Reducing energy usage and carbon dioxide emissions in construction

The process of civil engineering construction is a major user of energy in its own right, so it will be important to reduce this wherever possible. It is anticipated that studies of the whole-life energy usage of projects will become more common, requiring comparative assessment of the embodied energy in construction materials, energy usage to form a construction, and energy required during the operational life of a structure.

For example, Chau *et al.* (2006) have recently studied the embodied energies of various types of embedded retaining wall. They note that the manufacture of both steel and cement consumes a lot of energy, and they conclude that steel walls, using recycled steel, have much less embodied energy than concrete walls of similar stiffness. However, the use of virgin steel, now comparatively rare in European construction, entails much more embodied energy.

Reducing embodied energy in construction will lead to greater reuse of materials, the use of poorer, but locally available, fill materials, and greater use of many forms of ground improvement. Reuse of piled foundations has been studied at some length in the RUFUS project (Chapman *et al.*, 2004), and considerations of the use of fill and structural materials in construction on soft ground are discussed by O'Riordan (2007) on the basis of its importance to the design of the Channel Tunnel Rail Link in the UK.

The authors anticipate that practising geotechnical engineers will need to become proficient in assessing the energy requirements of construction, and this will also be an important aspect of research in coming years.

Alternative sources of energy

For many years, geotechnical engineers and geologists have been critically involved in the extraction of coal, oil and gas on land and sea, and in the related infrastructure. It is anticipated that this will continue, perhaps with renewed emphasis on coal as oil and gas potentially become more scarce. Interactions between mining engineering, geology, rock mechanics and soil mechanics will be essential. If, as seems likely, it is necessary to extract oil and gas from depleted or marginal fields, better understanding of the geomechanics of oil and gas reservoirs will be important. For example, problems such as 'sanding' due to overstressing of sandstone will require more investigation to give a better understanding of the rock and soil mechanics involved in the crushing of the material. Work on this is already underway at various research centres.

If oil and gas start to become more scarce, coal will probably grow in importance as a primary source of energy. In China, for example, there is currently a rapid expansion in deep coal mining, with mines typically extending to depths exceeding 1 km. Elsewhere, opencast working requires the formation of deep quarries, with the attendant problems of slope stability and water control. Although these forms of mining are generally seen as somewhat specialist, they require the basic disciplines of soil mechanics and, especially, rock mechanics, and the particular skills of combining theory, analysis, observation and precedent that characterise all forms of geotechnical engineering. Again, there are many potential opportunities for academic involvement.

As society tries to reduce its dependence on carbon fuels, and looks for alternative sources of energy, new geotechnical challenges arise: wind, tide and nuclear energy are considered here.

It is likely that very large numbers of wind turbines will be constructed in the coming years, both onshore and offshore. Many of these may require relatively conventional piled foundations, but the sheer number of them argues for careful review of the design process to optimise the balance of robustness and economy. Larger offshore turbines will probably use suction foundations, which are subject to continuing research and development (e.g. Houlsby & Byrne, 2005).

Harnessing tidal power is a challenge often considered, but realised on a significant scale only in the scheme at La Rance, France, which produces about 100 MW of power on average (EDF, 2007). This is the only tidal power scheme in Europe, and the largest in the world. The Sustainable Development Commission (2007) has reported that the UK has the potential to use tidal power as one of its major energy resources, particularly in the Severn Estuary, provided the environmental, social and economical impacts can be mitigated. Any development of tidal power requires the construction of very large barrages, probably earth dams, which has been uncommon in much of the world in the last few decades. It seems unlikely that many such schemes will be built, so their impact on the geotechnical industry as a whole may be fairly small. However, some revitalising of expertise will be needed, and such large projects could well spawn research opportunities.

A resurgence of construction for nuclear power seems likely. Apart from foundations for very heavy loads, this leads to the need, which in fact already exists, for deep disposal of nuclear waste, an issue with many geotechnical aspects and the focus of considerable current research. Suitable strata of relatively impermeable rock or clay have to be located and tested, requiring large investigations. Geological stability is particularly important, given that the repository may have to remain safe for much longer than the age of our present civilisation. The behaviour of the material and water in pores or fissures as temperature changes has to be understood, together with the grouts and clays used to surround and contain the radioactive waste; some of this material will initially be unsaturated. As some radio nuclides will inevitably escape their immediate containment, their adsorption onto the particle surfaces of surrounding materials will be important, as will a full understanding of their movement through the materials in groundwater. Geotechnical engineering and nuclear engineering will clearly overlap in these studies; the *Géotechnique* Symposium in Print of 2007 contains several relevant papers (Bernier *et al.*, 2007; Delage *et al.*, 2007; Gens *et al.*, 2007; Piriyakul *et al.*, 2007). Geotechnical engineering will also contribute to the safer construction of nuclear power plants in highly seismic zones such as Japan by more reliable prediction of the input motion to the structures and the seismic stability of the supporting ground, and by the design of foundation structures.

Mitigating effects of climate change

Potential effects of climate change vary around the world, including rising sea level, more severe rainfall events, and hotter, drier conditions. Needs for new or renewed structures for sea and river defence can be expected. Very major constructions might be needed, possibly funded on a global basis, if large areas of low-lying, populated land are not to be lost. Such structures, for example on the scale of the Oosterschelde Barrier, completed in the Netherlands in 1996, naturally lead to major advances in soil testing and research, and also in construction technology.

The stability of both existing and new slopes, cuttings and embankments could be challenged by heavier rainfall, requiring reassessment and modification. Similarly, foundation designs could be affected by higher or more variable water tables, or by more severe cycles of drying and wetting, especially in shrinkable clays. Current research in unsaturated soils, and into the effects of cycles of saturation and desaturation, will be important, and more will be needed, in combination with field studies, to understand and control this difficult area of ground behaviour.

Gas hydrates

Methane gas hydrates are solid, ice-like compounds formed from hydrogen-bonded cages of water molecules enclosing methane molecules, which can lead to large volumes of methane gas being stored within their structure (Sloan, 1998). These compounds are widespread in the top few hundred metres of sediment beneath continental margins in water depths greater than a few hundred metres (owing to the temperature and pressure restrictions under which hydrates form). This has led to gas hydrates being of international importance with regard to the following.

(*a*) The potential to be a future energy resource: 1 m^3 of hydrate is equivalent to 164 m^3 of methane gas at standard temperature and pressure. The estimated global volume of methane stored in hydrate ranges from 2 to 4×10^{16} m^3 at standard temperature and pressure (Kvenvolden, 1988, 1998; McDonald, 1990), which is twice the total reserves of hydrocarbon energy from all other fossil fuels. Exploration with a view to the development of hydrate reserves is currently under way in India, China, Japan, Malaysia and Korea. Geotechnical engineers will play an important role in this development.

(*b*) Their role in global warming: methane gas that is contained in gas hydrates is 20 times more potent as a greenhouse gas than carbon dioxide. It has been postulated that hydrate dissociation and the release of methane gas into the atmosphere would significantly increase global warming. Hydrate dissociation may have been the cause of some of the mass extinctions in the geological past (Haq, 1998). Geophysicists, supported by geotechnical engineers, are playing a key role in developing more effective ways of detecting hydrates.

(*c*) The possibility of being a geotechnical hazard: upon dissociation, ice-like gas hydrate converts back to its individual components of gas and water. This can lead to a reduction in strength of the sediment along with a change in stress conditions due to the over-pressurisation caused by the release of methane gas. This has been linked to large-scale slope failures on continental margins that have occurred in the geological past (Kayen & Lee, 1991; Popenoe *et al.*, 1993; Meinert *et al.*, 1998; Ashi, 1999; Berndt *et al.*, 2002). On a smaller scale, offshore operators are increasingly working on potentially hydrate-bearing sediments on continental slopes (e.g. the Ormen Lange development off Norway). It is feared that drilling activities might be capable of dissociating gas hydrates, giving rise to oil platform blowouts, casing failures and, at the extreme, platform subsidence (Bily & Dick, 1974; Franklin, 1980). Geotechnical engineers are involved in testing to examine the effects of dissociation on the properties of hydrate-bearing sediment, and in modelling these effects (e.g. by developing coupled thermo-chemo-mechanical numerical element modelling).

The challenges of determining the thermo-chemo-mechanical properties, through in situ and laboratory testing of sediments found at water depths of thousands of metres, and then modelling the impacts of hydrate mining, global warming, and hydrocarbon exploitation, are enormous, and are increasingly involving geomechanics researchers (e.g. Clayton *et al.*, 2005; Priest *et al.*, 2005; Soga *et al.*, 2007).

SOIL MODELLING

The development of stress–strain models for soils, mainly for use in finite element analysis, has been a major feature of research over the last 60 years. In practice, the commonly used models are relatively simple: elastic Mohr–Coulomb models, modified Cam-clay (Roscoe & Burland, 1968) and small extensions of it, and developments of the hyperbolic models of Duncan & Chang (1970). A few more complex models are in regular use in the UK (e.g. Jardine *et al.*, 1986; Simpson, 1992), but most models developed in research are rarely used, perhaps because they require many parameters that are difficult to evaluate.

The 2007 *Géotechnique* Symposium in Print served to illustrate the very great complexity of real soil behaviour (e.g. Simpson, 2008), and though its relevance to practical problems was apparent, there was still a clear gulf between what could be understood in principle and what could be applied in practice. Not only the deformation properties but also the strength of soils are poorly modelled in engineering practice (Hight *et al.*, 2007). Tatsuoka *et al.* (1991, 1997) and Siddiquee *et al.* (1999) have pointed out that failure to consider the effects of deposition, progressive failure and other effects in dense soils can lead to large overestimates of available shear strength, while failure to recognise the effects of grain size or to use the benefits of good compaction may lead to underestimates.

Further research is anticipated, mainly by academics, to capture the stress–strain behaviour of soil more accurately in order to develop more realistic models. The aims of the development will include

(*a*) more detailed understanding of stress–strain and hydraulic behaviour
(*b*) more general unified models
(*c*) modelling of soil particles and interparticle contacts for numerical methods that can directly simulate the particulate nature of soil (e.g. the discrete element method, DEM).

For the purposes of practical design, simple models will continue to be applied, but these will be most useful and reliable when the user is informed by the best possible understanding of real behaviour. Technology transfer from research to practice will remain important, though often difficult to achieve.

More detailed understanding of behaviour

The deformation properties of geomaterials are different from those of most man-made materials in several important respects.

(*a*) Their behaviour is essentially frictional, with or without interparticle bonding.
(*b*) They display an extremely wide range of strengths and stiffnesses.
(*c*) Their properties are non-constant: they readily change, controlled by such factors as density, confining pressure, stress history, rate effects, and ageing effects.
(*d*) Their properties are complex, with, for example, extremely high strain and stress non-linearity, large effects of stress history, dilatancy, anisotropy, and strain-softening associated with shear banding.

Because of the inherently difficult nature of geomaterials, their characterisation and modelling have developed from simplified pictures—essentially slip surface or continuum models—used in the early stages of geotechnical engineering.

The classical geomechanics theories of earth pressure, bearing capacity and slope stability provide a simple illustration of this point. Although based on different mathematical methods (upper- or lower-bound types or limit-equilibrium types) and different failure mechanisms, most of them assume isotropic, perfectly plastic soil properties with zero-thickness shear bands (i.e. shear surfaces). This simplified classical soil model ignores not only pre-peak non-linear stress–strain behaviour but also anisotropy, and strain-softening with shear bands having a thickness proportional to particle size (e.g. Finno *et al.*, 1997; Okuyama *et al.*, 2003). Because of this, realistic prediction of ground deformation and structural displacements by classical soil models is not possible (e.g. Siddiquee *et al.*, 1999), and continued research will be needed to include a wider range of features. Mathematical consistency or correctness alone, often sought in academic geotechnical studies, are not sufficient.

More accurate and reliable prediction of displacements during construction, in the long term, and due to loading such as earthquakes will continue to be challenging. Present trends in the prediction of displacements are moving

(*a*) towards complex non-linear analysis
(*b*) towards elasto-viscoplastic analysis
(*c*) towards the inclusion of cyclic loading, chemical effects, ageing effects and thermal effects
(*d*) towards increased consideration of coupling effects among these effects (probably important, but little understood at present)
(*e*) towards more developments in unsaturated soil mechanics.

Capabilities for the prediction of time-dependent ground deformation and residual structure displacements during long-term service are still very poor. For example, to predict the secondary consolidation of clay, both viscous and ageing effects, interacting with yielding properties, should be taken into account. Although the effects of ageing can be evaluated to some extent, our understanding of the actual process of ageing (usually by chemical and thermal effects) is very poor. This relates to both engineering and geological timescales, and is coupled with yielding, viscous and thermal effects.

Until quite recently, most soil modelling was for saturated soils or dry, coarse-grained materials. The behaviour of partially saturated soils is now attracting much attention, stimulated by construction in drier regions, the importance of fill materials in most regions, and growing interest in conditions around proposed deep repositories for nuclear waste (e.g. Gens *et al.*, 2007). Although this seems to be one more complex feature to add to soil behaviour, it is likely that reasonable approximations to the behaviour can be achieved without undue complexity. Climate change and the need for nuclear power are likely to drive this topic forward in the coming decades.

Future activities in the deep sea and on other planets may need new frameworks of theory to include the effects of very high pore water pressure, of stress–strain behaviour at very low pressure and at very low and high temperatures, and of pore fluids other than water. It will become necessary to decide the extent to which current ideas can be extended to these new conditions.

More general unified models

To achieve originality, researchers tend to specialise in narrow topics. For example:

(*a*) Researchers specialising either in saturated soft clay or in unbound granular materials develop different constitutive theories and models.
(*b*) Researchers specialising in soil mechanics and rock mechanics deal differently with discontinuities and heterogeneities in a given geomaterial mass.
(*c*) Geotechnical researchers quantify the physical properties of a given stratum, whereas geologists describe qualitatively the whole structure of a given stratum.

On the other hand, in geotechnical engineering practice engineers have to deal with a wide variety of geomaterials, requiring not only the detailed behaviour of individual specimens but also a qualitative description of the whole structure of a given stratum. Increased efforts to develop theories covering different types of geomaterial (bound and unbound, from soft clays, sands and gravels to rocks, saturated and unsaturated) would be very beneficial, since many real geomaterials fall between the various types. Nevertheless, it would be a mistake to lose important aspects of the modelling of one material by trying to make it fit a more general framework. In parallel, efforts to merge soil mechanics and rock mechanics as well as geotechnical engineering and applied geology may become more relevant and therefore be pursued, first in practice, then in education, and then in research and in learned societies.

Modelling soil particles and interparticle contacts: the DEM

Although soil is a particulate material, this nature is usually not directly taken into account in soil modelling. Recently, the discrete element method (DEM) has become popular because it can capture the behaviour of particulate materials better than the finite element method (FEM).

Many respondents to the questionnaire sent out by the authors suggested that the DEM would be a prominent feature of developments in analysis. Analyses to date have been limited to about 10^5 particles, generally circular (2D) or spherical (3D), but one respondent suggested that computing power and the use of parallel processing will allow analyses of 10^6 or even 10^9 particles. Cundall (2001) has suggested that 10^{11} particles could be available within 20 years. Although these numbers seem impressive, it is observed that in 1 m^3 there are 10^6 cm^3 and 10^9 mm^3, so real soils would have considerably larger numbers of particles per cubic metre, each particle being complex in shape. Hence the prospect of modelling most real boundary value problems on a particle-by-particle basis seems remote. It might be possible to model practical situations using much larger particles provided an appropriate scaling law is adopted, as suggested for a slope stability problem by Cundall (2001), but situations involving strain localisation into shear bands and pore water flow in narrow bands will be particularly difficult.

The DEM has been used as a ‘computer laboratory’ to

simulate direct shear tests or triaxial tests (e.g. Cui *et al.*, 2007; Zhang & Thornton, 2007) and to investigate soil liquefaction (El Shamy & Zeghal, 2005), and it could have other important applications in more fundamental work. For example, it has been used in studies of the relationships of interparticle friction, rolling and dilation to the angle of shearing resistance ϕ' (Thornton, 2000), and in studies of the significance of particle crushing in explaining the familiar log-linear compression behaviour of soils (e.g. McDowell, 2002; Cheng *et al.*, 2004). Other features of soil behaviour, including anisotropy of strength and stiffness, and creep, might also be studied in this way.

The DEM needs soil property characterisations that are totally different from those required for currently used geomaterial models. For more realistic simulations by DEM analyses of the stress–strain behaviour of geomaterial from small strains towards residual state (and also flow behaviour) in research and practice, development of the following aspects is anticipated

(*a*) simulations of the effects of particle shape, stiffness, strength, crushability, etc.
(*b*) a more realistic description of interparticle contacts (elasto-viscoplastic, with and without bonding)
(*c*) a significant increase in the particle numbers that can be simulated, so that shear banding in a large soil mass as encountered in field boundary value problems can be captured, and 3D simulations become feasible. Even then, for practical application, calibration of the model by laboratory stress–strain tests and/or full-scale field monitoring will always be necessary.

The essence of the DEM is to recognise the particulate nature of soils, in contrast to the basic assumptions of continuum mechanics. Even if this form of computation does not prove to be viable, it is likely that computations that model the non-continuous nature of soils, with discrete load paths, fractures and similar features, will become important and might provide a major step forward. A paradigm shift in thinking may be needed to apply the results in practical engineering.

Reducing the gap between research and practice

There are several reasons for the continuing use of classical soil mechanics in practice, including the following.

(*a*) It is still very difficult to capture the complete, accurate picture of a given soil stratum in the field with sufficient data to represent the important features discussed above. This is because of inherent heterogeneity due to the complicated histories of deposition and loading of soil strata.
(*b*) The best available soil models are usually highly complex, with many parameters not easily determined. The analyses may be very costly, though not very reliable, depending critically on small variations in some parameters.

At present, the gap between the academic soil models and those used in design is increasing, except, perhaps, in the design of special large projects. It is often impractical for a designer to develop a full understanding of a complex model, which might contain many features not relevant to the problem in hand. The development of simpler, more robust models that reproduce features relevant to a limited, but clearly delineated, range of situations will therefore remain important.

In Japan, because of the very high design seismic loads used in the design of infrastructure following the 1995 Kobe Earthquake, it has become necessary to use the peak strength as well as the residual strength, rather than a design shear strength that previously had been essentially the same as the residual shear strength. Possible effects of particle size are also taken into account if necessary.

Initial examples of this approach include

(*a*) modified Mononobe–Okabe dynamic earth pressure theory (Koseki *et al.*, 1998)
(*b*) modified Newmark method to evaluate seismic slip displacements (Okuyama *et al.*, 2003)
(*c*) limit equilibrium analysis using residual strength to obtain the safety factor for the critical failure plane that is sought in advance by limit equilibrium analysis using peak strength (Tatsuoka *et al.*, 1998: Leshchinsky, 2000).

New Japanese railway design codes incorporate part of the above (Railway Technical Research Institute, 1999, 2007; Koseki *et al.*, 2008).

GEOTECHNICAL INVESTIGATIONS

Information currently available for design is often insufficient to provide real confidence and economy, especially in the case of existing structures facing new threats such as natural disasters. Moreover, standards of design, construction and maintenance will continue to develop, requiring new datasets of ground information.

The development of cost-effective site investigation will continue, with the aim of capturing the overall, detailed and accurate properties of a given soil stratum. This will entail introducing new test methods, and improvements in software, hardware and data interpretation.

In situ testing

Geotechnical field investigation will always be important, and developments to be expected include those noted below.

(*a*) The use of existing sounding technologies, including the SPT, cone penetrometers, pressuremeters and dilatometers, will continue. Better correlations between their outputs and geomaterial behaviour will continue to be sought, based on the latest geomechanics knowledge and information. For example, most of the existing empirical correlations do not take into account anisotropy, strain and pressure level non-linearity, or the effects of stress or strain histories.
(*b*) More cost-effective, yet reliably accurate direct evaluations of soil properties (e.g. stiffness values at different strain levels and shear strength). In this respect, more development and use of non-destructive in situ testing (e.g. geophysical methods) can be expected, including remote sensing techniques to capture the elastic properties in an area much wider or in a larger geotechnical mass than the selected points where soundings are performed. Sounding methods and non-destructive methods evaluate different parts of the stress–strain range, so these two groups are complementary.
(*c*) Much faster field loading tests, such as the falling weight deflection (FWD) test and methods based on compaction machine dynamics, will become popular for real-time fast on-site compaction control of backfill. This type of loading test is needed to improve the current compaction control based on the apparent degree of compaction measured at a relatively small number of points.
(*d*) Reliable and cost-effective field investigations in very deep offshore situations will be required related to

natural resource exploration. Investigations in situations of very high or low pressures may even extend to planets other than the Earth, including the Moon and Mars.

Field monitoring

More cost-effective, accurate, reliable, robust (long-term) monitoring and measuring systems will register not only conventional geotechnical parameters, but also new components, such as chemical, biological and nuclear. Optical fibres will be widely used to monitor displacement and strain, and satellite photography will become more reliable and popular to monitor ground surface displacements accurately over large areas. 'Smart' structures monitored using wireless electronic pick-ups during construction and through the life of the project will become realistic, monitoring both ground and structures. Improved measurements will make observational methods more reliable and popular; it could be that every passive member, such as a tie, nail, strut or ground reinforcement, will report its state of loading and warn of overloads, allowing reductions in design factors of safety. These developments also will be important in preventing geotechnical failures related to earthquakes, landslides etc.

Sampling

In the last 60 years there have been very significant advances in the quality of soil sampling available. However, even the best samples are still disturbed, and it may be impossible to retain subtle microstructure, accurately reflecting the effects of ageing or recent stress or strain history. Further improvements in sampling would be very valuable, but the authors are not aware of any likely future developments that promise major gains.

Laboratory testing

Further developments are expected in the following

(*a*) more automation in operation, data acquisition and control of stress or strain, or both
(*b*) more flexibility in control of the stress and strain paths (i.e. the three principal stresses and their direction; pore water and air pressures; and the three principal strain rates and their directions) as well as loading histories in terms of stress, stain, strain rate (monotonic, sustained, cyclic and so on)
(*c*) more accurate control and measurements of stresses and strains (e.g. average values as well as local values inside a specimen as well as inside and outside shear bands)
(*d*) reduced cost in the production, maintenance and operation of the test equipment so that the developments described above can be enjoyed in as many laboratories as possible.

To achieve the above, innovations in both hardware and software (relevant test programme, test execution and data analysis and interpretation) are necessary.

For academics, it will necessary to review the basic assumptions of constitutive theories, and this will become feasible after more of the developments described above are made. For engineering practice, on the premise that meaningfully undisturbed samples become available and significant developments are made in laboratory testing, better data will be available from tests simulating typical field stress and strain paths and pre-loading history (creep etc.). However, to be practically useful and representative, large series of tests will generally be required, and the testing period will probably be too long for typical design programmes. Hence the results of high-quality test series carried out by universities for academic purposes or in relation to exceptional, large projects will continue to be important to engineering practice.

Even though new features of material behaviour are still being identified in laboratory tests, laboratory testing of geomaterials is becoming less popular than numerical analysis. This recent trend is due to the obvious applicability of the results of numerical analyses, whether they are right or wrong, and their apparent cost-effectiveness. To restore a better balance between testing and computing, it is hoped that the geotechnical community will strive to achieve

(*a*) a significant reduction in the cost of laboratory testing of geomaterials
(*b*) the development of design methods that use more realistic soil properties that can be obtained readily from laboratory tests
(*c*) the popular use of these tests to reduce the total construction cost (or the total lifetime cost).

Physical model testing

Centrifuge testing is now well established as a means to simulate a given prototype stress state, and its use will continue as a means of studying mechanisms of behaviour. The testing method has developed with time in terms of data accuracy, model scale and applicable loading history (including stage construction, monotonic and cyclic loading). Further improvements can be expected, such as better modelling of construction sequences, geotechnical processes, and inclusions such as geosynthetics and piles. However, it is not certain whether centrifuge testing will ever be able to tackle issues such as viscous behaviour and ageing, and issues of particle size remain problematic.

Databases

In many locations, especially urban, much construction is carried out in essentially similar ground conditions. Since geotechnical site investigation is considered expensive, databases of existing ground information, publicly accessible, are desirable. General acceptance and the involvement of policy makers are needed, and the process of developing comprehensive and reliable databases will probably take many years. Furthermore, it will never end, because continuous updating will be needed as new, and hopefully better, data become available.

The data of basic laboratory tests that are highly repeatable and reproducible among different laboratories could readily be included in a database. On the other hand, construction of a database of advanced laboratory tests would need careful quality control, or very discerning use. The information could be stored in geographic information systems, which could also be used as a store for real-time monitoring data, probably with restricted access initially.

Through the internet, databases could effectively be linked, making very large quantities of information available to every engineer. Highly developed computational processes will then be needed to find relevant data for a particular situation, to categorise their quality, and to interpret them.

Field reconnaissance

In addition to their use in deriving in situ parameters, geophysical methods such as 3D tomography are also relevant, and likely to be increasingly used, for site reconnaissance to identify stratification, boulders and voids. The enormous value of field inspections is well known to geologists and geotechnical engineers, but it is expensive, especially in global activities where sites may be very remote from the expert's base; at the limit, the site could even be on another planet. Developments in electronics and communications are likely to be very rapid, and future applications of these could be imagined. For example, a geological robot might be developed, delivered to site on a lorry, submarine or spaceship. It would be able to assemble and propel itself, take samples and carry out field tests such as geophysical and penetration tests, provide detailed photographs and micrographs, analyse rock jointing in exposures (perhaps by sending a flying probe up the rock face), and so on. All this might be controlled interactively from an office at great distance, making top-quality human expertise readily available anywhere in the world at acceptable cost.

DESIGN AND ANALYSIS

The geotechnical design process usually involves assimilation of data, a review of case history experience, codes of practice and empirical data, and some calculations, leading in combination to design decisions. This basic process is not expected to change, but features in which developments might be expected are noted below.

Numerical modelling

It is likely that finite element analysis will continue to be used, both for serviceability computations and for studies of failure mechanisms. This might largely replace the use of limit equilibrium methods, with advantages that features such as three dimensions, time dependence, anisotropy, progressive and post-peak behaviour could be included (e.g. Kovacevic *et al.*, 2007).

For design purposes, numerical modelling requires reliable soil constitutive models, reliable data on strata and soil properties, and adequate computational power. At present, computational power is outstripping soil modelling, with the prospect of relatively easy access to very large 3D analyses. However, as discussed above, the development of soil models is much more difficult, and continued research will be vital to practical use. Intelligent application of the limited capabilities of available models will be critically important, and possibilities of misuse may be an increasing danger as software becomes more accessible.

It is unlikely that DEM will become a practical tool for the analysis of geotechnical construction, but it is possible that it could be used to investigate some soil problems involving small details of behaviour. Examples might include the performance of augers and other forms of blade, or problems of sanding in oil wells, where a combination of reduction in pore pressure and high hydraulic gradient leads to crushing of sandstone and particle displacement (Li *et al.*, 2006).

Neural networks

Greater computing power enables engineers to carry out existing methods of analysis more quickly, and to extend them. The possibility of completely new modes of analysis and thought, which computers might make available, have been little explored to date, but could have a big impact in coming decades. Among these are the use of neural networks, which have been investigated as a new means of capturing essential features of soil behaviour to be used in analysis. Toll (2001) reviewed the use of neural networks in geotechnical engineering, and reported that the main use was in classification and parameter assessment and in foundation design. Recently, further use of neural networks has been reported in estimating the deflection of diaphragm walls (Kunga *et al.*, 2007) and tunnelling performance (Yoo and Kim, 2007). It is conceivable that these or some similar computer-based systems might provide superior brains, which can assimilate the very complex data currently available about soil behaviour, and use them to provide useful predictions. A disadvantage might be that the degree of understanding transmitted to human engineers might be reduced, which, regrettably, might be a common feature of future practice.

Statistical methods and risk analysis

In recent years, the specific identification of risks, for example in forming risk registers, has become a prominent part of the design process. Having a common language for risk, with clear understanding between clients, designers and constructors, is valuable and deserves more development. It is also desirable, where realistically possible, to quantify risk, in relation both to economic issues, such as delays in contracts, and to safety, which could relate to the derivation of factors of safety in codes of practice. The necessary balance in geotechnics between transparency of process, clarity of data, and the best use of experience and training has been discussed in texts such as *Degrees of Belief* (Vick, 2002) and in conferences such as the First International Symposium on Geotechnical Safety and Risk at Tongji University in 2007 (e.g. Faber, 2007).

In the geoenvironmental field, statistical analysis is common practice. However, for the design of conventional geotechnical structures, although statistical analyses can provide some useful indications, most researchers have found that the complexity of the geotechnical design process is such that it cannot be reduced to a computational process, so human judgement and experience, though often lacking complete transparency, are vital elements. It is expected that research in this area will continue, and will influence practice. With ever-more powerful computers, databases and logical algorithms, it may be that risks will be analysed with greater rationality and confidence than is possible at present. Related to this, and with major advances in monitoring systems, more use of the observational method can be expected, using reliability analysis to ensure that safety levels are not diminished.

Codes of practice

Assessment of ground properties, arguably the most important aspect of geotechnical design, is not easily reduced to a code of practice, so geotechnical codes tend to act as checklists, giving overall guidance about procedures together with some factors of safety. Recently some international codes, such as Eurocodes, have attempted to define the process of deriving design values of parameters in order to provide acceptable margins of safety and performance. At present the procedures rely heavily on human knowledge and expertise, but it could be that the development of information databases, together with major developments in artificial intelligence, will allow much of this process to be carried out by computer. Whether used by humans or computers, the past 60 years of *Géotechnique* and other journals is a vital part of the database.

There remains much to be debated about the relative importance of serviceability and ultimate limit states in geotechnical design, about where partial factors of safety should most logically be applied (Paikowsky, 2004; Schuppener, 2007; Simpson, 2007b), about the part that can be played by probability analysis, and about the role of performance-based design (Honjo & Nagao, 2007). In the short term there is a danger that, amidst this debate, much of the practical good sense of existing codes could be lost. Much work remains to be done to ensure that accumulated knowledge is used successfully with the relatively new safety concepts in current codes. In the longer term, it is reasonable to hope that ideas about the application of safety factors and margins will continue to converge. This will require a significant collaboration of the research community and practice, and undergraduate teaching about the purposes and magnitudes of safety margins in design will be needed to inform both users and future developers of the codes.

As noted earlier, civil engineering uses large amounts of energy, much of it embodied in the materials used. Economy in design will be essential, affected by the magnitudes of factors of safety, among other factors. Society will therefore need codes of practice to adopt the minimum acceptable factors of safety, which could lead to progressive refinement of the codes.

DISSEMINATION

Until recent years, technological understanding has been transmitted mainly in three ways: through the written word on paper, via lectures in teaching and conferences, and 'on the job' as knowledge is passed between colleagues. *Géotechnique* has been a leading organ of the first of these methods, and engineers have relied heavily on journals and textbooks. More recently, computer-based knowledge has become prominent, especially from the internet. Papers from journals can often be obtained from the internet, although this is not yet generally true of conference proceedings and textbooks. Young engineers searching for information are likely to appeal first to Google or Wikipedia, and then to papers on line, before finally resorting to paper publications if necessary. Commercial companies see a saving in this, because storage space for documents can be reduced. Issues of copyright on downloaded documents add an unfortunate complication for both publishers and users. On-line forums are active, and may become increasingly influential; geotechnical engineers have traditionally been generous in sharing their knowledge and information.

Peer-reviewed journals such as *Géotechnique* play an important role in providing readers with information that has been thoroughly vetted and so is likely to be reliable. The journals are also important in providing a controlled outlet for academic publications, which measure the achievements of researchers and so enable them to obtain funding for further work. It is important that these two roles continue in the future, although this does not necessarily require publication on paper; similar peer review could be applied to papers published mainly, or only, on the internet, and it is likely that development will proceed in this direction. Some means of guaranteeing the longevity of the publications will be needed. Internet journals will be somewhat different, containing direct links to other documents and data, and perhaps inviting direct communication with the authors.

Journals with an international readership are published mainly in English, but differences of language probably support the success of a range of journals in various languages devoted to the same subject. This can be frustrating for both authors and readers, impeding the spread of ideas. It is likely that one language will become even more ubiquitous in the coming decades, and it is also possible that computerised translation will become so powerful that readers can obtain a journal in any language they choose. This might lead to a reduction in the number of journals, but would enhance the transfer of technology.

Apart from publication, human interaction is important to dissemination. In order to identify reliable mentors, students and users of information need some personal connection to them, so lectures and conferences are likely to remain important. However, it could be that travel, especially internationally, will become more expensive, and various forms of video conferencing may take precedence. At present this is often unreliable and impersonal, but it can be expected that very rapid interaction, stimulating personal interchange, will be available in the near future, greatly reducing the need for travel. No doubt such a development will be considered both convenient and disappointing.

CONCLUDING REMARKS

The future of geotechnical construction is likely to be very active, involving the application of currently accepted processes and many developments aimed at reducing energy consumption and carbon dioxide emissions. These constraints apply both to construction processes, requiring the minimisation of embodied energy, and to the provision of constructions that will help society use less energy, or generate more energy from non-carbon sources. Geotechnical engineers will embrace expertise related to heat transfer in the ground, with important hydrogeological connections. Engineering of deep repositories for nuclear waste will involve ground materials at high stresses with changing temperatures and degrees of saturation, and consideration of the movement of radio nuclides in the ground.

More resourceful use of fill materials is envisaged, with ground improvement and geosynthetics playing important roles. The behaviour of soil is increasingly understood to be very complex, and continued research into this will be necessary. However, bringing the research into practical application remains very challenging, and is likely to be so for many years hence. Experience and empiricism will continue to be essential to geotechnical design.

Further rapid development of electronics and communications is expected to facilitate major advances in laboratory and field equipment, monitoring and observational approaches, and the possibility of remote site reconnaissance has been suggested.

Design processes will become more heavily reliant on computers as data sources and calculators. It is not clear to what extent artificial intelligence and computerised processes of risk analysis might replace the expertise of the human engineer. There is a danger that engineers will become deskilled, with no adequate replacement.

The future of journals such as *Géotechnique* is considered to be very important as a control on the quality of published information, which also helps to regulate the funding of research and the careers of researchers. However, the nature of journals could change considerably as they become mainly, if not entirely, electronic in nature.

The authors would like to wish *Géotechnique*, and the whole geotechnical community, a very successful next 60 years.

ACKNOWLEDGEMENTS

The authors would like to thank the many colleagues who have provided suggestions for this paper, by completing the questionnaire or by other means. They are grateful to Professor Chris Clayton and Dr Jeffrey Priest for assistance with text on gas hydrates, and to Geraldine Cheung and Martin Fairlie, who assisted with the preparation of some sections.

REFERENCES

Admiraal, J. B. M. (2006). A bottom-up approach to the planning of underground space. *Tunnelling and Undergroud Space Tech.* **21**, Nos 3–4, 464–465.

Aqil, U., Tatsuoka, F., Uchimura, T., Lohani, T. N., Tomita, Y. & Matsushima, K. (2005). Strength and deformation characteristics of recycled concrete aggregate as a backfill material. *Soils Found.* **45**, No. 4, 53–72.

Ashi, J. (1999) Large submarine-landslides associated with decomposition of gas hydrate. *Landslide News*, **12**, No. 12, 17–20.

Berndt, C., Mienert, J., Vanneste, M., Bünz, S. & Bryn, P. (2002). Submarine slope-failure offshore Norway triggers rapid gas hydrate decomposition. *Proc. 4th Int. Conf. on Gas Hydrates, Yokohama*, 71–74.

Bernier, F., Li, X.-L. & Bastiaens, W. (2007). Twenty-five years' geotechnical observation and testing in the Tertiary Boom Clay formation. *Géotechnique* **57**, No. 2, 229–237.

Bily, C. & Dick, J. W. L. (1974). Natural occurring gas hydrates in the Mackenzie delta, Northwest Territories. *Bull. Can. Pet. Geol.* **22**, No. 3, 340–352.

Brandl, H. (2006). Energy foundations and other thermo-active ground structures. *Géotechnique* **56**, No. 2, 81–122.

Chapman, T., Butcher, T. & Fernie, R. (2004). Re-using urban foundations. *Proc. Inst. Civ. Engrs, Civ. Engng* **157**, No. 1, 6.

Chau, C., Nicholson, D. & Soga, K. (2006). Comparison of embodied energy of four different retaining wall systems. *Proc. Int. Conf. Reuse Found. for Urban Sites (RuFus 2006)*, 277–285.

Cheng, Y. P., Bolton, M. D. & Nakata, Y. (2004). Crushing and plastic deformation of soils simulated using DEM. *Géotechnique* **54**, No. 2, 131–141.

Chow, B. (2006). Double-O-tube shield tunneling technology in the Shanghai Rail Transit Project. *Tunnelling and Underground Space Technology* **21**, No. 6, 594–601.

Clayton, C. R. I., Priest, J. A. & Best, A. I. (2005). The effects of disseminated methane hydrate on the dynamic stiffness and damping of a sand. *Géotechnique*, **55**, No. 6, 423–434.

Cundall, P. A. (2001). A discontinuous future for numerical modelling in geomechanics? *Proc. Inst. Civ. Engrs, Geotech. Engng* **149,** No. 1, 41–47.

Cui, L., O'Sullivan, C. O. & O'Neill S. (2007). An analysis of the triaxial apparatus using a mixed boundary three-dimensional discrete element model. *Géotechnique* **57**, No. 10, 831–844.

Delage, P., Le, T.-T., Tang, A.-M., Cui, Y.-J. & Li, X.-L. (2007). Suction effects in deep Boom Clay block samples. *Géotechnique* **57**, No. 2, 239–244.

Department of Trade and Industry (2003). *Our energy future: Creating a low carbon economy.* The Stationery Office.

Duncan, J. M. & Chang, C. Y. (1970). Nonlinear analysis of stress and strain in soils. *J. Soil Mech. Found. Div.* ASCE **96**, No. SM5, 1629–1653.

EDF (2007). *The Rance tidal power plant, power from the ocean.* http://www.edf.fr/html/en/decouvertes/voyage/usine/usine.html

El Shamy, U. & Zeghal, M. (2005). Coupled continuum-discrete model for saturated granular soil. *J. Engng Mech. ASCE* **131**, No. 4, 413–426.

Faber, M. H. (2007). Assessing and managing risks due to natural hazards. *Proc. 1st Int. Symp. on Geotechnical Safety and Risk, Tongji University*, **1**, 53–70.

Finno, R. J., Harris, W. W., Mooney, M. A. & Viggiani, G. (1997). Shear bands in plane strain compression of loose sand. *Géotechnique* **47**, No. 1, 149–165.

Franklin, L. J. (1980). In situ hydrates: a potential gas source. *Petroleum Engineer International*, November, 112–122.

Gens, A., Vaunat, J., Garitte, B. & Wileveau, Y. (2007). In situ behaviour of a stiff layered clay subject to thermal loading: observations and interpretation. *Géotechnique* **57**, No. 2, 207–228.

Haq, B. (1998). Natural gas hydrates: searching for the long-term climatic and slope stability records. In *Gas hydrates: Relevance to world margin stability and climate change* (eds J.-P. Henriet and J. Mienert), Special Publication 137, pp. 303–318. London: Geological Society.

Henkel, D. J. (1982). 22nd Rankine Lecture: Geology, geomorphology and geotechnics. *Géotechnique* **32**, No. 3, 175–194.

Hight, D. W., Gasparre, A., Nishimura, S., Minh, N. A., Jardine, R. J. & Coop, M. R. (2007). Characteristics of the London Clay from the Terminal 5 site at Heathrow Airport. *Géotechnique* **57**, No. 1, 3–18.

Hirosue, T. & Kagami, S. (2006). Recent challenges in underground projects in Japan. In *North American tunneling* (ed. L. Ozdemir). London: Taylor & Francis, 265–274.

Hofinger, H. & Adam, D. (2007). Tunnels as energy sources: practical applications in Austria. *Proc. 14th Eur. Conf. Soil Mech. Geotech. Engng, Madrid*, 907–912.

Honjo, Y. & Nagao, T. (2007). Development of a basic specific design code on performance based specification concept: The Technical Standards for Port and Harbor Facilities. *Proc. 1st Int. Symp. on Geotechnical Safety and Risk, Tongji University*, **1**, 105–116.

Houlsby, G. T. & Byrne, B. W. (2005). Design procedures for installation of suction caissons in clay and other materials. *Proc. Inst. Civil Engrs Geotech. Engng* **158,** No. 2, 75–82.

Intergovernmental Panel on Climate Change (2007). *Intergovernmental Panel on Climate Change: Fourth Assessment Reports.* Cambridge: Cambridge University Press.

Jardine, R. J., Potts, D. M., Fourie, A. B. & Burland, J. B. (1986). Studies of the influence of non-linear stress–strain characteristics in soil–structure interaction. *Géotechnique* **36**, No. 3, 377–396.

Kayen, R. E. & Lee, H. J. (1991). Pleistocene slope instability of gas hydrate-laden sediment on the Beaufort sea margin. *Marine Geotech.* **10**, Nos 1–2, 125–141.

Koseki, J., Tatsuoka, F., Munaf, Y., Tateyama, M. & Kojima, K. (1998). A modified procedure to evaluate active earth pressure at high seismic loads. *Soils Found.*, Special issue on geotechnical aspects of the January 17, 1995 Hyogoken-Nambu Earthquake, No. 2, 209–216.

Koseki, J., Tateyama, M., Watanabe, K. & Nakajima, S. (2008). Keynote lecture: Stability of earth structures against high seismic loads. *Proc. 13th Asian Regional Conf. Soil Mech. Geotech. Engng, Kolkata* **2**.

Kovacevic, N., Hight, D. W. & Potts, D. M. (2007). Predicting the stand-up time of temporary London Clay slopes at Terminal 5, Heathrow Airport. *Géotechnique* **57**, No. 1, 33–47.

Kunga, G. T. C., Hsiaob, E. C. L., Schusterb, M. & Juang, C. H. (2007). A neural network approach to estimating deflection of diaphragm walls caused by excavation in clays. *Comput. Geotech.* **34**, No. 5, 385–396.

Kvenvolden, K. (1988). Methane hydrate: a major reservoir of carbon in the shallow geosphere? *Chem. Geol.*, **71**, Nos 1–3, 41–51.

Kvenvolden, K. (1998). A primer on the geological occurrence of gas hydrates. In *Gas hydrates: Relevance to world margin stability and climate change* (eds J.-P. Henriet and J. Mienert), Special Publication 137, pp. 9–30. London: Geological Society.

Law, R. & Nicholson, D. P. (2006). Modelling of the chalk aquifer in central London for a proposed open cooling system. *Proc. 10th Int. Conf. on Thermal Energy Storage, Stockton College, New Jersey*, paper 11.

Leshchinsky, D. (2000). Design dilemma: use peak or residual strength of soil. *Geotextiles and Geomembranes*, **19**, No. 2, 111–125.

Li, L., Papamichos, E. & Cerasi, P. (2006). Investigation of sand production mechanisms using DEM with fluid flow. *Proceedings of EUROCK 2006: Multiphysics Coupling and Long Term Behaviour in Rock Mechanics* (eds A. Van Cotthem, R. Charlier, J.-F. Thimus and J.-P. Tshibangu), pp. 241–247. London: Taylor & Francis.

McDonald, G. (1990). The future of methane gas as an energy resource. *Ann. Rev. Energy* **15**, 53–83.

McDowell, G. R. & Harireche, O. (2002). Discrete element modelling of soil particle fracture. *Géotechnique* **52**, No. 2, 131–135.

Meinert, J., Posewang, J. & Baumann, M. (1998). Gas hydrates along the northeastern Atlantic margin: possible hydrate-bound margin instabilities and possible release of methane. In *Gas hydrates: Relevance to world margin stability and climate change* (eds J.-P. Henriet and J. Mienert), Special Publication 137, pp. 275–291. London: Geological Society.

Okuyama, Y., Yoshida, T., Tatsuoka, F., Koseki, J., Uchimura, T., Sato, N. & Oie, M. (2003). Shear banding characteristics of granular materials and particle size effects on the seismic stability of earth structures. *Proc. 3rd Int. Symp. on Deformation Characteristics of Geomaterials, Lyon*, 607–616.

O'Riordan, N. J. (2007). Keynote paper: Technical sustainability of construction on very soft ground. *Proceedings of the conference on soft ground engineering*, Portlaoise.

Paikowsky, S. G. (2004). *Load and resistance factor design for deep foundations*, NCHRP Report 507. Washington, DC: Transportation Research Board.

Piriyakul, K. & Haegeman, W. (2007). Void ratio function for elastic shear moduli for Boom Clay. *Géotechnique* **57**, No. 2, 245–248.

Popenoe, P., Schmuck, E. & Dillon, W. (1993). The cape fear landslide: slope failure associated with salt diapirism and gas hydrate decomposition. In *Submarine landslides: Selective studies in the US exclusive economic zone*, US Geological Survey Bulletin 2002, pp. 40–53. Reston, VA: US Geological Survey.

Priest, J. A., Best, A. I. & Clayton, C. R. I. (2005). A laboratory investigation into the seismic velocities of methane gas hydrate-bearing sand. *J. Geophys. Res. Solid Earth*, **110**, No. B4, paper B04102 (doi:10·1029/2004JB003259).

Railway Technical Research Institute (1999). *Seismic design standard for railway structures*, pp. 47 and 89–90. Tokyo: Maruzen (in Japanese).

Railway Technical Research Institute (2007). *Design standard for railway earth structures*, p. 257. Tokyo: Maruzen (in Japanese).

Roscoe, K. H. & Burland, J. B. (1968). On the generalised stress–strain behaviour of 'wet' clay. *Engineering plasticity* (eds J. Heymann and F. A. Leckie), pp. 535–609. Cambridge: Cambridge University Press.

Schuppener, B. (2007). Eurocode 7 and its latest developments. *Proc. 1st Int. Symp. on Geotechnical Safety and Risk, Tongji University*, **1**, 117–132.

Siddiquee, M. S. A., Tanaka, T., Tatsuoka, F., Tani, K. & Morimoto, T. (1999). Numerical simulation of the bearing capacity of strip footing on sand. *Soils Found.* **39**, No. 4, 93–109.

Simpson, B. (1992). 32nd Rankine Lecture: Retaining structures: displacement and design. *Géotechnique* **42**, No. 4, 539–576.

Simpson, B. (2007a). Use of underground space in urban areas. *Proc. 14th Eur. Conf. Soil Mech. Found. Engng, Madrid*, **5**.

Simpson, B. (2007b). Approaches to ULS design: the merits of Design Approach 1 in Eurocode 7. *Proc. 1st Int. Symp. on Geotechnical Safety and Risk, Tongji University*, **1**, 527–538.

Simpson, B. (2008). Keynote lecture: Engineering in stiff sedimentary clays. *Stiff Sedimentary Clays: Genesis and Engineering Behaviour*. London: Thomas Telford (forthcoming).

Sloan, E. D. Jr (1998). *Clathrate hydrates of natural gases*. New York: Marcel Dekker.

Soga, K., Lee, S. L., Ng, M. Y. A. & Klar, A. (2007). Characterisation and engineering properties of methane hydrate soils. In *Characterisation and engineering properties of natural soils* (eds T. S. Tan, K. K. Phoon, D. W. Hight, and S. Leroueil), Vol. 4, pp. 2591–2644. London: Taylor & Francis.

Sustainable Development Commission (2007). *Turning the tide: Tidal power in the UK*. London: Sustainable Development Commission.

Tatsuoka, F., Okahara, M., Tanaka, T., Tani, K., Morimoto, T. & Siddiquee, M. S. A. (1991). Progressive failure and particle size effect in bearing capacity of a footing on sand. *Proc. ASCE Geotech. Engineering Congress, Boulder*. ASCE Geotechnical Special Publication, Vol. 27, pp. 788–802.

Tatsuoka, F., Goto, S., Tanaka, T., Tani, K., & Kimura, Y. (1997). Particle size effects on particle size effects on bearing capacity of footing on granular material. *Proceedings of the International conference on deformation and progressive failure in geomechanics, IS Nagoya '97* (eds A. Asaoka, T. Adachi and F. Oka), pp. 133–138. Oxford: Pergamon Press.

Tatsuoka, F., Koseki, J., Tateyma, M., Munaf, Y. & Horii, N. (1998). Keynote lecture: Seismic stability against high seismic loads of geosynthetic-reinforced soil retaining structures. *Proc. 6th Int. Conf. on Geosynthetics, Atlanta* **1**, 103–142.

Tatsuoka, F., Hirawaka, D., Nojiri, M., Aizawah, H., Tateyama, M. & Watanabe, K. A. (2007). A new type integral bridge comprising of geosynthetic-reinforced soil walls. *Proc. 5th Int. Symp. on Earth Reinforcement (IS Kyushu)*, 803–809.

Thornton, C. (2000). Numerical simulations of deviatoric shear deformation of granular media. *Géotechnique* **50**, No. 1, 43–53.

Toll, D. G. (2001). Computers and geotechnical engineering: a review. In *Civil and structural engineering computing: 2001* (ed. B. H. V. Topping), pp. 433–458. Stirling: Saxe-Coburg Publications.

Vick, S. (2002). *Degrees of belief: Subjective probability and engineering judgment.* Reston, VA: ASCE.

Ward, L. (2008). Why Treasure Island is the super-green city of the future. *Popular Mechanics* **185**, No. 1.

Yoo, C. & Kim, J. (2007). Tunneling performance prediction using an integrated GIS and neural network. *Comput. Geotech.* **34**, No. 1, 19–30.

Zhang, L. & Thornton, C. (2007). A numerical examination of the direct shear test. *Géotechnique* **57**, No. 4, 343–354.

Salgado, R., Houlsby, G. T. & Cathie, D. N. (2008). *Géotechnique* **58**, No. 5, 369–375 [**doi:** 10.1680/geot.2008.58.5.369]

Contributions to *Géotechnique* 1948–2008: Foundation engineering

R. SALGADO*, G. T. HOULSBY† and D. N. CATHIE‡

Many of the important developments in the field of foundation engineering have been addressed in *Géotechnique* papers over the past 60 years. This paper briefly reviews some of these developments and related articles, particularly with respect to shallow and deep foundations. In the early days of *Géotechnique*, the power to perform sophisticated numerical analyses did not exist. Papers tended to focus on the solution of problems using simple models in which soil was modelled either as linear elastic or as perfectly plastic. Engineers sought simple closed-form analytical solutions for boundary-value problems. With the development of more powerful analytical, computational and experimental capabilities, and of more sophisticated pile installation technology (especially offshore), more recent papers have explored much more sophisticated approaches to a range of foundation problems, striving to achieve more realistic representation of working conditions. *Géotechnique* papers have attempted to solve the problems faced by the foundation engineering industry, with a strong emphasis on the underlying science; as a result, these papers have played a key role in the advancement of both the science and its applications in our discipline.

KEYWORDS: footings/foundations; historical review

Au cours des 60 dernières années, des communications de *Géotechnique* se sont penchées sur un grand nombre de développements importants dans le secteur de la technique des fondations. La présente communication se penche brièvement sur certains de ces développements et articles connexes, en particulier sur le plan des fondations superficielles et profondes. Au début de l'existence de *Géotechnique*, on ne disposait pas de la capacité d'effectuer des analyses numériques sophistiquées : les communications avaient tendance à se concentrer sur la solution de problèmes au moyen de modèles simples dans lesquels le sol était modélisé comme étant élastique linéaire ou parfaitement plastique. Les ingénieurs recherchaient de simples solutions analytiques de forme close pour des problèmes aux valeurs limites. Puis, avec le développement de moyens analytiques, de calcul et expérimentaux plus sophistiqués, et l'avènement d'une technologie plus sophistiquée pour l'installation de piles (notamment en mer), des communications plus récentes se sont penchées sur des méthodes beaucoup plus sophistiquées pour résoudre toute une série de problèmes de fondations, en s'efforçant de réaliser une représentation plus réaliste des conditions de travail. Des communications dans *Géotechnique* ont tenté de résoudre les problèmes affrontés par le secteur des techniques des fondations, en mettant l'accent sur la science sous-jacente ; en conséquence, ces communications ont joué un rôle essentiel dans les progrès réalisés tant sur le plan de la science que sur ses applications, dans notre discipline.

INTRODUCTION

Given that the worldwide body that represents much of the *Géotechnique* readership was, for many years, called the International Society for Soil Mechanics and Foundation Engineering, one might expect a substantial fraction of *Géotechnique* to be devoted to foundations. Somewhat surprisingly, this is not the case, with fewer papers than expected being devoted primarily to foundations. An interesting observation is that the most notable work on foundations does not appear in a few landmark papers, but as groups of papers dealing with distinct themes. Each theme was sustained over a number of years by groups of authors, and attracted a series of papers that collectively advanced the subject. These sustained contributions are an important demonstration of *Géotechnique*'s vitality. In this paper, we discuss these contributions collectively within themes rather than concentrating on individual papers.

The topic of foundations naturally divides into the study of shallow foundations and deep foundations (almost exclusively piles). Another division is into issues of capacity and deformation. Some papers present theories, while others report practical records, either at laboratory or at full scale. We have organised the discussion of papers based primarily on whether they deal with shallow or deep foundations.

In the early days of *Géotechnique*, the power to perform sophisticated numerical analyses did not exist, so problems were analysed using mainly simple models, in which soil was modelled either as linear elastic or as perfectly plastic. Simple closed-form analytical solutions for boundary-value problems were employed. Over the past 60 years, geotechnical engineers have gained a much better understanding of how soil responds to loading, and have developed numerical analyses that attempt much more realistic solutions to boundary-value problems. Significant progress has also been made in the technology used both to install and test deep foundations, and to address the specific requirements of offshore foundations—a challenge that hardly existed at the inception of *Géotechnique*.

We preface our review by some comments on offshore foundations. We view these as nothing other than shallow or deep foundations with specific design issues, related mainly to the magnitude of the loading, its cyclic nature, and the large horizontal component from environmental forces. The first explicit reference to offshore foundations in *Géotechnique* is the key paper by Bjerrum (1973), who introduced many of the major offshore foundation themes relevant to the emerging North Sea oil developments. The particular challenges that engineers faced in the North Sea drove the foundation engineering community to design much larger foundations than previously used. Open-ended pipe piles

Discussion on this paper closes on 1 December 2008, for further details see p. ii.
* School of Civil Engineering, Purdue University, USA
† Department of Engineering Science, Oxford University, UK.
‡ Cathie Associates SA/NV, Diegem, Belgium.

with much larger diameter and length than onshore piles required more powerful hammers with which to drive them. Gravity-based structures focused the attention of the profession on bearing capacity, with high cyclic horizontal loads and moments. In parallel, jack-up platforms used around the world introduced the challenge of estimating the penetration resistance of 'spudcan' foundations that were 10–20 m in diameter and penetrated up to two diameters in soft soil before achieving sufficient resistance for the stability of the mobile platform to be assured. Papers appearing in *Géotechnique* have captured these themes.

SHALLOW FOUNDATIONS

Bearing capacity

One of the first themes to emerge early in *Géotechnique* was that of the bearing capacity theory of foundations. The basis of the bearing capacity factor approach had of course already been set out by Terzaghi; it drew significantly on work on the indentation of metals, with appropriate modification for the frictional nature of soil, but much work remained to be done. An important early contribution, much cited later, was by Meyerhof (1951), who attempted a theoretical treatment of the bearing capacity of shallow foundations, accounting for the effects of foundation embedment depth. For many years the approach he pioneered, using shape and depth factors, became the standard approach for bearing capacity calculation. He also presented the results of field load tests for foundations in sand and clay. These results had a lasting impact because they provided data on bearing capacity that anchored theoretical calculations, providing engineers with a measure of confidence in their calculations, although by more exacting modern standards the amount of data presented in support of the theories was relatively slim.

At around the same time, an equally influential paper by Skempton (1951) on foundations in clay was published in a conference rather than *Géotechnique*!

Complementing the work of Meyerhof and Skempton was the early, and now famous, case study by Peck & Bryant (1953) of the bearing capacity failure of the Transcona grain elevator. In the same issue of *Géotechnique*, there is an interesting eyewitness account of the failure. The elevator was built on a raft foundation in Winnipeg, Canada, in 1911, on rather uniform, 10–15 m thick deposits of almost pure clay, overlying thinner layers of mixtures of clay and gravel and then limestone. The failure occurred in 1913 upon filling of the elevator. Calculations using the Skempton (1951) form of the bearing capacity equation for clay produced good agreement with the load at collapse, which was estimated from the known amount of grain stored at the time of failure, the dead weight of the structure, and the amount of pressure relief due to excavation of the soil for construction of the raft. Peck & Bryant recognised some of the difficulties in establishing a representative value of shear strength for the clay, including the possible effect of progressive failure and of the presence of closely spaced slickensides. In many ways, this paper still serves as a model of thoughtful back-analysis of a case history, although of course a more sophisticated analysis would now be possible.

A difficulty that Peck & Bryant did not mention is that, where strength increases with depth, averaging of shear strength values over certain depth ranges can produce misleading results. The elegant analysis by Davis & Booker (1973), who studied the bearing capacity of shallow foundations on soil with strength increasing with depth, allows a much more rigorous treatment of bearing capacity problems on clay. The Davis & Booker paper is a fine example of the contribution that numerical analysis began to make to the subject from the mid 1960s onwards.

Returning, though, to Meyerhof's contributions, although he contributed other papers to the early issues of *Géotechnique*, his highly influential paper on inclined and eccentric loading of foundations was published at the Third International Conference on Soil Mechanics and Foundation Engineering (Meyerhof, 1953). It was this paper that really set the pattern for bearing capacity calculations, with its use of inclination and eccentricity factors. Eccentric loading (or equivalently moment loading) was taken into account by considering an 'effective width' $B' = B - 2e$ (where e is the load eccentricity) and a corresponding 'effective area'. This simplification proves to be well founded, but the basis for a number of the other multiplicative factors that have been used in bearing capacity theory is rather less secure. This approach achieved its zenith in the work of Brinch Hansen (1970) and Vesić (1975), whose methods are both still much employed in practice. It is interesting that neither publication was in a mainstream journal, let alone *Géotechnique*.

Work in recent years has refined our understanding of some of the factors used in conventional bearing capacity analysis. For instance, Salgado *et al.* (2004) used modern numerical techniques to obtain more accurate shape and depth factors equations (see also Edwards *et al.*, 2005; Lyamin *et al.*, 2007). Other issues have been explored experimentally: for example, centrifuge tests by Kimura *et al.* (1985) showed that N_γ depends on the size of the footing.

An alternative approach to the calculation of the capacity of foundations has been developed, principally in the past 15 years. Rather than using a plethora of inclination, eccentricity and shape factors, the yield surface is considered directly as a function of the combinations of the vertical load V, lateral load H and moment M on the foundation leading to yield. *Géotechnique* has played a central role in publishing the papers that describe the development of the approach, but it should be noted that in fact it owes its origins to the work of Roscoe & Schofield (1957), and later Butterfield & Ticof (1979). It is interesting to speculate whether these ideas might have gained earlier acceptance had Roscoe & Schofield chosen to publish in *Géotechnique* rather than the *British Welding Journal*.

The most important driver for investigating the (V, H, M) yield surfaces has been their importance for offshore foundations. Various authors throughout the 1990s revisited the effective area concept for accounting for overturning moment and inclination factors for the effect of combined vertical and horizontal loads. Bransby & Randolph (1998) published the results of numerical analyses of the limit bearing capacity of undrained skirted foundations. They demonstrated that the yield locus for skirted foundations was not symmetric in H–V space owing to the deformation mechanisms that constrained failure. In fact, the issue of what exactly is the shape of the (V, H, M) yield surface has been the focus both of experimental work (e.g. Butterfield & Gottardi, 1994; Martin & Houlsby, 2000) and theoretical work (e.g. Bransby & Randolph, 1998; Taiebat & Carter, 2000; Gourvenec & Randolph, 2003; Randolph & Puzrin, 2003).

While the use of a (V, H, M) yield surface can be applied directly to bearing capacity calculations, it also serves as the starting point for the development of complete 'force resultant' or 'macro-element' plasticity models used to describe the entire response of a shallow foundation. This concept was first described by Nova & Montrasio (1991) in an important paper in *Géotechnique*, although the concept was anticipated by that of Schotman (1989). Since then, much work has been done on the development of force resultant models (e.g. Houlsby & Cassidy, 2002). These find particular applications to offshore applications, as they allow a complete and consistent numerical analysis of both foundation and structure. This is important, for instance, for dynamically sensitive structures, such as jack-up units or offshore wind turbine installations. One of the most important features of these models is that they offer a combined treatment of capacity and settlement, in much the same way that critical-state soil mechanics unified concepts of deformation and shear strength.

Jack-up foundations (or spudcans) have received particular attention. They are penetrated under a preload that is of the order of the maximum design axial load. The preload is then removed to provide the adequate safety factor on bearing capacity. This fascinating topic has motivated extensive research, particularly on the subject of penetration through a multilayered soil profile, and the strength (in V, H, M space) and stiffness of the preloaded foundation. The rotational stiffness of a spudcan foundation has an important effect on the structural integrity of the jack-up during extreme loading. Martin & Houlsby (2000, 2001) developed a macro-element model of the type described above to simulate the response of spudcan foundations to generalised, incremental (V, H, M) loading. This type of sophisticated foundation model is a considerable advance over the linear springs used hitherto, enabling calculation of a much more realistic load distribution in the jack-up structure.

Settlement

The importance of small-strain non-linearity is now well recognised, but this was not appreciated for much of the life of *Géotechnique*, and it was widely believed that deformation problems could be assessed with sufficient accuracy using linear isotropic elasticity theory. Linearity, of course, has the enormous advantage that superposition can be employed, so elastic solutions still play a vital role in assessing deformations, but their shortcomings are now better understood.

Geotechnical problems offered a whole range of new boundary-value problems to explore using elasticity theory, and *Géotechnique* has played a central and sustained role in the publication of these analyses, most of which relate to foundations. Among these are a number of papers from the group at Sydney University (especially E. H. Davis, H. G. Poulos and P. T. Brown), such as Davis & Poulos (1968). Much of this work led to the important reference work by Poulos & Davis (1974).

A key contribution was that by Gibson (1967); this contribution was so important that it receives the rare accolade that reference to a 'Gibson soil' is enough to define a soil with the shear stiffness increasing linearly with depth. This led to a whole new family of problems and solutions. Useful contributions to this were made by Awojobi, in a series of papers in the early 1970s (e.g. Awojobi, 1974). More recently, Gazetas and co-workers have made a number of contributions in elasticity. For example, Gazetas *et al.* (1985) proposed an equation that can be used in the calculation of settlement of a footing with any of a variety of shapes embedded in a homogeneous elastic soil.

The engineer using elastic analyses must: (a) match the engineering problem to an available, suitable solution to a similar boundary-value problem; and (b) properly define the soil properties. Assuming a solution is found that is applicable to the problem at hand, the problem reduces to one of establishing a soil profile and selecting suitable soil properties to use in calculations. This is not always so simple. For example, many solutions are available for a homogeneous elastic soil, when in fact the 'elastic' parameters of the soil, even for a relatively uniform deposit, will vary, with elastic modulus typically increasing with depth. An important decision is to choose suitable values for the elastic parameters, which requires identification of some depth of influence (depth over which most of the soil deformations will take place) and a suitable averaging algorithm. Of course, certain solutions more realistically capture specific problems: for instance, a 'Gibson soil' fits more closely the pattern of increasing stiffness with depth that is observed for many soft clay deposits.

In reality, footing settlement is not elastic except for exceptional cases in which displacements are extremely small. The linear elastic, perfectly plastic paradigm shaped much of the engineering thinking of the twentieth century, but soils are now known to yield very early in the loading process, so plasticity plays a role long before a complete plastic mechanism has formed. For long-term settlements in clay, for example, the role of consolidation (a plastic process) was recognised early, and Skempton & Bjerrum (1957) proposed an analysis to take three-dimensional consolidation into account when calculating these settlements. But plasticity asserts itself also for immediate settlements, depending on how far the combined loading action is from the (V, H, M) failure envelope. As long as plastic zones developing in the soil below and around the foundation remain confined by soil that is still in the elastic range, plastic deformations will be of the same order as elastic deformation, the role of plasticity is diminished, and the use of equivalent elastic parameters may be appropriate. As plastic zones begin to coalesce, the use of equivalent elastic parameters becomes increasingly inappropriate, and parameter values must, for sufficiently large settlements, be determined either from experiments or from more rigorous analyses that do take into account the plasticity of the soil in a sufficiently rigorous manner. An alternative path is to develop simplified analyses that take into account both the elastic and plastic responses of the soil, such as the interesting analysis proposed by Osman & Bolton (2005), in which they propose procedures to map from a representative stress–strain curve directly to the load–deformation response of a footing.

PILE FOUNDATIONS

Although many new techniques for pile installation have appeared over the last two decades, at the time of publication of the first issues of *Géotechnique*, piles were of one of two types: displacement piles (piles installed by displacing or pushing aside the soil occupying the space intended for the pile) and non-displacement piles (piles installed by first removing the soil from the space intended for the pile, then constructing the pile). Displacement piles were usually driven, and non-displacement piles were usually bored. Designing either displacement or non-displacement piles for axial loads using the working stress design approach consists in estimating the pile base resistance and the shaft resistance, adding these to obtain an ultimate load, and dividing the ultimate load by a suitable factor of safety.[§] This was the design practice 60 years ago. Then (and still in many cases at present), settlements were dealt with indirectly by using a factor of safety that experience showed to be sufficient to limit the settlement to a tolerable value. In the last two to three decades, analyses have appeared that allow explicit evaluation of settlement. The methods for design of laterally loaded piles, out of necessity, have focused more on deflection from the outset. Several papers appearing in *Géotechni-*

§ Variations of this procedure are used when load and resistance or partial factor design is employed.

que on pile foundations have started or contributed significantly to these trends. Pile installation technology has not been a meaningful part of the journal, but design innovations (piled raft design methods being a recent example) have. Most *Géotechnique* papers have approached pile analysis using the soil-property-based approach, in which the shaft and base resistances of axially loaded piles and lateral resistances in laterally loaded piles are estimated from soil properties (shear strength or stiffness) instead of from direct correlations with in situ test measurements.

Axially loaded piles

In the design of axially loaded piles, geotechnical engineers seek to prevent settlements that may endanger the superstructure or render it unserviceable. Originally, this was done by calculating a relatively loosely defined ultimate load (gradually, the load corresponding to a settlement of 10% of the pile diameter became the generally accepted definition for ultimate load) and dividing that load by a factor of safety that would keep settlements to tolerable levels. This required the capability to compute the shaft and base resistance of a pile that would correspond to this ultimate load. As an example, this was the focus of the work of Skempton (1959), who sought a correlation between the limit unit shaft resistance q_{sL} of bored piles installed in London Clay and the shear strength s_u of the clay. Skempton (1959) showed that the ratio $\alpha = q_{sL}/s_u$ is linked to aspects of construction techniques such as boring and concrete placement, and that it can be quite low (according to the paper, in the 0·3 to 0·6 range, with a typical value of 0·45) for piles installed in overconsolidated clay. This was one of the first studies attempting to link pile resistance to construction details. Although Skempton (1959) did not specifically identify one important construction-related factor (development of residual strength even before axial loading because of augering), he made important observations that allowed not only better estimation of shaft resistance in stiff clay but also an understanding of why the magnitude of this resistance was substantially less than the 'undisturbed', undrained shear strength of the clay as estimated from laboratory tests. Glossop (1968) noted in his Rankine lecture that there was a widespread belief (as late as 1950) that skyscrapers could not be built in London because the city rested on clay. The confidence in the use of large-diameter bored piles as foundations for tall buildings changed that perception, and Skempton's work contributed to that process.

The development of shaft resistance is of course more complex than contemplated by Skempton (1959), regardless of the pile type. This complexity is well illustrated by the work of Potts & Martins (1982), who analysed a section of a pile sufficiently removed from the pile top and base such that a one-dimensional finite element analysis, with the soil around the pile modelled as a disc of finite elements, suitably models the problem. The constraint was imposed that all nodes lying along a vertical line were tied together with respect to all degrees of freedom (directions). Application of this constraint guaranteed that shearing took place only in the vertical direction, and that there was no bending of the sides of any element. The pile itself was not modelled; instead, vertical displacements were applied to an internal cylindrical boundary to model pile loading. Potts & Martins (1982) made some important observations, including that the evolution of the stress state near the pile is not simple, potentially leading to substantial changes in the coefficient of lateral stress at the pile/soil interface as well as considerable shear strain localisation near the pile as a result of loading. The authors also stressed that the quality of the simulation of the response of soil near the pile, and thus of the shaft resistance, is very much dependent on the soil model, and on how completely this model describes the features of soil stress–strain response.

An important feature that any analysis must have is the ability to take into account the fact that materials that are not linear elastic tend to show strain localisation, as demonstrated also by Jardine *et al.* (1986) not only for piles but also for other types of foundation. As discussed by Randolph (2003) and White & Lehane (2004), the shaft resistance of driven piles and piles jacked into the ground using multiple jacking strokes is subject to degradation (this has become known as friction fatigue) owing to the changes in the state of the soil (including fabric, density and stress state) as the soil is subjected to multiple loading cycles during pile installation. Understanding and modelling this process is essential to development of full theoretical analyses for shaft resistance of these piles. A practical implication of this is that the presumed advantage of jacked piles over driven piles (because driven piles are subjected to many more loading cycles) is very much dependent on how many jacking strokes are used, and on what the shaft resistance against loading cycles asymptote is.

There have also been contributions in *Géotechnique* to the important problem of settlement analysis. The essence of the problem is to calculate the settlement corresponding to a given applied load at the pile head, which requires the intermediate step of calculating the load-transfer curve corresponding to that load. This is where the difficulty resides: how does one determine how much of the shaft and base resistance are mobilised for a given load (or given settlement, since the two go together)? As the applied load increases, we can expect the unit shaft resistance to increase gradually to a limit value. This limit will be reached first near the pile head, but will then be observed at cross-sections near the pile base as the load increases. If the load keeps increasing, end bearing resistance will be mobilised, and it too will increase towards its limit value, causing the pile to plunge if this limit is achieved. The papers that have proposed solutions for this problem (three examples from *Géotechnique*: Poulos & Mattes, 1969; Banerjee & Davies, 1978; Mylonakis, 2001) have assumed linear elastic soils, being more directly applicable to low load levels. Additional assumptions have been made to simplify the analyses: for example, Poulos & Mattes (1969) calculated the effects of shaft resistance by integration of Mindlin's equation, and Mylonakis (2001) assumed simplified stress and displacement fields in the soil. It is important to recognise that these methods are all continuum-based, but that they can be linked to the more accessible (but less rigorous) approach of replacing the soil by Winkler springs.

Laterally loaded piles

For laterally loaded piles, greater focus is placed on predicting lateral deflection and internal forces (shear forces and bending moments) than ultimate loads. Ultimate loads for very short piles may be reached because of the formation of shallow mechanisms in the soil, but, for piles with typical lengths, a pile would break or yield even before a mechanism formed in the soil, and the deflections would be too large before that happened.

Papers in *Géotechnique* have focused on modelling the soil as a continuum in the analysis of laterally loaded piles (e.g. Banerjee & Davies 1978; Randolph, 1981). There has been a resurgence of continuum-based analysis in recent years because better, more rigorous analyses, better soil models and greater computer power have become available; 20 to 30 years ago this was not true, and the p–y method, despite its numerous shortcomings, became the tool that practising engineers used in most laterally loaded pile analyses. This method replaces the soil with disjointed springs that have a non-linear force–deflection relation, a so-called p–y curve, p being the lateral load per unit length of pile and y being the lateral deflection. It is impossible to obtain theoretically rigorous p–y curves, but efforts have been made to make the process of obtaining them as rational as possible. One of the goals in this regard is to anchor them properly at large deflections, that is, to have a proper limit value for p, as that value naturally affects the values of p for smaller values of deflection. Randolph & Houlsby (1984) investigated the limit resistance of laterally loaded piles in clay using limit analysis, and showed that the resistance is less at small than at large depths (which, incidentally, can be critical in laterally loaded pile analyses), for which the ratio of the unit resistance to the undrained shear strength is of the order of 10.

Pile groups and piled rafts

The recognition that the load response of a pile in a group is influenced by neighbouring piles predates *Géotechnique*. Methods to take that interaction into account in a meaningful way in the prediction of the load response were only developed later, enabled by the computer, and papers in *Géotechnique* provided important contributions. Poulos (1968) studied the problem of pile groups by using Mindlin's equation, as done before by Poulos & Davis (1968) and later by Butterfield & Banerjee (1971) to account for the effect of loading throughout the length of a given pile on every other pile in the group. Poulos (1968) made the assumptions of no soil–pile slip, no bearing resistance against the pile cap, linear elasticity and incompressible piles (restricting the results to piles in comparatively weak soils). Poulos (1968) also introduced the useful concept of the interaction factor (the ratio of the additional settlement of a pile due to loading on an adjacent pile to the settlement of the pile under its own load), later also used by Randolph & Wroth (1979), and showed plots of the interaction factor against pile spacing and length. Randolph & Wroth (1979) also investigated pile group response, basing their pile group analysis on their previous work on the load–settlement response of single piles (Randolph & Wroth, 1978). Mylonakis & Gazetas (1998) showed that pile compressibility has an effect on the values of the interaction factors.

As in the Poulos (1968) solution, most solutions for pile groups neglected the raft action of the pile cap. Caps in which the resistance of the cap itself was taken into account became known later as piled rafts, a design concept that has received increasing attention in the last two decades or so, even if the concept first appeared in the literature in the 1950s and in practice much earlier. Perhaps a reason for the greater interest in this type of solution in recent years is that proper design of this type of foundation system is analysis-intensive, which means that, with the availability of better analyses and adequate computer power, the design of piled raft systems has become economically feasible in an increased number of projects. However, before engineers converge on piled rafts as the design solution for a particular project, simplified analyses are needed to assess whether performance-related gains more than compensate for the greater design cost. Poulos (2001) examined the entire design sequence, discussed analyses of various levels of sophistication that may be used to address the piled raft problem, and pointed to a number of key references covering them. Poulos showed, using his own analyses, that: (a) there is an optimal number of piles; (b) increasing raft thickness reduces differential settlements (but increases bending moments); (c) strategic placement of piles is required for optimal differential settlement reduction; and (d) modelling of column loads using an equivalent uniform loading is acceptable for estimation of total settlement, but not for prediction of differential settlements or bending moments. As Poulos (2001), Horikoshi & Randolph (1998) and Reul & Randolph (2003) also discussed the design process for piled rafts, using case histories (including centrifuge data) as illustrations.

Offshore piling

Although the pile capacities (and therefore pile sizes) needed for offshore structures are generally much larger than for land structures, it is surprising to find that the pile capacity estimation methods used today by the offshore industry have not been published in *Géotechnique*. However, the equally important question of the drivability of these piles is a theme that was addressed by Litkouhi & Poskitt (1980). Although the one-dimensional wave equation modelling of piledriving had been developed 20 years earlier, the dynamic resistance of typical stiff North Sea clays had not received much attention. Litkouhi & Poskitt demonstrated by simple laboratory tests that the viscous damping of these soils was highly non-linear. Most users of wave equation analysis software today use linear viscous damping (dynamic resistance proportional to pile velocity), and are blissfully unaware of the actual behaviour of the soil!

The theme of pile resistance to driving and the post-driving capacity of the plug in an open-ended pipe pile was a subject of research in the late 1980s, and Randolph *et al.* (1991) provided important insights into sand plug behaviour during and after driving. An understanding of the mechanisms of load transfer between the plug and the pipe during loading was developed further by De Nicola & Randolph (1997), but the implications of that work have yet to permeate industry practice.

Rankine Lectures

Finally, we note that two Rankine Lectures have been devoted to piling, both from Australia and both concentrating on the appropriate analysis techniques. Poulos (1989) examined the range of problems and solutions involving axially loaded piles, highlighting the need for a proper match between the level of rigour of an analysis and the problem to be solved, as well as the critical importance of knowing the operative values of soil properties. Randolph (2003) examined the science underlying piling engineering. He showed how it is possible with current knowledge to use effective stress analysis to understand and predict pile resistance, and called attention to important effects not widely recognised, such as shaft resistance degradation (friction fatigue) at a set depth as a pile is driven (or jacked in cycles) further and further beyond that depth. Randolph also examined the state of pile driving analysis, and of the analysis and design of piled rafts.

CONCLUSIONS

For 60 years, *Géotechnique* has published papers addressing the problem of how to best analyse, design and construct foundations. The papers in the journal show a clear evolution from simple, even rudimentary analyses to the relatively sophisticated ones that incorporate today's understanding of how soil responds to different types of loading, of modern numerical techniques for capturing soil response, and the sometimes complex displacement, strain and stress fields appearing in the boundary-value problems of foundation engineering, and the effects of foundation construction.

Many of the papers were drivers of innovation, expanding knowledge and establishing new trends in foundation design; virtually all were driven by the desire to solve the very real problems of foundations engineering.

The eighth Rankine Lecturer, R. Glossop, called attention to the fact that, throughout the nineteenth century, there was no systematic approach for foundation problems, and that there was indeed a reluctance to accept a scientific approach to foundation design. Although there continues to be a boundary between science and its application, that boundary is gradually becoming more permeable, thanks to the increasing quality and applicability of the science. *Géotechnique*, with its emphasis on proper science, and its recognition of new and improved design and technology, will continue to play a role in shaping the foundation engineering of the future.

REFERENCES

Awojobi, A. O. (1974). The invariance of Gibson's law for a stratum on a frictionless base. *Géotechnique* **24**, No. 3, 359–366.

Banerjee, P. K. & Davies, T. G. (1978). The behaviour of axially and laterally loaded single piles embedded in nonhomogeneous soils. *Géotechnique* **28**, No. 3, 309–326.

Bjerrum, L (1973). Geotechnical problems involved in foundations of structures in the North Sea. *Géotechnique*, **23**, No. 3, 319–358.

Bransby, M. F. & Randolph, M. F. (1998). Combined loading of skirted foundations. *Géotechnique* **48**, No. 5, 637–655.

Brinch Hansen, J. (1970). *A revised and extended formula for bearing capacity.* Bulletin No. 98, Danish Geotechnical Institute, Copenhagen, Denmark, 5–11.

Butterfield, R. & Banerjee, P. K. (1971). The problem of pile group and pile cap interaction. *Géotechnique* **21**, No. 2, 135–142.

Butterfield, R. & Gottardi, G. (1994). A complete three-dimensional envelope for shallow footings on sand. *Géotechnique* **44**, No. 1, 181–184.

Butterfield, R. & Ticof, J. (1979). The use of physical models in design: Discussion. *Proc. 7th Eur. Conf. Soil Mech. Found. Engng, Brighton* **4**, 259–261.

Davis, E. H. & Booker, J. R. (1973). The effect of increasing strength with depth on the bearing capacity of clays. *Géotechnique* **23**, No. 4, 551–563.

Davis, E. H. & Poulos, H. G. (1968). The use of elastic theory for settlement prediction under three-dimensional conditions. *Géotechnique* **18**, No. 1, 67–91.

De Nicola, A. & Randolph, M. F. (1997). The plugging behaviour of driven and jacked piles in sand. *Géotechnique* **47**, No. 4, 841–856.

Edwards, D. H., Zdravkovic, L. & Potts, D. M. (2005). Depth factors for undrained bearing capacity. *Géotechnique* **55**, No. 10, 755–758.

Gazetas, G., Dobry, R., Tassoulas, J. L. & O'Rourke, M. J. (1985). Elastic settlement of arbitrarily shaped foundations embedded in half-space. *Géotechnique* **35**, No. 3, 339–346.

Gibson, R. E. (1967). Some results concerning displacements and stresses in a non-homogeneous elastic half-space. *Géotechnique* **17**, No. 1, 58–67.

Glossop, R. (1968). The rise of geotechnology and its influence on engineering practice. *Géotechnique* **18**, No. 2, 107–150.

Gourvenec, S. & Randolph, M. F. (2003). Effect of strength non-homogeneity on the shape of failure envelopes for combined loading of strip and circular foundations on clay. *Géotechnique* **53**, No. 6, 575–586.

Horikoshi, K. & Randolph, M. F. (1998). A contribution to optimum design of piled rafts. *Géotechnique* **48**, No. 3, 301–317.

Houlsby, G. T. & Cassidy, M. J. (2002). A plasticity model for the behaviour of footings on sand under combined loading. *Géotechnique* **52**, No. 2, 117–129.

Jardine, R. J., Potts, D. M., Fourie, A. B. & Burland, J. B. (1986). Studies of the influence of non-linear stress–strain characteristics in soil–structure interaction. *Géotechnique* **36**, No. 3, 377–396.

Kimura, T., Kusakabe, O. & Saitoh, K. (1985). Geotechnical model tests of bearing capacity problems in a centrifuge. *Géotechnique* **35**, No. 1, 33–45.

Litkouhi, S & Poskitt, T. J. (1980). Damping constants for pile driveability calculations. *Géotechnique* **30**, No. 1, 77–86.

Lyamin, A., Salgado, R., Sloan, S. W. & Prezzi, M. (2007). Two- and three-dimensional bearing capacity of footings in sand. *Géotechnique* **57**, No. 8, 647–662.

Martin, C. M. & Houlsby, G. T. (2000). Combined loading of spudcan foundations on clay: laboratory tests. *Géotechnique* **50**, No. 4, 325–338.

Martin, C. M. & Houlsby, G. T. (2001). Combined loading of spudcan foundations on clay: numerical modelling. *Géotechnique* **51**, No. 8, 687–699.

Meyerhof, G. G. (1951). The ultimate bearing capacity of foundations. *Géotechnique* **2**, No. 4, 301–330.

Meyerhof, G. G. (1953). The bearing capacity of foundations under eccentric and inclined loads. *Proc. 3rd Int. Conf. Soil Mech. Found. Engng, Zurich* **1**, 440–445.

Mylonakis, G. (2001). Winkler modulus for axially loaded piles. *Géotechnique* **51**, No. 5, 455–461.

Mylonakis, G. & Gazetas, G. (1998). Settlement and additional internal forces of grouped piles in layered soil. *Géotechnique* **48**, No. 1, 55–72.

Nova, R. & Montrasio, L. (1991). Settlements of shallow foundations on sand. *Géotechnique* **41**, No. 2, 243–256.

Osman, A. S. & Bolton, M. D. (2005). Simple plasticity-based prediction of the undrained settlement of shallow circular foundation on clay. *Géotechnique* **55**, No. 6, 435–447.

Peck, R. B. and Bryant, F. G. (1953). The bearing-capacity failure of the Transcona elevator. *Géotechnique* **3**, No. 5, 201–208.

Potts, D. M & Martins, J. P. (1982). The shaft resistance of axially loaded piles in clays. *Géotechnique* **32**, No. 4, 369–386.

Poulos, H. G. (1968). Analysis of the settlement of pile groups. *Géotechnique* **18**, No. 4, 449–471.

Poulos, H. G. (1989). Pile behaviour: theory and application. *Géotechnique* **39**, No. 3, 363–415.

Poulos, H. G. (2001). Piled raft foundations: design and applications. *Géotechnique* **51**, No. 2, 95–113.

Poulos, H. G. & Davis, E. H. (1968). The settlement behaviour of single axially loaded incompressible piles and piers. *Géotechnique* **18**, No. 3, 351–371.

Poulos, H. G. & Davis, E. H. (1974). *Elastic solutions for soil and rock mechanics*. New York: Wiley.

Poulos, H. G. & Mattes, N. S. (1969). The behaviour of axially loaded end-bearing piles. *Géotechnique* **19**, No. 2, 285–300.

Randolph, M. F. (1981). The response of flexible piles to lateral loading. *Géotechnique* **31**, No. 2, 247–259.

Randolph, M. F. (2003). Science and empiricism in pile foundation design. *Géotechnique* **53**, No. 10, 847–875.

Randolph, M. F. & Houlsby, G. (1984). The limiting pressure on a circular pile loaded laterally in a cohesive soil. *Géotechnique* **34**, No. 4, 613–623.

Randolph, M. F. & Puzrin, A. M. (2003). Upper bound limit analysis of circular foundations on clay under general loading. *Géotechnique* **53**, No. 9, 785–796.

Randolph, M. F. & Wroth, C. P. (1978). An analysis of the deformation of vertically loaded piles. *Proc. ASCE J. Geotech. Engng Div.* **104**, No. GT12.

Randolph, M. F. & Wroth, C. P. (1979). An analysis of the vertical deformation of pile groups. *Géotechnique* **29**, No. 4, 423–439.

Randolph, M. F., Leong, E. C. & Houlsby, G. T. (1991). One-dimensional analysis of soil plugs in pipe piles. *Géotechnique* **41**, No. 4, 587–598.

Reul, O. & Randolph, M. F. (2003). Pile rafts in overconsolidated clay: comparison of in situ measurements and numerical analyses. *Géotechnique* **53**, No. 3, 301–315.

Roscoe, K. H. & Schofield, A. N. (1957). The stability of short pier foundations in sand, discussion. *British Welding Journal*, January, 12–18.

Salgado, R., Lyamin, A., Sloan, S. & Yu, H. S. (2004). Two- and three-dimensional bearing capacity of footings in clay. *Géotechnique* **54**, No. 5, 297–306.

Schotman, G. J. M. (1989). The effects of displacements on the stability of jack-up spud-can foundations. *Proc. 21st Offshore Technology Conf., Houston*, 15-524, OTC 6026.

Skempton, A. W. (1951). The bearing capacity of clays. *Building Research Congress, London* **1**, 180–189.

Skempton, A. W. (1959). Cast in-situ bored piles in London Clay. *Géotechnique* **9**, No. 4, 153–173.

Skempton, A. W. & Bjerrum, L. (1957). Contribution to the settlement analysis of foundations on clay. *Géotechnique* **7**, No. 4, 168–178.

Taiebat, H. & Carter, J. P. (2000). Numerical studies of the bearing capacity of shallow footings on cohesive soil subjected to combined loading. *Géotechnique* **50**, No. 4, 409–418.

Vesić, A. S. (1975). Bearing capacity of shallow foundations. In *Foundation engineering handbook* (eds H. F. Winterkorn and H.-Y. Fang), pp. 121–147. New York: Van Nostrand Reinhold.

White, D. J. & Lehane, B. M. (2004). Friction fatigue on displacement piles in sand. *Géotechnique* **54**, No. 10, 645–658.

Milligan, G. W. E., St John, H. D. & O'Rourke, T. D. (2008). *Géotechnique* **58**, No. 5, 377–383 [**doi:** 10.1680/geot.2008.58.5.377]

Contributions to *Géotechnique* 1948–2008: Retaining structures

G. W. E. MILLIGAN*, H. D. ST JOHN* and T. D. O'ROURKE†

Retaining structures and earth support systems form an important part of geotechnical engineering. In the first 60 years of *Géotechnique*, some 110 publications have appeared that related to such structures. These are discussed in this review under five headings: case histories and full-scale observations; ground movements associated with retaining structures; numerical analyses; compaction stresses acting on structures; and Symposia in Print. Individual comments are made on approximately half of the papers published, highlighting key contributions to the advancement of the understanding of retaining structures. Many of the papers have provided important information subsequently incorporated into standard reference documents for geotechnical engineers, notably three reports published by the Construction Industry Research and Information Association (CIRIA): *Temporary propping of deep excavations: Guidance on design*; *The observational method in ground engineering: Principles and applications*; and *Embedded retaining walls: Guidance for economic design*. In this way, and through the contributions of individual papers to other aspects of performance, *Géotechnique* has made a valuable contribution to the improved analysis, engineering design, construction and monitoring of retaining structures.

KEYWORDS: diaphragm and in situ walls; historical review; retaining walls

Les structures de soutien et les systèmes de terrassement représentent une partie importante de l'ingénierie géotechnique. Au cours des 60 premières années de *Géotechnique*, environ 110 communications ont été publiées sur des structures de ce genre. Dans le présent recueil, elles sont examinées dans les cinq catégories suivantes : études de cas et observations intégrales ; mouvements du sol relatifs à des structures de soutènement ; analyses numériques ; contraintes de compactage agissant sur les structures ; colloques publiés. Des commentaires individuels sont présentés sur environ la moitié des communications publiées, en mettant ainsi en valeur des contributions importantes pour l'avancement des connaissances sur les structures de soutènement. Un grand nombre de communications ont apporté des informations importantes, incorporées par la suite dans des ouvrages de référence pour ingénieurs géotechniciens, notamment trois rapports publiés par l'association Construction Industry Research and Information Association (CIRIA): *Soutènement Provisoire d'Excavations Profondes – Conseils Sur la Conception; Méthodes d'Observation Dans l'Ingénierie des Sols – Principes et Applications;* et *Murs de Soutien Encastrés – Conseils Pour une Conception Économique.* De cette façon, et à travers les contributions apportées par des communications individuelles sur d'autres aspects des performances, *Géotechnique* a été en mesure de contribuer de façon importante au perfectionnement des analyses, des études techniques, de la construction et du contrôle des structures de soutien.

INTRODUCTION

The total number of publications in the first 60 years of *Géotechnique* relating to retaining structures amounts to some 110 papers, technical notes and substantive correspondence or discussion. The exact figure depends on where the boundaries are drawn on subject matter to be included. For instance, a number of papers have been included that deal with the stability of slurry-filled trenches and associated ground movements, and others relate to the base stability of excavations in soft clay. The former are important in connection with the construction of diaphragm walls, both from an immediate practical point of view and also because of the influence of wall construction methods on the initial horizontal ground stresses affecting the design of the wall as well as soil displacements outside the perimeter of the excavation. The latter are important because an adequate margin of confidence against base failure is essential for safety, and because base failure and associated ground deformation will influence various aspects of the retaining structure performance, including wall penetration depths, strut loads, and soil displacements affecting adjacent buildings and facilities. Publications concerned with seepage pressures, and their influence on base stability and earth pressures, are included for the same reason. However, papers on the 'elastic' heave of excavations are not included, as being more a concern for foundations than for retaining structures.

There were 20 publications related to retaining structures in the first 20 years, then 16 in the subsequent decade. The fourth decade saw an increase to 20, boosted by the inclusion of a Symposium in Print relating to cantilever and propped walls in 1984. The decade from 1989 to 1998 was the most productive, with a total of 36 publications: again, this was helped by a number of papers in a Symposium in Print in 1994, this one on the observational method. The most recent decade has seen the number fall back to 18. A dozen of the papers are either of primarily historical interest or reviews of previously published information. Around 40 are concerned primarily with new theoretical developments, and about 30 primarily with each of model testing, numerical analysis or full-scale observations. Some papers have significant content under more than one of these headings.

CASE HISTORIES AND FULL-SCALE OBSERVATIONS

Some of the most interesting and useful papers in *Géotechnique* fall within this category. Three from the early years that have stood the test of time are those by Skempton & Ward (1952), Wu & Berman (1953) and Bjerrum & Eide (1956). The first of these is a classic case of a major excavation in soft clay with a multi-propped sheet pile retaining wall. It has all the right ingredients for a good case history: a detailed site investigation with high-quality data on soil conditions, including extensive testing on block

Discussion on this paper closes on 1 December 2008, for further details see p. ii.
* Geotechnical Consulting Group, UK
† Cornell University, USA

samples retrieved from a shaft excavated on the site; information on the construction sequence, and on all site events that could have influenced the observed performance; details of instrumentation used to measure forces transmitted from sheet piles to walings and within the props, including discussion on the reliability of the measurements; and an interpretation of the results in a way that provides useful information for future design.

Some of the results from the paper are shown in Fig. 1. Interestingly, the authors considered that the factual data on soil conditions and monitoring output were more important than their interpretation, recognising that methods of interpretation would develop with improved theory, while the data would always be useful for calibration of new analytical methods or as a contribution to empirical designs. One detailed point of interest was the increase in loads in the lowest level of props when piling was undertaken within the excavation, owing to disturbance (and perhaps partial failure) of the adjacent ground.

Wu & Berman (1953) present a somewhat similar case, for a very deep cut through soft to medium clay in Chicago. The work was for an extension of the metro system, from the early stages of which Peck had made extensive observations on prop loads, which had resulted in empirical design methods that were widely adopted. Wu & Berman refer back to the early work for the ground conditions, and then present earth pressure distributions derived from prop load measurements, as well as settlements measured on adjacent buildings. The pressure distributions are successfully related to classic design methods, and are interestingly different from Peck's earlier observations. Wu & Berman attribute this to the greater restraint at the base of the retaining walls in their case, which restricted the ground movements and altered the local pressure distribution.

Bjerrum & Eide (1956) also present information on excavations in soft clays, but have a rather different purpose. They were seeking confirmation of the 'reverse bearing capacity' analysis for base heave failure of deep excavations of limited width. To this end they collected information on 14 relevant cases, in most of which either total or partial failure had occurred. The data on each case are more limited than for a single case history, but include the critical information on excavation geometry, surcharge loading, and average soil strengths. The calculated factor of safety is shown to be below 1·0 for the cases where failure occurred, and just above for the cases where partial failure occurred, or no failure was observed. The paper had a profound influence on the evolution of design methods for braced excavations leading to the use of dimensionless base stability numbers in later semi-empirical procedures for predicting both ground movements and support loads (e.g. Peck, 1969; Clough & O'Rourke, 1990).

All three of these papers have provided information that has been incorporated into subsequent reviews and design advice, such as in CIRIA report C517 on *Temporary propping of deep excavations: Guidance on design* (Twine & Roscoe, 1999).

Other papers of interest involving measurements at full scale include Morgenstern & Amir-Tahmasseb (1965), Farmer & Attewell (1973) and Ng *et al.* (1999). All three are concerned with slurry-filled trench panels: Morgenstern & Amir-Tahmasseb with the question of stability following an observed failure, and the other two with ground movements determined from instrumented panels. Crofts *et al.* (1977) were concerned with the effects of trench excavation on parallel pipelines in the ground adjacent to the trench: the paper contains a useful collection of published data on

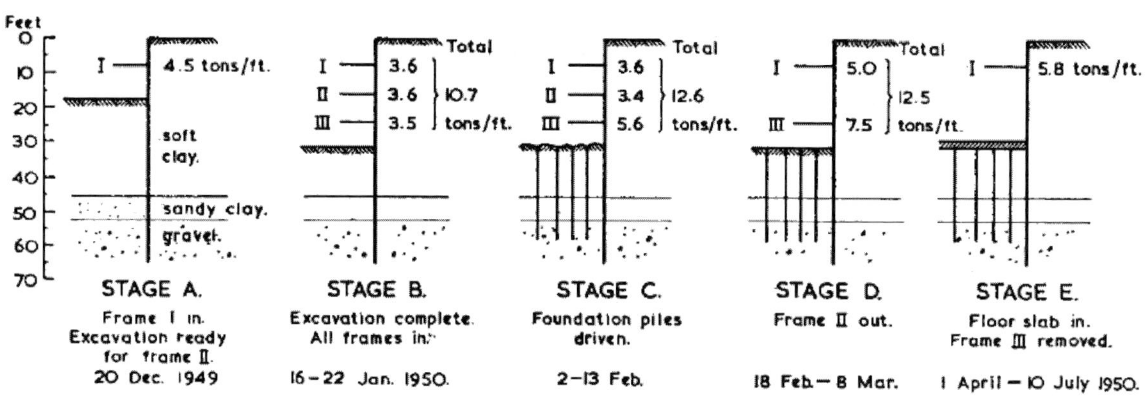

Waling loads at various stages of construction, as measured by the load gauges

Comparison of observed Waling Loads (tons per foot) with results calculated by Peck's Empirical Rule

	Stage B			Stage C			Stage D		
	Observed values	1·5 × Observed values	Peck's rule	Observed values	1·5 × observed values	Peck's rule	Observed values	1·5 × observed values	Peck's rule
Frame I ..	3·6	5·4	7·0	3·6	5·4	7·0	5·0	7·5	9·8
Frame II	3·6	5·4	5·7	3·4	5·1	5·7	—	—	—
Frame III	3·5	5·2	6·2	5·6	8·4	6·2	7·5	11·2	9·1
Total ..	10·7	16·0	18·9	12·6	18·9	18·9	12·5	18·7	18·9

Fig. 1. Observed waling loads from cofferdam in soft clay (from Skempton & Ward, 1952)

ground movements adjacent to excavations. Ou *et al.* (2000) provide information on ground displacements and strains due to excavation by a top-down construction method. Detailed measurements of changes in pore pressure and horizontal stress due to pile installation and then excavation for bored pile retaining walls are presented in a pair of papers by Richards *et al.* (2006, 2007). These are of particular interest, in that the ground conditions comprised Atherstone and Weald Clays, about which little information had previously been published. Blight & Dane (1989) provide some important information on the dangers of localised corrosion in metallic reinforcement in reinforced soil structures: they report detailed investigations of substantial structures that had to be demolished and replaced well before the end of their intended design life.

Both of the Symposia in Print in 1984 and 1994 included a number of papers concerned with full-scale observations. These are discussed further below, along with other related papers.

GROUND MOVEMENTS

The importance of ground movements has received increasing recognition, both because of the potential effects of the movements on the retaining structure itself or on nearby buildings and services, and also because the ground strength that can be mobilised may be limited by allowable deformations. An early *Géotechnique* paper on this subject of considerable practical importance was that by Rowe & Peaker (1965). They showed, by a series of high-quality model tests, that theoretical passive pressure coefficients based on peak strength values and high values of wall friction were unsafe. In loose sand this was because the full theoretical pressure could be achieved only after excessive wall movements, while in dense sand progressive failure meant that the average soil strength was less than the peak value. Some details of their apparatus and results of passive pressure measurements at various wall displacements are shown in Fig. 2.

From about this time a series of model tests were undertaken at Cambridge University, using an X-ray technique to study local strains within the soil mass associated with both active and passive modes of wall displacement. These investigations led to a series of publications, such as James & Bransby (1970, 1971), Bransby & Milligan (1975) and Milligan & Bransby (1976). Later, Bolton & Powrie (1987, 1988) measured similar strain fields in centrifuge tests and sought to develop a design approach using mobilised soil strengths based on simplified strain fields related to allowable wall deformations. Bica & Clayton (1998) made measurements of earth pressures and wall friction angles related to wall movements for the embedded length of cantilever walls. The results were consistent with previously published work, including that of Rowe & Peaker.

Of particular interest is the failure mode for cantilever walls observed by Bolton & Powrie (1987) whereby water filling a gap between the wall and adjacent clay generates hydrostatic pressure in the gap leading to collapse. This failure mode is taken into account in UK design methods (Gaba *et al.*, 2003), but was not recognised in wall design during upgrades of the New Orleans hurricane protection system. The collapse of I-walls during Hurricane Katrina in 2005, driven by similar gap formation, was a major contributor to levee failure that flooded substantial parts of Orleans Parish (Interagency Performance Evaluation Task Force, 2008).

Centrifuge testing has also been used to study a number of different aspects of retaining wall behaviour, providing information on failure modes, earth pressures, ground movements and structure bending moments. Examples of these are: Bolton & Stewart (1994), for a wall propped at excavation level and subject to rising groundwater; Richards &

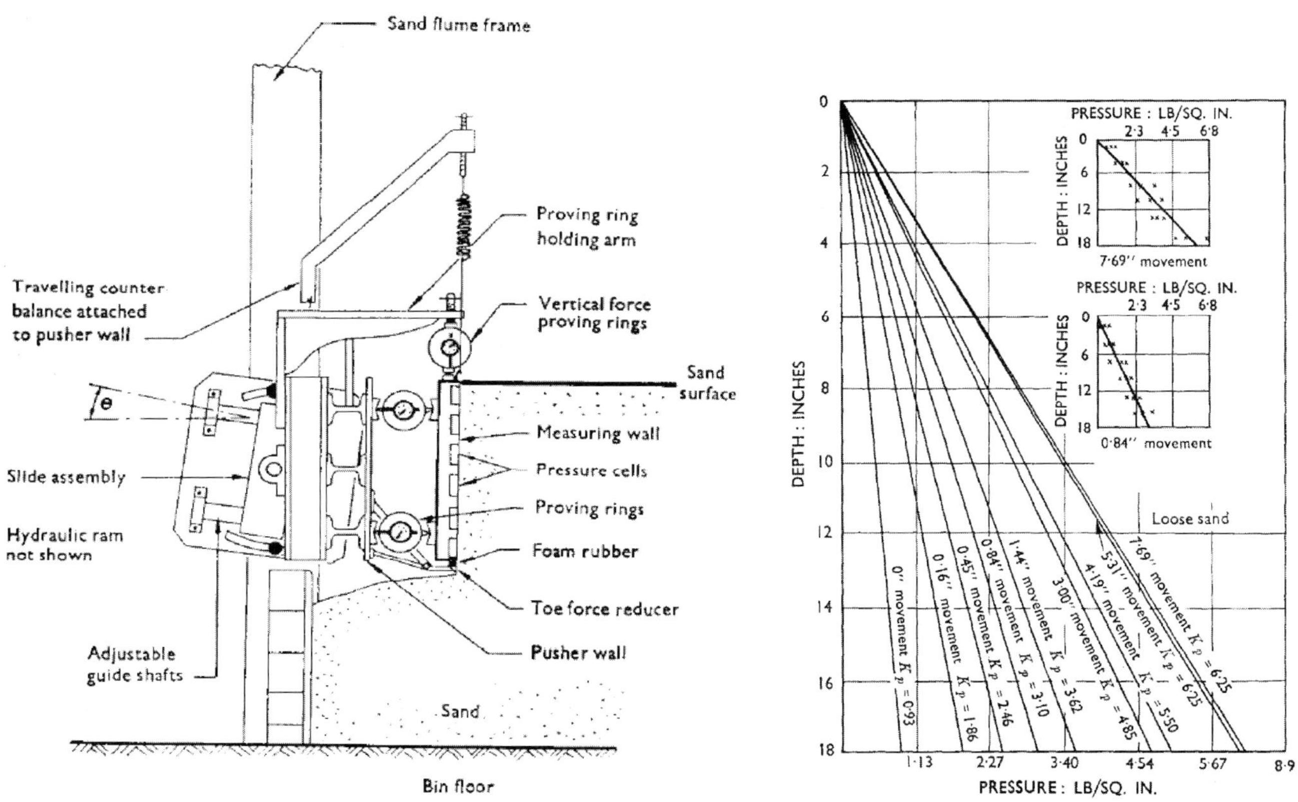

Fig. 2. Details of apparatus and typical measured passive pressure distributions varying with wall displacement (from Rowe & Peaker, 1965)

Powrie (1998), for a wall propped at the top and at excavation level; and Powrie & Daly (2002, 2007), for the use of berms and stabilising bases. Powrie & Kantartzi (1996) used centrifuge tests to study the pressures and movements in the ground around trenches filled with slurry or wet concrete. These studies have contributed to CIRIA report C580 (Gaba *et al.*, 2003) on *Embedded retaining walls: Guidance for economic design*, which is considered to be the standard industry document for embedded wall design in the UK.

NUMERICAL ANALYSES

Probably the most significant development in geotechnical design over the past 30 years has been the advance in the power and usefulness of numerical analysis. In theory, the application of finite element, finite difference or discrete element modelling should allow complete solution of complex real problems, for both collapse and serviceability conditions. In his Rankine Lecture, Potts (2003) reviewed the extent to which this is a reality, and highlighted the pitfalls and developments that are still required to fully achieve this aim.

Results of early attempts at numerical analyses were often disappointing, until the full extent of the variation of soil stiffness with strain was appreciated and incorporated into numerical models. Simpson *et al.* (1979) provided an early example of the application of non-linear behaviour in finite element analysis, with retrospective analyses of observed behaviour of retaining structures at New Palace Yard and Neasden Underpass. In his Rankine Lecture, Simpson (1992) presented a new model incorporating stress history as well as strain-level effects; he also compared measurements made at the excavation for the new British Library with calculations made with finite element analyses of varying levels of sophistication. The best analyses were shown to agree well with the measured wall deflections, as shown in Fig. 3.

Since the 1980s a number of aspects of retaining wall behaviour have been studied using finite element analyses. Examples of these are: Potts & Fourie (1984), investigating the effects of different types of construction and initial ground stresses for walls in overconsolidated clay; Potts & Fourie (1985), extending the previous paper by studying variations in wall stiffness; Powrie & Li (1991), analysing a wall propped at excavation level; Georgiadis & Anagnostopoulos (1998), on the effect of berms; and Powrie & Chandler (1998), on a wall with a stabilising platform. Gourvenec & Powrie (1999) modelled the installation effects of a diaphragm wall panel in three dimensions, obtaining ground movements and reductions in lateral stresses in the ground.

Until relatively recently, analyses with complex soil models have been restricted to two-dimensional analyses. With increasing computational power it has become more realistic to undertake fully three-dimensional analyses, but these are still comparatively time consuming and expensive. Zdravkovic *et al.* (2005) compared results from three-dimensional analyses of an excavation within diaphragm walls with various two-dimensional approximations. They demonstrated the conditions under which approximate two-dimensional analyses were adequate, in terms of both wall bending moments and ground movements. As with the model testing, results of the various numerical studies have been taken into account in CIRIA report C580 (Gaba *et al.*, 2003).

COMPACTION STRESSES

Where fill material is placed and compacted behind preformed retaining structures, the earth pressures against

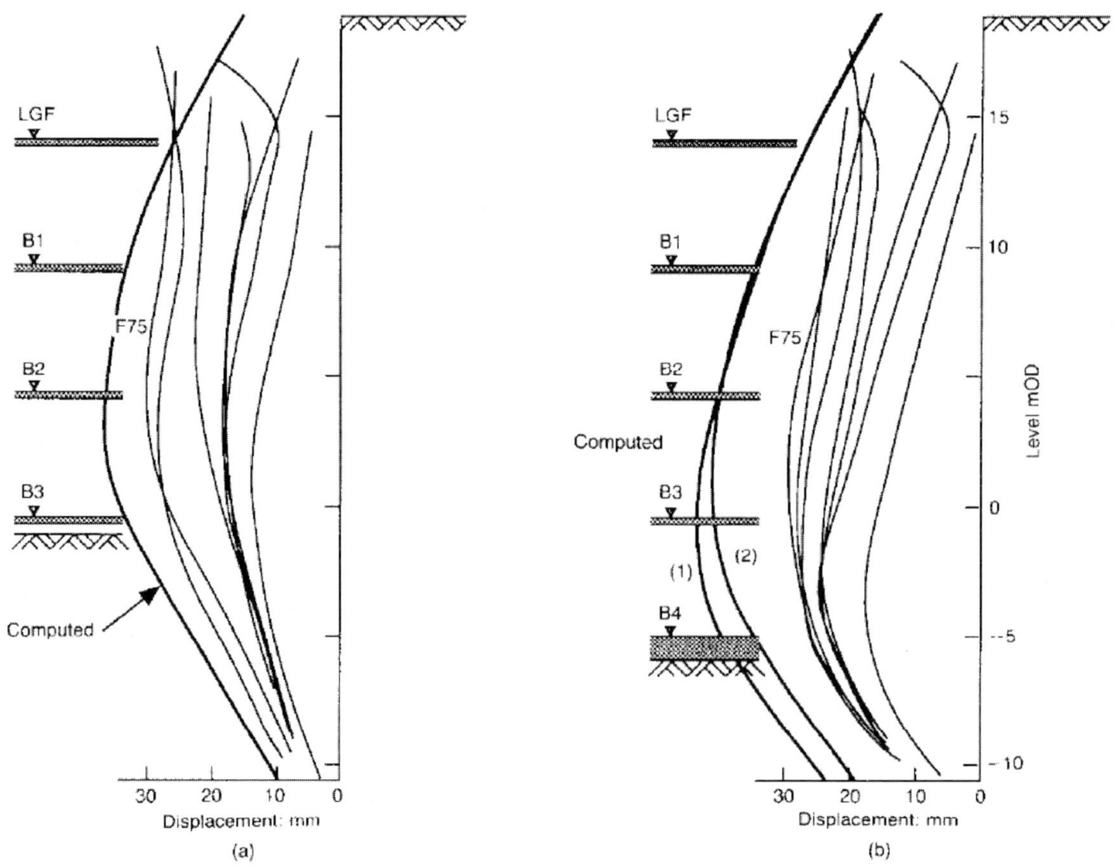

Fig. 3. Comparison of measured and computed displacements for retaining walls at British Library (from Simpson, 1992)

the wall may be expected to be different from, and higher than, those calculated from active pressure coefficients. The first *Géotechnique* paper to address this issue was that by Aggour & Brown (1974), from a theoretical point of view, while Jones & Sims (1975) reported measured pressures on bridge abutments. A significant contribution was made by Ingold (1979), who presented a relatively simple analytical method of determining stresses due to compaction, and compared calculated values with field observations, obtaining encouraging validation of the approach. Ingold's method has been widely adopted in practice.

Clayton & Symons (1992) extended the discussion to the stresses induced by compacted clay fill. Because of the effects of swelling and pore pressure dissipation the pressures can change significantly with time, and in high-plasticity materials can reach values higher than the passive pressures calculated using soil parameter values derived from standard compression tests.

Stresses behind abutments of integral bridges pose an additional problem. These are structures in which the abutment retaining walls are integral with the bridge deck. As the bridge expands and contracts, the fill behind is subject to alternating extension and compression. Broms & Ingelson (1971) report measurements of earth pressures against substantial abutments and their variation with movements of walls. High stresses were induced by compaction of the sand fill, which dropped to active values during outward movements of the walls, but rose to values equal to or higher than theoretical passive pressures during subsequent expansion of the bridge. The movements required to induce these pressures were smaller than expected.

Clayton *et al.* (2006) undertook laboratory tests to investigate the stress changes induced by the cyclic strains in the ground behind integral bridge abutments. They found that natural clay specimens did not show increases in horizontal stress due to the cyclic straining, but that sand specimens showed a progressive increase in maximum horizontal stress, approaching the theoretical passive pressure, consistent with the field measurements of Broms & Ingelson (1971).

SYMPOSIA IN PRINT

Papers submitted to two relevant Symposia in Print have been published in *Géotechnique*. In 1984 the Symposium was addressed to propped and cantilevered retaining walls. The 1994 Symposium was on the observational method in geotechnical engineering, and included four papers on applications of the observational method to retaining structures.

From the 1984 Symposium, two papers relate to retaining walls for the Bell Common cut-and-cover tunnel in London Clay, and two to retaining walls required in Gault Clay at Dunton Green, both arising from construction of the M25 motorway. For the former, Hubbard *et al.* (1984) describe the design of the propped walls. Questions arising from the conventional initial design led to a full soil–structure interaction analysis being undertaken using finite element methods, reported in this paper, and to the instrumentation and monitoring of construction, reported on by Tedd *et al.* (1984).

Garrett & Barnes (1984) report on the design of cantilever walls at Dunton Green, and also on measurements made during construction and with the walls in service. Clarke & Wroth (1984) present results of pressuremeter tests to measure lateral stresses in the ground, and derive values of soil parameters for use in finite element analyses, which gave results reasonably in accordance with the field measurements. These papers were of particular interest because at that stage there was very little detailed information on the Gault Clay, as previous developments had tended to avoid the clay formation owing to its unfavourable characteristics.

The final paper in this symposium was by Wood & Perrin (1984), and reported field measurements from a basement wall in London Clay with three levels of props. Measurements were made of wall movements, wall bending strains, ground movements and strut forces, and preliminary comparisons were made with designs based on conventional earth pressures.

As well as providing useful case histories, with data that has been incorporated into design guides such as CIRIA C580 (Gaba *et al.*, 2003), these publications did much to establish that proper design of embedded walls in heavily overconsolidated clays required the use of finite element analyses backed by further information on actual performance. Follow-up work is reported in a number of papers: Tedd & Charles (1985), on the strength of London Clay measured in both compression and extension at low stress levels relevant to retaining wall design; Carder & Symons (1989), with a further detailed case history of a wall in London Clay on the A329 at Reading; Symons & Tedd (1989), on the long-term performance of the walls at Bell Common; and a review by Symons & Carder (1992) of measurements from Bell Common, Reading, and further instrumented walls at Hatfield and Malden.

The 1994 Symposium contained four papers relating to retaining structures. Powderham (1994) provided an overview of the use of the observational method in cut-and-cover tunnel construction. Details of the major project at the Limehouse Link tunnel in East London are provided by Glass & Powderham (1994). Here the observational method was used to make considerable savings in propping costs. Young & Ho (1994) describe how a full-scale trial, finite element analyses and site instrumentation were combined in an observational approach to the construction of an anchored sheet-pile wall at the Channel Tunnel terminal in Folkestone. Ikuta *et al.* (1994) report on the construction of a large, deep basement excavation by a top-down procedure. Monitoring of the initial stages of excavation allowed the design of later stages to be rationalised and two arrays of temporary props included in the original construction method to be omitted.

All four of these papers are referenced in CIRIA report R185 on the *Observational method in ground engineering: Principles and application* (Nicholson *et al.*, 1999). This is now the standard guidance document for practical application of the observational method in UK construction.

FINAL COMMENTS

This review has mentioned individually fewer than half the papers published on the subject of retaining structures during the first 60 years of *Géotechnique*. Many other papers have made important theoretical contributions to subjects such as the determination of earth pressure coefficients or seepage pressures, and even the combination of the two. There are also some interesting individual publications that do not fit naturally into the general subject subdivisions used for this review. Examples are the paper by Lee *et al.* (1994) on major failures of high reinforced earth structures on a highway in Tennessee, and analyses by discrete element and finite element methods of drystone masonry walls, compared with measurements made on a full-scale test wall in Ireland in 1834, by Harkness *et al.* (2000) and Zhang *et al.* (2004).

As noted above, many of the papers selected have not only been of interest in themselves, but have also provided valuable feedstock for subsequent review papers that have become standard reference documents for geotechnical engineers. A prime example is the paper by Clough &

O'Rourke (1990) on construction-induced movements of in situ walls, while St John *et al.* (1993) review information on wall movements specific to London Clay. The three CIRIA reports on prop loads, wall design and the observational method are all important and useful practical documents, and include much that was first published in *Géotechnique*.

REFERENCES

Aggour, M. S. & Brown, C. B. (1974). The prediction of earth pressure on retaining walls due to compaction. *Géotechnique* **24**, No. 4, 489–502.

Bica, A. V. D. & Clayton, C. R. I. (1998). An experimental study of the behaviour of embedded lengths of cantilever walls. *Géotechnique* **48**, No. 6, 731–745.

Bjerrum, L. & Eide, O. (1956). Stability of strutted excavations in clay. *Géotechnique* **6**, No. 1, 32–47.

Blight, G. E. & Dane, M. S. W. (1989). Deterioration of a wall complex constructed of reinforced earth. *Géotechnique* **39**, No. 1, 47–53.

Bolton, M. D. & Powrie, W. (1987). The collapse of diaphragm walls retaining clay. *Géotechnique* **37**, No. 3, 335–353.

Bolton, M. D. & Powrie, W. (1988). Behaviour of diaphragm walls in clay prior to collapse. *Géotechnique* **38**, No. 2, 167–189.

Bolton, M. D. & Stewart, D. I. (1994). The effect on propped diaphragm walls of rising groundwater in stiff clay. *Géotechnique* **44**, No. 11, 111–127.

Bransby, P. L. & Milligan, G. W. E. (1975). Soil deformations near cantilever sheet pile walls. *Géotechnique* **25**, No. 2, 175–195.

Broms, B. B. & Ingelson, I. (1971). Earth pressure against the abutments of a rigid frame bridge. *Géotechnique* **21**, No. 1, 15–28.

Carder, D. R. & Symons, I. F. (1989). Long-term performance of an embedded cantilever retaining wall in stiff clay. *Géotechnique* **39**, No. 1, 55–75.

Clarke, B. G. & Wroth, C. P. (1984). Analysis of Dunton Green retaining wall based on results of pressuremeter tests. *Géotechnique* **34**, No. 4, 549–561.

Clayton, C. R. I. & Symons, I. F. (1992). The pressure of compacted fill on retaining walls. *Géotechnique* **42**, No. 1, 127–130.

Clayton, C. R. I, Xu, M. & Bloodworth, A. (2006). A laboratory study of the development of earth pressure behind integral bridge abutments. *Géotechnique* **56**, No. 8, 551–559.

Clough, W. & O'Rourke, T. D. (1990). Construction induced movements of in-situ walls. *Proceedings of the ASCE specialty conference on design and performance of earth retaining structures*, Cornell, ASCE SP15, pp. 439–470.

Crofts, J. E., Menzies, B. K. & Tarzi, A. I. (1977). Lateral displacement of shallow buried pipelines due to adjacent deep trench excavation. *Géotechnique* **27**, No. 2, 161–179.

Farmer, I. W. & Attewell, P. B. (1973). Ground movements caused by a bentonite-supported excavation in London clay. *Géotechnique* **23**, No. 4, 576–581.

Gaba, A. R., Simpson, B., Powrie, W. & Beadman, D. R. (2003). *Embedded retaining walls: Guidance for economic design*, C580. London: Construction Industry Research and Information Association.

Garrett, C. & Barnes, S. J. (1984). The design and performance of the Dunton Green retaining wall. *Géotechnique* **34**, No. 4, 533–548.

Georgiadis, M. & Anagnostopoulos, C. (1998). Effect of berms on sheet-pile wall behaviour. *Géotechnique* **48**, No. 4, 575–578.

Glass, P. R. & Powderham, A. J. (1994). Application of the observational method at the Limehouse Link. *Géotechnique* **44**, No. 4, 665–679.

Gourvenec, S. M. & Powrie, W. (1999). Three-dimensional analysis of diaphragm wall installation. *Géotechnique* **49**, No. 6, 801–823.

Harkness, R. M., Powrie, W., Zhang, X., Brady, K. C. & O'Reilly, M. P. (2000). Numerical modelling of full-scale tests on dry-stone masonry retaining walls. *Géotechnique* **50**, No. 2, 165–179.

Hubbard, H. W., Potts, D. M., Miller, D. & Burland, J. B. (1984). Design of the retaining walls for the M25 cut and cover tunnel at Bell Common. *Géotechnique* **34**, No. 4, 495–512.

Ikuta, Y., Maruoka, M., Aoki, M. & Sato, E. (1994). Application of the observational method to a deep basement excavated using the top-down method. *Géotechnique* **44**, No. 4, 655–664.

Ingold, T. S. (1979). The effects of compaction on retaining walls. *Géotechnique* **29**, No. 3, 265–283.

Interagency Performance Evaluation Task Force (2008). *Performance evaluation of the New Orleans and Southwest Louisiana Hurricane Protection System, Final and Interim Reports*, US Army Corps of Engineers, https://ipet.wes.army.mil/

James, R. G. & Bransby, P. L. (1970). Experimental and theoretical investigation of a passive earth pressure problem. *Géotechnique* **20**, No. 1, 17–37.

James, R. G. & Bransby, P. L. (1971). A velocity field for some passive earth pressure problems. *Géotechnique* **21**, No. 1, 61–83.

Jones, C. J. F. P. & Sims, F. A. (1975). Earth pressures against the abutments and wing walls of standard motorway bridges. *Géotechnique* **25**, No. 4, 731–742.

Lee, K., Jones, C. J. F. P. Sullivan, W. R. & Trolinger, W. (1994). Failure and deformation of four reinforced soil walls in eastern Tennessee. *Géotechnique* **44**, No. 3, 397–426.

Milligan, G. W. E. & Bransby, P. L. (1976). Combined active and passive rotational failure of a retaining wall in sand. *Géotechnique* **26**, No. 3, 473–494.

Morgenstern, N. & Amir-Tahmasseb, I. (1965). The stability of a slurry trench in cohesionless soils. *Géotechnique* **15**, No. 4, 387–395.

Ng, C. W. W., Rigby, D. B., Lei, G. H. & Ng, S. W. L. (1999). Observed performance of a short diaphragm wall panel. *Géotechnique* **49**, No. 5, 681–694.

Nicholson, D., Tse, C.-M. & Penny, C. (1999). *The observational method in ground engineering: Principles and applications*, R185. London: Construction Industry Research and Information Association.

Ou, C.-Y., Liao, J.-T. & Cheng, W.-L. (2000). Building response and ground movements induced by a deep excavation. *Géotechnique* **50**, No. 3, 209–220.

Peck, R. B. (1969). Deep excavations and tunneling in soft ground. *Proc. 7th Int. Conf. Soil Mech. Found. Engng, Mexico City*, State of the Art Volume, 225–290.

Potts, D. M. (2003). 42nd Rankine Lecture: Numerical analysis: a virtual dream or practical reality. *Géotechnique* **53**, No. 6, 533–573.

Potts, D. M. & Fourie, A. B. (1984). The behaviour of a propped retaining wall: results of a numerical experiment. *Géotechnique* **34**, No. 3, 383–404.

Potts, D. M. & Fourie, A. B. (1985). The effect of wall stiffness on the behaviour of a propped retaining wall. *Géotechnique* **35**, No. 3, 347–352.

Powderham, A. J. (1994). An overview of the observational method: development in cut and cover and bored tunnelling projects. *Géotechnique* **44**, No. 4, 619–636.

Powrie, W. & Chandler, R. J. (1998). The influence of a stabilizing platform on the performance of an embedded retaining wall: a finite element study. *Géotechnique* **48**, No. 3, 403–409.

Powrie, W. & Daly, M. P. (2002). Centrifuge model tests on embedded retaining walls supported by earth berms. *Géotechnique* **52**, No. 2, 89–106.

Powrie, W. & Daly, M. P. (2007). Centrifuge modelling of embedded retaining walls with stabilising bases. *Géotechnique* **57**, No. 6, 485–497.

Powrie, W. & Kantartzi, C. (1996). Ground response during diaphragm wall installation in clay: centrifuge model tests. *Géotechnique* **46**, No. 4, 725–739.

Powrie, W. & Li, E. S. F. (1991). Finite element analyses of an in situ wall propped at formation level. *Géotechnique* **41**, No. 4, 499–514.

Richards, D. J. & Powrie, W. (1998). Centrifuge model tests on doubly propped embedded retaining walls in overconsolidated kaolin clay. *Géotechnique* **48**, No. 6, 833–846.

Richards, D. J., Clark, J. & Powrie, W. (2006). Installation effects of a bored pile wall in overconsolidated clay. *Géotechnique* **56**, No. 6, 411–425.

Richards, D. J., Powrie, W., Roscoe, H. & Clark, J. (2007). Pore water pressure and horizontal stress changes measured during construction of a contiguous bored pile multi-propped retaining wall in Lower Cretaceous clays. *Géotechnique* **57**, No. 2, 197–205.

Rowe, P. W. & Peaker, K. (1965). Passive earth pressure measurements. *Géotechnique* **15**, No. 1, 57–78.

Simpson, B. (1992). Rankine Lecture: Retaining structures: displacement and design. *Géotechnique* **42**, No. 4, 541–576.

Simpson, B., O'Riordan, N. J. & Croft, D. D. (1979). A computer model for the analysis of ground movements in London Clay. *Géotechnique* **29**, No. 2, 149–175.

Skempton, A. W. & Ward, W. H. (1952). Investigations concerning a deep cofferdam in the Thames Estuary clay at Shellhaven. *Géotechnique* **3**, No. 3, 119–139.

St John, H. D., Potts, D. M., Jardine, R. J. & Higgins, K. G. (1993). Prediction and performance of ground response due to construction of a deep basement at 60 Victoria Embankment. In *Predictive soil mechanics* (eds G. T. Houlsby and A. N. Schofield), pp. 581–608. London: Thomas Telford.

Symons, I. F. & Carder, D. R. (1992). Field measurements on embedded retaining walls. *Géotechnique* **42**, No. 1, 117–126.

Symons, I. F. & Tedd, P. (1989). Behaviour of a propped embedded retaining wall at Bell Common Tunnel in the longer term. *Géotechnique* **39**, No. 4, 701–710.

Tedd, P. & Charles, J. A. (1985). The strength of London Clay in relation to the design of embedded retaining walls. *Géotechnique* **35**, No. 2, 199–204.

Tedd, P., Chard, B. M., Charles, J. A. & Symons, I. F. (1984). Behaviour of a propped embedded retaining wall in stiff clay at Bell Common. *Géotechnique* **34**, No. 4, 513–532.

Twine, D. & Roscoe, H. (1999). *Temporary propping of deep excavations: Guidance on design*, C517. London: Construction Industry Research and Information Association.

Wood, L. A. & Perrin, A. J. (1984). Observations of a strutted diaphragm wall in London Clay: a preliminary assessment. *Géotechnique* **34**, No. 4, 563–579.

Wu, T.-H. & Berman, S. (1953). Earth pressure measurements in open cut: Contract D-8 Chicago subway. *Géotechnique* **3**, No. 3, 248–258.

Young, D. K. & Ho, E. W. L. (1994). The observational approach to design of a sheet-piled retaining wall. *Géotechnique* **44**, No. 4, 637–654.

Zdravkovic, L., Potts, D. M. & St John, H. D. (2005). Modelling of a 3D excavation in finite element analysis. *Géotechnique* **55**, No. 7, 497–513.

Zhang, X., Koutsabeloulis, N. C., Hope, S. & Pearce, A. (2004). A finite element analysis for the stability of drystone masonry retaining walls. *Géotechnique* **54**, No. 1, 57–60.

Charles, J. A. & Bromhead, E. N. (2008). *Géotechnique* **58**, No. 5, 385–389 [**doi:** 10.1680/geot.2008.58.5.385]

Contributions to *Géotechnique* 1948–2008: Slope stability and embankment dams

J. A. CHARLES* and E. N. BROMHEAD†

The paper reviews the main contributions to geotechnical engineering in the allied fields of slope stability and embankment dams over the past sixty years of *Géotechnique*. Skempton's Rankine Lecture is probably the most important single contribution to the science of slope stability engineering, through the initial enunciation of the linked ideas of slip surfaces and residual strength. Embankment dam engineering papers do not have such a single identifiable seminal publication, but the contribution of Rankine Lecturers to the science of embankment dams is equally evident.

KEYWORDS: dams; embankments; historical review; slopes; soil stabilisation

La communication passe en revue les principales contributions à l'ingénierie géotechnique dans les domaines connexes de la stabilité des talus et les barrages à digue au cours des soixante années de l'existence de *Géotechnique*. La conférence de Rankine, dans Skempton, constitue probablement la contribution individuelle la plus importante à la science de la technique de la stabilité des talus, par le biais de l'énonciation des principes apparentés des surfaces de glissement et de la résistance résiduelle. Les communications techniques sur les barrages à digue ne font pas l'objet d'une publication fructueuse identifiable de ce genre, mais la contribution qu'apportent les conférences de Rankine à la science des barrages à digue est tout aussi évidente.

INTRODUCTION

In reviewing the most significant contributions to geotechnical science and practice that have been published in *Géotechnique* in the fields of slope stability and dam engineering, it is appropriate to start with the more general subject of slope stability, and then to progress to the particular case of embankment dams. By no means all the major advances in these subjects have appeared in *Géotechnique*, so this paper can only draw attention to a few important developments that have occurred during the last 60 years; it cannot form a comprehensive history of the subject.

SLOPE STABILITY ANALYSIS

Slope instability remains a major preoccupation for geotechnical engineers, and routine quantitative stability analyses have long been based on limit equilibrium methods, in which a factor of safety, defined as the ratio of the actual strength of the soil to that required to maintain limiting equilibrium, is calculated. Petterson (1955) reviewed the history of the early work at Gothenburg leading to the introduction, in 1916, of the use of circular arc sliding surfaces to analyse the stability of clay slopes.

The March and June 1955 issues of *Géotechnique* formed the Proceedings of the European Regional Conference on the Stability of Earth Slopes, which had been held at Stockholm in September 1954. The proceedings included the paper by Bishop (1955) on effective stress slip circle analysis. This type of limit equilibrium analysis is carried out by dividing the sliding mass into a number of vertical slices (Fig. 1), and Bishop examined the error introduced by simplifying assumptions concerning the inter-slice forces.

Little & Price (1958) programmed Bishop's method of analysis for the computer, which at the time was a major technical advance. The ability to carry out a large number of calculations very rapidly meant that it was possible to identify reliably the critical circular slip surface for which the factor of safety was a minimum. Probably from a notion that computers would never become widely available, several authors produced charts and stability coefficients from which the factor of safety of a slope could be obtained without laborious calculation (Bishop & Morgenstern, 1960; Spencer, 1967).

Over the years more complex methods of analysis appeared in *Géotechnique*. In many field situations, the critical slip surface deviates significantly from a circle or a plane, and a method of limit equilibrium analysis for surfaces of any shape was developed by Morgenstern & Price (1965). The method was sufficiently complicated to require the use of a computer; prior to that, even the most complicated methods of analysis were often carried out using 'hand' calculations. Sarma (1973) developed a method of analysis to determine the earthquake acceleration required to bring a slope to a state of limiting equilibrium, and suggested that this critical seismic acceleration was a more useful concept than the factor of safety. Papers on the analysis of the stability of slopes in three dimensions have on occasions appeared in *Géotechnique*, for example Hungr (1987), but three-dimensional analysis has not yet become the norm.

RESIDUAL STRENGTH AND PROGRESSIVE FAILURE

At the same time that improved methods of slope stability analysis were being published, understanding of the critical aspects of soil behaviour that control slope stability was also improving. Skempton's Rankine Lecture was probably the most seminal paper on the stability of slopes ever published in *Géotechnique*. It popularised the concept of residual strength, establishing the significance of ultimate large deformation shear strength for the stability of clay slopes (Skempton, 1964). Without in any way detracting from Skempton's achievement, proper recognition should be given to the major contribution that, a number of years previously, Haefeli (1951) had made to the subject of residual shear strength and progressive failure in his paper on the measurement of the shear strength of saturated cohesive soils, and the contributions to the concept made, *inter alia*, by Skempton's co-workers Hutchinson, Henkel and Early.

Discussion on this paper closes on 1 December 2008, for further details see p. ii.

* Building Research Establishment, UK
† Kingston University, UK

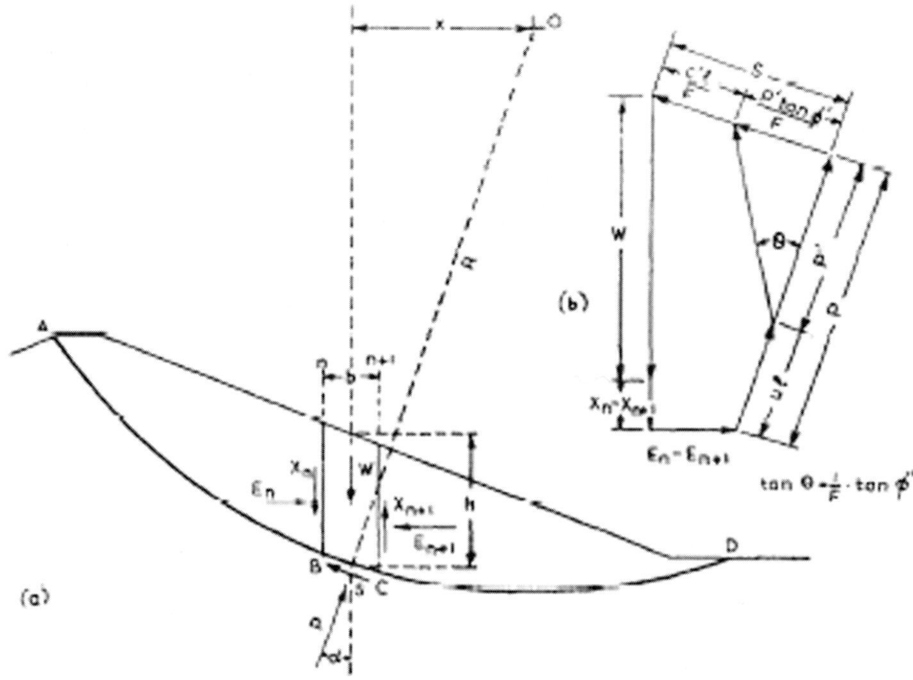

Fig. 1. Forces in the slices method of circular arc limit equilibrium stability analysis (Fig. 1 in Bishop, 1955)

The need for further development of these ideas was fulfilled by Skempton (1985) in a paper on the residual strength of clays in landslides, folded strata and the laboratory (Fig. 2), which was presented at the Institution of Civil Engineers on 6 June 1984, the day on which the Carsington embankment failed. With the recognition of the significance of residual strength came the need for more reliable laboratory determinations of this parameter, and Bishop *et al.* (1971) developed a new ring shear apparatus. The test results for a number of clay soils differed significantly from those obtained in multiple reversal direct shear box tests, and a critical examination was made of sources of error.

Back-analysis of the stability of the huge Folkestone Warren landslide by Hutchinson (1969) gave results that showed that the average strengths mobilised on the slip surfaces in the Gault Clay approximated to the residual strength. Deep-seated landslides that have been recorded during the last two centuries in the Warren have occurred during the winter period of high groundwater levels. The titles of two later papers by Hutchinson refer to *mudflows*, but doubtless it was the appreciation of the significance of basal shearing that led to his largely successful campaign to have such landslides called 'mudslides' rather than 'mudflows' (Hutchinson, 1970; Hutchinson & Bhandari, 1971).

Several different mechanisms have been suggested to explain why slopes cut in overconsolidated clay may become unstable many years after excavation. Field observations of pore pressure in an excavated slope were used by Vaughan & Walbancke (1973) to validate the belief that long-term pore pressure equilibration may be the principal mechanism controlling the delayed failure of cut slopes in London Clay. In an instrumented field trial to prove the concept, a progressive failure mechanism was induced in a 9 m deep cut slope in Gault Clay by pore pressure recharge (Cooper *et al.*, 1998).

The delayed collapse of cut slopes in stiff clay with an operational strength at failure significantly less than the peak strength has been analysed using advanced numerical techniques by Potts *et al.* (1997). A series of coupled finite element analyses using a strain-softening soil showed that progressive failure is considerable. The role of the climate-controlled hydraulic surface boundary condition also has an important role in controlling collapse.

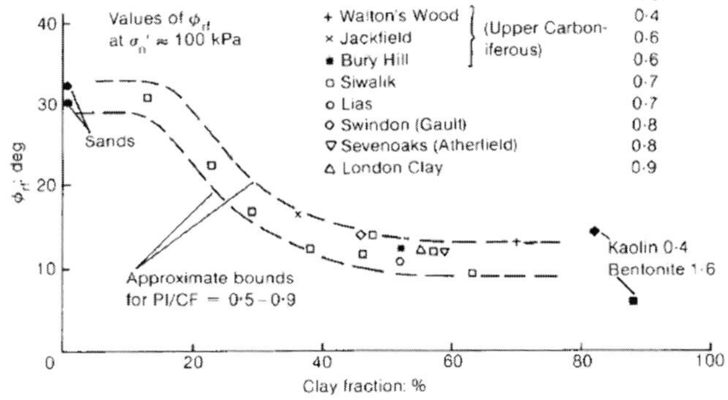

Fig. 2. Field residual strength and ring shear tests on sands, kaolin and bentonite (Fig. 18 in Skempton, 1985)

EMBANKMENT DAMS

The majority of British dams are earth embankments that were built before the first edition of *Géotechnique* was published! It is not surprising that the designers of these dams lacked our current understanding of soil behaviour. By the middle of the nineteenth century a fairly standard embankment design had been adopted, with an upstream slope of 1 vertical in 3 horizontal, a downstream slope of 1 vertical in 2·5 or 2·0 horizontal, and a narrow central core of puddle clay. Some embankments of this type were built as high as 30 m. In 1948 the construction of embankment dams in the United Kingdom still followed this traditional design, and largely did so for at least another decade, with the last puddle clay core dam being completed only in 1972.

From the 1950s the theories of modern soil mechanics were increasingly applied to the design and construction of embankment dams, and drainage of embankments to control seepage pressures became common. The most significant theoretical development was that slope stability was analysed as part of the design process using shear strength parameters for fill and foundations based on laboratory tests. Several of the *Géotechnique* papers that facilitated this development have been described in the previous sections.

With the knowledge that effective stress controlled shear strength came the recognition that the stability of a clay fill dam during construction depended critically on the dissipation of excess pore water pressures. Penman (1956) described a twin-tube hydraulic piezometer apparatus that was installed at Daer and Usk dams to monitor construction pore water pressures. The high pore pressures measured at Usk (Fig. 3) indicated that there would not be an acceptable factor of safety at the end of construction, and, on Skempton's advice, drainage layers were placed in the embankment shoulders to shorten drainage paths and increase the rate of consolidation.

At the time when geotechnical analysis was assuming a major role in the design of embankment dams, heavy earth-moving and compaction plant became widely available. This radically affected construction methods, and was a major factor in the change from puddle clay to rolled clay cores. The increased rate of embankment construction meant that the consolidation rate of clay fill became of major significance in dam design, because it indicated whether drainage layers were required to control pore pressures in the shoulders of a clay fill embankment. Construction pore water pressures are generated by the increasing weight as layers of fill are progressively placed and the drainage path lengthens, and therefore the mathematical treatment has to satisfy a condition on the pore water pressure at a moving boundary. In his paper on the progress of one-dimensional consolidation of a clay layer increasing in thickness with time, Gibson

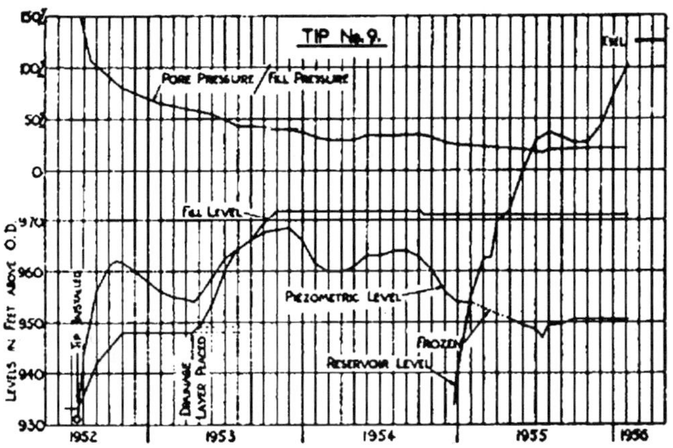

Fig. 3. Pore pressure results from piezometer tip No. 9 in downstream shoulder fill at Usk dam (Fig. 6 in Penman, 1956)

(1958) showed that, in the general case, the problem can be reduced to a linear integral equation, a solution of which was presented for fill placed at a constant rate. A numerical procedure for estimating the progress of consolidation during construction was illustrated for the Usk dam.

Any thought that modern developments in soil testing and analytical techniques had made slope instability in new embankment dams a thing of the past received a correction when the upstream slope of the 37 m high Carsington dam failed in 1984 with the embankment almost at full height. The dam had a rolled clay core with an upstream boot-shaped extension, and shoulders of compacted mudstone with drainage layers of crushed limestone. Skempton & Vaughan (1993) described the performance before failure, the extensive investigations that were conducted following the slip, and further information that was obtained when the dam was reconstructed. The slip sheared through the core, which contained shear surfaces due to rutting, and along a foundation layer of Yellow Clay, which contained solifluction shears (Fig. 4). Both materials were brittle, with low residual strengths. While the factor of safety based on measured peak strengths appeared acceptable, the effect of shear surfaces and progressive failure reduced it to unity. Potts *et al.* (1990) modelled the failure using a finite element technique in which strain-softening was reproduced and the role of progressive failure was confirmed.

The long-term performance of old embankment dams is of considerable significance for reservoir safety, and Dounias *et al.* (1996) used non-linear finite element techniques to analyse typical sections of old British puddle clay core embankment dams with shoulders of granular fill and strong foundations. Progressive failure was negligible during drained construction, operation and undrained drawdown. All the sections showed substantial arching of the puddle clay core between the stiffer supporting shoulders of the embankment, and this was even more pronounced in a clay-filled cut-off trench. The cores appeared to be vulnerable to cracking when subjected to full reservoir pressure on their upstream boundary, but very slow, drained impounding caused mean total stresses to rise so that fracture was not indicated. Hydraulic fracture was predicted in a deep cut-off trench filled with puddle clay. Cyclic drawdown and impounding increased stresses in the upstream side of the core, increasing the resistance to cracking.

The uncertainties surrounding the composition and behaviour of most old embankment dams are so great that a purely analytical approach to dam safety evaluation is unlikely to be adequate, and an observational approach is needed. Tedd *et al.* (1997) monitored the effect of reservoir drawdown and long-term consolidation on the deformation of a number of old dams. Where there was a central clay core, significant crest settlement was measured on reservoir drawdown, but this was not the case where there was an

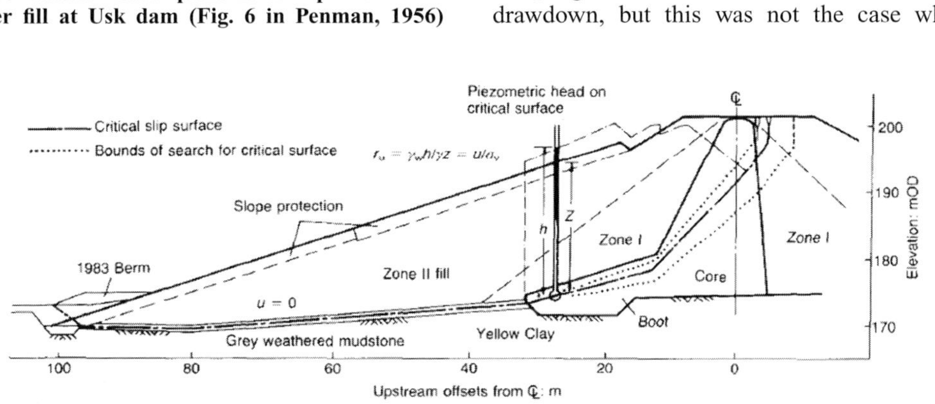

Fig. 4. Initial failure at chainage 725, Carsington dam (Fig. 23 in Skempton & Vaughan, 1993)

upstream watertight element. The field studies have provided a basis for interpreting measurements of crest settlement, thereby assisting in the identification of anomalous behaviour associated with deleterious processes, and in preventing normal behaviour associated with steady-state loading and reservoir fluctuations giving unnecessary cause for concern.

EARTHQUAKES AND DAMS

Earthquakes are not a great hazard in the United Kingdom, but *Géotechnique* is an international journal, and the Rankine Lectures of Newmark (1965) and Seed (1979) have presented major developments in predicting the performance of embankment dams in major seismic events. Newmark emphasised the importance of estimating the displacements that embankment slopes will undergo during earthquakes, rather than simply carrying out a pseudo-static stability check. Slope failure is initiated and movements begin when the inertia forces on a potential slide mass are large enough to overcome the yield resistance, and they stop when the inertia forces reverse. By integrating, as a function of time, the effective acceleration on the sliding mass in excess of the yield acceleration, velocities and displacements of the slide mass can be calculated. The analysis enables slope deformations to be related to the maximum ground acceleration during an earthquake.

Seed used studies of the field performance of embankment dams during seismic events to establish the factors to be considered in the earthquake-resistant design of dams, and the defensive measures that may be taken to mitigate the effects of these factors. The potential for earthquake-induced sliding is related to the different types of fill material and earthquake shaking intensities. The possibility of delayed failure, after the earthquake ground motions have stopped, due to pore water pressure redistribution within an embankment was examined. Methods for evaluating the stability and deformations of the slopes of a dam due to earthquake shaking were reviewed in relation to the uncertainties involved in the analytical procedures.

WASTE IMPOUNDMENTS

With the current emphasis on environmental protection, it is likely that the design and construction of waste impoundments and tailings lagoons will come under closer scrutiny in relation both to leakage and to the consequences of catastrophic breaching with the release of the impounded wastes. It may seem surprising that the safety of this type of structure has received less attention than that of the other major class of embankment dams, which retain water, although waste impoundments are as important as water-retaining embankment dams in terms of size of structure, economic significance and potential hazard to public safety.

The catastrophic failure of the Aznalcollar dam in April 1998 caused one of Spain's worst environmental disasters. A section of the 27 m high rockfill embankment with an upstream clay blanket slid bodily forward some 50 m without any prior warning, releasing millions of cubic metres of

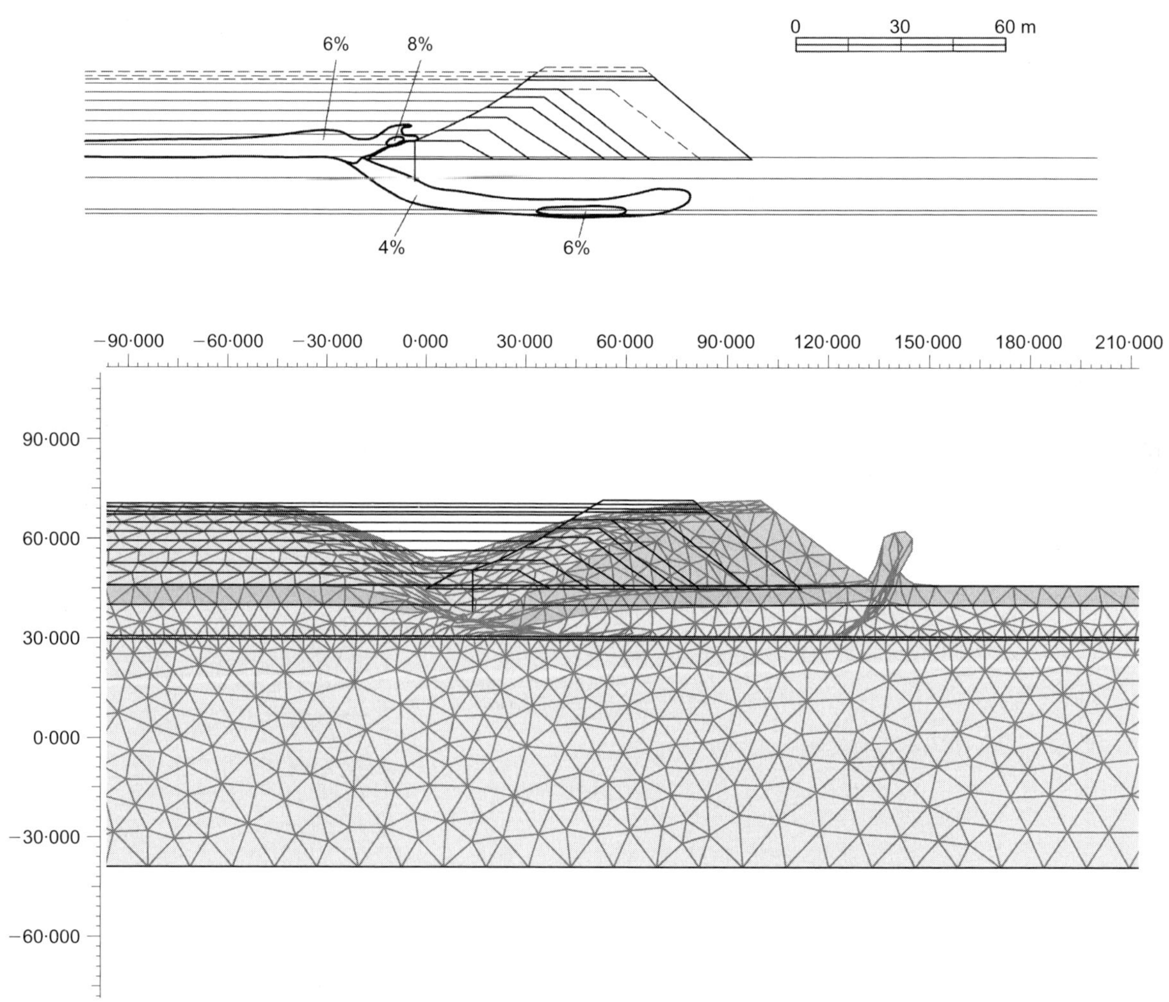

Fig. 5. Cross-section of slide at position of Profile 4 at Aznalcollar dam: (a) geometry after the slide; (b) reconstruction of position of sliding surfaces before failure (Fig. 17 in Alonso & Gens, 2006, Part 1)

fluidised acid mine tailings. There was no shearing of the impounded tailings or the rockfill; the deep translational slide was through the brittle, high-plasticity, foundation clay (Fig. 5). Olalla & Cuellar (2001) described the failure mechanism of this dam and, subsequently, Alonso & Gens (2006) have provided a comprehensive study of the failure in three papers dealing with, respectively, field observations and material properties, stability conditions and the failure mechanism, and the dynamics of the motion. Progressive failure is invoked as a significant factor, and the operational shear strength at the time of failure was midway between the peak and the residual angles of shearing resistance.

CONCLUSIONS

Techniques for the analysis of slope stability have progressed over the last 60 years from simple limit equilibrium methods to techniques using advanced finite element formulations. At the same time the understanding of soil behaviour as it affects slope stability has greatly progressed, and the Rankine Lectures of de Mello (1977), Penman (1986) and Vaughan (1994) have given many illustrations of geotechnical developments that affect the design and construction of embankment dams. Despite these advances, major slope and embankment failures have continued to occur. Vaughan concluded that the understanding of real behaviour is more important than accurate calculation, but that a great strength of modern numerical analysis is that it predicts mechanisms of behaviour that no longer need to be assumed in advance. However, there are still many situations where prediction has to be based directly on experience, and it continues to be vital that the engineering of slopes and dams is applied with an understanding of the mechanisms and materials involved.

REFERENCES

Alonso, E. E. & Gens, A. (2006). Aznalcollar dam failure—Part 1: Field observations and material properties; Part 2: Stability conditions and failure mechanism; Part 3: Dynamics of the motion. *Géotechnique* **56**, No. 3, 165–210.

Bishop, A. W. (1955). The use of the slip circle in the stability analysis of slopes. *Géotechnique* **5**, No. 1, 7–17.

Bishop, A. W., Green, G. E., Garga, V. K., Andresen, A. & Brown, J. D. (1971). A new ring shear apparatus and its application to the measurement of residual strength. *Géotechnique* **21**, No. 4, 273–328.

Bishop, A. W. & Morgenstern, N. R. (1960). Stability coefficients for earth slopes. *Géotechnique* **10**, No. 4, 129–150.

Cooper, M. R., Bromhead, E. N., Petley, D. J. & Grant, D. I. (1998). The Selborne cutting stability experiment. *Géotechnique* **48**, No. 1, 83–101.

de Mello, V. F. B. (1977). Seventeenth Rankine Lecture: Reflections on design decisions of practical significance to embankment dams. *Géotechnique* **27**, No. 3, 279–355.

Dounias, G. T., Potts, D. M. & Vaughan, P. R. (1996). Analysis of progressive failure and cracking in old British dams. *Géotechnique* **46**, No. 4, 621–640.

Gibson, R. E. (1958). The progress of consolidation in a clay layer increasing in thickness with time. *Géotechnique* **8**, No. 4, 171–182.

Haefeli, R. (1951). Investigation and measurements of the shear strengths of saturated cohesive soils. *Géotechnique* **2**, No. 3, 186–208.

Hungr, O. (1987). An extension of Bishop's simplified method of slope stability analysis to three dimensions. *Géotechnique* **37**, No. 1, 113–117.

Hutchinson, J. N. (1969). A reconsideration of the coastal landslides at Folkestone Warren, Kent. *Géotechnique* **19**, No. 1, 6–38.

Hutchinson, J. N. (1970). A coastal mudflow on the London Clay cliffs at Beltings, North Kent. *Géotechnique* **20**, No. 4, 412–438.

Hutchinson, J. N. & Bhandari, R. K. (1971). Undrained loading, a fundamental mechanism of mudflows and other mass movements. *Géotechnique* **21**, No. 4, 353–358.

Little, A. L. & Price, V. E. (1958). The use of an electronic computer for slope stability analysis. *Géotechnique* **8**, No. 3, 113–120.

Morgenstern, N. R. & Price, V. E. (1965). The analysis of the stability of general slip surfaces. *Géotechnique* **15**, No. 1, 79–93.

Newmark, N. M. (1965). Fifth Rankine Lecture: Effects of earthquakes on dams and embankments. *Géotechnique* **15**, No. 2, 139–159.

Olalla, C. & Cuellar, V. (2001). Failure mechanism of the Aznalcollar Dam, Seville, Spain. *Géotechnique* **51**, No. 5, 399–406.

Penman, A. D. M. (1956). A field piezometer apparatus. *Géotechnique* **6**, No. 2, 57–65.

Penman, A. D. M. (1986). Twenty-sixth Rankine Lecture: On the embankment dam. *Géotechnique* **36**, No. 3, 303–348.

Petterson, K. E. (1955). The early history of circular sliding surfaces. *Géotechnique* **5**, No. 4, 275–296.

Potts, D. M., Dounias, G. T. & Vaughan, P. R. (1990). Finite element analysis of progressive failure of Carsington embankment. *Géotechnique* **40**, No. 1, 79–101.

Potts, D. M., Kovacevic, N. & Vaughan, P. R. (1997). Delayed collapse of cut slopes in stiff clay. *Géotechnique* **47**, No. 5, 953–982.

Sarma, S. K. (1973). Stability analysis of embankments and slopes. *Géotechnique* **23**, No. 3, 423–433.

Seed, H. B. (1979). Nineteenth Rankine Lecture: Considerations in the earthquake-resistant design of earth and rockfill dams. *Géotechnique* **29**, No. 3, 215–263.

Skempton, A. W. (1964). Fourth Rankine Lecture: Long-term stability of clay slopes. *Géotechnique* **14**, No. 2, 77–102.

Skempton, A. W. (1985). Residual strength of clays in landslides, folded strata and the laboratory. *Géotechnique* **35**, No. 1, 3–18.

Skempton, A. W. & Vaughan, P. R. (1993). Failure of Carsington dam. *Géotechnique* **43**, No. 1, 151–173. Discussion: **45**, No. 4, 719–739.

Spencer, E. (1967). A method of analysis of the stability of embankments assuming parallel inter-slice forces. *Géotechnique* **17**, No. 1, 11–26.

Tedd, P., Charles, J. A., Holton, I. R. & Robertshaw, A. C. (1997). The effect of reservoir drawdown and long-term consolidation on the deformation of old embankment dams. *Géotechnique* **47**, No. 1, 33–48.

Vaughan, P. R. (1994). Thirty-fourth Rankine Lecture: Assumption, prediction and reality in geotechnical engineering. *Géotechnique* **44**, No. 4, 573–609.

Vaughan, P. R. & Walbancke, H. J. (1973). Pore pressure changes and the delayed failure of cutting slopes in overconsolidated clay. *Géotechnique* **23**, No. 4, 531–539.

Standing, J. R. & Potts, D. M. (2008). *Géotechnique* **58**, No. 5, 391–398 [**doi:** 10.1680/geot.2008.58.5.391]

Contributions to *Géotechnique* 1948–2008: Tunnelling

J. R. STANDING* and D. M. POTTS*

The response of the ground and nearby structures to tunnelling constitutes a complex soil–structure interaction boundary value problem. *Géotechnique* papers published over the past 60 years that have dealt with this subject are summarised and discussed here. The papers have been organised according to their subject matter, including historical perspectives, various forms of theoretical and numerical analyses, small-scale modelling, and full-scale field monitoring. The number of papers written has steadily accelerated with time, thus indicating the growing interest in tunnelling and its effects, particularly in the urban environment, and an increase in our understanding of the subject. However, our knowledge is far from complete, and there are still many challenges to be tackled, particularly concerning the effects of tunnelling on existing infrastructure.

KEYWORDS: ground movements; historical review; stress analysis; tunnelling

La réaction des structures au sol, et proches, à la construction de tunnels représente un problème complexe de la valeur limite de l'interaction sol–structure. Nous résumons et nous discutons ici de communications sur ce sujet, publiées dans *Géotechnique* au cours des soixante dernières années. Ces communications sont classées par sujet, y compris leurs perspectives historiques, les différentes formes d'analyses théoriques et numériques, la modélisation à petite échelle et des contrôles de champ à grande échelle. Le nombre de communications publiées a augmenté progressivement au fil du temps, en indiquant ainsi un intérêt croissant dans la construction de tunnels et ses effets, en particulier dans des environnements urbains, et le développement de nos connaissances sur ce sujet. Toutefois, nos connaissances sont loin d'être complètes, et nous avons encore de nombreux problèmes à résoudre, notamment en ce qui concerne les effets de la construction de tunnels sur l'infrastructure existante.

INTRODUCTION

Tunnels have been constructed since ancient times, and Kerisel (1956) was the first to take us back to the past. He mentioned the requirements stipulated by the great French military engineer of the early eighteenth century, the Maréchal de Vaubin, who, along with his enormous rampart-enclosed fortresses, also recognised the defensive value of tunnels, saying that with a crown thickness of 3 to 4 feet and ground cover of 5 to 8 feet they will resist bombs well!

Almost 60 *Géotechnique* papers have been written related to the subject of tunnelling in both soil and rock—that is, roughly one per year on average—although the number of papers has increased markedly in recent years. In the early years of *Géotechnique* there were few papers covering tunnels specifically. When papers started appearing they tended to be concerned with theoretical analyses of stresses and strains around tunnels. Of all the boundary value problems that geotechnical engineers have to deal with, perhaps understanding the conditions around tunnels and the soil–structure interaction between the lining and the ground is one of the most complex. It is probably for this reason that physical modelling and monitoring of full-scale tunnels is a frequently adopted approach to gain insight into mechanisms of deformation and failure of tunnels. More recently published papers using sophisticated numerical analyses have attempted to model tunnelling response, but there is almost always a need to implement some artificial device to achieve realistic results.

The complexity of tunnelling conditions increases further when considering cases where there are buildings above where tunnels are to be constructed. The ground movements are affected by the presence of the structure, and in particular its stiffness, foundations, and their interaction with the ground. In parallel with increasing sophistication of methods of analysis there have also been many advances in subjects such as machine technology and methods of tunnelling. As a consequence there are several papers that focus on specific elements of practice.

This précis of the papers published in *Géotechnique* that have discussed tunnelling is divided here into five broad sections: historical accounts and developments; theoretical analyses; numerical analyses; small-scale physical modelling; and full-scale field monitoring and case studies.

HISTORICAL ACCOUNTS AND DEVELOPMENTS

The Thames Tunnel chronologically follows the reference to Vaubin by Kerisel (1956). The construction of this tunnel (over the period 1825–1846), designed and overseen by Marc Brunel, later with his son I. K. Brunel, was described along with details of the geology by Skempton & Chrimes (1994). Of particular note is the first recorded use of a tunnelling shield and spiling (soil nailing of the tunnel face). Despite these developments, the project was dogged with problems, caused primarily by adverse ground conditions. It is a fascinating read. Reference is made to the earlier historical research by Glossop (1976), who provided a detailed account of the invention and early use of compressed air for tunnel construction. It is relevant to mention here because, as well as covering the inception of the technique patented by Cochrane in 1830, there is also an enlightening discussion as to why a proposition to Marc Brunel to use Cochrane's compressed air system to assist with the construction of the Thames Tunnel was turned down (because of Brunel's lack of confidence in the technique). Tunnelling using Cochrane's system was used successfully on a number of later projects, such as the Blackwall Tunnel, also beneath the Thames.

Kerisel (1975) in his Rankine Lecture, entitled 'Old structures in relation to soil conditions', made extensive reference to the effects of tunnelling on existing buildings, differentiating between old and new. He started this section of his paper with the statement that 'Soil engineers are

---

Discussion on this paper closes on 1 December 2008, for further details see p. ii.

* Department of Civil and Environmental Engineering, Imperial College London, UK.

asked more and more to predict the behaviour of buildings near to tunnels.' More than 30 years later this is still true, perhaps with even greater expectations about accuracy of 'predictions'. He posed some interesting ideas, and described a more simplistic approach than is used today (assuming that the settlement trough can be described by a circular arc), but with many insightful comments regarding issues concerning, for example, whether the building is hogging or sagging, its height-to-length ratio, and its construction details. Reference was made to the tunnelling parameter volume loss, often quoted in modern practice, which he referred to as the 'efficiency' with regard to building damage. He also tackled the difficult subject of the deformation and failure of old existing tunnels, making recourse to the behaviour of masonry arches and drawing on case studies. There is a great deal of interesting material in this paper.

It is very apt that the importance of engineering geology is highlighted in the first issue of *Géotechnique* by Von Moos (1948). He discussed this in a historical context of tunnelling, giving a brief summary of tunnel construction in Switzerland and mentioning case studies from the three principal areas: the Alps, the Mittelland and the Jura mountains. At that time some of the longest and deepest rock tunnels in the world had been constructed in the mountains there.

Grouting is a technique frequently associated with tunnelling, and Glossop (1961) in his second paper on the history of grouting gave a comprehensive account of its development from the mid-nineteenth century (the first paper covers mainly grouting of locks in the first half of the nineteenth century). In particular he drew attention to the grouting behind the first cast-iron-lined tunnel constructed beneath the Thames by Greathead in 1869, and to the machine subsequently devised by him in 1886 for the same purpose, which was still being used in the 1960s. Another interesting area is that of rock tunnelling, where grout was used to seal high-pressure tunnels for dams. Two other significant developments were also described by Glossop (1961): the use of 'aureoles' of grout to stabilise ground ahead of the tunnel heading (a series of concentric cones), and the invention of the 'tube à manchette' in the 1930s by Ischy. The latter is often associated with the more recent technique of 'compensation grouting' used to control tunnelling-induced settlements. A paper by Harris *et al.* (1994) describing one of the first uses of compensation grouting is discussed later.

THEORETICAL APPROACHES TO ANALYSING CONDITIONS IN AND AROUND TUNNELS

The papers in this section can be split into sub-groups, involving primarily

(*a*) solutions using elasticity
(*b*) lower- and upper-bound plasticity analyses
(*c*) other approaches, for example involving effective stress and seepage.

A major drawback with most of these methods is that it is necessary that very simplistic idealisations be made about the rock or soil, assuming for instance that the materials are linear elastic (obeying Hooke's law), homogeneous, isotropic, without macrofabric (fissures, joints), none of which are generally true in practice. Additionally they usually have no capability for accounting realistically for conditions at interfaces, say between the ground and the tunnel lining. However, for many years these were the only methods available, prior to the development of more sophisticated techniques of numerical analysis and laboratory soil testing, when only basic parameters (usually elastic) and constitutive models were used. Regardless of these limitations the methods have proved invaluable in assessing the stability of tunnels and in providing insight into overall trends in behaviour and mechanisms with respect to stresses and strains.

A classical elasticity approach was adopted by Terzaghi & Richart (1952) to investigate stresses in rock around cavities. They focused on the influence of the shape of the cavity (tunnel compared with spheroidal chamber) and identified that the correct vertical and horizontal stresses prior to excavation (and the ratio of horizontal to vertical stress, denoted here by K) are the critical factors that the analysis needs to establish. Distributions of stress components around tunnel and spheroidal cavities were presented for $K = 0{\cdot}25$.

Morgan (1961) used a similar approach to Terzaghi & Richart to calculate stresses in the ground around a tunnel, and then applied these to a tunnel lining to assess bending moments (assuming a frictionless interface and plane-strain conditions). This study was instigated following the introduction of concrete segments for tunnel linings, the flanges of which had failed in some cases. Overall stability of the lining was also assessed, as were the effects of consolidation of the ground around the lining. Muir Wood (1975) made a number of developments to the approach (and pointed out some shortcomings of Morgan's assumptions), accounting for a more general case of an elliptical lining, incorporating interface shear stresses between the lining and the ground and also groundwater pressures. The boundary conditions used in the previous elastic analyses were scrutinised by Pender (1980), who looked at three scenarios (stresses applied at a distant boundary, tunnel excavated in a prestressed medium, and interaction between a tunnel excavated in a prestressed medium and a tunnel lining) and their effect on calculated displacements. Also, Feng & Ding (1986) described a method for calculating the actions in circular flexible linings for rock tunnels (e.g. shotcrete) using a similar approach to Muir Wood (1975), although the interface between the lining and the rock was assumed to be frictionless. The earlier papers attracted discussion (e.g. Curtis, 1976), which usefully led to refinements to the proposed methods and provided insight into their limitations. The limitations of using elasticity and the various associated assumptions were acknowledged by the authors, but they suggested that there is benefit in applying the method: (a) to gain an initial insight into the overall problem prior to embarking on more advanced analyses; or (b) for cases where little is known about the ground conditions.

Sagaseta (1987) proposed a closed-form fluid mechanics approach for estimating soil deformations resulting from a 'point sink' within an elastic half-space from which a finite volume of soil is extracted under undrained conditions. The formulation is related to the 'strain path method', and so is different from the approaches described above. Strains are determined from the fact that the overall volume strain is zero for undrained conditions. Comparisons were made for the tunnelling scenario, between field data and results from the method, showing reasonable agreement. Once again the method was not intended to provide accurate quantitative results, but to give insight into mechanisms of behaviour. Useful comparisons were also made between Sagaseta's proposed method and the empirical error function approach in a subsequent discussion by Schmidt (1988). In both methods the magnitude of ground (volume) loss is required. Verruijt & Booker (1996) extended Sagaseta's work to provide again a surface settlement profile (and subsurface displacements), but for ground loss from either circular contraction or elliptical deformation of a tunnel in incompressible soil. Their work also extended Muir Wood's (1975) analysis, dealing with a more realistic elastic half-medium rather than an infinite space (of soil). Loganathan *et al.* (2000) referred to improvements made to Verruijt & Booker's work, providing more realistic narrower settlement troughs. This paper is discussed further in the small-scale physical modelling section.

Klar *et al.* (2005) investigated the problem of soil–pipe interaction due to tunnelling, focusing specifically on differences between analyses using Winkler springs (which rely on some form of 'subgrade modulus' and Hetenyi's (1946) elastic beam theory) and an elastic continuum solution. The main intention with the methods is to establish bending moments in the pipeline so that potential damage can be assessed. They concluded that it is safer to use their proposed continuum solution, as it more accurately models loading from tunnelling, and is more conservative than the analysis with Winkler springs.

The following papers generally fall into the second subgroup of analyses, involving the limit theorems of plasticity. The first three papers, discussed in chronological order, relate to the stability of shallow tunnels. Atkinson & Potts (1977) investigated the problem of circular tunnel stability in cohesionless soil (dry dense sand), describing small-scale gravitational and centrifuge plane-strain model tests. The results from these were obtained using upper- and lower-bound solutions, with the collapse mechanism for the former being obtained from the failure observed in the model tests. The normalised tunnel support pressures at failure plotted against normalised depth of cover, obtained from both 1*g* and centrifuge test data, were very well bracketed between the calculated upper and lower bounds, providing evidence of the applicability of the bound theorems. Davis *et al.* (1980) investigated almost the same problem, but this time involving clay under undrained conditions. A number of collapse mechanisms were considered for the upper-bound analyses. Stability against depth of cover for both the plane-strain case and also a tunnel heading were considered and assessed in terms of tunnel stability number N (overburden plus surcharge less tunnel pressures, all divided by undrained shear strength). They reported that there is good agreement between centrifuge model test data and the analytical solutions, thus providing a method for estimating the tunnel pressure required for stability. Leca & Dormieux (1990) extended the case with granular (frictional) material, but looking specifically at conditions at the face (not considered by Atkinson & Potts). They assessed three quite complex three-dimensional failure mechanisms for upper-bound analysis, two for collapse and one for blow-out (details were given in extensive appendices), and compared the results with lower-bound solutions from another study, which were reported to be very similar to those of Davis *et al.* (1980) mentioned above. The critical support load was expressed in terms of factors similar to those used in estimating foundation bearing capacity. The results from centrifuge tests simulating the problem were found to fall within the bounds, but generally being closer to the upper bound, thus suggesting that critical face pressures can be reasonably estimated using the proposed approach.

This is a good point to introduce a paper by Mair *et al.* (1993), which, although not theoretical, provides a very useful empirical formulation that can be applied to the error function (Gaussian) approach for assessing the shape of subsurface settlement troughs that become narrower and deeper (for a constant volume loss) as the distance to the tunnel crown reduces. The formulation is based on data from both small-scale model tests and field observations, and is often used for assessing tunnelling-induced settlements of subsurface structures such as existing tunnels and services. Comparisons were made between predictions from the Mair *et al.* (1993) formulation and the method proposed in the next paper to be discussed.

Osman *et al.* (2006a) considered plane-strain tunnel stability and surface and subsurface deformations in undrained clay. They did this first by adopting a plastic deformation mechanism where deformations of the soil within the zone are compatible with empirical Gaussian distributions (i.e. error functions) while outside the soil was assumed to be rigid. Modified formulations based on the error function were derived for surface and subsurface vertical and horizontal displacements. An upper-bound approach was used for assessing stability, equating virtual plastic work in shearing along the boundaries of the proposed mechanism and virtual potential energy loss combined with work done by tunnel pressure and surcharge, from which a modified stability number N was formulated. Good agreement was shown when comparing results from the analysis with centrifuge and field data and also other upper-bound solutions (similar to those of Davis *et al.*, 1980). A second paper by Osman *et al.* (2006b) continued with the theme of the first, extending it to calculate ground displacement fields at various stages of reduction in tunnel support pressure, and hence volume loss, using an upper-bound style of calculation, representing the soil by a strain-hardening plastic soil model. The results from the analyses were compared with those from centrifuge tests (in kaolin), showing reasonable agreement in most cases. A closed-form solution was also proposed for predicting maximum surface settlement for a deep tunnel situation.

The case of groundwater loading on permeable and impermeable tunnel, sewer or pipe linings was investigated by Atkinson & Mair (1983). The approach used was to consider total stresses bearing on the lining, accounting for effective stresses, pore pressures and the influence of seepage. Some simplifying assumptions were made, such as a smooth tunnel interface, and radial flow into the permeable tunnel (with straight flow lines), and tunnels close to the ground surface where they might be affected by other seepage flow conditions were not considered. The main conclusion from the paper was that long-term loads acting on tunnel linings below the water table are unaffected by whether the lining is permeable or impermeable. (A numerical study by Shin *et al.* (2002a), discussed more fully later, also considered this problem, and showed that the above conclusions are valid only for certain cases. They concluded that the ground loading on, and hoop thrust in, an impermeable lining in highly permeable soil are not much higher than those for a permeable lining. In contrast, an impermeable lining in a low-permeability soil attracts significantly higher ground loading and hoop thrust than a permeable lining.)

The final paper within this heading, by Merritt & Mair (2008), proposed a theoretical model for the mechanics of tunnelling machine screw conveyors. This recent paper indicated the level of sophistication involved with modern closed-face tunnelling machine design. The theory, developed from consideration of the kinematics and equilibrium of an element of soil flowing through a screw conveyor, offers the possibility of predicting the pressure gradient and torque of a tunnel-boring machine (TBM) screw conveyor with respect to screw geometry and rotational speed, soil flow rate and soil properties. The method was validated against the results from field and comprehensive model tests (discussed later in a paper by the same authors).

ASSESSING CONDITIONS IN AND AROUND TUNNELS USING NUMERICAL ANALYSIS

Numerical analysis is generally associated with either finite element or finite difference methods (FEM or FDM). These methods of analysis allow complex boundary value problems to be investigated using a variety of material constitutive models, interfaces and boundaries. Such analyses provide information on the soil and tunnel displacements, tunnel stability, and interaction with other structures. The papers involving numerical analysis date back to just over the past 10 years: the technique has developed rapidly in recent years with increased computing power.

Despite the sophistication of the constitutive models available, it is still generally accepted by experienced numerical analysts that it is not possible to model accurately a field-monitored greenfield response resulting from straightforward tunnel construction, especially for tunnels installed in stiff soils with a high K_0 value. In such cases several approaches are available, but generally the surface settlement trough predicted is too wide and not deep enough. Numerous reasons for this have been postulated, such as three-dimensional effects, construction technique and method of imposing volume loss. Achieving a realistic simulation without resorting to some controlling artifice (i.e. some 'empirical factor') is still a 'holy grail' in numerical analysis. This reflects the complexity of the tunnelling boundary value problem.

The first paper in *Géotechnique* discussing the modelling of tunnel construction using numerical analysis, by Addenbrooke *et al.* (1997), which predated subsequent papers by five years, investigated potential reasons for the discrepancies mentioned above. The focus was the influence of pre-failure stiffness, using a variety of isotropic and anisotropic linear and non-linear elastic perfectly plastic models. The conditions for the analyses, which were run using plane-strain conditions, were based on those at St James's Park, where twin tunnels were constructed and where detailed field monitoring results were available for comparison. The overall conclusion for a single tunnel was that the model used does influence the accuracy of prediction, but for the scenario considered it was still not possible to replicate field measurements unless an unrealistic stiffness was adopted. Franzius *et al.* (2005), drawing on the work of Addenbrooke *et al.*, also sought reasons for the differences between FEM predictions and field measurements (again from St James's Park). This time the researchers concentrated on the influence of soil anisotropy and K_0, and also the effect of modelling using three-dimensional analyses, which were possible by this time with improved computing power (for the complex constitutive models, boundary conditions and coupled analyses implemented). The overall conclusion was that, even with a high degree of anisotropy, the transverse settlement trough was still too wide and shallow for tunnels constructed in high-K_0 soils. In such soils, longitudinal settlement troughs were also assessed and found to extend too far, making it difficult to establish steady-state settlement conditions behind the tunnel face. Steady-state conditions could be achieved only using an unreasonably high degree of anisotropy, which led to excessive volume losses.

Shin *et al.* (2002b) described an alternative approach for analysing three-dimensional conditions using a Fourier series aided finite element method, which can be run more cheaply and efficiently than conventional three-dimensional analyses. This form of analysis takes into account factors such as geometry, boundary conditions and material properties that vary in the circumferential sense around the tunnel axis (e.g. to model the ground surface). It has the benefit that 'empirical factors' usually involved in conventional two-dimensional analyses are not required. The conditions modelled related to NATM tunnelling in decomposed granite as part of the Seoul Subway system in Korea. The tunnel heading was also modelled. Results from the analysis were found to compare well with conventional two-dimensional modelling of the same problem and the field data available.

Two papers considered the problem of long-term ground response to tunnelling, with respect to dissipation of excess pore pressures and consequent consolidation (or swelling), changes in groundwater regime, and the permeability of the tunnel lining. Shin *et al.* (2002a) made comparisons between behaviour in a low-permeability soil (London Clay) and a more free-draining soil (weathered granite relevant to Seoul) for a common tunnel geometry scenario. An approach for modelling finite lining permeability ranging from impermeable to fully permeable was presented. In the case of the permeable lining two flow regimes were considered, one with drawdown and one with a maintained phreatic surface. A number of conclusions relating to the two ground conditions were given, for example surface settlements and lining loads, with particular emphasis on the effects of lining permeability and hydraulic boundary conditions. Wongsaroj *et al.* (2007) analysed the conditions around the tunnels at the St James's Park site (London Clay), focusing on the subsurface response, where they identified zones involving rigid-body, consolidation and swelling movements. Comparisons were made between the extensive subsurface monitoring data and the numerical analysis. It was concluded that good predictions can be achieved only when stiffness anisotropy, variation in permeability with the London Clay sub-units and non-uniform drainage in the lining are taken into account.

The final three papers in this section all described complex soil–structure interaction problems involving elements or structures as well as the tunnel itself. Jenck & Dias (2004) used a three-dimensional FDM to model the interaction between a concrete building with beams and floors supported on a raft foundation and a shallow tunnel (with pressurised face). The analyses led to a building stiffness threshold being established for both transverse and longitudinal profiles, below which surface settlements are exacerbated and above which they are reduced. The analysis confirmed that, with a raft foundation, there are only negligible horizontal strains, and so using greenfield predictions will be very conservative in terms of assessing damage. The paper also identified areas where refinements in the analyses are required (e.g. modelling of the tunnelling process, interface between the structure and the soil). Dias *et al.* (2002) investigated the benefit of reinforcing a tunnel face by spiles using the same FDM software as the previous paper. Two techniques for modelling the reinforced face were considered. In one case the effect of the spiles was modelled by using a homogenisation technique that represents their influence by adopting global anisotropic material properties. For this situation a two-dimensional analysis can be used. The results from this approach were compared with those from a conventional three-dimensional analysis. It was concluded that deformations obtained from the two approaches are comparable, although plastic rupture zones between the reinforcing elements cannot be modelled using the homogenisation method unless limiting spile/soil friction is taken into account. Schroeder *et al.* (2004) investigated the complex problem of pile group loading on existing tunnels using FEM analyses. Initially three-dimensional analyses were used to justify the applicability of using a two-dimensional approach to model the loading of pile rows (i.e. as walls). The rows were then loaded on one or two sides of a tunnel to assess the effects on the tunnel. The results were used to formulate design guidelines relating to distance between tunnel and piles and maximum allowable tunnel deformations, in terms of both tunnel distortions and global movements.

SMALL-SCALE PHYSICAL MODELLING

Laboratory modelling with small models provides a means for investigating tunnelling-related boundary value problems under controlled conditions. The tunnel can be modelled as a whole, or perhaps components of the lining or TBM or other associated processes such as grouting.

Centrifuge modelling of the tunnelling scenario allows various configurations to be investigated at prototype stress levels. Schofield (1980) in his Rankine Lecture discussed the extensive Cambridge University centrifuge modelling of tunnels in soft ground, explaining how the mechanisms observed were used to develop theoretical formulations for the tunnel stability using plasticity solutions (e.g. Davis *et al.*, 1980, mentioned above). Atkinson (1975) gave a brief report on centrifuge tests with tunnels in dense sand; further details were also given by Atkinson & Potts (1977), as discussed in the previous section. In most of these tests a steadily increasing tunnelling volume loss was imposed by deflating a membrane representing the tunnel lining. As tunnelling in the urban environment becomes more common, previously unexplored problems have been investigated. Loganathan *et al.* (2000) used centrifuge tests to model the two-dimensional effects of tunnelling at various depths on adjacent piles and pile groups in clay. Extensive instrumentation was used to monitor the axial force and bending moment along the length of the piles. The results from the tests were compared with ground movement predictions from empirical and closed-form analytical solutions, which compared reasonably well. Mitigating tunnelling-induced ground movements can be a serious consideration in urban environments. Bilotta (2008) used centrifuge models to investigate the potential of using diaphragm walls to modify the two-dimensional displacement field from tunnelling in clay to safeguard buildings and services. The offset, depth, thickness and roughness of the wall were varied to assess their effects. The experimental results, supplemented with numerical analysis, indicated that the embedment length of the wall and its roughness are key variables. An efficiency parameter was introduced to assess the degree to which the wall modifies the reference settlement profile (i.e. without a wall).

Modelling testing at 1*g* was adopted to investigate the effect of tunnelling on existing tunnels by Kim *et al.* (1998). Two apparatus and a miniature tunnelling machine were developed to model tunnel construction parallel and perpendicular to existing tunnels in clay. Jacking forces and redistribution of stresses from liner deformation and volume loss were key factors controlling interaction effects for the perpendicular and parallel tunnels respectively. Significant liner deformations and bending moments were also observed during tunnel installation.

The problem of mud 'pumping', where high-speed trains passing through tunnels constructed in readily eroding soils led to increases in pore pressure and hence fluidisation of fine material, causing flow out of weep holes and loss of tunnel support, was investigated by Hayashi & Shahu (2000). The 'erosive soil' was discussed and characterised. Initially a small-scale model, representing conditions from the top of the walls to just below invert level, was used that was able to simulate cyclic loading from the trains, water seepage through the soil, and tunnel pressure. Tests indicated that the problem could be solved by relocating the weep holes at a higher elevation within the tunnel wall in conjunction with a suitable filter. Two full-scale model tests were then set up to verify the solution.

The influence of radial rock bolts on the failure mode of a tunnel opening in soft rock was studied by Indraratna (1993) using small-scale 'segments' of the ground. The segment represents one eighth of a plane perpendicular to the tunnel axis and ground (i.e. a 45° sector of an axisymmetric 'slice'). The segments were cast from gypstone, a synthetic rock, sometimes with discontinuities incorporated, with model bolts installed in the face of the tapered end. Each segment was tested under plane-strain conditions in a triaxial apparatus with Teflon interfaces to reproduce isotropic conditions. It was shown that using a high density of bolts results in minimal propagation of slip-lines and risk of failure close to the tunnel boundary.

Komiya *et al.* (2001) and Soga *et al.* (2004) reported model tests performed to look into the effects of compensation grouting in clay to reduce tunnelling-induced ground movements. Komiya *et al.* (2001) presented results from a grout jacking field trial, where injection was from a newly constructed tunnel. These highlighted that, although the grouting produced heave, it was followed by settlements from consolidation and grout shrinkage. Model testing was then undertaken to assess the main controlling factors. A grouting efficiency parameter was introduced, being the ratio of heaved to injected volume, and the laboratory study identified that efficiency improves with overconsolidated clay and injection volume. The results were supplemented with numerical analysis. Soga *et al.* (2004) presented findings from tests where grout was injected at single and multiple points into a confined cylindrical kaolin sample. Both fracture and compaction grouting were investigated (the latter using inflated balloons), and numerical analysis was used to gain a better understanding of conditions within the apparatus. It was confirmed that a better efficiency is achieved with overconsolidated soils, with compaction grouting (as compared with fracture grouting), and that for normally consolidated clay it could be improved by reducing spacing and time between injections.

Control of the face pressure in closed-face TBMs is directly linked to the operation of the screw conveyor that removes the spoil from the plenum chamber, and it is clear that face pressure controls volume loss and ground movements. Merritt & Mair (2006) described tests performed with a sophisticated tenth-scale instrumented model screw conveyor in which a number of soils were tested, sometimes with soil conditioning agents, investigating variables such as spoil strength, screw speed and screw geometry. The tests provide new insight into the behaviour of soil flow through screw conveyors, and have led to the proposed theoretical model discussed earlier (Merritt & Mair, 2008).

FULL-SCALE FIELD MONITORING AND CASE STUDIES

Field monitoring of tunnel construction has been reported in a significant number of papers, with a particular emphasis on tunnels in London Clay. This clearly reflects the importance of field observations both to control the works and understand mechanisms of behaviour for future projects. It seems appropriate to start by referring to the Second Rankine Lecture, given by Cooling (1962), concerning 'Field measurements in soil mechanics'. He drew attention to the fact that, in lined tunnel construction and operation, small movements can have large importance. At that time, and for many years later, extensive high-quality measurements in and around tunnels on a number of important projects were made by the Building Research Station (BRS). Details were given of devices for measuring changes in tunnel diameter and the design of instrumented tunnel linings, which was significant with respect to the adoption of precast concrete wedge-block linings (particularly for tunnels transporting water under pressure). Mention was made of the distortion of the Bakerloo line tunnels from removing overburden above them, during construction of the Shell Centre basement, causing ovalisation, with the vertical diameter increasing by about 0·2 in over 12 ft, and the horizontal diameter decreasing slightly.

Uncertainty of ground conditions, especially with respect

to water, is one of the main reasons why the observational method was conceived. It is particularly relevant to the complex conditions of tunnelling, which are not easy to analyse, as discussed earlier. Peck (1969), in his Rankine Lecture, discussed the advantages and limitations of the method, citing one case study where two tunnels were to be constructed in very close proximity to the foundations of a building with deep basements. A very expensive protection scheme was designed, but the feeling was that movements would not be detrimental. Comprehensive monitoring in advance of the tunnels reaching the structure indicated that the surface settlements were small, and it would not be necessary to implement the mitigation measures. Thus great savings were made with the tunnels successfully constructed. Twenty-five years later, Powderham (1994) gave an overview of the observational method with respect to three major tunnelling projects: the Channel Tunnel, the Docklands Light Railway, and the Limehouse Link (the latter was also discussed by Glass and Powderham (1994), but cut-and-cover tunnels have not been included in this review). Refinements to the technique were given, such as the 'progressive modification' approach, and its use with respect to contractual issues and implications and 'value engineering' were discussed.

A number of the earlier field monitoring papers are concerned with tunnelling in rock. Terzaghi (1962) discussed the measurement of stresses in rock, particularly with reference to the design of permanent linings of tunnels and mines. The importance of the ratio of the initial horizontal to vertical stress (denoted here by *K*) was raised in his earlier paper concerning the analysis of stresses around cavities (Terzaghi & Richart, 1952). There is therefore benefit in measuring initial rock stresses accurately. It was concluded that stress conditions near to newly exposed surfaces tend to be erratic and unreliable, and it is necessary to make measurements from drill holes extending beyond the disturbed zone that are 'delicate, time-consuming and expensive' operations that need to be justified economically. Mayer (1963) in the Third Rankine Lecture, on 'Recent work in rock mechanics', referred to how testing of rock samples can enable the conditions under which a given rock will flow or rupture to be predicted, along with where cracks are likely to develop. With this information, potential rock falls can be prevented, for example by the appropriate installation of rock bolts. This method was applied on a large scale during the driving of the Mont Blanc tunnel, which is described briefly. Mayer (1953) also discussed the measurement of the mechanical properties of rocks (e.g. Young's modulus), although this related primarily to galleries in rock.

'Ground support for tunnels in weak rocks' was the title of Ward's Rankine Lecture in 1978. His name is associated with many extensive tunnel monitoring projects undertaken by the BRS (as mentioned by Cooling, 1962), and this paper covered one aspect of his work. It did not seek to define how to design supports for tunnels in weak rocks, but drew attention to the many factors that influence support loading and convergence, and how they operate. The importance of monitoring was emphasised, and it was pointed out that tunnelling in weak rock is much more of an art than soft ground tunnelling. The final conclusion was that the skilled engineer, by observing the behaviour of the ground, can cause it to work for him by controlling its yield and hence its ability to support load. A case study that implements this final conclusion was described by Ward *et al.* (1983) with reference to the Kielder experimental tunnel, which was built to assess design and construction requirements for the proposed 32 km long aqueduct tunnel. It was established that the load carried by the lining, deduced from strain measurements, was dependent on how much the rock was allowed to yield. Thus by reducing the stiffness of supports, which allowed some movement of the rock, loads carried by them were also greatly reduced. To assist with design of the main tunnel, empirical characteristic lines (ground-reaction curves) demonstrating the interaction between support loading and rock displacement were produced. The dominating influence of rock discontinuities was also highlighted.

The extensive monitoring carried out within a trial excavation for underground caverns at Dinorwig power station was described by Douglas *et al.* (1983). Detailed measurements of rock mass displacement and stress changes were made, as well as measurements of loads in primary and secondary reinforcement and strains in sprayed concrete. Observed behaviour was compared with results from theoretical studies. It was reported that, overall, significant economies were made with the construction of the subsequent caverns, particularly with respect to increasing rates of excavation and reducing the quantity of required support.

The first of the papers involving tunnels in London Clay concerns the Bakerloo line tunnels mentioned 26 years earlier by Cooling (1962). Burford (1988) gave an update of their response, showing excellent long-term monitoring data that clearly indicate that the tunnels were continuing to heave (by up to about 50 mm) with little decrease in the rate of movement 27 years after the excavation above them was made. Some years later, quite close by at Waterloo, monitoring of ground and structural movements was carried out as part of compensation grouting operations, instigated to offset settlements from tunnel construction for new passageways and a new escalator at the station, as described by Harris *et al.* (1994). The observational method was implemented for this project, such that field measurements were checked against defined limits of acceptable behaviour, for which there were planned contingencies in case these were approached. This was important because of the sensitivity of a number of overlying structures. Monitoring of displacements during tunnelling and compensation grouting was by precise levelling at the surface and by using rod extensometers and electrolevel beams in boreholes subsurface. Excellent results are presented from the different instruments. The work was successfully completed with a maximum settlement of less than 15 mm and no damage to the structures above. Standing & Burland (2006) investigated unexpected volume losses from tunnelling for the Jubilee Line Extension just across the river Thames in Westminster. Volume losses south of the lake in St James's Park were about twice those north of it. A number of causes were identified relating primarily to construction method (over-advanced unsupported tunnel heading) and geology (different Thames terrace deposits either side of the lake, and the position of the London Clay divisions relative to the tunnels).

Extensive monitoring was undertaken during construction of the Heathrow Express (HE) tunnels in London Clay (by NATM and shield tunnelling techniques). Cooper *et al.* (2002) assessed the movements of the existing Piccadilly line tunnels, from the HE tunnel construction beneath them, in terms of settlements, rotations and distortions. Various responses were presented and discussed, including transient tunnel rotation and ovalisation, asymmetry of the settlement troughs when tunnels were constructed close to each other (e.g. either side of the concourse tunnel), and long-term increases of maximum settlements. Minor damage to the existing concrete segment linings was also reported. Van Der Berg *et al.* (2003) also reported on extensive monitoring of ground response from construction of the HE tunnels, but further along their route (at Terminal 4). In particular, longitudinal displacements were measured ahead of the advancing NATM tunnel by installing borehole deflectometer and inclinometer devices from existing tunnels either side of the tunnel to be constructed and the surface. Insight was gained into the magnitude and zones of displacement and how they developed, and about the instruments themselves (especially the grout used to install them). The measurements allowed the proportion of the volume loss from the 'face-take' to be established. Clayton *et al.* (2006) extended the previous work, this time concentrating on in-tunnel displacements and surface settlements, both transverse and longitudinal to the tunnel under construction. In both papers there is a thorough discussion of the instruments and their accuracy. On the subject of tunnel monitoring, Clayton *et al.* (2002) discussed the performance of pressure cells for sprayed concrete tunnel linings. Numerical and physical experiments were performed to study the response of typical cells, and their field response was assessed by installing and monitoring cells within the new HE tunnels. Various factors governing their response were identified and discussed, such as cell-action effects, installation, temperature changes and shrinkage. These need to be taken into account for realistic measurement of lining stresses.

The settlements during tunnelling in soft alluvial clays using compressed air, and the effects after switching off the air pressure, were considered by Glossop & Farmer (1979). A case study was described involving a tunnel driven in Belfast sleech, with measurements transverse to the tunnel of pore pressure and surface and subsurface settlements. A two-stage process was observed, with initially a typical undrained 'volume loss' response. After removal of air pressure, deformation patterns and reduced piezometric levels indicated compression of the silty clay associated with drainage into the tunnel and primary consolidation around it.

Two papers described the ground response to slurry shield tunnelling as part of the Lyon Metro D in complex beds of silts, sands and clays. Both papers used comprehensive surface and subsurface field observations to develop and test numerical models and analysis. Benmebarek *et al.* (1998) described the tunnelling machine and geology, and presented vertical and horizontal displacements at different depths: generally movements were very small, with the largest being less than 10 mm. Face losses on this particular project were small because of the slurry pressure, but face loss from previous projects, along with the annular void between the lining and the ground, created by passage of the shield, were identified as primary sources of short-term settlement. Following from these observations, a new procedure for two-dimensional FDM numerical analysis was proposed. Bernat *et al.* (1999) presented and used data from the same monitoring sections to examine soil–tunnelling interactions using an empirical 'deconfinement factor', which represents stress reductions around the tunnel during construction, implemented in an FEM analysis. Good comparisons, for both surface and subsurface settlements, were obtained as long as accurate account was taken of the stratigraphy and deconfinement factor.

SUMMARY

Tunnelling in the ground represents a complex soil–structure interaction boundary value problem—one of the most difficult the geotechnical engineer has to deal with. A wide range of papers connected with tunnelling in soils and rocks have been published in *Géotechnique* during the past 60 years. Their variety, ranging from theoretical and numerical analysis through small-scale physical models to full-scale field monitoring and associated case studies, reflects the interest and importance of the subject. It is particularly relevant to the urban environment, where tunnelling-induced settlements can potentially cause damage to existing infrastructure, services and buildings. There is still a great deal to be learnt and understood, and the realistic modelling of this three-dimensional boundary value problem is still a challenge. It seems very likely that there will be papers submitted to *Géotechnique* for many years to come, continuing the progress of tunnelling knowledge.

REFERENCES

Addenbrooke, T. I., Potts, D. M. & Puzrin, A. M. (1997). The influence of pre-failure soil stiffness on the numerical analysis of tunnel construction. *Géotechnique* **47**, No. 3, 693–712.

Atkinson, J. H. (1975). The stability of a shallow circular tunnel in dense sand during surface excavation or filling. *Géotechnique* **25**, No. 3, 591–592.

Atkinson, J. H. & Mair, R. J. (1983). Loads on leaking and watertight tunnel linings, sewers and buried pipes due to groundwater. *Géotechnique* **33**, No. 3, 341–344; Gunn, M. J. & Taylor, R. N.; Hungr, O. (1985). Discussion. **35**, No. 1, 73–77.

Atkinson, J. H. & Potts, D. M. (1977). Stability of a shallow circular tunnel in cohesionless soil. *Géotechnique* **27**, No. 2, 203–215.

Benmebarek, S., Kastner, R., Ollier, C. & Charif, A. (1998). Auscultation et modélisation numérique du processus de creusement á l'aide d'un tunnelier. *Géotechnique* **48**, No. 6, 801–818.

Bernat, S., Cambou, B. & Dubois, P. (1999). Assessing a soft soil tunnelling numerical model using field data. *Géotechnique* **49**, No. 4, 427–452.

Bilotta, E. (2008). Use of diaphragm walls to mitigate ground movements induced by tunnelling. *Géotechnique* **58**, No. 2, 143–155.

Burford, D. (1988). Heave of tunnels beneath the Shell Centre, London, 1959–1986. *Géotechnique* **38**, No. 1, 135–137.

Clayton, C. R. I., Van Der Berg, J. P., Heymann, G., Bica, A. V. D. & Hope, V. S. (2002). The performance of pressure cells for sprayed concrete tunnel linings. *Géotechnique* **52**, No. 2, 107–115.

Clayton, C. R. I., Van Der Berg, J. P. & Thomas, A. H. (2006). Monitoring and displacements at Heathrow Express Terminal 4 station tunnels. *Géotechnique* **56**, No. 5, 323–334; .

Cooling, L. F. (1962). Field measurements in soil mechanics. *Géotechnique* **12**, No. 2, 77–104.

Cooper, M. L., Chapman, D. N., Rogers, C. D. F. & Chan, A. H. C. (2002). Movements in the Piccadilly Line tunnels due to the Heathrow Express construction. *Géotechnique* **52**, No. 4, 243–257.

Curtis, D. J. (1976). Discussion: 'The circular tunnel in elastic ground'. *Géotechnique* **26**, No. 3, 231–237.

Davis, E. H., Gunn, M. J., Mair, R. J. & Seneviratnes, H. N. (1980). The stability of shallow tunnels and underground openings in cohesive material. *Géotechnique* **30**, No. 4, 397–416.

Dias, D., Kastner, R. & Jassionnesse, C. (2002). Sols renforcé par Boulonnage: Etude numérique et application au front de taille d'un tunnel profound. *Géotechnique* **52**, No. 1, 15–27.

Douglas, T. H., Richards, L. R. & Arthur, L. J. (1983). Trial excavation for underground caverns at Dinorwig power station. *Géotechnique* **33**, No. 4, 407–431.

Feng, Z. L. & Ding, D. P. (1986). The interaction between flexible linings and surrounding rock masses. *Géotechnique* **36**, No. 4, 587–591.

Franzius, J. N., Potts, D. M. & Burland, J. B. (2005). The influence of soil anisotropy and K_0 on ground surface movements resulting from tunnel excavation. *Géotechnique* **55**, No. 3, 189–199.

Glass, P. R. & Powderham, A. J. (1994). Application of the observational method at Limehouse Link. *Géotechnique* **44**, No. 4, 665–679.

Glossop, R. (1961). The invention and development of injection processes. Part II: 1850–1960. *Géotechnique* **11**, No. 4, 255–279.

Glossop, N. H. & Farmer, I. W. (1979). Settlement associated with removal of compressed air pressure during tunnelling in alluvial clay. *Géotechnique* **29**, No. 1, 67–72.

Glossop, R. (1976). The invention and early use of compressed air to exclude water from shafts and tunnels during construction. *Géotechnique* **26**, No. 2, 253–280.

Harris, D. I., Mair, R. J., Love, J. P., Taylor, R. N. & Henderson, T. O. (1994). Observations of ground and structure movements for compensation grouting during tunnel construction at Waterloo station. *Géotechnique* **44**, No. 4, 691–713.

Hayashi, S. & Shahu, J. T. (2000). Mud pumping problem in tunnels on erosive soil deposits. *Géotechnique* **50**, No. 4, 393–408.

Hetenyi, M. (1946). *Beams on elastic foundations.* Ann Arbor: University of Michigan Press.

Indraratna, B. (1993). Effect of bolts on failure modes near tunnel openings in soft rock. *Géotechnique* **43**, No. 3, 433–442.

Jenck, O. & Dias, D. (2004). Analyse tridimensionnelle en différences finies de l'interaction entre une structure en béton et le creusement d'un tunnel à faible profondeur. *Géotechnique* **54**, No. 8, 519–528.

Kerisel, J. (1956). Historique de la méchanique des sols en France jusqu'au 20e siècle. *Géotechnique* **6**, No. 4, 151–166.

Kerisel, J. (1975). Old structures in relation to soil conditions. *Géotechnique* **25**, No. 3, 433–483.

Kim, S. H., Burd, H. J. & Milligan, G. W. E. (1998). Model testing of closely spaced tunnels in clay. *Géotechnique* **48**, No. 3, 375–388.

Klar, A., Vorster, T. E. B., Soga, K. & Mair, R. J. (2005). Soil–pipe interaction due to tunnelling: comparison between Winkler and elastic continuum solutions. *Géotechnique* **55**, No. 6, 461–466.

Komiya, K., Soga, K., Akagi, H., Jafari, M. R. & Bolton, M. D. (2001). Soil consolidation associated with grouting during shield tunnelling in soft clayey ground. *Géotechnique* **51**, No. 10, 835–846; Shirlaw, J. N. (2003). Discussion. **53**, No. 4, 447–448.

Leca, E. & Dormieux, L. (1990). Upper and lower bound solutions for the face stability of shallow circular tunnels in frictional material. *Géotechnique* **40**, No. 4, 581–606.

Loganathan, N., Poulos, H. G. & Stewart, D. P. (2000). Centrifuge testing of tunnelling-induced ground and pile deformations. *Géotechnique* **50**, No. 3, 283–294.

Mair, R. J., Taylor, R. N. & Bracegirdle, A. (1993). Subsurface settlement profiles above tunnels in clay. *Géotechnique* **43**, No. 2, 315–320; Heath, G. R. (1995). Discussion. **45**, No. 2, 361–362.

Mayer, A. (1953). Les propriétés mécaniques des roches. *Géotechnique* **3**, No. 8, 329–338.

Mayer, A. (1963). Recent work in rock mechanics. *Géotechnique* **13**, No. 2, 99–120.

Merritt, A. S. & Mair, R. J. (2006). Mechanics of tunnelling screw conveyors: model tests. *Géotechnique* **56**, No. 9, 605–615.

Merritt, A. S. & Mair, R. J. (2008). Mechanics of tunnelling machine screw conveyors: a theoretical model. *Géotechnique* **58**, No. 2, 79–94.

Morgan, H. D. (1961). A contribution to the analysis of stress in a circular tunnel. *Géotechnique* **11**, No. 1, 37–46; Engelbreth, K. (1961). Discussion. **11**, No. 3, 246–248.

Muir Wood, A. M. (1975). The circular tunnel in elastic ground. *Géotechnique* **25**, No. 1, 115–127.

Osman, A. S., Mair, R. J. & Bolton, M. D. (2006a). On the kinematics of 2D tunnel collapse in undrained clay. *Géotechnique* **56**, No. 9, 585–595.

Osman, A. S., Bolton, M. D. & Mair, R. J. (2006b). Predicting 2D ground movements around tunnels in undrained clay. *Géotechnique* **56**, No. 9, 597–604.

Peck, R. B. (1969). Advantages and limitations of the observational method in applied soil mechanics. *Géotechnique* **19**, No. 2, 171–187.

Pender, M. J. (1980). Elastic solutions for a deep circular tunnel. *Géotechnique* **30**, No. 2, 216–222; Gumbel, J. E. (1981). Discussion. **31**, No. 3, 434–436.

Powderham, A. J. (1994). An overview of the observational method: development in cut and cover and bored tunnelling projects. *Géotechnique* **44**, No. 4, 619–636.

Sagaseta, C. (1987). Analysis of undrained deformation due to ground loss. *Géotechnique* **37**, No. 3, 301–320.

Schmidt, B. (1988). Discussion: 'Analysis of undrained deformation due to ground loss'. *Géotechnique* **38**, No. 4, 647–649.

Schofield, A. N. (1980). Cambridge Geotechnical Centrifuge operations. *Géotechnique* **30**, No. 3, 227–268.

Schroeder, F. C., Potts, D. M. & Addenbrooke, T. I. (2004). The influence of pile group loading on existing tunnels. *Géotechnique* **54**, No. 6, 351–362.

Shin, J. H., Addenbrooke, T. I. & Potts, D. M. (2002a). A numerical study of the effect of groundwater movement on long-term tunnel behaviour. *Géotechnique* **52**, No. 6, 391–403.

Shin, J. H., Potts, D. M. & Zdravkovic, L. (2002b). Three-dimensional modelling of NATM tunnelling in decomposed granite soil. *Géotechnique* **52**, No. 3, 187–200; Knill, J. L.; Shirlaw, J. N. (2003). Discussion. **53**, No. 7, 685–687.

Skempton, A. W. & Chrimes, M. M. (1994). Thames Tunnels: geology, site investigation and geotechnical problems. *Géotechnique* **44**, No. 2, 191–216.

Soga, K., Au, S. K. A., Jafari, M. R. & Bolton, M. D. (2004). Laboratory investigation of multiple grout injections into clay. *Géotechnique* **54**, No. 2, 81–90.

Standing, J. R. & Burland, J. B. (2006). Unexpected tunnelling volume losses in the Westminster area, London. *Géotechnique* **56**, No. 1, 11–26.

Terzaghi, K. (1962). Measurement of stresses in rock. *Géotechnique* **12**, No. 2, 105–124.

Terzaghi, K. & Richart, F. E. (1952). Stresses in rock about cavities. *Géotechnique* **3**, No. 2, 57–90.

Van Der Berg, J. P., Clayton, C. R. I. & Powell, D. B. (2003). Displacements ahead of an advancing NATM tunnel in the London clay. *Géotechnique* **53**, No. 9, 767–784.

Verruijt, A. & Booker, J. R. (1996). Surface settlements due to deformation of a tunnel in an elastic half plane. *Géotechnique* **46**, No. 4, 753–756; Sagaseta, C. (1998). Discussion. **48**, No. 5, 709–713.

Von Moos, A. (1948). Engineering geology in Switzerland: a review of the literature. *Géotechnique* **1**, No. 1, 40–48.

Ward, W. H. (1978). Ground supports for tunnels in weak rocks. *Géotechnique* **28**, No. 2, 133–171.

Ward, W. H., Tedd, P. & Berry, N. S. M. (1983). The Kielder experimental tunnel: final results. *Géotechnique* **33**, No. 3, 275–291.

Wongsaroj, J., Soga, K. & Mair, R. J. (2007). Modelling of long-term ground response to tunnelling under St James's Park, London. *Géotechnique* **57**, No. 1, 75–90.

Phear, A. G. & Harris, S. J. (2008). *Géotechnique* **58**, No. 5, 399–404 [**doi:** 10.1680/geot.2008.58.5.399]

Contributions to *Géotechnique* 1948–2008: Ground improvement

A. G. PHEAR* and S. J. HARRIS†

Ground improvement comprises approximately 30 different methods of ground treatment, including modification, chemical alteration, reinforcement with steel or geosynthetic, strengthening by drainage, densification by vibration or consolidation, the use of electro-osmosis, and the use of the observational technique. This paper attempts to highlight influential papers published in *Géotechnique* since 1947, and urges the reader to seek these out. Descriptions are given of those papers considered to be most influential, ranging from the densification of soils at Cape Kennedy in the USA to the construction of a 41 m high reinforced soil wall in South Africa. The use of vibro-compaction and vibro-replacement techniques has featured significantly, as has consolidation using vertical drains. The importance of observational techniques and their relationship to ground improvement has also received great attention in *Géotechnique.* A summary of papers is also presented in a table to provide a quick reference tool.

KEYWORDS: consolidation; ground improvement; historical review; observational method; reinforced soil; vertical drains; vibro-compaction

L'amélioration du sol comprend une trentaine de méthodes de traitement du sol, y compris la modification, le traitement chimique, le renforcement par l'acier, le raidissement géosynthétique par le drainage, la densification par la vibration ou la consolidation, l'utilisation de l'électro-osmose, ou encore l'emploi de la technique d'observation. Dans la présente communication, on s'efforce de mettre en lumière des communications qui font autorité, publiées dans *Géotechnique* depuis 1947, et nous conseillons vivement au lecteur de consulter; on y trouve également des descriptions de communications jugées particulièrement importantes, de la densification des sols à Cape Kennedy, aux Etats-Unis, jusqu'à la construction d'un mur de sol renforcé, de 41 m de haut, en Afrique du Sud. L'utilisation de techniques de compactage et de remplacement par vibration occupe une place de choix, de même que la consolidation au moyen de drains verticaux. De même, l'importance des techniques d'observation et leur rapport avec le renforcement du sol est évoquée très souvent dans *Géotechnique.* Enfin, un récapitulatif de communications est présenté sous forme de tableau, afin de faciliter les recherches.

INTRODUCTION

Ground improvement techniques have developed greatly over the 60 years for which *Géotechnique* has been in existence. There are now over 30 ground improvement techniques. For most of these, there are variations that extend the versatility of the basic technique. Also, many are used in combination, and the scope for innovation is huge. Ground improvement is used both for temporary works, and for short-term and long-term applications.

Mitchell & Jardine (2002) observed that

> The term, ground improvement, is open to different interpretations. First, it is an intention or objective, not the process of achieving it, although the term is often used in that sense. Secondly, improvement is a relative condition as to which aspect and to what degree there is improvement.

Ground improvement has featured as the main topic in three *Géotechnique* Symposia in Print. The first Symposium in Print (SiP) was published in March 1975, and was on the subject of 'Ground treatment by deep compaction'. The subject of soil treatment was again addressed in March 1981, when the third SiP was published, on 'Vertical drains'.

The eleventh SiP was published in June 2000 and focused on 'Ground and soil improvement'. The theme for that issue was restricted to the mechanical improvement of the ground using vibro-compaction, vibro-replacement, dynamic compaction, soil mixing, compaction grouting, surcharging, deep drainage, and other improvement techniques used to consolidate or reinforce the soil.

Less obviously, the ninth SiP, published in 1994, on the 'Observational method in geotechnical engineering', also contained several papers on ground improvement. Its complementary use is discussed later in this paper.

The aim of this paper is to highlight papers published in *Géotechnique* in the last 60 years that have been the most influential in the development of the practice of ground improvement techniques, and to urge readers to seek them out. This is inevitably subjective, as the authors started with papers that they use regularly, or which are well known. However, this was benchmarked by a review of the references quoted in the main British enabling documents on ground improvement: the Construction Industry Research and Information Association reports on vertical drains (Holtz *et al.*, 1991), on ground treatment (Charles & Watts, 2002; Mitchell & Jardine, 2002) and on the observational method in ground engineering (Nicholson *et al.*, 1998). The state-of-the-art paper on dynamic compaction methods by Greenwood & Kirsch (1983) was also reviewed for references from *Géotechnique*.

Charles (2002) wrote that ground improvement has developed largely in an empirical, contractor-led fashion, and, for the most part, it can be described as an experience-based technology. In contrast, *Géotechnique* addresses mainly the scientific aspects of ground engineering. Ground improvement is a part of ground engineering where practice usually precedes theory, and research follows development, not just in trying to explain why a technique works but, more importantly, in trying to establish rational, rather than empirical, design methods and to see how the technique can be improved and its limitations identified.

Table 1 includes a range of papers published in *Géotechnique* in the last 60 years on ground improvement topics.

Discussion is presented on those topics that in the authors' experience are commonly used or considered for use in current engineering practice.

Discussion on this paper closes on 1 December 2008, for further details see p. ii.

* Ove Arup & Partners, Solihull, UK.
† Ove Arup & Partners, Newcastle Upon Tyne, UK.

Table 1. Summary of selected ground improvement papers published in *Géotechnique* 1947–2007

	Decade					
	1947–1956	1957– 1966	1967–1976	1977–1986	1987–1996	1997–2007
Electro-osmosis	Casagrande (1949)		Bjerrum *et al.* (1967)			Su & Wang (2003)
Reinforced soil					Blight & Dane (1989) Lee *et al.* (1994)	Zornberg (2002) Rowe (2002)
Lime/cement stabilisation	Sherwood (1957)					Boardman *et al.* (2001) Larsson *et al.* (2005)
Surcharging	Tomlinson (1956)		Tomlinson & Wilson (1973)	Watson *et al.* (1984)	Choa (1994)	
Vertical drains				Davies & Humpheson (1981) Hansbo *et al.* (1981) McGown & Hughes (1981) Nicholson & Jardine (1981)	Teh & Houlsby (1991) Choa (1994)	Almeida *et al.* (2000) Long & O'Riordan (2001) Nash & Ryde (2001) Hird & Moseley (2000)
Observational method			Peck (1969)		Hammond & Thorn (1994) Choa (1994)	
Vibro-replacement			Hughes *et al.* (1975) McKenna *et al.* (1975)			Renton-Rose *et al.* (2000)
Dynamic compaction			Menard & Broise (1975)		Hansbo (1996)	
Groundwater control					Roberts & Preene (1994)	
Vacuum consolidation						Almeida *et al.* (2000)
Explosive compaction						Gohl *et al.* (2000)
Lime/cement columns						Larsson *et al.* (2005)

VIBRATORY TECHNIQUES

Vibro stone columns

The formation of columns of compacted stone in the ground using a heavy vibrating poker to displace the in situ ground and to compact the imported material is referred to as vibro stone columns or vibro-replacement (Watts, 2000). There are several ways to design stone columns. The currently used design methods for bearing capacity are mostly refinements of work carried out about 30 years ago by Baumann & Bauer (1974) and Hughes & Withers (1974). For settlement calculations Baumann & Bauer, and also Priebe (1995), are widely used. At the time of the first Symposium in Print, the state of knowledge of how deep compaction techniques work was relatively primitive, and several of the papers addressed this (Charles, 2002). Two of these are considered influential in the development of the design and testing of stone columns, and are discussed below.

Hughes *et al.* (1975) described a case history of the plate loading of an isolated stone column in soft clay on Canvey Island in Essex. The soft clay was 9 m thick and overlay a medium dense silty sand. The stone columns were 10 m long and penetrated through the soft clay into the much stiffer silty sand. This paper is important, because the behaviour of the column in this field trial proved to be in accordance with the behaviour predicted by the theory proposed by Hughes & Withers (1974). As noted above, this theory is still widely used for the design of stone columns.

McKenna *et al.* (1975) described an apparently unsuccessful application of ground improvement. It is a case history of stone columns supporting a widespread load—in this case a trial embankment. This was built as part of the design studies for a section of the M5 motorway in Somerset. To investigate the effectiveness of stone columns in reducing the settlement of high embankments built on soft alluvium, vibro-replacement stone columns constructed using the wet process were installed under one end of a trial embankment. The columns were about 11 m long, but did not penetrate the full thickness of the alluvium. The embankment was up to nearly 8 m high. The foundations were instrumented, and a comparison of the improved and unimproved ground showed that the columns apparently had no effect on the amount or rate of settlement of the embankment. This paper stimulated discussion at the symposium and further research into the reasons for the poor performance of the ground treatment. It was concluded that the reasons postulated were probably incorrect, and that the explanation was more complicated. The columns had probably performed as rigid friction piles, which had been punched into the clay because it had been remoulded to a very low strength by the column installation process. The widespread load contained the clay stratum, preventing column bulging at any depth and so allowing stress transfer down the columns (Greenwood, 1991).

These two case histories of a single loaded column and a group of columns under widespread load demonstrate the importance of the containment of the soil on the column behaviour. As a result, plate load tests on single isolated columns do not represent the performance of columns under widespread loading. This is because a single column fails primarily because of bulging at shallow depths, but under widespread load columns may fail by bulging or can be like a rigid pile, according to the circumstances.

It is now recognised that plate load tests on single columns should primarily be used only to check the quality and consistency of workmanship, and zone tests are required to demonstrate that vibro stone column treatment has achieved the specified performance criteria. This approach to quality control and performance testing is adopted by BRE in their document *Specifying vibro stone columns* (Watts, 2000).

Vibro-compaction

Vibro-compaction is a technique for increasing the density of granular soils by the insertion to depth of a heavy vibrating poker. This method was first developed in Germany in the 1930s. A major development in the vibro market worldwide over the last 25 years has been the treatment of soils to resist the effects of liquefaction during earthquakes. It had previously generally been considered that vibro compaction is effective only where the fines (silt plus clay) content does not exceed 15%. Slocombe *et al.* (2000) importantly show how densification of granular soils using vibro has progressed with the development of improved depth vibrators, and better knowledge of how soils respond to both vibration and displacement, and they also show the importance of good workmanship. As a result, sands with significantly higher fines content can now be compacted using vibro. This is an example of the key role of equipment development in the evolution of vibro techniques. Another important aspect of this paper—and of several others in the eleventh SiP—is that its emphasis is much more on testing to confirm the post-treatment performance than was the case in the first SiP 25 years earlier. This illustrates how ground improvement technology has matured, but it is also a reflection of the general move within the construction industry towards quality assurance and performance specifications.

Dynamic compaction

Dynamic compaction (DC) is the process of systematically tamping the ground surface with a heavy weight dropped from height. A major limitation of the method is the limited depth to which surface impact loading can densify the soil. The uncertainty of the depth to which treatment is effective is also a problem. Menard & Broise (1975) stated that the depth that could be compacted, H (in metres), could be determined from the relationship $H < (Mh)^{0 \cdot 5}$, where a weight of mass M tonnes is dropped from a height of h metres. This relationship has dominated analyses of dynamic compaction for the last 25 years.

ADDING LOAD OR INCREASING VERTICAL EFFECTIVE STRESS

Precompression (and surcharging)

Precompression is the deliberate act of compressing the ground under an applied pressure prior to placing or completing the permanent load. It is described as surcharging where the stress intensity from the preloading is greater than the intensity from the permanent loading (Mitchell & Jardine, 2002). Precompression is used to induce settlements that have three recognisable phases: immediate; primary consolidation; and secondary consolidation (or creep).

One of the earliest recorded uses of precompression ground improvement in the UK is given by Tomlinson (1956). This was supervised by Thomas Telford's resident engineer, and was during the construction of the Clachnacharry Locks on the Caledonian Canal near Inverness in Scotland in 1811. The locks needed to be built 400 yards (~370 m) out from the shore of Loch Beauley on soft mud 55 feet (~17 m) thick. The elasticity of the mud prevented the usual process of piledriving, because the piles rebounded after each blow of the piling hammer. It was therefore decided to precompress the mud using over 20 feet (~6 m) of clay dug from the adjacent hill, topped off by a quantity of stone (which was later used to build the locks). This load was left in place for six months. After that time it had compressed the mud by about 11 feet (~3 m), with settlement being monitored by a spirit level from a mark on the shore. When it was felt that no further sinking would take place (as the load was much greater than that of the future lock pit), the lock was excavated and the masonry chamber was constructed.

Vertical drains

The five papers published in the third SiP, on 'Vertical drains', in 1981 are all considered to be influential, and two are described below.

Hansbo *et al.* (1981) described a design theory that can be used for vertical drainage, which is still used today; their paper also contained six well-documented case records, related to different drain types, from sites in Sweden, Italy and the Netherlands. Experience gained from these case records showed that in many cases, even if the overall performance appeared successful, there were aspects of soil/drain behaviour that could not be explained on the basis of the previously existing theories. It was also noticed that experienced engineering judgement is needed with respect to uncertainties in parameter selection, design and drain performance, to avoid disappointing results.

Davies & Humpheson (1981) presented a case history of the behaviour of vertical drains in Belfast, where one section of a new link road crossed soft estuarine deposits, known locally as 'sleech', on embankments up to 8 m high. Vertical drains were considered to be the most appropriate solution. A site trial was carried out to compare vertical sand drains and band drains installed beneath trial embankments. Both performed well, and the results showed good agreement between the laboratory and field data. This paper demonstrates the relationship between the coefficient of consolidation (c_h) and the vertical effective stress for soft clays. At stresses less than the preconsolidation pressure (p'_c) the values of c_h were almost an order of magnitude higher (i.e. faster) than those above p'_c: see Fig. 1 (Fig. 14 in the paper). The c_h values back-calculated from the trial embankments and those obtained in the laboratory were in good agreement, probably because of the lack of preferential drainage paths in the sleech. It was also noted that, in order to obtain a reasonable prediction of the dissipation of excess pore pressures with time, account must be taken of the variation of c_h with vertical effective stress.

The ability to derive increased confidence in values of the coefficient of consolidation measured in situ from piezocone dissipation tests was examined by Almeida *et al.* (2000). In their description of an embankment built on very soft clay in Rio de Janeiro they compared the coefficients of horizontal consolidation determined from special laboratory tests and piezocone dissipation tests. They referred to the methods of Teh & Houlsby (1991) to derive the coefficient of consolidation from such dissipation tests.

Vertical drains are often used in combination with pre-

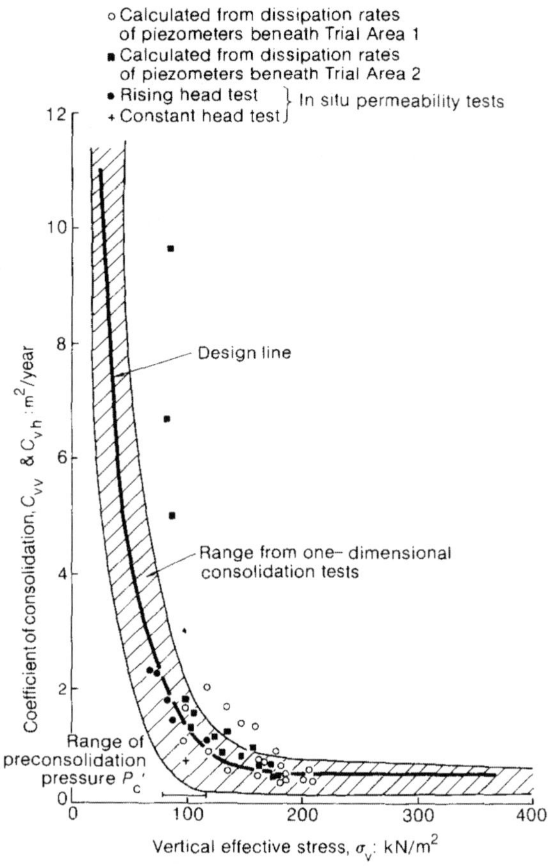

Fig. 1. Variation of coefficient of consolidation with vertical effective stress (originally Fig. 14 in Davies & Humpheson, 1981)

compression and/or temporary stabilising berms, as described by Long & O'Riordan (2001). This is a case history of the construction of embankments up to 8 m high over very soft and highly compressible clay for the new bypass at Athlone in the Republic of Ireland. Long & O'Riordan also described investigative work using a piezocone to determine the increase in undrained shear strength in soft clay beneath an embankment nine years after its construction. They observed an increase approximately equal to 0·22 × vertical effective stress (which is consistent with the work published by Mesri, 1989).

Vacuum preloading

The principle of vacuum preloading is the application of atmospheric pressure to form a temporary surcharge for soft clays while applying a vacuum to the surface of the soil beneath a membrane. It was developed in Sweden in the 1950s. Vacuum preloading provides a good example of the interaction of theory and practice, because it requires both a good understanding of the soil behaviour, and appropriate tools and techniques (Charles, 2002). The paper by Chu *et al.* (2000) is a case history of a 50 ha site in Tianjin in China, where vacuum preloading was used to precompress a 20 m thick deposit of soft clay by 1 m.

THE OBSERVATIONAL METHOD IN GROUND IMPROVEMENT

An observational approach has long been applied to piling and ground improvement works, although it had never been formally recognised as the observational method. The first formal recognition of this was by Peck (1969) in his Rankine Lecture. It was subsequently developed in *Géotechnique* in the ninth SiP in 1994, and in the CIRIA report by Nicholson *et al.* (1998). It is particularly applicable to ground improvement by deep compaction, but is also suitable for embankment and other construction on soft foundations where techniques such as precompression and vertical drains are adopted.

The paper by Peck (1969) is considered to be one of the most influential in the practical development of geotechnical engineering practice and philosophy over the last 35 years or so. Two of the case studies therein address ground improvement topics. The first is the Cleveland Ore Yard in Ohio, USA, where Terzaghi first used a 'best-way-out' observational method. The second is the unusual but nonetheless very interesting case history of the Cape Kennedy causeway in Florida, USA. This was the 4 mile long (~6·5 km) causeway built to permit transport of Saturn rockets from their assembly building to one of the launching pads. The rockets were carried in a vertical position on a platform supported on caterpillar tracks. These ran on a 3 ft thick (0·9 m) stabilised platform over hydraulically placed uniform fine sand. The ground surface was only a few feet above sea level. Construction of the second causeway started before the ground investigation was complete, and unfortunately this found several zones with an SPT N value of less than 2. This raised the possibility of liquefaction failure under the passage of the fully loaded transporter, so some form of ground improvement was required. Conventional techniques would have been costly and would have taken too long, so it was decided to use an observational method and compress the sand by repeated passes of the transporter itself with increasingly heavy loads. This project was carried out very carefully, with the input of three expert consultants.

Examples of the application of the observational method are given by Nicholson *et al.* (1998). There may also be scope for combining techniques to enhance the treatment scheme, such as vibro-compaction plus piling, as described by Raison (1996) for the North Morecambe Gas Terminal.

The observational method is equally applicable to the construction of embankments on soft clays and for land reclamation works. Choa (1994) describes the application of the observational method to the Changi Airport reclamation in Singapore and the Tianjin Port development in China. The settlement and piezometers records were reviewed against predictions. Modification plans to be triggered, if necessary, included controlling the consolidation settlement by varying the surcharge preloading and the ground treatment method, such as vertical drains. A contingency plan was developed to maintain the stability of the Tianjin East Pier. This used vacuum dewatering to provide an additional surcharge and thus accelerate the construction programme.

REINFORCED SOIL

A well-documented case history of the corrosion problems encountered with a complex system of reinforced soil walls up to 41 m high in South Africa was presented by Blight & Dane (1989). The writers described the importance of using a free-draining backfill around the galvanised strips, and the compounded detrimental effects of using salt water in the compaction process. Their paper includes a detailed description of their investigations, and methods of measurement of the stresses in the steel strips, and of the lateral deflections of the wall. The paper is accompanied by an interesting discussion of the issues raised.

Lee *et al.* (1994) presented case histories of failure

investigations for four reinforced soil walls, curved in plan, built in Tennessee, USA. The paper describes the design and investigations of the walls, and highlights many design issues, including: the effect of using poorly compacted coarse granular fill; three-dimensional deformation; stresses induced in the reinforcement due to either foundation or fill settlement; and the validity of applying empirical design methods to walls with unprecedented features. The paper was also accompanied by an interesting discussion of the issues raised.

CONCLUSIONS

There has been a wealth of papers over the last 60 years describing various ground improvement techniques, which have charted the developments made in techniques. Perhaps not surprisingly, the authors of this review have been drawn to those papers that concentrate on case histories, rather than research-based publications.

The papers selected demonstrate the commitment required to obtain convincing evidence of the efficacy of a ground improvement technique. Similarly, where failures have occurred, extensive work has been demonstrated to provide credible explanations that can stand scrutiny. The lack of papers published for some techniques, such as lime and or cement stabilisation, arguably could be due to the inherent difficulties in obtaining such evidence, particularly where more prescriptive improvement methods are utilised, where extensive instrumentation is not used.

It should be acknowledged that the existence of a significant number of technical journals gives many opportunities to submit and publish papers, and the practical nature of ground improvement might not immediately encourage authors to submit to *Géotechnique*. In addition, of course, Thomas Telford Ltd commenced publication of the *Ground Improvement* journal in 2001.

The long-term post-construction performance of treated ground appears to be seldom documented in case histories in *Géotechnique* in the last 60 years. The aim of ground improvement is, after all, to provide foundations that have adequate bearing capacity and settlement characteristics to assure satisfactory performance throughout the design life of the structure or embankment that is to be built on them. Charles (2002) noted that further research is needed to, *inter alia*, gain a more realistic appreciation of what can be achieved by ground treatment through the preparation and study of well-documented case histories of long-term performance. Hopefully some of these will be published in *Géotechnique* in the near future.

REFERENCES

Almeida, M. S. S., Santa Maria, P. E. L., Martins, I. S. M., Spotti, A. P. & Coelho, L. B. M. (2000). Consolidation of a very soft clay with vertical drains. *Géotechnique* **50**, No. 6, 633–643; Discussion. **52**, No. 2, 148–154.

Baumann, V. & Bauer, G. E. A. (1974). The performance of foundations on various soils stabilised by the vibro-compaction method. *Can. Geotech. J.* **3**, No. 11, 509–530.

Bjerrum, L., Moum, J. & Eide, O. (1967). Application of electro-osmosis to a foundation problem in a Norwegian quick clay. *Géotechnique* **17**, No. 3, 214–235; Discussion. **18**, No. 2, 276–278.

Blight, G. E. & Dane, M. S. W. (1989). Deterioration of a wall complex constructed of reinforced earth. *Géotechnique* **39**, No. 1, 47–53; Discussion. **39**, No. 3, 567–570.

Boardman, D. I., Glendinning, S. & Rogers, C. D. F. (2001). Development of stabilisation and solidification in lime-clay mixes. *Géotechnique* **50**, No. 6, 533–543.

Casagrande, L. (1949). Electro-osmosis in soils. *Géotechnique* **1**, No. 3, 159–177.

Charles, J. A. (2002). Ground improvement: the interaction of engineering science and experience-based technology. *Géotechnique* **52**, No. 7, 527–532.

Charles, J. A. & Watts, K. S. (2002). *Treated ground: Engineering properties and performance*, CIRIA Report C572. London: Construction Industry Research and Information Association.

Choa, V. (1994). Application of the observational method to hydraulic fill reclamation projects. *Géotechnique* **44**, No. 4, 735–745.

Chu, J., Yan, S. W. & Yang, H. (2000). Soil improvement by the vacuum preloading method for an oil storage station. *Géotechnique* **50**, No. 6, 625–632; Discussion. **52**, No. 2, 148–154.

Davies, J. A. & Humpheson, C. (1981). A comparison between the performance of two types of vertical drain beneath a trial embankment in Belfast. *Géotechnique* **31**, No. 1, 19–31.

Gohl, W. B., Jefferies, M. G., Howie, J. A. & Diggles, D. (2000). Explosive compaction: design, implementation and effectiveness. *Géotechnique* **50**, No. 6, 657–665; Discussion. **52**, No. 8, 622–623.

Greenwood, D. A. (1991). Assurance of performance of stone columns. In *Deep foundation improvements: Design, construction and testing*, ASTM STP 1089, pp. 148–171. ASTM International.

Greenwood, D. A. & Kirsch, K. (1983). Specialist ground treatment by vibratory and dynamic measures. *Proceedings of the international conference on advances in piling and ground treatment*, London, pp. 17–45.

Hammond, A. J. & Thorn, M. R. (1994). A methodology for the investigation and treatment of strata affected by gold mining in Johannesburg. *Géotechnique* **44**, No. 4, 715–725.

Hansbo, S. (1996). Dynamic consolidation of mixed fill: a cost-effective alternative to piling: a case record. *Géotechnique* **46**, No. 2, 351–355.

Hansbo, S., Jamiolkowski, M. & Kok, L. (1981). Consolidation by vertical drains. *Géotechnique* **31**, No. 1, 45–66.

Hird, C. C. & Moseley, V. J. (2000). Model study of seepage in smear zones around vertical drains in layered soil. *Géotechnique* **50**, No. 1, 89–97.

Holtz, R. D., Jamiolkowski, M. B., Lancellotta, R. & Pedroni, R. (1991). *Prefabricated vertical drains: Design and performance*. CIRIA Ground Engineering Report. Oxford: CIRIA/Butterworth-Heinemann.

Hughes, J. M. & Withers, N. J. (1974). Reinforcing of soft cohesive soils with stone columns. *Ground Engng* **7**, No. 3, 42–44, 47–49.

Hughes, J. M. O., Withers, N. J. & Greenwood, D. A. (1975). A field trial on the reinforcing effect of a stone column in soil. *Géotechnique* **25**, No. 1, 31–44.

Larsson, S., Stille, H. & Olsson, L. (2005). On horizontal variability in lime-cement columns in deep mixing. *Géotechnique* **55**, No. 1, 33–44.

Lee, K., Jones, C. J. F. P., Sullivan, W. R. & Trolinger, W. (1994). Failure and deformation of four reinforced soil walls in eastern Tennessee. *Géotechnique* **44**, No. 3, 397–426; Discussion: *Géotechnique* **45**, No. 4, 749–752.

Long, M. M. & O'Riordan, N. J. (2001). Field behaviour of very soft clays at the Athlone embankments. *Géotechnique* **51**, No. 4, 293–309.

McGown, A. & Hughes, F. H. (1981). Practical aspects of design and installation of vertical drains. *Géotechnique* **31**, No. 1, 3–17.

McKenna, J. M., Eyre, W. A. & Wolstenholme, D. R. (1975). Performance of an embankment supported by stone columns in soft ground. *Géotechnique* **25**, No. 1, 51–59.

Menard, L. & Broise, Y. (1975). Theoretical and practical aspects of dynamic compaction. *Géotechnique* **25**, No. 1, 3–18.

Mesri, G. (1989). A re-evaluation of $s_{\mathrm{umob}} \approx 0{\cdot}22\sigma'_{\mathrm{p}}$ using laboratory shear tests. *Can. Geotech. J.* **26**, No. 1, 301–309.

Mitchell, J. M. & Jardine, F. M. (2002). *A guide to ground treatment*, CIRIA Report C573. London: Construction Industry Research and Information Association.

Nash, D. F. T. & Ryde, S. J. (2001). Modelling consolidation accelerated by vertical drains in soils subject to creep. *Géotechnique* **51**, No. 3, 257–273.

Nicholson, D. P., Tse, C. M. & Penney, C. (1998). *The observational method in ground engineering: principles and applica-*

tions, CIRIA Report 185. London: Construction Industry Research and Information Association.

Nicholson, D. P. & Jardine, R. J. (1981). Field behaviour of vertical drains at Queenborough Bypass. *Géotechnique* **31**, No. 1, 67–90.

Peck, R. B. (1969). Ninth Rankine Lecture: Advantages and limitations of the observational method in applied soil mechanics. *Géotechnique* **19**, No. 2, 171–187.

Priebe, H. J. (1995). The design of vibro replacement. *Ground Engng* **28**, No. 12, 31–37.

Raison, C. A. (1996). Discussion. In *The observational method in geotechnical engineering*, pp. 175–179. London: Thomas Telford.

Renton-Rose, D., Bunce, G. C. & Finlay, D. W. (2000). Vibro-replacement for industrial plant on reclaimed land, Bahrain. *Géotechnique* **50**, No. 6, 727–737; Discussion. **52**, No. 10, 764.

Roberts, T. O. L. & Preene, M. (1994). The design of groundwater control systems using the observational method. *Géotechnique* **44**, No. 4, 727–734.

Rowe, R. K. (2002). Behaviour of reinforced embankments on soft rate-sensitive soils. *Géotechnique* **52**, No. 1, 29–40.

Sherwood, P. T. (1957). The stabilisation with cement of weathered and sulphate-bearing clays. *Géotechnique* **7**, No. 4, 179–191.

Slocombe, B. C., Bell, A. L. & Baez, J. L. (2000). The densification of granular soils using vibro methods. *Géotechnique* **50**, No. 6, 715–725.

Su, J. Q. & Wang, Z. (2003). The two-dimensional consolidation theory of electro-osmosis. *Géotechnique* **53**, No. 8, 759–763.

Teh, C. I. & Houlsby, G. T. (1991). An analytical study of the cone penetration test in clay. *Géotechnique* **41**, No. 1, 17–34.

Tomlinson, M. J. (1956). Telford and soil mechanics. *Géotechnique* **6**, No. 3, 99–105.

Tomlinson, M. J. & Wilson, D. M. (1973). Preloading of foundations by surcharge on filled ground. *Géotechnique* **23**, No. 1, 117–120.

Watson, G. H., Crooks, J. H. A., Williams, R. S. & Yam, C. C. (1984). Performance of pre-loaded and stage-loaded structures on soft soils in Trinidad. *Géotechnique* **34**, No. 2, 239–257.

Watts, K. S. (2000). *Specifying vibro stone columns*. Garston: Building Research Establishment.

Zornberg, J. G. (2002). Discrete framework for limit equilibrium analysis of fibre-reinforced soil. *Géotechnique* **52**, No. 8, 593–604.

Zdravkovic, L. & Carter, J. (2008). *Géotechnique* **58**, No. 5, 405–412 [**doi:** 10.1680/geot.2008.58.5.405]

Contributions to *Géotechnique* 1948–2008: Constitutive and numerical modelling

L. ZDRAVKOVIC* and J. CARTER†

A review of the first 60 years of *Géotechnique* publications shows clearly how the subject of soil mechanics has evolved. In terms of constitutive and numerical modelling of soil, early forms of numerical analysis involved hand calculations of ultimate states applying classical methods of analysis: limit equilibrium, limit analysis or stress field solutions. Consequently, the soil was considered to behave as a rigid plastic material, and to follow one of the two basic failure laws of classical soil mechanics, namely the Tresca or Mohr–Coulomb failure criteria. For assessing the deformation of structures, soil was normally considered to be linear elastic. The foundations of modern numerical analysis and constitutive modelling were laid in the early to mid 1960s, with the development of the finite element method and the postulation of the critical state framework of soil behaviour respectively. Clearly, the continuous advancement of computer power has been essential in applying new developments to modern geotechnical analysis. This paper reviews some of the main milestones in the evolution of geotechnical analysis in the past 60 years, commenting, where appropriate, on what problems still lie ahead.

KEYWORDS: constitutive relations; historical review; numerical modelling and analysis; theoretical analysis

Un examen des 60 premières années de publications de *Géotechnique* démontre clairement l'évolution suivie par la discipline de la mécanique des sols. En ce qui concerne la modélisation constitutive et numérique des sols, les premières formes d'analyse numérique comportaient des calculs manuels d'états limites avec l'application de méthodes d'analyse classiques : équilibre limite, analyse limite ou solutions sur place pour les contraintes. C'est ainsi que l'on estimait que le sol se comportait comme une matière plastique rigide en suivant une des deux lois de base sur la rupture de la mécanique des sols classique, à savoir les critères de rupture de Tresca ou de Mohr–Coulomb. Pour l'évaluation de la déformation des structures, le sol était normalement considéré comme étant élastique linéaire. C'est vers le début ou la moitié des années soixante que l'on posa les fondations de l'analyse numérique et de la modélisation constitutive modernes, avec le développement de la méthode aux éléments finis et la postulation du cadre de l'état critique respectivement dans le comportement des sols. Le progrès continu de l'informatique a joué manifestement un rôle essentiel dans l'application de nouveaux développements dans l'analyse géotechnique moderne. La présente communication passe en revue les principales étapes de l'évolution de l'analyse géotechnique au cours des 60 dernières années, en se penchant, le cas échéant, sur les problèmes que l'on devra affronter dans l'avenir.

THE BEGINNINGS

From reviewing *Géotechnique* issues published in the past 60 years, it is evident that the first 10 years of the journal reported mainly case studies, ground investigations, and the development of experimental techniques. The first numerical procedure reported in *Géotechnique* was Bishop's method of slices calculation for slope stability, with a circular slip surface (Bishop, 1955). After the method of Fellenius developed 20 years earlier, this was the starting point for the development of future similar procedures that are still used in engineering practice. Any calculations at that time were performed by hand, and were therefore time consuming. The first account of this, as well as of the first use of a computer, is reported in relation to Bishop's method of slices by Little & Price (1958), who state that for 200 circles, each with about 30 slices, for a 150 ft high dam 'a team of four (people) took more than 4 weeks to complete the calculation.' 'With the powerful new weapon of the electronic computer' it took less than half an hour for all 200 circles, and the cost of the analysis per circle was reduced from £4 to 2 shillings (10 pence in new money)! At the same time, however, the authors warn that with any future use of computers 'it will be necessary to guard against over-enthusiasm for complicated methods of analysis where there is a danger that limitations of accuracy in measuring soils and other properties will be forgotten.' A far-reaching message that, unfortunately, still applies!

In terms of constitutive modelling, early accounts of the analyses of boundary value problems considered the soil to be either linear elastic or rigid plastic, following the basic failure criteria of Tresca or Mohr–Coulomb. The first realistic framework that had an enormous influence on the development of modern constitutive modelling, and which continues to do so today, is the critical state framework, introduced by Roscoe *et al.* (1958). This paper shows the analysis of a consistent set of clay data, from both drained and undrained experiments, that forms a framework for the yielding of clays. For the first time in *Géotechnique* we read about the concept of the 'critical voids ratio', which answers the authors' question as to whether the loading path ends at any specific point. The paper also shows an isometric view of the yield surface for Weald clay, which combines Hvorslev and what will later become Roscoe surfaces. Experiments by Henkel (1960) on the same clay show further evidence of the unique relationship between the mean effective stress, shear strength and water content at failure, contributing further to the validity of the critical state concept. This framework is still the most dominant one in constitutive modelling of soils.

In terms of numerical analysis, the early 1960s saw the development of finite difference procedures to look at consolidation and seepage through elastic soil (e.g. Gibson, 1958; Tomlin, 1966). Numerical analysis of geotechnical

Discussion on this paper closes on 1 December 2008, for further details see p. ii.

* Imperial College London, UK.
† The University of Newcastle, Australia.

problems started to develop with the development of finite element techniques, first reported in *Géotechnique* by Zienkiewicz *et al.* (1968). Having published his first book on finite elements in 1967, in this paper for the first time a discretisation of a soil domain is shown, applying triangular finite elements. The aim was to determine zones of tensile principal stress in an elastic rock mass around tunnel excavations or in foundations underneath a dam. As the material cannot take tension, the paper describes a possible method for modelling stress transfer from zones where tension occurs. Dealing with 'no tension' in soils is still not straightforward, and current computer programs employ different procedures and conditions to account for tension.

Other solutions of geotechnical problems in this period have mainly considered the soil to be linear elastic. Typically, parametric analyses were performed, on the basis of which suitable design charts or interaction diagrams were produced to aid engineering design. For example, Milovic *et al.* (1970), using the finite element method, produced charts for stress distributions in an isotropic elastic medium underneath rigid and rough strip footings, due to an applied inclined load on the footing at a certain eccentricity. Poulos & Davis (1968) and Poulos (1968) produced comprehensive studies on settlement of a single incompressible pile and then on settlement interaction between two identical piles and pile groups in an elastic soil, utilising Mindlin's equations. Results were presented in terms of charts that plot load distribution in a single pile, load and settlement distribution between the piles, all with respect to pile length to diameter ratio and for different values of the Poisson's ratio. These studies formed the basis of subsequent developments in pile interaction analyses, some of which are in use even today.

Another important advance in the application of the linear elastic soil model to foundation problems was first published by Gibson (1967), with a later, more comprehensive, summary provided in his Rankine Lecture (Gibson, 1974). In this body of work the soil, while being represented as an elastic continuum, had the important distinction of an elastic modulus that varied linearly with depth below the soil surface. In this way an attempt was made to provide a more realistic model of natural soils, where this type of depth effect is well known, and where it is generally acknowledged that soil stiffness does increase with depth. In particular, it was shown by Gibson that for the case of an incompressible half space the loaded surface settles by an amount proportional to the local intensity of applied pressure, where the constant of proportionality (effectively the coefficient of subgrade reaction) is independent of the size or shape of the loaded area and equal numerically to twice the rate of increase of Young's modulus with depth. While it is likely that this work was initially inspired by the desire to provide more reliable predictions of the behaviour of foundations on soils such as London Clay, it is worth noting that the model has much wider application.

Clearly, in any numerical analysis, such as the finite element method, a linear elastic behaviour does not require special incremental solution procedures. The stress–strain relationship is linear, and therefore the stiffness matrix in the system of equations does not change over a load step. This is not the case when soil behaviour is considered to be non-linear elasto-plastic. In such a situation the stiffness matrix varies along the loading path, and the solution of the problem must be determined in an incremental manner. For the first time in *Géotechnique* a solution of an elasto-plastic problem was reported by Smith (1970), who used the initial strain method for incremental solution. The problem analysed was a thick cylinder loaded uniformly over its inner perimeter, for which there is an analytical solution that is used for comparison with the numerical results. At the same time the research group at Swansea developed an elastic viscoplastic approach for soil behaviour, where classical soil plasticity was represented with viscoplastic, rather than purely plastic, strains. They also developed an algorithm for the incremental solution as a viscoplastic problem, and Zienkiewicz *et al.* (1975) presented a viscoplastic formulation of some constitutive models commonly used in geotechnical finite element analysis (Mohr–Coulomb and modified Cam Clay). They also showed analysis results for some boundary value problems simulated with these models, which were solved with a viscoplastic non-linear solution algorithm. However, at present, almost 30 years later, there is still no unified opinion as to which non-linear solver is the most efficient and accurate in solving geotechnical problems. Existing computer programs use either viscoplastic or tangent stiffness, or modified and full Newton–Raphson approaches.

Apart from constitutive models and solution procedures, the finite element method of analysis requires the truncation of a continuum to a domain of a manageable size for implementation in the analysis. Using examples of idealised slopes Smith & Hobbs (1974) investigated the effect of the size (i.e. distance to far boundaries) and coarseness of the mesh on analysis results. They concluded that the far boundaries should be sufficiently far away from the problem and that the mesh should be finer in the areas of high stress concentration. This advice still applies in non-linear elasto-plastic geotechnical analysis. The authors also performed plane-strain analyses of centrifuge model slopes, using the von Mises constitutive model, and obtained higher failure heights than those observed in experiments. Without explaining why, the authors state that 'if the Tresca criterion were used, rather than the von Mises, lower collapse loads would result.' A possible reason for this is that the von Mises circle is circumscribed around the Tresca hexagon in the deviatoric plane (in stress space), with the input strength being fixed to that in triaxial compression. As the problem involves failure under plane-strain conditions, the von Mises circle mobilises a higher strength than that given by the Tresca hexagon, and hence enables higher failure heights. It is important to note that this problem is not fully recognised by some numerical analysts even today. Although the von Mises and Drucker–Prager yield surfaces are easier for implementation in a finite element code, owing to the absence of corners, care must be taken when choosing input parameters. Further to this, Smith & Hobbs (1976) showed an extension of the finite element method to incorporate consolidation in the soil, coupling its mechanical and hydraulic behaviour. Biot's theory of consolidation was implemented in a finite element program and applied to the analysis of embankments on soft ground, where the predicted settlement of the foundation soil with time agreed well with observations. A further parametric study investigated the effect of embankment stiffness on pore pressures generated in the foundation soil. This development was an important step towards general application of numerical analysis in geotechnics, as most real situations involve transient soil behaviour, rather than purely drained or purely undrained conditions, which are possible to analyse without recourse to a coupled formulation.

MODERN CONSTITUTIVE MODELLING

Clays

In terms of constitutive models, after postulation of the critical state framework, the development of the most widely used critical state model, the modified Cam clay (MCC) model, in 1968 was not reported first in *Géotechnique*. However, several subsequent extensions to the MCC model reported in *Géotechnique* papers tried to improve the behaviour of the model below the yield surface, which is assumed to be elastic in the original model. It was recognised that real soils do not behave in this manner, and, for example, Pender (1978) made an assumption that undrained stress paths below yield are of a parabolic shape, therefore being non-linear instead of linear as in the original model. Van Eekelen & Potts (1978) introduced several improvements to the model:

(*a*) a cut-off surface on the dry side to reduce the mobilised peak strength of overconsolidated soils
(*b*) a Mohr–Coulomb yield surface in the deviatoric plane to better represent soil failure
(*c*) non-associated plasticity in the deviatoric plane, with circular plastic potential surface
(*d*) a single parameter cyclic model below the yield surface to enable modelling of cyclic processes.

In general, it is an established procedure even today to improve simple elasto-plastic constitutive models (of Mohr–Coulomb or MCC type) by coupling them with simple non-linear models, which then replace the elastic behaviour of the elasto-plastic model.

As laboratory testing became more advanced in the 1980s, more was understood about the small-strain stiffness behaviour of soils in the range of up to 1% strain. Simple models were developed to reproduce stiffness degradation from the initial elastic plateau. However, they did not have a firm theoretical framework, but were simple curve-fitting techniques that could be coupled with plastic models. In this category Simpson *et al.* (1979) presented a model for London Clay (the LC model) developed in strain space, which can account for stiffness variation with both stress and strain level. The authors demonstrated for the first time in *Géotechnique* the importance of modelling small-strain stiffness non-linearity in soil–structure interaction problems at serviceability limit states, by back-analysing some excavations in London Clay. Subsequently, Simpson (1992), talking about the design of retaining walls in his Rankine Lecture, presented his ideas of a 'brick' model, in which soil stiffness behaviour is represented as that of a brick on strings. The stiffness degradation curve was presented in a stepwise manner, each step corresponding to a string length (strain), and proportion of stiffness compared with the maximum one. Further to this, Jardine *et al.* (1986) presented a different empirical model for small-strain stiffness degradation with stress and strain level and applied it in the analysis of a number of boundary value problems (footing, strutted excavation, axially loaded pile). They demonstrated that results of soil–structure interaction problems using linear (instead of non-linear) elasticity can be misleading. All these developments led to a very positive outcome for geotechnical practice in the UK and elsewhere, as the necessity of modelling small-strain non-linearity at serviceability limit states is today very well understood, and designers are trying to incorporate these models in their finite element programs. The appeal of these simple models is that they need only a stiffness degradation curve for calibrating input parameters, and are then coupled with a suitable plastic model.

The next major step in the development of a sound constitutive framework was the introduction of kinematic yield surfaces below a critical state bounding surface, reported in *Géotechnique* first by Mroz *et al.* (1979). The bounding surface has the same elliptical shape as the MCC yield surface in the deviatoric stress–mean effective stress plane, but the behaviour below it is elasto-plastic, and this surface now presents a gross yield locus. A small kinematic surface, which is of the same shape as the bounding surface, encompasses truly elastic soil behaviour, and as the stress state engages this surface and moves on its way to the bounding surface, elasto-plastic behaviour is invoked. Both isotropic and kinematic hardening exist, and certain rules must be satisfied for the movement and interaction of these surfaces. With this concept, stress path non-linearity and plasticity, as well as small-strain stiffness degradation below gross yield, can be simulated. Although it is capable of simulating some of the most important aspects of real soil behaviour, the application of this modelling framework has become increasingly popular only in the past 10 years, as the importance of simulating small-strain behaviour became more appreciated. Many such model formulations appeared in *Géotechnique*, starting with Stallebrass & Taylor (1997), who presented a model with two kinematic surfaces: a yield surface, which is a boundary of elasticity; and a history surface. Simulating laboratory experiments on kaolin, the authors demonstrated the model's capability in dealing with the effects of recent stress history and stress path direction. Various further similar models introduced the effects of structure and de-structuration in soils (e.g. Kavvadas & Amorosi, 2000; Rouainia & Muir Wood, 2000; Baudet & Stallebrass, 2004). At present, the kinematic surface framework appears to be a way forward in the future constitutive modelling of clays, but a number of issues still have to be resolved for its use in successful modelling of overconsolidated clays.

Other aspects of clay behaviour, such as strength anisotropy, creep and partial saturation, have been addressed in a number of new models based on a critical state framework, in which the shape of the yield or the bounding surface is still the ellipse of the MCC model. Of these, arguably the first significant breakthrough in modelling unsaturated clays using a rigorous applied mechanics approach came from Alonso *et al.* (1990), with the formulation of the Barcelona Basic Model (BBM). This model utilises hardening plasticity and two independent stress variables: the excess of total stress over air pressure, and the suction. It accounts for the stiffness changes of the soil induced by suction change, reproduces the irreversible response of the soil on stress and suction reversals, and simulates collapsible soil behaviour. The model is relatively simple because of its need for only a modest amount of experimental data to determine the input parameters, and it is suitable for unsaturated soils of moderate to low plasticity. This was the starting framework for future similar model developments, for example those of Wheeler & Sivakumar (1995) and Bolzon *et al.* (1996). However, it should be noted that others, e.g. Khalili & Kabaz (1998), contend that the effective stress approach for unsaturated soils originally proposed by Bishop has some advantages over the use of two independent stress variables.

Whittle (1993) presented a critical state bounding surface plasticity model for clays, the MIT-E3 model, which can simulate non-linearity and plasticity below gross yield, similar to the kinematic surface models. In addition, one of the hardening rules of this model allows the rotation of the bounding surface in general stress space, and the model can therefore simulate both inherent and induced anisotropy of clays. This is arguably the most advanced, but also the most complex, model that can deal with this aspect of soil behaviour. Borja & Kavazanjian (1985) reported for the first time in *Géotechnique* a constitutive model for time-related soil behaviour. Here, the total strain consists of elastic and plastic time-independent strains, evaluated from the MCC yield surface, and time-dependent plastic strains, evaluated from creep rate expressions empirically derived from triaxial tests. However, it is the model of Yin & Graham (1996), which introduces the equivalent time concept, that makes a step forward in modelling creep. Although this paper showed model development for one-dimensional consolidation only (a complete model was published later, but not in *Géotechnique*), it assumed that the total strain consists of elastic and viscoplastic parts. The use of equivalent time allows the model to have stress–strain-equivalent time states independent of stress path (i.e. total strain rate is equal to creep strain rate). The model also introduces the limit time line, which helps to model soils that do not experience creep: that is, if the equivalent time is set to be very large (infinity), the creep rate will be equal to zero.

Sands

Modelling of sands was significantly advanced by the formal introduction of the state parameter concept by Been & Jefferis (1985), although the 'distance' between the state of the sample and the ultimate critical state was first mentioned in *Géotechnique* by Wroth & Bassett (1965). This is a fundamental physical concept for describing the behaviour of sands, as it combines the influence of void ratio and stress level with respect to the critical state conditions. A constitutive model that makes use of this concept for the first time in *Géotechnique* is that of Manzari & Dafalias (1997). This model consists of three conical surfaces that have the shape of a smooth triangle in the deviatoric plane: they are the dilatancy, critical state and bounding surfaces. It also includes a kinematic yield surface, which is a circular cone and which limits the elastic zone in the material. If the initial value of the state parameter is positive, this implies that the sand will dilate, passing through phase transformation upon reaching the dilatancy surface, engaging peak on the bounding surface, and then slowly approaching the critical state surface. If this value is negative, the sand will contract, failing along the critical state surface (which now coincides with the bounding surface) and never engaging the dilatancy surface. Because of the small elastic yield surface, the model also mobilises plasticity before the gross yield, and can simulate cyclic loading of sands. Several extensions of this model (not published in *Géotechnique*) improve its cyclic behaviour, in particular, but the main concept is maintained.

Modelling of granular materials has also been attempted through the Cosserat approach, which looks at the microstructure of granular materials within the continuum approach. An example of such an approach was published in *Géotechnique* by Muhlhaus & Vardoulakis (1987), who developed a two-dimensional Cosserat theory for a granular material consisting of rods of the same diameter, adopting aspects of slip and rotation between individual rods. To be able to predict the dimensions of the shear band, the grain size must be introduced into the constitutive model, which is not the case in continuum models. The results show that the width of the shear band depends only on the grain size, and not on the geometry of the soil body, which could be useful when trying to analyse progressive failure in soils in general.

Another, more widespread, development in modelling granular materials is the concept of distinct element modelling, where the behaviour of individual grains and intergranular contacts is introduced, instead of treating the soil as a continuum. The first such application was presented by Cundal & Strack (1979), who described a numerical procedure for simulating the behaviour of assemblies of discs and spheres. The method utilises an explicit solution scheme, where the motion of particles is described by a dynamic equation, taking account of the inertia of the discs. The computer program BALL was written, and numerical results were compared with photoelastic tests on an assembly of discs, showing good agreement of force chains through particle contacts between numerical and experimental results. Thornton (1979), using discrete element modelling, examined the strength of a close-packed regular (face-centred cubic) array of uniform rigid spheres, allowing for the first time three degrees of rotational freedom for the system of particles. This work showed agreement with Lade's experimental failure envelopes for sand.

Clearly, modelling grains as circular discs and spheres is quite idealistic, and attempts have since been made to account for more realistic grain shapes. Among these, Lin & Ng (1997) presented three-dimensional discrete element modelling using random arrays of elastic ellipsoidal particles, with a new numerical algorithm for contact detection of these particles. By comparing the results from single-sized spherical and single-sized ellipsoidal particles, they were able to reproduce some important effects of particle shape: ellipsoidal particles achieve lower porosity under the same consolidation procedure; in triaxial compression they also achieve higher strength, larger initial modulus, more dilation and less particle rotation, which are all aspects observed when comparing experimental results from rounded and angular sand grains. Latham *et al.* (2001) went even further, developing tetrahedral particle shapes, with which they were able to investigate the effects of particle orientation, as this influences the interlocking potential of the granular structure. Finally, Cheng *et al.* (2003) described the use of distinct element modelling in simulating the behaviour of crushable materials, utilising spherical particles. This is achieved by modelling a grain as an agglomerate of bonded spheres.

The main use of distinct element modelling has so far been in simulating laboratory experiments and understanding material response from the micromechanical point of view (e.g. Thornton, 2000; Cui & O'Sullivan, 2006; Zhang & Thornton, 2007).

All constitutive model developments reviewed so far have been performed within the concepts of classical plasticity. *Géotechnique* also reported the work of Collins & Kelly (2002), who applied the principles of thermomechanics in developing a systematic procedure for deriving constitutive models for soils. The paper develops a thermodynamically consistent set of variables that describe the state of the soil. It shows the formulation of plasticity models using fundamental physical concepts of work, energy and dissipation. Analysing a number of basic well-known models, such as the Drucker–Prager and MCC models, the paper highlights their shortcomings in terms of satisfying basic thermomechanical principles, and proposes a family of new models where these principles are satisfied. Although this is not yet a widespread framework for modelling soils, it is fundamentally consistent, and could prove more applicable in the future.

MODERN NUMERICAL ANALYSIS

After the initial, mainly elastic, numerical analyses of geotechnical problems, the 1980s saw much greater application of numerical analysis to real geotechnical problems. These applications range from investigating the behaviour of laboratory and field experiments to understanding the mechanisms of behaviour of real geotechnical structures.

Laboratory and field applications

In terms of laboratory experiments, the work of Potts *et al.* (1987) reported on the use of finite element analysis in understanding stress and strain non-uniformities in the material tested in a direct shear box apparatus and what effect these have on the interpretation of shear box experiments. Arroyo *et al.* (2006) showed three-dimensional finite difference analyses of axial wave propagation through a triaxial soil sample, thus simulating bender element tests for measuring elastic soil stiffness. The results have significant implications for laboratory testing, as they indicate that the size of the sample affects the propagating signal owing to reflections from the lateral boundaries of the sample. Errors in shear velocity estimates could therefore be introduced by the slender geometry of a sample, high wavelengths, reflecting boundaries and low material attenuation. The implication of the results is that certain experimental details need to be carefully considered before bender element testing can be taken into consideration with more certainty.

The trapdoor problem is often considered suitable for investigating problems that involve active or passive failure in the soil, such as arching around tunnels in the former, or the uplift force of buried anchors in the latter case. Koutsabeloulis & Griffiths (1989), using a non-associated Mohr–Coulomb model, investigated the two modes of trapdoor failure in both plane-strain and axisymmetric conditions and presented the results in the form of non-dimensional influence charts of a range of soil properties, trapdoor sizes and layer thickness. Sloan *et al.* (1990), using finite element limit analysis (i.e. upper and lower bounds), investigated the undrained stability of a trapdoor problem (i.e. Tresca soil), bracketing the exact solutions to within 10% or better for a range of trapdoor geometries. This is the first application of this hybrid method of analysis reported in *Géotechnique*.

Teh & Houlsby (1991), utilising the strain path method and large-displacement finite element analysis, investigated the quasi-static penetration of a cone penetrometer into clay, which was modelled as a von Mises soil. The findings emphasised the primary importance of the soil rigidity index I_r and the horizontal stress in the ground in influencing the cone factor N_{kt}. The authors therefore proposed an equation relating N_{kt} to soil parameters and cone roughness. Lu *et al.* (2004) investigated cone penetration into a Tresca soil using large-displacement finite element analysis, but introducing a novel approach of 'arbitrary Lagrangian–Eulerian' large-displacement analysis. An expression for the cone factor N_{kt} was also proposed in terms of soil rigidity, cone roughness and stresses in the ground, which compared well with the previous proposal of Teh & Houlsby, giving confidence in its application in practice. In a similar way, Houlsby & Carter (1993) made use of a parametric FE study to investigate the effects of the geometry of the self-boring pressuremeter on the interpretation of this field test. The effects of length-to-diameter ratio (L/D) and penetration depth-to-diameter ratio (H/D) were examined, together with the soil rigidity index. The results were presented as a chart of correction factors for strengths derived from pressuremeter tests, which can be applied in practice.

Shallow foundations

In terms of real geotechnical problems, the behaviour of foundations, both shallow and deep, has been extensively studied and reported in *Géotechnique*. The bearing capacity of shallow foundations is normally calculated in practice using Terzaghi's bearing capacity equation, which takes account of footing geometry, embedment depth and load inclination. However, most of the load factors in this equation are empirical, and several different expressions for each of these have been adopted in practice over the years. Griffiths (1982) showed how the FE method can be applied for obtaining theoretically accurate values of bearing capacity factors N_c, N_q and N_γ, utilising the Mohr–Coulomb constitutive model. It was shown that the numerical values are closer to the lower empirical values, and also that N_γ depends on the footing roughness. More recently, using the FE limit analysis approach, Salgado *et al.* (2004) performed a comprehensive study of the undrained bearing capacity of general foundation shapes. Full three-dimensional analyses of embedded square, circular and rectangular foundations, as well as two-dimensional analyses of embedded strip foundations were performed, with proposed finite values of shape and depth factors. Embedment depths of five times the footing width/diameter (B) were considered. The maximum difference between upper- and lower-bound solutions was not more than 12%, except for the circular footing with the maximum embedment depth of $5B$, for which it reached 25%. Consequently, the solutions can be taken as sufficiently accurate for engineering practice. This approach was recently extended to footings on sand (Lyamin *et al.*, 2007), and a different form of the bearing capacity equation was proposed that does not rely on the traditional approximation that the effects of foundation shape and depth can be considered separately for each of the effects of soil self-weight and surcharge (embedment).

Since the early 1990s, foundation design has seen a move from the classical bearing capacity equation towards the consideration of individual loading components (i.e. horizontal force H, vertical force V and moment M). This has seen the formulation of a three-dimensional loading surface in H–V–M stress space that is a boundary to possible and impossible loading combinations for a particular footing. The first numerical evidence of such formulation published in *Géotechnique* is that of Bransby & Randolph (1998), who showed the results and mechanisms from general undrained loading of a surface strip footing on a Tresca soil, which allowed tension to develop between the footing and the soil. The authors proposed equations for various cross-sections of the loading surface that can be directly applied in practice. In a similar study, Taiebat & Carter (2000) investigated failure envelopes for general undrained loading of a surface circular footing on a Tresca soil, which also mobilised full tension in the interface with the soil.

Martin & Houlsby (2001) took the concept of a three-dimensional loading surface to formulate a macro model of a footing and soil. They described the load–displacement response of a spudcan footing on clay as a plasticity model with three degrees of freedom (vertical, rotational and horizontal). This model, named 'model B', has a three-dimensional loading surface as a yield surface, as well as a flow rule, both derived from laboratory experiments of a model footing on kaolin. The behaviour inside the yield surface is described by elastic stiffness factors determined from FE analyses, while upon reaching the yield surface the model can harden or soften according to a hardening rule. The idea is that such a macro model can be used in a simplified structural analysis of a jack-up rig, for example, where the footing and the soil at the end of jack-up legs can be represented with this macro model. Houlsby & Cassidy (2002) developed a similar concept, called 'model C', for footings on sand.

Adopting a similar approach, Zdravkovic *et al.* (2001) investigated the pull-out capacity of suction caissons in soft clay, utilising the advanced constitutive model MIT-E3, which can take account of soil strength anisotropy. Realistic anisotropy, from laboratory experiments in the hollow cylinder apparatus, was input into the model. Similar to the studies described above, the authors derived a predictive equation and interaction charts for determining the pull-out capacity of suction caissons. In addition, for the first time in *Géotechnique* the effect of soil strength anisotropy, in this case on the pull-out capacity, was demonstrated and quantified. It shows that the pull-out capacity of a suction caisson can be significantly overestimated if the soil is considered isotropic.

Piles

Investigations of piled foundations began by phenomenological investigations of pile installation effects on lateral stresses and pore water pressures in the ground. Randolph *et al.* (1979) used the FE method to simulate pile installation in clay as a cavity expansion problem, and Potts & Martins (1982) investigated pile loading by applying increments of vertical displacement along the boundary of the pile shaft. Clearly, these were the first attempts in looking at single pile behaviour, but piles are normally placed in a group, and the main issues become their interaction, and transfer of the load between individual piles and the pile cap. In recent years insight into these issues and guidance for practical design have been offered by Horikoshi & Randolph (1998) and then by Reul & Randolph (2003). In the former publication the authors investigated for the first time in *Géotechnique* the concept of piled raft foundations. They used a relatively simple approach of a plate on piles, utilising Mindlin's solution, and validated it against the case study from the centrifuge. The optimum design was established in terms of reducing differential settlements, with the following guidelines.

(*a*) Piles should be distributed over the central area of the raft.
(*b*) The stiffness of the pile group should be approximately equal to the stiffness of the raft alone.
(*c*) The total pile capacity should be designed for between 40% and 70% of the design load.

In the latter publication the authors performed three-dimensional FE back-analyses of well-instrumented piled raft foundations in overconsolidated clay. Having obtained good agreement between numerical analyses and field measurements, the authors performed further studies with altered pile distributions below the raft, confirming that the central distribution of piles below the raft reduces both the settlement and the differential settlement, as well as increases the raft's participation in carrying the load.

Retaining structures

The finite element method has been used to examine conventional retaining wall design in the work of Potts & Fourier (1984), showing, among other things, that classical limit equilibrium design underestimates structural forces for excavations in front of a single propped retaining wall in high-K_0 soils (such as London Clay). In a similar way, Powrie & Li (1991) investigated a type of retaining wall that is propped at formation level, in terms of soil/wall/prop stiffness and initial K_0 in the soil. *Géotechnique* has also reported work on three-dimensional modelling of the diaphragm wall panel installation procedure, where the soil is first excavated for the panel, and then bentonite pressure is applied on the excavation surfaces before concreting (Gourvenec & Powrie, 1999; Ng & Yan, 1999). The effect of panel length on horizontal ground movements and changes in stress distribution in the ground was investigated. More recently, Zdravkovic *et al.* (2005) investigated modelling issues related to a retaining wall in a three-dimensional analysis. The authors demonstrated that the current approach of modelling the wall with isotropic properties in all coordinate directions is not realistic, and predicts ground and wall movements much smaller than those that are likely to occur in reality. These walls normally cannot transmit any substantial bending, and transmit only very small axial force in the horizontal direction. They also investigated issues of rectangular excavations being analysed as plane-strain or axisymmetric problems, and how such solutions apply to a general three-dimensional geometry.

Tunnelling

In terms of tunnelling, the main issue has been the prediction of realistic settlement troughs above tunnels, in particular for high K_0 soils. Lee & Rowe (1989) reported that elastic isotropic models predict settlement troughs that are too shallow and too wide. Using the FE method for simulating tunnel excavation, their results indicate that elastic soil anisotropy should be taken into account and that, in particular, care should be given to selection of the ratio of G_{vh} to E_v. Franzius *et al.* (2005) investigated this further in three-dimensional analyses of tunnel excavations in London Clay (a high-K_0 material), utilising non-linear elastic small-strain anisotropy and comparing the results for the transversal settlement trough in a cross-section along the tunnel with the results from a plane-strain analysis. The authors showed that three-dimensional modelling has negligible effects on the transverse surface settlement trough, but that longitudinal settlements do not develop a steady state. Transverse settlements improved with the use of an unrealistically high degree of stiffness anisotropy, but longitudinal settlements reached steady state only when, in addition to this, a low K_0 was introduced in the vicinity of the tunnel. However, both assumptions were unreasonable for London Clay, and resulted in the analysis predicting a volume loss that was too high. In general, simulation of tunnelling in high-K_0 soils still needs further investigation, and is a headache for many analysts. Shin *et al.* (2002) proposed a novel method for modelling the coupled behaviour of a tunnel lining, in an attempt to investigate the effects of tunnel lining permeability on ground movements and on load distribution on the lining. The method involves coupling of beam elements (to represent the structural properties of the lining) and thin solid elements (on the soil side of the lining, to represent its hydraulic properties). Cases of high- and low-permeability soils were considered, with flow regimes that either do or do not draw down the phreatic surface, and in which tunnel lining is either fully permeable or impermeable. The results showed significant interaction between the soil and lining permeabilities and, consequently, a significant effect on ground movements and loads in the lining in the long term, agreeing well with limited observations from low-permeability soils such as London Clay. Schroeder *et al.* (2004) looked at another practical problem of the effect of new piles, installed around existing tunnels, on tunnel loading and deformation. This is a very common problem in cities such as London, where new developments have to conform to very strict rules imposed by London Underground Ltd for their infrastructure tunnels. The authors produced design guidelines that relate tunnel deformation to the minimum distance between the first row of piles and the existing tunnel.

Embankments and slopes

The role of progressive failure in general slope stability was quantified for the first time in *Géotechnique* in the paper by Potts *et al.* (1990), which described the failure of Carsington dam in the UK. Classical limit equilibrium analysis of this dam showed a sufficient factor of safety for its stability. However, the dam failed during construction, and finite element analysis performed by the authors, utilising a non-linear strain-softening elasto-plastic material model, revealed that progressive failure of a brittle clay layer in the dam foundations was the cause of dam failure. Potts *et al.* (1997) applied a similar method of analysis to explain the delayed collapse of cut slopes in stiff clays, with particular reference to London Clay. By considering various typical heights and inclinations of slopes, and also surface hydraulic boundary conditions, the authors were able to predict the time to failure. The findings from this paper have significant implications for the stability of cut slopes in London Clay along motorways and railway lines.

For embankments on soft, normally consolidated soil, the issue for design is often the choice of undrained strength for the foundation soil. As the potential slip surface engages stress states ranging from plane-strain compression through simple shear to plane-strain extension, design practice often recommends that a simple shear strength be taken for calculating failure height. Zdravkovic *et al.* (2002) investigated this issue for the case study of a real embankment brought to failure, using finite element analysis and an anisotropic soil constitutive model, MIT-E3. By comparing the results from isotropic and anisotropic analyses, the authors demonstrated that the back-analysed soil strength at embankment failure, using an isotropic constitutive model, is not necessarily applicable for further design or redesign of an embankment if the soil is highly anisotropic. The new design can be largely conservative or non-conservative, depending on the shape and size of the new embankment. Consequently, if the isotropic soil strength has to be used, which is most likely in a design office scenario, it needs to be averaged in a certain way.

CONCLUSIONS

This review of numerical analysis and constitutive model developments published in *Géotechnique* has revealed several important points.

(*a*) Numerical analysis and constitutive modelling clearly involve significant use of mathematical techniques. Despite this, many major geotechnical developments in this area have been published in *Géotechnique* and not in numerical methods journals.

(*b*) It is evident that the development of constitutive models has advanced significantly more than their application in routine numerical analysis. The most advanced constitutive models (such as MIT-E3, kinematic surface models, and unsaturated models) exist only in research software, and are still a long way from being used commercially. One of the reasons for this is that they require significantly more input parameters, which in turn require special testing, and are therefore not suitable for everyday design.

(*c*) It is also evident that, in general, the numerical analyses of geotechnical problems presented in *Géotechnique* have almost always investigated practical issues, and have aimed to provide guidance and improved methods for design. Modern numerical analysis has certainly enabled geotechnical practice to understand mechanisms of soil behaviour underneath structures and construction processes that are of increasing complexity.

Potts (2003), in his Rankine lecture, posed a very pertinent question about what is at present the real role of numerical analysis and constitutive modelling in geotechnical practice. Through various examples he demonstrated the advantages and pitfalls of using numerical analysis, recognising that numerical analysis will increasingly become the preferred design tool in the future. However, going back to the warning message quoted at the beginning of this review, Potts also warned that useful numerical analysis can be performed only if one fully understands both the numerical tool and the real soil behaviour!

REFERENCES

Alonso, E. E., Gens, A. & Josa A. (1990). A constitutive model for partially saturated soils. *Géotechnique* **40**, No. 3, 405–430.

Arroyo, M., Muir Wood, D., Greening, P. D., Medinas, L. & Rio, J. (2006). Effects of sample size on bender-based axial G_0 measurements. *Géotechnique* **56**, No. 1, 39–52.

Baudet, B. & Stallebrass, S. (2004). A constitutive model for structured clays. *Géotechnique* **54**, No. 4, 269–278.

Been, K. & Jefferies, M. G. (1985). A state parameter for sands. *Géotechnique* **35**, No. 2, 99–112.

Bishop, A. W. (1955). The use of the slip circle in the stability analysis of slopes. *Géotechnique* **5**, No. 1, 7–17.

Bolzon, G., Schrefler, B. A. & Zienkiewicz, O. C. (1996). Elastoplastic soil constitutive laws generalised to partially saturated states. *Géotechnique* **46**, No. 2, 279–289.

Borja, R. I. & Kavazanjian, E. (1985). A constitutive model for the stress–strain-time behaviour of 'wet' clays. *Géotechnique* **35**, No. 3, 283–298.

Bransby, M. F. & Randolph, M. F. (1998). Combined loading of skirted foundations. *Géotechnique* **48**, No. 5, 637–655.

Cheng, Y. P., Nakata, Y. & Bolton, M. D. (2003). Discrete element simulation of crushable soil. *Géotechnique* **53**, No. 7, 633–641.

Collins, I. F. & Kelly, P. A. (2002). A thermomechanical analysis of a family of soil models. *Géotechnique* **52**, No. 7, 507–518.

Cui, L. & O'Sullivan, C. (2006). Exploring the macro- and micro-scale response of an idealised granular material in the direct shear apparatus. *Géotechnique* **56**, No. 7, 455–468.

Cundal, P. A. & Strack, O. D. L. (1979). A discrete numerical model for granular assemblies. *Géotechnique* **29**, No. 1, 47–65.

Franzius, J. N., Potts, D. M. & Burland, J. B. (2005). The influence of soil anisotropy and K_0 on ground surface movements resulting from tunnel excavation. *Géotechnique* **55**, No. 3, 189–199.

Gibson, R. E. (1958). The progress of consolidation in a clay layer increasing in thickness with time. *Géotechnique* **8**, No. 4, 171–182.

Gibson, R. E. (1967). Some results concerning displacements and stresses in a non-homogeneous elastic half-space. *Géotechnique* **17**, No. 1, 58–67.

Gibson, R. E. (1974). The analytical method in soil mechanics. *Géotechnique* **24**, No. 2, 115–140.

Griffiths, D. V. (1982). Computation of bearing capacity factors using finite elements. *Géotechnique* **32**, No. 3, 195–202.

Gourvenec, S. M. & Powrie, W. (1999). Three-dimensional finite-element analysis of diaphragm wall installation. *Géotechnique* **49**, No. 6, 801–823.

Henkel, D. J. (1960). The relationships between the effective stresses and water content in saturated clays. *Géotechnique* **10**, No. 2, 41–54.

Horikoshi, K. & Randolph, M. F. (1998). A contribution to optimum design of piled rafts. *Géotechnique* **48**, No. 3, 301–317.

Houlsby, G. T. & Carter, J. P. (1993). The effects of pressuremeter geometry on the results of tests in clay. *Géotechnique* **43**, No. 4, 567–576.

Houlsby, G. T. & Cassidy, M. J. (2002). A plasticity model for the behaviour of footings on sand under combined loading. *Géotechnique* **52**, No. 2, 117–129.

Jardine, R. J., Potts, D. M., Fourier, A. B. & Burland, J. B. (1986). Studies of the influence of non-linear stress–strain characteristics in soil–structure interaction. *Géotechnique* **36**, No. 3, 377–396.

Kavvadas, M. & Amorosi, A. (2000). A constitutive model for structured soils. *Géotechnique* **50**, No. 3, 263–273.

Khalili, N. & Khabbaz, M. H. (1998). A unique relationship for the determination strength of unsaturated soils. *Géotechnique* **48**, No. 5, 681–687.

Koutsabeloulis, N. C. & Griffiths, D. V. (1989). Numerical modelling of the trap door problem. *Géotechnique* **39**, No. 1, 77–89.

Latham, J.-P., Lu, Y. & Munjiza, A. (2001). A random method for simulating loose packs of angular particles using tetrahedra. *Géotechnique* **51**, No. 10, 871–879.

Lee, K. M. & Rowe, R. K. (1989). Deformations caused by surface loading and tunnelling: the role of elastic anisotropy. *Géotechnique* **39**, No. 1, 125–140.

Lin, X. & Ng, T.-T. (1997). A three-dimensional discrete element model using arrays of ellipsoids. *Géotechnique* **47**, No. 2, 319–329.

Little, A. L. & Price, V. E. (1958). The use of an electronic computer for slope stability analysis. *Géotechnique* **8**, No. 3, 113–120.

Lu, Q., Randolph, M. F., Hu, Y. & Bugarski, I. C. (2004). A numerical study of cone penetration in clay. *Géotechnique* **54**, No. 4, 257–267.

Lyamin, A. V., Salgado, R., Sloan, S. W. & Prezzi, M. (2007). Two- and three-dimensional bearing capacity of footings in sand. *Géotechnique* **57**, No. 8, 647–662.

Manzari, M. T. & Dafalias, Y. F. (1997). A critical state two-surface plasticity model for sands. *Géotechnique* **47**, No. 2, 255–272.

Martin, C. M. & Houlsby, G. T. (2001). Combined loading of spudcan foundations on clay: numerical modelling. *Géotechnique* **51**, No. 8, 687–699.

Milovic, D. M., Touzot, G. & Tournier, J. P. (1970). Stresses and displacements in an elastic layer due to inclined and eccentric load over a rigid strip. *Géotechnique* **20**, No. 3, 231–252.

Mroz, Z., Norris, V. A. & Zienkiewicz, O. C. (1979). Application of an anisotropic hardening model in the analysis of elasto-plastic deformation of soils. *Géotechnique* **29**, No. 1, 1–34.

Muhlhaus, H.-B. & Vardoulakis, I. (1987). The thickness of shear bands in granular materials. *Géotechnique* **37**, No. 3, 271–283.

Ng, C. W. W. & Yan, R. W. M. (1999). Three-dimensional modelling of a diaphragm wall construction sequence. *Géotechnique* **49**, No. 6, 825–834.

Pender, M. J. (1978). A model for the behaviour of overconsolidated soil. *Géotechnique* **28**, No. 1, 1–25.

Potts, D. M. (2003). Numerical analysis: a virtual dream or practical reality? *Géotechnique* **53**, No. 6, 535–573.

Potts, D. M. & Fourier, A. B. (1984). The behaviour of a propped retaining wall: results of a numerical experiment. *Géotechnique* **34**, No. 3, 383–404.

Potts, D. M. & Martins, J. P. (1982). The shaft resistance of axially loaded piles in clay. *Géotechnique* **32**, No. 4, 369–386.

Potts, D. M., Dounias, G. T. & Vaughan, P. R. (1987). Finite element analysis of the direct shear box test. *Géotechnique* **37**, No. 1, 11–23.

Potts, D. M., Dounias, G. T. & Vaughan, P. R. (1990). Finite element analysis of progressive failure of Carsington embankment. *Géotechnique* **40**, No. 1, 79–101.

Potts, D. M., Kovacevic, N. & Vaughan, P. R. (1997). Delayed collapse of cut slopes in stiff clay. *Géotechnique* **47**, No. 5, 953–982.

Poulos, H. G. (1968). Analysis of settlement of pile groups. *Géotechnique* **18**, No. 4, 449–471.

Poulos, H. G. & Davis, E. H. (1968). The settlement behaviour of single axially loaded incompressible piles and piers. *Géotechnique* **18**, No. 3, 351–371.

Powrie, W. & Li, E. S. F. (1991). Finite element analysis of an in situ wall propped at formation level. *Géotechnique* **41**, No. 4, 499–514.

Randolph, M. F., Carter, J. P. & Wroth, C. P. (1979). Driven piles in clay: the effect of installation and subsequent consolidation. *Géotechnique* **29**, No. 4, 361–393.

Reul, O. & Randolph, M. (2003). Piled rafts in overconsolidated clay: comparison of in situ measurements and numerical analyses. *Géotechnique* **53**, No. 3, 301–315.

Roscoe, K. H., Schofield, A. N. & Wroth, C. P. (1958). On the yielding of soils. *Géotechnique* **8**, No. 1, 22–53.

Rouainia, M. & Muir Wood, D. (2000). A kinematic hardening constitutive model for natural clays with loss of structure. *Géotechnique* **50**, No. 2, 153–164.

Salgado, R., Lyamin, A. V., Sloan, S. W. & Yu, H. S. (2004). Two- and three-dimensional bearing capacity of foundations in clay. *Géotechnique* **54**, No. 5, 297–306.

Schroeder, F. C., Potts, D. M. & Addenbrooke, T. I. (2004). The influence of pile group loading on existing tunnels. *Géotechnique* **54**, No. 6, 351–362.

Shin, J. H., Addenbrooke, T. I. & Potts, D. M. (2002). A numerical study of the effect of groundwater movement on long-term tunnel behaviour. *Géotechnique* **52**, No. 6, 391–403.

Simpson, B. (1992). Retaining structures: displacement and design. *Géotechnique* **42**, No. 4, 541–576.

Simpson, B., O'Riordan, N. J. & Croft, D. D. (1979). A computer model for the analysis of ground movements in London Clay. *Géotechnique* **29**, No. 2, 149–175.

Sloan, S. W., Assadi, A. & Purushothaman, N. (1990). Undrained stability of a trapdoor. *Géotechnique* **40**, No. 1, 45–62.

Smith, I. M. (1970). Incremental numerical solution of a simple deformation problem in soil mechanics. *Géotechnique* **20**, No. 4, 357–372.

Smith, I. M. & Hobbs, R. (1974). Finite element analysis of centrifuge and built-up slopes. *Géotechnique* **24**, No. 4, 531–559.

Smith, I. M. & Hobbs, R. (1976). Biot analysis of consolidation beneath embankments. *Géotechnique* **26**, No. 1, 149–171.

Stallebrass, S. E. & Taylor, R. N. (1997). The development and evaluation of a constitutive model for the prediction of ground movements in overconsolidated clay. *Géotechnique* **47**, No. 2, 235–253.

Taiebat, H. A. & Carter, J. P. (2000). Numerical studies of the bearing capacity of shallow foundations on cohesive soil subjected to combined loading. *Géotechnique* **50**, No. 4, 409–418.

Teh, C. I. & Houlsby, G. T. (1991). An analytical study of the cone penetration test in clay. *Géotechnique* **41**, No. 1, 17–34.

Thornton, C. (1979). The conditions for failure of a face-centred cubic array of uniform rigid spheres. *Géotechnique* **29**, No. 4, 441–459.

Thornton, C. (2000). Numerical simulation of deviatoric shear deformation of granular media. *Géotechnique* **50**, No. 1, 43–53.

Tomlin, G. R. (1966). Seepage analysis through zoned anisotropic soils by computer. *Géotechnique* **16**, No. 3, 220–230.

Van Eekelen, H. A. M. & Potts, D. M. (1978). The behaviour of Drammen clay under cyclic loading. *Géotechnique* **28**, No. 2, 173–196.

Wheeler, S. J. & Sivakumar, V. (1995). An elasto-plastic critical state framework for unsaturated soil. *Géotechnique* **45**, No. 1, 35–53.

Whittle, A. J. (1993). Evaluation of constitutive model for overconsolidated clays. *Géotechnique* **43**, No. 2, 289–313.

Wroth, C. P. & Bassett, R. H. (1965). A stress–strain relationship for the shearing behaviour of a sand. *Géotechnique* **15**, No. 1, 32–56

Yin, J.-H. & Graham, J. (1996). Elastic visco-plastic modelling of one-dimensional consolidation. *Géotechnique* **46**, No. 3, 515–527.

Zdravkovic, L., Potts, D. M. & Jardine, R. J. (2001). A parametric study of the pull-out capacity of bucket foundations in soft clay. *Géotechnique* **51**, No. 1, 55–67.

Zdravkovic, L., Potts, D. M. & Hight, D. W. (2002). The effect of strength anisotropy on the behaviour of embankments on soft ground. *Géotechnique* **52**, No. 6, 447–457.

Zdravkovic, L., Potts, D. M. & St John, H. D. (2005). Modelling of a 3D excavation in finite element analysis. *Géotechnique* **55**, No. 7, 497–513.

Zhang, L. & Thornton, C. (2007). A numerical examination of the direct shear test. *Géotechnique* **57**, No. 4, 343–354.

Zienkiewicz, O. C., Humpheson, C. & Lewis, R. W. (1975). Associated and non-associated visco-plasticity in soil mechanics. *Géotechnique* **25**, No. 4, 671–689.

Zienkiewicz, O. C., Valliappan, S. & King, I. P. (1968). Stress analysis of rock as a 'no tension' material. *Géotechnique* **18**, No. 1, 56–66.

White, D. J. (2008). *Géotechnique* **58**, No. 5, 413–421 [**doi:** 10.1680/geot.2008.58.5.413]

Contributions to *Géotechnique* 1948–2008: Physical modelling

D. J. WHITE*

This paper reviews the major contributions to *Géotechnique* that relate to physical modelling, including developments in modelling technology, important experimental observations, and the resulting advances in geotechnical engineering. An increasing proportion of the papers published by this journal involve physical modelling, conducted either at 1*g* or in a geotechnical centrifuge. Over the 60 years since *Géotechnique* was first published, experimental techniques have advanced significantly, improving the realism of small-scale simulations, and raising the quality and detail of the measurements that can be made. These techniques are reviewed, and some of the consequent advances in relation to foundations, tunnels, retaining walls and slopes are highlighted, as reported in the pages of *Géotechnique*.

KEYWORDS: centrifuge modelling; historical review; model tests

La présente communication passe en revue les principales contributions à *Géotechnique* portant sur une modélisation physique, y compris des développements dans la technologie de la modélisation, d'importantes observations expérimentales ainsi que les progrès résultants en ingénierie géotechnique. Une proportion croissante de communications publiées dans le présent journal comporte une modélisation physique, effectuée à 1 g ou dans une centrifuge géotechnique. Au fil des 60 années qui se sont écoulées depuis le premier numéro de *Géotechnique*, les techniques expérimentales ont effectué des progrès considérables, en optimisant le réalisme de simulations à petite échelle, et en renforçant la qualité et le détail des mesures qui peuvent être effectuées. Ces techniques sont passées en revue, en mettant l'accent sur certains des progrès qui en découlent, sur le plan des fondations, des tunnels, des murs de soutènement et des pentes et talus, qui ont été reportés dans les pages de *Géotechnique*.

INTRODUCTION

Since the birth of *Géotechnique*, physical modelling has matured as an experimental technique relevant to geotechnical engineering. The key milestones of this development are described in the pages of *Géotechnique*, which has been chosen by many involved in physical modelling as the repository for their best work. In this paper, the major contributions to the development of geotechnical physical modelling are highlighted, and some of the resulting advances in the theory and practice of geotechnical engineering are described.

A total of about 200 papers, representing approximately 6% of the *Géotechnique* archive, are concerned primarily with physical modelling, and many others make reference to this body of work. However, during the first 20 years of *Géotechnique*, from 1948 to 1968, only ~10 papers described physical modelling: that is one every second year, representing 1–2% of the journal. Most of these early contributions describe model tests conducted in large tanks—generally of sand—which aimed to establish the forces on retaining walls and piles. These models were not intended to replicate any particular field-scale equivalent; they were aimed at understanding generic modes of behaviour.

In 1970, the Rankine Lecture delivered by Roscoe (1970) included a description of the 5 m radius geotechnical centrifuge that had recently been commissioned in Cambridge, UK—a machine described as 'terrifying' by de Josselin de Jong, in his vote of thanks. Roscoe showed how the progressive failure of a kaolin slope could be simulated in the centrifuge. Later that year Lyndon & Schofield (1970) published the results from a similar experiment conducted using the geotechnical centrifuge at UMIST in Manchester, using London Clay. Their post-failure cross-sectional view—drawn with the dimensions multiplied up to the field-scale equivalent—is at first sight indistinguishable from the many cross-sections of field scale slope failures found in the early volumes of this journal (Fig. 1). The challenge set out by Roscoe was that 'the only satisfactory way of truly modelling to scale a prototype problem, in which the self-weight of the soil is significant, is to use a centrifuge.'

Over the following 40 years around 90 papers on centrifuge modelling have been published in *Géotechnique*—23 in the past 5 years. Many early developments in centrifuge techniques took place in the UK, in Cambridge and Manchester. *Géotechnique* contains many of the key publications emerging from this work, together with numerous contributions from the international centrifuge modelling community.

However, Roscoe's intermediate clause—'in which the self-weight of the soil is significant'—should not be forgotten. Significant contributions to *Géotechnique* also include physical modelling of geo-environmental processes and small in situ testing tools, which can be simulated in conditions that replicate field-scale behaviour without the inconvenience of an inhospitable centrifuge environment. Furthermore, as described later, many important aspects of geotechnical behaviour have been elucidated through small-scale model tests conducted at 1*g*—taking advantage of the easier control of events compared with the centrifuge.

Two key developments have advanced the art of geotechnical physical modelling over the past 50 years. The development of the centrifuge in the 1970s allowed the realism of physical modelling to be enhanced, through the correct modelling of self-weight stresses. The subsequent development of miniaturised electronics and microcomputers has led to enhanced methods of data acquisition, control, and image analysis. The refinement of these techniques continues to yield dramatic improvements in the utility of physical modelling. More realistic simulations can be conducted, and more detailed observations can be gathered.

Discussion on this paper closes on 1 December 2008, for further details see p. ii.

* Centre for Offshore Foundation Systems, University of Western Australia.

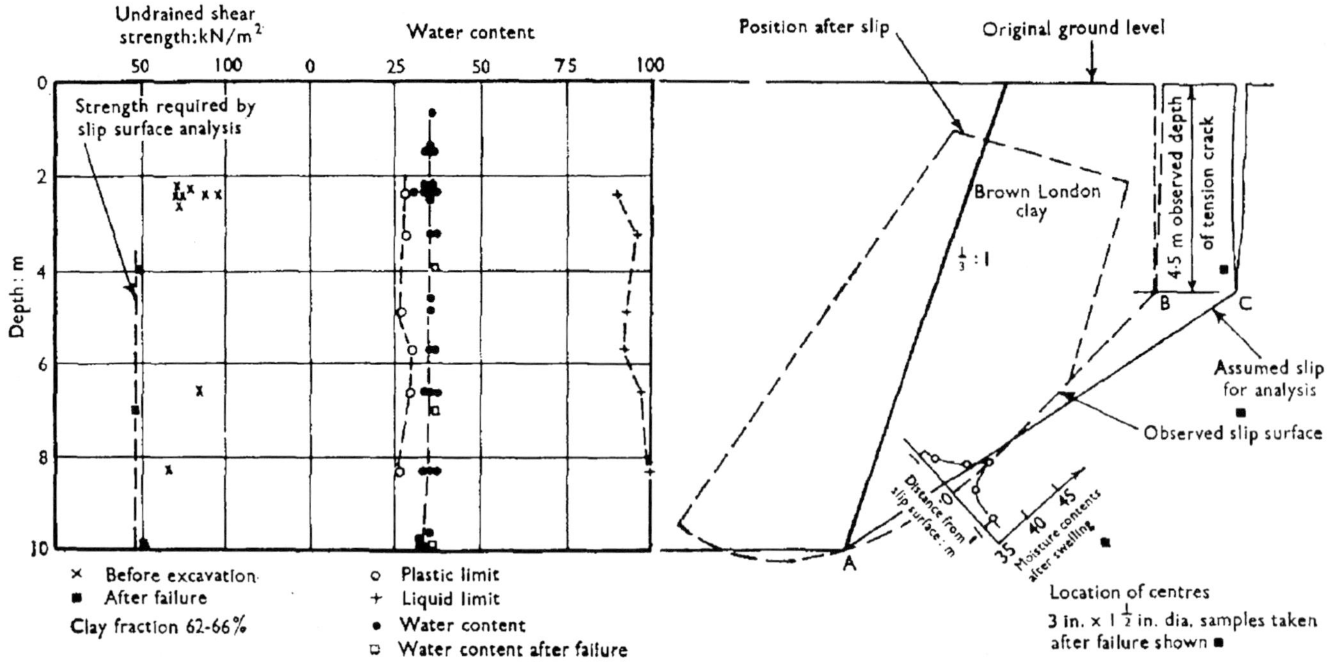

Fig. 1. Early physical modelling of slopes (Lyndon & Schofield, 1970)

PHYSICAL MODELLING TECHNIQUES

Geotechnical centrifuge development

Approximately half of the ~200 physical modelling contributions to *Géotechnique* make use of a centrifuge in order to ensure that the stress levels in the model are comparable to field-scale conditions. The majority of these papers describe research conducted in Cambridge or Manchester, in the groups led by Professor Andrew Schofield and Professor Peter Rowe respectively. Schofield and Rowe pioneered the use of the geotechnical centrifuge in Europe, in parallel with developments in Japan and following earlier work in the USSR, which was at that time unknown in the West. The idea of using a centrifuge to correctly model civil engineering structures in which self-weight forces are significant can be traced back to the French engineer Édouard Phillips in the nineteenth century, as described in *Géotechnique* by Craig (1989).

The earliest mention of centrifuge modelling in the pages of *Géotechnique* is the final section of Roscoe's Rankine Lecture, delivered in 1970. Despite leading a research group focused on the development of theoretical models for soil behaviour, he argued boldly that 'with the centrifuge it is possible to obtain answers immediately to full-scale problems without having to appeal to, or wait for the development of, any theory' (Roscoe, 1970). In a letter to *Géotechnique*, Golder (1971) relates a more light-hearted attempt to test soil using centrifugal force, which was conducted on the lawn outside the UK Building Research Establishment in 1936.

Rowe's Rankine Lecture, given in 1972, was concerned with the identification of soil fabric during site investigations, but concluded with a description of the second geotechnical centrifuge built in Manchester—at the (then) Victoria University of Manchester (Rowe, 1972). Unlike most physical modelling, Rowe's work relating to site-specific situations frequently involved using intact samples of natural soil, which were built into models placed within the centrifuge. This approach followed rationally from his conclusion that strength and consolidation testing of natural soil elements in the laboratory should be conducted in cells sufficiently large to accommodate representative amounts of the natural fabric—leading to the consolidation device now known as the 'Rowe cell'. Applying this logic to the centrifuge and his particular interest in earth embankment dams led him to design a machine sufficiently large to accommodate soil models that are 1 m × 2 m in plan.

Some of Rowe's most significant centrifuge work, conducted with Craig, contributed to the development of the large gravity platforms deployed in the North Sea in the 1970s and the Oosterschelde storm surge barrier. These studies influenced the final form of these structures, and provided the necessary performance data to support the design (Smith, 1997; Craig, 2002).

Ten years after Roscoe's Rankine Lecture, his successor, Schofield, delivered the Rankine Lecture (Schofield, 1980), focusing on centrifuge operations in Cambridge. Twenty-six years later, in 2006, Professor Robert Mair—Schofield's successor—delivered the Rankine Lecture (Mair, 2008), and also described extensive centrifuge modelling studies conducted in Cambridge. Working alongside Schofield and later Mair, Professor Malcolm Bolton, another strong proponent of centrifuge modelling, has made more than 40 contributions to *Géotechnique*, many of which are concerned with centrifuge modelling.

The research conducted by the groups in Cambridge and Manchester, and the resulting sequence of Rankine Lectures, provides the backbone of the *Géotechnique* archive of centrifuge modelling research, but many seminal contributions come from elsewhere. During the past 10 years, *Géotechnique* has featured centrifuge modelling articles from research groups in Japan, Singapore, France, Germany, the USA and Australia in addition to the UK.

Modern experimental methods

Modern geotechnical physical modelling, in parallel with other branches of experimental mechanics, has benefited from digital and robotic technology, which has allowed improved control and monitoring. In early physical model tests, such as the classic experiments on piles and walls by Marsland (1953), Whitaker (1957) and Hanna (1963), external loads were applied by modified strength-testing machines, and ground movements were monitored by dial gauges—or in Marsland's case by eye through a microscope.

In early centrifuge tests load was imposed by self-weight alone, owing to the inability to provide control within the centrifuge environment.

Recent editions of *Géotechnique* include examples of the more sophisticated experimental methods that represent the evolving state of the art. Many ingenious devices have been developed to replicate construction activity at small scale, often in a centrifuge. New techniques have been developed to simulate excavation and backfilling in-flight. These include the simple approach of draining heavy fluid to simulate the reduction in stress during excavation or tunnelling (Davis *et al.*, 1980; Bolton & Powrie, 1988), which has been augmented by techniques for in-flight 'concreting' of diaphragm walls (Powrie & Kantartzi, 1996), insertion of props (Richards & Powrie, 1998, Fig. 2(a)), loading of adjacent piles (Choy *et al.*, 2007), and deterioration of sewer linings (Spasojevic *et al.*, 2007).

To install foundations in a realistic manner, miniaturised systems for hammer-driving (De Nicola & Randolph, 1997) and suction pumping (Gaudin *et al.*, 2006; Chen & Randolph, 2007, Figs 2(c), 2(d)) have been devised. Model test beds have been improved, with dynamic compaction (Merrifield & Davies, 2000), miniature vertical drains (Hird & Moseley, 2000) and stone columns (Al-Khafaji & Craig, 2000; Muir Wood *et al.*, 2000). Robots have been developed to construct sand compaction piles (Lee *et al.*, 2004) and conduct deep mixing (Lee *et al.*, 2006) in the centrifuge.

Earthquake loading has been simulated on shaking tables, although this research is poorly represented in *Géotechnique*, featuring only as a small section of Newmark's (1965) Rankine Lecture. Earthquakes have been modelled in the centrifuge (Scott, 1987; Lee & Hushmand *et al.*, 1988; Lee & Schofield, 1988; Kutter & James, 1989) using special containers developed to reduce boundary effects (Zeng & Schofield, 1996; Teymur & Madabhushi, 2003) and artificial pore fluids to ensure correct scaling of inertia and consolidation (Dewoolkar *et al.*, 1999). Liquefaction from wave loading has also been simulated (Sassa & Sekiguchi, 1999).

Servo-controlled actuators have been developed to allow arbitrary sequences of load and displacement to be imposed

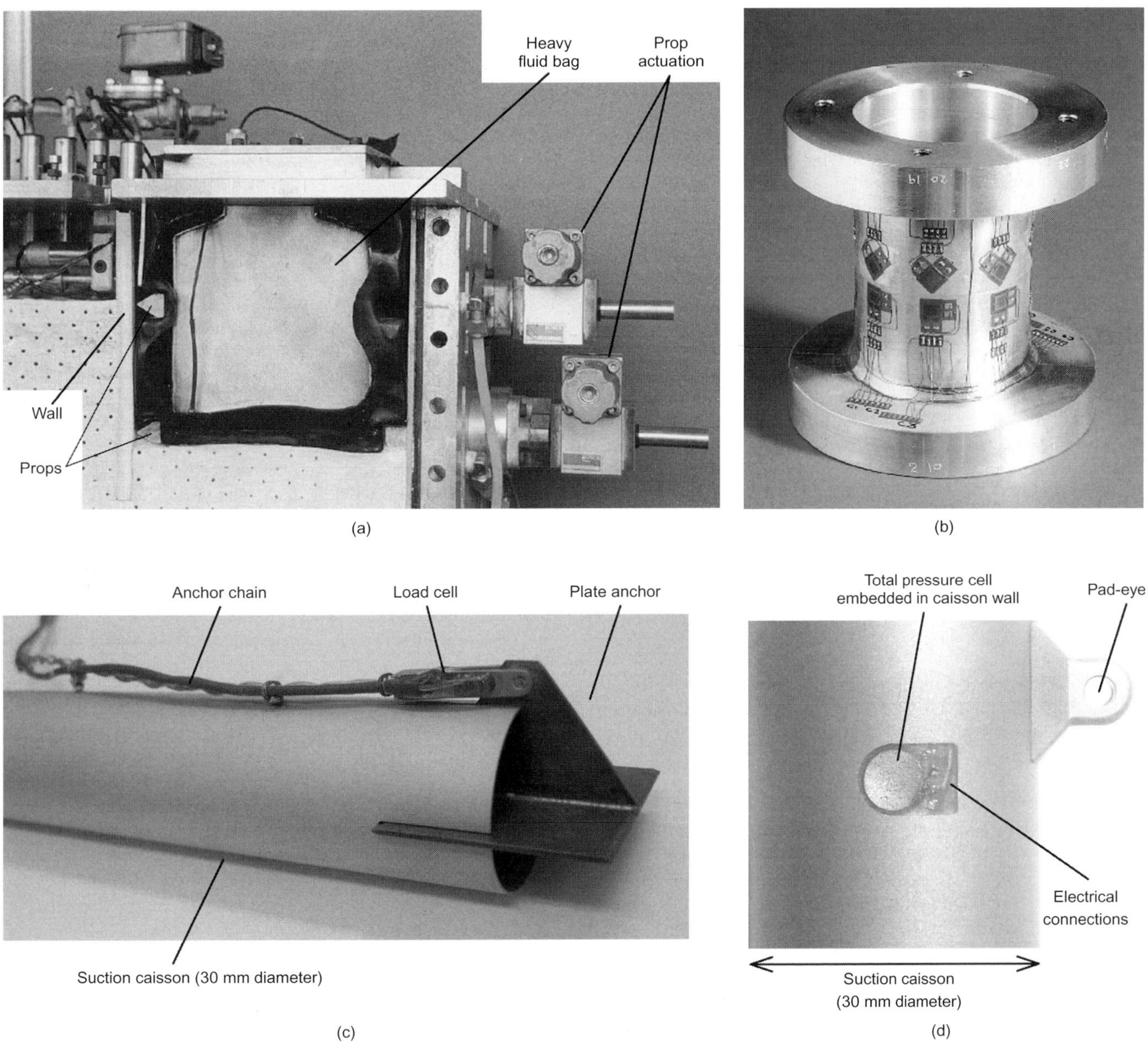

Fig. 2. Examples of modern physical modelling technology: (a) in-flight construction of a double-propped retaining wall (Richards & Powrie, 1998); (b) six-component load cell (Bienen *et al.*, 2006); (c) suction-embedded plate anchor (Gaudin *et al.*, 2006); (d) instrumented suction caisson (Chen & Randolph, 2007)

on model structures and foundations. Martin & Houlsby (2000) describe a foundation loading system with full control of three axes—vertical, horizontal and rotational. Bienen *et al.* (2006) describe a miniature Stewart platform (Stewart, 1965) that provides control of all six degrees of freedom—three translational and three rotational (Fig. 2(b)).

To allow proper back-analysis of a physical modelling event, it is necessary to conduct the miniature equivalent to a ground investigation in order to characterise the test bed. For this purpose, miniature vane shear and cone penetration test devices have been devised, and adapted for in-flight use in the centrifuge (Davies & Parry, 1982; Bolton *et al.*, 1999). These devices have been used to illustrate the repeatability that can be achieved, as evidenced through CPT tip resistance profiles recorded in the same type of sand tested at six different European centrifuge laboratories: a variation of 10% was found (Bolton *et al.*, 1999). In a reversal of the centrifuge testing philosophy of miniaturising reality, an enlarged version of the original T-bar penetrometer, which was first developed for use in the centrifuge (e.g. Stewart *et al.*, 1994; Horikoshi & Randolph, 1996), has become popular in the field as an in situ test for characterising soft sediments (Randolph *et al.*, 1998; Kolk & Wegerif, 2005).

To measure displacements within an exposed plane of a soil model, Butterfield *et al.* (1970) and Andrawes & Butterfield (1973) described a technique based on stereophotogrammetry, which provided remarkable accuracy. By manually measuring particle movements, as seen in stereo pair photographs, displacements as small as a fraction of a grain size could be detected over a ~0·5 m field of view. The recent introduction of digital technology has removed the need for painstaking manual measurements, and pre-failure deformations can now be detected using digital imaging combined with particle image velocimetry (PIV) and close-range photogrammetry (White *et al.*, 2003). Photographic techniques are limited to the observation of external surfaces, but Borsic *et al.* (2005) describe how electrical impedance tomography can reveal the internal density distribution of a soil model. Miniature transducers have been developed to measure stress (Garnier *et al.*, 1999) and pore pressure (Take & Bolton, 2003) within soil masses, and earth pressures on foundations and piles (Klotz & Coop, 2001; White & Lehane, 2004; Chen & Randolph, 2007, Fig. 2(d); Choy *et al.*, 2007).

Each of the following sections focuses on a particular type of geotechnical construction. Some of the most significant developments that have emerged from physical modelling are highlighted.

SHALLOW FOUNDATIONS

The load–displacement response of a shallow foundation remains a significant area of geotechnical engineering research, and has been the subject of more than 20 papers in *Géotechnique* during the past five years. Early work by Meyerhof (1951), Hanna (1963) and De Beer (1963, 1970) described extensive model tests and limit equilibrium solutions, which established the general bearing capacity expressions that feature in every undergraduate textbook. Meyerhof, De Beer and Hanna all recognised the difficulty of selecting an appropriate friction angle to use in the bearing capacity equation. This difficulty arises because peak friction angle varies with stress level, and a range of stress levels exist within the failing soil beneath a footing. Bjerrum, in a special lecture that was never delivered because of his sudden death, but which was published in *Géotechnique* in 1973, highlighted the implications of this uncertainty in relation to the design of the first concrete gravity structure installed in the North Sea.

> A variation in friction angle of only 2° may result in a variation in the value of N_γ of 50%. Most of the [existing expressions for N_γ] are derived in a semi-empirical way, being based partly on the results of loading tests. The loading tests are, however, in most cases carried out with model footings of very small size, the dimensions generally being of the order of inches, or at the most one or two feet. Extrapolation of the results to the structures in the North Sea having dimensions of about 330 ft (100 m) therefore requires careful consideration of the scale effect. (Bjerrum, 1973)

With the advent of centrifuge modelling in the early 1970s, it became possible to simulate large-scale footings in controlled and repeatable soil conditions, eliminating the need for extrapolation. The classic parametric studies by Ovesen (1975) and Kimura *et al.* (1985) clarified the variation of N_γ with footing size, while confirming, through 'modelling of models', that small centrifuge tests were free from unwanted errors associated with grain size (Steenfelt, 2006). Ovesen's classic study draws on his own tests conducted in Florida, combined with data gathered by Mikasa & Takada (1973) in Japan. In these tests, the observed unit bearing capacity was consistent for a given prototype (i.e. field scale) footing size, regardless of whether the model was 10 mm or 30 mm in diameter—representing successful modelling of models. In contrast, as the prototype footing diameter increased, the unit bearing capacity decreased. Ovesen's study involved small models and relatively low acceleration levels. During the same period, Rowe & Craig (1979), working in Manchester, were simulating gravity platforms up to 100 m in diameter, to support the early oil and gas developments in the North Sea.

Bjerrum's concern with establishing a value for the vertical bearing capacity factor N_γ did not stem from any concern that the 100 m diameter Ekofisk tank would sink vertically, but was because the resistance to inclined loading (resulting from wave action) was assessed by applying a reduction factor to the capacity under purely vertical load.

An alternative approach to describe the capacity of a foundation under combined loading is to consider the capacity in terms of an envelope in vertical, horizontal and moment (V-H-M) load space. The first mention of this approach in *Géotechnique* is on the second page of Roscoe's Rankine Lecture, in which he relates how Sir John Baker asked him to design the foundations of a portal frame, to support the V, H and M loads given to him by the structural engineer. Roscoe & Schofield (1956) plotted the capacity of the resulting foundations as an envelope in combined load space. However, the pages of *Géotechnique* portrayed combined loading in terms of inclination and eccentricity factors for a further 35 years, until Nova & Montrasio (1991) proposed a return to the yield envelope approach, which they coupled with a work-hardening plasticity theory to describe the general footing response. The theory was compared with results from a programme of model tests, and was able to calculate the footing displacements at yield, and subsequent hardening or softening of the footing response. This form of plasticity model 'treats in a unifying conceptual framework both displacements under working loads and failure conditions' (Nova & Montrasio, 1991).

Based on these and other model tests, Butterfield & Gottardi (1994) suggested that the failure surface approach 'might replace, in a simple and more useful form, the plethora of load inclination and eccentricity factors currently used to predict such failure loads'. Further papers in *Géotechnique* describe the highly sophisticated physical model tests that have underpinned the development and calibration of these plasticity 'macro-element' models for foundation behaviour (Montrasio & Nova, 1997; Gottardi *et al.*, 1999; Martin & Houlsby, 2000, Fig. 3(a); Byrne & Houlsby, 2001; Cassidy *et al.*, 2004). Centrifuge model tests validated a more simple approach for incorporating the benefit of rotational fixity into the analysis of the 'spudcan' foundations of jack-up drilling rigs (Dean *et al.*, 1998). Centrifuge modelling studies have also provided guidance on other aspects of the behaviour of spudcan foundations, including punch-through failure in sand-over-clay conditions (Craig & Chua, 1990), bearing capacity and soil backflow during deep penetration (Hossain *et al.*, 2005, Fig. 3(b)) and the increased extraction resistance due to consolidation during operation (Purwana *et al.*, 2005).

These physical model tests have validated many aspects of the analysis techniques that are used in practice, and are found in international design codes for offshore structures (SNAME, 2002; ISO, 2008). For onshore design, centrifuge model tests have also been used to validate simplified approaches for calculating foundation settlement, accounting for soil non-linearity (Atkinson, 2000).

TUNNELLING

The stability of tunnel headings and the ground movements associated with tunnel construction have been widely investigated through physical model tests. A classic series of model tests accompanied by plasticity limit analysis led to the development of calculation methods for the collapse of tunnels in sand (Atkinson, 1975; Atkinson & Potts, 1977) and in clay (Davis *et al.*, 1980). These tests, and other more recent studies, have been used to calibrate simplified methods for predicting tunnelling-induced ground movements, which closely match recent field measurements (Mair *et al.*, 1993; Loganathan *et al.*, 2000, Fig. 4; Osman *et al.*, 2006a). The early tests have been revisited recently to calibrate techniques to link tunnel support pressure to surface settlement (Osman *et al.*, 2006b). Physical modelling is particularly valuable to the understanding of tunnel behaviour because numerical modelling is unable to match observed settlement troughs, even when using a highly sophisticated constitutive model and including three-dimensional effects and anisotropy (Franzius *et al.*, 2005).

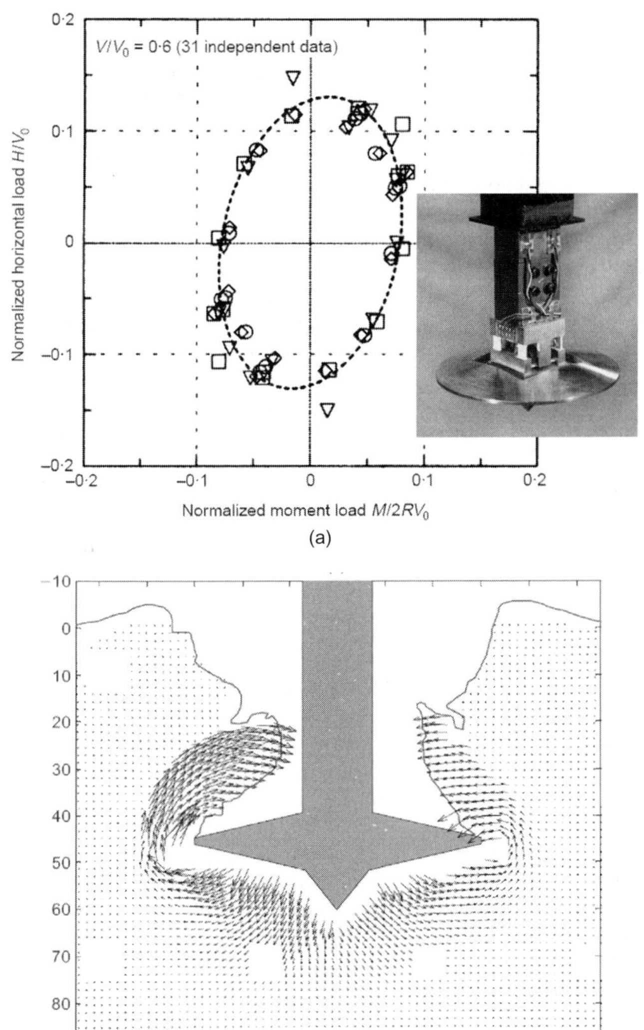

Fig. 3. Spudcan foundations: (a) capacity under combined loading (Martin & Houlsby, 2000); (b) backflow during vertical penetration (Hossain *et al.*, 2005)

RETAINING WALLS

Physical model tests published in *Géotechnique* have been used to assess the validity of theoretical analyses for the limiting pressures on retaining walls (e.g. Rowe & Peaker, 1965, Fig. 5(a); James & Bransby, 1970; Powrie, 1996; Bica & Clayton, 1998). These tests have also been used to identify the resulting soil deformation mechanisms and therefore the nearby settlement and pre-failure wall movements (Bransby & Milligan, 1975; Milligan & Bransby, 1976; Bolton & Powrie, 1988). These observations inspired simple kinematic mechanisms for the prediction of wall and ground movement during excavation. These mechanisms provide a link between soil strain and wall movement, at least for a relatively rigid wall. Mechanisms of this kind allow an assumed soil stress–strain response to be used to select a wall embedment that is sufficient to limit ground movements to a specified serviceability limit (Bolton & Powrie, 1988), and can be found in modern textbooks (Powrie, 2004; Wood, 2004).

In their centrifuge tests of unpropped diaphragm walls in stiff clay, Bolton & Powrie (1987) observed the formation of a flooded tension crack on the retained side of the wall (Fig. 5(b)). They argued that designers 'should always be aware of this possibility: stability under these conditions might be viewed as a minimum requirement for any wall', although if the retained area is paved over or built on, tension cracks may be prevented. However, cantilever flood defence walls, such as those that failed when Hurricane Katrina struck New Orleans in 2005, are vulnerable to this mechanism. During the subsequent investigation, physical model tests were con-

Fig. 4. Tunnelling (Loganathan *et al.*, 2000)

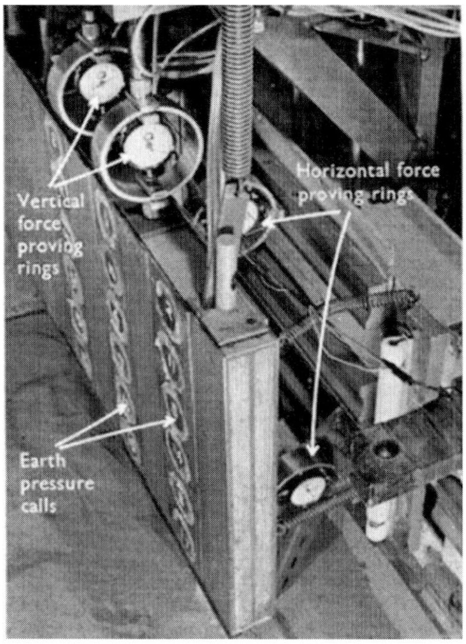

(a)

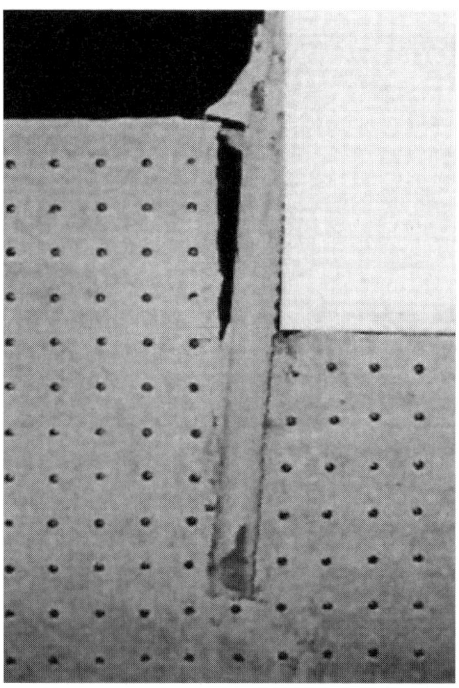

(b)

Fig. 5. Retaining walls: (a) early studies of earth pressure (Rowe & Peaker, 1965); (b) diaphragm wall failure due to tension cracking (Bolton & Powrie, 1987)

ducted using the US Army geotechnical centrifuge to identify the modes of failure. It was established that 'a key factor in the failure was the formation of a gap between the wall and the levee fill on the canal side of the wall, allowing water pressure to act on the wall below the surface of the levee' (Steedman, 2006; IPET, 2007). This possibility was not considered during the design of the wall. An additional destabilising mechanism identified by these tests was the creation of high uplift pressures in the sand beneath the embankment by the water flowing down the gap. This same uplift mechanism has been observed in centrifuge model tests, as reported in two *Géotechnique* papers that have identical titles (and the same last author) (Hird *et al.*, 1978; Padfield & Schofield, 1983).

SLOPES

In his Rankine Lecture on slope behaviour, Leroueil (2001) described the Selborne cut slope experiment (Cooper *et al.*, 1998) as 'the first time that the development of progressive failure up to generalised failure has been observed'. This comment is strictly true only if referring to field observations. Quantitative measurements of progressive slope failure feature in *Géotechnique* as early as Roscoe's Rankine Lecture. His study—the first centrifuge modelling published in *Géotechnique*—includes results from early tests that 'clearly show that the rupture develops progressively upwards from the toe' (Roscoe, 1970). Tension cracking at the surface followed, matching the mechanism observed at Selborne. Subsequent papers describe similar observations, enjoying the advantage over field-scale studies of a full view of the slope cross-section through a window in the model container (Lyndon & Schofield, 1970; Endicott, 1974). The analysis by Smith & Hobbs (1974) of many centrifuge slope tests conducted in Manchester is the first comparison between finite element analysis and centrifuge modelling to appear in *Géotechnique*, and these authors highlighted the complementary roles of the two modelling techniques, accompanied by field evidence.

These early slope tests were very simple, with failure being initiated by the self-weight of the slope and equilibration of pore pressures. Real slopes are subject to seasonal variations in the hydraulic boundary condition at the free surface, which can drive progressive failure. Take & Bolton (2004, 2008) revisited the issue of progressive failure that was first tackled in the centrifuge by Roscoe (see also Take, 2003). They subjected their slopes to successive wet and dry seasons within a humidity-controlled chamber equipped with a rainfall simulator (Fig. 6). Meanwhile, the patterns of movement within the slope cross-section were tracked to micrometre-level accuracy using image analysis (White *et al.*, 2003) and the pore pressure response within the slope was measured using miniature tensiometers (Take & Bolton, 2003). In these tests, the stress history of the soil, the geometry of the slope, and the imposed changes in humidity, temperature and rainfall were all known and controlled, and the resulting pore pressure and the detailed ground movements were continuously monitored. These tests demonstrated the important role that seasonal cycles of pore pressure have in the progressive degradation of a slope.

CONCLUDING REMARKS

The sophistication of modern physical modelling is epitomised by the study described in the previous paragraph. In many respects, the information available from this centrifuge model is more complete than can ever be gathered in a field test. Other recent papers in *Géotechnique* describe similarly advanced physical modelling studies.

This level of sophistication means that modern physical modelling experiments should be regarded as case studies of value comparable to those undertaken in the field. Field studies have the important benefit of incorporating natural soil properties and variability, whereas physical modelling allows better control of events and ground conditions, and provides more detailed measurements of the resulting behaviour. Unlike a field trial, an experiment can be quickly repeated at will with controlled changes to the soil and boundary conditions. The more controlled conditions in a physical model test compared with the field provide a better basis for establishing the validity of theoretical and numerical analyses.

Many of the model test observations and events described in this paper lie beyond current constitutive and numerical

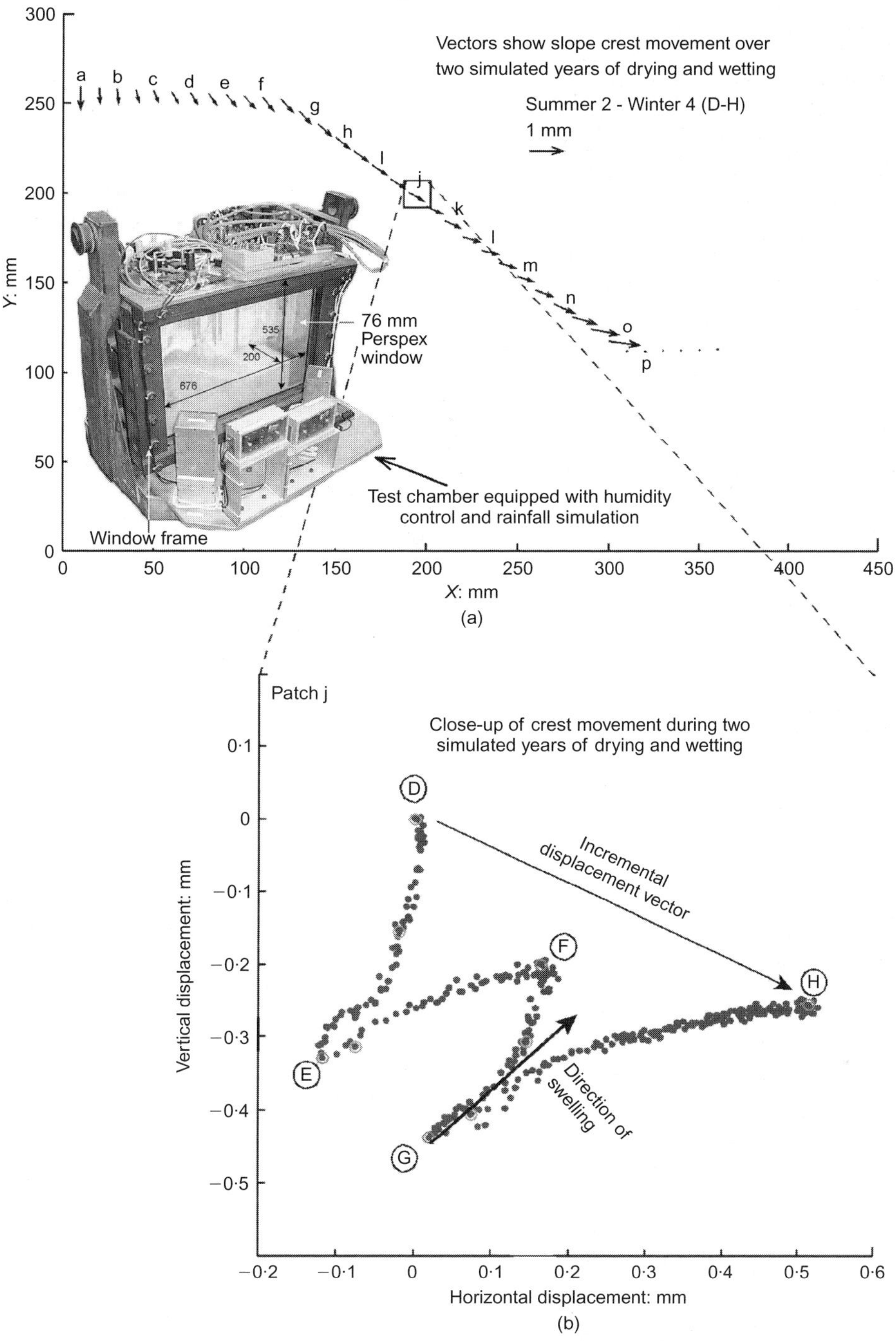

Fig. 6. Recent physical modelling of slopes (Take & Bolton, 2008)

modelling capabilities. Physical model testing in its various forms is therefore to be regarded as a powerful tool that is complementary to numerical modelling and field investigations. Each has a distinct role within the research and practice of geotechnical engineering.

Géotechnique has published many of the key papers that document the evolution of physical modelling over the past 60 years. These papers have revealed a variety of important geotechnical phenomena, and have validated analyses that underpin many aspects of geotechnical practice. As modelling techniques advance, the sophistication of physical models, and the detail of the resulting measurements, will continue to increase. The best of this work will appear in the pages of *Géotechnique*, and should be keenly anticipated.

REFERENCES

Al-Khafaji, Z. A. & Craig, W. H. (2000). Drainage and reinforcement of soft clay tank foundation by sand columns. *Géotechnique* **50**, No. 6, 709–713.

Andrawes, K. Z. & Butterfield, R. (1973). The measurement of planar displacements of sand grains. *Géotechnique* **23**, No. 5, 571–576.

Atkinson, J. H. (1975). The stability of a shallow circular tunnel in dense sand during surface excavation or filling. *Géotechnique* **25**, No. 3, 591–592.

Atkinson, J. H. (2000). Non-linear soil stiffness in routine design. *Géotechnique* **50**, No. 5, 487–508.

Atkinson, J. H. & Potts, D. M. (1977). Stability of a shallow circular tunnel in cohesionless soil. *Géotechnique* **27**, No. 2, 203–215.

Bica, A. V. D. & Clayton, C. R. I. (1998). An experimental study of the behaviour of embedded lengths of cantilever walls.

Géotechnique **48**, No. 6, 731–746.

Bienen, B., Byrne, B. W., Houlsby, G. T. & Cassidy, M. J. (2006). Investigating six degree of freedom loading of shallow foundations on sand. *Géotechnique* **56**, No. 6, 367–379.

Bjerrum, L. (1973). Geotechnical problems involved in foundations of structures in the North Sea. *Géotechnique* **23**, No. 3, 319–358.

Bolton, M. D. & Powrie, W. (1987). The collapse of diaphragm walls retaining clay. *Géotechnique* **37**, No. 3, 335–353.

Bolton, M. D. & Powrie, W. (1988). Behaviour of diaphragm walls in clay prior to collapse. *Géotechnique* **38**, No. 2, 167–189.

Bolton, M. D., Gui, M. W., Gamier, J., Corte, J. F., Bagge, G., Laue, J. & Renzi, R. (1999). Centrifuge cone penetration tests in sand. *Géotechnique* **49**, No. 4, 543–552.

Borsic, A., Comina, C., Foti, S., Lancelotta, R. & Musso, G. (2005). Imaging heterogeneities with electrical impedance tomography: laboratory results. *Géotechnique* **55**, No. 8, 539–548.

Bransby, P. L. & Milligan, G. W. E. (1975). Soil deformations near cantilever sheet-pile walls. *Géotechnique* **25**, No. 2, 75–195.

Butterfield, R. & Gottardi, G. (1994). A complete three-dimensional failure envelope for shallow footings on sand. *Géotechnique* **44**, No. 1, 181–184.

Butterfield, R., Harkness, R. M. & Andrawes, K. Z. (1970). A stereo-photogrammetric technique for measuring displacement fields. *Géotechnique* **20**, No. 3, 308–314.

Byrne, B. W. & Houlsby, G. T. (2001). Observation of footing behaviour on loose carbonate sand. *Géotechnique* **51**, No. 5, 463–466.

Cassidy, M. J., Byrne, B. W. & Randolph, M. F. (2004). A comparison of the combined load behaviour of spudcan and caisson foundations on soft normally consolidated clay. *Géotechnique* **54**, No. 2, 91–106.

Chen, W. & Randolph, M. F. (2007). External radial stress changes and axial capacity for suction caissons in soft clay. *Géotechnique* **57**, No. 6, 499–511.

Choy, C. K., Standing, J. R. & Mair, R. J. (2007). Stability of a loaded pile adjacent to a slurry-supported trench. *Géotechnique* **57**, No. 10, 807–819.

Cooper, M. R., Bromhead, E. N., Petley, D. J. & Grant, D. I. (1998). The Selborne cutting stability experiment. *Géotechnique* **48**, No. 1, 183–191.

Craig, W. H. (1989). Édouard Phillips (1821–89) and the idea of centrifuge modelling. *Géotechnique* **39**, No. 4, 697–700.

Craig, W. H. (2002). The Seven Ages of centrifuge modelling. *Proceedings of the workshop on constitutive and centrifuge modelling: two extremes*, Monte Verita, pp. 165–174.

Craig, W. H. & Chua, K. (1990). Deep penetration of spud-can foundations on sand and clay. *Géotechnique* **40**, No. 4, 541–556.

Davies, M. C. R. & Parry, R. H. G. (1982). Determining the shear strength of clay cakes in the centrifuge using a vane. *Géotechnique* **32**, No. 1, 59–62.

Davis, E. H., Mair, R. J., Gunn, M. & Seneviratne, N. (1980). The stability of shallow tunnels and underground openings in cohesive material. *Géotechnique* **30**, No. 4, 397–416.

de Beer, E. E. (1963). The scale effect in the transposition of the results of deep-sounding tests on the ultimate bearing capacity of piles and caisson foundations. *Géotechnique* **13**, No. 1, 39–75.

de Beer, E. E. (1970). Experimental determination of the shape factors and the bearing capacity factors of sand. *Géotechnique* **20**, No. 4, 387–411.

De Nicola, A. & Randolph, M. F. (1997). The plugging behaviour of driven and jacked piles in sand. *Géotechnique* **47**, No. 4, 841–856.

Dean, E. T. R., James, R. G., Schofield, A. N. & Tsukamoto, Y. (1998). Drum centrifuge study of three-leg jack-up models on clay. *Géotechnique* **48**, No. 6, 761–785.

Dewoolkar, M. M., Ko, H. Y. & Pak, R. Y. S. (1999). Centrifuge modelling of models of seismic effects on saturated earth structures. *Géotechnique* **49**, No. 2, 247–266.

Endicott, L. J. (1974). The determination of stress fields from plane strain data. *Géotechnique* **24**, No. 4, 517–530.

Franzius, J. N., Potts, D. M. & Burland, J. B. (2005). The influence of soil anisotropy and K_0 on ground surface movements resulting from tunnel excavation. *Géotechnique* **55**, No. 3, 189–199.

Garnier, J., Ternet, O., Cottineau, L.-M. & Brown, C. J. (1999). Placement of embedded pressure cells. *Géotechnique* **49**, No. 3, 405–414.

Gaudin, C., O'Loughlin, C. D., Randolph, M. F. & Lowmass, A. C. (2006). Influence of the installation process on the performance of suction embedded plate anchors. *Géotechnique* **56**, No. 6, 381–391.

Golder, H. Q. (1971). Letter. *Géotechnique* **21**, No. 3, 269–270.

Gottardi, G., Houlsby, G. T. & Butterfield, R. (1999). The plastic response of circular footings on sand under general planar loading. *Géotechnique* **49**, No. 4, 453–470.

Hanna, T. H. (1963). Model studies of foundation groups in sand. *Géotechnique* **13**, No. 4, 334–351.

Hird, C. C. & Moseley, V. J. (2000). Model study of seepage in smear zones around vertical drains in layered soil. *Géotechnique* **50**, No. 1, 89–97.

Hird, C. C., Marsland, A. & Schofield, A. N. (1978). The development of centrifugal models to study the influence of uplift pressure on the stability of a floodbank. *Géotechnique* **28**, No. 1, 85–106.

Horikoshi, K. & Randolph, M. F. (1996). Centrifuge modelling of piled raft foundations on clay. *Géotechnique* **46**, No. 4, 741–752.

Hossain, M. S., Hu, Y., Randolph, M. F. & White, D. J. (2005). Limiting cavity depth for spudcan foundations penetrating clay. *Géotechnique* **55**, No. 9, 679–690.

Hushmand, B., Scott, R. F. & Crouse, C. B. (1988). Centrifuge liquefaction tests in a laminar box. *Géotechnique* **38**, No. 2, 253–262.

Interagency Performance Evaluation Taskforce (IPET) (2007). *Performance evaluation of the New Orleans and Southeast Louisiana Hurricane Protection System. Final report of the Interagency Performance Evaluation Taskforce. Volume V: The performance—Levees and floodwalls.* Washington, DC: US Army Corps of Engineers.

ISO (2008). *Petroleum and natural gas industries—Site-specific assessment of mobile offshore units—Part 1: Jack-ups*, ISO 19905-1 (draft). Geneva: International Organization for Standardization.

James, R. G. & Bransby, P. L. (1970). Experimental and theoretical investigations of a passive earth pressure problem. *Géotechnique* **20**, No. 1, 17–37.

Kimura, T., Kusakabe, O. & Saitoh, K. (1985). Geotechnical model test of bearing capacity problems in a centrifuge. *Géotechnique* **35**, No. 1, 33–45.

Klotz, E. U. & Coop, M. R. (2001). An investigation of the effect of soil state on the capacity of driven piles in sands. *Géotechnique* **51**, No. 9, 733–751.

Kolk, H. J. & Wegerif, J. (2005). Offshore site investigation: new frontiers. *Proc. 1st Int. Symp. Frontiers in Offshore Geotechnics, Perth*, 145–161.

Kutter, B. L. & James, R. G. (1989). Dynamic centrifuge model tests on clay embankments. *Géotechnique* **39**, No. 1, 91–106.

Lee, F. H. & Schofield, A. N. (1988). Centrifuge modelling of sand embankments and islands in earthquakes. *Géotechnique* **38**, No. 1, 45–58.

Lee, F. H., Juneja, A. & Tan, T. S. (2004). Stress and pore pressure changes due to sand compaction pile installation in soft clay. *Géotechnique* **54**, No. 1, 1–16.

Lee, F. H., Lee, C. H. & Dasari, G. R. (2006). Centrifuge modelling of wet deep mixing processes in soft clay. *Géotechnique* **56**, No. 10, 677–691.

Leroueil, S. (2001). Natural slopes and cuts: movement and failure mechanisms. *Géotechnique* **51**, No. 3, 197–243.

Loganathan, N., Poulos, H. G. & Stewart, D. P. (2000). Centrifuge model testing of tunnelling induced ground and pile deformations. *Géotechnique* **50**, No. 3, 283–294.

Lyndon, A. & Schofield, A. N. (1970). Centrifugal model test of a short-term failure in London Clay. *Géotechnique* **20**, No. 4, 440–442.

Mair, R. J. (2008). Tunnels and geotechnics: new horizons. *Géotechnique* **58**, forthcoming.

Mair, R. J., Taylor, R. N. & Bracegirdle, A. (1993). Subsurface settlement profiles above tunnels in clays. *Géotechnique* **43**, No. 2, 315–320.

Marsland, A. (1953). Model experiments to study the influence of seepage on the stability of a sheeted excavation in sand. *Géotechnique* **3**, No. 6, 223–241.

Martin, C. M. & Houlsby, G. T. (2000). Combined loading of spudcan foundations on clay: laboratory testing. *Géotechnique* **50**, No. 4, 325–338.

Merrifield, C. M. & Davies, M. C. R. (2000). A study of low energy dynamic compaction: field trials and centrifuge modelling. *Géotechnique* **50**, No. 6, 675–683.

Meyerhof, G. G. (1951). The ultimate bearing capacity of foundations. *Géotechnique* **2**, No. 4, 301–332.

Mikasa, M. & Takada, N. (1973). Significance of centrifugal model test in soil mechanics. *Proc. 8th Int. Conf. Soil Mech., Moscow* **1·2**, 273–278.

Milligan, G. W. E. & Bransby, P. L. (1976). Combined active and passive rotational failure of a retaining wall in sand. *Géotechnique* **26**, No. 3, 473–494.

Montrasio, L. & Nova, R. (1997). Settlement of shallow foundations on sand: geometrical effects. *Géotechnique* **47**, No. 1, 49–60.

Muir Wood, D., Hu, W. & Nash, D. F. T. (2000). Group effects in stone column foundations: model tests. *Géotechnique* **50**, No. 6, 689–698.

Newmark, N. M. (1965). Effects of earthquakes on dams and embankments. *Géotechnique* **15**, No. 2, 139–159.

Nova, R. & Montrasio, L. (1991). Settlements of shallow foundations on sand. *Géotechnique* **41**, No. 2, 243–256.

Osman, A. S., Mair, R. J. & Bolton, M. D. (2006a). On the kinematics of 2D tunnel collapse in undrained clay. *Géotechnique* **56**, No. 9, 585–595.

Osman, A. S., Bolton, M. D. & Mair, R. J. (2006b). Predicting 2D ground movements around tunnels in undrained clay. *Géotechnique* **56**, No. 9, 597–604.

Ovesen, N. K. (1975). Centrifugal testing applied to bearing capacity problems of footings on sand. *Géotechnique* **25**, No. 2, 394–401.

Padfield, C. J. & Schofield, A. N. (1983). The development of centrifugal models to study the influence of uplift pressure on the stability of a flood bank. *Géotechnique* **33**, No. 1, 57–66.

Powrie, W. (1996). Limit equilibrium analysis of embedded retaining walls. *Géotechnique* **46**, No. 4, 709–723.

Powrie, W. (2004). *Soil mechanics: Concepts and applications*, 2nd edn. London: Spon Press.

Powrie, W. & Kantartzi, C. (1996). Ground response during diaphragm wall installation in clay: centrifuge model tests. *Géotechnique* **46**, No. 4, 725–739.

Purwana, O. A., Leung, C. F., Chow, Y. K. & Foo, K. S. (2005). Influence of base suction on extraction of jack-up spudcans. *Géotechnique* **55**, No. 10, 709–719.

Randolph, M. F., Hefer, P. A., Geise, J. M. & Watson, P. G. (1998). Improved seabed strength profiling using a T-bar penetrometer. *Proceedings of the international conference on offshore site investigation and foundation behaviour*, London, pp. 221–235.

Richards, D. J. & Powrie, W. (1998). Centrifuge model tests on doubly propped embedded retaining walls in overconsolidated kaolin clay. *Géotechnique* **48**, No. 6, 833–846.

Roscoe, K. H. (1970). The influence of strains in soil mechanics. *Géotechnique* **20**, No. 2, 129–170.

Roscoe, K. H. & Schofield, A. N. (1956). Stability of short pier foundations in sand. *Br. Weld. J.* **3**, No. 8, 343–354.

Rowe, P. W. (1972). The relevance of soil fabric to site investigation practice. *Géotechnique* **22**, No. 2, 195–300.

Rowe, P. W. & Peaker, K. (1965). Passive earth pressure measurements. *Géotechnique* **15**, No. 1, 57–78.

Rowe, P. W. & Craig, W. H. (1979). Applications of models to the prediction of offshore gravity platform foundation performance. *Proceedings of the international conference on offshore site investigation*, pp. 261–281. London: Graham & Trotman.

Sassa, S. & Sekiguchi, H. (1999). Wave-induced liquefaction of beds of sand in a centrifuge. *Géotechnique* **49**, No. 5, 621–638.

Schofield, A. N. (1980). Cambridge geotechnical centrifuge operations. *Géotechnique* **30**, No. 3, 227–268.

Scott, R. F. (1987). Failure. *Géotechnique* **37**, No. 4, 423–466.

Smith, I. M. & Hobbs, R. (1974). Finite element analysis of centrifuged and built-up slopes. *Géotechnique* **24**, No. 4, 531–559.

Smith, I. M. (1997). Obituary: Peter Walter Rowe 1922–1997. *Géotechnique* **47**, No. 6, 1083–1086.

SNAME (2002). *Commentaries to guidelines for site specific assessment of mobile jack-up units*, Technical and Research Bulletin 5-5A. Jersey City, NJ: Society of Naval Architects and Marine Engineers.

Spasojevic, A. D., Mair, R. J. & Gumbel, J. E. (2007). Centrifuge modelling of the effects of soil loading on flexible sewer liners. *Géotechnique* **57**, No. 4, 331–342.

Steedman, R. S. (2006). Investigating New Orleans. *Ingenia, Magazine of the Royal Academy of Engineering* **28**, 24–28.

Steenfelt, J. S. (2006). Obituary: Niels Krebs Ovesen. *Géotechnique* **56**, No. 3, 223–224.

Stewart, D. (1965). A platform with six degrees of freedom. *Proc. Instn Mech. Engrs* **180**, No. 15, 371–384.

Stewart, D. P., Jewell, R. J. & Randolph, M. F. (1994). Design of piled bridge abutments on soft clay for loading from lateral soil movements, *Géotechnique* **44**, No. 2, 277–296.

Take, W. A. (2003). *The influence of seasonal moisture cycles on clay slopes.* PhD dissertation, University of Cambridge, UK.

Take, W. A. & Bolton, M. D. (2003). Tensiometer saturation and the reliable measurement of matric suction. *Géotechnique* **53**, No. 2, 159–172.

Take, W. A. & Bolton, M. D. (2004). Identification of seasonal slope behaviour mechanisms from centrifuge case studies. *Proc. Skempton Memorial Conference, Advances in Soil Mechanics*, Vol. 2, pp. 992–1004. London: Thomas Telford .

Take, W. A. & Bolton, M. D. (2008). Progressive seasonal softening leading to first time failure of stiff clay slopes. *Géotechnique* (submitted for publication).

Teymur, B. & Madabhushi, S. P. G. (2003). Experimental study of boundary effects in dynamic centrifuge modelling. *Géotechnique* **53**, No. 7, 655–663.

Whitaker, T. (1957). Experiments with model piles in groups. *Géotechnique* **7**, No. 4, 147–167.

White, D. J. & Lehane, B. M. (2004). Friction fatigue on displacement piles in sand. *Géotechnique* **54**, No. 10, 645–658.

White, D. J., Take, W. A. & Bolton, M. D. (2003). Soil deformation measurement using particle image velocimetry (PIV) and photogrammetry. *Géotechnique* **53**, No. 7, 619–631.

Wood, D. M. (2004). *Geotechnical modelling*. Abingdon: Spon Press.

Zeng, X. & Schofield, A. N. (1996). Design and performance of an equivalent-shear-beam (ESB) container for earthquake centrifuge modelling. *Géotechnique* **46**, No. 1, 83–102.

Coop, M. & Schnaid, F. (2008). *Géotechnique* **58**, No. 5, 423–428 [**doi:** 10.1680/geot.2008.58.5.423]

Contributions to *Géotechnique* 1948–2008: Laboratory and field testing

M. COOP* and F. SCHNAID†

A review is presented of the development of laboratory and field testing as viewed through the pages of *Géotechnique*, and from the personal standpoint of two authors whose careers started after many of the most important apparatus and techniques had already been invented. The review highlights the shifting tides of fashion to and fro, from laboratory to field tests, but emphasises the interdependence of the two. While the early years of *Géotechnique* witnessed the more obviously dramatic developments, more recent years have seen refinements in apparatus and techniques, particularly for small strains, and also analyses, in which the effects of more complex aspects of soil behaviour have been accounted for, such as structure, stress and strain history, and time-related effects. Although more subtle in nature, these developments have still had immense influence on engineering practice.

KEYWORDS: field instrumentation; historical review; in situ testing; laboratory equipment; laboratory tests; soil classification

La présente communication passe en revue le développement d'essais en laboratoire et sur le terrain, recueillis dans les pages de *Géotechnique*, et du point de vue personnel de deux auteurs dont la carrière a débuté alors que les appareils et techniques les plus importants avaient déjà été inventés. Ce tour d'horizon souligne l'évolution des tendances, alternant entre essais en laboratoire et sur le terrain, en insistant toutefois sur l'interdépendance de ces deux formes d'essais. Alors que dans les premières années de *Géotechnique*, on assista aux développements les plus fulgurants, les années plus récentes ont été marquées par un raffinement non seulement de l'appareillage et des techniques, en particulier pour des variétés limitées, mais également des analyses, avec la reconnaissance des effets d'aspects plus complexes du comportement des sols, comme la structure, l'historique des déformations et des sollicitations, et les effets relatifs au temps. De nature plus subtils, ces développements ont eu néanmoins une influence considérable sur la pratique de l'ingénierie.

INTRODUCTION

A review of the early years of soil mechanics as revealed through the pages of *Géotechnique* is inevitably a sobering experience. The strength of views and ideas of the great soil mechanics of that time, not to mention their succinctness of expression and often elegant use of English (or French), contrasts starkly with the output of some modern research, which, with the exception of course of *Géotechnique* (!) by comparison often seems incremental, somewhat vapid, and politically correct (within a received soil mechanics agenda). Perhaps the difference is simply the result of the maturing of a science, although their era was driven not by numbers of papers, or by some measure of superficial and immediate uptake, but by real and lasting impact. Let anyone who is in any doubt about this spend a morning in their library perusing the dusty pages of Volume 2 (or perhaps more realistically and sadly with a pdf download), which contains the Proceedings of the Conference on the Measurement of Shear Strength of Soils in Relation to Practice. The fireworks of discussion to be found therein makes us wonder whether a living science and passionate debate are compatible with the presidential-style addresses of modern conferences, seemingly submerged in the gimmickry of presentational software.

We would like to pay our personal tribute to the authors of the wonderful papers that we have read in compiling this review—not only the visionary researchers whose work has outlasted their own careers, and even their lifetimes, but also those whose direction might seem in the modern context to have been 'wrong'; after all, everyone has 20–20 hindsight. Of course, the papers we have chosen to include represent a very personal choice, with absolutely no rigorous criteria, and by no means represent any sort of score, grade or prize—things with which we all seem to have become obsessed.

LABORATORY TESTS

From the standpoint of contemporary soil mechanics, the development of our apparatus, for all their defects, seems to be one of intelligent design, albeit by numerous creators, rather than one of Darwinian evolution. But this would be to ignore all the intriguing twists and turns of development, and the dead ends followed, that fill many pages of *Géotechnique*. For example, the conference proceedings in Volume 2 (De Beer, 1950; Geuze & Tan Tjong Kie, 1950; Skempton & Bishop, 1950; Kjellman, 1951; Mayer, 1951) make for fascinating reading. It is clear that the triaxial is already starting to become the dominant laboratory apparatus, but the papers do reveal early divergence between the geotechnical communities of different countries, with a preference for torsional shear as well as triaxial extension and direct shear in France, and a simple shear device from Sweden (although the current commonly used version of the apparatus did not appear until 1966; Bjerrum & Landva, 1966). To some extent, the cracks were papered over by the claim that the different geologies of different countries required different tests. But of greatest historical interest is the heated debate between the champions of the triaxial test (mostly British) and those of the ingenious 'cell-test' as used in Holland and Belgium. We can only be grateful that there was not some turgid European Community committee at that time to fudge some sort of awful compromise!

The cell apparatus, illustrated in Fig. 1, had a sample enclosed in a membrane and placed within a water-filled pressure chamber. An increment of axial stress was applied, and then the cell fluid was allowed to escape, drip by drip, until the sample was deemed to be 'about to fail'. The axial

Discussion on this paper closes on 1 December 2008, for further details see p. ii.
* Imperial College London, UK.
† Federal University of Rio Grande do Sul, Porto Alegre, Brazil.

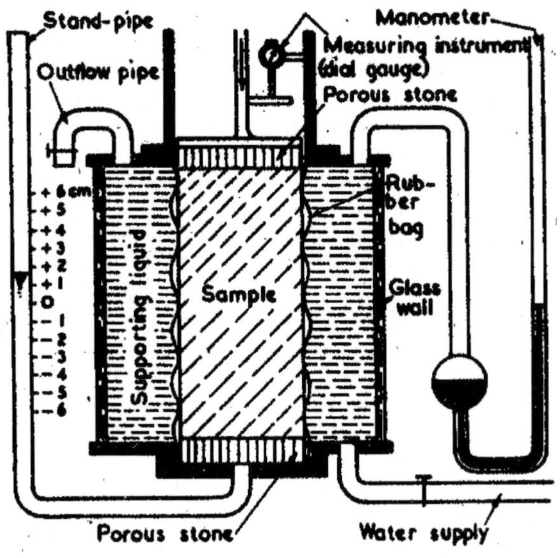

Fig. 1. The cell apparatus (from De Beer, 1950)

stress was then increased in numerous steps and the new cell pressure at failure determined for each. The combinations of the axial stress and cell pressure allowed a Mohr's circle to be drawn for each stage, with the advantage over the triaxial that the entire Mohr–Coulomb failure envelope could apparently be traced out with a single test. Tests could be undrained (or quick) or drained depending on whether time was allowed for consolidation of the sample at each step.

The cell-test advocates criticised the triaxial because of the rate dependence of the strengths measured, and because they noted that the test was not one of pure shear because, as the axial stress increases, both the shear stresses and the normal stresses within the sample change—criticisms that seem a little unfair, since they would also apply to the cell-test! However, in the absence of modern concerns for the validity of a multi-stage test on soils that might well have a structure, the debate for the triaxial advocates centred on several series of comparative tests, run by both camps, which showed that the cell-test gave lower strengths than the triaxial, and that for a quick undrained test it gave the unlikely result that ϕ_u was not equal to zero for a saturated soil. This Bishop attributed to the fact that the mobilised strength was dependent on the strain level (even if the differences between peak, critical and residual states were not well understood at that time); the cell-test was simply not applying a sufficiently large strain to the sample to reach the peak strength. Perhaps even more damningly for the future of the cell-test was the fact that Bishop realised that, since the strength envelope depended on the mobilised axial strain, it would be dependent on the dimensions of the apparatus and, in particular, on the flexibility of the chamber. Nevertheless, it remains an intriguing technique.

As described in the conference proceedings, neither the cell apparatus nor the triaxial had pore pressure measurement during undrained tests, even if a series of wonderfully imaginative mechanical devices is described in papers around this time to address this key issue of soils testing (e.g. Bishop & Eldin, 1950; Penman, 1953; Glynn & Spence, 1961). There was even a very early mid-height pore pressure probe in 1954 (Hanrahan, 1954). However, all of these devices became forlornly obsolete with the advent of modern electronic pressure transducers, which we believe first appeared in *Géotechnique* in Richardson & Whitman (1963), although the design of a mid-height pore pressure probe in direct contact with the sample came much later (Hight, 1982). The other key measurement in the laboratory that required some ingenuity to datalog was volumetric strain, as measured through the displacement of water from saturated samples. From the comfort of a modern laboratory, many of the electro-mechanical devices can appear rather 'Heath Robinson' in style, but that would be to neglect the amazing inventiveness of the authors, which we wonder how many of us would now have. Some were based on burettes (e.g. Lewin, 1971), some on mercury pots on springs (Rowlands, 1972; Darley, 1973), and some on levers (Klementev, 1974). But perhaps the key development was that of Menzies (1975), who utilised a rolling 'Bellofram' seal around a piston, so turning a volumetric strain into a linear motion that was more easily measured and datalogged; this remains the basis of several current popular designs.

In parallel with the automation of measurements for tests that were becoming ever longer, there was a need to provide stable pressure sources, which was initially achieved through the mercury pot system of Bishop & Henkel (1953). At this point academe started to diverge in its requirements from industry as the next 'Holy Grail' was to automate the pressure control so that ever-more complex tests could be carried out. Again, the ingenuity is amazing, from mercury pots running up and down their rails through motorized winches (Bishop & Wesley, 1975), which would be enough to give any present-day safety officer apoplexy, to ball bearings dropping one by one into a loading bucket (Lewin & Burland, 1970), and the most brilliant mechanical analogue of the Pi-plane used by Ko & Scott (1967) to achieve a rotation of the principal stresses on a true triaxial—all again sadly obsolete in our rather grey electronic age. Although servo-hydraulics were used for dynamic testing (Cullingford *et al.*, 1972) the two most successful systems for monotonic loading were ultimately based on motorised screw systems (Menzies *et al.*, 1977) or motorised air pressure controllers, which was the route chosen by Imperial College and City University (Atkinson, 1985).

An underlying theme through the years is the mechanical genius of Bishop, which is clear in the rather complex ring shear he developed in conjunction with NGI (Bishop *et al.*, 1971), with its great advantage of shearing on the sample mid-plane, and also in the elegant and simple design of the hydraulic triaxial apparatus, which was first published by Bishop & Wesley (1975). This had been based on versions for larger samples at Imperial College (Atkinson, 1973) as well as even earlier apparatus developed by other students of Bishop for investigating creep. For many this remains the standard research triaxial apparatus; the use of hydraulic loading through an axial piston sealed with rolling Bellofram facilitates the application of more complex stress paths, and obviates the need for expensive loading frames.

For us it was perhaps surprising how early it was realised that the differences in behaviour seen in various apparatus were in part attributable to different modes of loading, and how quickly the limitations of the axisymmetry in the triaxial and the uncontrolled rotation of principal stresses in the simple shear were addressed. Initially comparisons focused on the differences between triaxial and plane-strain apparatus (Cornforth, 1964; Roscoe, 1970), but cubical true triaxial cells were soon developed that allowed independent application of principal stresses on three pairs of flexible faces supported by a metal frame (Ko & Scott, 1967; Arthur & Menzies, 1972). An intriguing variant was the independent stress control apparatus (Bishop, 1966; Green & Bishop, 1969; Green & Reades, 1975), which used a cuboidal sample in a triaxial chamber, but had a belt on two of the horizontal faces to control the stress or strain in that direction. A common criticism of the flexible-faced true triaxial was that the frame restricted strain in the corners of the sample. This was overcome by the Cambridge type of true triaxial with independently moving rigid faces (Hambly, 1969; Roscoe, 1970), which is an apparatus that almost defies description in its ingenuity. Similarly, a rigid-faced simple shear apparatus was developed, but although the strains were uniform, the stresses were not, as investigated by Roscoe using multiple load cells in the platens. The debate over flexible versus rigid loading boundaries applied equally to consolidation tests with the development of the Rowe consolidation cell (Rowe & Barden, 1966), in which the vertical stress was applied through a membrane, allowing the key advantage of being able to control drainage and apply a back-pressure to the sample.

True triaxial apparatus are limited to applying principal stresses on the sample faces, and apparatus soon appeared that allowed a rotation of the principal stresses, so allowing a fuller investigation of the effects of anisotropy. The directional shear cell (Arthur *et al.*, 1977) is a plane-strain device that allows independent control of the normal and shear stresses on two pairs of faces of a cubical element. Although there had been several previous attempts to build torsional shear devices, the hollow cylinder apparatus developed by Hight *et al.* (1983) finally allowed a complete control of both the intermediate principal stress and the rotation of the major principal stress relative to the sample axis.

In the early 1980s the number of papers appearing in *Géotechnique* about laboratory apparatus suddenly declined, perhaps in part because of a changing emphasis to the use of all the apparatus developed to investigate soil behaviour, but also perhaps as a result of the introduction of the *Geotechnical Testing Journal* by ASTM in 1978. Perhaps the two key developments during the later issues of *Géotechnique* were the design of apparatus for partially saturated soils, and the improvement of the measurement of soil stiffness at small strains. An important innovation in the former sphere was the development of osmotic control systems for suction (Kassiff, & Shalom, 1971), although the actual measurement of suctions during a test rather than through indirect techniques such as the pressure plate or filter paper came somewhat later, with the design of the 'suction probe' tensiometer (Ridley & Burland, 1993).

The 1980s and early 1990s also saw an increasing realisation that the behaviour of soils at small strains was highly non-linear, and that much of the difference between dynamic and monotonically measured stiffnesses could be accounted for by the strain level. Jardine *et al.* (1984) was a highly influential paper in emphasising the need for local strain measurement in laboratory tests. Although the importance of local strain measurement had already been realised by Brown & Snaith (1974) and Boyce & Brown (1976) for measuring permanent strains in cyclic loading, it was now apparent that very accurate strain measurement was also required for monotonic loading if realistic predictions of ground displacements were to be made. This led to a rejuvenation in laboratory testing, which was better able to resolve the small strains required than field tests, and a range of local strain-measuring devices was quickly devised, from inclinometer types (Burland & Symes, 1982; Ackerley *et al.*, 1987), to proximity transducers (Hird & Yung, 1987), Hall effect transducers (Clayton & Khatrush, 1986), and strain-gauged strips for axial and radial strain (Kim *et al.*, 1994; El-Ruwayih, A. A., 1976; Matyas & Radhakrishna, 1968). But ultimately perhaps the most accurate has proven to be the LVDTs, as used by Brown & Snaith (1974) for axial strains and Menzies (1976) on a radial strain belt. The continued improvement to the precision of these devices eventually allowed soil stiffnesses to be measured at strain levels similar to those applied by dynamic tests. However, few developments of dynamic laboratory tests actually appear in *Géotechnique*, with the exception of bender elements, which first appeared in Bates (1989). Although this non-destructive technique could be used in most laboratory apparatus, they were typically mounted in triaxial cells and initially with an axially mounted arrangement. Pennington *et al.* (1997) introduced laterally mounted elements, which then allowed all three shear moduli to be measured for a cross-anisotropic soil, even if the uncertainty in the interpretation of bender elements seems to grow in proportion to the number of researchers using them! Combined with local measurement of strains during monotonic probes this then permitted the complete set of elastic parameters to be obtained.

IN SITU TESTING

In the minds of many engineers there has always been an ongoing intellectual battle between those working in the laboratory and those working in the field. To some extent this battle represents the different perspectives to problem solving embraced by researchers and practitioners, or the prejudices of those who tend to rely on experience rather than knowledge. But we do not have to conform to these preconceived views. In fact, the historical and main reason as to why in situ testing has consolidated itself as an established area of knowledge is because of the strong integrative process established between the laboratory and the field, where conceptual frameworks developed in the laboratory have induced the technological and theoretical developments of field model tests. Once a new level of insight and understanding produced by laboratory tests proved to be effective, it quickly inspired the development of model in situ tests designed to capture particular information in the field by a straightforward procedure. Some of these field model tests proved to be remarkably robust, but no matter how novel an instrument was, it was never totally independent of previous theoretical and experimental knowledge.

In browsing through the pages of *Géotechnique* the solidarity between the areas of laboratory and field tests becomes immediately clear. The early laboratory techniques also described marked well-defined breakthroughs in the development of our field, and framed the gradual progress of in situ testing geomechanics. The following major developments should be highlighted:

(*a*) the testing and standardisation of ground investigation tools used both for preliminary site exploration and for assessment of soil properties from semi-empirical correlations
(*b*) the development of interpretative analytical methods based on bearing capacity and cavity expansion theories
(*c*) the gradual growth of numerical methods based on constitutive models framed by critical-state soil mechanics and reinforced by the strain path method
(*d*) the recognition that small-strain stiffness is a major feature of natural soils, which has prompted the use of in situ geophysical shear wave measurements in geomechanics.

The first 10 years of *Géotechnique* record the birth of a new scientific field by setting the background to these developments, and to the main postulates currently used in modern soil mechanics. It is a science of everyday work, addressing the fundamental questions that every engineer has wondered about when dealing with soils and rocks. Recognition in the mid 1940s that procedures of sampling and testing in clay were not always reliable to obtain accurate soil properties prompted the development of the vane test. Skempton (1948) acknowledged the need to seek methods other than laboratory tests, preferably ones in which the strength is measured in situ. He describes the construction of an apparatus similar in principle to that developed by the Royal Swedish Geotechnical Institute, recognising the need for proper calibration procedures, and introducing sound analytical interpretation of the rotation of the blades. After the key contributions from Brinch Hansen (1950), Skempton & Northey (1952) and Bjerrum (1954), the vane has quickly become accepted as a test that measures the undrained shear strength and sensitivity of clays in situ.

The need to provide field measurements of pore water pressure to take full advantage of the principles of effective stress prompted the development of a field piezometer apparatus by Penman (1956). The device was a tip containing a porous stone buried in the soil and connected by plastic tubing to a Bourdon gauge. Only decades later did the measurements of pore pressure became routine in site characterisation. In parallel, the importance of geophysical methods was already recognised. Brown & Robertshaw (1953) reported the in situ measurement of Young's modulus for rock by dynamic methods. The authors even introduced a correlation between the modulus and the longitudinal wave velocity that is not entirely different from today's practice. A few years later, Jones (1958) described a method for obtaining a dynamic shear modulus of soils by a mass vibrating perpendicular to the surface. In these pioneering contributions not only the experimental tools but also the main theoretical principles of soil dynamics have been consistently described.

Penetration tests started to gain importance in *Géotechnique* much later. Originating in the 1930s, the cone penetration test (CPT) used in parts of Europe and the standard penetration test (SPT) used in North and South America appeared in *Géotechnique* in 1970. In the following years there were several publications devoted to penetration tests, comparing SPT and CPT tests on the basis of semi-empirical correlations, and discussing the effects of geometry, procedures and penetration rates on the accuracy of predicted soil properties. In truth it was already evident in these first publications that standardisation was a key element in obtaining reliable and comparable results. Ireland *et al.* (1970) described the significant differences in drilling techniques, equipment and test procedures adopted in the SPT, and the consequent variations in the observed blow count N. A decade later Skempton (1986) observed the essential need to correct the N value to the value that would have been measured using a specified rod energy. A recommended value, which became recognised internationally, is 60% of the free-fall energy of the standard hammer weight and drop. Nevertheless, the criticism of the SPT remains.

A major breakthrough appeared in the 1970s with the realisation that improved understanding and interpretation of pressuremeter tests could be obtained by application of cavity expansion theory. The probe contains sufficient geometric symmetries to reduce the problem to a relatively simple form, so that the measured pressure–displacement curve can be used to back-calculate the mechanical properties of the tested soil, adopting cavity expansion solutions combined with constitutive models capable of representing soil behaviour. An elegant study made of the problem of interpreting the results of Ménard pressuremeter tests was presented by Palmer (1972). The author showed that for an elastic perfectly plastic soil the results of a single test in saturated clay, under undrained conditions, were enough to determine a complete stress–strain relation in plane strain. Five years later, Hughes *et al.* (1977) extended the solution of cavity expansion to sands with allowance for the volume change that occurs in shear. Within the assumptions that the stress–dilatancy theory is valid, and that sand fails with a constant ratio of principal stresses, the authors introduced a method of determining the angle of internal friction ϕ' and the angle of dilation ψ from self-boring pressuremeter test data. Values of shear modulus G and in situ horizontal stress σ_{ho} were also measured and reported. Unquestionably, the theoretical bases provided by cavity expansion theory enhanced the value of the self-boring pressuremeter test as a site investigation tool devised to obtain accurate soil properties in situ. Around this time, the conceptual insights of the penetration mechanism of piles and cones were given by Randolph *et al.* (1979) in a detailed study of the installation of model piles using large-strain cylindrical cavity expansion analysis of incompressible soils. These major innovative concepts were recognised by Wroth (1984) in his Rankine Lecture, which framed the knowledge of the time by describing the relationships between soil properties determined by in situ tests.

Understanding of the fundamental mechanics of deep penetration problems was a more complex challenge. The slow progress in our ability to describe this mechanism started with Baligh's recognition that the process was not a simple variation on bearing capacity problems, but a steady-state motion in which the continuity of steady penetration could be described within the scope of plasticity theory (Baligh & Scott, 1976). The proposed solution to penetration problems (Baligh, 1986a, 1986b) accounted for the undrained steady penetration of a solid pile in a homogeneous isotropic elastic-perfectly plastic and incompressible clay that is initially subjected to isotropic stresses. In the assumed boundary conditions along the soil/pile boundary interface it was considered that the deformation and strains caused by penetration were independent of the shearing behaviour, which implies that the penetration problem becomes totally strain controlled. The strain path method has been used ever since to predict soil velocities, shear stresses and pore pressure fields in undrained penetration, and has supported new forms of interpretation of the piezocone test.

The theoretical framework of both cavity expansion theory and the strain path method supported most developments related to the interpretation of in situ tests in recent decades. Some important contributions have been selected here to illustrate these developments. Yu & Houlsby (1991) extended the solutions of cavity expansion by the introduction of dilation to the analysis of large strain of both cylindrical and spherical cavities. The dilatancy of the soil is accounted for by adopting a Mohr–Coulomb criterion with a non-associated plastic flow rule. The importance of the analysis is that no restrictions are imposed on the magnitude of the deformations, which is an important feature of soil behaviour when modelling problems such as cone penetration. Also in the early 1990s, Teh & Houlsby (1991) combined the strain path method with finite element calculations to model the consolidation process around a penetrating cone, where penetration is viewed as a steady flow of an elastic-perfectly plastic isotropic, homogeneous material that is followed by an uncoupled consolidation process to calculate the dissipation around a stationary cone. In spite of the approximate nature of this solution, it is sufficiently realistic to capture the essential elements of the dissipation process.

There are several other important contributions that would have deserved to be mentioned in this brief review—for example recent published papers dealing with the in situ measurement of the small-strain shear modulus from geophysical tests. However, these papers do not form a body of collective work in *Géotechnique*, and in writing this review we made the choice of selecting papers not only because they can be seen as important single individual contributions, but because they can be viewed as part of a combined effort, developing to form a rational framework that, through the pages of *Géotechnique*, has had a significant impact on the way engineering science and engineering practice are perceived today.

THE FUTURE

The pages of *Géotechnique* reveal the profound impact that the journal has had on the development of laboratory and in situ tests, even if most of the key 'inventions' came during the first few decades, and recent years have tended rather to see improvements of the techniques, the quality of measurements and the means of analysis. It would certainly be beyond foolishness to predict the next 60 years, but confining ourselves to 20 years, which is about the hoped-for span of our own careers, it is most likely that this process of refinement will continue, and that the sort of paradigm shift beloved of academics will have to wait a little longer. The longevity of the basic soils apparatus is certainly a testament to the genius of the great soil mechanics of the early decades of *Géotechnique*. Specialised and rather odd apparatus will of course continue to be built for specific research purposes: for example, we will need a suite of tests to characterise the properties of individual soil particles to relate these to the macroscopic behaviour. But while it is likely that particulate approaches will continue to inform our understanding of continuum models, it is most unlikely that we shall be using them for routine design within that period to the extent that we will find rubbish skips full of triaxial apparatus and self-boring pressuremeters. In the spirit of *Géotechnique*, we should be delighted if the authors of a review of laboratory and in situ tests for its eightieth birthday were able to lament the lack of foresight of Messrs Coop and Schnaid.

REFERENCES

Ackerley, S. K., Hellings, J. E. & Jardine, R. J. (1987). Discussion: 'A new device for measuring local axial strains on triaxial specimens'. *Géotechnique* **37**, No. 4, 414–415.

Arthur, J. R. F., Chua, K. S. & Dunstan, T. (1977). Induced anisotropy in a sand. *Géotechnique* **27**, No. 1, 13–30.

Arthur, J. R. F. & Menzies, B. K. (1972). Inherent anisotropy in a sand. *Géotechnique* **22**, No. 1, 115–128.

Atkinson, J. H. (1973). *The deformation of undisturbed London Clay*. PhD thesis, University of London.

Atkinson, J. H. (1985). Simple and inexpensive pressure control equipment for conventional and stress path triaxial testing of soils. *Géotechnique* **35**, No. 1, 61–63.

Baligh, M. M. & Scott, R. F. (1976). Analysis of wedge penetration in clay. *Géotechnique* **26**, No. 1, 185–208.

Baligh, M. M. (1986a). Undrained deep penetration, I: Shear stresses. *Géotechnique* **36**, No. 4, 471–485.

Baligh, M. M. (1986b). Undrained deep penetration, II: Pore pressures. *Géotechnique* **36**, No. 4, 487–501.

Bates, C. R. (1989). Dynamic soil property measurements during triaxial testing. *Géotechnique* **39**, No. 4, 721–726.

Bishop, A. W. (1966). Sixth Rankine Lecture: The strength of soils as engineering materials. *Géotechnique* **16**, No. 2, 91–130.

Bishop, A. W. & Eldin, G. (1950). Undrained triaxial tests on saturated sand and their significance to the general theory of shear strength. *Géotechnique* **2**, No. 1, 13–32.

Bishop, A. W. & Henkel, D. J. (1953). A constant-pressure control for the triaxial-compression test. *Géotechnique* **3**, No. 8, 339–344.

Bishop, A. W. & Wesley, L. D. (1975). A hydraulic triaxial apparatus for controlled stress path testing. *Géotechnique* **25**, No. 4, 657–670.

Bishop, A. W., Gren, G. E., Garga, V. K., Andersen, A. & Brown, J. D. (1971). A new ring shear apparatus and its application to the measurement of residual strength. *Géotechnique* **21**, No. 4, 273–328.

Bjerrum, L. (1954). Geotechnical properties of Norwegian marine clays. *Géotechnique* **4**, No. 2, 49–69.

Bjerrum, L. & Landva, A. (1966). Direct simple shear tests on a Norwegian quick clay. *Géotechnique* **16**, No. 1, 1–20.

Boyce, J. R. & Brown, S. F. (1976). Measurement of elastic strain in granular material. *Géotechnique* **26**, No. 4, 637–640.

Brinch Hansen, J. (1950). Vane tests in a Norwegian quick-clay. *Géotechnique* **2**, No. 1, 58–63.

Brown, P. D. & Robertshaw, J. (1953). The in situ measurement of Young's modulus for rock by dynamic method. *Géotechnique* **3**, No. 7, 283–286.

Brown, S. F. & Snaith, M. S. (1974). The measurement of recoverable and irrecoverable deformations in the repeated load triaxial test. *Géotechnique* **24**, No. 2, 255–259.

Burland, J. B. & Symes, M. (1982). A simple axial displacement gauge for use in the triaxial apparatus. *Géotechnique* **32**, No. 1, 62–65.

Clayton, C. R. I. & Khatrush, S. A. (1986). A new device for measuring local axial strains on triaxial specimens. *Géotechnique* **36**, No. 4, 593–597.

Cornforth, D. H. (1964). Some experiments on the influence of strain conditions on the strength of sand. *Géotechnique* **14**, No. 2, 143–167.

Cullingford, G., Lashine, A. K. F. & Parr, G. B. (1972). Servo-controlled equipment for dynamic triaxial testing of soils. *Géotechnique* **22**, No. 3, 529.

Darley, P. (1973). Discussion: 'Apparatus for measuring volume change suitable for automatic logging'. *Géotechnique* **23**, No. 1, 140–141.

De Beer, E. E. (1950) The cell-test. *Géotechnique* **2**, No. 2, 162–172.

El-Ruwayih, A. A. (1976). Design, manufacture and performance of a lateral strain device. *Géotechnique* **26**, No. 1, 215–216.

Geuze, E. C. W. A. & Tan Tjong Kie (1950). The shearing properties of soils. *Géotechnique* **2**, No. 2, 141–161.

Glynn, T. E. & Spence, R. A. (1961). An automatic control for pore-pressure measurements. *Géotechnique* **11**, No. 3, 183–185.

Green, G. E. & Bishop, A. W. (1969). A note on the drained strength of sand under generalized strain conditions. *Géotechnique* **19**, No. 1, 144–149.

Green, G. E. & Reades, D. W. (1975). Boundary conditions, anisotropy and sample shape effects on the stress–strain behaviour of sand in triaxial compression and plane strain. *Géotechnique* **25**, No. 2, 333–356.

Hambly, E. C. (1969). A new true triaxial apparatus. *Géotechnique* **19**, No. 2, 307–309.

Hanrahan, E. T. (1954). An investigation of some physical properties of peat. *Géotechnique* **4**, No. 3, 108–123.

Hight, D. W. (1982). A simple piezometer probe for the routine measurement of pore pressure in triaxial tests on saturated soils. *Géotechnique* **32**, No. 4, 396–401.

Hight, D. W., Gens, A. & Symes, M. J. (1983). The development of a new hollow cylinder apparatus for investigation of the effects of principal stress rotation in soils. *Géotechnique* **33**, No. 4, 355–383.

Hird, C. C. & Yung, P. (1987). Discussion: 'A new device for measuring local axial strains on triaxial specimens'. *Géotechnique* **37**, No. 4, 413–414.

Hughes, J. M. O., Wroth, C. P. & Windle, D. (1977). Pressuremeter tests in sands. *Géotechnique* **27**, No. 4, 455–471.

Ireland, H. O., Moreto, O. & Vargas, M. (1970). The dynamic penetration test: a standard that is not standardized. *Géotechnique* **20**, No. 2, 185–192.

Jardine, R. J., Symes, M. J. & Burland, J. B. (1984). The measurement of soil stiffness in the triaxial apparatus. *Géotechnique* **34**, No. 3, 323–340.

Jones, R. (1958). In situ measurement of the dynamic properties of soil by vibration methods. *Géotechnique* **8**, No. 1, 1–21.

Kassiff, G. & Shalom, A. B. (1971). Experimental relationship between swell pressure and suction. *Géotechnique* **21**, No. 3, 245–255.

Kim, Y.-S., Tatsuoka, F. & Ochi, K. (1994). Deformation characteristics of sedimentary soft rocks by triaxial compression tests. *Géotechnique* **44**, No. 3, 461–478.

Kjellman, W. (1951). Testing the shear strength of clay in Sweden. *Géotechnique* **2**, No. 3, 225–231.

Klementev, I. (1974). Lever-type apparatus for electrically measuring volume change. *Géotechnique* **24**, No. 4, 670–671.

Ko, H.-Y. & Scott, R. F. (1967). A new soil testing apparatus. *Géotechnique* **17**, No. 1, 40–57.

Lewin, P. I. (1971). Use of servo mechanisms for volume change measurement and K_0 consolidation. *Géotechnique* **21**, No. 3, 259–262.

Lewin, P. I. & Burland, J. B. (1970). Stress-probe experiments on saturated normally consolidated clay. *Géotechnique* **20**, No. 1, 38–56.

Matyas, E. L. & Radhakrishna, H. S. (1968). Volume change characteristics of partially saturated soils. *Géotechnique* **18**, No. 4, 432–448.

Mayer, A. (1951). Les measures de cisaillement dans les laboratoires Francais. *Géotechnique* **2**, No. 3, 236–243.

Menzies, B. K. (1975). A device for measuring volume change. *Géotechnique* **25**, No. 1, 133–134.

Menzies, B. K. (1976). Discussion: 'Design manufacture and performance of a lateral strain device'. *Géotechnique* **26**, No. 3, 542–544.

Menzies, B. K., Sutton, H. & Davies, R. E. (1977). A new system for automatically simulating K_0 consolidation and K_0 swelling in the conventional triaxial cell. *Géotechnique* **27**, No. 4, 161–179.

Palmer, A. C. (1972). Undrained plane-strain expansion of a cylindrical cavity in clay: a simple interpretation of the pressuremeter test. *Géotechnique* **22**, No. 3, 451–457.

Penman, A. D. M. (1953). Shear characteristics of a saturated silt measured in triaxial compression. *Géotechnique* **3**, No. 8, 312–328.

Penman, A. D. M. (1956). A field piezometer apparatus. *Géotechnique* **6**, No. 2, 57–65.

Pennington, D. S., Nash, D. F. T. & Lings, M. L. (1997). Anisotropy of G_0 stiffness in Gault Clay. *Géotechnique* **47**, No. 3, 391–398.

Randolph, M. F., Carter, J. P. & Wroth, C. P. (1979). Driven piles in clay: the effects of installation and subsequent consolidation. *Géotechnique* **29**, No. 4, 361–393.

Richardson, A. M. & Whitman, R. V. (1963). Effect of strain-rate upon undrained shear resistance of a saturated remoulded fat clay. *Géotechnique* **13**, No. 4, 310–324.

Ridley, A. M. & Burland, J. B. (1993). A new instrument for the measurement of soil moisture suction. *Géotechnique* **43**, No. 2, 321–324.

Roscoe, K. H. (1970). Tenth Rankine Lecture: The influence of strains in soil mechanics. *Géotechnique* **20**, No. 2, 129–170.

Rowe, P. W. & Barden, L. (1966). A new consolidation cell. *Géotechnique* **16**, No. 2, 162–170.

Rowlands, G. O. (1972). Apparatus for measuring volume change suitable for automatic logging. *Géotechnique* **22**, No. 3, 525–526.

Skempton, A. W. (1948). Vane tests in the alluvial plain of the River Forth near Grangemonth. *Géotechnique* **1**, No. 2, 111–124.

Skempton, A. W. (1986). Standard penetration test procedures and the effects in sands of overburden pressure, relative density, particle size, ageing and overconsolidation. *Géotechnique* **36**, No. 3, 425–447.

Skempton, A. W. & Bishop, A. W. (1950). The measurement of the shear strength of soils. *Géotechnique* **2**, No. 2, 90–108.

Skempton, A. W. & Northey, R. D. (1952). The sensitivity of clays. *Géotechnique* **3**, No. 1, 30–53.

Teh, C. I. & Houlsby, G. T. (1991). An analytical study of the cone penetration test in clay. *Géotechnique* **41**, No. 1, 1–15.

Wroth, C. P. (1984). 24th Rankine Lecture: The interpretation of in situ soil tests. *Géotechnique* **34**, No. 4, 449–489.

Yu, H. S. & Houlsby, G. T. (1991). Finite cavity expansion in dilatant soil: loading analysis. *Géotechnique* **41**, No. 2, 173–183.

Jefferies, M., Delage, P. & Cotecchia, F. (2008). *Géotechnique* **58**, No. 5, 429–433 [**doi:** 10.1680/geot.2008.58.5.429]

Contributions to *Géotechnique* 1948–2008: Soil behaviour

M. JEFFERIES*, P. DELAGE† and F. COTECCHIA‡

Soil behaviour (that is, strains induced by changes in stresses, suction, or physico-chemical conditions) has been a recurring theme of contributions to *Géotechnique* throughout its 60 years. Focusing on experimental work, this essay illustrates the developing understanding of soil mechanics through brief reviews of 47 contributions. These contributions were chosen for their insight or influence, necessarily involving subjective judgement on our part. Some work comprises now well-understood wisdom. Other work provides fundamental, but not necessarily widely appreciated, insight—and continues to be relevant to developments in geomechanics today. Most interestingly, a striking feature in these various contributions is the commonality in findings across the many soils and loading paths investigated. From a practical perspective, this body of work allows assessment of soil behaviour, under a wide range of circumstances, using surprisingly few tests. More generally, these cited contributions provide a history of the subject—and suggest where the subject may go over the next few decades.

KEYWORDS: compressibility; creep; elasticity; historical review; partial saturation; plasticity; shear strength; stiffness

Le comportement du sol (c'est à dire les déformations induites par des variations de contraintes, de succion ou de conditions physico-chimiques) est un thème que l'on retrouve souvent dans les communications publiées dans *Géotechnique*, depuis maintenant 60 ans. Le présent article se base sur les travaux expérimentaux publiés pour illustrer l'évolution des connaissances de la mécanique des sols, par le biais d'un examen rapide de quarante-cinq communications. Ces contributions ont été choisies pour leur intérêt ou pour l'impact qu'elles on eu, ce qui implique nécessairement de notre part une certaine part de subjectivité. Certains comprennent des connaissances désormais bien connues, d'autres apportent des éléments fondamentaux, qui ne sont pas toujours très connus, mais qui sont toujours pertinents dans les développements actuels en géomécanique. On relève avec intérêt qu'un des aspects saillants des différentes communications est une convergence des conclusions sur les sols les plus variés et des chemins de contraintes examinés. Sur un plan pratique, cet ensemble d'travaux permet d'évaluer le comportement des sols dans tout un éventail de circonstances, en n'utilisant qu'un nombre étonnamment limité de tests. D'un point de vue plus général, ces communications fournissent un historique du sujet, et contiennent des indications sur de son évolution probable au cours des prochaines décennies.

INTRODUCTION

Soil behaviour, as discussed here, is taken as the strains induced either by changes in the imposed hydromechanical conditions (stresses and/or suction) or by changes in the physico-chemical conditions. Soil behaviour depends on the nature of the soil constituents (particle size distribution, particle roundness, mineralogy, pore water chemistry) and on the soil state (defined by internal variables including the specific volume, the microstructure, and the degree of saturation). Starting perhaps a century ago, the influence of these properties and state factors on soil behaviour has been much investigated. Most often, investigation used reconstituted soils under axisymmetric stress conditions. The extension to soil behaviour under general stress conditions (including the effects of principal stress rotation, intermediate stress, and unloading), to undisturbed as opposed to reconstituted soils, and to considerations about discontinuous behaviour (related to the effects of fissures, joints, strain localisation or grain crushing) has been less considered, even though practically important. Some common behaviours and frameworks have been found across the range of soils, the practical utility of which is that relatively few 'good' tests allow an engineer to make a reasoned assessment over a surprisingly wide range of conditions.

Discussion on this paper closes on 1 December 2008, for further details see p. ii.

* Golder Associates Vancouver, Canada.
† Ecole des Ponts, Paris (Université Paris-Est, CERMES UR Navier), France.
‡ Technical University of Bari, Italy.

In this essay, with apologies to authors of many important but neglected papers (because of space limitations), we look at the history of the developing understanding of soil behaviour as seen through 60 years of contributions to *Géotechnique*—and which largely matches most of the developments in the subject.

KEY CONTRIBUTIONS TO UNDERSTANDING OF SOIL BEHAVIOUR

The influence of water content on undrained clay strength had been known since the mid 1930s. Bishop & Eldin (1950) used undrained triaxial tests on reconstituted sand samples to validate the effective stress principle, and show that undrained shear strength is uniquely related to porosity, regardless of the subsequent changes in total confining stress. By testing a granular material, the apparent cohesion indicated by $\phi_u = 0°$ is purely a consequence of induced excess pore pressures caused by the imposed boundary conditions impeding drainage and resulting in an unchanged effective stress state—an elegant and fundamental result from well-conceived experiments.

During the 1950s, basic clay behaviour was investigated using Weald and London Clays, reconstituted samples being used for repeatable, controlled experiments. Henkel (1956) explored the effect of overconsolidation on the triaxial compression undrained strength. Simple, systematic trends were found. Then, again using triaxial tests but now a range of stress paths, Henkel (1960) found water content to be uniquely related to effective stress for normally consolidated clays, irrespective of drainage conditions or stress path, introducing specific volume as an internal variable controlling soil behaviour. Along similar lines, Parry (1960) moved to stress–strain behaviour of overconsolidated clays and found the stress–strain behaviour (not just strength) could be normalised in terms of overconsolidation ratio. Together these three papers are a historically important body of work, substantiating earlier suggestions by Rendulic and Hvorslev, and providing the basis for what has become known as critical-state soil mechanics.

Most practical work involves undisturbed rather than remoulded soils. Skempton & Northey (1952) introduced the liquidity index as a normalising state variable for the influence of geological and compositional factors on clay behaviour. Noting the difference between the normal consolidation of clay during natural sedimentation and accretion in situ and the normal consolidation behaviour of the same clay reconstituted in the laboratory, they showed that the differing structure of the clay directly influenced the clay's compressibility. This work highlighted how compressibility and undrained shear strength depend on structure, and in particular how undrained shear strength does not depend only on water content. Data were also given showing the important effects of post-sedimentation processes, such as thixotropy and leaching, on clay behaviour; this is still topical today.

Particularly interesting aspects of site characterisation and the response of natural stiff clays were provided by Ward *et al.* (1959), who tested block samples of London Clay obtained from across a large area of London. On a practical level, this work provides a compendium of high-quality data on London Clay, with the first quantification of deficiencies in conventional site investigation procedures. Perhaps more importantly, the participation of seven companies in this work showed what could be achieved collaboratively—a forerunner to the now widely appreciated CIRIA organisation. Further understanding, in effective stress terms, of undisturbed London Clay was presented by Bishop *et al.* (1965), also based on extensive testing of high-quality block samples.

Although the importance of the soil fabric (arrangement of the soil particles) and structure (also accounting for the links between elementary units) to soil behaviour had been indirectly recognised 50 years ago, by the 1970s technical advances in scanning electron microscopy technology allowed this facet of natural soils to be fully examined. Collins & McGown (1974) used this technology to investigate various recent normally consolidated or lightly overconsolidated natural soils (clays, silts and sand). They found very few occurrences of single clay layer arrangements, and this showed that clay aggregations were a relevant elemental unit linking microstructure to macroscopic behaviour.

Clay fabric is not restricted to particle arrangements, because natural clays, and stiff clays in particular, contain fissures or joints. These discontinuities are weaker than the intact material, and as the discontinuities have a characteristic spacing (intensity), fissured clay strength depends on the size of the sample. Lo (1970) introduced the operational strength in situ, a strength that is intermediate between the intact clay strength and the strength on the fissure planes themselves. A strength–size relationship was found and, taking anisotropy and time to failure into account, Lo showed that it was possible to use conventional $\phi = 0$ analysis for short-term (undrained) stability analyses of stiff fissured clays. This is a useful, simple, practical approach for an otherwise difficult aspect of soil strength. At the same time, the residual strength along fissure planes after large sliding movement was being studied within slope stability investigations. Lupini *et al.* (1981) reported a compilation of many ring shear tests to show the mechanistic basis by which the nature of the particles forming the clay controls the clay's residual friction angle.

Developing knowledge about clay structure prompted investigations relating clay behaviour to specific structural features, with ideas from plasticity theory (and critical-state soil mechanics in particular) providing a context for experimental investigations. Wong & Mitchell (1975) pioneered this approach with an investigation of the elasto-plastic behaviour of a Champlain Clay from the Ottawa area using experimental determination of yield surface evolution and corresponding plastic deformations at yield. The peculiar behaviour of this sensitive clay was related to the interparticle bonding. Extending these ideas to a wider range of soils, Leroueil & Vaughan (1990) compiled an impressive number of published experimental data on natural soils and porous weak rocks from various continents (soft clays, residual soils, cemented sands, porous mudstone, tuff). This compilation allowed detailed description of yield along various stress orientations for these soils, with common behavioural features being then apparent. This unified view of yielding behaviour confirmed that, for saturated soils, the clay structure was an additional internal variable to include in constitutive modelling. The investigation of Bothkennar clay by Smith *et al.* (1992) provided further detailed information about yield processes, made possible by very precise local strain measurements. Recalling the thixotropic aspects of clay behaviour identified 30 years earlier by Skempton & Northey (1952), but now including the concept of yield surface evolution, Graham *et al.* (1983) reviewed how strain rate affected clay behaviour, using data on Canadian, Irish, Norwegian and Thai clays, with new tests then extending the framework to that of yield envelopes scaling with strain rate.

With a common pattern of clay yield apparent, and clay structure clarified as a necessary internal variable, how to quantify this internal variable becomes the next step. Burland (1990) looked back to the framework of Henkel (1956, 1960) and Parry (1960) and used the remoulded first compression behaviour to normalise in situ clay strength and stiffness, in essence further developing the framework of Skempton & Northey (1952), but changing from their liquidity index to a void index. Following the same trend, Cotecchia & Chandler (2000) considered the effects of structure during sedimentation, and related different responses in compression to different geological histories and structures. They confirmed the difference between preconsolidation pressure and yield pressure for clays experiencing structural changes after sedimentation. They showed how sensitivity could be a parameter to quantify the effects of structure on the size of the state boundary surface. These contributions, following the critical-state framework, unified what at first sight appeared to be a wide spectrum of clay behaviour.

Following the masterly introduction to unsaturated soil concepts and experimental techniques provided by Croney (1952), various attempts were made at the end of the 1950s to extend the effective stress principle to unsaturated soils. Based on considerations about stress and suction distributions in unsaturated sands with menisci located at the intergrain contacts, most attempts led to expressions basically similar to the Bishop (1959) expression.[§] Jennings & Burland (1962) presented a detailed analysis of the applicability of effective stress in saturated and unsaturated soils, and demonstrated experimentally that the collapse phenomenon under soaking clearly showed that a unique effective stress like Bishop's suggestion could not properly account for the 'changes in volume and shearing strength of a soil due to changes in effective stress', mainly as a result of microstructural considerations. This led to the understanding

§ $\sigma' = (\sigma - u_a) - \chi(u_a - u_w)$, where u_a and u_w are the air and water pressures respectively, $s = u_a - u_w$ the suction, and χ is a parameter between 0 and 1, depending on the degree of saturation.

that two independent stress variables were needed, as suggested by Coleman (1962) in a landmark contribution.

Based on the use of two independent stress variables, Matyas & Radhakrishna (1968) described the compression behaviour of unsaturated soils in an experimental investigation using suction-controlled isotropic compression tests. The change of state of the unsaturated samples tested was, for the first time, fully described in terms of changes in volume and degree of saturation through two state surfaces. The relevance of critical-state concepts to unsaturated soils was demonstrated by Wheeler & Sivakumar (1995), Maâtouk *et al.* (1995) and Cui & Delage (1996). The last study showed the basis of suction hardening and the anisotropic nature of compacted soils (giving an inclined yield surface). Together, these three contributions confirmed the extension of the critical-state framework to unsaturated soils proposed by Alonso *et al.* (1990). Recently, in the topical and very important context of high-level nuclear waste disposal, Lloret *et al.* (2003) extended the understanding into heavily compacted swelling soils, with Romero *et al.* (2003) adding insight into the effect of temperature.

In the late 1970s sands came into focus worldwide, partly because sand liquefaction was becoming a major component of earthquake hazard reduction programmes—possibly the only example where soil behaviour has been a focus of government policy. In many ways, sand testing followed clays, with initial work being to normalise experimental data in terms of state variables. Of course, a common state variable for sands was relative density (or relative void index), but this suffers from two fundamental deficiencies that can be exposed by a trivial amount of testing. First, sand behaviour changes at constant relative density with changes in stress level. Second, even at constant stress level and fixed relative density, behaviour can be very different from one sand to another—something especially marked for the practical situation of sands with traces of silt. Relative density is a misleading state variable. To address the effect of stress level on sand behaviour, Bolton (1986) introduced the relative dilatancy index, an index that accounts for dilatancy reduction by confining stress while retaining the familiar aspects of relative density. However, the relative dilatancy index does not address the deficiency of dissimilar behaviour at the same index value between different soils (mineralogy, angularity, fines content etc.).

The work on clays discussed above shows a trend towards using the critical state as a fundamental framework for understanding their behaviour. Historically, the concept of a critical state originated in the 1930s for sands. Building on the identification by Parry (1958) that peak dilatancy for soils denser than critical was proportional to the distance from the critical state, and including the concept of stress–dilatancy, the peak strength and dilatancy of sands were found to be apparently uniquely related to the void ratio difference from the sand's current critical state—the void ratio difference being called the state parameter for the soil. This state parameter is a simple, practical soil behaviour index, and the approach is most eloquently articulated in Been & Jefferies (1986). As data have accumulated, with till-like soils in particular, it has become apparent that Parry's trend seems universal, but there is a coefficient of proportionality that varies slightly from one soil to another, depending on particle size distribution and mineralogy.

The state parameter was introduced as an appropriate normalising index for cohesionless soil behaviour. Of course, adopting the critical state as a reference condition requires that the critical state can be measured. The critical state of clays has been measured for some 50 years, and the measurement is an established technology. Measuring the critical state of sand proved more controversial. Concerns developed during the period 1970–1990 about how sample preparation and stress path affected the measured critical state, with various workers suggesting that the critical state was not unique, and therefore not suitable for normalising soil behaviour. Been *et al.* (1991) explored these arguments in considerable detail, providing experimental evidence that the critical state is sensibly a true and unique property of sand within a narrow range.

Both Bolton and Been & Jefferies identified apparently fundamental trends in the strength and dilatancy of cohesionless soils across a large body of data, but this raises the question: how might useful empirical trends become formally linked to full understanding? Two sets of independent studies, both using the isotropic compression of sand, considered this question. In a comprehensive series of tests, Coop (1990) found very compressible sands tended to a unique normal compression line as the stress level increased. This tendency idealises as a limiting compression locus, and affirms a similar framework for compressible sand-like soils as was found for remoulded clays 40 years earlier. By contrast, Jefferies & Been (2000) tested a quartz sand at more usual engineering stress levels and found a spectrum of non-unique normally consolidated compression behaviour. Such non-unique normal compression behaviour makes the overconsolidation ratio and the state parameter independent variables, each of which influences soil behaviour. And it is as independent state variables that the overconsolidation ratio and state parameter appear, pervasively, in advanced plasticity models of soil behaviour. Fundamentally, this non-unique normal compression behaviour appears related to the creation of additional contacts between soil particles as the soil compresses, with different rates for the contact creation, depending on the specific volume.

The formation of grain contacts is separate from grain crushing, the latter being a further work dissipation process. Grain crushing underlies the limiting compression locus, and becomes increasingly important as the stress level increases. An important contribution to the understanding of grain crushing effects was provided by McDowell & Bolton (1998), who found that fractal-like successive fracturing of soil particles, which was controlled by the particle tensile strength, provided a physical basis for the normal compression line. This work might provide a glimpse of the future, in which the easily measured particle size distribution of a soil, combined with knowledge of the particle mineralogy and morphology (and hence the properties), predicts the soil's critical state and thus its detailed mechanical behaviour.

Soil stiffness received less attention than strength until the 1990s, when it was recognised that design should refer to the serviceability state far more. The serviceability state is a domain in which it is essential to assess the elastic stiffness. The elasticity of a standard reference sand (Ticino) was extensively investigated by Bellotti *et al.* (1996) using seismic waves on reconstituted samples in a very large test cell. Many data are presented, and this remains a landmark study on sand elasticity. A very different insight was provided by Cuccovillo & Coop (1997), who tested intact samples of two lightly cemented sands and measured the effect of shearing in breaking cemented bonds—a fascinating study, and a lovely contrast with the behaviour of reconstituted sand. For clays, Viggiani & Atkinson (1995) reported on the very small strain shear stiffness of four soils using predominantly reconstituted samples. Using both bender elements and local strain measurements, a most interesting result was the dependence of shear modulus on mean stress across a wide range of anisotropic stress states. More generally, the effect of overconsolidation ratio was established, and the overall framework has much in common with that of sand elasticity. At the same time, Pennington *et al.* (1997) used multiple-direction bender element testing to measure the elasticity of remoulded and undisturbed Gault Clay. They found anisotropic elastic shear stiffness to be much greater in the undisturbed natural samples; anisotropy in shear modulus might be viewed as an implicit characterisation of soil structure.

Plastic yielding is generally the reason for non-linear soil behaviour, and can arise surprisingly early in a loading. Atkinson *et al.* (1990) used reconstituted clays to show that even the most recent stress history—either a sudden change in the direction of the stress path or a period of time at a constant stress state—affects the soil response. The stiffness near the start of a loading path was found to change by as much as an order of magnitude following different rotations of the loading path. Such behaviour falls in the framework of kinematic plastic hardening. This form of behaviour was reported for sand by Jefferies (1997), who showed recovery of apparent plastic strain in unloading. Going further into these more general aspects of plastic behaviour, Kuwano & Jardine (2007) explored kinematic yielding of sand, by stress-probing with high-resolution equipment, to link the evolution of macro-scale plastic yield with underlying micromechanical processes. These behaviours are not yet accounted for in practical engineering, but will become so as they are necessary for accurate modelling of geotechnical works across a spectrum of situations, ranging from stiffness reduction during liquefaction to soil–structure interactions in NATM tunnelling. There is a lot more to soils than their response in first loading.

The soil behaviour discussed so far is that under triaxial conditions. Of course, triaxial conditions are not especially representative of much engineering reality, with plane strain being far more common in constructed engineering works. For constant principal stress direction, Cornforth (1964) reported a comparison between the results of triaxial compression, extension and plane strain tests on sand over a range of densities. Any practical engineer can quickly appreciate how strain conditions affect soil strength and stiffness from this elegant work and presentation. Also using a fixed principal stress direction, but with more general strain states, Wood (1975) tested reconstituted isotropically consolidated kaolin in 'true triaxial' shear in which the proportion of all three applied stresses and strains could be varied independently. Data for three groups of stress paths were presented. Today these two contributions, even some 40 years later, provide important validation cases for constitutive models in modern finite element codes. For a natural stiff clay, Callisto & Rampello (2002) provide experimental evidence for the shape of the strength envelope in the general stress space, again using true triaxial tests.

It was not until the late 1970s that the possibly dominant effect of principal stress rotation was discerned. In part, this insight followed from soil properties being commonly anisotropic because of their geological deposition. But there is much more to the effect of principal stress rotation on soil behaviour than that caused by anisotropic properties. Wong & Arthur (1986) found that rotating the principal stress direction radically alters soil behaviour from that seen under constant directions of principal stress, with proportionately much less dilation, even in very dense sand. Principal stress rotation causes a different particle movement mode from shear deformation, with densification towards the minimum specific volume rather than dilation to the critical state: this is a possible explanation for some of the very dense sands found in nature. More recently, Zdravkovic & Jardine (1997) investigated the stiffness of dense silt under principal stress rotation and confirmed the importance of such rotational loading for soil behaviour. The practical consequence of understanding the effect of principal rotation includes engineering earthquake-resistant foundations and engineering storm-resistant offshore structures. The application to foundation engineering for wind-generated electricity is obvious and topical. Now known for some 40 years, principal stress rotation remains an important but largely neglected loading mechanism of soil.

In laboratory testing, the tested sample is considered as a continuum 'soil element'. Yet even the simplest triaxial test on dense sand shows localised zones of more intense shearing, eventually leading to a slip plane. Understanding of soil behaviour is incomplete without understanding localisation. A comprehensive study by Desrues *et al.* (1996) used X-ray-based tomography to detect the inception and subsequent development of localisation during triaxial tests on dense and loose sand. The complex geometrical localisation patterns were measured as well as void ratio evolution in the localisation zones. The local void ratio in the shear zones evolved to a limit value (a local critical state), but this limit was often quite different from the global sample void ratio. The real importance of this work, however, lies in its value as validation data for advanced constitutive models that, as yet, do not produce the localised strain patterns seen in the simplest triaxial test on dense sand by any engineer. This is a widely cited and influential paper, and it is easy to appreciate why—a remarkable contribution and a fitting point to finish this overview of influential and insightful contributions on soil behaviour to be found in the pages of *Géotechnique*.

CLOSURE

Focusing on soil behaviour, this essay is a history of experimental contributions. These contributions give high-quality soil property data in itself (and from a possibly surprising range of soils worldwide). But looking only to soil properties misses much of what is offered. A striking feature is the commonality of findings across many contributors and many soils investigated: a critical-state framework is applicable from sands to undisturbed clays, including saturated and unsaturated conditions, and allows the effects of various soil constituents to be captured in simple state variables. Plastic yielding is pervasive, and the yield surfaces are related to soil structure. Despite the pervasive nature of yielding, elasticity also exists, and the framework for it has been defined. Soil behaviour under plane strain, yield in unloading, and principal stress rotation have all been investigated and reported. From a practical perspective, the findings allow assessment of a soil's behaviour, under a wide range of circumstances, using few (but high-quality) tests. In modern parlance, there is a lot of 'bang for buck' on offer, and most is easy to read and understand. Of course, as we noted at the start, we have neglected many contributions: soil behaviour has been a recurring theme in *Géotechnique* throughout its six decades, and this essay is short. We have presented a brief history of how the subject has developed, and have looked to early work (now part of accepted understanding), to fundamental insights, and to situations found in engineering works. Hopefully, we have raised your curiosity in reading this wonderful, international journal in greater depth—there is much to be found of both basic interest and practical relevance.

REFERENCES

Alonso, E. E, Gens, A. & Josa, A. (1990). A constitutive model for partially saturated soils. *Géotechnique* **40**, No. 3, 405–430.

Atkinson, J. H., Richardson, D. & Stallebrass, S. E. (1990). Effect of recent stress history on the stiffness of overconsolidated soil. *Géotechnique* **40**, No. 4, 531–540.

Been, K. & Jefferies, M. G. (1986). Discussion: 'A state parameter for sands'. *Géotechnique* **36**, No. 1, 127–132.

Been, K., Jefferies, M. G. & Hachey, J. (1991). The critical state of sands. *Géotechnique* **41**, 365–381

Bellotti, R., Jamiolkowski, M., Lo Presti, D. C. F. & O'Neil, D. A. (1996). Anisotropy of small strain stiffness in Ticino sand. *Géotechnique* **46**, No. 1, 115–131.

Bishop, A. W. (1959). The principle of effective stress. *Teknisk Ukeblad* **106**, No. 39, 859–863.

Bishop, A. W. & Eldin, G. (1950). Undrained triaxial tests on saturated sands and their significance in the general theory of shear strength. *Géotechnique* **2**, No. 1, 13–32.

Bishop, A. W., Webb, D. L. & Lewin, P. I. (1965); Undisturbed samples of London Clay from the Ashford common shaft: strength–effective stress relationships. *Géotechnique* **15**, No. 1, 1–31.

Bolton, M. D. (1986). The strength and dilatancy of sands. *Géotechnique* **36**, No. 1, 65–78.

Burland, J. B. (1990). On the compressibility and shear strength of natural clays. *Géotechnique* **40**, No. 3, 329–378.

Callisto, L. & Rampello, S. (2002). Shear strength and small-strain stiffness of a natural clay under general stress conditions. *Géotechnique* **52**, No. 8, 547–560.

Coleman, J. D. (1962). Stress–strain relations for partially saturated soils. *Géotechnique* **12**, No. 4, 348–350.

Collins, K. & McGown, A. (1974). The form and function of microfabric features in a variety of natural soils. *Géotechnique* **24**, No. 2, 223–254.

Coop, M. R. (1990). The mechanics of uncemented carbonate sands. *Géotechnique* **40**, No. 4, 607–626.

Cornforth, D. H. (1964). Some experiments on the effect of strain condition on the strength of sand. *Géotechnique* **14**, No. 2, 143–167.

Cotecchia, F. & Chandler, R. J. (2000). A general framework for the mechanical behaviour of clays. *Géotechnique* **50**, No. 4, 431–448.

Croney, D. (1952). The movement and distribution of water in soils. *Géotechnique* **3**, No. 1, 1–16.

Cuccovillo, T. & Coop, M. R. (1997). Yielding and pre-failure deformation of structured sands. *Géotechnique* **47**, No. 3, 491–508.

Cui, Y. J. & Delage, P. (1996). Yielding and plastic behaviour of an unsaturated compacted silt. *Géotechnique* **46**, No. 2, 291–311.

Desrues, J., Chambon, R., Mokni, M. & Mazerolle, F. (1996). Void ratio evolution inside shear bands in triaxial sand specimens studied by computed tomography. *Géotechnique* **46**, No. 3, 529–546.

Graham, J., Crooks, J. H. A. & Bell, A. L. (1983). Time effects on the stress–strain behaviour of natural soft clays. *Géotechnique* **33**, No. 3, 327–340.

Henkel, D. J. (1956). The effect of overconsolidation on the behaviour of clays during shear. *Géotechnique* **6**, No. 4, 139–150.

Henkel, D. J. (1960). The relationships between the effective stresses and water content in saturated clays. *Géotechnique* **10**, No. 2, 41–54.

Jennings, J. E. B. & Burland, J. B. (1962). Limitations to the use of effective stresses in partly saturated soils. *Géotechnique* **12**, No. 2, 125–144.

Jefferies, M. G. (1997). Plastic work and isotropic softening in unloading. *Géotechnique* **47**, No. 5, 1037–1042.

Jefferies, M. G. & Been, K. (2000). Implications for critical state theory from isotropic compression of sand. *Géotechnique* **50**, No. 4, 419–429.

Kuwano, R. & Jardine, R. J. (2007). A triaxial investigation of kinematic yielding in sand. *Géotechnique* **57**, No. 7, 563–579.

Leroueil, S. & Vaughan, P. R. (1990). The general and congruent effects of structure in natural soils and weak rocks. *Géotechnique* **40**, No. 3, 467–488.

Lo, K. Y. (1970). The operational strength of stiff fissured clays. *Géotechnique* **20**, No. 1, 57–74.

Lloret, A., Villar, M., Sanchez, M., Gens, A., Pintado, X. & Alonso, E. E. (2003). Mechanical behaviour of compacted bentonite under high suction changes. *Géotechnique* **53**, No. 1, 27–40.

Lupini, J. F., Skinner, A. E. & Vaughan, P. R. (1981). The drained residual strength of cohesive soils. *Géotechnique* **31**, No. 2, 181–213.

Maâtouk, A., Leroueil, S. & La Rochelle, P. (1995). Yielding and critical state of a collapsible unsaturated silty soil. *Géotechnique* **45**, No. 3, 465–477.

Matyas, E. L. & Radhakrishna, H. S. (1968). Volume change characteristics of partially saturated soil. *Géotechnique* **18**, No. 4, 432–448.

McDowell, G. R. & Bolton, M. D. (1998). On the micromechanics of crushable aggregates. *Géotechnique* **48**, No. 5, 667–679.

Wood, D. M. (1975) Explorations of principal stress space with kaolin in a true triaxial apparatus. *Géotechnique* **25**, No. 4, 783–797.

Parry, R. H. G. (1958). Correspondence: 'On the yielding of soils'. *Géotechnique* **8**, No. 3, 184–186.

Parry, R. H. G. (1960). Triaxial compression and extension tests on remoulded saturated clay. *Géotechnique* **10**, No. 4, 166–180.

Pennington, D. S., Nash, D. F. T. & Lings, M. L. (1997). Anisotropy of G_0 shear stiffness in Gault Clay. *Géotechnique* **47**, No. 3, 391–398.

Romero, E., Gens, A. & Lloret, A. (2003). Suction effects on a compacted clay under non-isothermal conditions. *Géotechnique* **53**, No. 1, 65–81.

Skempton, A. W. & Northey, R. D. (1952). The sensitivity of clays. *Géotechnique* **3**, No. 1, 30–53.

Smith, P. R., Jardine, R. J. & Hight, D. W. (1992). The yielding of Bothkennar clay. *Géotechnique* **42**, No. 2, 257–274.

Viggiani, G. & Atkinson, J. H. (1995). Stiffness of fine-grained soil at very small strains. *Géotechnique* **45**, No. 2, 249–265.

Ward, W. H., Samuels, S. G. & Butler, M. E. (1959) Further studies of the properties of London Clay. *Géotechnique* **9**, No. 2, 33–58.

Wheeler, S. J. & Sivakumar, V. (1995). An elasto-plastic critical state framework for unsaturated soil. *Géotechnique* **45**, No. 1, 35–54.

Wong, P. K. K. & Mitchell, R. J. (1975). Yielding and plastic flow of sensitive cemented clay. *Géotechnique* **25**, No. 4, 763–782.

Wong, R. K. S. & Arthur, J. R. F. (1986). Sand sheared by stresses with cyclic variations in direction. *Géotechnique* **36**, No. 2, 215–226.

Zdravkovic, L. & Jardine, R. J. (1997). Some anisotropic stiffness characteristics of a silt under general stress conditions. *Géotechnique* **47**, No. 3, 407–437.

Powrie, W. (2008). *Géotechnique* **58**, No. 5, 435–439 [**doi:** 10.1680/geot.2008.58.5.435]

Contributions to *Géotechnique* 1948–2008: Groundwater

W. POWRIE*

This paper reviews work on the subject of groundwater published in *Géotechnique* over the past 60 years. Contributions relating to the effects of groundwater on geotechnical structures, measurement of pore water pressure using piezometers, analysis of groundwater flow, groundwater control or construction dewatering, estimating hydraulic conductivity, and passive vertical drains are summarised and discussed. Three major areas of significant impact are identified: (a) the realisation of the dominant role of pore water pressure equilibration in the delayed failure of cut slopes; (b) determination of the hydraulic conductivity of soils by means of rising- or falling-head tests in piezometers and boreholes; and (c) understanding the role of soil fabric in groundwater control, and using it to extend the operational range of pumped groundwater control techniques to soils of low permeability.

KEYWORDS: drainage; groundwater; historical review; permeability

La présente communication passe en revue les travaux sur le sujet de la nappe phréatique, publiés dans *Géotechnique*, au cours des 60 dernières années. Des contributions sur les effets de la nappe phréatique sur les structures géotechniques, la mesure de la pression interstitielle au moyen de piézomètres, l'analyse de l'écoulement de la nappe phréatique, le contrôle de la nappe phréatique ou le drainage pour la construction, l'estimation de la conductivité hydraulique, ainsi que des drains verticaux passifs, sont résumées et discutées. On identifie trois principaux aspects qui ont eu un impact significatif. Il s'agit des suivants : (a) réalisation du rôle dominant de l'équilibrage de la pression interstitielle dans la rupture retardée de talus ; (b) détermination de la conductivité hydraulique des sols au moyen de tests de montée ou abaissement de la hauteur de refoulement dans des piézomètres et des forages ; et (c) connaissance du rôle de l'assemblage du sol dans le contrôle de la nappe phréatique, et son utilisation pour le renforcement de la plage opérationnelle des techniques de contrôle de la nappe phréatique pompée dans des sols à faible perméabilité.

INTRODUCTION

Without groundwater, geotechnical engineering would be much simpler—but also less of a challenge and much less interesting. A result of this is that many of the papers published in *Géotechnique* over its first 60 years feature groundwater in some way: for example, its effects on the behaviour of slopes, retaining walls, foundations and other geotechnical structures. In selecting those papers on groundwater that have particularly influenced the theory and practice of geotechnical engineering, I have focused on those that have groundwater as their main theme. Most of these address issues of groundwater flow and control, the characterisation of groundwater regimes through the measurement of pore water pressures, and the influence of ground conditions and soil fabric. Some papers concerned with the effects of groundwater on geotechnical structures are included, if this is the principal or only influence addressed.

I have consulted with researchers and practising engineers in the field to find out which *Géotechnique* papers they use and consider important, and I have tried to provide some reasonably objective evidence or explanation for the influence and importance of each of the selected papers. Ultimately, however, the selection is a personal one. In assessing its impact, a paper should be viewed in the context of geotechnical engineering knowledge and practice at the time. From a modern perspective, the contribution made by a paper might seem obvious because it long ago became assimilated into the accepted canon of knowledge and understanding—but this might be argued to be the highest test of its lasting value.

Discussion on this paper closes on 1 December 2008, for further details see p. ii.

* School of Civil Engineering and the Environment, University of Southampton, UK.

GROUNDWATER EFFECTS ON GEOTECHNICAL STRUCTURES

The generally more onerous loading and potentially damaging effects of a high groundwater level on a geotechnical structure were, at least empirically, as well known in 1948 as now. While its effects were well documented (e.g. the retaining wall failures at Cricklewood in 1943, 41 years after its construction, and at Wembley Hill in 1918, 13 years after its construction, both reported by Skempton, 1946), the rather more subtle and insidious process of pore pressure change following excavation in low-permeability soils, as a result of the transient flow of groundwater to a new equilibrium state, was rather less universally accepted and understood. By the late 1950s it was realised by some that effective, rather than total, stress analysis could be the key to understanding the delayed failure of cut slopes and retaining walls in clay soils. However, alternative theories based on a deterioration in the drained strength over time or the release of strain energy due to a reduction of the strength of interparticle bonds due to weathering were still current up to the early 1970s.

Vaughan & Walbancke (1973) used observations of pore pressure changes in a cut slope over a period of nine years to demonstrate that the delayed failure of such structures could be explained entirely by the gradual recovery of pore pressures towards the new equilibrium state, without the need to invoke any significant fall in the drained strength over time. Thus Vaughan & Walbancke (1973) settled a debate that had been raging for 20 years or more. It is now universally accepted that increases in pore water pressure towards a new equilibrium condition by transient groundwater flow following an unloading event such as an excavation can cause an initially stable trench, slope or retaining wall to fail in the medium to long term. It is accepted that this is primarily a groundwater effect, and that stability and serviceability checks must be carried out for such structures under the estimated long-term conditions. The problem that remains, however, is the reliable determination of how long the transition from short-term, undrained, to long-term equilibrium conditions will take. Also, other mechanisms such as ratcheting due to cyclic seasonal effects are again beginning to be investigated, to explain longer-term failures of infrastructure slopes beyond the timescale for excess pore water pressure equilibration.

MEASUREMENT OF PORE WATER PRESSURE

Over a period of more than 30 years, a number of papers were published in *Géotechnique* that addressed the related issues of the time lag between a change in pore water pressure occurring in the soil and the registration of that change in a piezometer measuring it (the so-called equalisation time), and the measurement of hydraulic conductivity by means of constant- or falling-head tests in piezometers. All of these developed, and sought to improve, the work of Hvorslev (1951), who had shown that for a piezometer in an incompressible soil the progress of the pore water pressure u measured by a piezometer from an initial value u_0 to a final value u_∞ (the latter being in equilibrium with the pore water pressure in the ground) was given by

$$U = \frac{u_\infty - u}{u_\infty - u_0} = e^{-\frac{Fkt}{V\gamma_w}} \quad (1)$$

where F is a shape factor (sometimes known as the intake factor), k is the soil hydraulic conductivity, t is the elapsed time, V is a volume factor, and γ_w is the unit weight of water. The volume factor is defined as the volume of water that must flow into or out of the piezometer for it to register a unit change in pore water pressure: for a standpipe of cross-sectional area A, $V = A/\gamma_w \approx 0{\cdot}1A$ m^3/kPa. The shape factor F is defined as

$$F = q/kH \quad (2)$$

where q is the steady-state flowrate of water into or out of the piezometer for a head difference H. F can be determined analytically only for a spherical piezometer of radius a, for which $F = 4\pi a$; Hvorslev (1951) determined an approximate shape factor for a cylindrical piezometer by representing it as an inscribed prolate spheroid.

Gibson (1963) derived dimensionless curves relating the normalised pore water pressure U to μT for falling-head tests in spherical piezometers in compressible soils, where $\mu = 4\pi a^3 m_v \gamma_w / A$, $T = c_v t/a^2$, m_v is the coefficient of volume compressibility of the soil, c_v is the consolidation coefficient $= k/m_v\gamma_w$, and the other symbols are as already defined. Gibson also derived a relationship between the flowrate of water out of (or into) the piezometer and time in a constant-head test, again taking into account the hydraulic conductivity and compressibility of the soil. In 1966, Gibson published a technical note in which he extended his solution for the constant-head test to take into account the effect of a piezometer tip (e.g. a high air entry value ceramic) that might be of similar or smaller hydraulic conductivity than the soil (Gibson, 1966). By solving the equations numerically, Wilkinson (1968) developed solutions for the constant-head in situ permeability test in a compressible soil in which the hydraulic conductivity and the compressibility varied during the test. He also extended the solution to cylindrical (rather than spherical) piezometers, and to account for the effects of imperfections such as air or gas in the piezometer pocket, soil smear during installation, and leakage past the grout seal. Gibson (1970) developed solutions for constant-head in situ hydraulic conductivity tests in compressible but inelastic soils.

More than a decade later, Brand and Premchitt published a series of papers and a technical note based on an extensive programme of research carried out at the Asian Institute of Technology. Brand & Premchitt (1980) presented a detailed review of previous work on shape factors for cylindrical piezometers in incompressible soils. They then proposed a modified version of Hvorslev's (1951) expression, based on experimental data from a liquid electric analytical model. Premchitt & Brand (1981) addressed the issue of pore pressure equalisation in compressible soils. They noted that 'since its publication, Gibson's important [1963] solution for pore pressure equalisation has received little attention.' They confirmed Gibson's (1963) analysis for a spherical piezometer by means of a series of laboratory tests, and developed an 'equalisation chart' to facilitate the use of Gibson's analysis to determine the hydraulic conductivity and compressibility of the soil from pore water pressure equalisation tests (i.e. falling- or rising-head tests in boreholes or piezometers). Brand & Premchitt (1982) developed a similar analysis and presented a similar chart for cylindrical piezometers, eliminating the need to model these as spheres of an equivalent radius. More than a decade later still, Bromhead (1996) used numerical analyses to quantify the effect on a constant-head test of installing the piezometer into a soil pocket of finite, rather than infinite, extent.

Gibson's (1963) method of analysis for an in situ hydraulic conductivity test is incorporated into the current UK code of practice for site investigations, BS 5930 (BSI, 1999). However, the expression used for the shape factor for a cylindrical piezometer is that given by Al-Dhahir & Morgernstern (1969), rather than Brand & Premchitt (1980). The effect of installing a piezometer into a soil lens of finite extent is not considered in BS 5930, although Bromhead (1996) notes that the way in which the results are interpreted will not cause the derived hydraulic conductivity to vary by a factor of more than two or three.

ANALYSIS OF GROUNDWATER FLOW

The fundamental principles of groundwater flow and the determination of the soil parameter—hydraulic conductivity or permeability—that governs it were well established by 1948. For example, Darcy's law had been formulated by 1856, and Dupuit's classic analysis of the steady-state flow of groundwater to an individual well was published in 1863. Theis's analysis of non-equilibrium flow towards wells was published in 1935. By this time it was well understood that the flow of groundwater through a porous medium followed the same basic physical laws as the flow of heat or electricity through a conductor (Muskat, 1937), and the solution of groundwater flow problems by the technique of flownet sketching was established (e.g. Casagrande, 1937). Further landmark publications in the analysis of individual wells and idealised flows were published just before the establishment of *Géotechnique* (e.g. Cooper & Jacob, 1946), or just afterwards but in other journals (e.g. Jacob & Lohman, 1952; Hall, 1955). Nonetheless, key contributions were published in *Géotechnique* by Chapman (1957, 1960), whose charts and equations for determining the height of the seepage face that develops when groundwater flows through an open vertical surface and the associated flowrates, with and without capillary effects, are still in regular use in the design analysis of wellpoint systems.

Many of the developments in the analysis of groundwater flow that have been made by papers published in *Géotechnique* have been based on numerical modelling, using the computing power that has become increasingly available to geotechnical engineers and researchers, and routinely so since the late 1980s. An early example was Tomlin (1966), who reported on the development of a computer-based method of analysing seepage through zoned anisotropic soils. Powrie, Roberts & Moghazi (1989) used a numerical method to explore the effects of high-permeability lenses on the performance of a wellpoint dewatering system. Their conclusion, that these would be significant if their hydraulic conductivity were more than about two-and-a-half orders of magnitude (i.e. a factor of about 400) greater than the surrounding soil, seems to have some generality in seepage problems. Griffiths & Fenton (1993) used numerical methods to investigate seepage beneath water-retaining structures founded on a spatially random soil. They found that flow-rates and pore water pressures were affected significantly by the standard deviation and the correlation structure (scale of fluctuation) of the hydraulic conductivity of the soil. Stochastic analyses have yet to replace traditional deterministic methods—perhaps because many engineers find it problematic enough to assign even a mean value to a parameter as difficult to determine as hydraulic conductivity—but this paper does show what can be achieved, as well as illustrating the effects of the variations in parameter values that geotechnical engineers must expect.

By modelling groundwater control systems around excavations as a single equivalent well, Powrie & Preene (1992) identified appropriate simple models (i.e. essentially plane or radial flow) for excavations of different aspect ratio and the distance to the recharge boundary. They also developed a dimensionless chart for the rapid estimation of flowrates, which is reproduced in current industry guidance CIRIA Report C515 (Preene *et al.*, 2000).

GROUNDWATER CONTROL (CONSTRUCTION DEWATERING)

The aim of groundwater control is to reduce pore water pressures in the vicinity of an excavation below the natural water table, so that the excavation will remain stable and workably dry. Modern groundwater control is generally pre-emptive, in that wells are sunk and pumped to intercept the water before it can flow into the excavation itself (a technique pioneered by Robert Stephenson during construction of the Kilsby tunnel on the London to Birmingham railway, in the 1830s). By the 1950s, jetted wellpoints and deep wells pumped using electric submersible pumps were well established for groundwater control in soils of medium to high hydraulic conductivity (sands and sandy gravels).

The focus on relatively permeable soils, and indeed the term 'dewatering' itself, tended to divert attention away from the fact that the key issue was the reduction of pore water pressure rather than the removal of groundwater as such. That the presence of high pore water pressures in strata of relatively low hydraulic conductivity could be immensely damaging and potentially dangerous was elegantly demonstrated in a classic case history by Ward (1957). This concerned a ~6 m deep sewer trench excavated at Millbrook, Southampton, in 1949, through Test valley gravel into the underlying Bracklesham Beds—a notoriously variable Eocene deposit comprising at this location a layer of sandy clay overlying laminated silty clay and fine sand. The sides of the trench were supported by sheet pile walls, but no other groundwater controls were envisaged. The method of construction was to cast a 150 mm thick concrete slab onto the trench floor, and then to lay, level and joint the 0·75 m diameter precast concrete sewer pipes.

Major problems were encountered with uplift of the formation and the concrete slab, particularly when the trench was left open overnight, and settlement and cracking of the pipes after backfilling. Ward diagnosed the problem as being due to the delamination of the fine sand and silty clay deposits, as a result of the reduced overburden and unrelieved pore water pressure, which caused the layers of silty clay to separate and 'open up like a concertina'. The problem was solved by the installation of simple passive relief wells, bored using a post-hole auger from about 0·9 m above formation level and extending to about 1·8 m below it. Ward's paper is an economical but complete description of the problem and its solution. It includes detailed measurements of the movements that occurred and the piezometric levels in the laminated deposits both before and after the installation of the relief wells, and some superb photographs of split samples showing very clearly the structure of the material. The important influence of soil fabric on groundwater flow and pore water pressures was later expounded by Rowe (1972) in his Rankine Lecture. This is a landmark paper, which demonstrated beyond doubt, and with reference to numerous case studies and practical examples, the importance of scale and effective sample size in hydraulic conductivity tests.

Nearly 40 years after the Milbrook sewer, construction of the A55 North Wales Coast Road Conwy Crossing in 1987 required three large excavations up to 15 m deep to be made in similarly structured glacial lake deposits, and to remain stable for a period of well over a year. The technique adopted was a system of vacuum wells pumped by devices known as ejectors. This large project to control pore water pressures in a soil of such low hydraulic conductivity (about 4×10^{-7} m/s horizontally) sparked a programme of associated research into appropriate methods of analysis for groundwater control in fine soils. Some of the results were published in *Géotechnique*, including papers on steady-state and time-drawdown behaviour (Preene & Powrie, 1993; Powrie & Preene, 1994).

These papers identified analytical methods and assumptions that could be used to predict reasonably reliably the performance of groundwater control systems in fine soils. The second paper established beyond doubt that groundwater control in a low-permeability soil was a consolidation process; pore pressure reduction would therefore take some time to achieve, which must be allowed for in a construction programme. Also, fine soils are generally less stiff than sands and gravels, so that settlements induced by groundwater control measures in such strata might also be a consideration. The papers gave guidance on the selection of the key parameters of hydraulic conductivity, effective soil modulus and consolidation coefficient on the basis of the test results likely to be available to the designer of a pore pressure control system. The work finds regular application in practice through its incorporation into CIRIA Report C515, *Groundwater control: theory and practice* (Preene *et al.*, 2000)

The practical experience of the Conwy Crossing project, together with the associated research, gave confidence in the ability of ejector systems to control, predictably, pore pressures in low permeability where previously groundwater control would not have been attempted. Since Conwy, ejectors have been used widely throughout the UK and elsewhere, on projects including the Jubilee Line Extension, the Docklands Light Railway under the Thames to Lewisham, and the Channel Tunnel Rail Link (CTRL). Audaciously, the technique was used on CTRL Phase 1 to control pore pressures in the Weald Clay by acting through the fissures (Roberts *et al.*, 2007); and in an application that Bill Ward would have appreciated, to preserve the integrity of the structure of the Bracklesham Beds by preventing delamination during basement excavation and construction of the raft foundation for the Skandia Life building in Southampton.

Other well-documented case studies of groundwater control and construction dewatering systems, in addition to Ward (1957), reported in *Géotechnique* include the Mangla Dam (Starr *et al.*, 1969) and Sizewell B nuclear power station (Knight *et al.*, 1996). Both of these were designed with the aid of an electrical resistance network analogue model (Herbert & Rushton, 1966)—a technique now rendered virtually obsolete by numerical analysis using modern computers. Roberts & Preene (1994) refer to the groundwater control systems at Kilsby Tunnel, the Tees Barrage and the Pontsticill water treatment works in relation to the use of the observational method in design. Powrie & Roberts (1995) describe an unusual dewatering and groundwater recharge system in Chalk at Winchester, in which geochemical effects caused major difficulties towards the end of the project.

ESTIMATION OF HYDRAULIC CONDUCTIVITY

In many natural soil deposits, it is the effects of inhomogeneity, anisotropy and soil fabric that make the effective hydraulic conductivity of the ground so difficult to determine. Even in more uniform deposits, disturbance during sampling can lead to discrepancies between hydraulic conductivities measured in laboratory tests and those that operate in the field. For this reason, correlations between simple physical properties such as a representative particle size (e.g. D_{10}) and hydraulic conductivity (k) have been developed. Perhaps the simplest and best known is Hazen's (1892) formula for clean filter sands,

$$k\ (m/s) \approx 0{\cdot}01(D_{10})^2 \text{ (with } D_{10} \text{ in mm)} \tag{3}$$

This was the approach taken by Loudon (1952), who carried out a large number of simple but careful and thorough tests on a range of sands, and used the results to develop and verify empirical equations relating the hydraulic conductivity to particle size, angularity and porosity. The method may not be an adequate substitute for direct hydraulic conductivity testing, but it certainly provides a useful additional check for soils approximating to the clean sands for which it was intended.

A broader and more pragmatic approach was taken by Preene & Powrie (1993), who considered the steady-state performance of 30 pore water pressure control systems in low-permeability deposits. Hydraulic conductivities had been determined using a range of methods, from correlations with particle size to in situ pumping tests. By back-analysis of the flow data, they drew up guidelines for estimating the operational hydraulic conductivity of the ground on the basis of the particular data available—for example, maximum values from in situ falling-head tests in boreholes or piezometers. These should enable the actual flowrate to be predicted to within a factor of three—a range over which the groundwater control system would ideally be designed to operate. Assessing the effective or operational hydraulic conductivity of the ground clearly remains a challenge.

VERTICAL DRAINS

March 1981 saw the publication of the third *Géotechnique* Symposium in Print, on vertical drains. The symposium comprises five papers, two of which are concerned with analysis and design while three describe and discuss in some detail a number of case studies. In particular, Nicholson & Jardine (1981) present a detailed case record of a scheme at Queenborough, Kent, in which they compare the performance of two types of vertical drain with calculations of pore pressure dissipation rates and settlements based on laboratory parameters. In general, the agreement was close—a useful result, giving some confidence in the methods of soil testing and analysis used. Both this paper and the symposium as a whole continue to represent a valuable practical guide and reference.

CONCLUDING COMMENTS

In this review, I have focused on three main groundwater-related areas in which papers published in *Géotechnique* over the past 60 years have made significant contributions. These are

(*a*) realisation of the dominant role of pore water pressure equilibration in the delayed failure of cut slopes, culminating in and represented by Vaughan & Walbancke (1973)
(*b*) determination of the hydraulic conductivity of soils from tests in piezometers and boreholes, including shape factors for cylindrical piezometers and analysis of tests in compressible soils
(*c*) realisation of the important effects of soil fabric on groundwater flow and its implications for groundwater control, and the exploitation of soil fabric to extend the operational range of pumped pore pressure control techniques to soils of relatively low permeability.

Impacts have generally resulted from series of papers by one or more groups of workers published over a period of time, rather than a single pivotal paper. Perhaps the clearest example of this is the body of work on hydraulic conductivity tests in piezometers and boreholes, starting with Gibson (1963) and ending with Brand & Premchitt (1982) or even Bromhead (1996). Interestingly, of these only the paper by Gibson (1963) has found its way into the British Standard (BS 5930, 1999), even though the work by Brand and Premchitt would clearly warrant inclusion. Research on pumped pore pressure control systems in low-permeability soils by Preene, Powrie and Roberts has impacted on and is used in practice through its incorporation into CIRIA Report C515 (Preene *et al.*, 2000), helped to some extent by the common authorship of the various documents. Perhaps it is inevitable that, as the state of geotechnical knowledge advances so that the number or people with a real understanding of the latest developments in a particular area becomes relatively few, the impact of papers on practice will be indirect through interpretation in codes and guidance documents rather than directly as a source reference. Well-documented case studies papers will always be instructive, and should be an exception to this.

As a researcher, I have generally subscribed to the view attributed to Lord Rayleigh that 'in science, by a fiction as remarkable as any to be found in law, what has once been published . . . is spoken of as *known*, and it is too often forgotten that the rediscovery in the library may be a more difficult and uncertain process than the first discovery in the laboratory.' At times, when preparing this review and uncovering a large number of apparently worthy and useful papers covering a wide range of material that seem to have become forgotten, I have felt tempted to side instead with Professor Bob Ham, who has said words to the effect that 'three years in the field can save . . . maybe three weeks in the library.' I have even briefly thought that if I were familiar with all of the papers that had been published in *Géotechnique* over the past 60 years, I would not need to write another research proposal. But of course this is not the case. Knowledge generally advances in small steps, and even papers with the most promising of titles usually leave many questions unanswered. Add to this the effect of developments in scientific understanding, construction and measurement technology, and the finite half-life of knowledge, and there will always be a need for discovery and new research.

So what of the future? Predicting the future is even more difficult than determining hydraulic conductivity, and if there are two types of futurologist—those who don't know, and those who don't know they don't know—I would place myself firmly in the first of these groups. But it is possible to select some pointers from papers published recently, and some not so recently, in *Géotechnique*. Numerical analysis will become increasingly complex, and one day stochastic methods as espoused by Griffiths & Fenton (1993) may become the norm. Brandl (2006), in the published version of his 2001 Rankine Lecture, considers the importance of groundwater flow and hydraulic conductivity in assessing the performance of 'energy foundations', used in ground source heat pump systems and for the storage of excess building heat for later retrieval. Hardisty *et al.* (1998) consider the flow of immiscible liquid contaminants in fractured aquifers: two-phase flow and dual porosity systems are both areas, either separately or together, for future theoretical, model and practical development. Smethurst *et al.* (2006) consider the effects of climate and vegetation on the behaviour of a road cutting slope. There is a strong environmental theme here; with impacts both on and of the environment becoming increasingly important, I would expect all of these areas to feature in papers on groundwater in future issues of *Géotechnique*. I do not expect the pace of change to appear rapid to those who drive it, and Darcy's law will still apply. However, I do expect that when someone (probably someone else, but you never know) carries out a similar review in 2068, significant advances—probably practice-led—will have been made.

ACKNOWLEDGEMENT

I am grateful to Dr Martin Preene for his contribution to the genesis of this paper, both in the discussion of initial ideas and for his thoughtful comments on an earlier draft.

REFERENCES

Al-Dhahir, Z. A. & Morgernstern, N. R. (1969). Intake factors for cylindrical piezometer tips. *Soil Sci.* **107**, No. 1, 17–21.

BSI (1999). *Code of practice for site investigations*, BS 5930. London: British Standards Institution.

Brand, E. W. & Premchitt, J. (1980). Shape factors of cylindrical piezometers. *Géotechnique* **30**, No. 4, 369–384.

Brand, E. W. & Premchitt, J. (1982). Response characteristics of cylindrical piezometers. *Géotechnique* **32**, No. 3, 203–216.

Brandl, H. (2006). Energy foundations and other thermo-active ground structures. 41st Rankine Lecture. *Géotechnique* **56**, No. 2, 81–122.

Bromhead, E. N. (1996). Interpretation of constant head in situ permeability tests in soil zones of finite extent. *Géotechnique* **46**, No. 1, 133–143.

Casagrande, A. (1937). Seepage through dams. *J. New England Water Works Assoc.*, **51**, No. 2, June, 131–172.

Chapman, T. G. (1957). Two-dimensional groundwater flow through a bank with vertical faces. *Géotechnique* **7**, No. 1, 35–40.

Chapman, T. G. (1960). Capillary effects in a two-dimensional groundwater flow system. *Géotechnique* **10**, No. 2, 55–61.

Cooper, H. H. & Jacob, C. E. (1946). A generalised graphical method for evaluating formation constants and summarizing well field history. *Trans. Am. Geophys. Union* **27** , No. 4, 526–534.

Darcy, H. (1856). *Les fontaines publiques de la ville de Dijon*. Paris: V. Dalmont.

Dupuit, J. (1863). *Etudes theoretiques et practiques sur les mouvements des eaux dans les canaux decouverts et à travers les terrains permeable.* Paris: Dunod.

Gibson, R. E. (1963). An analysis of system flexibility and its effect on time-lag in pore pressure measurements. *Géotechnique* **13**, No. 1, 1–11.

Gibson, R. E. (1966). A note on the constant head test to measure soil permeability in situ. *Géotechnique* **16**, No. 3, 256–259.

Gibson, R. E. (1970). An extension to the theory of the constant head in situ permeability test. *Géotechnique* **20**, No. 2, 193–197.

Griffiths, D. V. & Fenton, G. A. (1993). Seepage beneath water retaining structures founded on spatially random soil. *Géotechnique* **43**, No. 4, 577–587.

Hall, H. P. (1955). An investigation of steady state flow toward a gravity well. *Houille Blanche* **10**, No. 8, 8–35.

Hardisty, P. E., Wheater, H. S., Johnston, P. M. & Bracken, R. A. (1998). Behaviour of light immiscible liquid contaminants in fractured aquifers. *Géotechnique* **48**, No. 6, 747–760.

Hazen, A. (1892). *Physical properties of sands and gravels with reference to their use in filtration*. Report to the Massachusetts State Board of Health.

Herbert, R. & Rushton, K. R. (1966). Ground water flow studies by resistance networks. *Géotechnique* **16**, No. 1, 53–75.

Hvorslev, M. J. (1951). Time lag and soil permeability in ground water observations. *US Waterways Experimental Station Bulletin.* Vicksburg: US Army Corps of Engineers.

Jacob, C. E. & Lohman, S. W. (1952). Nonsteady flow to a well of constant drawdown in an extensive aquifer. *Trans. Am. Geophys. Union* **33** , No. 4, 559–569.

Knight, D. J., Smith, G. L. and Sutton, J. S. (1996). Sizewell B foundation dewatering: system design, construction and performance. *Géotechnique* **46**, No. 3, 473–490.

Loudon, A. G. (1952). The computation of permeability from simple soil tests. *Géotechnique* **3**, No. 4, 165–183.

Muskat, M. (1937). *The flow of homogeneous fluids through porous media*. New York: McGraw-Hill.

Nicholson, D. P. & Jardine, R. J. (1981). Performance of vertical drains at Queenborough bypass. *Géotechnique* **31**, No. 1, 67–90.

Preene, M. & Powrie, W. (1993). Steady-state performance of construction dewatering systems in fine soils. *Géotechnique* **43**, No. 2, 191–205.

Preene, M., Roberts, T. O. L., Powrie, W. & Dyer, M. (2000). *Groundwater control: Design and practice*, CIRIA Report C515. London: Construction Industry Research and Information Association.

Premchitt, J. & Brand, E. W. (1981). Pore pressure equalization of piezometers in compressible soils. *Géotechnique* **31**, No. 1, 105–123.

Powrie, W. & Preene, M. (1992). Equivalent well analysis of construction dewatering systems. *Géotechnique* **42**, No. 4, 635–639.

Powrie, W. & Preene, M. (1994). Time-drawdown behaviour of construction dewatering systems in fine soils. *Géotechnique* **44**, No. 1, 83–100.

Powrie, W. & Roberts, T. O. L. (1995). Case history of a dewatering and recharge system in Chalk. *Géotechnique* **55**, No. 4, 599–609.

Powrie, W., Roberts, T. O. L. and Moghazy, H. E.-D. (1989). Effects of high permeability lenses on efficiency of wellpoint dewatering. *Géotechnique* **39**, No. 3, 543–547.

Roberts, T. O. L. and Preene, M. (1994). The design of groundwater control systems using the observational method. *Géotechnique* **44**, No. 4, 727–734.

Roberts, T. O. L., Roscoe, H., Powrie, W. & Butcher, D. E. (2007). Controlling clay pore pressures for cut-and-cover tunnelling. *Proc. Instn Civ. Engrs Geotech. Engng* **160**, No. 4, 227–236.

Rowe, P. W. (1972). The relevance of soil fabric to site investigation practice. *Géotechnique* **22**, No. 2, 195–300.

Skempton, A. W. (1946). *Principles and applications of soil mechanics. Lecture II: Earth pressure and stability of slopes.* London: Institution of Civil Engineers.

Smethurst, J. A., Clarke, D. & Powrie, W. (2006). Seasonal changes in pore water pressure in a grass-covered cut slope in London Clay. *Géotechnique* **56**, No. 8, 523–537.

Starr, M. R., Skipp, B. O. & Clarke, D. A. (1969). Three-dimensional analogue used for relief well design in the Mangla dam project. *Géotechnique* **19**, No. 1, 87–100.

Theis, C. V. (1935). The relation between the lowering of the piezometric surface and the rate and duration of discharge of a well using groundwater storage. *Trans. Am. Geophys. Union* **16**, Part 2, 519–524.

Tomlin, G. R. (1966). Seepage analysis through zoned anisotropic soils by computer. *Géotechnique* **16**, No. 3, 220–230.

Vaughan, P. R. & Walbancke, H. J. (1973). Pore pressure changes and the delayed failure of cutting slopes in overconsolidated clay. *Géotechnique* **23**, No. 4, 531–539.

Ward, W. H. (1957). The use of simple relief wells in reducing pore water pressure beneath a trench excavation. *Géotechnique* **7**, No. 3, 134–139.

Wilkinson, W. B. (1968). Constant head in situ permeability tests in clay strata. *Géotechnique* **18**, No. 2, 172–194.

Soga, K. & Jefferis, S. A. (2008). *Géotechnique* **58**, No. 5, 441–448 [**doi:** 10.1680/geot.2008.58.5.441]

Contributions to *Géotechnique* 1948–2008: Soil science and interdisciplinary aspects of geotechnical engineering

K. SOGA* and S. A. JEFFERIS†

It has long been recognised that soils are not simply assemblages of inert mineral grains, but complex systems within which multiple phenomena can occur. Our objective in this overview is to highlight the key papers and concepts in mineralogy, pollution and the environment that that have been published in *Géotechnique* over the last 60 years, and to consider the future directions for progress in these areas of geotechnical engineering. To provide a framework for our review we have adopted the concept of coupled phenomena, and we have considered six processes (mechanical, hydrological, chemical, electrical, thermal and biological). By reviewing the major technical achievements in the areas of chemical, electrical, thermal and biological effects on soil systems, the importance of a sound understanding of these phenomena in geotechnical and geoenvironmental engineering is highlighted. We conclude that, although much excellent work has been published in *Géotechnique*, there are still many opportunities and challenges offered by soil processes, and especially low-intensity chemical and biological processes.

KEYWORDS: chemical properties; historical review; mineralogy; pollution migration/control; temperature effects

On reconnaît depuis longtemps que les sols ne sont pas simplement des assemblages de grains minéraux inertes, mais plutôt des systèmes complexes au sein desquels de multiples phénomènes peuvent survenir. Dans la présente communication, notre objectif est de mettre en lumière les principaux concepts et communications dans le secteur de la minéralogie, de la pollution et de l'environnement qui ont été publiés par *Géotechnique* au cours des soixante dernières années, et de nous pencher sur l'orientation à suivre dans l'avenir pour l'avancement dans les domaines de l'ingénierie géotechnique. Afin de délimiter la portée de notre examen, nous avons adopté le principe des phénomènes accouplés, et nous avons examiné six procédés (mécanique, hydrologique, chimique, électrique, thermique et biologique). En examinant les principales réalisations techniques dans les secteurs chimique, électrique, thermique et biologique sur les systèmes de sol, nous soulignons l'importance d'une bonne connaissance de ces phénomènes en ingénierie géotechnique et géoenvironnementale. La conclusion que nous tirons est que bien que nous ayons publié un grand nombre de travaux excellents en *géotechnique*, les procédés des sols présentent encore un grand nombre d'opportunités et de défis, en particulier les procédés chimiques et biologiques de faible intensité.

INTRODUCTION

Our objective in this overview is to highlight the key papers and concepts on mineralogy, pollution and the environment that have been published in *Géotechnique* over the last 60 years, and to consider future directions for progress in these areas of geotechnical engineering. We have subtitled the review 'Interdisciplinary aspects of geotechnical engineering', as the required knowledge base has often had to be 'imported' into geotechnical engineering from other fields of science, and other engineering disciplines. However, this has not been solely a process of knowledge transfer. A recurring problem has been how to analyse the many complex simultaneous interactions that can occur in a soil, and to provide engineering solutions for processes that may occur at scales ranging from micrometres (clay particles) to kilometres (geomorphological processes).

Consolidation provides a simple example of the interaction of two processes—the dissipation of pore pressure in a porous medium, and soil deformation. We now refer to these interactions as coupled phenomena, and geotechnical engineers have keenly embraced the concepts of coupling, especially as their interests have expanded into areas of the geo-environment. Other examples of coupled phenomena include quick clays (chemical and mechanical coupling), waste containment (chemical, hydraulic/hydrological and mechanical), pollution transport (chemical and hydrological), electro-osmosis (electrical, hydrological and mechanical), electro-kinetics (electrical, chemical and hydrological), energy foundations (thermal, hydrological and mechanical) and bio-fouling (biological and hydrological).

A variety of coupled phenomena have been examined in the past, but the significance of coupling was not always explicitly recognised until the complex array of coupled phenomena was elegantly elaborated by Mitchell in his Rankine Lecture in 1991 (Mitchell, 1991). Table 1 from his paper sets out key coupled phenomena. His lecture was a landmark presentation, and the related paper must be regarded as a defining contribution in this area of soil mechanics. It provides the fundamental logical framework for the analysis of coupled phenomena in soils.

In this overview we will use the concept of coupled phenomena to provide a framework for the discussion. The various coupled problems that have been investigated in the past are listed in the upper-right half of Table 2. Biological phenomena have been added to the list of phenomena presented by Mitchell, although it is recognised that defining coupling coefficients will be a challenge (e.g. Mitchell & Santamarina, 2005). We also identified key papers that deal explicitly with coupling problems; these are listed in the lower-left half of Table 2.

At the outset it must be recognised that for many coupled problems, especially those involving chemical, biological and electrical phenomena, there has been considerable publication in journals and conferences aside from *Géotechnique*. A key contribution of *Géotechnique* has been to provide a medium for the publication of authoritative, geotechnically focused papers in areas that may have been

Discussion on this paper closes on 1 December 2008, for further details see p. ii.
* Department of Civil Engineering, University of Cambridge, UK.
† Environmental Geotechnics Ltd, UK.

Table 1. Direct and coupled flow phenomena (Mitchell, 1991)

Flow	Gradient			
	Hydraulic head	Temperature	Electrical	Chemical concentration
Fluid	Hydraulic conduction: Darcy's law	Thermo-osmosis	Electro-osmosis	Chemical osmosis
Heat	Isothermal heat transfer	Thermal conduction: Fourier's law	Peltier effect	Dufour effect
Current	Streaming current	Thermoelectricity: Seebeck effect	Electric conduction: Ohm's law	Diffusion and membrane potentials
Ion	Streaming current	Thermal diffusion of electrolyte: Soret effect	Electrophoresis	Diffusion: Fick's law

initially reported and developed elsewhere—often in disciplines quite distinct from geotechnical engineering. Coupled hydraulic–mechanical phenomena (H–M) such as pore pressure and effective stress and consolidation, although strongly represented in *Géotechnique*, are outside the scope of this review.

CHEMICAL EFFECTS

A convenient starting point for the review is to consider papers focusing on mineralogy—understanding the basic physical and chemical properties of soils that are the fundamental source of many of the coupling phenomena. It is pleasing to find an early paper in *Géotechnique* by Ralph Grim, an expert on clay mineralogy. In his paper 'Mineralogical composition in relation to properties of certain soils' (Grim, 1949) he presented data on nine soils from around the world, and drew the following conclusions.

> The analyses of these soils and of others that have somewhat unusual properties permit a listing of certain factors of composition that are apt to give a soil unusual properties ... Also they cause a soil to have properties that are quite likely to change as a consequence of changes in soil environment, such as changes in groundwater circulation or changes due to construction. The factors are as follows:
>
> (*a*) clay mineral composition containing montmorillonite minerals
> (*b*) clay mineral composition containing halloysite minerals
> (*c*) high base-exchange capacity
> (*d*) sodium as an important exchangeable base
> (*e*) high soluble salt content
> (*f*) concentration of component particles in fine silt size grade.

These conclusions have stood the test of time, and they are still the reference points for 'unusual soil properties' that we would now recognise as swelling or particularly sensitive soils.

In the early 1950s *Géotechnique* published several papers describing a process of salt leaching of glacial marine clays by fresh water as the land elevated over geological time (Skempton & Northey, 1952; Rosenqvist, 1953; Bjerrum, 1954). Leaching of saline soils is a good example of how pore fluid chemistry (including cation type and concentration and pH) can influence the mechanical behaviour of clays, and the processes involved are well summarised in Bjerrum's Rankine Lecture of 1967 (Bjerrum, 1967). The reduced concentration of ions in the pore fluid as a result of leaching changes a stable flocculated structure to a metastable one in which dispersion is preferred (as indicated by the reduction in Atterberg limits when remoulded) and so leads to a significant reduction in the undrained shear strength and a large increase in compressibility when the soil structure is disturbed. The leached clay becomes 'sensitive' to disturbance, and clays that behave like a liquid on remoulding were named 'quick clays'. This cycle of work was an early example, in *Géotechnique*, of soil mechanics meeting chemistry at large scale.

The effect of pore fluid composition on the properties of clays, including compressibility and shear strength, has been a regular topic for research. It has long been recognised that clay particles carry a net negative charge as a result of lattice imperfections, and that as a result they exhibit a cation exchange capacity. They also can show an anion exchange capacity as a result of the broken edges of particles, although the nature of the binding and the anion capacity is sensitive to the anion species and the pore water pH. Clay–pore fluid interactions are further complicated by the fact that, when a water-saturated clay interacts with an electrolyte, it can behave as a partial semi-permeable membrane. The solute ions (electrolytes) are inhibited from diffusing inwards, but solvent can flow inwards or outwards (depending on the concentration gradient), causing compression or swelling by pore pressure changes at particle contacts (osmotic pressure).

The diffuse double layer theory, developed by Gouy (1910) and Chapman (1913), has proved a popular point of reference throughout the years for authors seeking to explain the effects of interparticle forces on the behaviour of clay–water systems. The balance of interparticle forces, including electrostatic interactions, van der Waals attraction and Born repulsion, at a particle contact scale has been used to explain chemical effects on soil fabric (such as dispersed structures, face-to-face aggregated systems and edge-to-face card-house structures) and the corresponding mechanical behaviours. The paper by Bolt in 1956 is one of the early papers that illustrate the difference in the compression curves of sodium montmorillonite, calcium montmorillonite and sodium illite when prepared at different salt concentrations (Bolt, 1956). Work on homoionised clay and the applicability of the double layer theory is taken up in many other papers. Olson (1962) provides an early description of tests on homoionised clays, and Mesri & Olson (1970, 1971) and Sridharan and his co-workers (e.g. Sridharan & Venkatappa Rao, 1973, 1979; Sridharan *et al.*, 1986) examined the consolidation and shear behaviour of sodium and calcium montmorillonite and kaolinite with water at different electrolyte concentrations, and with non-polar organic liquids as pore fluids. Moore & Mitchell (1974) provided further experimental and theoretical investigations to highlight the importance of including van der Waals forces in predictions of macroscopic properties such as soil strength from the microscopic properties of soil–water–ion systems.

Basic double layer theory, although a very useful tool, is no longer sufficient to explain the behaviour of soil–water–

Table 2. Coupling processes and papers in *Géotechnique*

Processes	Mechanical (M)	Hydrological (H)	Chemical (C)	Electrical (E)	Thermal (T)	Biological (B)
Mechanical		Effective stress Consolidation Unsaturated soils Soil cracking	Clay mineralogy Quick clays Residual strength Chemical weathering Grouting Clay liners Cement/lime stabilisation Soil heaving	Electro-osmotic consolidation	Desiccation cracks Ground freezing Energy pipelines Energy foundations Permafrost thawing Nuclear waste Oil sands Methane hydrates	Lithification Biological weathering Biogenesis of gassy soils
Hydrological	Many publications		Waste containment Chemical and colloid migration Osmotic processes	Geophysics Environmental geophysics	Drying Ground freezing	Bio-fouling Bio-clogging
Chemical	Grim (1949) Bolt (1956) Sherwood (1957) Olson (1962) Hawkins & Pinches (1987) and others	Rowe & Booker (1984, 1986) Mitchell (1991) Manassero & Dominijanni (2003) Rowe (2005) and others		Electro-kinetics	Heat generating wastes Thermally enhanced remediation	Permeable reactive bio-barriers and impermeable bio-barriers
Electrical	Casagrande (1949) Bjerrum *et al.* (1967) Mitchell (1991)	See left	Mitchell (1991) Yeung & Mitchell (1993) Acar *et al.* (1994)		Thermo-electric effect Peltier and Seebeck effects	Biologically mediated redox environments Electro-stimulation
Thermal	Ward & Sewell (1950) Morgenstern (1981) Blight (1997) Brandl (2006) Gens (2008) and others	Mitchell (1991) Blight (1997)				Geo-thermal impacts
Biological	Maclean & Smart (1978) Sills *et al.* (1991)	Paksy *et al.* (1998) Rowe (2005)				

electrolyte systems. Factors not considered in the theory, such as pH, ion size/type, ion redistributions, anion adsorption, the influence of large, natural organic molecules, and the stacked state of clay platelets, must be considered. Furthermore, it is important to note that soil structure is often an aggregation of particles, and that a complex arrangement of particle clusters and inter- and intra-aggregate pores also influences their mechanical behaviour. The limitations of the double layer theory are discussed in Güven (1992) and Mitchell & Soga (2005), and in texts on colloid science.

Lupini *et al.* (1981), a well-cited paper, describes how the drained residual strength of cohesive soils is influenced by soil plasticity. They quote the work by Kenney (1977) illustrating that the residual strength in the 'sliding' shear mode of failure is influenced by the pore water chemistry in addition to the soil plasticity effect (i.e. the pore water chemistry can change the soil plasticity). This failure mode occurs when behaviour is dominated by platy, low-friction particles, and a low strength shear surface of strongly orientated platy particles then develops. This led to a number of papers in *Géotechnique* that discussed the effect of pore fluid chemistry on residual strength. The general consensus is that, under constant confining stress, the residual shear strength of micaceous minerals and montmorillonite/bentonite increases with increasing concentrations of ions in the pore fluid, and with cations of higher valence. In these materials there are significant physico-chemical forces, and the increase in strength is considered to be due to the increase in true effective interparticle stresses, which vary with pore fluid electrolyte concentration as well as cation exchange (Moore, 1991; Di Maio & Fenelli, 1994). The paper by Moore & Brunsden (1996) provides extensive field data—information at a scale that is too rarely available—on the effects of salts and cation exchange on coastal soils.

The papers mentioned above, and a number of others, were and are key to our understanding of the impact of the local chemical environment and its prior evolution on the physical properties of clay soils. However, the present state of the art is such that concepts such as double layer theory are generally more useful for the explanation of observed responses of clay systems to chemical change than the *a priori* prediction of the effects of chemicals on soil-mechanical properties.

The effect of chemicals naturally present in soils has been addressed by several authors. An early paper (Sherwood, 1957) discussed the stabilisation of soils containing sulphate with cement, and showed that sulphates could significantly damage the strength of stabilised soils. The paper did not specifically address the swelling of lime- or cement-stabilised soils. However, this and other work by Sherwood has been fundamental in highlighting the problems of heave of soils containing sulphur species when stabilised with lime and/or cement—although sadly the literature has not been sufficient to prevent recurring problems of heave and damage.

The heave of lime/cement-stabilised soils is not the only source of heave problems in soils containing sulphur species, and this may be linked to another major area of importance to geotechnical engineers—redox potential. Although seldom explicitly recognised, construction activity can have dramatic impacts on soil redox. For example, flooding by excluding air can lead to reduction in redox, whereas dewatering can raise the local redox. Change of redox can change the speciation of soil chemicals. For example, Hawkins & Pinches (1987) showed that sulphur species can be oxidised to sulphate and acid, and hence that the presence of oxidisable sulphur species can increase the aggressivity of soils to concrete. It may be noted that the crystallisation of sulphates as gypsum in the soil can also cause heave.

Hydrological–chemical (H–C) interaction brings in the whole arena of pollution transport, and here it is appropriate to note that there was a *Géotechnique* Symposium in Print on contaminated land in 1992.

The papers by Rowe & Booker (1984, 1986) provide a sound theoretical basis for the analysis of pollution migration under the combined effects of hydraulic and chemical potential (H–C) coupling, which was further developed and finely dissected with regard to the prevention of pollution migration by barrier systems by Rowe in his Rankine Lecture (Rowe, 2005). An understanding of diffusion processes is fundamental to the analysis of pollution migration, and it is interesting to note that in soil mechanics Fick's law of diffusion is usually taken as sufficiently valid for engineering processes, whereas in chemical engineering the issues of mass transfer and the dependence of the apparent diffusion coefficient on the concentration of the diffusing species may have to be considered (Coulson *et al.*, 1990). This may be an area for future research. Manassero & Dominijanni (2003) provided a further significant development in chemical migration theory for clay barriers with a detailed formulation of osmotic processes including osmotic efficiency, the ability of clays to restrict the flow of ions.

ELECTRICAL EFFECTS

When an electrical potential is applied across a liquid, including a soil pore water, cations move to the cathode and anions to the anode. However, in soils there will also be a movement of the mobile part of the diffuse double layer towards the cathode, as this will carry a net positive charge owing to the negative charge of fine soil particles. If the water arriving at the cathode is drained away, and no water is allowed to enter at the anode, then consolidation of the soil and hence a strength gain around the anode occurs owing to the decrease in pore pressure.

This coupled electrical–hydraulic phenomenon (called electro-osmosis) has been addressed in several *Géotechnique* papers, springing from the work in the 1930s by L. Casagrande, who pioneered the application of electro-osmosis for dewatering and slope stabilisation. In 1949, he published a comprehensive paper on work conducted at the UK Building Research Station, and commented that

> this work has not led the author to change his earlier opinion that the field of practical application of this method is likely to be limited by economic factors. The author considers that for economic use the rate of direct removal of water is too small to permit the consolidation of large soil masses within the short period which is usually available in building practice. However, the original use of the method for overcoming difficulties in excavations in waterlogged ground is both practicable and economic. (Casagrande, 1949)

In the 1960s the Norwegian Geotechnical Institute conducted extensive research on electro-osmosis as a way to obtain a permanent increase in the shear strength of clays around deep excavations. An excellent field study of the combined effects of consolidation and electrochemical reactions was presented by Bjerrum *et al.* (1967). By applying the electro-osmotic consolidation technique at a quick clay site to a depth of 10 m over an area of 10 by 20 m, they demonstrated an increase in strength from reduction in water content and consolidation. However, interestingly, in addition to the increase in strength from consolidation, they found an increase in Atterberg limits of the clay as a result of electrochemical reactions, which contributed additional strength gain. The steel anodes (reinforcing bars) used for the project were observed to be electrolytically dissolved during the treatment, thus contributing iron for ion exchange and potentially clay cementation.

Electro-chemical effects during electro-osmotic consolidation can be beneficial or detrimental. Such effects include ion diffusion, ion exchange, development of variable pH gradients, desiccation by heat generation around the electrodes, chemical reactions, mineral decomposition, precipitation of salts or secondary minerals, and soil fabric changes. As discussed above, electro-chemical strengthening of a quick clay was demonstrated by Bjerrum *et al.* (1967). Mitchell (1991) describes the possible use of electro-osmosis to stabilise the leaning Tower of Pisa. However, a field trial conducted at the site experienced a great amount of heat and gas generation, which reduced system efficiency, and the application proved unsuccessful (Mitchell & Soga, 2005).

Although there was limited work reported on electric–hydraulic–chemical coupling during the 1970s and 1980s, possible new applications in waste containment barriers and the remediation of contaminated sites stimulated new interest in the 1990s. Coupling between electrical and hydraulic flows can generate electrokinetic phenomena in fine-grained soils. The transport of dissolved and suspended constituents into and out of the soil by electrokinetics and electrophoresis (movement of charged particles) and electro-chemical reaction can be utilised to remove contaminants from fine-grained soils, to introduce treatment agents, and to create electrokinetic barriers to contaminant transport. Yeung & Mitchell (1993) described the theoretical background and showed experimental results on the migration of cations and anions in compacted clay under the influences of hydraulic, electrical and chemical gradients imposed simultaneously.

Acar *et al.* (1994) demonstrated cadmium removal for the decontamination of soils at laboratory scale. They highlighted the importance of acid/base profile between the electrodes, which significantly affects the electrokinetic processes. When electric current is applied across a soil sample, the acid front generated at the anode will advance towards the cathode and neutralise the base generated at the cathode. As a result of the pH gradients and the prevailing chemistry during the process, there will be complex chemical reactions such as dissolution of available salts/clay minerals, adsorption/desorption and complexation reactions, and formation of cementitious products, which in turn result in time-dependent changes in the structure and hence the engineering characteristics of the soil deposit. Acar *et al.*'s careful consideration of chemistry during electro-osmosis provided further insight into the findings made by Bjerrum *et al.* (1967) described earlier.

THERMAL EFFECTS

Thermal effects include thermally driven moisture flows in unsaturated soils, ground freezing, frost heave, and thaw consolidation and weakening. Because of the three-phase nature of soils (solid, water, gas), heat flow through soils occurs by several mechanisms: (a) conduction, (b) convection by water and/or air flow in the voids, (c) radiation across voids, (d) moisture and vapour migration, (e) vaporisation and condensation, and (f) freezing and thawing. Heat transfer in soils occurs mainly by conduction, and by convection for high-permeability soils, although advective transport is likely to dominate if there is any significant groundwater flow. If there is a phase change of water or other fluids in the voids, the latent heat process will require significant heat flow. Except for exposed surface soils (see Blight, 1997), radiation usually contributes only negligibly to heat transfer; its effect in sand is less than 1% of the overall heat transfer (Rees *et al.*, 2000; quoted in Brandl, 2006).

Several papers dealing with heat flow effects have been published in *Géotechnique*. One of the early papers was by Ward & Sewell (1950), who examined the problem of insulating ground beneath kilns, blast furnaces and other high-temperature installations, and of the cooling and possible freezing of soil beneath cold-storage warehouses. The work conducted at the Building Research Station was based on investigations of damaged industrial plant, where the foundations had moved appreciably following the shrinking and heaving effects of heat and frost on the ground. A further theoretical paper is that by Scott (1961), who provided a discussion on the prediction of the depth of summer thawing in perennially frozen ground.

A thermal gradient in the ground will induce water migration towards the colder region. Thermally driven water flows in saturated soils are rather small—a few tenths of a centimetre water head per degree Celsius. However, when the ground freezes, water is drawn to the frost front owing to the latent heat from the phase change from water to ice. The coupling of the heat and mass flow adds considerable complexity to the theory of frost heave, as was described in detail by Morgenstern (1981) in his Rankine Lecture.

Morgenstern (1981) developed the procedure for forecasting the heave of a chilled buried gas pipeline laid on unfrozen ground. His analysis showed that the rate of water intake is limited on the cold side of the warmest ice lens: therefore the warm-side temperature alone affects the value of the suction at the frost front. This is because an ice lens acts like an impermeable barrier with regard to water migration in the frozen soil. This was confirmed by field studies of monitoring the movement of a test pipeline subjected to in situ frost heave.

The reverse problem, thawing of ice-rich frozen ground resulting in large settlements and reduced bearing capacity, was also considered by Morgenstern (1981). If thawing proceeds under fully drained conditions, the time-dependent settlement occurs with the progress of the thaw front with time. If thawing proceeds too quickly for the pore pressures generated to dissipate, settlement is impeded but the shear strength is reduced owing to the excess pore pressures. Hence, for analysis, coupling of the thaw process with the consolidation process is necessary. Morgenstern (1981) developed a concept of 'residual stress', which the soil retains after thaw under undrained conditions, and illustrated how it influences pore pressures, settlements associated with thaw, and the undrained strength of a soil mass. He further argued that studies of thaw-induced soil–structure interaction problems are not yet well developed, and testing to obtain the appropriate deformation parameters is in its infancy. Although written in 1981, this remains true today and may become progressively more significant. For example, as highlighted by Gens in his Rankine Lecture (Gens, 2008), the thawing of permafrost due to global warming has the potential for substantial impacts on civil engineering infrastructure.

In unsaturated soils thermally driven moisture flows are very important because of the contribution of vapour phase flow, which is driven by gradients of temperature and moisture content. Moisture is driven away from the heat source; this may cause shrinkage in the warm zone and expansion in the cold zone. The practical implications are subgrade instability, settlements of building basements founded on swelling soils, and limited performance of ground source heat pump coils and buried cables. Blight (1997) in his Rankine Lecture presented an overview of the soil–water balance occurring in an unsaturated soil between the ground surface and water table. The paper included an excellent discussion on evapotranspiration processes and the associated cracking and fissuring of soils as a hydro-mechanical coupling process, a subject that requires further study. Thomas & Li (1997) presented a paper that describes a theoretical formulation for coupled heat and water flow in non-deforming unsaturated soil for both steady-state and transient state conditions. In contrast to earlier papers published elsewhere, they took vapour transfer characteristics fully into account. Continuity of both liquid water and water vapour was considered by means of a conservation of mass equation. The energy conservation equation was based on heat transfer by means of conduction, convection and latent heat of vaporisation. Physical model heating experiments were performed, and the data compared well with the results of the numerical simulations.

The early work by Geraminegad & Saxena (1986) considered the coupling effects of moisture, heat and gas flow on soil skeleton volume changes. Thomas & He (1995) extended this work by proposing a theoretical model that considers the so-called state surface approach to relate volumetric strain not only to total stress but also to suction and temperature changes. More recently, Gens (2008) in his Rankine Lecture presented a more advanced theoretical thermo-hydro-mechanical (THM) model and demonstrated its use by describing the results of numerical simulations of chilled buried cable (as the freezing case) and high-level radioactive waste disposal (as the heating case).

High-level nuclear waste disposal is an area in which thermal effects are challenging, as a swelling clay such as bentonite, compacted in an unsaturated state, may be used as part of the engineered barrier. After burial, the clay will first be subjected to elevated temperature in the region near the radioactive waste, and then eventually water saturation from the recovery of the groundwater table. The clay can therefore experience significant deformation due to moisture content changes. Thermal expansion of pore water increases the pore water pressure, and consequently decreases the effective stress of the soil. Furthermore, an increase in temperature may cause cementation and a loss of the swelling characteristic of the bentonite and other mineralogical, chemical and microbial changes. Since the 1980s this problem has been the driver for the development of detailed THM coupled models. However, the addition of chemical and biological processes to achieve the THMCB models necessary to back-analyse large-scale laboratory tests and/or predict full-scale behaviour could yet require a radical recasting of current models.

Resource recovery is another application of THM models. Morgenstern (1981) in his Rankine Lecture introduced in situ thermal enhancement of heavy oil recovery from unconsolidated materials, which has generated intensive research more recently as the oil price has increased such that the recovery method has become economically viable. The in situ recovery process involves a massive injection of steam in order to reduce the viscosity of the bitumen. The efficiency of subsequent withdrawal is influenced by fluid/gas flows and the strength and compressibility of oil sands at elevated temperatures. Extraction of methane gas from methane hydrate soils is another example. These soils exist under low-temperature and high-pressure conditions such as permafrost and deep seabed regions. By decreasing the pressure or increasing the temperature, the hydrate dissociates and methane gas is generated. This is associated with ground deformation as the solid hydrate, which provided cohesion between soil particles, is lost, so the soil loses strength (e.g. Soga *et al.*, 2007).

Closed loops of high-density polyethylene piping can be embedded in underground structures (such as piles, tunnel linings and retaining walls). A heat carrier fluid (water, water with antifreeze, or a saline solution) then can be circulated through the system to exchange energy between ground and buildings. These 'energy foundations' represent an innovative technology that can contribute to environmental protection and provide substantial long-term cost savings and minimised maintenance. Brandl (2006), based on his Rankine Lecture in 2001, provides a comprehensive summary of the theories, design and construction details of energy foundations, including several useful case histories.

Brandl (2006) states that the most economical and environmentally friendly condition for energy foundations is seasonal operation with an overall energy balance across a year. This means heating in winter (i.e. heat extraction from the ground) and cooling in summer (i.e. heat sinking/recharging into the ground). To achieve this he recommends a low-permeability soil and groundwater with only a low hydraulic gradient and a temperature differential, ΔT, limited to 2°C between absorber fluid inflow and return flow from the primary circuit. His experience has been that geothermal heating/cooling system-based energy foundations may save up to two-thirds of conventional heating/cooling costs.

Thermo-hydro-mechanical effects such as the change in shaft resistance and base capacity of energy piles and movements of diaphragm walls and tunnels may be of concern. Based on site measurements, Brandl (2006) found that these mechanical properties are not affected by the heat exchange processes to a statically relevant magnitude. However, he warns that exposure of the surface of deep foundations to temperatures below freezing should be avoided, and advises caution when considering energy foundations in fine-grained soils with a high content of active clay minerals, which are susceptible to large temperature-induced volume change, as discussed above.

BIOLOGICAL EFFECTS

Microbiological life can be expected in all environments where there is or has been liquid water, and thus the presence of microbial life including viruses, bacteria, fungi and larger life forms must be expected in soils. In *Géotechnique*, microbiology has been discussed in relation to the oxidation of sulphur species in soils to sulphates, the behaviour of peats and waste and landfill. However, there is little formal discussion of the influence of biological processes on the properties of the mineral elements of soils or the binding of soil grains.

The brief technical note by Maclean & Smart (1978) shows that clay particles can be bound to microorganisms perhaps by electrostatic charges or exo-polysaccharides. From this work it seems that biological materials may be able to flocculate clay-sized material settling from suspension, and thus influence the nature of the final settled material. In this context it is appropriate to note that some theories of the origins of life suggest that clay mineral and/or quartz surfaces may have been the key to the development of life by sorption and concentration organic molecules from a primordial soup (Bernal, cited in Brown, 2005, and Bohn *et al.*, 1985). Bernal, as cited in Brown (2005), considered both clays and quartz (quartz could impart chirality—handedness—seen in biological molecules). Bohn *et al.* (1985) considered only sorption onto clays.

There is a useful *Géotechnique* literature on gassy soils such as may be produced biogenically. Sills *et al.* (1991) considered the behaviour of an offshore soil containing gas bubbles. Wichman (2000) developed a finite strain theory for a gassy sludge, and detailed laboratory investigations are reported by Sills & Gonzalez (2001).

Biogenic slimes can cause clogging of landfill drainage systems, and in *Géotechnique* this was first addressed by Paksy *et al.* (1998), who subjected laboratory columns containing granular drainage material to varying rates of volatile fatty acid (VFA) application (VFAs are formed during the biodegradation of organic wastes). Landfill drainage clogging was also analysed by Rowe (2005) in his Rankine Lecture, where a predictive model was derived based on the chemical/biochemical reactions within the landfill.

The exploitation of the ground as a heat source/sink also can lead to bio-fouling problems. These are relatively easy to manage in closed-circuit systems. However, in open systems where water from an aquifer is passed through a heat exchanger or heat pump and then returned to the ground there is potential for bio-fouling in the wells and heat exchangers. Furthermore, the precipitation of minerals and hence clogging may occur as a result of temperature changes and/or as a result of redox changes—for example precipitation of iron manganese hydroxides on exposure to air, or the input of dissolved oxygen to the aquifer (redox-related precipitation is likely to be biologically mediated).

With sustainability becoming an increasingly important driver for all activities, including the geotechnical, it is to be expected that there will be considerable developments in research activity on low-energy-intensity biological manipulation/management of soil properties.

FUTURE OPPORTUNITIES

Geotechnical engineers work on multidisciplinary problems, and their work must encompass a wide range of processes and still provide engineering solutions. In the early years of geotechnical engineering researchers tackled problems on hydro-mechanical interactions such as consolidation and chemical–mechanical interaction in relation to clay mineralogy and pore fluid chemistry, including quick clays and residual strength. These may be considered to be classical 'soil science/soil mechanics' problems.

During the late 1980s and early 1990s the management of contaminated sites and landfills became a driving force for the development of environmental geotechnics, in which we have established conceptual frameworks for the fate and transport of chemicals in the ground, and have regularly exploited these at full scale. The early years of the present century could be classified as the years of sustainability, with ever more opportunities for geotechnical engineers in the field of sustainable development, waste management, effective utilisation of natural resources and energy, and maintenance and repair of ageing infrastructure (e.g. National Research Council, 2006).

The geotechnical engineer now has the basic engineering and scientific knowledge to become more proactive rather than merely reactive to the opportunities offered by soil processes, and perhaps especially low-intensity chemical and biological processes. Furthermore, in relation to biological processes, we must recognise that life is supported by soil systems, and the geotechnical engineer must become versed in soil protection, including the problems that flow from surface sealing by hard development.

SUMMARY AND CONCLUSIONS

This review of 60 years of papers published in *Géotechnique* demonstrates that it has long been recognised that soils are not simply inert mineral grains, but complex systems within which multiple phenomena occur. To provide a framework for our review we have adopted the concept of coupled phenomena. However, we recognise that only some aspects of the many possible phenomena have been addressed. For example, for the electrical–hydraulic (E–H) phenomenon, several detailed papers have been published on electro-osmosis, but the inverse phenomenon of H–E coupling, streaming potential, has received much less attention and remains an area of opportunity. Furthermore, reference to Table 2 shows that five coupled phenomena, C–T, C–B, E–T, E–B and T–B, have yet to receive attention in *Géotechnique* and remain areas of opportunity for research and perhaps blue-sky applications.

This review has highlighted the importance of a sound understanding of mineralogy, pollution and the environment to geotechnical engineering, and has shown that, although much excellent work has been published in *Géotechnique*, there are still many opportunities and challenges. However, without the invigoration of new research and new applications, this area of geotechnics risks becoming overly introspective, lacking the diversity that flourished when geotechnics embraced geo-environmental engineering.

REFERENCES

Acar, Y. B., Hamed, J. T., Alshawabkeh, A. N. & Gale, R. J. (1994). Removal of cadmium (II) from saturated kaolinite by the application of electrical current. *Géotechnique* **4**, No. 2, 239–254.

Bjerrum, L. (1954). Geotechnical properties of Norwegian marine clays. *Géotechnique* **4**, No. 2, 49–69.

Bjerrum, L. (1967). Engineering geology of Norwegian normally-consolidated marine clays as related to settlements of buildings. *Géotechnique* **17**, No. 2, 81–118.

Bjerrum, L., Moum, J. & Eide, O. (1967). Application of electro-osmosis to a foundation problem in a Norwegian quick clay. *Géotechnique* **17**, No. 3, 214–235.

Blight, G. E. (1997). Interactions between the atmosphere and the Earth. *Géotechnique* **47**, No. 4, 713–767.

Bohn, H., McNeal, B. & O'Connor, G. (1985). *Soil Chemistry*, 2nd edn. Chichester: Wiley Interscience.

Bolt, G. H. (1956). Physico-chemical analysis of the compressibility of pure clays. *Géotechnique* **6**, No. 2, 86–93.

Brandl, H. (2006). Energy foundations and other thermo-active ground structures. *Géotechnique* **56**, No. 2, 81–122.

Brown, A. (2005). *The sage of science*. Oxford: Oxford University Press.

Casagrande, L. (1949). Electro-osmosis in soils. *Géotechnique* **1**, No. 3, 159–177.

Chapman, D. L. (1913). A contribution to the theory of electro-capillarity. *Phil. Mag.* **25**, No. 6, 475–481.

Coulson, J. M., Richardson, J. F., Backhurst, J. R. & Harker, J. H. (1990). *Chemical engineering*, Vol. 1, 4th edn. Oxford: Pergamon.

Di Maio, C. & Fenelli, G. B. (1994). Residual strength of kaolin and bentonite: the influence of their constituent pore fluid. *Géotechnique* **44**, No. 2, 217–226.

Gens, A. (2008). Soil–environment interactions in geotechnical engineering. *Géotechnique* (to be published).

Geraminegad, M. & Saxena, S. K. (1986). A coupled thermoelastic model for saturated-unsaturated porous media. *Géotechnique* **36**, No. 4, 539–550.

Gouy, G. (1910). Sur la constitution de la charge électrique à la surface d'un électrolyte. *J. Phys. Théor. Appl* **4**, No. 9, 457–468.

Grim, R. E. (1949). Mineralogical composition in relation to the properties of certain soils. *Géotechnique* **1**, No. 3, 139–147.

Güven, N. (1992). Molecular aspects of clay–water interactions. In *Clay–water interface and its rheological implications* (eds N. Güven and R. M. Pollastro), CMS workshop lectures, 4, pp. 2–79. Clay Minerals Society.

Hawkins, A. B. & Pinches, G. M. (1987). Sulphate analysis on black mudstones. *Géotechnique* **37**, No. 2, 191–196.

Kenney, T. C. (1977). Residual strengths of mineral mixtures. *Proc. 9th Int. Conf. Soil Mech. Found. Engng, Tokyo* **1**, 155–160.

Lupini, J. F., Skinner, A. E. & Vaughan, P. R. (1981). The drained residual strength of cohesive soils. *Géotechnique* **31**, No. 2, 181–213.

Maclean, P. A. & Smart, P. (1978). Effect of bacteria on soil structures. *Géotechnique* **28**, No. 1, 122–123.

Manassero, M. & Dominijanni, A. (2003). Modelling the osmosis

effect on solute migration through porous media. *Géotechnique* **53**, No. 5, 481–492.
Mesri, G. & Olson, R. E. (1970). Shear strength of montmorillonite. *Géotechnique* **20**, No. 3, 261–270.
Mesri, G. & Olson, R. E. (1971). Consolidation characteristics of montmorillonite. *Géotechnique* **21**, No. 4, 341–352.
Mitchell, J. K. (1991). Conduction phenomena: from theory to geotechnical practice. *Géotechnique* **41**, No. 3, 299–340.
Mitchell, J. K. & Santamarina, J. C. (2005). Biological considerations in geotechnical engineering. *J. Geotech. Geoenviron. Engng ASCE* **131**, No. 10, 1222–1233.
Mitchell, J. K. & Soga, K. (2005). *Fundamentals of soil behavior*. New Jersey: John Wiley & Sons.
Moore, C. A. & Mitchell, J. K. (1974). Electromagnetic forces and soil strength. *Géotechnique* **24**, No. 4, 627–640.
Moore, R. (1991). The chemical and mineralogical controls upon the residual strength of pure and natural clays. *Géotechnique* **41**, No. 1, 35–47.
Moore, R. & Brunsden, D. (1996). Physico-chemical effects on the behaviour of a coastal mudslide. *Géotechnique* **46**, No. 2, 259–278.
Morgenstern, N. R. (1981). Geotechnical engineering and frontier resource development. *Géotechnique* **31**, No. 3, 305–365.
National Research Council (2006). *Geological and geotechnical engineering in the new millennium: Opportunities for research and technological innovation*. Washington, DC: National Academies Press.
Olson, R. (1962). The shear strength properties of calcium illite. *Géotechnique* **12**, No. 1, 23–43.
Paksy, A., Powrie, W., Robinson, J. P. & Peeling, L. (1998). A laboratory investigation of anaerobic microbial clogging in granular landfill drainage media. *Géotechnique* **48**, No. 3, 389–401.
Rees, S. W., Adjali, M. H., Zhou, Z., Davies, M. and Thomas, H. R. (2000). Ground heat transfer effects on the thermal performance of earth-contact structures. *Renewable and Sustainable Energy Review* **4**, No. 3, 213–265.
Rosenqvist, I. Th. (1953). Considerations on the sensitivity of Norwegian quick-clays. *Géotechnique* **3**, No. 5, 195–200.
Rowe, R. K. (2005). Long-term performance of contaminant barrier systems. *Géotechnique* **55**, No. 9, 631–678.
Rowe, R. K. & Booker, J. R. (1984). The analysis of pollutant migration in a non-homogeneous soil. *Géotechnique* **34**, No. 4, 601–612.
Rowe, R. K. & Booker, J. R. (1986). A finite layer technique for calculating three-dimensional pollutant migration in soil. *Géotechnique* **36**, No. 2, 205–214.
Scott, R. F. (1961). Heat transfer in soil involving change of state. *Géotechnique* **11**, No. 2, 144–153.
Sherwood, P. T. (1957). The stabilization with cement of weathered and sulphate-bearing clays. *Géotechnique* **7**, No. 4, 179–191.
Sills, G. C. & Gonzalez, R. (2001). Consolidation of naturally gassy soft soil. *Géotechnique* **51**, No. 7, 629–639.
Sills, G. C., Wheeler, S. J., Thomas, S. D. & Gardner, T. N. (1991). Behaviour of offshore soils containing gas bubbles. *Géotechnique* **41**, No. 2, 227–241.
Skempton, A. W. & Northey, R. D. (1952). The sensitivity of clays. *Géotechnique* **3**, No. 1, 30–53.
Soga, K., Lee, S. L., Ng, M. Y. A. & Klar, A. (2007). Characterisation and engineering properties of methane hydrate soils. In *Characterisation and engineering properties of natural soils*, Vol. 4, pp. 2591–2642. London: Taylor & Francis.
Sridharan, A. & Venkatappa Rao, G. (1973). Mechanisms controlling volume change of saturated clays and the role of the effective stress concept. *Géotechnique* **23**, No. 3, 359–382.
Sridharan, A. & Venkatappa Rao, G. (1979). Shear strength behaviour of saturated clays and the role of the effective stress concept. *Géotechnique* **29**, No. 2, 177–193.
Sridharan, A., Rao, S. M. & Murthy, N. S. (1986). Compressibility behaviour of homoionized bentonites. *Géotechnique* **36**, No. 4, 551–564.
Thomas, H. R. & He, Y. (1995). Analysis of coupled heat, moisture and air transfer in a deformable unsaturated soil. *Géotechnique* **45**, No. 4, 677–689.
Thomas, H. R. and Li, C. L. W. (1997). An assessment of a model of heat and moisture transfer in unsaturated soil. *Géotechnique* **47**, No. 1, 113–131.
Ward, W. H. & Sewell, E. C. (1950). Protection of the ground from thermal effects of industrial plant. *Géotechnique* **2**, No. 1, 64–81.
Wichman, B. G. H. M. (2000). A finite strain theory for gassy sludge. *Géotechnique* **50**, No. 1, 35–41.
Yeung, A. T. & Mitchell, J. K. (1993). Coupled fluid, electrical and chemical flows in soil. *Géotechnique* **43**, No. 1, 121–134.

Harrison, J. P. (2008). *Géotechnique* **58**, No. 5, 449–455 [**doi:** 10.1680/geot.2008.58.5.449]

Contributions to *Géotechnique* 1948–2008: Engineering geology, rock mechanics and rock engineering

J. P. HARRISON*

Throughout its history *Géotechnique* has regularly published papers on the fundamentals of engineering geology, the core concepts of rock mechanics, and the practice of rock engineering. In totality these papers provide a concise history of the development of these areas of geotechnics. Here, a review of many of these papers is presented in terms of the specific themes of engineering geology, rock slope instability, behaviour of intact rock and discontinuities in rock, field measurements in rock mechanics, discontinuous rock masses, numerical modelling, and rock engineering. Papers covering a wide range of issues within these themes are outlined, together with an indication of how they have influenced later developments. The review clearly demonstrates how *Géotechnique* has been at the forefront of rock mechanics, rock engineering and engineering geology for the whole of its 60-year existence.

KEYWORDS: historical review; rock mechanics; rocks

Tout au long de son histoire, *Géotechnique* a publié, à des échéances régulières, des communications sur les principes fondamentaux de la géologie technique, les concepts essentiels de la mécanique des roches et la pratique de l'ingénierie des roches. L'ensemble de ces communications fournit une histoire abrégée du développement de ces domaines de la géotechnique. Nous présentons ici un recueil d'un grand nombre de ces communications sur le plan des thèmes spécifiques de la géologie de l'ingénieur, de l'instabilité des pentes rocheuses, du comportement de roches intactes et discontinuités dans la roche, des mesures sur le terrain en mécanique des roches, des masses rocheuses discontinues, de la modélisation numérique, et de la technique des roches. Nous présentons des communications portant sur tout éventail de questions dans le cadre de ces thèmes, en formulant une indication sur l'influence qu'elles ont eu sur des développements ultérieurs. L'étude démontre clairement la façon dont *Géotechnique* s'est maintenue à l'avant-garde dans la mécanique des roches, l'ingénierie des roches et la géologie de l'ingénieur tout au long des 60 années de son existence.

INTRODUCTION

The 1960s saw many journals commence publication on the subjects of rock mechanics and engineering geology. Examples include the *Quarterly Journal of Engineering Geology* (Geological Society of London), *Felsmechanik und Ingenieurgeologie* (International Society for Rock Mechanics, but superseding the Springer publication *Geologie und Bauwesen*, which ran from 1929 to 1962), *Bulletin of Engineering Geology and the Environment* (the then International Association of Engineering Geology), *Engineering Geology* (Elsevier) and the *International Journal of Rock Mechanics and Mining Sciences* (Pergamon Press, latterly Elsevier). These journals continue publication in one form or another, and have developed enviable reputations. Perhaps, then, it is no surprise that *Géotechnique* is often perceived as a 'soils' journal. However, on perusing all 60 volumes of *Géotechnique* it is clear that this perception is incorrect: the journal contains many tens of contributions dealing with a wide range of issues in engineering geology and rock mechanics, and a good number of these are now seen to have been seminal.

This short review brings together a number of these contributions in a way that aims to show how *Géotechnique*, throughout its history, has covered the entire range of geotechnical engineering in rock, from the fundamentals of engineering geology, through the core concepts of rock mechanics, to the practice of rock engineering.

ENGINEERING GEOLOGY

The Europe of 1948 was very different from the Europe of 2008—in geopolitics as well as in geotechnics. In the immediate aftermath of the Second World War a key concern was to rebuild the scientific and engineering links that had been broken by the conflict, and so Volume 1 of *Géotechnique* was able to carry a paper that reviewed engineering geology in Switzerland in the context of the isolation enforced by the War (Von Moos, 1948). Not only is engineering geology as we still recognise it laid out in this contribution, but the paper also underscores just how much Europe has changed since the 1940s. Although being cast as a review, the paper is largely a presentation of case histories, a format that has continued down the years for papers concerning engineering geology and rock engineering.

Although issues such as weathering and alteration of rock are nowadays firmly regarded as being within the realm of engineering geology, Mayer (1963), in the Third Rankine lecture, discussed these under the title 'Recent work in rock mechanics'. In particular, Mayer described the development of methods for measuring the 'alterability of a rock', and suggested that not only were there relations between 'alteration of rock' and the elastic modulus and compressive and tensile strengths of rock, but that changes of rock texture—which could be both chemically and microscopically observed and thereby measured—might be one of the principal physical properties of rock. Indeed, Mayer made the point that there was no great benefit to be gained from expensive in situ tests if no effort was made to understand the microstructure of the rock. Engineering geologists are now fully aware of these concepts, and it is therefore fair to say that Mayer's dream of being able to 'look forward to the day when a series of standard tests in the laboratory and parallel tests in situ would make it possible to classify rock for civil engineering work' has been realised.

Other important engineering geology case histories within *Géotechnique* are those by Terzaghi (1962a), in which the effects of sheeting joints in granite on the stability and

Discussion on this paper closes on 1 December 2008, for further details see p. ii.

* Imperial College London, UK.

construction of the foundation of the Mammoth Pool Dam in California are reported, by Price & Knill (1967) with their discussion of the rock slope instability found at Edinburgh Castle Rock, and by Fookes *et al.* (1977) with their description of the site investigation, engineering geology and rock slope design at Fowey, Cornwall. While all three of these contributions are exemplary in the way they combine techniques in order to develop a thorough understanding of the geomechanics of the sites, that by Terzaghi is particularly noteworthy for the following concluding comment:

> ... on many projects, the results of the application of rock and soil mechanics to the solution of the problems of subsurface engineering will continue to be very disappointing until it becomes universal practice to maintain intimate cooperation between the designers of the projects and the resident engineers during the entire period of construction ... If, for geographical reasons, continuous cooperation between the designers and the construction organization cannot be maintained, it is essential that at least one of the resident engineers should be as thoroughly familiar with the reasoning on which the project was based as the designers themselves.

Almost half a century after this was written, can we say with certainty that this tenet is being universally applied, or are we continuing to encounter difficulties that are largely avoidable?

The principles of engineering geological mapping, together with recommendations for field practice, were set out by Fookes (1969) in the context of the work undertaken at Mangla Dam, Pakistan. This paper introduced a theme that continues to be developed in engineering geology: it underpins Fookes' 1997 Glossop lecture, in which the importance of developing a complete geological model was stressed, and is arguably a precursor to the 'total engineering geology approach' advocated by Baynes *et al.* (2005). These key concepts of developing a geological model in order to understand the geotechnical behaviour of the ground have featured elsewhere in the journal. For example, Fookes & Denness (1969) investigated the fissure and fracture patterns in Cretaceous sediments (both chalk and clay) of South-East England, in work that has become a key paper on the geomechanics of natural fractures in weaker sediments, and continues to be a reference source to researchers. Similarly, in the 16th Rankine Lecture, Meigh (1976) showed how the geotechnical properties and behaviour of the Triassic rocks of the UK could be understood in terms of their various depositional environments. This contribution was built upon by Cripps & Taylor (1981) in their important paper on the engineering properties of mudrocks. A recent elegant contribution carried by *Géotechnique* on understanding the geotechnical behaviour of rock in terms of its constituent minerals and diagenesis is the work of Burton *et al.* (2001), who examined strength variation in young reef limestones. The authors showed clearly how the transformation over time of the mineral composition of coral from aragonite into calcite affects the mechanical properties of these materials, and suggested that classification based on coral species rather than the more usual age and/or diagenesis may be most appropriate for engineering purposes.

ROCK SLOPE INSTABILITY

The importance of understanding rock slope instability in the context of engineering geology features often in the journal. Terzaghi (1962c) reviewed slope stability in what he termed 'hard unweathered rock', in a paper that has become highly cited (in for example, Jaeger's 1971 Rankine lecture and the 1995 benchmark work by Schmidt & Montgomery). In a discussion contribution, Bjerrum & Jorstad (1963) offered additional case history data from Norway, some of which valuably linked elevation, time of year, and days of first and last temperatures above freezing and frost. As we now attempt a fuller understanding of the effects of global warming on geotechnics, this work may have gained new significance.

De Freitas & Watters (1973) presented a detailed review of toppling instability in rock slopes, together with a number of field examples, and concluded that this is a common phenomenon, which can be found at a variety of scales and in a wide range of rocks. The key nature of this paper in encapsulating the phenomenon of toppling is indicated by the fact that this work continues to be cited by researchers in the field. Hamel & Adams (1974) added valuable field examples from North America in materials that had not been affected by the Pleistocene glaciations, and raised the interesting hypothesis that rock creep—and by extension soil creep—may be a form of toppling instability. Given the current interest in numerical analysis using distinct element methods in soil mechanics, this hypothesis may deserve further testing.

Of the many papers carried by *Géotechnique* on the subject of natural slope stability, one is extraordinary: the 1997 paper of Voight & Elsworth that considers the mechanisms behind, and the analysis of, instability of volcano slopes. Such slopes are unique in that the instability derives from processes within the volcano, and the authors showed how the intrusion of magma leads to a complex system of interacting effects such as over-steepened slopes and increased pore pressures due to retrograde boiling and hydrothermal systems. This is without doubt an exotic branch of geotechnics, but one that fits well within the remit of *Géotechnique*. The work is an object lesson in tackling a complex problem from first principles, and although few geotechnical engineers will have need to reference this work, its significance for geoscience cannot be understated.

BEHAVIOUR OF INTACT ROCK AND DISCONTINUITIES IN ROCK

The early years of rock mechanics were, not surprisingly, concerned with the requirement to understand the behaviour of rock material, leading to the publication of many papers on the development of testing techniques and the analysis of the ensuing results. An early example of such work is a report on the geotechnical properties of brown coal (Trollope *et al.*, 1965), which concluded that, in many ways (and principally because of the effect of moisture content), the behaviour of this material is similar to that of overconsolidated clay. This similarity suggests a unified framework for rock and soil behaviour, and some advances have been made in this direction using critical-state concepts (e.g. Gerogiannopoulos & Brown, 1978; Novello & Johnston, 1995; Cuccovillo & Coop, 1999; Cuss *et al.*, 2003). However, it seems that there is more work to do both in understanding the behaviour of rock in this context and in developing wider appreciation of critical-state concepts within the rock mechanics community. A few years later, the anisotropy of strength and stiffness of granite was explored through physical tests and microfabric analysis (Douglass & Voight, 1969) in a truly benchmark publication, supporting such key papers in the field of rock mechanics and structural geology as those by Kranz (1979), Kranz *et al.* (1979) and Moore & Lockner (1995).

Perhaps because of its eminence in the area of soil behaviour, *Géotechnique* has attracted a number of contributions concerning the behaviour of weak rocks. Thus Elliott & Brown (1985) presented detailed laboratory test results on the yield and strength of Bath stone, and used an analysis of plastic strain vectors at yield to demonstrate that an associated plastic flow rule was appropriate for this material. Importantly, yield stress was found to be independent of stress path. This work continues to be a significant resource, including influential work such as that on the effects of structure in natural soils and weak rocks by Leroueil & Vaughan (1990). Dobereiner & deFreitas (1986) characterised weak sandstones using mineral composition, packing density, grain contact area and grain size distribution as a means of predicting geotechnical properties. These authors developed an index parameter for assessing the strength and deformability of these rocks in terms of their saturated moisture content, thereby demonstrating once more how an appreciation of concepts from both soil and rock mechanics is required for these materials.

Work to improve our understanding of brittle fracture processes in rock continues, and one recent contribution to this is the concept of the relative mobilisation of cohesive and frictional components of strength in these materials. Using case history data of breakouts around circular tunnels in the Lac du Bonnet granite of Manitoba, Canada, Hajiabdolmajid *et al.* (2003) proposed that these components mobilise independently within the fractured rock that develops in these failures. This is a credible, but currently controversial, concept, and it will be interesting to see how it develops in coming years.

The mechanical behaviour of discontinuities has also been thoroughly and authoritatively explored in the pages of *Géotechnique*. In one of the earliest reports of direct shear testing on discontinuities, Krsmanovic (1967) presented results obtained using 40 cm square specimens of sandstone and conglomerate. The large size of these specimens (relative to those routinely used nowadays) and the massive size of the testing rig means these data are still of fundamental importance. The vital problem of the shear strength of rock containing discontinuous joints was studied two years later by Lajtai (1969) through a series of careful direct shear tests. The supporting analysis concluded that compressive, shear and tensile failure mechanisms occur in these conditions, a point that is perhaps not as widely recognised by geotechnical engineers today as it might be.

Field tests to determine the shear strength of discontinuities are awkward to perform, and as a result the analysis of the so-called tilt test has received significant attention in the rock mechanics community. In this vein, Cawsey & Farrar (1976) described a simple tilt test apparatus and presented friction angle results for natural discontinuities in the Upper Chalk. These results suggest that the initial friction angle for these features is about 37°, decreasing to about 34° after a few repetitions of sliding, but then increasing to about 40° after many sliding repetitions. As a result of visual inspection of the surfaces the authors concluded that the initial reduction is due to the surfaces becoming smoother owing to both damage and infilling with rock flour, with the subsequent increase being due to angular fragments of chalk jamming between the surfaces of the discontinuity. A detailed discussion contribution by Hencher (1976) provided further results for stronger rocks (Portland limestone, Delabole slate and Darleydale sandstone), and showed that while the residual friction angle was some 7° lower than the peak, a shear displacement of more than 1·5 m was required for this to develop. Unfortunately these valuable contributions seem to have been ignored by the rock mechanics community in its subsequent development of tilt tests. Furthermore, the principle that friction angle may reduce as the shear displacement continues to increase up to very large values is probably—and erroneously—not accounted for in the majority of analyses undertaken by geotechnical engineers.

Recognising that conventional linear direct shear tests cannot induce shear displacements that are sufficiently large to enable the residual strength of discontinuities to be developed, Xu & deFreitas (1988) developed a rotary shear machine as a means of overcoming this limitation. This was used to determine the strength of artificial discontinuities in rock, with a later report (de Toledo & de Freitas, 1993) dealing with the behaviour of a synthetic discontinuity infilled with various thicknesses of Gault clay. In this, and a later discussion contribution by Papaliangas *et al.* (1995), the complex mechanical conditions found in filled discontinuities were explored in a preliminary fashion; a conclusion has not yet been reached on the fundamental mechanisms, as the recent work by Indraratna *et al.* (2005) demonstrates.

FIELD MEASUREMENTS IN ROCK MECHANICS

Some of the earliest work reported in *Géotechnique* on field measurements in rock mechanics was on geophysical methods to determine rock mass elastic properties. In particular, Brown & Robertshaw (1953) used p-wave velocity to determine the elastic modulus of rock in foundations, and Evison (1956) used p- and s-wave velocities to determine both Young's modulus and Poisson's ratio. Terzaghi (1962b) reported in situ stress measurements in Swedish mines, and presented results that showed how the principal circumferential stresses determined in a tunnel in 'sound, granitic gneiss' demonstrated significant variability of both magnitude and direction. We now realise that such variability is inherent in rock, and is the result of the presence of discontinuities and material heterogeneity and anisotropy. Indeed, the importance of Terzaghi's findings was recognised by Amadei (1996) in his influential work on the importance of anisotropy when estimating and measuring in situ stresses in rock.

Of all the field measurements undertaken in rock mechanics, determination of the in situ stress state is perhaps both the most common and the most difficult. Experience gained over many years has confirmed that there are no general relations that can be applied in advance of a stress determination campaign at any given location to estimate the magnitudes of the principal stress components. Prediction of the principal stress directions (with, perhaps, the exception that one component may act in a vertical direction) is equally difficult. Development of an instrument that would allow determination of the complete stress tensor from a single installation was thus a crucial requirement from the outset of modern rock mechanics, and such a device was described in *Géotechnique* by Rocha & Silverio (1969). This elegant yet sophisticated device took the form of a solid epoxy resin cylinder, in the surface of which were embedded a number of electrical resistance strain gauge rosettes: in use it was cemented into a borehole and then overcored while the gauges were monitored, with the complete stress tensor being calculated from the strain changes. This instrument was the first overcoring gauge capable of giving the complete stress tensor from a single installation, but as the cylinder was solid and of different elastic properties from the rock in which it was installed it was possible for strains to be induced during overcoring that were sufficient to debond the gauges. To overcome this problem the device was made into a thin-walled cylinder, with this critical development being reported independently by Rocha *et al.* (1974) and Worotnicki & Walton (1976). This arrangement remains essentially unchanged in the commercially available hollow inclusion cells available currently.

DISCONTINUOUS ROCK MASSES

Although early work in rock mechanics tended to consider rock as a continuum, by the mid 1960s the profound effect of discontinuities on its properties began to be acknowledged. Development of techniques for the measurement and analysis of discontinuity geometry then came to the fore, with one of the first such pieces of work being the seminal paper 'Sources of error in joint surveys' by Ruth Terzaghi (1965). This work commenced with what could now be regarded as an introduction to the basics of hemispherical projections, but then moved on to the classical analysis on orientation bias. Importantly, by introducing the concept of 'blind zones' in sampling, the paper concluded by setting out the fundamental prerequisites for conducting minimally biased discontinuity surveys on outcrops, in tunnels and on borehole core. These principles have immense practical significance, and for these alone the paper is deserving of every geotechnical engineer's acquaintance. The significance of this work in the development of rock mechanics and rock engineering cannot be underestimated: it is cited by over 100 research papers, including such significant works as those of Priest & Hudson (1976), the ISRM suggested method (Barton, 1978) and Einstein & Baecher (1983).

By the 1980s the effect of discontinuities on the mechanical behaviour of rock was becoming much better understood, and thus Hoek, in the 23rd Rankine Lecture (Hoek, 1983), was able to elucidate the strength of jointed rock masses. Although essentially a distillation of work published over the preceding decade on empirical peak strength criteria, the shear strength of discontinuities, the strength of rock containing inclined planes of weakness and the strength of fractured rock masses, the paper is a concise collection of principles that remain central to the practice of rock engineering. Indeed, the concepts and material first collected together in this paper have gone on to form the basis of a number of important papers and books, including the influential work by Hoek and Brown on practical techniques for assessing the strength of discontinuous rock masses (Hoek & Brown, 1997).

The hydraulic properties of discontinuous rock masses had been under investigation at this time, with one particularly noteworthy development being that of the crack tensor (Oda, 1985). The crack tensor in this work resulted from integration over a volume of material of the geometry of circular discs that represent cracks of uniform aperture, and was used to obtain the permeability tensor of a fractured mass of impermeable rock. The mathematical development revealed that the principal axes of the crack and permeability tensors were coincident, and a comparison of permeability results obtained from the crack tensor with those obtained by numerical analysis both supported this and demonstrated the utility of the approach. The crack tensor concept continues to be key to developments in a wide range of problems in both the strength and hydraulic behaviour of discontinuous rock masses, and is likely to continue to be so as rock mechanics research extends our understanding of coupled hydro-mechanical properties of rock masses for applications such as nuclear waste isolation.

The deformability and deformation of jointed rock masses was explored in a reproduction of the Eighth Laurits Bjerrum Memorial Lecture (Barton, 1986). This compilation paper discussed aperture and conductivity of discontinuities, compaction and depletion in fractured reservoirs, hydraulic shearing, the modelling of shearing and dilation, in situ stress distributions, stability of underground openings, and fluid flow through fractured rock masses. Despite empiricism being a key part of the work in this paper, it nevertheless forms the basis of many analytical and design techniques used to the present day. Arguably, this indicates that there is a great deal of fundamental research still to be undertaken on the deformability and deformation of discontinuous rock masses.

Although a certain maturity had been reached in the understanding of the geotechnical behaviour of discontinuous rock masses by the early 1990s, limit equilibrium analysis of complex block shapes was hampered by their awkward geometry. As a means of overcoming this difficulty, the powerful analytical technique of block theory had been undergoing steady development for a number of years, and so it was in 1995 that the 35th Rankine lecture was given by Goodman, one of its principal developers (Goodman, 1995). This synthesis paper gave a clear and straightforward introduction to block theory, and through the use of various examples showed how it could be applied to good effect in rock engineering. A particular strength of block theory is in helping to produce hemispherical projection diagrams that clearly and concisely show how the mechanical behaviour of a rock mass depends on the orientation of the engineered structure, and is perhaps the single most important reason why geotechnical engineers should be familiar with it. Uniquely for *Géotechnique*, and probably any geotechnical journal, the paper included the words and music to Rankine's 'The three-foot rule', together with an additional sixth verse penned by Goodman that, in support of his conclusions related to the strength of discontinuous rock masses, exhorted engineers to adopt 'a 3-d rule'! Unfortunately, despite its obvious utility in analysing complex discontinuous rock masses, block theory is not as widely used as it should be, and one can therefore conclude that Goodman's exhortations are unfulfilled.

Géotechnique continues to publish influential papers on rock mass geometry. One of the most recent is that by Zhang *et al.* (2002), in which stereological principles were used to derive a relation between observed trace length and size distribution of elliptical discontinuities. Additionally, the authors gave a procedure for determining major axis orientation, aspect ratio and the probability density function of the major axis length for elliptical discontinuities from trace length sampling on different sampling windows. This is a fundamental piece of work that will become an essential reference for other workers in this field.

NUMERICAL MODELLING

The development of numerical analysis for rock mechanics is well represented in *Géotechnique*. One of the earliest contributions was by Zienkiewicz *et al.* (1968), who made a significant development in non-linear finite element stress analysis by considering rock as a material with zero tensile strength. A number of surface and subsurface examples were presented in the paper that clearly show how stress redistribution takes place to leave a tension-free zone that is very different from the zone of tensile stress developed with an elastic solution. With the current preponderance of continuum-based numerical analyses in rock engineering, this is a point that needs to be restated. Effectively, this work utilised a 'ubiquitous joint' material. Such a material is applicable when the location and development of particular fractures is not important, but if this is not the case then the fracture propagation process must be explicitly captured. To this end, Scavia (1995) introduced a two-dimensional boundary element formulation that used linear elastic fracture mechanics to model the propagation of discontinuities in rock, and used it to show how pre-existing cracks in a rock slope could extend, coalesce and lead to instability. Being able to simulate progressive breakdown convincingly in this way is a critical step in understanding the behaviour of rock, and as a result the approach has found notable success in modelling both fracture coalescence (Bobet & Einstein, 1998) and the initiation and progressive failure of natural rock slopes (Eberhardt *et al.*, 2004).

The end result of intense fracture propagation is a discontinuous rock mass, and distinct element methods are usually used for the analyses of such material. Of particular interest is the flow of fluid through discontinuous rock masses, and Harper & Last (1989) showed how a distinct element code could be used to interpret changes to the conductivity of a fractured rock mass caused by fluid injection. This work showed that the response is highly nonlinear, a finding of practical significance for all rock engineering applications in fractured rock that involve injection or abstraction of fluids. However, our knowledge in this area is still limited (Jing, 2003).

An extreme case of discontinuous materials is given by granular media, and as a means of studying these Cundall & Strack (1979) described the distinct element method for assemblages of circular discs. This is probably the first report of such a development, and has gone on to become the most highly cited paper in *Géotechnique*.

Not all numerical techniques reported in the journal have attracted the acclaim they deserve. One such example is that by Jones & Bellamy (1973), in which a frictionless, laminated, transversely isotropic elemental model for prediction of mining subsidence was proposed. The authors demonstrated the efficacy of the approach by reference to two case histories from UK coal mines, and while the technique seems to be a great advance on the National Coal Board's empirical method, it was probably developed too late to overcome the widespread use of the latter in the UK. That said, a great deal of longwall coal extraction still takes place in various countries of the world, and one wonders whether geotechnical engineers in these locations would be better served applying the work of Jones and Bellamy than attempting to make the NCB method fit their local circumstances.

ROCK ENGINEERING

One aim of a deeper understanding of rock mechanics principles is improved rock engineering, and a number of papers appearing in *Géotechnique* tackle this issue. One of the earliest looked at the matter of stresses induced around cavities of various shapes (Terzaghi & Richart, 1952), and although the analyses were limited to those conditions for which closed-form solutions could be obtained, this work is immediately recognisable as presenting the fundamentals of stress analysis around underground openings in a manner that continues to be used in almost all rock mechanics textbooks. This pioneering paper dealt with fundamental issues, and as such has been a source of reference for much other influential material such as Ward's 18th Rankine lecture on tunnel support in weak rock (see below), and the important work on stresses in anisotropic rocks developed by Amadei *et al.* (1987).

Two Rankine lectures from the 1970s deal extensively with different aspects of rock engineering. First, Jaeger (1971)—in the 11th Rankine Lecture—produced a particularly broad contribution in a paper that began with the strength of intact rock and the influence of testing machine stiffness, moved through testing methods and the strength of discontinuities, and concluded with the stability of rock slopes and the strength of fractured rock masses. Much of the material in this paper now forms the basis of textbooks on the fundamentals of rock mechanics and rock engineering, and the work on the frictional behaviour of both intact rock and rock fractures influenced the significant publication on rock friction by Byerlee (1978). This work remains a source of reference to the rock physics community.

The 18th Rankine Lecture (Ward, 1978) dealt with the support of tunnels in weak rock, through a detailed case history of the Kielder experimental tunnel in north-east England. As an adjunct to the Kielder water scheme, this was a series of short 3·3 m diameter headings driven through limestone, mudstone and sandstone at a depth of up to 100 m. The paper defined 'weak' as 'needing support', and as such it encompassed both weak intact rock and strong heavily fractured rock. These materials usually involve different failure mechanisms and hence require different analytical techniques. Thus, among many others, this paper is cited by both Brown *et al.* (1983) in their classic paper on ground-support interaction, and Goodman in his 35th Rankine Lecture (see above). This case history is perhaps unique in that it describes the progressive collapse of a section of tunnel that was deliberately permitted to fail. Such field experiments are rare indeed, and geotechnical engineers should become reacquainted with this paper for this report alone.

The Kielder experimental tunnel yielded another rock engineering case history, this one presenting final results (Ward *et al.*, 1983). This paper included high-quality monitoring data that decisively showed the beneficial effect on stability of sprayed concrete linings, and of the detrimental effect of delaying the introduction of support into a tunnel. Although these concepts are now universally recognised, given the advances made in both numerical modelling and rock strength models, and the lack of more recent data, the monitoring results themselves are probably more useful now than at the time of publication. A similarly detailed case history is that by Douglas *et al.* (1983) regarding the trial excavation for the machine hall at Dinorwig pumped storage power station. This involved the careful instrumentation of the roof of a 24·5 m span cavern in slate containing dolerite dykes, and a reconciliation of the observed and computed displacements. The paper (and the geotechnical work behind it) is exemplary, and should be required reading for all engineers involved in the design of large-span caverns. The paper emphasised the importance of extensively instrumenting such a structure and then using the results to both ratify and optimise the support and reinforcement design. It is a pity that, while we now have tools for numerical analysis that far surpass those available at the time of the Dinorwig project, we seldom implement anything like as extensive an instrumentation scheme as was used there; somehow we seem to have embarked upon an approach that regards numerical analysis as self-sufficient, when in fact it must always be supported by rigorous field observations.

CONCLUSIONS

This review clearly demonstrates that publications in *Géotechnique* have been at the forefront of rock mechanics, rock engineering and engineering geology for the whole of its 60-year existence. Many of the contributions on these subjects published in the journal have proved with the passage of time to be either key contributions in their own right, or to be the source of fruitful lines of development. It is apparent that this continues to be the case: many papers published in *Géotechnique* in recent years are likely to be similarly influential, and it will be interesting to watch the development of the concepts they propose.

However, one unsettling aspect is clear: a number of the papers highlighted in this review present facts and concepts that are crucial to high-quality geotechnical engineering, but these same papers—and hence the crucial facts and concepts they contain—have become lost to much of the geotechnical community. Hopefully, through this review engineers will be able to reacquaint themselves with their valuable contents.

REFERENCES

Amadei, B. (1996). Importance of anisotropy when estimating and measuring in situ stresses in rock. *Int. J. Rock Mech. Min. Sci. Geomech. Abs.* **33**, No. 3, 293–325.

Amadei, B., Savage, W. Z. & Swolfs, H. S. (1987). Gravitational stresses in anisotropic rock masses. *Int. J. Rock Mech. Min. Sci. Geomech. Abs.* **24**, No. 1, 5–14.

Barton, N. (1978). Suggested methods for the quantitative description of discontinuities in rock masses. *Int. J. Rock Mech. Min. Sci.* **15**, No. 6, 319–368.

Barton, N. R. (1986). Deformation phenomena in jointed rock. *Géotechnique* **36**, No. 2, 147–167.

Baynes, F. J., Fookes, P. G. & Kennedy, J. F. (2005). The total engineering geology approach applied to railways in the Pilbara, Western Australia. *Bull. Eng. Geol. Env.* **64**, No. 1, 67–94.

Bjerrum, L. & Jorstad, F. (1963). Correspondence: Stability of steep slopes on hard unweathered rock. *Géotechnique* **13**, No. 2, 171–173.

Bobet, A. & Einstein, H. H. (1998). Numerical modeling of fracture coalescence in a model rock material. *Int. J. Fracture* **92**, No. 3, 221–252.

Brown, E. T., Bray, J. W., Ladanyi, B. & Hoek, E. (1983). Ground response curves for rock tunnels. *J. Geotech. Engng ASCE* **109**, No. 1, 15–39.

Brown, P. D. & Robertshaw, J. (1953). The in-situ measurement of Young's modulus for rock by a dynamic method. *Géotechnique* **3**, No. 7, 283–286.

Burton, C. L., Waltham, A. C. & McLaren, S. J. (2001). Strength variation in young reef limestones. *Géotechnique* **51**, No. 10, 887–889.

Byerlee, J. (1978). Friction of rocks. *Pure Appl. Geophys.* **116**, No. 4–5, 615–626.

Cawsey, D. C. & Farrar, N. S. (1976). A simple sliding apparatus for the measurement of rock joint friction. *Géotechnique* **26**, No. 2, 382–386.

Cripps, J. C. & Taylor, R. K. (1981). The engineering properties of mudrocks. *Q. J. Engng Geol.* **14**, No. 4, 325–346.

Cuccovillo, T. & Coop, M. R. (1999). On the mechanics of structured sands. *Géotechnique* **49**, No. 6, 741–760.

Cundall, P. A. & Strack, O. D. L. (1979). A discrete numerical model for granular assemblies. *Géotechnique* **29**, No. 1, 47–65.

Cuss, R. J., Rutter, E. H. & Holloway, R. F. (2003). The application of critical state soil mechanics to the mechanical behaviour of porous sandstones. *Int. J. Rock Mech. Min. Sci.* **40**, No. 6, 847–862.

de Freitas, M. H. & Watters, R. J. (1973). Some field examples of toppling failure. *Géotechnique* **23**, No. 4, 495–514.

de Toledo, P. E. C. & de Freitas, M. H. (1993). Laboratory testing and parameters controlling the shear strength of filled rock joints. *Géotechnique* **43**, No. 1, 1–19.

Dobereiner, L. & deFreitas, M. H. (1986). Geotechnical properties of weak sandstones. *Géotechnique* **36**, No. 1, 79–94.

Douglas, T. H., Richards, L. R. & Arthur, L. J. (1983). Trial excavation for underground caverns at Dinorwig power station. *Géotechnique* **33**, No. 4, 407–431.

Douglass, P. M. & Voight, B. (1969). Anisotropy of granites: a reflection of microscopic fabric. *Géotechnique* **19**, No. 3, 376–398.

Eberhardt, E., Stead, D. & Coggan, J. S. (2004). Numerical analysis of initiation and progressive failure in natural rock slopes: the (1991) Randa rockslide. *Int. J. Rock Mech. Min. Sci.* **41**, No. 1, 69–87.

Einstein, H. H. & Baecher, G. B. (1983). Probabilistic and statistical-methods in engineering geology: specific methods and examples. 1. Exploration. *Rock Mech. Rock Engng* **16**, No. 1, 39–72.

Elliott, G. M. & Brown, E. T. (1985). Yield of a soft, high porosity rock. *Géotechnique* **35**, No. 4, 413–423.

Evison, F. F. (1956). The seismic determination of Young's modulus and Poisson's ratio for rocks in situ. *Géotechnique* **6**, No. 3, 118–123.

Fookes, P. G. (1969). Geotechnical mapping of soils and sedimentary rock for engineering purposes with examples of practice from the Mangla Dam project. *Géotechnique* **19**, No. 1, 52–74.

Fookes, P. G. (1997). Geology for engineers: the geological model, prediction and performance. *Q. J. Engng Geol. Hydrogeol.* **30**, No. 4, 293–424

Fookes, P. G. & Denness, B. (1969). Observational studies on fissure patterns in Cretaceous sediments of south-east England. *Géotechnique* **19**, No. 4, 453–477.

Fookes, P. G., Reeves, B. J. & Dearman, W. R. (1977). The design and construction of a rock slope in weathered slate at Fowey, South-west England. *Géotechnique* **27**, No. 4, 533–556.

Gerogiannopoulos, N. G. & Brown, E. T. (1978). Critical state concept applied to rock. *Int. J. Rock Mech. Min. Sci.* **15**, No. 1, 1–10.

Goodman, R. E. (1995). Block theory and its application. *Géotechnique* **45**, No. 3, 383–423.

Hajiabdolmajid, V., Kaiser, P. & Martin, C. D. (2003). Mobilised strength components in brittle failure of rock. *Géotechnique* **53**, No. 3, 327–336.

Hamel, J. V. & Adams, R. (1974). Discussion: 'Some field examples of toppling failure'. *Géotechnique* **24**, No. 4, 691–692.

Harper, T. R. & Last, N. C. (1989). Interpretation by numerical modelling of changes of fracture system hydraulic conductivity induced by fluid injection. *Géotechnique* **39**, No. 1, 1–11.

Hencher, S. R. (1976). Correspondence: 'A simple sliding apparatus for the measurement of rock joint friction'. *Géotechnique* **26**, No. 4, 641–644.

Hoek, E. (1983). Strength of jointed rock masses. *Géotechnique* **33**, No. 3, 187–223.

Hoek, E. & Brown, E. T. (1997). Practical estimates of rock mass strength. *Int. J. Rock Mech. Min. Sci.* **34**, No. 8, 1165–1186.

Indraratna, B., Welideniya, H. S. & Brown, E. T. (2005). A shear strength model for idealised infilled joints under constant normal stiffness. *Géotechnique* **55**, No. 3, 215–226.

Jaeger, J. C. (1971). Friction of rocks and stability of rock slopes. *Géotechnique* **21**, No. 2, 97–134.

Jing, L. (2003). A review of techniques, advances and outstanding issues in numerical modelling for rock mechanics and rock engineering. *Int. J. Rock Mech. Min. Sci.* **40**, No. 3, 283–353.

Jones, C. J. F. P. & Bellamy, J. B. (1973). Computer prediction of ground movements due to mining subsidence. *Géotechnique* **23**, No. 4, 515–530.

Kranz, R. L. (1979). Crack growth and development during creep of Barre granite. *Int. J. Rock Mech. Min. Sci.* **16**, No. 1, 23–35.

Kranz, R. L., Frankel, A. D., Engelder, T., & Scholz, C. H. (1979). Permeability of whole and jointed Barre Granite. *Int. J. Rock Mech. Min. Sci.* **16**, No. 4, 225–234.

Krsmanovic, D. (1967). Initial and residual shear strength of hard rocks. *Géotechnique* **17**, No. 2, 145–160.

Lajtai, E. Z. (1969). Strength of discontinuous rocks in direct shear. *Géotechnique* **19**, No. 2, 218–233.

Leroueil, S. & Vaughan, P. R. (1990). The general and congruent effects of structure in natural soils and weak rocks. *Géotechnique* **40**, No. 3, 467–488.

Mayer, A. (1963). Recent work in rock mechanics. *Géotechnique* **13**, No. 2, 99–118.

Meigh, A. C. (1976). The Triassic rocks, with particular reference to predicted and observed performance of some major foundations. *Géotechnique* **26**, No. 3, 391–452.

Moore, D. E. & Lockner, D. A. (1995). The role of microcracking in shear-fracture propagation in granite. *J. Struct. Geol.* **17**, No. 1, 95–114.

Novello, E. A. & Johnston, I. W. (1995). Geotechnical materials and the critical state. *Géotechnique* **45**, No. 2, 223–235.

Oda, M. (1985). Permeability tensor for discontinuous rock masses. *Géotechnique* **35**, No. 4, 483–495.

Papaliangas, T., Hencher, S. R. & Lumsden, A. C. (1995). Discussion: 'Laboratory testing and parameters controlling the shear strength of filled rock joints'. *Géotechnique* **45**, No. 1, 175–183.

Price, D. G. & Knill, J. L. (1967). The engineering geology of Edinburgh Castle Rock. *Géotechnique* **17**, No. 4, 411–432.

Priest, S. D. & Hudson, J. A. (1976). Discontinuity spacings in rock. *Int. J. Rock Mech. Min. Sci.* **13**, No. 5, 135–148.

Rocha, M. & Silvério, A. (1969). A new method for the complete determination of the state of stress in rock masses. *Géotechnique* **19**, No. 1, 116–132.

Rocha, M., Silverio, A., Pedro, J. O. & Delgado, J. S. (1974). A new development of the LNEC stress tensor gauge. *Proc. 3rd Congr. ISRM, Denver* **1**, 464–467.

Scavia, C. (1995). A method for the study of crack propagation in rock structures. *Géotechnique* **45**, No. 3, 447–463.

Schmidt, K. M. & Montgomery, D. R. (1995). Limits to relief. *Science* **270**, No. 5236, 617–620.

Terzaghi, K. (1962a). Dam foundation on sheeted granite. *Géotechnique* **12**, No. 3, 199–208.

Terzaghi, K. (1962b). Measurement of stresses in rock. *Géotechnique* **12**, No. 2, 105–124.

Terzaghi, K. (1962c). Stability of steep slopes on hard unweathered rock. *Géotechnique* **12**, No. 4, 251–270.

Terzaghi, K. & Richart, F. E. (1952). Stresses in rock about cavities. *Géotechnique* **3**, No. 2, 57–90.

Terzaghi, R. (1965). Sources of error in joint surveys. *Géotechnique* **15**, No. 3, 287–304.

Trollope, D. H., Rosengren, K. J. & Brown, E. T. (1965). The mechanics of Brown Coal. *Géotechnique* **15**, No. 4, 363–386.

Voight, B. & Elsworth, D. (1997). Failure of volcano slopes. *Géotechnique* **47**, No. 1, 1–31.

Von Moos, A. (1948). Engineering geology in Switzerland: a review of the literature. *Géotechnique* **1**, No. 1, 40–48.

Ward, W. H. (1978). Ground supports for tunnels in weak rocks. *Géotechnique* **28**, No. 2, 133–171.

Ward, W. H., Tedd, P. & Berry, N. S. M. (1983). The Kielder experimental tunnel: final results. *Géotechnique* **33**, No. 3, 275–291.

Worotnicki, G. & Walton R. J. (1976). Triaxial 'hollow inclusion' gauges for the determination of rock stress in situ. *Proceedings of the ISRM symposium on investigation of stress in rock and advances in shear measurement*, Sydney, Supplement, pp. 1–8.

Xu, S. & deFreitas, M. H. (1988). Use of a rotary shear box for testing the shear strength of rock joints. *Géotechnique* **38**, No. 2, 301–309.

Zhang, L., Einstein, H. H. & Dershowitz, W. S. (2002). Stereological relationship between trace length and size distribution of elliptical discontinuities. *Géotechnique* **52**, No. 6, 419–433.

Zienkiewicz, O. C., Valliappan, S. & King, I. P. (1968). Stress analysis of rock as a 'no tension' material. *Géotechnique* **18**, No. 1, 56–66.

Lancellotta, R. & Pappin, J. W. (2008). *Géotechnique* **58**, No. 5, 457–460 [**doi:** 10.1680/geot.2008.58.5.457]

Contributions to *Géotechnique* 1948–2008: Dynamics

R. LANCELLOTTA* and J. W. PAPPIN†

Although the home of *Géotechnique* is not in an area of high seismicity, the journal has played an important part in enabling its readership and the geotechnical profession to keep abreast of world developments in the important and often difficult area of geotechnical dynamics. This note highlights contributions published in *Géotechnique* that have been influential in developing not only conceptual approaches but also engineering practice. Topics covered range from soil behaviour, earth structures to foundations and earth retaining walls. While seismic actions are mainly discussed many of the concepts are also relevant to machine base design.

KEYWORDS: centrifuge modelling; dynamics; earthquakes; historical review; in situ testing; liquefaction; retaining walls; slopes

Bien que le siège de *Géotechnique* ne se trouve pas dans une zone fortement sismique, le journal a joué un rôle important en permettant à ses lecteurs et à la profession de la géotechnique de se tenir à jour dans les développements mondiaux du secteur important et souvent difficile de la dynamique géotechnique. La présente communication se penche sur des contributions publiées dans *Géotechnique*, qui ont joué un rôle important dans le développement non seulement de méthodes conceptuelles, mais également de la pratique de cette discipline. Les sujets traités vont du comportement du sol et de structures terrestres jusqu'aux fondations et aux murs de soutènement. Bien que l'on discute principalement des actions sismiques, un grand nombre de concepts sont également pertinents pour l'étude de socles de machine.

INTRODUCTION

The aim of this note is to highlight contributions published in *Géotechnique* that have been influential in developing not only conceptual approaches, but also engineering practice in the area of dynamics. This area covers a wide range of topics, spanning from soil behaviour to soil–structure interaction in the presence of seismic actions. Although by necessity subjective, it comments on contributions collectively, concentrating on those themes within which *Géotechnique* papers may have driven cultural shifts or have promoted advances in engineering design.

SEISMIC DESIGN OF EARTH STRUCTURES

Traditionally, civil and geotechnical engineers have generally felt comfortable with the use of safety factors, based on past experience, against the attainment of a collapse limit state while guaranteeing the serviceability and the performance of the structures under working loads. However, in the presence of seismic actions most structures exhibit a behaviour that is dominated by displacements, so that a displacement-based approach gives more reliable indication of the real performance of geotechnical structures. This approach, based on a simple pseudo-static analysis and with a factor of safety less than 1 at the peak ground acceleration, was pioneered by Newmark (1965) in the fifth Rankine Lecture, and represented a major step forward in the philosophy of evaluating the performance of embankments during earthquakes. Further refinements to account for pore pressure build-up and the variation in acceleration throughout the embankment were suggested by Ambraseys & Sarma (1967) and Sarma (1975) respectively.

Later, Seed (1979), in the 19th Rankine Lecture, demonstrated that this method was an adequate design procedure for those soils that do not build up large pore pressure, or cause significant strength loss due to associated displacements. Seed's lecture also highlighted the fact that, in fine cohesionless soils, critical conditions may be attained after the shaking has stopped, so that seismic analysis must consider post-seismic effects. Significant advances in random vibration theory have also promoted the application of this approach to earth dams. Gazetas *et al.* (1981), although based on a one-dimensional model, may be intended to offer an insight into this approach.

In the 33rd Rankine Lecture, Ishihara (1993) remarked how, traditionally, one of the major concerns in the early development of soil mechanics was associated with settlements and stability of cohesive soils, whereas cohesionless soils have long been recognised as stable and free from major distress such as failures or intolerable settlements. This concern is reversed when considering dynamic loading, such as earthquake, where the sudden loss of strength experienced by saturated, loosely packed cohesionless soils, broadly called liquefaction, is a major aspect that must be investigated. The lecture discusses in detail procedures used for liquefaction analysis based on field sounding tests, and introduces new index parameters to forecast the undrained behaviour of sands, within the context of state parameter philosophy. Interestingly, many publications discussing the liquefaction phenomenon rely heavily on the paper by Been & Jefferies (1985), which introduced a state parameter for sands that provides a key understanding of the likelihood of liquefaction for a particular soil. Recently, Pietruszczak *et al.* (2003) have provided an interesting study in making soil less likely to liquefy by reducing the degree of saturation by a small amount, and they show numerical and experimental evidence to evaluate this effect.

EARTH-RETAINING STRUCTURES

Even though there has been considerable debate in the literature about the use of force-based approaches (Steedman & Zeng, 1990; Zeng & Steedman, 1993), current practice for computing earth thrust on earth-retaining structures in the presence of seismic action still relies on an extension of the Coulomb solution, due to Okabe (1924) and Mononobe & Matsuo (1929), and referred to in the literature as the Mononobe–Okabe approach (Kramer, 1996). This approach considers the Coulomb wedge acted upon by horizontal and

Discussion on this paper closes on 1 December 2008, for further details see p. ii.

* Politecnico di Torino, Italy.
† Arup, Hong Kong.

vertical pseudo-static forces (in addition to forces acting under static conditions), whose magnitude is related to the mass of the wedge, through the pseudo-static accelerations.

The Mononobe–Okabe approach is conceptually simple, and most researchers agree that the estimated total lateral thrust is approximately correct. The vertical distribution of the dynamic incremental force requires further scrutiny, however, owing to the fact that the Mononobe–Okabe approach assumes a uniform acceleration field throughout the backfill, whereas phase changes can have an influence on the actual distribution. Tajimi (1973) first investigated this point, showing that the earth stress distribution is a function of a dimensionless parameter $\omega H/v_s$, where H is the wall height, v_s is the shear wave velocity, and ω is the angular frequency of the base shaking. Steedman & Zeng (1990) have explored this point further, reaching the conclusion that the resultant thrust acts above one third of the height of the wall. Furthermore, since the maximum thrust and the maximum moment are in phase with the acceleration at the mid height, this acceleration may be the most appropriate value for the design.

It is well known that, when wall friction is present, a non-uniform stress field arises as well as a non-planar failure surface. This renders the problem of computing exact values of passive earth resistance rather complex, and standard codes suggest solutions provided by limit equilibrium methods with a curved (typically log spiral) surface (see Caquot & Kerisel, 1948). However, since these procedures are essentially of a kinematical nature, they are not conservative. For this reason, it is of interest to rely on solutions based on the lower-bound theorem of plasticity, as in the solution recently provided by Lancellotta (2007), the novelty of which lies in a transformation of axes, and in a properly scaled gravity body force, which allows the problem of seismic passive resistance to be solved using the same stress field equations as the usual static case.

A strength of the pseudo-static approach is that it allows estimation of the seismic thrust on the retaining wall, so that the ultimate limit state can be investigated in a rather simple and convenient way. But, as expected, more information is required when the overall performance of the wall has to be assessed, and this is related to permanent displacements that the wall can suffer. These displacements can be predicted by using the approach pioneered by Richards & Elms (1979), based on the sliding block procedure originally proposed by Newmark (1965), and subsequently improved by Whitman & Liao (1985).

Although these aspects related to gravity retaining walls have been long explored in papers published elsewhere, relatively few contributions are related to the behaviour of tie-back walls during earthquakes. One exception is the paper by Zeng & Steedman (1993), in which, by using centrifuge model tests as described by Schofield & Lee (1988), the consequences of loss of strength and stiffness below the ground surface (which may produce a degradation of the anchor system) highlight the need for more complex physical and numerical investigations of these aspects.

CENTRIFUGE TESTING

Centrifuge model testing has proved very useful in validating mathematical models that describe the behaviour of soil subjected to earthquake ground motion. Hushmund *et al.* (1988), for example, report on liquefaction testing carried out in a laminar box. Zeng & Steedman (1998) extended these tests to study the bearing capacity and overturning of shallow foundations, and show a complex behaviour that cannot be captured by the use of simple pseudo-static methods. Nevertheless, pseudo-static methods, such as those presented by Sarma & Iossifelis (1990) and Budhu & Al-Karni (1993), will continue to be used, and they are satisfactory for design purposes in circumstances where a significant margin against failure can be demonstrated.

PILES AND PILED FOUNDATIONS

Géotechnique has also contained a variety of papers for pile design in dynamic situations. Dobry & Gazetas (1988) present a useful method for deriving the dynamic stiffness and damping of floating pile groups relevant to soil structure interaction problems. Nikolaou *et al.* (2001) present an overview of the effects of ground distortion in earthquakes and the effects on pile bending moments. In addition, the paper by Bhattacharya *et al.* (2004) provides insight into the very difficult problem of determining when piles may be detrimentally affected by passing through layers of liquefied soil.

SEISMIC TECHNIQUE FOR SOIL INVESTIGATION

An important aspect of dynamics in geotechnical engineering is the determination of soil properties that are appropriate to modelling soil behaviour under dynamic loading. It is well known that soils have high small-strain stiffness G_{max}, and that this stiffness reduces with increasing shear strain amplitude. As described by Abbiss (1981), G_{max} is directly related to shear wave velocity, which can be directly measured in situ, making this an important site investigation technique. The most significant recent development in shallow seismic exploration for geotechnical engineering application is the so-called SASW method (spectral analysis of surface waves), pioneered at the University of Texas at Austin (see Stokoe *et al.*, 1994). In this method, a source at the surface generates Rayleigh surface waves, detected by at least two receivers located at a known distance apart from the source. If we consider a layered half-space, the velocity of propagation of a surface wave depends on the frequency (or wavelength) of the wave. This dependence is called dispersion, and arises because waves of different wavelengths sample different portions of the soil. Therefore, by using surface waves with a range of wavelengths, or frequencies, all layers can be sampled, and the stiffness profile can be determined by using an inversion algorithm.

Papers published in *Géotechnique* on this subject do not represent the collective body of work, but have to be considered as individual contributions that deserve some consideration. Notably, the first shallow exploration method was the steady-state Rayleigh method, suggested by Jones (1958), with application to road pavements (Jones, 1962). In these experiments Jones obtained the dispersion curve by moving the receiver along a straight line from the source, working at a given fixed frequency, and looking for a position such that the source and the receiver were in phase. The distance between any two adjacent receivers is the wavelength at that frequency. The process was then repeated for different frequencies and the dispersion curve obtained. This original idea is currently used in a more efficient way, since information over a broad range of frequencies is obtained at one time by using an impulse source and processing the associated signal (Foti, 2003).

A further application of seismic techniques is related to the potential of detecting soil porosity. In fine-grained soils, porosity is generally inferred from laboratory measurements on undisturbed samples, but for coarse-grained soils sophisticated freezing techniques are required to obtain undisturbed samples. Therefore there is a need to search for alternative procedures, and a promising one derives the porosity from measurements of seismic wave velocities, as discussed by

Foti *et al.* (2002) and Foti & Lancellotta (2004), using concepts of poroelasticity developed by Biot (1956).

A useful laboratory technique for determining wave velocities in laboratory tests is the use of bender elements. Papers published in *Géotechnique* on this subject (Viggiani & Atkinson, 1995; Jovicic *et al.*, 1996; Pennington *et al.*, 1997; Arroyo *et al.*, 2003, 2006) collectively give a perspective of the most relevant aspects related to the interpretations of these tests, including near-field effects, and the influence of sample size and of anisotropic soil behaviour.

Traditionally, however, there was reluctance to make direct use of these G_{max} values in static situations because of concern that the apparently high G_{max} values were largely a result of the high frequency associated with the testing. Bolton & Wilson (1989) categorically demonstrated, by laboratory testing, that the shear modulus values of cohesionless soils were purely a function of shear strain amplitude, and were independent of frequency. By combining in situ shear wave velocity testing to derive G_{max} with an understanding of the stiffness reduction with shear strain amplitude, it is now becoming routine to determine appropriate stiffness values for dynamic applications, including seismic (e.g. Papastamatiou *et al.*, 1993) and static (e.g. Clayton & Heymann, 2001).

In addition to the measurement of shear wave and compression wave velocities, there are many useful papers in *Géotechnique* discussing field and laboratory testing to derive an understanding of soil properties appropriate to dynamic problems. Konrad (1998) and Schnaid & Yu (2007), for example, provide valuable insight into determination of the state parameter for sands from static and seismic cone penetrometer tests. Similarly, Qadimi & Coop (2007) discuss the behaviour of saturated carbonate sand when subjected to cyclic triaxial tests, and give insight as to when particle crushing affects the generation of pore water as compared with conventional sand materials.

SUMMARY

Although the home of *Géotechnique* is not in an area of high seismicity—indeed some would say it is in an area of low to trivial seismicity—it is evident from the above discussion that it has, at the very least, played an important part in enabling its readership and the geotechnical engineering profession to keep abreast of world developments in the important and often difficult area of geotechnical dynamics. While the area of machine base foundations has not been mentioned, many of the developments above, including geophysical field and laboratory testing, an understanding of pore pressure generation and dynamic soil structure interaction, are very relevant to these problems. It is hoped that *Géotechnique* will continue to attract and publish such high-quality papers over the broad spectrum of this topic.

REFERENCES

Abbiss, C. P. (1981). Shear wave measurements of the elasticity of the ground. *Géotechnique* **31**, No. 1, 91–104.

Ambraseys, N. N. & Sarma, S. K. (1967). The response of earth dams to strong earthquakes. *Géotechnique* **17**, No. 3, 181–213.

Arroyo, M., Muir Wood, D. & Greening, P. D. (2003). Source near-field effects and pulse tests in soil samples. *Géotechnique* **53**, No. 3, 337–345.

Arroyo, M., Muir Wood, D., Greening, P. D., Medina, L. & Rio, J. (2006). Effects of sample size on bender-based axial G_0 measurements. *Géotechnique* **56**, No. 1, 39–52.

Been, K. & Jefferies, M. G. (1985). A state parameter for sands. *Géotechnique* **35**, No. 2, 99–112.

Biot, M. A. (1956). Theory of elastic waves in fluid-saturated porous solid. *J. Acoust. Soc. Am.* **28**, 168–191.

Bolton, M. D. & Wilson, J. M. R. (1989). An experimental and theoretical comparison between static and dynamic torsional soil tests. *Géotechnique* **39**, No. 4, 585–599.

Bhattacharya, S., Madabushi, S. P. G. & Bolton, M. D. (2004). An alternative mechanism of pile failure in liquefiable deposits during earthquakes. *Géotechnique* **54**, No. 3, 203–213.

Budhu, M. & Al-Karni, A. (1993). Seismic bearing capacity of soils. *Géotechnique* **43**, No. 1, 181–187.

Caquot, A. & Kerisel, J. (1948). *Tables for the calculation of passive pressure, active pressure and bearing capacity of foundations*. Paris: Gauthier-Villars.

Clayton, C. R. I. & Heymann, G. (2001). Stiffness of geomaterials at very small strains. *Géotechnique* **51**, No. 3, 245–255.

Dobry, R. & Gazetas, G. (1988). Simple method for dynamic stiffness and damping of floating pile groups. *Géotechnique* **38**, No. 4, 557–574.

Foti, S., Lai, C. G. & Lancellotta, R. (2002). Porosity of fluid-saturated porous media from measured seismic wave velocities. *Géotechnique* **52**, No. 5, 359–373.

Foti, S. & Lancellotta, R. (2004). Soil porosity from seismic velocities. *Géotechnique* **54**, No. 8, 551–554.

Gazetas, G., Debchaudhury, A. & Gasparini, D. A. (1981). Random vibration analysis for the seismic response of earth dams. *Géotechnique* **31**, No. 2, 261–277.

Hushmund, B., Scott, R. F. & Crouse, C. B. (1988). Centrifuge liquefaction tests in a laminar box. *Géotechnique* **38**, No. 2, 253–262.

Ishihara, K. (1993). Liquefaction and flow failure during earthquakes. *Géotechnique* **43**, No. 3, 351–415.

Jones, R. B. (1958). In situ measurements of the dynamic properties of soils by vibration methods. *Géotechnique* **8**, No. 1, 1–21.

Jones, R. B. (1962). Surface wave technique for measuring the elastic properties and thickness of roads: theoretical development. *Br. J. Appl. Phys.* **13**, No. 1, 21–29.

Jovicic, V., Coop, M. R. & Simic, M. (1996). Objective criteria for determining G_{max} from bender element tests. *Géotechnique* **46**, No. 2, 357–362.

Konrad, J. M. (1998). Sand state from cone penetration tests: a framework considering grain crushing stress. *Géotechnique* **48**, No. 2, 201–215.

Kramer, S. L. (1996). *Geotechnical earthquake engineering*. Upper Saddle River, NJ: Prentice Hall.

Lancellotta, R. (2007). Lower-bound approach for seismic passive earth resistance. *Géotechnique* **57**, No. 3, 319–321.

Mononobe, N. & Matsuo, H. (1929). On the determination of earth pressure during earthquake. *Proc. 2nd World Conf. Earthquake Engineering, Tokyo* **9**, 177–185.

Newmark, M. N. (1965). Effects of earthquakes on dams and embankments. *Géotechnique* **15**, No. 2, 139–160.

Nikolaou, S., Mylonakis, G., Gazetas, G. & Tazoh, T. (2001). Kinematic pile bending during earthquakes: analysis and field measurements. *Géotechnique* **51**, No. 5, 425–440.

Okabe, S. (1924). General theory of earth pressure and seismic stability of retaining wall and dam. *J. Jpn Soc. Civ. Engrs* **10**, No. 6, 1277–1330.

Papastamatiou, D., Pappin, J. W., Richards, J. A. & Sweeney, M. (1993). Earthquake response at Grange mouth. *Géotechnique* **43**, No. 4, 537–553.

Pennington, D. S., Nash, D. F. T. & Lings, M. L. (1997). Anisotropy of G_0 shear stiffness in Gault clay. *Géotechnique* **47**, No. 3, 391–398.

Pietruszczak, S., Pande, G. N. & Oulapour, M. (2003). A hypothesis for mitigation of risk of liquefaction. *Géotechnique* **53**, No. 9, 833–838.

Qadimi, A. & Coop, M. R. (2007). The undrained cyclic behaviour of a carbonate sand. *Géotechnique* **57**, No. 9, 739–750.

Richards, R. & Elms, D. (1979). Seismic behaviour of gravity retaining walls. *J. Geotech. Div. ASCE* **105**, No. 4, 449–464.

Sarma, S. K. (1975). Seismic stability of earth dams and embankments. *Géotechnique* **25**, No. 4, 743–761.

Sarma, S. K. & Iossifelis, I. S. (1990). Seismic bearing capacity factors of shallow strip footings. *Géotechnique* **40**, No. 2, 265–273.

Schofield, A. N. & Lee, F. H. (1988). Centrifuge modelling of sand embankments and islands in earthquakes. *Géotechnique* **38**, No. 1, 45–58.

Schnaid, F. & Yu, H. S. (2007). Interpretation of the seismic cone test in granular soils. *Géotechnique* **57**, No. 3, 265–272.

Seed, H. B. (1979). Considerations in earthquake-resistant design of earth and rockfill dams. *Géotechnique* **29**, No. 3, 215–263.

Steedman, R. S. & Zeng, X. (1990). The influence of phase on the calculation of pseudo-static earth pressure on a retaining wall. *Géotechnique* **40**, No. 1, 103–112.

Stokoe, K. H., Wright, S. G., Bay, J. A. & Roesset, J. M. (1994). Characterization of geotechnical sites by SASW method. *Proc. 13th Int. Conf. Soil Mech. Found. Engng, New Delhi, Special volume on geophysical characterization of sites*, 15–34.

Tajimi, H. (1973). Dynamic earth pressures on basement wall. *Proc. 5th World Conf. Earthquake Engineering, Rome* **2**, 1560–1569.

Viggiani, G. & Atkinson, J. H. (1995). Interpretation of bender element tests. *Géotechnique* **45**, No. 1, 149–154.

Whitman, R. V. & Liao, S. (1985). *Seismic design of retaining walls*, Miscellaneous Paper GL-85. Vicksburg, MS: US Army Engineers Waterways Experiment Station.

Zeng, X. & Steedman, R. S. (1993). On the behaviour of quay walls in earthquakes. *Géotechnique* **43**, No. 3, 417–431.

Zeng, X. & Steedman, R. S. (1998). Bearing capacity failure of shallow foundations in earthquakes. *Géotechnique* **48**, No. 2, 235–256.

Skempton, A. W. & Bishop, A. W. (1950). *Géotechnique* **2**, No. 2, 90–108

First Technical Session

***Monday morning*, 5 *June*, 1950**

Mr L. F. Cooling, M.Sc., in the Chair

The following Paper was introduced by the Authors and submitted for discussion. On the motion of the Chairman, the thanks of the Conference were accorded to the Authors.

THE MEASUREMENT OF THE SHEAR STRENGTH OF SOILS

by

A. W. Skempton, D.Sc., and A. W. Bishop, M.A.

SYNOPSIS

Tests are described for determining the shear strength parameters c and ϕ of the principal soil types under controlled conditions of pore pressure dissipation. The significance of the results is discussed in terms of their application in stability analyses.

The presentation is necessarily a personal one, but the Authors believe that the Paper is not unrepresentative of the views and of the practice of other soil laboratories in Great Britain.

Des essais sont décrits pour la détermination des paramètres c et ϕ de résistance au cisaillement des types principaux des sols dans des conditions contrôlées de dispersion de la pression poreuse. La signification de ces résultats est discutée en fonction de leur application dans les analyses de stabilité.

La présentation en est nécessairement personnelle, mais les auteurs sont d'avis que l'article représente dans une certaine mesure les vues et la pratique des autres laboratoires d'analyse de sol en Grande Bretagne.

FUNDAMENTAL CONCEPTS

EFFECTIVE STRESS

The fundamental principle underlying all work on the shear strength of soils is that the mechanical properties are controlled by the intergranular forces. If σ denotes the total normal stress acting on any plane in a soil and if u the pressure in the pore space, then, to a close approximation, the effective stress σ' is given by the equation

$$\sigma' = \sigma - u \qquad (1).$$

In this expression σ' is the sum of the intergranular forces resolved normal to the plane, per unit area of the plane.

An examination of this equation is given in a recent paper by Bishop and Eldin (1950), but here it is necessary only to note that equation (1) has been substantiated experimentally by Terzaghi (1936) for sand clay and concrete, by Rendulic (1937) and Taylor (1944) for clays, and by Bishop and Eldin for sands. There is, therefore, no reason to doubt the general validity of the principle of effective stress.

TRUE COHESION AND INTERNAL FRICTION

In 1773 Coulomb suggested that the shear strength characteristics of a soil could be represented by the expression

$$s = c + \sigma \tan \phi \qquad (2),$$

90

where, in modern nomenclature,

s denotes shear strength
c ,, apparent cohesion
σ ,, total pressure normal to shear plane
ϕ ,, angle of shearing resistance

In Coulomb's equation c and ϕ are empirical constants the values of which, for any given soil, depend upon the conditions of test or the conditions under which the soil is stressed in the field ; in particular, whether or not the soil is allowed to drain during the shearing process.

In a more fundamental form, first put forward by Hvorslev in 1937, Coulomb's equation is rewritten

$$s = c_e + (\sigma - u) \tan \phi_e \quad . \quad . \quad . \quad . \quad . \quad . \quad . \quad . \quad . \quad (3),$$

where c_e denotes true cohesion
ϕ_e ,, true angle of internal friction.

The cohesion and internal friction will, in general, depend upon the void ratio of the soil at the instant of shear failure.

For purposes of simplicity the rate of strain is taken as an independent variable and its effect on the shear strength will be considered separately. Existing data do not justify any more elaborate treatment, and, so far as the authors are aware, no serious error is thereby introduced.

In any isotropic soil the cohesion is a non-directional property and it may be regarded as the resultant of the physico-chemical forces acting between the particles. In clean sands these forces are negligibly small, whilst in clays they are important. Indeed, there is evidence to suggest that in bentonite the shear strength is composed almost wholly of true cohesion (Skempton, 1948*a*).

Internal friction is derived principally from the actual friction of grain on grain. It is, however, also taken as including the resistance to shear developed as a result of the work which has to be done when the soil changes volume during shear. Internal friction is not itself a directional property, but in the general case of an element under unequal principal stresses the shear resistance along different planes will vary in accordance with the variation in σ, and hence the internal friction imparts directional properties to the soil.

Whether or not this view of the physics of shear strength is wholly acceptable, the Coulomb-Hvorslev equation has been confirmed experimentally, at least as a first approximation, by Hvorslev (*loc. cit*), Peynircioglu (1939) and recent research at Imperial College. In the Authors' opinion it must form the basis of any fundamental consideration of shear strength.

ANGLE OF SHEARING RESISTANCE

The measurement of c_e and ϕ_e is, however, a matter of some difficulty in many soils and, moreover, an analysis of stability based on these basic properties is likely to be rather elaborate in practical use. Ultimately such analyses will be carried out, as has been done already in a provisional way for certain relatively simple cases by Skempton (1948*b*) and Hansen and Gibson (1949). But at the present time the usual method of procedure is :

(i) To determine the values of c and ϕ in Coulomb's equation under definite conditions of water-content change during shear ; namely, no water-content change (*undrained test*), or full water-content change (*drained test*).

(ii) To use these values of c and ϕ as if they denoted the cohesion and internal friction of the soil : where practical conditions approximate to zero water-content change the undrained test parameters are used, with respect to total stresses : where practical conditions approximate to full water-content change the drained test parameters are used, with respect to effective stresses : where practical conditions cannot be even approximately represented by either drained or undrained tests, both tests * should be made and the appropriate strength deduced from a knowledge of the degree of consolidation under field conditions.

This procedure is semi-empirical ; its justification must be judged by three criteria :

(*a*) simplicity ;
(*b*) reliability in practice ;
(*c*) small errors as compared with a more rigorous analysis based on c_e and ϕ_e and a knowledge of the pore pressure u.

Of these (*a*) is undoubted, and there is some evidence for (*b*) ; but principally from that type of problem which has been shown by Skempton and by Hansen and Gibson to conform also to (*c*)—namely those problems where a saturated clay is stressed without any change in water content and $\phi = 0$ with respect to applied stresses. One of the most urgent needs in the subject at the present time is thus for a theoretical examination, and practical results, concerning the reliability of the present procedure in those cases where a water content change precedes the stage for which it is desired to calculate the factor of safety (typical cases include the foundation stability of slowly constructed earth dams, rapid draw-down, and the stability of roads and runways).

But even in the absence of such results the present conventional procedure will probably continue to be used ; and the purpose of the more advanced theoretical examination will be to provide information on the corrections which must be applied in the conventional analyses. Consequently the Authors will consider the various tests by which the values of c, the apparent cohesion, and ϕ, the angle of shearing resistance, can be measured on samples of the principal soil types.

SAMPLING

No matter how correct a theory of shear strength may be, and no matter how carefully the laboratory tests are carried out, the results will be misleading unless samples have been taken which are closely representative of the soil as it exists in Nature. Consequently it is of the utmost importance to obtain undisturbed samples, or in the case of fills and sub-grades to predict with reasonable accuracy the state of the soil during and after construction.

Unless this is possible, the testing programme is likely to yield results of very limited value. For example, it appears that it is difficult, or even impossible, to obtain undisturbed samples of sensitive silty clays from depths of more than about 50 feet (Carlson, 1948 ; Skempton, 1948*c*). For such cases the vane test has been developed (see Section VI) and it would be inadvisable to carry out any elaborate laboratory shear tests to determine the in-situ strength. As a second example, the problem of predicting the moisture contents in a clay subgrade some years after construction is still a matter of practical difficulty, although recent work (Croney, Lewis, and Coleman, 1950) at the Road Research Laboratory shows great promise in this direction. In highway and airfield work, this is a far more important matter than any refinements of testing technique or stability analysis.

Nevertheless there is a wide range of practical problems where the conditions are known with sufficient accuracy and adequate samples can be obtained. It is with these cases that this Paper is mainly concerned. The techniques of sampling lie outside the scope of the present discussion.

PRINCIPAL SOIL TYPES

From the point of view of studying shear strength, soils may be classified into five principal types. Any one of these types may be either fully or partially saturated ; and its properties have to be considered in either the undisturbed or remoulded states.

* Or a series of undrained tests can be carried out in which the pore pressures are measured. Such tests are, however, not easily made and can hardly be regarded as part of a normal laboratory routine.

Broadly speaking, undisturbed soils beneath ground-water level are fully saturated ; and partially saturated when above water level. Soils used in the construction of dams, embankments, or subgrades are generally remoulded and, at least initially, only partially saturated.

Sands and gravels.—In this type of soil only a small percentage of the particles are smaller than 0·06 millimetre. The true cohesion is negligible and the permeability is relatively great. Most sands and gravels, unless in a loose state of packing, exhibit positive dilatancy during shear.

Clay-gravels.—This type of soil contains sufficient silt and clay fractions to impart appreciable cohesion, but not so much of these finer fractions as seriously to reduce the angle of shearing resistance under normal conditions. Hoggin forms a familiar example. But in many boulder clays, the clay content is so high as to place the soil in the class of " clays."

Silts.—In its most characteristic form this type of soil has a liquid limit of less than 35 and contains not more than about 15 per cent. clay fraction (less than 0·002 millimetre). Unless very soft, silts exhibit positive dilatancy.

Non-fissured clays.—This class of soil is plastic and smooth to the touch, with a liquid limit typically greater than 35, and without any obvious macro-structure. The clay fraction is usually in excess of 15 per cent. Unless highly over-consolidated most clays exhibit negative dilatancy.

Fissured clays.—This type of soil is identical with that described above, except for the presence of a network of fissures. These fissures can be readily seen if a sample is broken apart. They are of greatest significance in over-consolidated clays, and require special consideration in predicting the effects of softening.

THE BASIC SHEAR TESTS

GENERAL

The outstanding importance of the drainage conditions during shear has been recognized since 1932, when Terzaghi published the results of two series of triaxial tests on a saturated clay. In the first of these no drainage was allowed and the angle of shearing resistance ϕ was found to be approximately zero ; whilst in the second series the clay was allowed to consolidate during the test, when the value of ϕ was found to be 23 degrees.

At the present time the tests normally carried out may be classified as follows (after Casagrande 1941) : *

No drainage during shear—

Undrained (Quick) Test : the samples are placed in the testing apparatus in any given state (undisturbed, compacted to a specified density, etc.), subjected to an applied cell pressure under conditions of no drainage and then sheared, also under conditions of no drainage. At least three samples must be used, each being tested under a different cell pressure.

Consolidated-undrained (Cons.-Quick) Tests : several (not fewer than three) samples are allowed to consolidate, or soften, under a defined pressure † and are then subjected to undrained tests ; each sample being under a different cell pressure.‡ The procedure

* The nomenclature given is that adopted in the Civil Engineering Code of Practice No 1 (1950) " Site Investigation," shortly to be issued by the Institution of Civil Engineers on behalf of the Civil Engineering Codes of Practice Joint Committee. The nomenclature given in parentheses is also used ; particularly in the U.S.A.

† The shear strength will depend on whether the sample is consolidated under an isotropic cell pressure (as is usual in triaxial testing procedure) or under an anisotropic stress system (which is generally more representative of the conditions in the ground) (Hansen and Gibson, 1949).

‡ For saturated clays, in which $\phi = 0$ in undrained tests, it is sufficient to test the sample only under its consolidation pressure ; thus only one sample need be set up under any given consolidation pressure.

is then repeated on further sets of samples, each set being consolidated under different pressures.

Full drainage during shear—

Drained (Slow) Test : the samples are allowed to consolidate, or soften, under a defined pressure and are then sheared under conditions of no excess pressure in the pore space, the cell pressure (or vertical pressure in the shear box) being held constant.

RATE OF STRAIN

Recent research by Casagrande and co-workers has shown that the strength of a sand is virtually uninfluenced by the rate of strain but that, on the other hand, the strength of a clay depends to an appreciable extent on this factor.

Undrained tests are usually completed within a period ranging from 5 minutes to 30 minutes. But in the subgrade of a road or runway, for example, the shear stresses are applied at a rate of the order of 1,000 or even 10,000 times greater than that used in the laboratory test. With special equipment Casagrande and Shannon (1948) have carried out shear strength measurements under such transient rates of loading and they find that the results may be 30 to 60 per cent., or in extreme cases 100 per cent. greater than those obtained in the normal tests. Clearly some allowance for this effect should logically be made in subgrade design.

In contrast, the rate of loading in most constructional operations is measured in terms of weeks or months : a rate of the order of 10,000 times smaller than that in the routine tests. Taylor (1943) and Casagrande and Wilson (1949), have investigated the influence of very low rates of strain on clays and it appears that the shear strength typically falls to values about 15 to 30 per cent. lower than that normally measured. Here again some allowance for this effect should be made in an analysis which attempts to be thoroughgoing. From the theoretical considerations put forward by Hansen and Gibson (1949), it is probable, however, that the $\phi = 0$ analysis contains an implicit factor of safety of the same order as the decrease in strength due to slow rates of loading. It may well be that the two effects tend to balance out and there is, at any rate, a considerable body of evidence pointing to the reasonable accuracy of the simple $\phi = 0$ analysis, as shown by Skempton and Golder (1948) and Cadling and Odenstad (1950).

So far as laboratory practice is concerned the variation in rates of loading in undrained tests are of little consequence. A considerable number of experiments in the Authors' laboratory indicate that the differences in strength obtained in tests carried out 5 times faster or 5 times slower than the standard rate (about 10 minutes to failure) do not amount to more than about 5 or, in extreme cases, 10 per cent ; even at water contents equal to the liquid limit. But in drained tests on clays the time of loading varies from about 8 hours to 3 days ; and consequently for research work it is necessary to apply corrections to those results before comparing them with undrained tests.

LABORATORY TESTING PROCEDURES

GENERAL

The choice of apparatus for carrying out a shear test is primarily determined by the conditions of drainage under which it is desired to carry out the test.

Table 1 indicates the procedure which is generally followed in testing the main soil types. For example, the use of the small shear box for undrained and consolidated-undrained tests is limited to clays of low permeability tested at a rapid rate of shear, owing to the relatively short drainage paths in the apparatus.

Other factors also have to be considered, such as whether samples can be satisfactorily prepared (in very soft, friable or coarse grained soils), and the degree of operational skill required to carry out the test.

Tests are normally carried out under a controlled rate of strain, as the conditions at failure are then completely specified.

TRIAXIAL APPARATUS

The triaxial compression test is carried out under a constant cell pressure on a sample 3 inches long × 1½ inch in diameter in the apparatus shown diagrammatically in Figs 1 and 2 (for sand, silts and clays) or 9 inches long × 4 inches diameter (for gravelly soils). The cell pressure is maintained constant in long duration tests * by the mercury control which, once it has been set at the correct level, automatically compensates for any leakage of water out of the triaxial apparatus. For short duration tests (normally about 10 to 20 minutes) the compressed air in the cylinder is sufficient to maintain a constant pressure.

Table 1

Apparatus used for shear-strength measurements in relation to main soil types

Soil type	Test conditions		
	Undrained	Consolidated-undrained	Drained
Sand	—	Triaxial	Triaxial or shear box, deep sounding in situ
Gravel	—	—	Large shear box
Clay-gravel	Triaxial or large shear box (if of low permeability)	Large shear box or triaxial, depending on grading and permeability	Large shear box or triaxial, depending on grading
Silt	Triaxial vane in situ	Triaxial	Triaxial or shear box
Soft clay (non-fissured)	Triaxial or unconfined compression, or vane, vane in situ	Triaxial (shear box) oedometer + triaxial	Triaxial or shear box
Peat	Triaxial	Triaxial	Triaxial or shear box
Fissured clay ..	Triaxial	Triaxial or oedometer + triaxial	Triaxial

In an undrained test on a cohesive soil the specimen is simply placed between two solid end pieces (of duralumin or " Perspex ") and enclosed in a thin rubber membrane which is sealed against the end pieces by rubber rings. The rubber membranes in general use in Britain are about 0·2 millimetre in thickness, and it has been found that these given an apparent increase in compression strength of about 0·8 lb. per square inch. At Imperial College, when very soft clays are being tested, a membrane only about 0·1 millimetre in thickness is used; this being made in the laboratory from self-vulcanizing latex. The correction is then about 0·3 lb. per square inch.

In the consolidated-undrained test, for any type of soil, the sample is placed on a porous disc, which is connected through the base of the cell to a burette. This permits consolidation to take place under the applied pressure. When consolidation is complete † the valve to the burette is closed, and the sample is sheared under undrained conditions with any desired cell

* The mercury control maintains a constant pressure without any difficulty for tests lasting a week or more.

† I.e. when pore water ceases to be expelled from the sample into the burette.

pressure. When clays are being tested, side drains made of thin strips of filter paper are employed to increase the rate of consolidation.

The procedure is similar in the case of the drained test, except that the valve to the burette remains open throughout the test : and in testing clays the shearing process must not be completed in less than about 1 to 3 days (using side drains).

In the case of saturated non-cohesive sands, the rubber membrane is surrounded by a former while the sample is placed in it, and the cap sealed in place. The burette is then lowered to apply a small negative pressure sufficient to hold the sample rigid while the former is removed, the cell assembled, and the lateral pressure applied. A consolidated-undrained or drained test is then carried out.

The same method is applied in the case of dry sands (Fig. 2), where a lowering of the mercury level in the burette (by the screw control) sets up a negative pore pressure while the cell is being assembled. Volume changes are measured during the test by maintaining atmospheric pressure in the pore space (indicated by the oil manometer), which means that the volume of air in the system is constant, and hence changes in the volume of the

Fig. 1

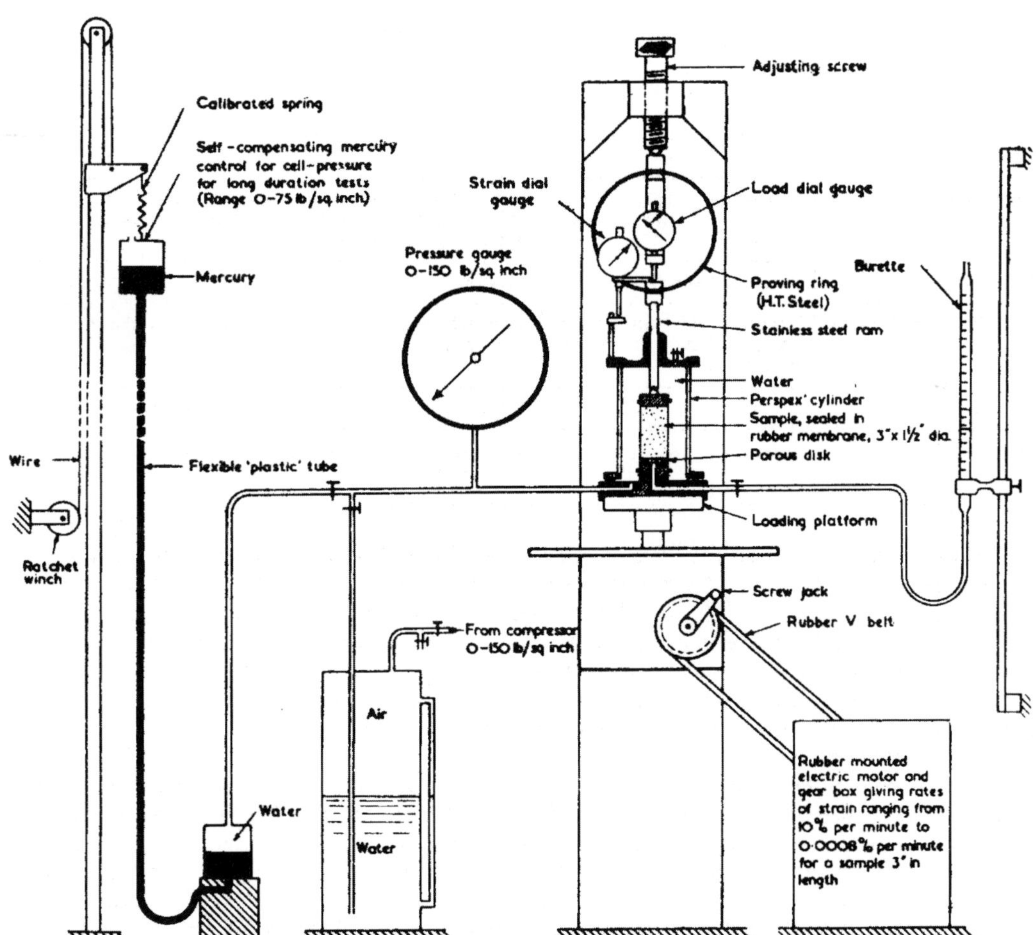

Diagrammatic layout of triaxial apparatus for testing saturated or partly-saturated soils

sample are indicated by changes in mercury level in the burette. This procedure is, of course, applicable only to drained tests.

SPECIAL CASES OF THE TRIAXIAL TEST

Unconfined compression test.—This is the limiting case of the undrained triaxial test, in which the lateral pressure is zero. The rubber membrane and cell may be dispensed with, and in this form it provides a test which can be simply and rapidly carried out either in the laboratory or in the field (Cooling and Golder, 1940).

In the case of intact saturated clays it provides all the information required for a " $\phi = 0$ analysis," but the results may be misleading in the case of fissured clays. In these the unconfined compression strength may be 30 per cent. below the value as measured in an undrained triaxial test under the overburden pressure. The unconfined test is also misleading in dilating or partly saturated soils, in which the angle of shearing resistance is not equal to zero. (Bishop and Eldin, 1950).

Fig. 2

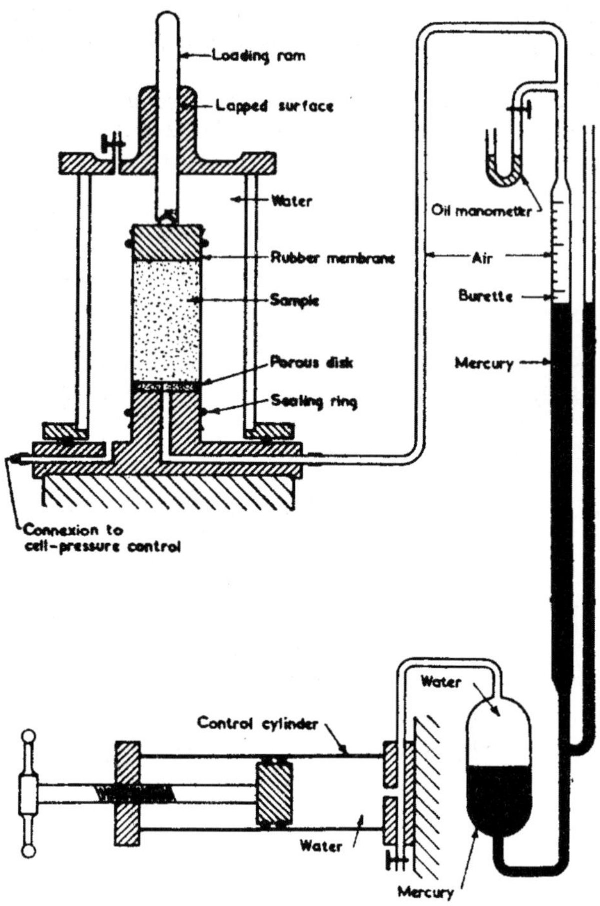

Layout of triaxial apparatus for measuring volume changes in dry sand samples

Samples consolidated in the oedometer.—In certain cases it is convenient to consolidate the sample (or allow it to swell) in the oedometer before cutting from it a specimen for an undrained triaxial test (Bishop, 1948*b*). This approximates to a special case of the anisotropically-consolidated-undrained test and is useful in certain practical foundation and runway problems where the corresponding test carried out wholly in the triaxial apparatus is technically rather complicated.

SHEAR BOX

The shear-box test is normally carried out on samples 6 centimetres square × 2 centimetres in thickness or 12 inches square × 6 inches in thickness. The 6-centimetre shear box is illustrated in Fig. 3, and details of the large shear box have been given by Bishop (1948*a*).

The use of the shear box is normally limited to drained tests, as, owing to the short drainage paths, undrained and consolidated-undrained tests can only be carried out on clays of low permeability at high testing rates. In any case the high stress gradients within the sample make the test unsuitable for a study of rheological properties, as internal redistribution of the pore water will take place even if drainage is prevented.

For drained tests on clays, using the 6-centimetre box, the time of shearing should not be less than about ½–1 day. With the limited data at present available it appears that the results are in close agreement with those obtained from drained tests in the triaxial apparatus.

The large shear box enables gravels with a particle size of up to 1 inch diameter to be tested (or up to 1½ inch diameter if well graded).

Figs 4

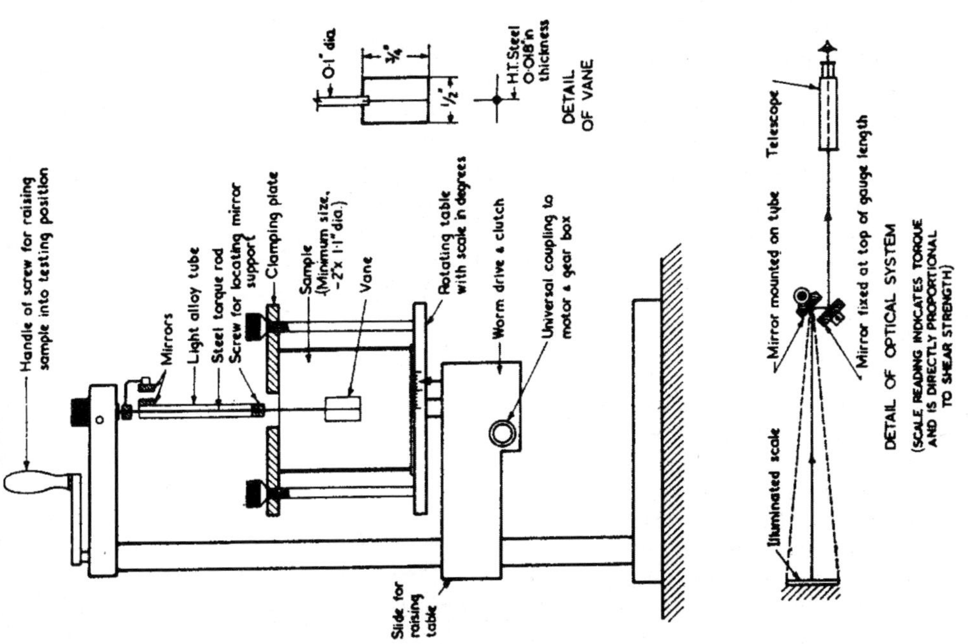

Laboratory vane apparatus

Figs 3

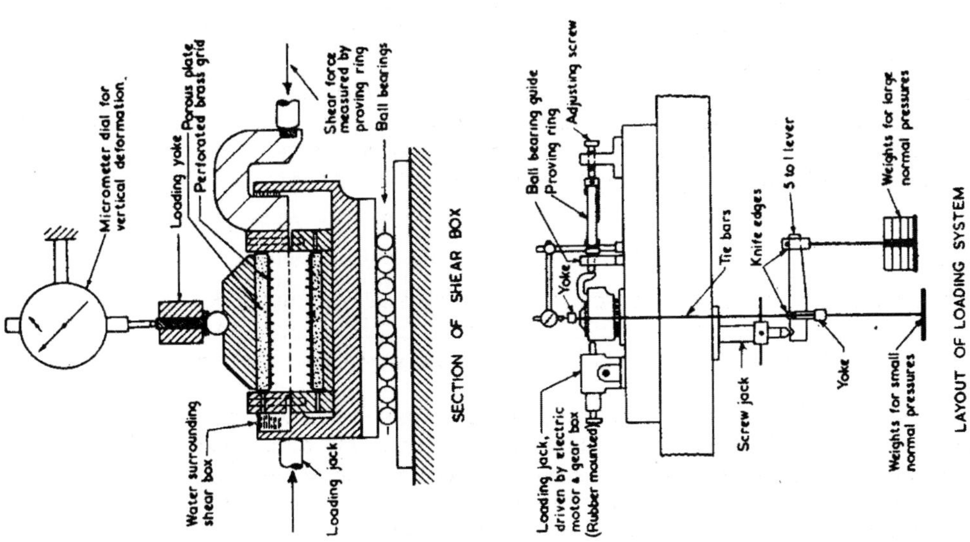

The 6-centimetre shear box

VANE TEST

The vane test as used in the laboratory * at Imperial College is illustrated in Fig. 4. The torsion set up in the thin steel rod on which the vane is mounted, due to the relative rotation of the sample and the vane, is measured optically. This gives a very sensitive measurement of the shear strength mobilized on the cylindrical surface bounding the vane, and is entirely free from frictional errors in the apparatus. Strengths as low as 0·03 lb. per square inch may be measured with reasonable accuracy.

Tests on seven different saturated non-fissured clays have shown good correlation between unconfined compression strength and the strength as measured with the vane; using the expression

$$\text{torque} = c \, . \, \pi \left(\frac{d^2h}{2} + \frac{d^3}{6}\right)$$

where d and h denote the diameter and height of the vane, and c denotes the shear strength of the clay.

The test is primarily used as an undrained test on very soft undisturbed and remoulded clays from which compression specimens are not readily prepared. It is of interest to note that the value of c at the liquid limit is of the order of 0·1 to 0·2 lb. per square inch for most clays.

PENETRATION TESTS

The type of penetration test which has figured most widely in laboratory procedure is the cone test, for use on cohesive soils where $\phi_u = 0$, as a measure of consistency or shear strength.

As it appeared to have particular advantages in the case of very soft samples, a careful study of its value as a measure of shear strength has been made at Imperial College. The apparatus used for carrying out the test at controlled rates of penetration is illustrated in Fig. 5.

The force required to cause penetration can be expressed as $P = k \, . \, c \, . \, \pi \tan^2 B/2 \, . \, p^2$

where c denotes shear strength ($= \frac{1}{2}$ unconfined compression strength)
B ,, cone angle (90 degrees in our tests)
p ,, penetration
k is the cone factor

It was found, however, that the value of the conversion factor k varied with water content for any given clay and varied widely from one clay to another, as shown in Table 2.

Table 2

Clay	LL	PL	Cone factor
Shellhaven (undisturbed)	96	32	6·9–7·2
,, (remoulded)	,,	,,	4·6–6·0
London Clay	73	26	7·0–7·5
Silent Valley	46	20	5·0–5·1
Horten	29	16	2·8–3·4

The cone test can thus be considered only as a very rough guide to the undrained shear strength of a clay. Even as a measure of the sensitivity of any particular clay it may be misleading, as the cone factor (which depends on the ratio of Young's Modulus to the shear strength of the sample) will in general be different in the undisturbed and remoulded states.

* The first description of a laboratory vane test is that by Evans and Sherratt (1948).

Other penetration tests, such as the California Bearing Ratio test and the Proctor Needle are empirical consistency tests and are not intended as measurements of shear strength. They will therefore not be considered here.

IN-SITU TESTS

VANE TEST

The vane test was first used as a means of measuring the undrained shear strength of deep clay strata by Carlson (1948) * in Sweden. Another form has been developed in England by the Authors to overcome the difficulty of penetrating hard layers and sand or gravel strata overlying the clay ; and to permit undisturbed samples to be taken from the same bore-hole (Skempton, 1948c).

Two factors have been given particular consideration :—

(*a*) the location of the vane at a depth below the bore-hole casing where the clay is free from any appreciable disturbance or change in stress ;

(*b*) the choice of the dimensions of the apparatus so that corrections due to adhesion of the clay to the vane spindle and due to friction in the bearings shall be only a small proportion of the total result.

Fig. 5

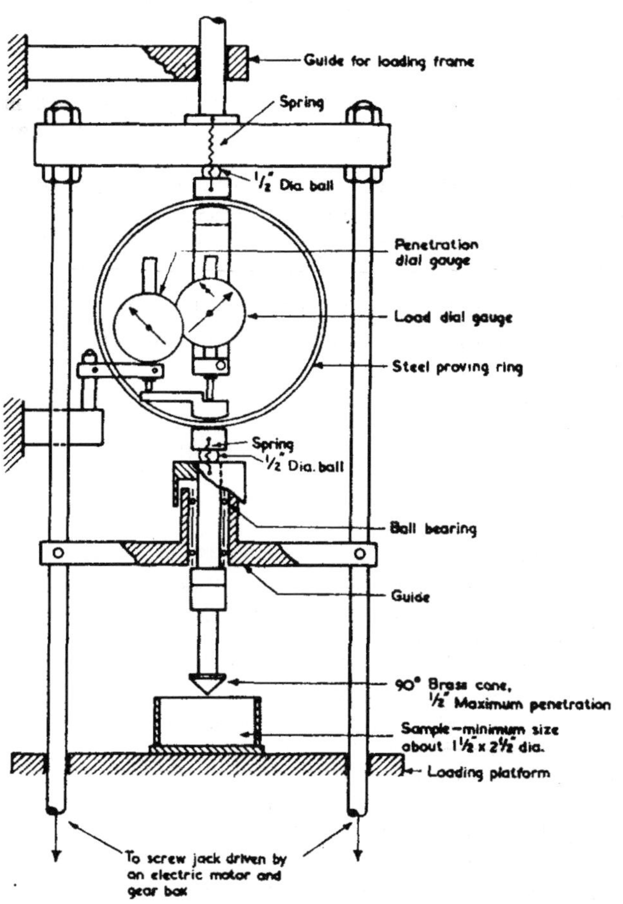

Cone penetration test

The apparatus finds its main use in deep beds of soft clay or silt, particularly when high sensitivity makes undisturbed sampling difficult or unreliable. In a simple hand-operated form the vane also provides a convenient means of measuring shear strength in the upper few feet of a soil (Smith, 1945 ; Evans, 1950).

DEEP SOUNDING TEST

This test, which was originally developed in the Netherlands, is in effect a field penetration test at a controlled rate of penetration, in which the end bearing pressure is measured separately from the side friction. It is useful qualitatively in locating the boundaries of hard and soft strata, and quantitatively in measuring the bearing capacity of sand and gravel strata (Vermeiden, 1948).

In problems relating to the design of piled foundations, for which it has been most widely used, the measured ultimate bearing capacity of the cone is taken to be that of the pile point. This procedure has been confirmed by full-scale loading tests (Plantema,

* A more recent form of the Swedish type of vane is described by Cadling and Odenstad (1950).

1948). The test may, however, be considered as a drained shear test in the case of sands, and the angle of internal friction calculated, at least approximately, from bearing capacity theory.

TYPICAL RESULTS AND THEIR PRACTICAL SIGNIFICANCE

DESCRIPTION OF SOILS

Before describing the results of shear tests and discussing their significance it may be useful to outline the other results which, in the Authors' opinion, are a minimum requirement for the description of any soil.

In-situ Cohesionless Soils: Particle-size distribution, in-situ density and water content, upper and lower porosity limits (Kolbuszewski, 1948), degree of saturation.

Compacted Cohesionless Soils: as above, but with Proctor density * and optimum water content instead of the in-situ values.

In-situ Cohesive Soils: Natural water content, Atterberg limits, degree of saturation, particle-size distribution, sensitivity to remoulding, pre-consolidation load.

Compacted Cohesive Soils: Atterberg limits, Proctor density * and optimum water content, particle-size distribution, natural water content, and density of material in the borrow pits.

SENSITIVITY

Many normally or lightly over-consolidated clays and silts show a marked drop in strength on remoulding: Terzaghi (1944) defines the "sensitivity" as the ratio of the undisturbed strength c to the remoulded strength c_r, and a sensitivity of between 2 and 4 is commonly encountered. In clays which may be classified as "sensitive" this ratio lies between 4 and 8, whilst in "extra-sensitive" clays the ratio is greater than 8 (Terzaghi and Peck 1948).

In some problems the fact that a clay may be sensitive or extra-sensitive is as important as its undisturbed strength; and for this reason a statement of the sensitivity is an essential part of the description of a cohesive soil.

The "remoulding loss" may be defined as $\frac{c - c_r}{c}$ and, apart from extra-sensitive clays, it is roughly equal to the liquidity index $\frac{w - PL}{LL - PL}$. Most clays which are heavily over-consolidated show almost no remoulding loss; for example, London Clay, boulder clays, etc.

UNDRAINED TESTS

These tests are carried out with no water-content change. Consequently they can be applied to those problems in which the soil is stressed at constant water content, and in which the soil can be sampled, or formed into samples (in the case of compacted fills) that can be tested immediately in the triaxial apparatus, or by the vane.

The soils included within the scope of these tests are clays, silts, and clay-gravels † and the problems include foundation of structures and dams, provided that consolidation during construction is negligible, and the design of walls and cuttings where the factor of safety is required for the condition immediately after construction.

Stability analyses are carried out with respect to total stresses and with the shear parameters c_u and ϕ_u defined as in Fig. 6. The results fall into two classes: (*a*) those in which ϕ_u is not zero and (*b*) those in which ϕ_u is zero.

The result $\phi_u = 0$ is obtained in all saturated clays and in some saturated silts. Golder

* The Authors consider the "modified Proctor" compaction test to be applicable only in special work and they deprecate its use as a standard test.

† The vane cannot, of course, be used in clay-gravels.

and Skempton (1948) have published data for twenty saturated clays in which ϕ_u is either zero or not greater than 1 degree and in which c_u varies from 1½ lb. per square inch to 45 lb. per square inch under cell pressures covering a range from zero (unconfined compression test *) to 60 lb. per square inch. Many other similar results have been published. Characteristic variations in c_u with depth in a normally consolidated and in an overconsolidated clay are shown in Figs 7. In the former case the strengths were used in a $\phi = 0$ analysis of a steel

Figs 6

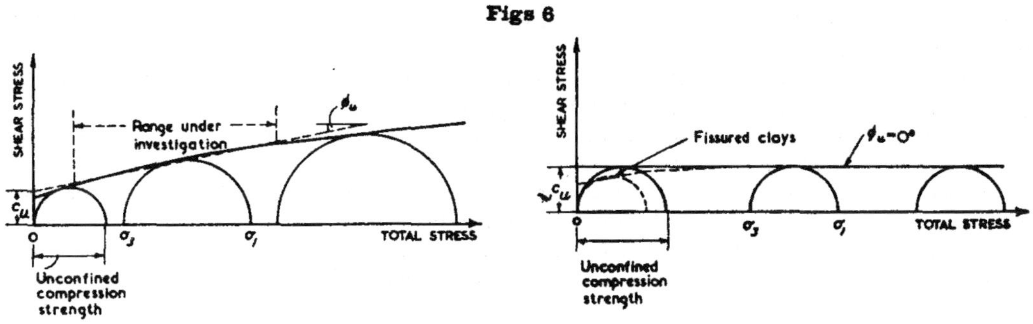

(a) UNDRAINED TEST (TRIAXIAL) – GENERAL CASE (TYPICAL OF PARTLY SATURATED OR DILATING SOILS)

(b) UNDRAINED TEST (TRIAXIAL) $\phi_u - 0°$ (TYPICAL OF SATURATED SOILS)

Typical results of undrained tests

sheet pile cofferdam 35 feet deep, 50 feet wide and 250 feet long, whilst in the latter case they were used in a $\phi = 0$ analysis of the foundation stability of an earth dam 60 feet high. In both examples consolidation during construction could be neglected without any appreciable error. It is to be noted that it is this fact which provides the justification of the $\phi = 0$ assumption (together with the $\phi_u = 0$ result in the tests), and that the $\phi = 0$ assumption in analysis is wholly unaffected by the gain in strength with depth.

The angle of shearing resistance ϕ_u is not zero in partially saturated clays, silts, and clay gravels, and in some saturated (dilating) silts. Also it may be accepted as a definition of a clay-gravel that, even when saturated, it has a value of ϕ_u greater than zero ; if ϕ_u does equal zero the material should be classified as a gravelly clay. Many boulder clays fall into this latter category, and they are essentially clays : the sand and gravel not being present in sufficient proportion to impart any marked positive dilatancy. Clay shales and siltstones (i.e. intensely over-compacted clays and silts) also fall into the class of soils with ϕ_u not equal to zero. In most soils in which ϕ_u is not zero the Mohr envelope is slightly curved, Fig. 6 (*a*), but it may be taken as linear over the range of pressure under investigation. Golder and Skempton quote data on nine partially saturated clays, five saturated silts (LL less than 35) and seven clay shales or siltstones. The values of ϕ_u typically lie between 15 and 30 degrees. In a saturated clay gravel, containing 65 per cent. by weight of particles retained on the ¼-inch sieve and 15 per cent. clay fraction, ϕ_u was found to be 24 degrees.

CONSOLIDATED-UNDRAINED TESTS

In these tests the soil is allowed to soften or consolidate under a specified pressure, and is then sheared under conditions of no water-content change. All soil types can be tested by this procedure, but of the " sands and gravels " category only very fine silty sands are likely,

* Some of the saturated fissured clays have a slightly lower strength in unconfined than in undrained triaxial tests where σ_3 is equal to or greater than the overburden pressure (see Fig. 6*b*). The unconfined test should not be used for fissured clays.

in practice, to involve shear at constant water content. The problems in which the test results can be directly applied include the design of earth dams for the case of rapid draw-down and the design of road and runway bases and subgrades.

In the former problem the samples are allowed to come to equilibrium under the effective stresses existing in the dam before drawdown and are then sheared under various cell pressures under conditions of constant water content. In the latter problem the samples are brought to the water content and density estimated as being those which will exist under the pavement, and are then tested as mentioned above (or, frequently, in the C.B.R. apparatus). The stability calculations are carried out with respect to changes in total stresses, using the shear parameters obtained from the tests.

A similar procedure is adopted for determining the strength to be used in calculating the initial stability of " impervious " rolled fill in embankments and dams, since during the placing of the layers appreciable softening can take place due to rain.

Figs 7

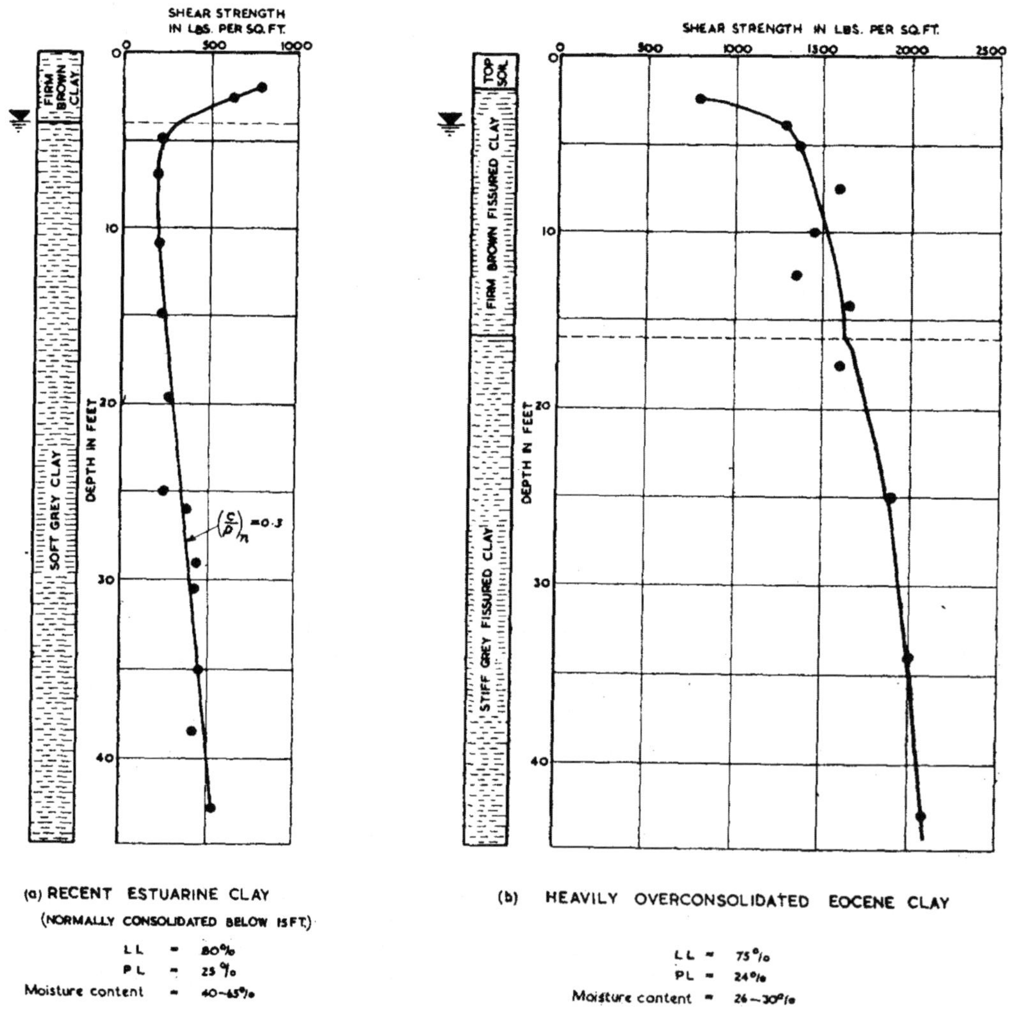

Variation of strength with depth for two typical clays

Figs 8

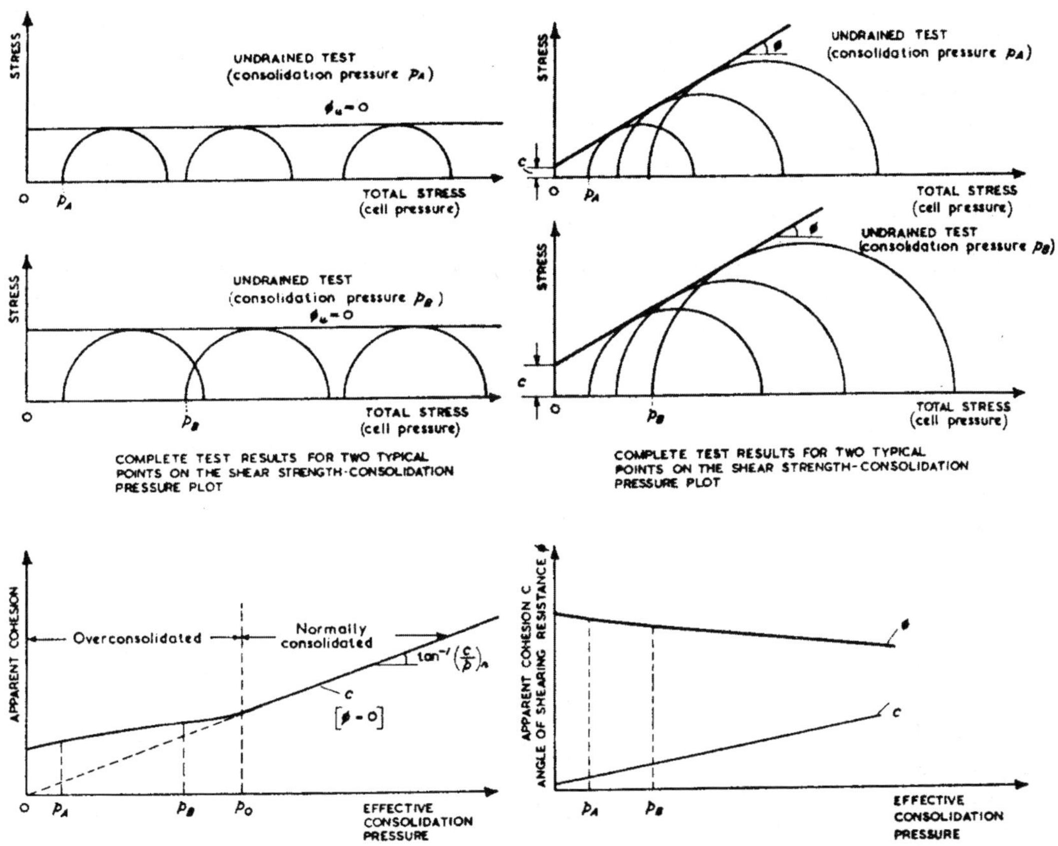

(a) UNDISTURBED SATURATED SAMPLE OF CLAY (TYPICAL OF SATURATED COMPRESSIBLE SOILS)

(b) COMPACTED CLAY–GRAVEL (TYPICAL OF DILATING OR PARTLY SATURATED SOILS)

Consolidated-undrained tests

In all saturated clays $\phi = 0$, and in some cases even a clay which was partially saturated initially will become fully or almost fully saturated after consolidation or swelling, and will then show $\phi = 0$. In clay gravels and dilating silts ϕ will not be zero. The results are most conveniently plotted as shown in Figs 8. The values of c and ϕ to be used in the analysis can be read off the graph for any given consolidation pressure.

A considerable number of tests have been carried out on clays in which $\phi = 0$ after consolidation or softening, and a typical set of results has been published by Bishop (1948*b*) relating to undisturbed London Clay. Similarly, in the design of a runway subgrade, tests have been made on samples of a silty clay which were compacted to a specified density and water content and then allowed to soften under various pressures corresponding to different thicknesses of construction. When tested under undrained conditions it was found that $\phi = 0$ and a curve could readily be drawn relating shear strength to pressure.

The Authors have had only one occasion to carry out tests on a material which shows ϕ not equal to zero after consolidation. This was a clay gravel (55 per cent. retained on the $\frac{1}{4}$-inch sieve and 20 per cent. combined silt and clay fraction).

The design of cuttings and retaining walls in stiff fissured clays is still a matter of uncertainty, but a provisional method has been suggested (Skempton, 1948*d*) based on a knowledge of the original unsoftened strength c_1 and the final fully softened strength c_0. The strength c_1 is obtained from undrained tests on undisturbed samples, while c_0 is the strength as measured in undrained tests on samples which have been allowed to soften under zero pressure. These samples are wrapped in filter paper, with solid end pieces, and immersed in water for several days. This technique is mentioned since it seems to provide a possible method of tackling the problem, but it is not considered to be any more than a tentative solution. In foundation problems there is, of course, no need to allow the clay to soften before testing, since the fissures cannot open in such cases.

In research, especially on saturated clays, the relation between undrained shear strength and consolidation pressure is a useful parameter, and especially in the range of pressure where the clay is normally consolidated. Here the ratio is written $\left(\frac{c}{p}\right)n$, as in Fig. 8 (*a*). Terzaghi and Peck (1948) * use the symbol ϕ_{cq} where $\tan \phi_{cq} = \left(\frac{c}{p}\right)n$. It should be emphasized that the ratio of c to p measured by tests on undisturbed samples may be appreciably greater than that obtained from field evidence in sensitive clays consolidated under their own overburden pressure (Terzaghi, 1947). This is probably due to the lateral pressures in the ground, in normally consolidated clays, being less than the vertical pressure : whereas in the laboratory tests the sample is consolidated under an isotropic pressure (Hansen and Gibson, 1949).

DRAINED TESTS

These tests are carried out with full water content change (i.e. no excess pore pressures) both in consolidation and in shear. Consequently the results can be applied directly to those problems in which the soil is stressed at zero excess pore pressure.

All soil types can be tested under fully drained conditions, and the problems include all stability analyses in sands and gravels, except the case of rapid drawdown or flow slides in loose silty sands. In clays the drained test can be used directly in analysing the long-term stability of cuttings, retaining walls and the downstream slope of earth dams. And the test can be used in problems concerning partial dissipation of pore pressures.

Stability analyses are carried out with respect to effective stresses and with the shear parameters c_d and ϕ_d defined as in Figs 9. The results fall into two classes : (*a*) those in which c_d is zero and (*b*) those in which c_d is not zero.

The result $c_d = 0$ is obtained in all sands and gravels. Many test results have been published giving values of ϕ_d for these soils, and these can be summarized approximately as in Table 3.

Table 3

Approximate limits of ϕ_d in cohesionless soils
(Pressure range 0–3 tons per square foot)

Soil	ϕ_d (degrees)	
	Loose	Dense
Uniform sands	30	40
Well-graded sands	32	45
Sandy gravels	35	50

* Casagrande (1941) uses the symbol ϕ_{cq} in the sense :

$$\sin \phi_{cq} = \frac{(c/p)_n}{1 + (c/p)_n}$$

Since effective stresses must be used in stability calculations it is often necessary to construct a flow net as the first step in an analysis in these soils.

The apparent cohesion c_d is, in general, not zero in clay-gravels, silts, and clays, unless they are normally consolidated. Not many results of drained tests on undisturbed samples of cohesive soils have been published. TheVicksburg Triaxial Shear Report (U.S. War Department, 1947) is the chief source of information.*

Figs 9

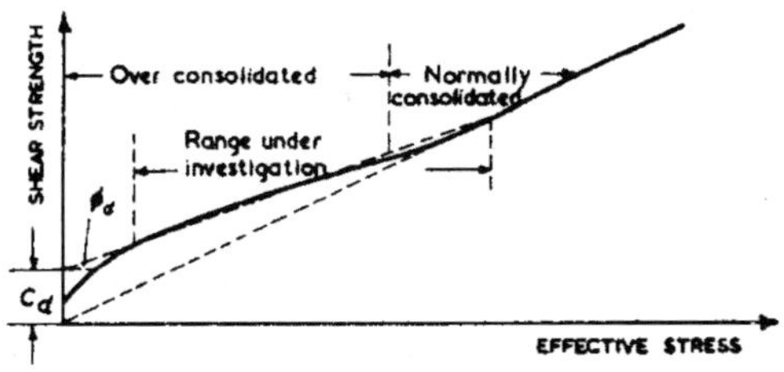

(a) DRAINED TEST – GENERAL CASE

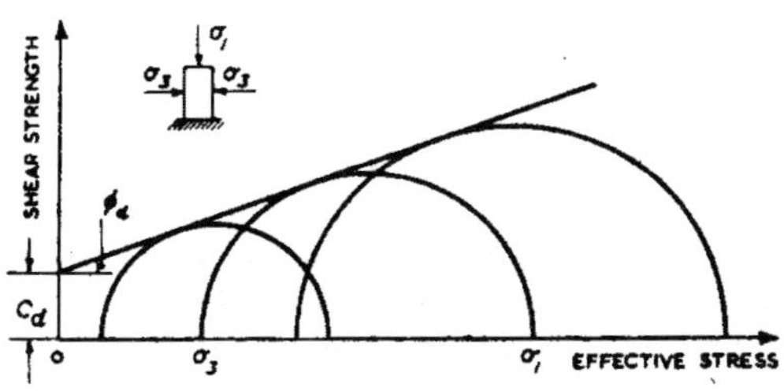

(b) DRAINED TEST — TRIAXIAL TEST RESULTS

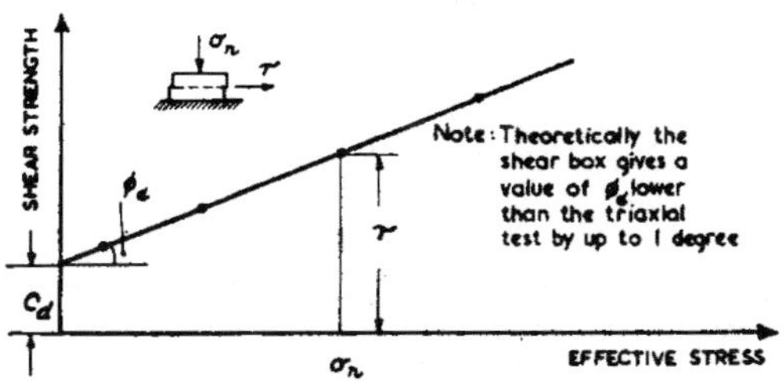

(c) DRAINED TEST – SHEAR BOX RESULTS

Drained tests

PARTIALLY DRAINED CONDITION

In a clay stratum beneath an earth dam there may be appreciable, but not full, consolidation during construction. No specific test corresponds to this condition, but in such cases a conservative procedure is to take the values of shear strength as found from undrained tests on specimens consolidated to the vertical effective pressure in the clay as

* These tests were carried out at Harvard.

calculated from Terzaghi's theory of consolidation. This method has been described in some detail (Bishop, 1948*b*).

In the Authors' opinion an analysis more representative of the actual stress change in the clay layer could be evolved using the drained as well as the consolidated-undrained test. From these two tests it is possible to estimate the effect of dissipation of pore pressures set up by shear stresses in addition to those due to the vertical pressures (weight of bank).

REFERENCES

Bishop, A. W. (1948*a*). A large shear box for testing sands and gravels. *Proc. Second Int. Conf. Soil Mech.* 1 : 207–211.

Bishop, A. W. (1948*b*). Some factors involved in the design of a large earth dam in the Thames Valley. *Proc. Second Int. Conf. Soil Mech.* 2 : 13–18.

Bishop, A. W., and Eldin, G. (1950). Undrained triaxial tests on saturated sands and their significance in the general theory of shear strength. *Géotechnique.* 2 : 13–32.

Cadling, L., and Odenstad, S. (1950). The vane borer. *Proc. Roy. Swedish Geotechnical Inst. No.* 2.

Casagrande, A. (1941). *Harvard Triaxial Report No.* 3.

Casagrande, A., and Shannon, W. L. (1948). Research on stress-deformation and strength characteristics of soils and soft rocks under transient loading. *Pub. Harvard Univ. Grad. Sch. Eng. No.* **447.** 132 pp.

Casagrande, A., and Wilson, S. D. (1949). Final Report to U.S. Waterways Experiment Station on investigation of effect of long-time loading on the strength of clays and shales at constant water content. *Harvard University.* 77 pp.

Carlson, L. (1948). Determination in situ of the shear strength of undisturbed clay by means of a rotating auger. *Proc. Second Int. Conf. Soil Mech.* 1 : 265–270.

Cooling, L. F., and Golder, H. Q. (1940). Portable apparatus for compression tests on clay soils. *Engineering.* 149 : 57–58.

Croney, D., Lewis, W. A., and Coleman, J. D. (1950). Calculation of the moisture distribution beneath structures. *Civil Eng. (L.).* 45 : 103–106.

Evans, I. (1950). The measurement of the surface bearing-capacity of soils in the study of earth-crossing machinery. *Géotechnique.* 2 : 46–57.

Evans, I., and Sherratt, G. G. (1948). A simple and convenient instrument for measuring the shearing resistance of clay soils. *J. Scientific Instruments.* 25 : 411–414.

Golder, H. Q., and Skempton, A. W. (1948). The angle of shearing resistance in cohesive soils for tests at constant water content. *Proc. Second Int. Conf. Soil Mech.* 1 : 185–192.

Hansen, J. B., and Gibson, R. E. (1949). Undrained shear strengths of anistropically consolidated clays. *Géotechnique.* 1 : 189–204.

Hvorslev, M. H. (1937). Über die Festigkeitseigenschaften gestoerter bindiger Boeden. (On the physical properties of disturbed cohesive soils.) *Ingeniorvidenskabelige Skrifter* A No. 45. 159 pp.

Kolbuszewski, J. J. (1948). An experimental study of the maximum and minimum porosities of sands. *Proc. Second Int. Conf. Soil Mech.* 1 : 158–165.

Peynircioglu, H. (1939). Über die Scherfestigkeit bindiger Boeden. (Shearing resistance of cohesive soil.) *Veroeffentlichungen des Instituts der Deutschen Forschungsgesellschaft fuer Bodenmechanik an der Technischen Hochschule, Berlin.* No. 7 : 28–52.

Plantema, G. (1948). Results of a special loading-test on a reinforced concrete pile, a so-called pile sounding ; interpretation of the results of deep-soundings, permissible pile loads and extended settlement observations. *Proc. Second Int. Conf. Soil Mech.* 4 : 112–118.

Rendulic, L. (1937). Ein Grundgesetz der Tonmechanik und sein experimenteller Beweis. (The fundamental law of clay mechanics and the experimental proof.) *Bauingenieur.* 18 : 459–467.

Skempton, A. W. (1948*a*). A study of the immediate triaxial test on cohesive soils. *Proc. Second Int. Conf. Soil Mech.* 1 : 192–196.

Skempton, A. W. (1948*b*). The effective stresses in saturated clays strained at constant volume. *Proc. Seventh Int. Cong. Appl. Mech.* 1 : 378–392.

Skempton, A. W. (1948*c*). Vane tests in the alluvial plain of the River Forth near Grangemouth. *Géotechnique.* 1 : 111–124.

Skempton, A. W. (1948*d*). The rate of softening in stiff fissures clays, with special reference to London clay. *Proc. Second Int. Conf. Soil Mech.* 2 : 50–53.

Skempton, A. W., and Golder, H. Q. (1948). Practical examples of the $\phi = 0$ analysis of stability of clays. *Proc. Second Int. Conf. Soil Mech.* 2 : 63–70.

Smith, A. H. V. (1945). Some preliminary trials with the A.O.R.G. vane apparatus. *Army Op. Res. Group Memo.* 540.

Taylor, D. W. (1943). *M.I.T. Triaxial Report No.* 9.

Taylor, D. W. (1944). *M.I.T. Triaxial Report No.* 10.

Taylor, D. W. (1948). Fundamentals of soil mechanics. *Wiley, New York.* 700 pp.

Terzaghi, K. (1932). Tragfaehigkeit der Flachgruendungen. (Bearing capacity of shallow foundations.) *Prelim. Pub. First Cong. Int. Ass. Bridge Struct. Eng.* 659–672.

Terzaghi, K. (1936). Simple tests determine hydrostatic uplift. *Eng. News Rec.* **116 : 872–875.**

Terzaghi, K. (1944). Ends and means in soil mechanics. *Eng. J.* 27 : 608–615. *Publ. Harvard Univ. Grad. Sch. Eng. No.* 402. 8pp. (1944).
Terzaghi, K. (1947). Shear characteristics on quicksand and soft clay. *Proc. Seventh Texas Conf. Soil Mech.* 10 pp.
Terzaghi, K., and Peck, R. B. (1948). Soil mechanics in engineering practice. *Wiley, New York.* 566 pp.
Vermeiden, J. (1948). Improved soundings apparatus, as developed in Holland since 1936. *Proc. Second Int. Conf. Soil Mech.* 1 : 280–287.
U.S. War Department (1947). Triaxial shear research and pressure distribution studies on soils. *U.S. Waterw. Expt. Sta. Vicksburg.* 332 pp.

DISCUSSION

The Chairman, in opening the discussion, said that he felt that the Paper was representative of views widely held in Great Britain at the present time. He emphasized the point made by Dr Skempton that the usual procedure for the analysis of stability problems was to use values of c and ϕ in Coulomb's equation which had been determined from samples subjected to laboratory test-conditions compatible with the conditions under which the soil would be stressed in the field. He felt that the fundamental work being carried out by Dr Skempton on the shear strength of soils was of considerable importance. On the question of silts, experience in Great Britain was not extensive and the present position was uncertain, especially in relation to their behaviour under field conditions. Whilst it was possible to obtain a value of ϕ equal to zero with respect to total stresses in undrained triaxial tests, it was doubtful whether silts ever behaved as ($\phi = 0$) materials in the field. It was known that their true cohesion was very small and that their strength must therefore be derived from the frictional quality. When it was possible in the practical case to measure pore-water pressures, he thought that an analysis using c_e and ϕ_e, based on effective pressures, would be preferable. He congratulated the Authors on the ingenuity displayed in their development of laboratory testing equipment.

Mr H. B. Sutherland said that he had received his introduction to soil mechanics from Mr Cooling at the Building Research Station, having spent some time working at the laboratory there. Subsequently he had been two years at Harvard University working under Professor Arthur Casagrande. He had therefore had the interesting experience of acquiring his initial knowledge of the shear strength of soils in Britain and had then come into contact with some rather different ideas held in the United States during his stay there. The tests used in Great Britain for the determination of shear strength values were modifications of the tests developed and used in the United States, but whilst those tests were the same in principle there were some differences in interpretation. From the point of view of practical application the shear strength of the soil under any one particular condition was the most important factor. To what extent c and ϕ contributed to the shear strength was relatively unimportant : c and ϕ could vary with different drainage conditions for the same soil, and once that was fully recognized many details of the interpretation of shear tests faded into insignificance.

The work of Hvorslev had been preceded by the researches of A. Casagrande and Albert (1930) into some of the most important aspects of the characteristics of the shear strength of soils, particularly the effect of the time-rate of shear-load application upon the results of shear tests.

He was of the opinion that the terms " quick " and " slow " tests should be retained as the standard nomenclature in shear-strength testing. They had originally been suggested by Casagrande, were widely used in America, and had been adopted by Rutledge in the Triaxial Shear Research Review published in 1947. He thought that it would be a mistake to use the terms " undrained " and " drained " in the new British Code of Practice, as that would tend to destroy uniformity in nomenclature.

He believed that the effect of the rate of strain on the shear strength of soils was somewhat larger than that indicated by Dr Skempton. Tests carried out by Casagrande and Wilson (1949) showed that in one instance, for a clay-shale, the strength obtained from a 1 month test was only 38 per cent. of the 1 minute test strength ; for a bentonite the 1 month strength was 56 per cent. of the 1 minute test strength, and for clay the 1 month strength was 76 per cent. of the 1 minute test strength. Those tests were undertaken as part of an investigation of the effect of long-time loading on compressive strength at constant water content.

He would like to ask Dr Skempton the following questions :

(1) What his opinions were regarding the relative merits of the triaxial and shear-box tests ?

(2) How could one be certain that full drainage occurred in a constant rate of strain test ?

In conclusion, he would stress his appreciation of the very excellent work on the shear strength of soils being carried out by Dr Skempton and his staff.

Mr W. Kjellman was of the opinion that the λ-theory due to Dr Skempton was the most important contribution to the understanding of the fundamentals of shear-strength theory that had appeared recently. More experimental work should be done to establish the validity of the relations between strain and effective stress change upon which that theory was based. It would be necessary to introduce a further parameter expressing the effect of pure shear, as had been done by Odenstad in a recent Paper. That theory of Odenstad's, moreover, was in agreement with the experimental findings of Professor Taylor. Had Dr Skempton any experimental results which might be used as a check for such theories ? He was not in disagreement with anything in the present Paper, but he wished to make the following comments :

(1) He suggested that the vane should be used only for undrained tests and only in materials which did not exhibit dilatancy. When the vane was used, samples should be taken as a check.

(2) When the validity of the cone test was being investigated at the Imperial College and the cone conversion-factor was found to vary with the soil type, what test was used as a standard to measure the soil strength ? So far as he was aware, the only method of measuring the strength of very sensitive clays was with the vane.

(3) He wished to know why dilatancy increased the strength of a soil.

(4) The method of determining the porosity of sands by the use of air pressure and volume measurements, described by Mr Bishop, was not new. There was, in fact, a Swiss patent for it.

M. J. Florentin dit qu'il veut remercier tout d'abord les conférenciers du remarquable exposé relatif aux méthodes expérimentales utilisées en Angleterre pour la mesure de la résistance au cisaillement des sols. Toutefois, il y a un point qui le trouble et qui l'amène à poser la question suivante : les laboratoires anglais de mécanique du sol, ont-ils abandonné la mesure de la résistance au cisaillement avec des appareils de torsion ? Si oui, pourquoi ?

Il pense que la comparaison entre les différents appareils actuellement utilisés aurait été plus complète si l'on y avait inclus la mesure du cisaillement par torsion. Il pense, comme beaucoup d'autres, que le cisaillement par torsion est le seul moyen de produire le cisaillement sans effets parasites. A son avis, la torsion sous des compressions différentes constitue un bon moyen de contrôle de la loi de Coulomb, puisque les résultats expérimentaux se traduisent par la détermination d'un effort tangentiel t correspondant à une compression n. Dans les essais de compression, (sans ou avec étreinte latérale) on fait appel, en plus, à la théorie de Mohr sur la rupture des corps. Ce que l'on tire en définitive de ces essais, ce sont les cercles construits sur les contraintes principales extrêmes atteintes au cours de la rupture. Or, il croit se souvenir que la théorie de la plasticité (et l'on ne peut nier que les matériaux essayés soient en général plastiques), conduit à la conclusion que la rupture d'un corps plastique n'est pas complètement indépendante de la contrainte intermédiaire.

En résumé, il lui semble que sur le plan théorique, l'essai de torsion n'implique aucune hypothèse supplémentaire à la loi de Coulomb, alors que les essais de compression supposent en plus que le domaine de rupture est délimité par l'enveloppe des cercles de Mohr ayant les contraintes extrêmes comme diamètre.

Mais ces divergences théoriques peuvent à la rigueur être considérées comme négligeables dans la mesure ou il est beaucoup plus pratique de procéder à des essais de compression (avec ou sans étreinte latérale) ; d'ailleurs, il ne crois pas qu'il soit question, pour le moment, de reconsidérer la question de la rupture plastique, dont les résultats peuvent être considérés comme formant la loi de base.

Toutefois, il voudrait se permettre d'insister sur quelques avantages pratiques de l'essai de torsion. M. Langer avait construit aux laboratoires du Bâtiment et des Travaux Publics avant la guerre (la construction date de 1936–1937) des appareils de torsion qui diffèrent de ceux de M. Hvorslev qui sont bien connus de tous les chercheurs. Ces appareils permettent d'essayer des échantillons cylindriques (et non en forme de couronne comme dans ceux de M. Hvorslev) ; ils présentent donc l'avantage de permettre l'essai d'échantillons intacts. Toutefois, l'objection que l'on a fait à ces appareils, et qui est d'ailleurs contenue implicitement dans la conception de M. Hvorslev, était que l'on ne connait pas la distribution des contraintes tangentielles dans le plan de cisaillement. La forme théorique est celle de la figure 10, mais pratiquement on peut l'assimiler à la distribution de la figure 11.

Fig. 10

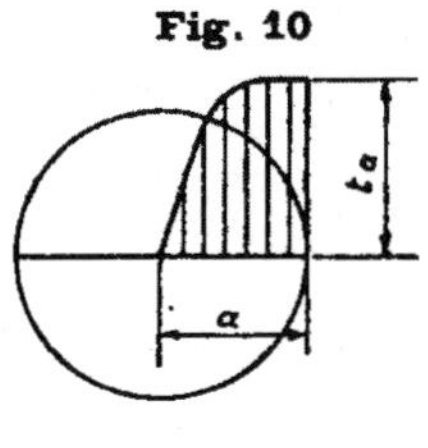

Fig. 11

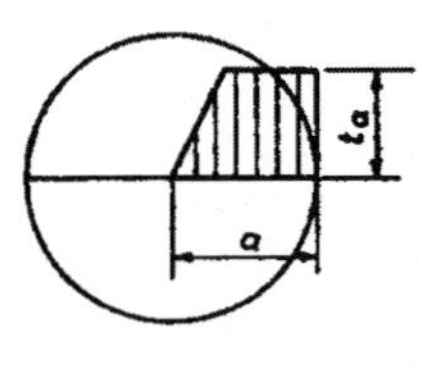

Dans chacun des deux cas, évidemment le moment de la surface hachurée par rapport au centre doit équilibrer le couple de torsion M. Si l'on appelle θ l'angle de rotation, M. Nadai (voir Plasticity 1931, page 125–128) a montré qu'il existe entre la contrainte de cisaillement maximum t_a atteinte sur le bord de l'échantillon et les différentes grandeurs géométriques et mécaniques de l'essai, la relation suivante :—

$$t_a = \frac{1}{2\pi a^3}\left(\theta\frac{dM}{d\theta} + 3M\right)$$

Cette formule appelle les remarques suivantes :—

(1) Au moment de la rupture $\frac{dM}{d\theta}$, est nul et tout se passe comme si la contrainte t est constante tout le long d'un même diamètre.

(2) Il est possible, en relevant la courbe $\frac{dM}{d\theta}$, de déterminer à chaque instant, au cours de l'essai, la contrainte maximum t_a et même une loi simplifiée de distribution des contraintes analogue à celle que suppose la figure 2. Il suffit, en effet, de tenir compte de l'équation ci-dessus de M. Nadai (ce qui donne t_a) et du fait que M est équilibré à chaque instant par le moment des efforts de cisaillement par rapport au centre de l'échantillon.

L'objection de l'indétermination de la distribution des contraintes dans l'appareil de torsion de M. Langer étant écartée, il lui parait nécessaire d'insister sur certains avantages pratiques de cet appareil, avantages partagés avec celui de M. Hvorslev, avec la différence qu'à moins d'utiliser des carottiers spéciaux il est plus délicat d'opérer avec celui-ci sur des échantillons intacts.

L'appareil permet de réaliser un certain nombre de conditions pratiques de rupture, ce qui permet une application assez systématique de la méthode semi-empirique attribuant à ϕ et à c des valeurs de paramètres expérimentaux et non de constantes physiques.

Pour l'étude de glissement du talus du poste de Genissiat (voir Etude des Glissements des Talus Argileux, par J. Florentin. Publication de l'Institut Technique de Bâtiment et des Travaux Publics, série D, No. 23), il a cherché, d'une part à déterminer les causes de ce glissement et, d'autre part, à savoir quelle pouvait être l'évolution du coéfficient de sécurité dont on disposerait, une fois ces causes supprimées. Il a opéré, d'une part sur des échantillons intacts à la teneur en eau naturelle, en déterminant la résistance au cisaillement maximum et la résistance au cisaillement résiduel après rupture. Il a ensuite cherché l'influence de l'arrivée d'eau dans le plan de cisaillement, ce qui est très fréquent dans un glissement. Enfin, il a cherché à étudier la résistance au cisaillement dont on dispose le long d'un plan de glissement réel, afin d'estimer la sécurité du talus, après enlèvement des matériaux qui y avaient été mis en remblai, et surtout une fois prises certaines précautions permettant d'assurer le drainage du talus.

Pour ce dernier cas, qui techniquement est le plus important, il a opéré, avec des charges normales de l'ordre de grandeur de celles qui existent le long de la courbe de glissement déjà observée, par torsion d'échantillons préalablement coupés suivant le plan de cisaillement. Il a constaté que la résistance au cisaillement est proportionnelle à la pression, ce qui conduit à définir un angle de glissement, mais il a également constaté que cet angle est plus grand si le plan de glissement est drainé. Enfin, il a obtenu, sur échantillons préalablement coupés, un angle de glissement supérieur à l'angle de cisaillement observé sur les échantillons intacts, ce qui conduisit, dans ce cas, à concevoir que la résistance au cisaillement est la somme de deux termes : l'un qui dépend du frottement du matériau et qui, d'après la loi de Coulomb, serait proportionnel à la pression, l'autre qui dépend de l'adhérence crée par la charge normale entre les deux plages d'un ancien plan de glissement. Les essais montrent que cette force d'adhésion était également proportionnelle à la pression.

Il s'excuse d'avoir été si long sur cet exemple pratique et personnel. Il l'a fait parceque l'étude numérique a montré que le coéfficient de sécurité avec les caractéristiques déterminées au laboratoire, reste voisin de 1 tant que l'on ne prend pas de précautions pour drainer le talus. Effectivement, avant que le drainage soit efficace et après enlèvement de remblai qui surchargeait le talus, des petits mouvements ont été observés. Actuellement, avec le drainage par puits verticaux et par puits horizontaux qui s'ajoute à l'effet d'une plantation d'acacias, on peut dire que les mouvements sont stoppés.

Mr B. Jakobson agreed with Dr Skempton that c and ϕ could not generally be regarded as fundamental parameters of a clay. Nevertheless he felt that they could be regarded in that way when they were deduced with respect to effective stresses. If, in a practical problem, the effective stresses were known, then a correct analysis could be made. Unfortunately, the drained test was rather difficult and time-consuming as a routine test.

He would like to mention, in connexion with the difference between the triaxial test and the shear-box test, that the sum of the principal stresses was different both initially and throughout the test in both cases. Moreover, in the shear-box test the directions of the principal stresses rotated during the test. That would lead to the water content at failure being different in the two tests for a given consolidation pressure. The sum of the principal stresses was probably greater in the box test than in the triaxial test. He suggested that a most important experiment to be carried out was one in which the sum of the principal stresses was maintained constant, a shear stress being applied and the variation in the pore-water pressure being measured.

Mr Ivor Evans said that he did not wish to comment on the Paper, but would like to present the following additional results on some cone tests which he had carried out on a number of clays, the shear strength used to calculate the cone factor k being found from laboratory vane tests.

Liquid limit	Plastic limit	k
86·5	25·0	7·6
67·0	27·7	6·3
58·0	21·8	5·6
54·2	22·2	7·1
50·0	27·7	6·2
46·6	26·6	5·0
39·3	17·7	5·2
39·0	18·8	5·0

Those results indicated that the factor k was rather low for silts. That could be partly explained by the fact that, as the cone was pressed into a silt, the surface tended to be depressed round the cone, the true bearing-area being, in consequence, rather less than that computed on the assumption that the surface remained horizontal throughout the test. If a logarithmic plot of cone movement against load were drawn, the slope was equal to 2 if no depression of the surface occurred. It was found in silts, however, that the slope was often somewhat higher ($2\frac{1}{2}$).

Further, it had been found possible to express the bearing capacity q in the form :

$$q = cN_c + \gamma RN_\gamma$$

where N_c and N_γ denoted the Terzaghi bearing-capacity factors for an infinite strip, c denoted the undrained triaxial shear strength, and R denoted the radius of the cone.

Mr E. C. W. A. Geuze said that he was concerned with an exact definition of the word " fundamental " in the reference to " fundamental parameter " or " fundamental property " in the Paper. If a pattern, both self-consistent and in accord with experimental results, could be generated from a number of concepts or variables, then he submitted that those concepts or variables could be, in a sense, regarded as " fundamental." He agreed that, generally, those fundamental variables could not be measured directly, and, moreover, it would be very difficult to use them directly for the solution of a practical problem. An empirical method was usually all that was necessary.

He was rather surprised that the Authors had said that the rate of strain had a very small effect on the strength of sands. His own experience was contrary to that. He was in agreement with M. Florentin that the torsion test was best from a theoretical point of view, but its successful execution was extremely difficult.

Dr A. W. Skempton, in reply, thanked the Chairman for his kind remarks and said that many soil mechanics specialists in Britain had been brought up in Mr Cooling's laboratory, or in close contact with it, and thus a certain measure of agreement had been obtained on the methods to be used for determining the shear strength of soils and on other basic problems in soil mechanics.

In many practical problems, partial drainage occurred. But in the laboratory tests must be carried out under definite limiting conditions, namely, no drainage or full drainage. It would be difficult, or even impossible, to attempt to simulate partial-drainage conditions in a laboratory test ; yet, from the results of both drained and undrained tests, the shear strength corresponding to partial drainage could be obtained.

Mr Sutherland's contribution was interesting since it brought the American point of view forward in the Conference to some extent. With regard to the question of nomenclature, he understood that there was an increasing tendency in the United States of America to adopt the terms " drained " and " undrained " instead of " slow " and " quick." The shear-box test was used in Britain especially for drained tests since it was rather more convenient for those than was the triaxial test. A comparison of the results of drained and undrained tests on London Clay carried out in the triaxial apparatus and the shear box by Mr Henkel and Mr Gibson at the Imperial College had shown that for practical purposes the difference was small, although for research purposes the triaxial test was to be preferred.

He was of the opinion that Professor Casagrande's finding, quoted by Mr Sutherland, that the strength in one extremely slow undrained test was only 38 per cent. of the strength in a 1-minute test, was important. He doubted, however, if it was at all typical for most clays ; otherwise earthworks which had been designed upon the results of normal-rate undrained laboratory tests, with a factor of safety certainly less than 3, would have failed. Such failures, had not, however, taken place.

In reply to Mr Sutherland's question concerning the time necessary to carry out a drained constant-rate-of-strain test, he would say that a preliminary calculation based upon a value of the coefficient of consolidation, and using Terzaghi's theory, would be sufficient to determine the order of time necessary to give, say, 95 per cent. dissipation of pore-water pressure. A few tests at slower rates could be carried out in addition to confirm that. Mr Henkel had, in fact, calculated that, for London Clay, a drained triaxial test using side filter-drains and a specimen 3 inches long and 1½ inch in diameter could be carried out in 2–3 days, giving 95 per cent. dissipation of pore-water pressure. That estimate had been subsequently confirmed experimentally.

He had been interested in Mr Kjellman's contribution to the discussion, and agreed that both vane-tests and undisturbed samples should be taken, but he felt, at least for clays encountered in Britain, that the emphasis should be placed first on obtaining undisturbed samples ; if that could not be done satisfactorily, then vane-tests should be carried out. In the laboratory, the cone-tests had been compared both with compression tests and with vane-tests. A Paper was being prepared on those experiments.

Dr Skempton said that he was in agreement with many of M. Florentin's comments on the torsion test but felt that that test was not as useful in practice as the shear-box test or the triaxial test. On Mr Jakobson's point, he had to say that if the true cohesion, true angle of internal friction, and pore-water pressures were known, then the answer to the problems had been obtained. The λ-theory and other similar theories permitted, in principle, the calculation of the pore-water pressure and, hence, of the effective stresses in samples; but it was difficult either to calculate or to measure the pore-water pressure in the field. Thus he believed that the more practical procedure was that outlined in the Paper. The results of the cone tests carried out by Mr Evans were in accord with the results of similar tests that had been done at the Imperial College. He would agree that it was important to make a correction for the depression of the surface of the sample round the cone, but he had found that, even after such a correction had been made, the variation in the factor k with different clays was still very large (4·5–7), and he had therefore been compelled to reject the cone test for research purposes.

He thanked Mr Geuze for his remarks, with which he was in agreement. The statement in the Paper on the small effect of the rate of strain on the strength of sands was based on the result given by Casagrande and Shannon in their monograph.[1]

Mr A. W. Bishop, in reply, said that he had found, from direct shear-box tests carried out on sands at the Imperial College, that, unless the thickness of the sample was at least one-quarter of the length of the sample, no unique measure of the strength could be obtained.

[1] CASAGRANDE, A., and SHANNON, W. L., 1948. Research on stress-deformation and strength characteristics of soils and soft rocks under transient loading. *Pub. Harvard Univ. Grad. Sch. Eng. No.* 447, 132 pp.

It appeared that, in many of the shear-box tests in America that condition was not satisfied and he was of the opinion that this had led to wrong conclusions being drawn.

Research work had been conducted at the Imperial College with the object of distinguishing the various factors contributing to the shear strength of sands. An important step had been to separate the work done in overcoming internal friction and that required to cause the sample to dilate against the applied pressure. Fig. 12 showed the results of a direct shear test carried out on a dense sand.

If σ_n denoted the normal stress, τ the shear stress, and τ_d the shear stress necessary to cause the sample to dilate against the normal stress, then for an incremental displacement $\delta\Delta$ the work done against the normal stress was :

$$\sigma_n \delta V = \tau_d \delta\Delta$$

$$\therefore \quad \frac{\tau_d}{\sigma_n} = \frac{\delta V}{\delta\Delta}$$

$$\frac{\text{Frictional strength}}{\text{Normal stress}} = \frac{(\tau - \tau_d)_{\max.}}{\sigma_n} = \left(\frac{\tau}{\sigma_n} - \frac{\delta V}{\delta\Delta}\right)_{\max.}$$

The maximum value of $\frac{\tau}{\sigma_n} - \frac{\delta V}{\delta\Delta}$ was called the tangent of the "angle of internal friction."

Fig. 12

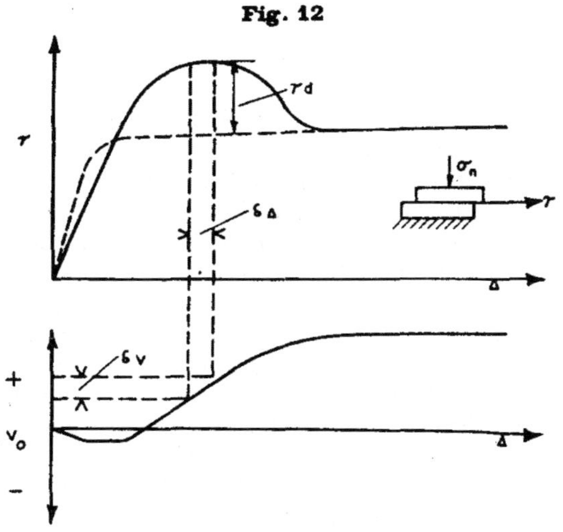

σ_n denotes normal stress.
τ ,, shear strength.
τ_d ,, shear stress necessary to cause sample to dilate against the normal stress.

For displacement $\delta\Delta$, work done against normal stress is :

$$\sigma_n \cdot \delta V = \tau_d \cdot \delta\Delta$$

Hence $$\frac{\tau_d}{\sigma_n} = \frac{\delta V}{\delta\Delta}$$

$$\frac{\text{Frictional strength}}{\text{Normal stress}} = \frac{\tau - \tau_d}{\sigma_n}$$

$$= \frac{\tau}{\sigma_n} - \frac{\delta V}{\delta\Delta} = \tan\phi_f$$

Shear-box test on sand. Frictional and dilatancy components of shear strength

Figs 13

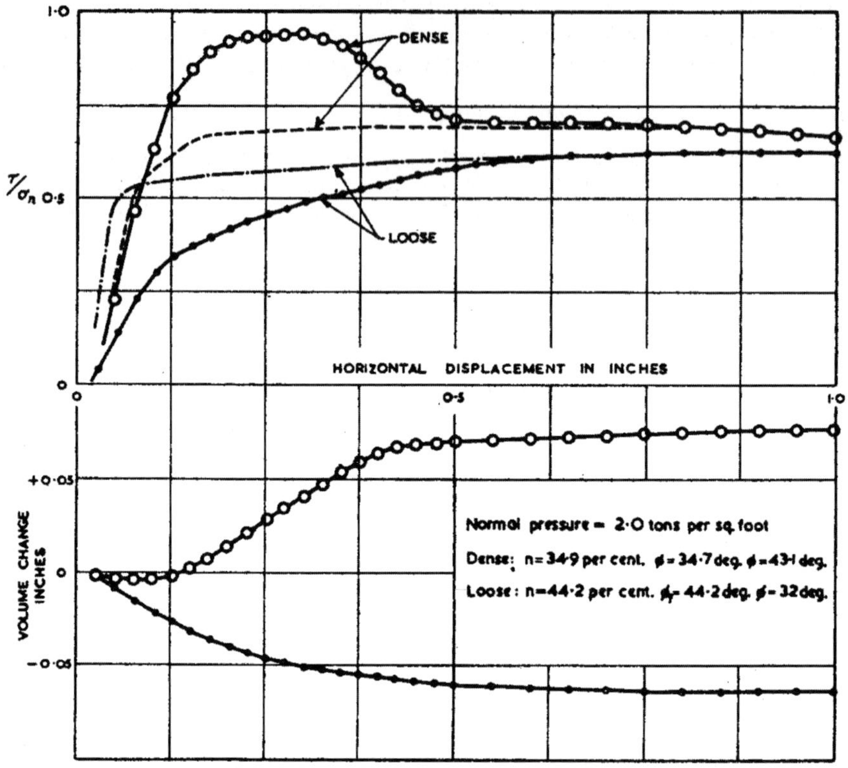

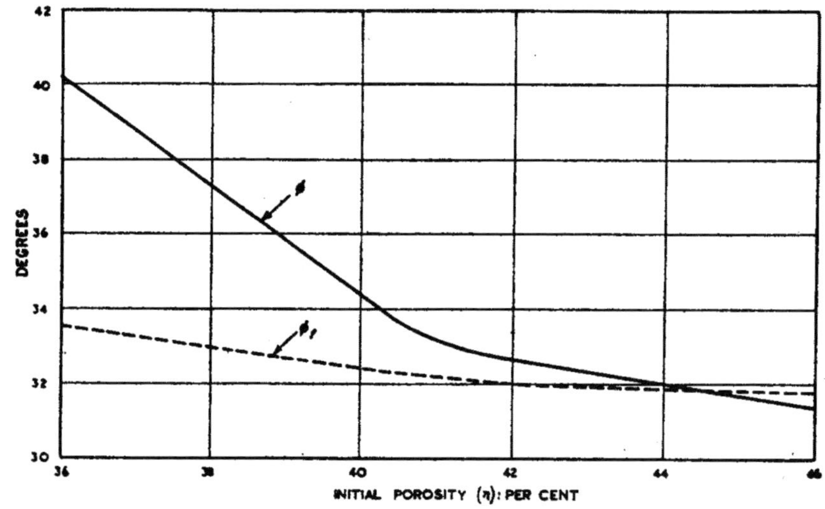

Shear tests on Stone Court sand

The graph of τ and of volume change against horizontal displacement was shown in Fig. 12, from which the dotted curve of $(\tau - \tau d)$ against Δ had been calculated.

In Fig. 13 the results of shear tests on a sand in a dense state of packing were compared with those for a loose state of packing. The lower diagram in Fig. 13 showed the values of ϕ $\left(\text{where } \tan \phi = \frac{\tau_{max.}}{\tau_n}\right)$ and of ϕ_f $\left(\text{where } \tan \phi_f = \frac{(\tau - \tau_d)_{max.}}{\sigma_n}\right)$ plotted against the initial porosity of the sample. Those results indicated that the increase in shear strength (as measured by ϕ) with density was due primarily to the increased tendency of the sample to dilate, and that the work done in overcoming frictional forces (measured by ϕ_f) was almost unchanged.

Skempton, A. W. & Northey, R. D. (1952). *Géotechnique* **3**, No. 1, 30–53

THE SENSITIVITY OF CLAYS

by

A. W. SKEMPTON, D.Sc., and R. D. NORTHEY, Ph.D.

SYNOPSIS

The sensitivity of clays is defined as the ratio of their undisturbed and remoulded strengths, and varies from about 1·0 for heavily over-consolidated clays to values of over 100 for the so-called extra-sensitive or "quick" clays. As a result of experimental work it is shown that thixotropy can account for low or medium sensitivity but not for high sensitivity. Both laboratory and field evidence, however, show that a reduction in the salt concentration in the pore water, caused by leaching clays of marine or estuarine origin, can result in high sensitivities. Moreover, of the small number of clays so far investigated, those which have not been leached are found to exhibit only low or medium sensitivities of the order of magnitude explicable in terms of thixotropy. But it is, nevertheless, not suggested that leaching is the sole cause of high sensitivity and other possibilities are briefly discussed. Reasons are given, supported by experimental work, for the insensitivity of heavily over-consolidated clays with water contents approximating to the plastic limit.

A result of particular interest from the scientific point of view is that the water content and undisturbed strength of a clay remain unaltered by leaching, although the liquid limit and remoulded strength are appreciably reduced. It is suggested that this implies effectively "solid" contact between the particles of an undisturbed clay. Finally some evidence is given which indicates that the fissured structure present in many clays may be due to syneresis.

La sensitivité des argiles est définie comme étant le rapport entre leur résistance à l'état brut et leur résistance après remoulage, et varie de 1,0, environ, pour les argiles trop fortement tassées, jusqu'à des valeurs de plus de 100 pour des argiles dites extra-sensitives ou "sables mouvants". A la suite de travaux d'expérience, on montre que la thixotropie peut entre en ligne de compte pour les basses ou moyennes de sensitivités, mais non pas pour les fortes sensitivités. Cependant, les travaux de laboratoire et sur les lieux montrent qu'une réduction de la concentration en sel dans l'eau de porosité, causée par le lavage des argiles d'origine maritime ou alluvionaire, peut provoquer de fortes sensitivités. Qui plus est, du petit nombre d'argiles qui ont fait l'objet de recherches jusqu'à maintenant, celles qui n'ont pas été lavées n'ont présenté seulement que des sensitivités faibles ou moyennes de l'ordre de grandeur explicable en fonction de la thixotropie. Cependant, il n'est pas suggéré que le lavage soit la seule cause de haute sensitivité et d'autres possibilités font l'objet d'une discution rapide. On donne des raisons, s'appuyant sur des travaux expérimentaux, pour le manque de sensitivité des argiles trop fortement tassées dont la teneur en eau approche la limite plastique.

Un résultat d'un intérêt tout particulier du point de vue scientifique est que la teneur en eau et la résistance à l'état brut restent inchangées par le lavage, alors que la limite liquide et la résistance après remoulage soient sensiblement réduites. Il est suggéré que ceci signifie qu'il y ait un contact "solide" effectif entre les particules d'une argile à l'état brut et s'oppose à la théorie communément acceptée que les particules sont séparées par des couches relativement épaisses d'eau adsorbée. Finalement il est donné des preuves qui indiquent que la structure fissurée qui se rencontre dans beaucoup d'argiles puisse être due à une synérèse.

INTRODUCTION

Most clays, except those which have been heavily over-consolidated, lose a proportion of their strength when remoulded. In Fig. 1 are shown the stress/strain curves for a sample of post-glacial clay from the Thames estuary in its undisturbed and remoulded state; the water content being identical in both tests. In this sample the undisturbed shear strength is 5·3 lb. per square inch and the remoulded shear strength is 0·7 lb. per square inch. Terzaghi (1944) has defined the sensitivity of clays as the ratio

$$S_t = \frac{\text{undisturbed strength}}{\text{remoulded strength}} = \frac{c}{c_r}$$

and, consequently, for the clay in Fig. 1 the sensitivity is 7·6. The remoulding loss is $c - c_r$ and the percentage remoulding loss is defined by the expression $\left(\frac{c - c_r}{c}\right)100$.

From the point of view of their sensitivity to remoulding, clays may conveniently be classified as in Table 1 :—

Table 1

$S_t = c\,.\,1{\cdot}0$	:	insensitive clays
$S_t = 1$ to 2	:	clays of low sensitivity
$S_t = 2$ to 4	:	clays of medium sensitivity
$S_t = 4$ to 8	:	sensitive clays
$S_t = >8$	:	extra-sensitive clays
$S_t = >16$	:	quick-clays

Fig. 1

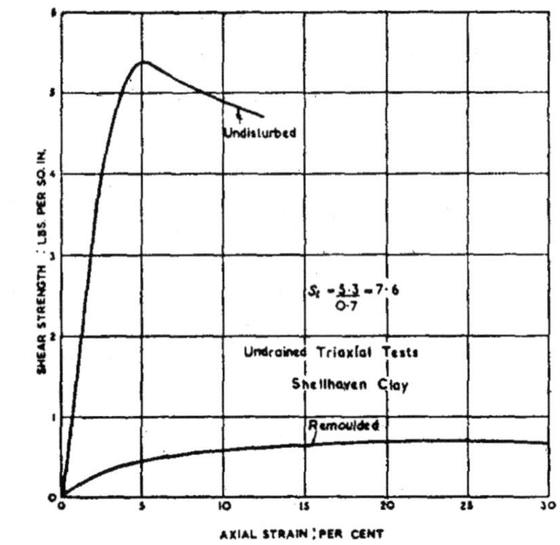

Stress/strain curves for typical sensitive clay

Clays which have been heavily over-consolidated during their geological history (e.g. London Clay) are insensitive. So also are most boulder clays. There seem to be few examples of clays of low sensitivity, but sensitivities of 2 to 4 are very common among normally consolidated clays, and sensitivities of 4 to 8 are quite frequently encountered. The practical significance of a decrease in strength such as that illustrated in Fig. 1 is obviously considerable. Sampling has to be very carefully carried out in such clays if truly undisturbed specimens are required, while the effects of disturbance, caused for example by pile-driving, may be serious. Even the extra-sensitive clays are not rare and, indeed, in parts of Scandinavia and Canada they are met almost as commonly as sensitive clays. So far, however, very few examples have been reported from other countries. A special category of extra-sensitive clays, and found typically in the same localities, is the group of the so-called ' quick-clays ' which become so extremely soft on remoulding as to appear practically fluid. The most extreme case which has been investigated relates to a clay from St. Thuribe, near Quebec (Peck et. al. 1951), with a sensitivity of about 150. This material was responsible for the extraordinary flow slide of 1898 when some 3,500,000 cubic yards of earth ran out of the land adjacent to the River Blanche in 3 or 4 hours, and dammed the valley to a depth of fully 25 feet over a length of 2 miles (Sharpe 1938).

It is not difficult to see that sensitivity presents several interesting scientific problems. How is it that a clay can be quite firm in its undisturbed state and yet, without any change in water content or porosity, loose 90 per cent. or even 99 per cent. of its strength after remoulding ? Why are some clays entirely unaffected by remoulding ? Are the factors responsible for medium sensitivity the same as those controlling high sensitivity or are some other factors involved in extra-sensitive clays ? Is the mineralogy of the clays one of these factors ? Can the phenomenon be explained in terms of thixotropy ? Have the conditions of deposition and subsequent geological history an important influence ? In this article an attempt is made to answer some of these questions. But it must be made clear, at the outset, that no finality is claimed. Insufficient field data and laboratory results have so far been obtained to enable more than a few tentative conclusions to be drawn. Yet a stage has perhaps been reached where the problem of sensitivity can be usefully surveyed and some working hypotheses suggested for future research.

POSSIBLE FACTORS

From general considerations and from the available experimental results, it may be inferred that at least four factors should be considered :—

(*a*) Mineralogy : with special reference to the mineralogy of the " clay fraction."

(*b*) Particle size distribution.

(*c*) Thixotropy : this being the term used to define a " process of softening, caused by remoulding, followed by a time-dependent return to the original, harder state " (Burgers and Scott Blair 1948). It is essentially a reversible effect, and consequently any permanent changes, such as alterations in chemical composition or in electrolyte concentration in the pore water of a clay, are excluded. Among others,* Moretto (1948) and the Authors† have carried out quantitative investigations on thixotropy in clays. Several papers have been published in the geological literature of Germany and Britain, but they are not quantitative in the way required in the present study.

(*d*) Micro-structural stability : a term used to imply the existence in an undisturbed clay of a more or less well-developed structural arrangement of the particles (micro-structure) which is broken down by remoulding ; a process which is largely irreversible. Qualitative hypotheses for the development of such meta-stable structures have been suggested by several writers, notably Casagrande (1932) and Terzaghi (1941). In 1946, however, Rosenquist described some experiments showing that the effect of a partial leaching out of the salt from the pore water of clays was greatly to increase their sensitivity. He appears to have considered this as being unrelated to structural effects, but a critical experiment by the Authors has shown that the action of leaching is, essentially, that of developing a meta-stable structure. Consequently although the leaching of clays (i.e. decreasing the salt concentration in the pore water) might be considered as a separate (and fifth) factor, it is more logically taken as one of the causes of the development of a meta-stable structure. There may be other causes leading to an organized structure in natural clays (subject to irreversible breakdown on remoulding), such as those suggested by Casagrande and Terzaghi, but they have not yet been investigated in any detail.

Table 2

Clay	Mineralogy of clay fraction (in order of predominance)	Clay fraction (<2 micron)	Activity $\frac{PI}{\text{clay frac.}}$	Undisturbed shear strength lb./sq. ft.	Sensitivity S_t
(1) St. Thuribe	Quartz, Mica, trace of Montmorillonite	36 per cent	0·33	800	*c.* 150
(2) Detroit I	Mica, Illite, Calcite, Quartz, Montmorillonite	36 ,,	0·36	360	2·5
(3) Horten	Mica, Quartz, Illite, Montmorillonite	40 ,,	0·42	850	17
(4) Beauharnois	Mica, Illite, Calcite, Quartz, Montmorillonite	79 ,,	0·52	380	14
(5) Gosport	Illite, Halloysite	*c.*55 ,,	0·88	600	2·2
(6) Shellhaven	Illite, Kaolinite, some organic colloids	41 ,,	1·33	760	7·6
(7) Mexico City	Montmorillonite (mainly)	*c.*90 ,,	*c.*4·5	960	5·3

References : (1) Peck and others (1951). (2) Moretto (1948). (3) Hansen (1950) and Northey (1950). (4) Moretto (1948). (5) Skempton (1948 a). (6) Skempton (1951 a). (7) Rutledge (1944).

* Quantitative data on thixotropic regain in clays were given as early as 1937 by Hvorslev, but he used the cone test now known to be inaccurate (Skempton and Bishop 1950).

† All the results of the thixotropy experiments and of the leaching tests are given in detail by Northey (1950).

MINERALOGY AND PARTICLE SIZE

There is at the present time sufficient data to show that neither mineralogy or particle size distribution have any major significance in the problem of sensitivity, beyond the fact that the phenomenon is restricted to clays and silts. Thus Rosenquist (1946) reports that no essential mineralogical differences could be found between several Norwegian "quick-clays" and other clays of medium sensitivity; nor was there any essential difference in the particle size distribution. The information collected together in Table 2 strongly supports these conclusions. The seven clays in this Table have the same order of undisturbed strength, but very different sensitivities. Yet there seems to be no correlation between sensitivity and mineralogy except that the three extra-sensitive clays contain minerals other than the true clay minerals in their clay fraction. But this is, in itself, no indication of sensitivity; as a comparison of the Beauharnois and Detroit clays will show. In these two clays the mineral assemblage is very similar, the activities* are of the same order, and the undisturbed strengths are almost equal; yet the sensitivities are 14 and 2·5 respectively.

THIXOTROPY

The properties of an ideal purely thixotropic plastic material are illustrated in Figs 2 (a) and 3 (a). In its undisturbed state, when tested at a given rate of shear, a shear strength c will be measured. When tested at the same rate of shear immediately after having been remoulded, the shear strength is c_r. If the material is then allowed to remain under constant external conditions and without any change in porosity, the strength after a time interval t, as measured at the same rate of shear, will be c_t where $c_t > c_r$. Remoulding the material will reduce the strength again to the value c_r. By repeating the experiment with different intervals between remoulding and testing a hardening curve such as that shown in Fig. 3 (a) is obtained and, for a purely thixotropic material, the original strength c will ultimately be regained.

The natural sensitivity of the material is $S_t = \frac{c}{c_r}$, the acquired sensitivity after a hardening period t is $AS_t = \frac{c_t}{c_r}$, the thixotropic regain is $c_t - c_r$ and the percentage thixotropic regain, after this same period t, is $\left(\frac{c_t - c_r}{c_r}\right) \times 100$. The actual values of c, c_r and c_t will depend upon the rate of shear used in the test, owing to the presence of plastic viscosity; see Fig. 2 (a). But the ratios of c/c_r and c_t/c_r will not vary to any important extent with the rates of shear normally used.

In a thixotropic fluid material, the resistance to shear is, in the simplest (Newtonian) case, directly proportional to the rate of shear: see Fig. 2 (c).

* Activity is defined as the ratio of plasticity index to clay fraction content, and is related to mineralogy (Skempton 1950).

Figs 2

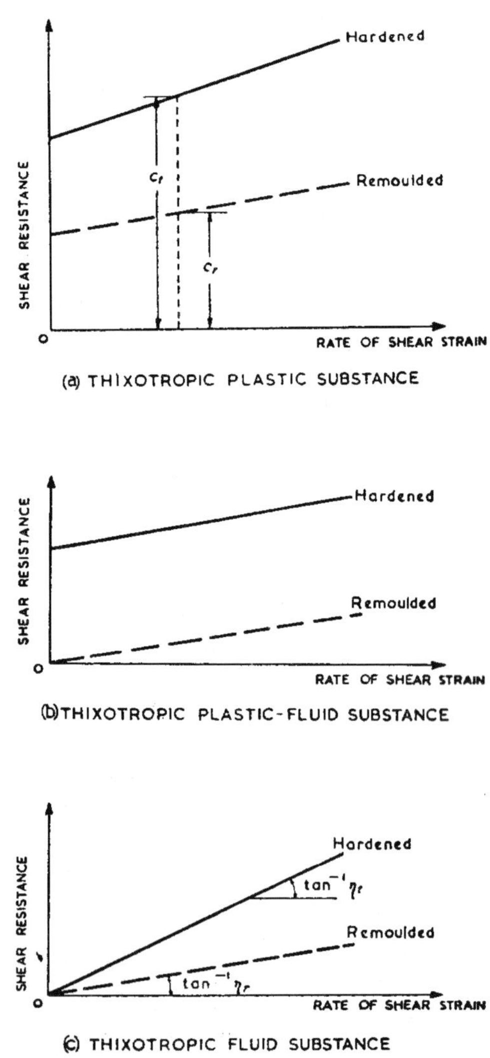

Thixotropic systems

Figs. 3

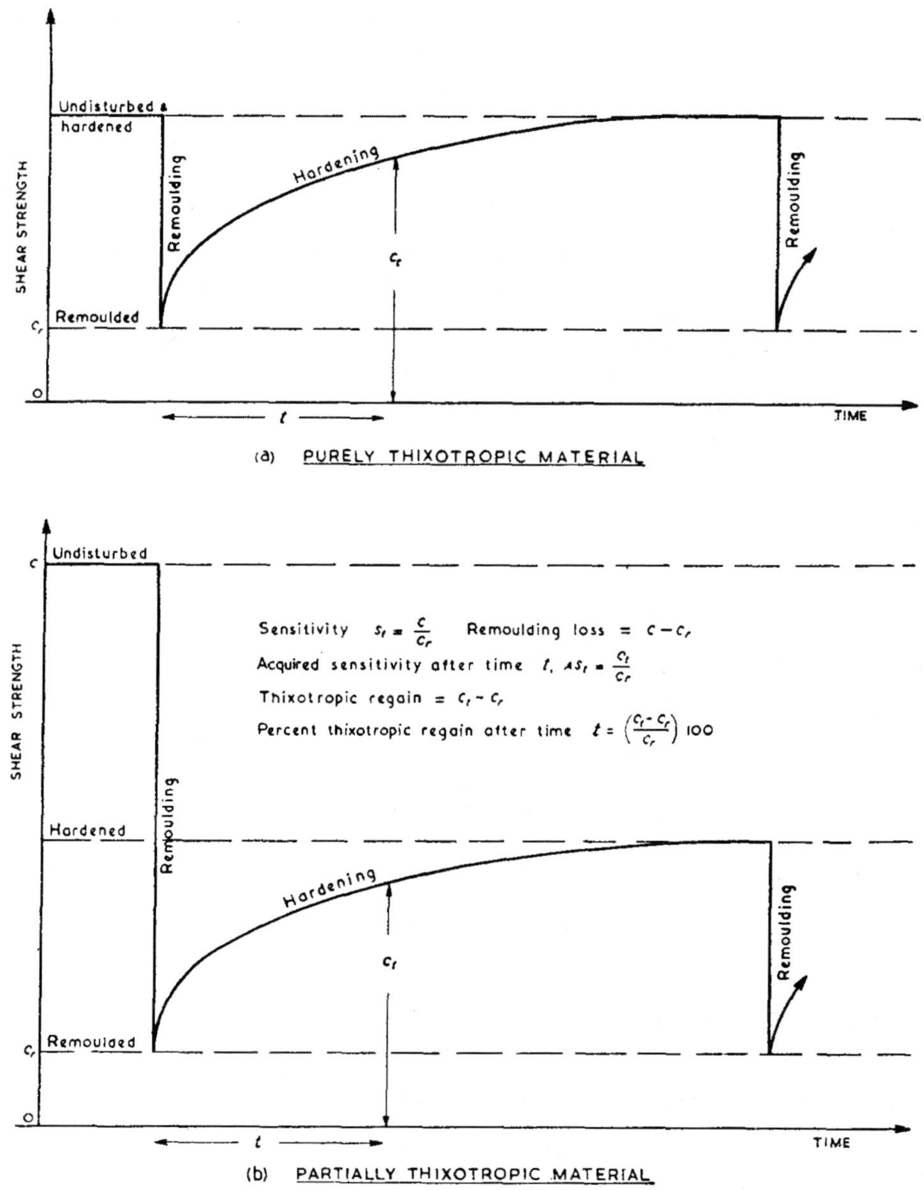

Full and partial thixotropic regain

The slope of these lines is, of course, the coefficient of viscosity and, in close analogy with plastic materials, the natural sensitivity of a fluid can be defined as the ratio η/η_r; and the acquired sensitivity after a hardening period t would be given by the ratio η_t/η_r. A more complex special case* is that of a material which is plastic in the undisturbed or hardened condition, and fluid when remoulded see Fig. 2 (b).

* Some Authors, notably Boswell (1949), would restrict the term thixotropy to this case, which conforms with the original definition (see Freundlich, 1935). If the more general definition used in this article is unacceptable a less specific term such as " age-hardening " may be substituted for " thixotropy."

With clays, however, at the range of water contents usually encountered, plastic behaviour is exhibited even in the remoulded state. This is illustrated in Fig. 4 where the shear-strength/rate-of-shear curves are plotted for freshly remoulded London clay at water contents equal to and well above the Atterberg "liquid limit." In the latter state the clay appears visually to behave almost as a fluid but, as will be seen from Fig. 4, it is actually a very soft plastic material, with only a small viscous component. In this Figure, and throughout the remainder of the article, water content is expressed as the weight of water per 100 grammes of dry clay. But it is essential to note that the water content is of little significance until related to the Atterberg Limits. This relation is established by the liquidity index: defined (Terzaghi 1936) by the expression

$$LI = \frac{w-PL}{LL-PL}$$

Fig. 4

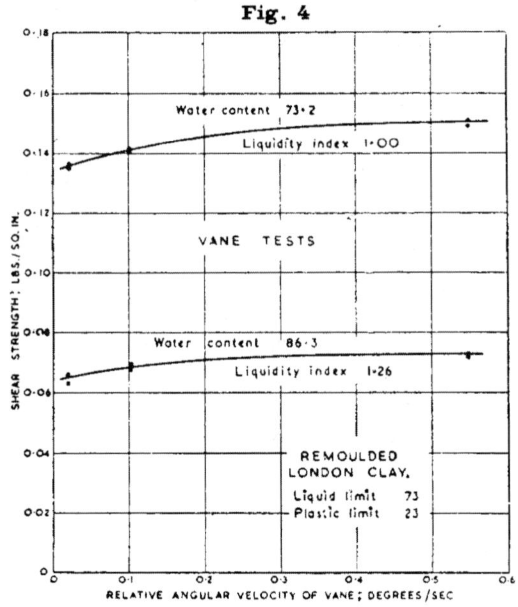

Strength variation with rate of testing

where w denotes the water content, and LL and PL denote the liquid and plastic limits of Atterberg (see Casagrande 1932). When the water content equals the liquid limit, $LI=1{\cdot}0$, and when the water content equals the plastic limit, $LI=0$.

Fig. 4 is typical of many clays both in their undisturbed, remoulded, and thixotropically hardened states; and there is no doubt that clays behave as plastic materials over the full range of liquidity index encountered in practice. But, as will be seen later, most natural clays are not purely thixotropic materials. That is to say the natural undisturbed strength is not fully regained by thixotropic hardening, and in some of the clays tested the evidence is that only a small fraction of the undisturbed strength is regained, as illustrated in Fig. 3 (b). Thus the conclusion is reached that whereas clays exhibit thixotropy this is, in general, not sufficient to account wholly for the phenomenon of sensitivity.

The results on which this statement is based have been obtained by Moretto (1948) and the Authors, and are summarized in Figs. 5 to 9. In these experiments a clay was thoroughly mixed with water, usually at the liquid limit, and then placed in a number of containers so arranged that the samples could be stored for long intervals without change in volume or water content. In the Authors' tests the containers consisted of Perspex cylinders about 3 inches in diameter and 2½ inches long. These were completely filled with clay and stored under water with filter papers and a glass plate, somewhat larger than the cylinder, at each end. On the upper plate a glass jar was placed containing lead shot. In this way drying out of the clay was impossible, and swelling was prevented. At various times a sample was removed from store and tested by means of the laboratory vane apparatus (Skempton and Bishop 1950). The clay was then remoulded and again tested. This remoulded strength should be equal to that of the remoulded clay at the time of preparing the specimens. Any change in remoulded strength would indicate some permanent alteration in the clay. In some cases such alterations were found to occur after 2 or 3 months, owing to the outward diffusion of dissolved salts, and these tests had therefore to be rejected. Moretto stored his samples in paraffin wax; a technique which eliminated the difficulty with diffusion but was not always successful in preserving a truly constant water content. He used the unconfined compression test for measuring strength.

Fig. 5

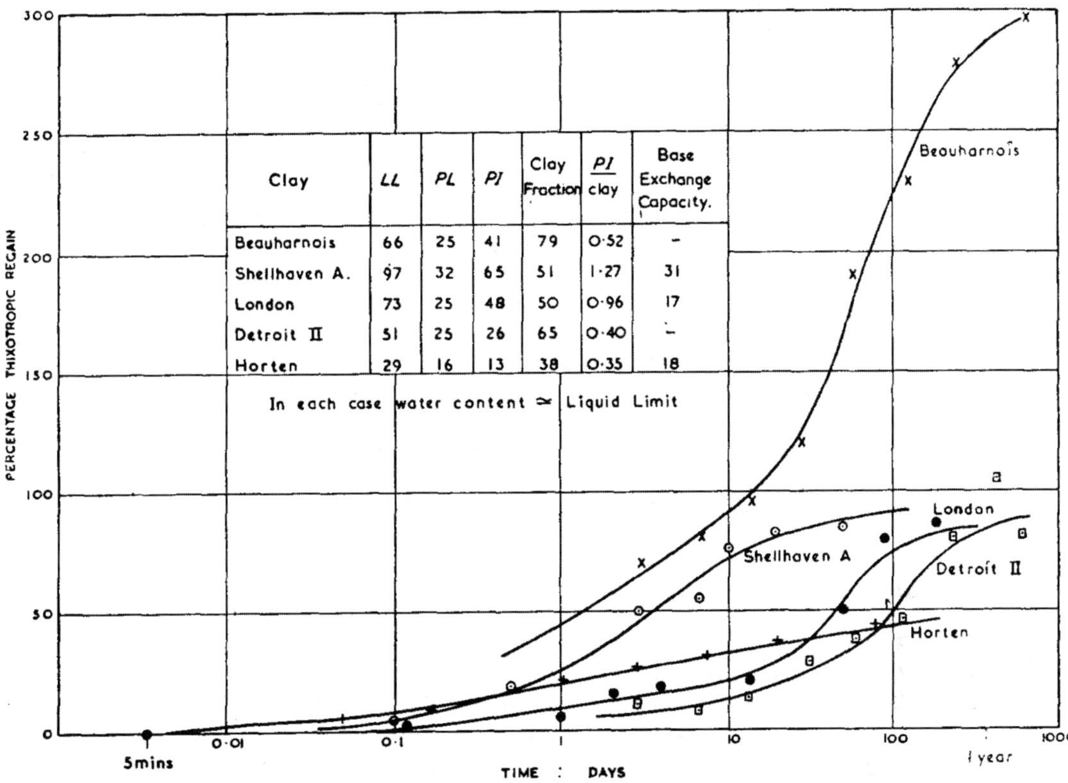

Clay	LL	PL	PI	Clay Fraction	PI/clay	Base Exchange Capacity.
Beauharnois	66	25	41	79	0·52	-
Shellhaven A.	97	32	65	51	1·27	31
London	73	25	48	50	0·96	17
Detroit II	51	25	26	65	0·40	-
Horten	29	16	13	38	0·35	18

Thixotropic regain in some typical clays

Fig. 6

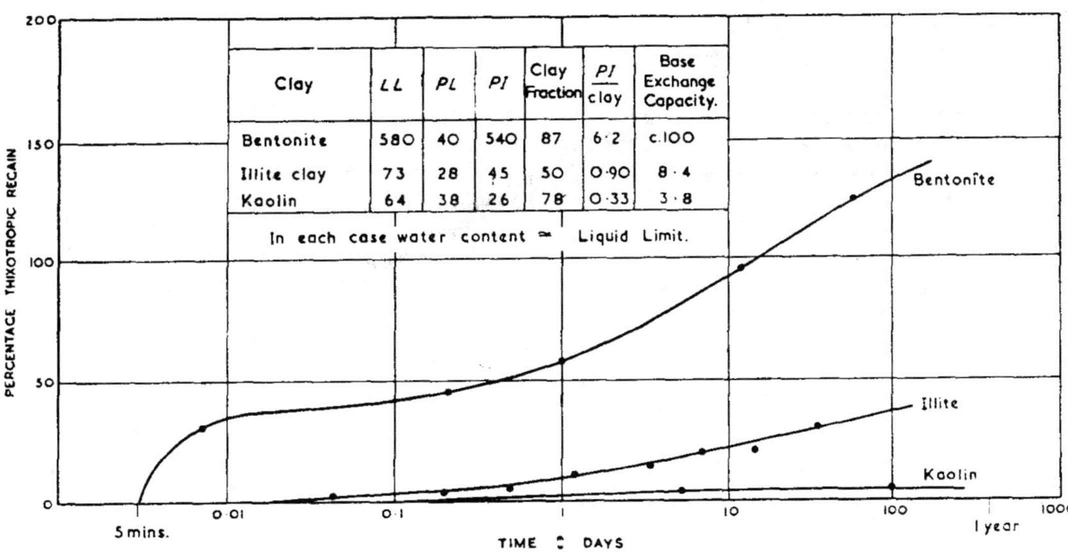

Clay	LL	PL	PI	Clay Fraction	PI/clay	Base Exchange Capacity.
Bentonite	580	40	540	87	6·2	c.100
Illite clay	73	28	45	50	0·90	8·4
Kaolin	64	38	26	78	0·33	3·8

Thixotropic regain in three clay minerals

Fig. 7

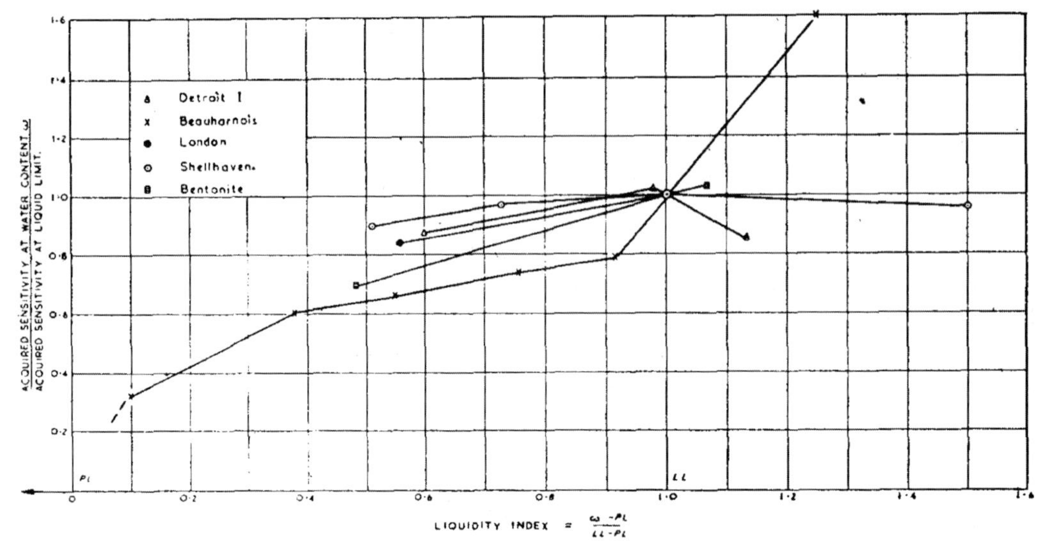

Effect of water content on thixotropic regain

The percentage thixotropic regains for five clays are plotted in Fig. 5, with a logarithmic time base. Very little regain occurred in the first few hours but after 1 day the clays had increased in strength by the order of 10 to 40 per cent. There is an indication that hardening was practically completed after about 1 year, and it appears that the London,* Shellhaven,* and Detroit† clays would never show a regain of more than 200 per cent (or an acquired sensitivity of more than 3). Similarly the upper limits for the Horten* and Beauharnois† clays would appear to be about 100 per cent and 400 per cent respectively.

In Fig. 6 results are given, from the Authors' tests, for kaolin, illite clay, and bentonite. Kaolin shows almost no thixotropy and illite shows only a small effect. In contrast, the bentonite shows a remarkable regain at very short time intervals, and it is not possible to suggest an upper limit for this material since the strength continued to increase throughout the experiment.

The data in Figs. 5 and 6 relate to clays with water contents equal to their liquid limit. The influence of water content on thixotropic regain is shown in Fig. 7. The acquired sensitivity at 100 days has been used for the comparisons. In all five materials the thixotropic regain decreases with decreasing water content below the liquid limit, and there is a suggestion that thixotropy may be zero at water contents at or close to the plastic limit. This is in agreement with the properties of the boulder clays with which the Authors are acquainted. These materials were formed under conditions corresponding to complete or almost complete remoulding, and have been left in position for many thousands of years. Yet they are insensitive—even those with a moderate or high clay content and activity. The explanation is presumably to be sought in the fact that these boulder clays typically have a liquidity index of about zero; and their thixotropy has therefore (by inference from Fig. 7) been inhibited.

At water contents greater than the liquid limit the evidence is conflicting. At such water contents the Detroit and Shellhaven clays are slightly less thixotropic, but the Beauharnois

* Tested by the Authors.

† Tested by Moretto. The percentage thixotropic regains plotted in Fig. 5. differ slightly from those given in Moretto's paper since there the remoulded strength is taken as that at 10 per cent. axial strain (in the unconfined compression test) whereas the Authors take the maximum stress.

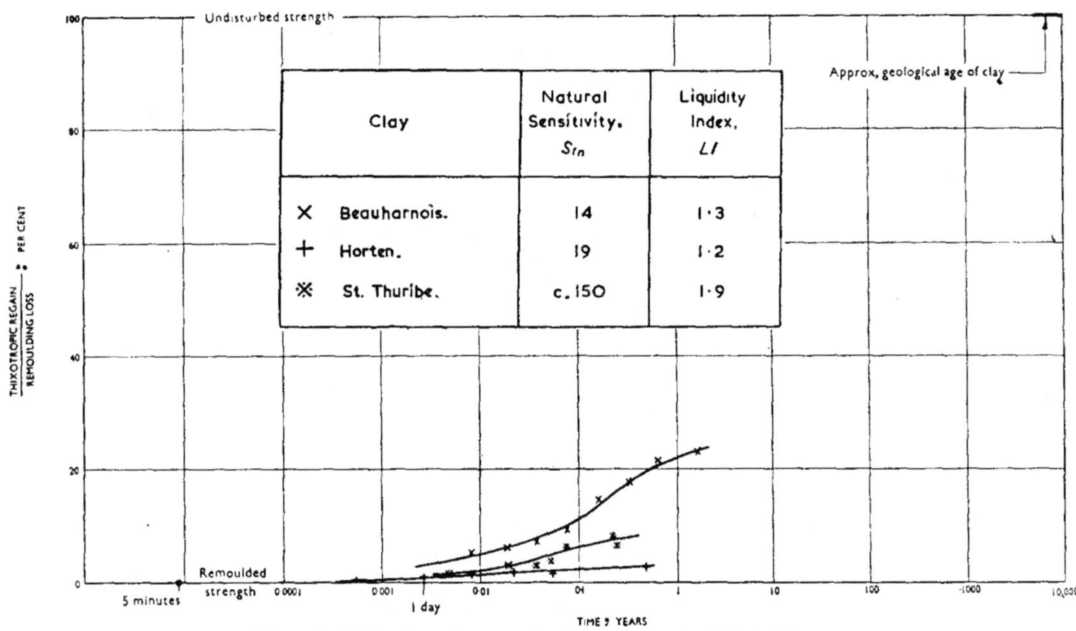

Fig. 8 Thixotropic regain in extra sensitive clays

clay shows an increased thixotropy. It might be expected that thixotropy* would be more marked the higher the liquidity index, in accordance with the Beauharnois result, yet there is no reason to doubt the validity of the Shellhaven and Detroit tests, and the elucidation of this point must await further research.

Having summarized the laboratory tests on thixotropy in clays it is now necessary to compare the results with the natural sensitivities of the various materials. This is possible with six clays ; the laboratory results for four of these are given in Fig. 5, and for the other two references can be made to the papers by Moretto (1948) and Peck *et al* (1951). Three of the clays are extra-sensitive, while the other three have sensitivities of less than 8.

In Figs. 8 and 9 the results have been plotted on a logarithmic time scale and with an ordinate scale extending from the remoulded strength to the undisturbed strength ; this range being taken as 100 per cent. Thus if the natural sensitivity of a clay is 14, as in the Beauharnois clay for example, and if the acquired sensitivity by hardening over a period of 1 year is 3·9, then the ratio of thixotropic regain to remoulding loss at this time is $(3{\cdot}9-1)/(14-1) = 0{\cdot}22$ or 22 per cent. Now so far as the Authors are aware all those six clays are of late glacial or early post-glacial age and were deposited about 5,000 to 10,000 years ago (Zeuner 1946). This order of time has therefore been available for hardening in nature and obviously, in estimating whether the full undisturbed strength could be regained thixotropically, a very great extrapolation is required. But owing to the shape of the regain curves the task is not so difficult as to prevent any such estimate from being made. Thus it appears to be impossible that the Horten clay could ever regain its natural undisturbed strength, but, in contrast, it is possible that the Detroit I clay could do so.

An examination of Figs. 8 and 9 leads to the conclusions that whereas in clays of medium sensitivity it is possible that sensitivity may be due to thixotropy, it is very improbable that this could be the case in sensitive or extra-sensitive clays. Although quoting no quantitative results Rosenquist (1946) draws a similar deduction from his studies on Norwegian clays : namely that the quick-clays " usually will recover very little of their original firmness."

* A discussion of the physics of thixotropy lies outside the scope of the present article. But in a very general way it may be taken that thixotropy results from the gradual rearrangement of the particles, under the action of bonding forces, into positions of increasing mechanical stability. With the greater freedom for movement at higher water contents it might consequently be expected that thixotropy would be greater.

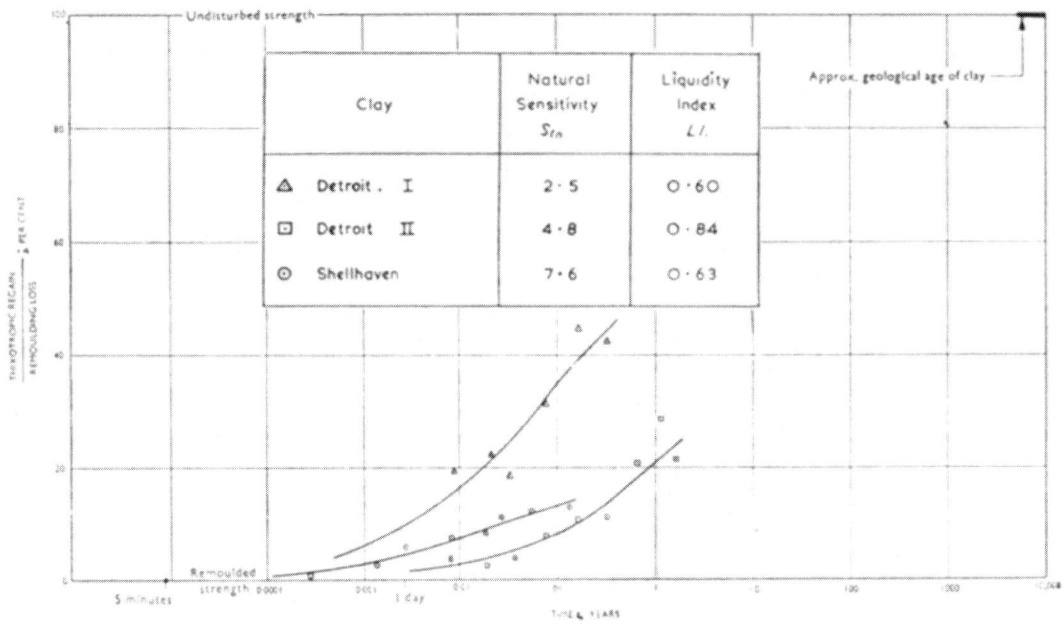

Fig. 9 Thixotropic regain in clays of moderate sensitivity

It is, of course, admitted that these conclusions cannot be proved; but with regard to the extra-sensitive clays it would be far more difficult to sustain the opposite view, namely that sensitivity can be accounted for solely by thixotropy. Some other cause of high sensitivity must therefore be sought, and this problem will be considered later in the article.

FISSURED MACRO-STRUCTURE

Before leaving the thixotropic tests brief mention should be made of the observed occurrence of fissures in the samples of London and Shellhaven clays after storage for periods of more than 3 months. The appearance of these fissures in one of the London clay samples is shown in Fig. 10. The clay had been remoulded at its liquid limit and stored under water as previously described; in this particular case for 1 year. The " chunky " nature of the clay is readily seen and the contrast with the same sample after remoulding was very marked. The fissures are similar to those found in nature in the Shellhaven clay, and although on a smaller scale, to those in many stiff fissured clays of which London clay is a well-known example.

Fig. 10

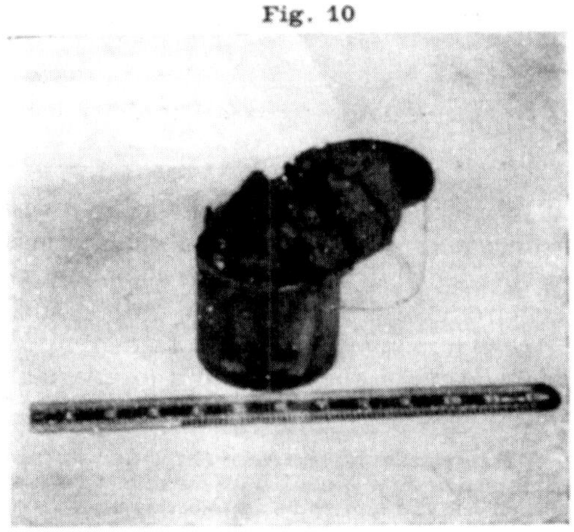

Now it was found that the water content of the clay in the immediate vicinity of the fissures was slightly lower than that in the bulk of the clay. It is therefore suggested that the fissures developed as a result of syneresis, a colloidal process whereby the particles draw themselves together under the action of attractive forces and expel some of the pore water. This phenomenon may take place in natural clay deposits and, if so, it provides an explanation of the fissured macro-structure alternative to that put forward by L. Casagrande (1948), in which the influence of electric streaming potentials was invoked.

Fig. 11

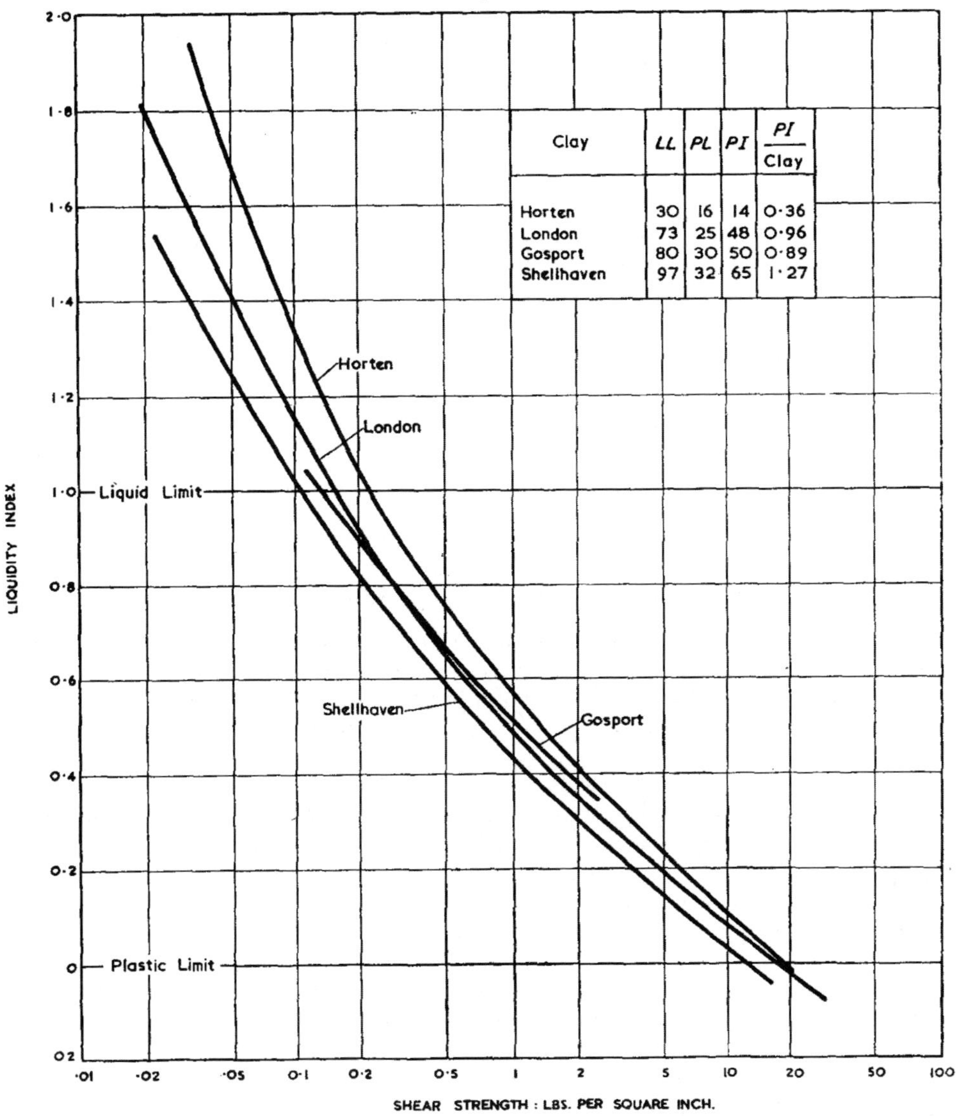

Relation between liquidity index and shear strength of moulded clays

MICRO-STRUCTURAL STABILITY

In order to study the influence on sensitivity of the structural stability of the particle arrangement, consider two normally consolidated clays existing under the same effective overburden pressure; one with a medium sensitivity and the other with a high sensitivity. The clays from Gosport (Skempton **1948** a) and Horten (Hansen **1950**) may be taken as examples.

Now it is known* that the ratio of undisturbed strength c to effective overburden pressure p depends upon the liquid limit (or plasticity index) of the clay, and that in general the higher the liquid limit the greater the ratio $(c/p)_n$. For Gosport clay $(c/p)_n = 0{\cdot}30$ and for Horten

* The relationship between $(c/p)_n$ and liquid limit was first published by Skempton (1948 b). Results for only five clays were then available. Since that time many more results have been obtained, but the relation has not had to be appreciably altered. A paper on this subject is in preparation.

clay $(c/p)_n = 0{\cdot}17$. Thus if the overburden pressure is 15 lb. per square inch ($= 1{\cdot}1$ kg./sq. cm.) the undisturbed shear strengths are 4·5 lb. per square inch and 2·5 lb. per square inch respectively. In terms of practical problems this difference in strength is important but, when viewed in the light of the very great differences in sensitivity of these clays, Gosport $S_t = 2{\cdot}5$ and Horten $S_t = 17$, the differences in strength are not very significant. Indeed the extreme range of $(c/p)_n$ for all the clays so far studied is from 0·16 to 0·43, and thus the *order* of the undisturbed shear strength is the same for all normally consolidated clay sat the same overburden pressure.

Considering, again, the Gosport and Horten clays, the remoulded strengths of the two samples are $4{\cdot}5/2{\cdot}5 = 1{\cdot}8$ lb per square inch and $2{\cdot}5/17 = 0{\cdot}15$ lb. per square inch respectively. It will at once be seen, then, that the Horten clay is so much more sensitive than the Gosport clay not because it has a higher undisturbed strength, but because it has a far lower remoulded strength. Moreover this result is generally true : high sensitivity implies an exceptionally low remoulded strength. Yet the results of tests on many clays have shown that the remoulded strength of a clay depends principally upon the liquidity index. Curves relating remoulded shear strength to liquidity index are given in Fig. 11 for four clays, and they show that the order of the remoulded strength depends only on the liquidity index. Hence it follows that the liquidity index of an extra-sensitive clay must be very considerably greater than that of a medium or low sensitivity clay existing under the same overburden pressure. In the particular example being considered the liquidity index of Gosport clay corresponding to a remoulded strength of 1·8 lb. per square inch is 0·41, and the liquidity index of Horten clay corresponding to a remoulded strength of 0·15 lb. per square inch is 1·14. The essential difference between these two clays may therefore be expressed by the fact that, although both have been normally consolidated under the same overburden pressure, one has a water content rather nearer to the plastic than the liquid limit, while the other has a water content still well above the liquid limit. And the higher sensitivity of the latter as compared with the former is a direct consequence of this fact.

This important conclusion is true not only for the two particular clays considered above, but appears to be generally valid. In Fig. 12, curves* (or points, where the data is limited) have been plotted relating liquidity index to effective consolidation pressure, for 16 clays ; and a study of the position of these curves with respect to the sensitivities of the clays, as given in the accompanying Table, reveals a definite correlation between sensitivity and the liquidity index under any given pressure.

The point may be made in another way by comparing the in-situ values of liquidity index, at a given consolidation pressure, with the liquidity index of the same clay after consolidation in the oedometer from a slurry with a high initial water content. To avoid undue confusion in Fig. 12 the "slurry" curves only for Gosport, Horten, and London clays are given, but these immediately show the profound difference between the medium sensitive and the extra-sensitive clay. The in-situ liquidity index for Gosport clay is only slightly greater (for a given pressure) than that obtained in the slurry test. In contrast, the in-situ liquidity index of the Horten clay is very far removed from that in the slurry test.

Now the oedometer slurry tests correspond to the remoulded state, and these curves represent approximately the values of liquidity index which would be attained if the natural clay were remoulded and allowed to consolidate, still under its natural overburden pressure. Thus it can be seen that the Horten clay in its natural state possesses an extremely " metastable " structure since, although the particle arrangement is stable under the overburden pressure yet, after remoulding, the clay would consolidate (still under the same pressure) to the " slurry " curve. In the example given previously the in-situ liquidity index of 1·14

* Some of the curves in Fig. 12 can be determined only approximately, especially with regard to their slope, but the possible errors in no way detract from the validity of the reasoning based on this graph. It may be noted that all the clays are normally consolidated with the exception of St. Thuribe (lightly over-consolidated) and London (heavily over-consolidated). In these two latter clays, liquidity index has been plotted against pre-consolidation load.

Fig. 12

Relation between liquidity index and consolidation pressure

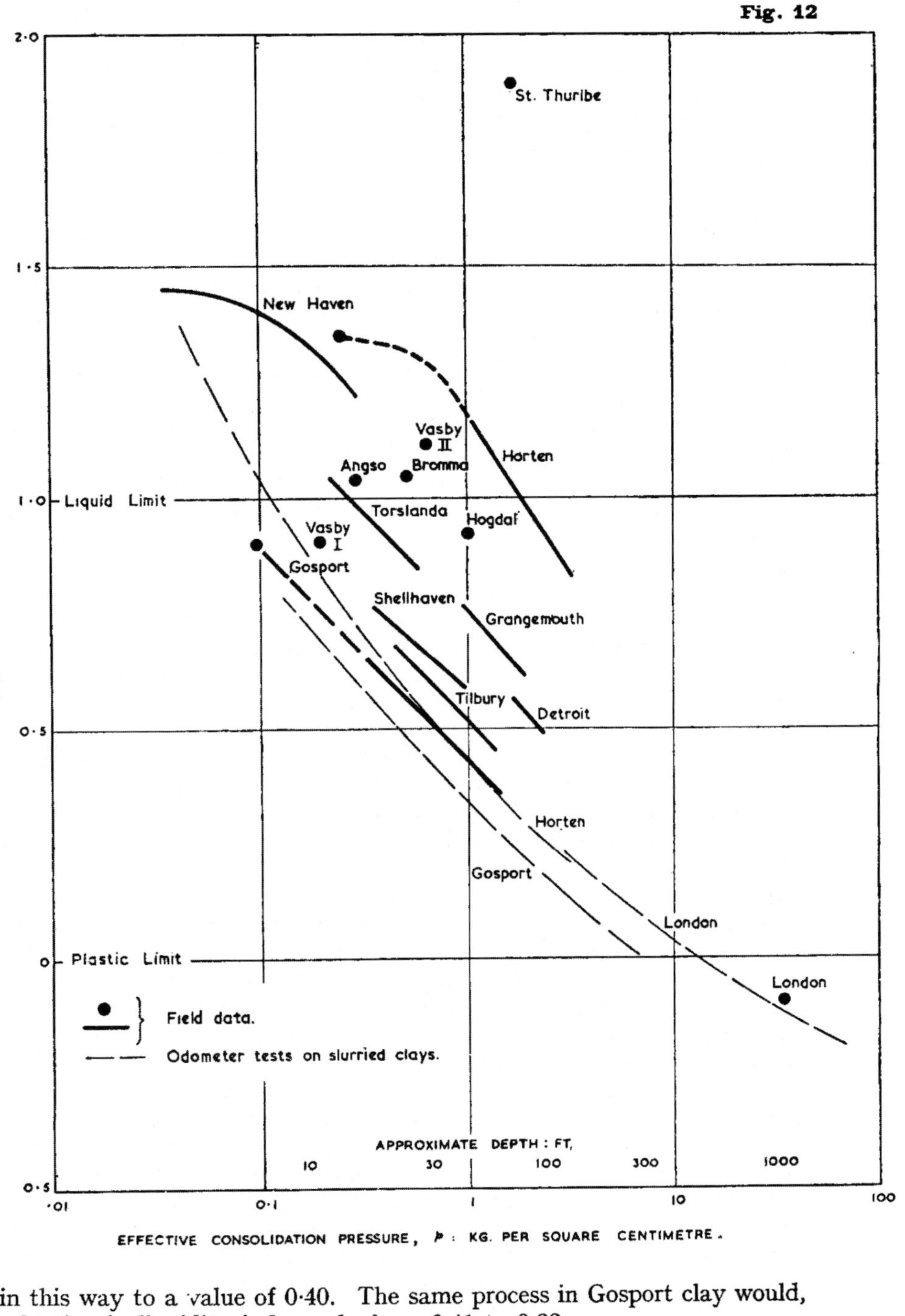

would be reduced in this way to a value of 0·40. The same process in Gosport clay would, however, cause a reduction in liquidity index only from 0·41 to 0·32.

Summarizing the foregoing discussion, it has been shown that high sensitivity implies exceptionally low remoulded strengths, and this implies the existence in the natural undisturbed clay of a meta-stable structure. In order successfully to explain sensitivity it is therefore necessary to account for this meta-stable structural arrangement of the particles.

Location		Average values				Reference
		Sensitivity	Liquid limit	Plastic index	Liquidity index	
St. Thuribe		c.150	33	12	1·9	Peck *et al.* (1951)
Horten, Norway		17	32	16	1·12	Hansen (1950)
New Haven, Conn.		c.10	93	43	1·4	Terzaghi (per. comm.)
Hogdal	Sweden	9·7	64	39	0·92	Cadling and Odenstad (1950) and Kjellman (per. comm.)
Vasby II	Sweden	9·0	79	54	1·12	
Bromma	Sweden	8·0	69	38	1·05	
Angso	Sweden	8·0	87	29	1·07	
Torslanda	Sweden	6·0	67	43	0·95	
Vasby I	Sweden	5·6	125	83	0·91	
Shellhaven	Gt. Britain	5·5	105	73	0·59	Skempton(1951)a
Gosport-mud	Gt. Britain	4·4	80	50	0·90	Skempton(1948)a
Grange-mouth	Gt. Britain	4·3	41	22	0·64	Skempton(1948)b
Tilbury	Gt. Britain	3·2	100	70	0·56	Skempton(1951)b
Gosport-deep	Gt. Britain	2·4	80	50	0·48	Skempton(1948)a
Detroit		2·3	31	13	0·56	Cummings *et al.* (1950)
London		c.1·0	77	50	−0·10	Cooling & Skempton (1942)

LEACHING EXPERIMENTS

Although there may well be several possible causes of the development of a meta-stable structure in clays one cause has now been established, by laboratory experiment, and confirmed to a limited degree by field evidence. This is the process whereby the salt concentration in the pore water of a marine or estuarine clay is reduced, subsequent to deposition, by the leaching action of fresh water. Attention to this effect seems first to have been given in Norway. Rosenquist (1946) mixed a clay with a 3 per cent solution of NaCl and allowed the remoulded material to stand for 5 days. He does not quote the acquired sensitivity at this time, but it is unlikely that it would have been at all considerable (see previous discussion on thixotropy). The greater part of the NaCl was then removed from the clay by dialysis for 15 days. After this treatment the strength increased slightly but, on remoulding, it was found that the sensitivity was very high.* Rosenquist then discusses the geological conditions under which leaching out of some of the salts in the pore water of a clay could take place, and adds the extremely important statement that he has investigated several quick-clays which are known to have been deposited in sea water and yet at the present time " not the slightest trace of chloride ion could be ascertained by means of silver nitrate."

The Authors' first experiments on leaching were designed to show whether the sensitivity of a natural undisturbed sample of clay could be increased by the reduction of the salt concentration in the pore water. Three specimens were cut from an undisturbed sample of Shellhaven clay. On the first of these, determinations were made of the shear strength, sensitivity, water content, Atterberg Limits, and the salt concentration in the pore water. A second specimen was placed in a Perspex cylinder, each end covered with a porous disc of greater diameter than the cylinder, and stored under water with a load on the upper disc. In this way, as in the thixotropy experiments, the clay could neither swell or consolidate. But this specimen was subjected to leaching by a percolating flow of distilled water for a period of 14 days. To act as a control the third specimen was similarly stored but it was not leached. At the end of 14 days both specimens were tested to determine the various properties mentioned above, and the results are summarized in Table 3.

It will be seen from the control specimen results that mere storage under water produces no significant changes in any of the properties. But the effect of leaching is to reduce the salt concentration and the liquid limit, and the remoulded strength, without any corresponding change in the water content or undisturbed strength. Quantitatively, as a result of reducing the salt concentration in the pore water by about one-half, the sensitivity of the clay is nearly doubled.

This experiment was evidently in agreement with Rosenquist's work and provided direct evidence for the influence of leaching. But it was also clear that both Rosenquist and the

* Rosenquist used the cone penetration test which is now known not to be accurate: with this test the sensitivity of the dialysed sample was measured as about 100.

Table 3

Leaching Test on Specimen of Undisturbed Shellhaven Clay Sample B

Condition	Water content	Atterberg limits LL	Atterberg limits PL	Liquidity index	Shear strength lb./sq. in. c	Shear strength lb./sq. in. c_r	Sensitivity	Conc. of salt in pore water gm./litre
Natural state	55	60	25	0·86	3·3	0·33	10	26
Control specimen after 14 days without leaching	56	60	25	0·88	3·1	0·27	11	27
After leaching for 14 days	55	56	25	0·97	2·8	0·15	18	12

Authors had arranged their tests in such a way that the clay was not under load during the leaching process. There was, therefore, a most serious restriction on the generality of the results. It had been shown that, under the condition of the test, leaching reduced the remoulded strength and, moreover, that it did so essentially by reducing the liquid limit of the clay without changing the water content. That is to say, the liquidity index was increased. But if the clay was under a consolidation pressure while being leached, as must be the case in nature, then the effect of leaching might be to lower not only the liquid limit but also the water content. In other words the clay might consolidate during leaching and the liquidity index might consequently either remain unaltered or, at any rate, it might increase far less than the amount observed in the Authors' test.

Since this problem was one of critical significance, experiments were carried out to investigate all the effects of leaching on a clay while actually under a consolidation pressure. Shellhaven and Horten clays were used in these tests.

Shellhaven clay (Sample A) was made up with distilled water to a water content of 172 (liquidity index = 2·15) and, after thorough mixing, the material was placed in two identical oedometers made of Perspex, specially designed for the consolidation of slurries under low pressures. Over a period of about 1 month the samples were consolidated under increasing load until finally they were in equilibrium under a pressure of 0·55 lb. per square inch (= 0·039 kg. per square centimetre). One sample was then leached with distilled water for 21 days, while remaining under this pressure. The other sample was also left for this period of time, and under the same pressure, but with no leaching. The samples were then removed from the oedometers and their shear strength both in the " undisturbed " and remoulded states

Table 4

Combined Leaching and Consolidation Tests, Remoulded Shellhaven Clay. Sample A

Test: Consolidation pressure lb./sq. in.	Test: Condition	Water content	Atterberg limits LL	Atterberg limits PL	Liquidity index	Shear strength (a) lb./sq. in. c	Shear strength (a) lb./sq. in. c_r	Sensitivity	Conc. of salt in pore water gm./litre
0 to 0·55	No leaching	107	93	32	1·23	0·15	0·055	2·7	12
0 to 0·55	Leached, after consolidation	106	90	32	1·28	0·18	0·043	4·2	11
0 to 4·5	No leaching	72	91	32	0·68	1·08	0·45	2·4	11
0 to 4·5	Leached during consolidation from 0·55 to 4·5 lb./sq. in.	72	89	32	0·70	1·13	0·31	3·6	10

Note : (a) all specimens tested 50 days after start of test.

measured by the laboratory vane test. Water contents, Atterberg Limits, and salt concentration were also determined. The results are given in Table 4 and show that although the salt concentration was reduced by leaching only from 12 to 11 grammes per litre the sensitivity was increased from 2·7 to 4·2 and, moreover, this change was accompanied by no appreciable alteration in water content.

The experiment was then repeated, but after consolidating to 0·55 lb. per square inch both samples were further consolidated up to a pressure of 4·5 lb. per square inch (=0·32 kg. per square centimetre). One sample was continuously leached during this consolidation from 0·55 to 4·5 lb. per square inch and the other served as a control test in which no leaching was carried out. The results are again to be found in Table 4, and are similar to those obtained in the earlier experiment.* Here, then, was direct evidence that leaching increased the sensitivity, and the liquidity index, but caused no appreciable change in the water content or undisturbed strength. The experiment was not altogether satisfactory, however, owing to the very low permeability of the samples which restricted the amount of leaching that could occur in a reasonably short time.

This defect was overcome in the tests on Horten clay. These were carried out in the same manner as those previously described, except that the clay was initially sedimented from a 1 per cent salt solution. This concentration was sufficient to cause flocculation and the clay settled out of suspension, with a water content of about 88 (liquidity index = 5·1). This material was then consolidated to 1·29 lb. per square inch (= 0·09 kg. per square centimetre) in the two oedometers. While under this pressure, one sample was leached with distilled water for 24 days, and the other was maintained under load without leaching, as a control. The results of tests on the samples after removal from the oedometers are given in Table 5.

Table 5

Combined Leaching and Consolidation Tests, Remoulded Horten Clay

Test		Water content	Atterberg limits		Liquidity index	Shear strength (a) lb./sq. in.		Sensitivity	Conc. of salt in pore water gm./litre
Consolidation pressure lb./sq. in.	Condition		*LL*	*PL*		c	c_r		
0 to 1·29	No leaching	31·6	28·6	15·6	1·23	0·145	0·133	1·1	12·6
0 to 1·29	Leached after consolidation	31·2	24·5	15·6	1·75	0·144	0·039	3·7	2·2

Note: (a) both specimens tested 47 days after start of test.

It will be seen from Table 5 that with the Horten clay leaching caused a substantial reduction in salt concentration in the pore water and that the sensitivity was increased from 1·1 (in the unleached sample) to 3·7. The increase in sensitivity was due entirely to a reduction in the remoulded strength—the " undisturbed " strength remaining unchanged—and the reduction in remoulded strength was in turn a consequence of a considerable increase in liquidity index, following upon a reduction in liquid limit without any significant reduction in water content; even though the clay was under load during the leaching process. This last point, it will be remembered, was the chief concern in these combined leaching and consolidation tests.

The sensitivity of the specimens which were not leached, namely about 2·4 for Shellhaven and 1·1 for Horten, are approximately the values that would be expected purely as a result

* The changes in salt content were too small to show whether or not there was a difference between the effects of leaching during, or after, consolidation.

Figs 13

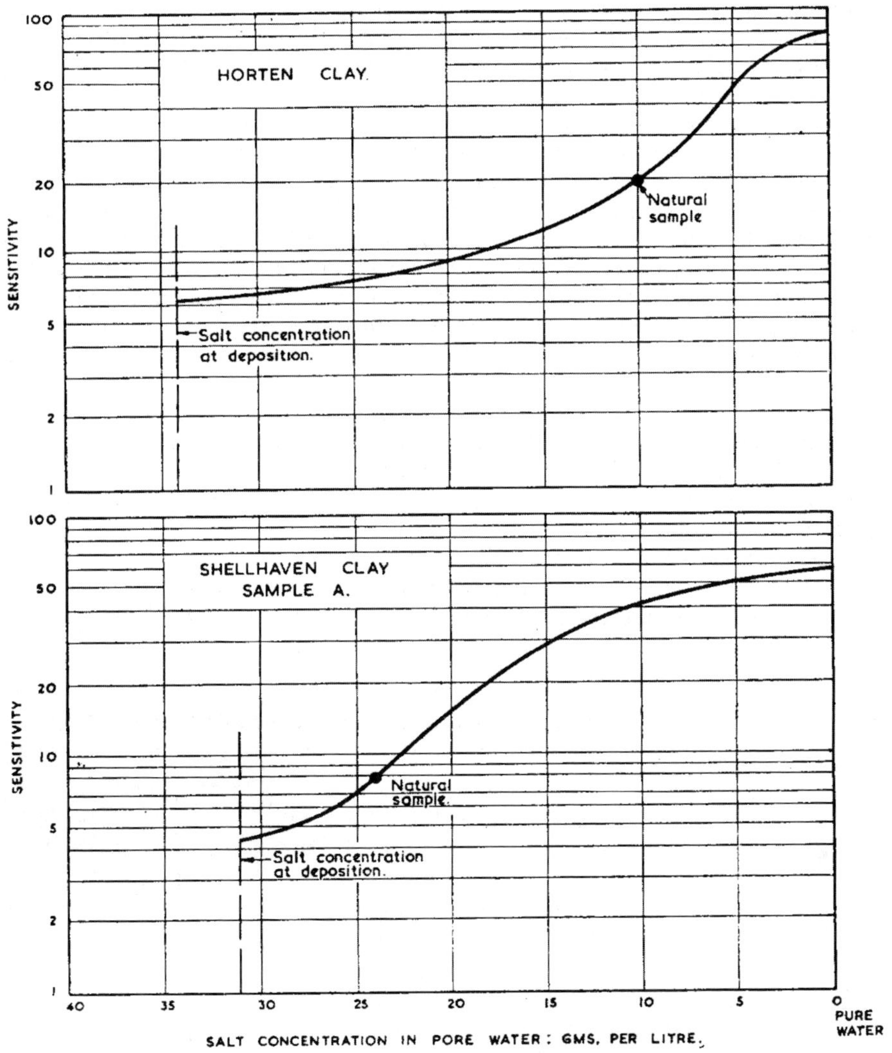

Calculated increase in sensitivity due to decrease in salt concentration

of thixotropic regain during the period of about 50 days which had elapsed from the start of the test.

In summary, the effects of reducing the salt concentration in the pore water of a clay by leaching are :—

(1) to reduce the liquid limit and the remoulded strength,

(2) to leave unaltered the water content and the undisturbed strength,

(3) consequently, to increase the liquidity index and the sensitivity.

It is particularly to be noted that as a result of leaching the clay is left with a meta-stable structure, in the sense previously discussed and illustrated in Fig. 12.

ESTIMATION OF SENSITIVITY CONSEQUENT UPON LEACHING

On the basis of experimental data it is possible to estimate the changes in sensitivity brought about by leaching in the Shellhaven and Horten clays. The necessary experimental work and the calculations are outlined* in the Appendix and the results are presented in Figs 13. It will be seen that according to these calculations the sensitivities of the two clays before any leaching had taken place were about 5. When account is taken of the degree of approximation inherent in the calculations, it is possible that this initial sensitivity might be the result solely of thixotropic hardening.

With progressive decrease in the salt concentration the sensitivity increases, and if leaching were to be carried to completion the calculations indicate that both clays would have sensitivities of about 60 to 80. Here, again, no great accuracy can be given to the calculations, but there seems to be little doubt of the general trend of the curves in Figs. 13 and of the conclusion that very high sensitivity can result from substantial reductions in the salt concentrations in the pore water of clays of marine origin.

THE META-STABLE STRUCTURE CAUSED BY LEACHING

The effect of leaching in producing a meta-stable structure in marine clays may tentatively be visualized as follows. In the original state, after deposition and consolidation, the particles exist in an open structure in effectively solid contact established through the interpenetration of the films of " bound " or " rigid " water: see Fig. 14 (a). These films are considered to contain within their thickness only a few water molecules (Grim and Cuthbert 1945). Owing to the concentration of salt in the pore water (equal initially to that of sea water) the particles are also surrounded by rather thick layers of adsorbed water, far less closely associated with the particles than the very thin " bound " films, but denser and more viscous than the " free " (or normal) water existing beyond the boundary of the absorbed layers. On remoulding the clay the " solid " bonds are broken and the strength is reduced. This is the account of clay structure and of remoulding given by Terzaghi (1941) and it seems to provide a reasonable and simple picture. To introduce the influence of leaching it is postulated that a reduction in the salt concentration reduces the thickness of the adsorbed layers and, returning now to the clay in its undisturbed state, but after partial leaching has occurred, the state may be represented as in Fig. 14 (b). The particles are still in effective contact and their packing has not been changed. The water content and shear strength therefore remain unaltered. But the proportion of " free " water is now greatly increased. Consequently on remoulding there is more of this free water available than before leaching and the remoulded strength is lower in the leached clay. Also the liquid limit of the clay would be appreciably lower, with the thinner adsorbed layers, and the liquidity index would therefore be higher since the water content has not been altered by leaching.

Figs 14

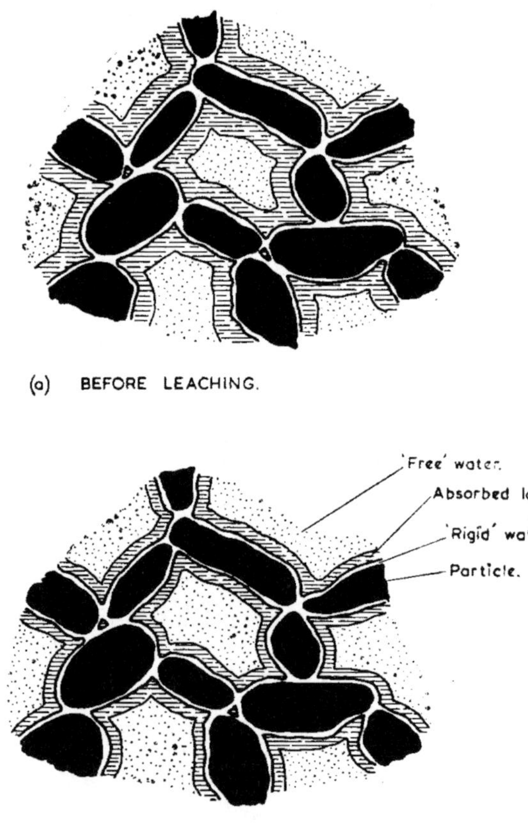

Effect of leaching on undisturbed marine clay

The foregoing hypothesis is undoubtedly over-

* Details are given by Northey (1950).

simplified ; but it is in accordance with the experimental facts, at least in a rough and ready way, and it illustrates what is meant by the term "meta-stable" structure. Since there is some controversy* as to whether clay particles are effectively in contact or separated by relatively large distances, it is of some interest to mention that, in the Authors' opinion, the former case is more probable in view of the strength and water content remaining constant during leaching. On the latter hypothesis it would seem inevitable that a change in the thickness of the adsorbed layers would be accompanied by a change in strength and porosity of packing.

FIELD EVIDENCE ON SENSITIVITY AND LEACHING

It is clear that sensitivity can be due to leaching only in clays of marine or estuarine origin. The sensitivity of fresh-water clays and of marine clays which have not been leached, must be due either to thixotropy or some other cause other than a reduction in salt concentration. Probably the most common case in which leaching can occur is that of a late- or post-glacial marine or estuarine clay which has been subjected to isostatic uplift, subsequent to deposition, and its surface brought above sea level. A slow seepage of fresh water from rainfall or ground water can then be set up which in the course of time may remove some or all of the dissolved salts in the pore water. Another possibility, for clays whether above or below sea level, is that leaching may occur as a result of an upward flow of fresh water from sand or gravels beneath the clay. Rosenquist (*loc. cit.*) suggests that, even with no flow of the fresh water, leaching may take place by slow diffusion of the salts from the clay into the underlying fresh water.

The evidence, known to the Authors, relating senstitivity and leaching is assembled in Table 6. The clays from Norway and Horten were deposited in the sea, and the Shellhaven

Table 6

Sensitivity of Normally Consolidated Clays, Related to Conditions of Deposition and Present Salt Concentration in the Pore Water

Location	Conditions of deposition: Environment	Conditions of deposition: Salt conc. gm./litre (approximate)	Present salt conc. in pore water gm./litre	Sensitivity	Authority
Norway	Marine	34 (a)	nil	high	Rosenquist (1946)
Horten	Marine	34 (a)	10	19	Authors of present article
Shellhaven B	Estuarine	31 (b)	12 (c)	18	,, ,, ,,
Shellhaven A	Estuarine	31 (b)	24	8	,, ,, ,,
Gosport	Marine	34 (a)	30	2·4	,, ,, ,,
Rosyth	Estuarine	30	29	1·8	,, ,, ,,
Gloucester	River Water	small	3	2·1	,, ,, ,,

Notes : (a) salt concentration in sea water taken as 34 gm./litre.
(b) present salt concentration in water adjacent to location of clay.
(c) leached in laboratory.

clay was formed near the mouth of the Thames estuary. They have been leached, as proved by the present salt concentration in the pore water, and they are sensitive or extra-sensitive clays. The Gosport and Rosyth clays were deposited, respectively, under marine and estuarine conditions, but the salt concentration in the pore water is, in each case, approximately equal to that of the water in which they were formed. They are clays of medium sensitivity and may owe their sensitivity entirely to thixotropy. The Gloucester clay is a flood plain deposit

* See for example, a discussion on " The Chemistry and Rheology of Clay " reported in *Nature*, vol. 160, p. 511 (1947).

of the River Severn. It contains little salt in the pore water, but it was formed in fresh water. Consequently it has not been leached; and it also is a clay of medium sensitivity.

The field evidence quoted above is therefore in accordance with the conclusions:

(1) sensitive and extra-sensitive clays can result from the complete or partial leaching of marine and estuarine clays; although it is, of course, not logical to assume that leaching is the only cause of high sensitivity.

(2) of the clays which have so far been investigated, those in which no leaching has occurred exhibit only a medium sensitivity, of the order known to be developed by thixotropy.

OTHER POSSIBLE CAUSES OF SENSITIVITY

As stated above, there is no reason to assume that all sensitive and extra-sensitive clays have been leached. It is possible that other factors may exist which can satisfactorily account for the phenomenon. One such factor has been postulated by Casagrande (1932) and Terzaghi (1941, 1947) who introduce various suggestions for the way in which an organized structural arrangement of the particles can be built up during the very slow process of formation of clays in nature. This structure is considered to be more " open " or meta-stable than that developed in the relatively very rapid consolidation process necessarily used in laboratory tests. Consequently it seems that these suggestions cannot be checked by direct experiment. But field evidence for the existence of a more open structure in natural clays, as compared with remoulded clays, has been given in Fig. 12, and although this evidence can be explained by leaching (e.g. in the Horten clay) it is possible that some of the sensitive clays in this graph have not been leached and may owe their high liquidity index to structural developments of the types envisaged by Casagrande and Terzaghi.

At the present time these considerations must remain largely speculative. It is, however, perhaps worth mentioning a feature, noticeable in some normally consolidated clays, which may be relevant to this problem; namely the existence of a zone near the surface of such clays in which the strength and liquidity index remain approximately constant over an appreciable range of depth; see Figs 15. In this figure Zone III is the region where drying

Figs 15

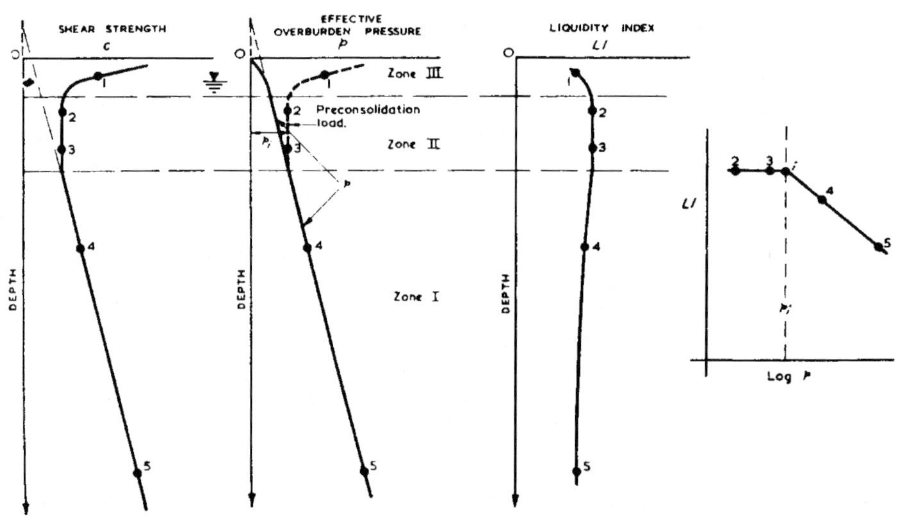

Variations in properties with depth in normally consolidated clays (with surface drying in Zone III)

has taken place—the so-called " crust " which is common in clays which have been uplifted subsequent to deposition. Zone II is the region at present under discussion, while in Zone I the clay is normally consolidated : the strength increasing, and the liquidity index decreasing, with depth. The constant strength and liquidity index in Zone II appear to be related to a constant pre-consolidation load p_i, although the effective overburden pressure p is increasing with depth. Consequently if liquidity index be plotted against p a horizontal line is obtained, as shown in Figs 15, for values of p less than p_i. If, however, the liquidity index in Zone II is plotted against p_i a single point is obtained (point i in Figs 15). The curves of liquidity index against effective pressure in Fig. 12 have been plotted in the former manner and the approximately constant portions, corresponding to Zone II, are seen in the New Haven and Horten clays. Now in some cases the pressure p_i may simply be the result of intermittent drying during formation of the clay (Casagrande and Fadum 1944), but other cases are known in which this explanation is most unlikely.* The problem is one to which little attention has so far been devoted. Where drying cannot have occurred, it has been suggested by A. W. Bishop that over-consolidation in the upper layers could result from cyclic pore pressure changes due either to tidal or temperature variations ; but alternatively it must be tentatively admitted that the pressure p_i may be some measure of the action of inter-particle forces bonding the clay into a structure sufficiently strong to resist any appreciable consolidation by the overburden pressure ; so long as this does not exceed p_i. Under the action of overburden pressures greater than p_i the clay consolidates in a normal manner. If the foregoing hypothesis is correct it seems reasonable to suppose that, since consolidation has been inhibited in Zone II, the liquidity index at any given depth in Zone I would be greater than if consolidation had proceeded normally in Zone II. In this way it may be possible in some clays to account for the " open " or meta-stable structure referred to previously.

The Authors have records of some six clays in which p_i is approximately known (the values ranging from 400 to 1,200 lb. per square foot) and there seems to be a rough correlation between sensitivity and p_i. More than this cannot be claimed on the basis of the present data, but nevertheless the Authors feel justified in drawing attention to this aspect of the problem of sensitivity since it may prove a fruitful subject for future research.

THE INSENSITIVITY OF HEAVILY OVER-CONSOLIDATED CLAYS

Mention has been made previously of the small or even negligible thixotropy exhibited by clays at water contents approximating to the plastic limit. The fact that heavily over-consolidated clays are insensitive is probably also related to the breakdown of any organized structure by the great pressures and deformations to which such clays have been subjected. London Clay is a typical example. It has in the past been consolidated under about 1,000 feet of overburden, and its water content was then reduced below the plastic limit (Cooling and Skempton 1942). Even if the clay was sensitive when still under small pressures and with a moderate or high liquidity index (e.g. similar to Shellhaven clay), it is difficult to see how any meta-stable structure could be maintained when the particles had been packed so closely together under this great pressure, and the liquidity index was rather less than zero.

In the case of London Clay an examination of Fig. 12 will show that in fact no such structure can exist, for the point representing the undisturbed clay lies only just above the oedometer test " slurry " consolidation curve. That is to say the undisturbed and the remoulded clay must have practically identical structures. There is, moreover, a suggestion in Fig. 12 that for all clays the curves relating in-situ liquidity index and consolidation pressure are tending to converge with the " slurry " curves at high pressures and this tendency must, of course, be accompanied by a corresponding decrease in sensitivity.

CONCLUSIONS

(1) From experimental work it has been shown that thixotropy can account for low or medium sensitivity, but it appears to be unable to account for high sensitivity.

* Personal communication from Professor Terzaghi.

(2) Both laboratory and field evidence show that high sensitivity can be developed in marine or estuarine clays in which the original concentration of salt in the pore water has been reduced by leaching.

(3) Of the few clays so far investigated those in which leaching has not occurred have only a medium sensitivity, the magnitude of which could be attributed to thixotropy.

(4) There is no reason to assume that leaching is the sole cause of high sensitivity, and much further research will be necessary before this phenomenon can be fully understood.

(5) Heavily over-consolidated clays, with water contents typically equal to about the plastic limit, are insensitive probably for the following reasons : (*a*) thixotropy is negligible at such water contents, and (*b*) any meta-stable micro structure that may have existed in the clay at an earlier stage in its geological history, when still under comparatively low overburden pressures, will have been broken down by the intense loading and deformation to which the clay has been subjected.

(6) It is possible that the fissured macro-structure present in many clays may be the result of syneresis.

ACKNOWLEDGEMENTS

The Authors are grateful to Professor K. Terzaghi, Mr W. Kjellman and Professor R. B. Peck for valuable data and for helpful discussions, both verbal and written. Professor R. E. Grim supplied the sample of Illite clay and very kindly carried out mineralogical analyses of the London, Shellhaven, and Horten clays. The samples of Horten clay were given to the Authors by Mr J. Brinch Hansen to whom they are also indebted for the translation from the Norwegian of Dr I. Rosenquist's paper. The experimental work was carried out in the Civil Engineering Department of Imperial College, University of London, with the aid of a National Research Scholarship granted to the junior Author by the New Zealand Government.

REFERENCES

BOSWELL, P. G. H., 1949. A preliminary examination of the thixotropy of some sedimentary rocks. *Quart. J. Geol. Soc.* 104 : 499.

BURGERS, J. M., and SCOTT BLAIR, G. W., 1948. Report on the principles of rheological nomenclature. Joint Committee on Rheology of the International Council of Scientific Unions. *Proc. Int. Rheol. Cong.* (Amsterdam).

CADLING, L., and ODENSTAD, S., 1950. The vane borer. *Proc. Roy. Swedish Geotech. Inst.* No. 2. (Stockholm).

CASAGRANDE, A., 1932 (a). The structure of clay and its importance in foundation engineering. *J.Boston. Soc. C.E.* 19 : 168.

CASAGRANDE, A., 1932 (b). Research on the Atterberg Limits of soils. *Publ. Roads.* 13 : 121.

CASAGRANDE, A., and FADUM, R. E., 1944. Applications of soil mechanics in designing building foundations. *Tr.Am.Soc.C.E.* 109 : 383.

CASAGRANDE, L., 1948. Structures produced in clays by electric potentials and their relation to natural structures. *Nature.* 160 : 470.

COOLING, L. F., and SKEMPTON, A. W., 1942. A laboratory study of London clay. *J.Inst.Civ.Engrs.* 17 : 251.

CUMMINGS, A. E. KERKOFF, and PECK, R. B., 1950. Effect of driving piles into soft clay. *Tr. Am. Soc. C.E.* 115 : 275.

FREUNDLICH, H., 1935. Thixotropy. *Hermann et Cie*, Paris.

GRIM, R. E., and CUTHBERT, F. L., 1945. Some clay-water properties of certain clay minerals. *J. Am. Ceramic Soc.* 28 : 90.

HANSEN, J. B., 1950. Vane tests in a Norwegian quick clay. *Géotechnique.* 2 : 58.

HVORSLEV, M. J., 1937. Uber die festigkeitseigenschaften gestorter bindiger boden. *Ingeniorvidenskabelige Skr.* No. 45 (Copenhagen).

MORETTO, O., 1948. Effect of natural hardening on the unconfined compression strength of remoulded clays. *Proc. Second Int. Conf. Soil Mech.* 1 : 137.

NORTHEY, R. D., 1950. An experimental study of the structural sensitivity of clays. Ph.D. Thesis (Faculty of Science). *University of London.*

PECK, R. B., IRELAND, H. O., and FRY, T. S., 1951. Studies of soil characteristics ; the earth flows of St. Thuribe, Quebec. *Dept. Civ. Eng. Univ. of Illinois. Soil Mech. Series.* No. 1.

ROSENQUIST, I. T., 1946. Om de norske kvikkleirers egenskaper og mineralogiske sammensetning. *N.I.M. Forhandlingar.* 10 : 1. (*Stockholm.*)

RUTLEDGE, P. C., 1944. Relations of undisturbed sampling to laboratory testing. *Tr. Am. Soc. C.E.* 109 : 1155.

SHARPE, C. F. S., 1938. Landslides and related phenomena. *Columbia Univ. Press, New York.*
SKEMPTON, A. W., 1948 (a). The geotechnical properties of a deep stratum of post-glacial clay at Gosport. *Proc. Second Int. Conf. Soil Mech.* 1 : 145.
SKEMPTON, A. W., 1948 (b). Vane tests in the alluvial plain of the River Forth, near Grangemouth. *Géotechnique.* 1 : 111.
SKEMPTON, A. W., 1950. Soil mechanics in relation to geology. *Proc. Yorkshire Geol. Soc.*
SKEMPTON, A. W., 1951 (a). Unpublished report on soil tests at Shellhaven, to John Mowlem & Co., Ltd.
SKEMPTON, A. W., 1951 (b). Unpublished report on soil tests at Tilbury, to Sir Alexander Gibb & Partners.
SKEMPTON, A. W., and BISHOP, A. W., 1950. The measurement of the shear strength of soils. *Géotechnique.* 2 : 90.
TERZAGHI, K., 1936. Stability of slopes of natural clay. *Proc. First Int. Conf. Soil Mech.* 1 : 161.
TERZAGHI, K., 1941. Undisturbed clay samples and undisturbed clay. *J. Boston Soc. C.E.* 28 : 211.
TERZAGHI, K., 1944. Ends and means in soil mechanics. *Eng. Journ. (Canada).* 27 : 608.
TERZAGHI, K., 1947. Shear characteristics of quicksands and soft clay. *Proc. Second Texas Conf. Soil Mech.*
ZEUNER, F. E., 1946. Dating the past: An introduction to geochronology. *Metheun, London.*

APPENDIX

ESTIMATION OF SENSITIVITY CONSEQUENT UPON LEACHING

On the basis of experimental data it is possible to estimate the sensitivity of the Horten and Shellhaven clays for any given concentration of salt in the pore water. In outline these tests and calculations are as follows :

(1) The variations in liquid and plastic limits with changes in salt content were determined. Some samples of the clay were leached for different periods of time, while to other NaCl was added, and the Atterberg Limits and the salt concentration were then measured. In both clays the plastic limit was found to be almost independent of the salt content but the liquid limit decreased very appreciably as the salt content decreased.

(2) The remoulded strengths of both clays were measured over a wide range of water content, with different liquid limits resulting from different salt contents. It was found that, in spite of the large variation in liquid limit, the remoulded shear strength of any one of the clays was a unique function of liquidity index. That is to say all the results fell on or close to the curves relating remoulded strength and liquidity index, for the particular clay, previously given in Fig. 11.

(3) For the Shellhaven and Horten clays the remoulded strength of the natural sample is known, together with the salt concentration and liquidity index. The experimental evidence given in the article shows that the water content of a clay remains practically unaltered by a change in the salt concentration. Thus from the relation between salt content and liquid limit, the change in liquidity index in the undisturbed clay, following a change in salt concentration, can be calculated. But, from the relation between remoulded strength and liquidity index, the change in remoulded strength with change in salt concentration can also be predicted.

(4) The leaching experiments indicated that the undisturbed strength of a clay remained practically unchanged by a decrease in salt content. In these tests, however, the alterations were not great and there is field evidence (Skempton 1948 b) that the strength of normally consolidated clays, under a given overburden pressure, is less for low than for high liquid limit clays. In the calculations allowance has been made for this effect, since it tends to reduce the estimated influence of leaching on sensitivity. But it should be emphasized that the evidence for the tendency for undisturbed strength to be less with lower liquid limits is derived from a comparison of different clays. It may be that, for any given clay, the effect of decreasing liquid limit, caused by a decreasing salt concentration after deposition, is considerably

less than that deduced from comparisons of different clays. If this were the case then the changes in salt concentration would be even more pronounced than those estimated here, and illustrated in Figs 13.

Now the undisturbed shear strengths of the Shellhaven and Horten clays are known, and also the liquid limits and salt concentrations. For any other salt concentration the liquid limit is also known, from the experiments mentioned under (1) above, and the corresponding changes in undisturbed strength can then be found from the relation given by Skempton and referred to previously in this section.

(5) The variation in both remoulded and undisturbed strengths with changes in salt concentration are now known for both clays and it is immediately possible to calculate the sensitivity for each clay at any given concentration of salt in the pore water. The results are given in Figs 13.

CORRESPONDENCE

The Secretary,
The Institution of Civil Engineers.

DEAR SIR,

The article on "The sensitivity of clays" (*Géotechnique,* vol. 3, p. 30) by Drs A. W. Skempton and R. D. Northey is a significant and thought-provoking contribution.

Reference is made in this article to the extremely sensitive clays from St Thuribe, Canada. These are known to be post-glacial clays laid down in marine waters that invaded the St Lawrence River valley. Therefore the initial concentration of salt in the porewater should have been essentially the same as that in sea water. The salt concentration at the present time is approximately 2·2 grammes per litre. This would seem to indicate a large amount of leaching and is in good agreement with the other data presented by the Authors.

Yours faithfully,

R. B. PECK.

NEWS ITEMS

Recent Publications

Soil Mechanics for Road Engineers ; D.S.I.R. H.M. Stationery Office, June 1952.

The Suction of Moisture held in Soil and other Porous Materials, by D. Croney ; Road Research Tech. Paper No. 24. D.S.I.R., H.M. Stationery Office, May 1952.

The Application of Electro-osmosis to practical problems in foundations and earthworks, by L. Casagrande ; Building Research Tech. Paper No. 30, D.S.I.R., H.M. Stationery Office, May 1952 (1st pub. 1947).

Triaxial Testing of Soils and Bituminous Mixtures ; A.S.T.M., Philadelphia. Special Tech. Pub. No. 106, 1951.

The Resistance of Piles to Penetration, by R. V. Allen ; E. and F. N. Spon, London, July 1952 (2nd edition).

International Society of Soil Mechanics and Foundation Engineering : British Section

Annual General Meeting.—The Annual General Meeting of the British Section will be held at the Institution of Civil Engineers at 5.30 p.m. on Wednesday, 22nd October, 1952. Tea will be served at 5.0 p.m.

Informal Meeting.—There will be an Informal Meeting of the British Section at the Institution of Civil Engineers immediately after the Annual General Meeting, at which the article on "Shellhaven" by Dr Skempton and Mr Ward, published in the September issue of *Géotechnique*, will be discussed.

Croney, D. (1952). *Géotechnique* **3**, No. 1, 1–16

THE MOVEMENT AND DISTRIBUTION OF WATER IN SOILS

by

D. Croney, B.Sc.(Eng.)

SYNOPSIS

Moisture plays a particularly important role in the soil mechanics of shallow foundations, because of the comparatively large seasonal changes occurring in the moisture condition of the soil close to the earth's surface. In particular, methods of estimating the changes which occur in the moisture condition when soil is paved are required in the design of roads and airfields.

This article discusses the fundamental factors which give rise to moisture movements in soil and, on the basis of this discussion, methods are outlined for estimating the moisture distribution with depth beneath impervious pavements under equilibrium conditions. Differences in the soil moisture suction (or reduced pressure at which water is retained in the soil structure) provide the "potential" causing water to move within a soil mass. The movements which occur can only be expressed in terms of moisture content if the relationship between soil moisture suction and moisture content for the soil is known. The nature of this relationship and the manner in which it can be used to estimate equilibrium moisture distributions are discussed in relation to incompressible, partially compressible, and fully compressible soils.

Some experimental confirmation of the theories involved has been obtained from laboratory and field investigations at present in progress.

L'humidité joue un rôle particulièrement important dans la mécanique des sols des fondations peu profondes, à cause des changements saisonniers d'une ampleur relativement grande, qui se produisent dans l'humidité du sol juste au-dessous de la surface du terrain. En particulier, pour l'étude des routes et des aérodromes, on a besoin d'un moyen d'évaluer l'ampleur des changements d'humidité lorsque le sol est pourvu d'un revêtement.

Cet article traite des facteurs fondamentaux qui donnent lieu aux mouvements de l'humidité dans le sol ; sur la base de cette discussion, l'auteur donne une indication des méthodes pour évaluer la répartition de l'humidité à une profondeur donnée au-dessous des revêtements imperméables dans des conditions d'équilibre. Les différences dans la succion d'humidité du sol (c'est-à-dire la basse pression à laquelle la structure du sol peut retenir l'eau) donnent lieu au "potentiel" qui provoque des mouvements de l'eau dans le sol. Ces mouvements ne pourront être exprimés en termes de teneur en eau que si le rapport entre la succion et la teneur en eau du sol est connu. L'auteur discute le caractère de ce rapport et la manière dans laquelle on peut s'en servir pour évaluer ces répartitions équilibres d'humidité par rapport aux sols incompressibles, partiellement compressibles, et entièrement compressibles.

Les théories en cause ont reçu certaines confirmations expérimentales par suite des recherches en laboratoire et en chantier qui sont en cours.

INTRODUCTION

Since the moisture condition of soil close to the surface is subject to large variations, the influence which moisture has on the mechanical properties of soil is of considerable importance in soil mechanics applied to shallow foundations. In particular, the road engineer requires to know what changes are likely to take place in the moisture condition beneath a road surface, and what effect these changes will have on the stability of the pavement.

It is usual engineering practice to assess the moisture condition of soil, and the effect which moisture has on soil properties, in terms of the moisture content expressed as a percentage of the dry weight. This is perhaps unfortunate since a difference of moisture content does not provide the "potential" causing water to move in soil. Not only may two soils in contact be in a state of moisture equilibrium when they have different moisture contents, but it is possible that adjacent parts of the same soil may be at appreciably different moisture contents without any transfer of moisture occurring.

The more fundamental property of soil water, which is the primary cause of moisture movements in the liquid phase, is *soil moisture suction*. This is defined as the reduction of pressure (below atmospheric pressure) present in the soil water, which arises from the forces by which the moisture is retained in the soil structure. If a change of moisture content occurs at any point in a mass of soil initially in a state of moisture equilibrium, the soil moisture suction at that point is changed, and moisture movements occur throughout the soil until a new condition of suction equilibrium is established. The changes of moisture content which will occur during the re-distribution process can only be estimated if the relationships between suction and moisture content for the soil or soils comprising the soil mass are known.

During the last few years an investigation has been in progress at the Road Research Laboratory into the suction properties of soils and the relationship between suction and moisture content. The primary object of the work is to enable an estimate to be made of the moisture distribution with depth reached in the soil beneath impervious road and airfield pavements, and hence to assess the changes which will occur when uncovered soil is paved.

CLASSIFICATION OF SOIL MOISTURE

Soil moisture is usually classified in terms of, (a) gravitational water which moves towards the water-table under the action of gravity, (b) ground water, contained below the level of the water-table, and, (c) held water which is retained above the water-table after all movement of gravitational water has ceased. The water-table is defined as the level in the soil at which the soil water is at atmospheric pressure, and it corresponds therefore to the equilibrium level to which water would rise in a borehole put down in the soil.

Below the water-table the pressure of the water contained in the soil pores increases linearly with depth, in the same manner as the pressure increases below a free-water surface. Above the water-table the pressure decreases, and the pore-water pressure (measured from atmospheric pressure as zero) is negative. In soil which is in a state of moisture equilibrium, i.e. in which there is no tendency for water to move up from, or down to, the water-table, the pore-water pressure decreases linearly with height above the water-table, the pressure x feet above the water-table being less than atmospheric pressure by an amount equivalent to a head of x feet of water.

The surface tension and adsorptive forces by which water is held in the soil above the water-table are responsible for the reduction of pressure. These forces give rise to what may be termed the moisture suction, or more briefly the suction characteristics of the soil. It is these characteristics which largely determine the way in which water distributes itself in soil above the water-table. Other factors which may be of importance under certain conditions are temperature gradients, and the effect which such gradients have on the vapour pressure of soil water.

SOIL MOISTURE SUCTION

Because the pressure of the water contained in a small sample of soil removed from above the water-table is less than atmospheric pressure, the sample will absorb moisture if it is placed in contact with free-water. The soil suction is the difference between the pressure of the water in the sample and atmospheric pressure.

The presence of soil suction can be detected, and for suctions less than 1 atmosphere the magnitude can be determined, by the apparatus shown in Fig. 1. A small water-filled reservoir is connected to a horizontal flow tube, so that changes in the volume of water contained in the system can be detected by observing the movement of the water meniscus in the flow tube. The upper surface of the reservoir consists of a flat sintered glass membrane, the pores of which are sufficiently small to prevent air from entering the system, whilst allowing the passage of water. When a sample of soil is placed on the membrane the water meniscus moves towards the reservoir, indicating that water is being drawn into the soil. If a vacuum pump is

Fig. 1

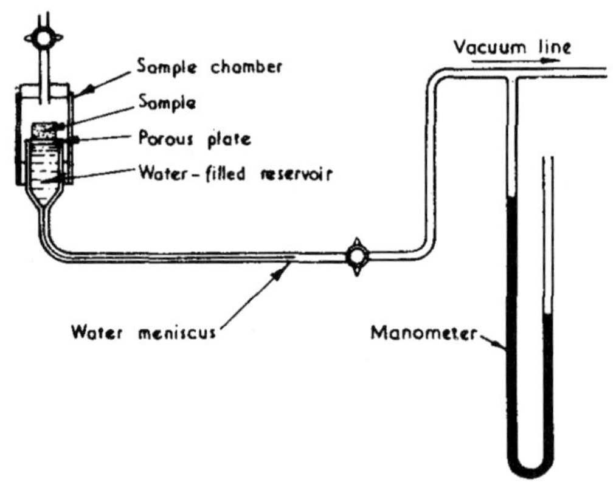

Method of measuring soil suction

connected to the open end of the flow tube, the reduction of pressure necessary to arrest movement of water into the soil is equal to the suction exerted by the soil.

The suction of soil increases as the soil becomes drier. By definition the suction of a sample in equilibrium with free-water is zero, but as moisture is removed from the soil the suction rises, until at oven-dryness it may be equivalent to many thousands of atmospheres. The rapid increase of suction as soil dries makes it convenient, when considering a wide moisture range, to express soil suction on a logarithmic scale. On the *pF* scale (Schofield, 1935) the common logarithm of the suction expressed in centimetres of water is equivalent to the *pF* value of the soil moisture.

Using Kelvin's equation relating the pressure deficiency across a water meniscus to the depression of vapour pressure at the surface of the water, it is possible to express soil suction in terms of the relative humidity of the atmosphere with which the soil is in vapour equilibrium. This relationship, and other aspects of the thermodynamics of soil moisture, are considered in greater detail elsewhere (Croney and Coleman, 1948). Table 1 shows the relationship between the *pF* scale, the equivalent negative hydraulic head, the equivalent negative pressure expressed in lb. per square inch, and the corresponding relative humidity.

Table 1

The *pF* scale in terms of the equivalent negative hydraulic head, the equivalent suction in lb. per square inch and relative humidity

pF	Equivalent negative hydraulic head		Suction (lb./sq. in.)	Relative humidity (per cent)
	cm.	ft.		
0	1	$3{\cdot}28 \times 10^{-2}$	$1{\cdot}42 \times 10^{-2}$	
1	10	$3{\cdot}28 \times 10^{-1}$	$1{\cdot}42 \times 10^{-1}$	
2	10^2	3·28	1·42	99·99
3	10^3	$3{\cdot}28 \times 10$	$1{\cdot}42 \times 10$	99·92
4	10^4	$3{\cdot}28 \times 10^2$	$1{\cdot}42 \times 10^2$	99·27
5	10^5	$3{\cdot}28 \times 10^3$	$1{\cdot}42 \times 10^3$	93·00
6	10^6	$3{\cdot}28 \times 10^4$	$1{\cdot}42 \times 10^4$	48·43
7	10^7	$3{\cdot}28 \times 10^5$	$1{\cdot}42 \times 10^5$	00·07
$-\infty$	0	0	0	100

THE RELATIONSHIP BETWEEN SOIL SUCTION AND MOISTURE CONTENT

Although the suction of soil increases with decreasing moisture content, the relationship for any material is not in general unique. A single method cannot be used for exploring the variation of suction with moisture content over the moisture range from saturation to oven-dryness, and in practice at least three methods have to be employed.

Over the suction range *pF* zero to *pF* 3, i.e. up to suctions approaching 1 atmosphere, a suction plate method similar in principle to that shown in Fig. 1 can be used. The sample, initially saturated in the presence of free-water, is placed on a saturated porous plate, the pressure within the pores of which can be adjusted to any value between atmospheric pressure and about 3 centimetres of mercury. The suction on the plate is increased by stages, the sample being allowed to reach moisture equilibrium with the plate between each change of suction. From the series of wet weights corresponding to each suction, and the dry weight of the sample determined at the conclusion of the tests, the variation of moisture content with suction can be obtained. A similar test can be carried out with the sample " wetting ", if an initially dry sample is used and the suction is reduced in steps from its maximum value.

For suctions between *pF* 3 and *pF* 4·5 a centrifuge technique is used. The sample under test is placed on the surface of a porous cylinder mounted in the centrifuge cup. The lower end of the cylinder is in contact with free-water maintained at constant level, so that in the earth's field a suction of a few centimetres of water is exerted on the sample. This suction is increased n-fold in a field of ng, so that with a sample 5 centimetres above the water-table, and a field of 5,000g, suctions of the order of 25,000 centimetres of water (pF4·4) can be obtained.

Above pF 4·5, indirect methods are used, the simplest of which is the vacuum desiccator. This method involves the relationship between suction and relative humidity already mentioned. The relative humidity of the atmosphere, in vapour equilibrium with many aqueous solutions, has been related accurately to the concentration of the solutions. This provides a ready means of obtaining a known humidity within an enclosure. Sulphuric acid solutions can be conveniently used to give humidities corresponding to the suction range pF 4·5 to pF 7. The sample under test is suspended in an enclosure containing acid of the required concentration, and moisture transfer occurs in the vapour phase until the sample is in suction equilibrium with the humidity of the acid. Evacuation to a pressure approximating to the vapour pressure of the acid accelerates the transfer process.

Although the relationship between suction and moisture content is broadly similar for all porous materials, there are important differences associated with the particular physical characteristics of each material. These differences are best appreciated by considering the relationship between suction and moisture content for the four groups into which the materials with which the civil engineer is concerned can be classified. These are, (1) materials of rigid structure, such as chalk, plaster, cement products, etc., (2) incompressible soils, such as clean sands and gravels, (3) saturated compressible soils, such as heavy clays, and, (4) partially compressible soils such as sandy and silty clays.

Materials of rigid structure. Fig. 2 shows the suction/moisture-content relationships for samples of soft chalk and hard chalk, between saturation at zero suction and oven-dryness. The difference between the saturation moisture contents of the two samples reflects the effect of consolidation pressure during the formation of the chalk beds.

For both samples, as the suction applied to the initially saturated material is increased, no appreciable change in moisture content occurs until a suction of about pF 2·8 is reached, when the water menisci at the surface of the material break down allowing air to enter. A steady decrease of moisture content occurs as the suction is further increased until the samples are virtually dry at pF 4·5. The suction at which air first enters the structure is rather lower for the soft chalk than for the hard variety, confirming the difference between the average pore size of the two materials. At suctions above pF 3·6 the curves are practically identical for both samples. In this range the whole of the water present is probably contained in the "breathing" channels perforating the shell-like material of which the chalk is constituted. The size of these channels would be unaffected by the consolidation process, and similar curves would therefore be expected for both materials in this range.

For both samples of chalk there is a marked hysteresis between the curves corresponding to the wetting and drying conditions. This hysteresis probably arises from the degree of control which the smaller pores exert on the suction at which the larger pores release their water as the suction is increased. Thus water is retained in large pores, during the drying process, at suctions much greater than those at which the same pores will re-fill during the wetting process.

As a result of the hysteresis there is no unique relationship between suction and moisture content. For the soft chalk, at a suction of pF 3 the moisture content would lie between 16·4 and 25 per cent, the actual value depending on the the previous moisture history of the material. If the range over which the suction varies is limited, the hysteresis is smaller. For the soft chalk (Fig. 2), the loop ABCD represents the limiting moisture content condition when the suction varies between the limits pF 3 and pF 3·5.

Fig. 3 shows the relationship between suction and moisture content for samples of plaster of Paris made to different water/plaster ratios. Drying curves only are shown (saturation to air-dryness), and the suction is in this instance expressed on a linear scale. The effect of the water/plaster ratio on the pore size of the mix is reflected in, (a) the increase of saturation moisture content, and, (b) the decrease in the suction at which air enters the structure, both of which occur as the water/plaster ratio is increased.

Incompressible soils. Since the volume of an incompressible soil is not affected by changes of moisture content, the suction/moisture-content relationship for a sample of such a soil is similar to that for a material of rigid structure. Consideration of the incompressible soil

Fig. 2

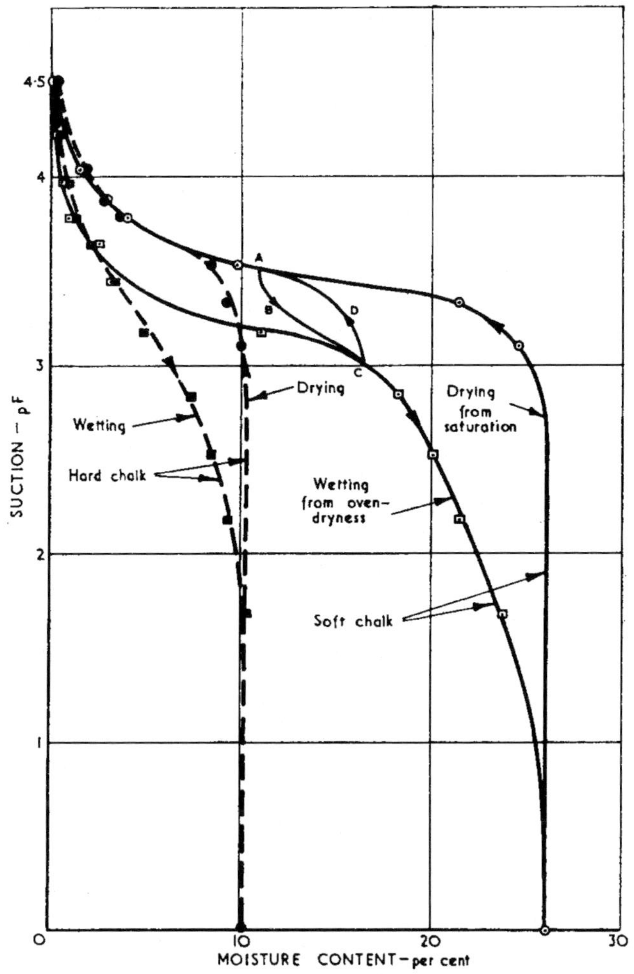

Relationships between suction and moisture content for samples of hard and soft chalk

case is, however, complicated by the fact that for each soil there is a family of relationships corresponding to different dry densities of the soil.

As the dry density of the soil is increased, the saturation moisture content decreases and the size of the pores decreases. The general effect of increasing the dry density, therefore, on the suction/moisture-content relationship is similar to that of decreasing the water/plaster ratio as shown in Fig. 3. A complete study of the suction characteristics of incompressible soils has not yet been carried out at the Road Research Laboratory. Fig. 4 shows wetting and drying curves for a well graded sand at one dry density. The grading of the sand is shown inset on the figure.

*

Fig. 3

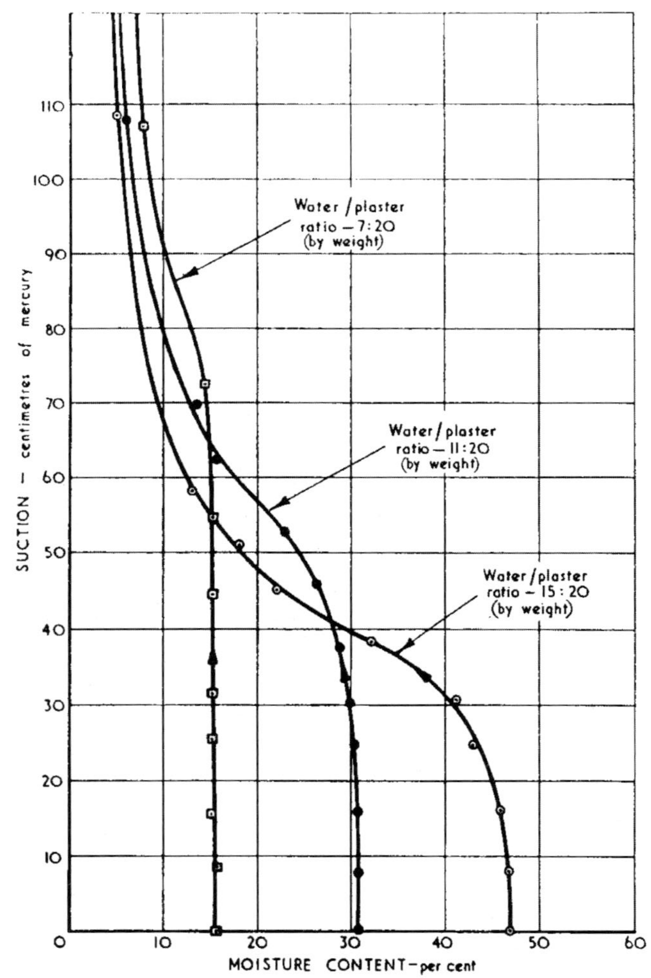

Fig. 4

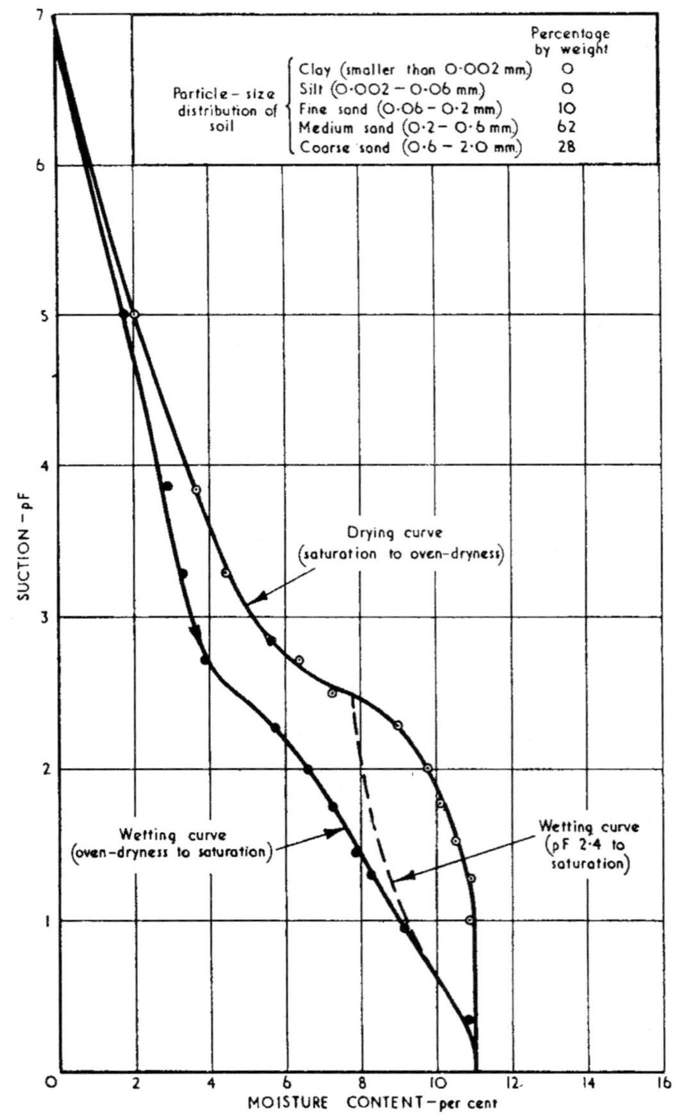

Above: Relationship between suction and moisture content for three mixes of plaster of Paris—drying condition

Right: Relationship between suction and moisture content for samples of sand at a dry density of 133 lb./cu. ft.

Compressible soils. The relationship between suction and moisture content for compressible soils is more complex. Since such soils swell and shrink with changing moisture content, the relationship between suction and moisture content does not represent a constant condition of dry density. A continuous increase in the dry density occurs as the suction is increased from zero, at least until the shrinkage limit is reached. Furthermore, during the drying process some relative re-orientation of the particles is likely to occur so that the soil does not revert, on re-wetting, to its original density. These points can be seen more clearly from Fig. 5, which shows the results of a series of tests carried out on a heavy clay soil, at the Road Research Laboratory.

An " undisturbed " sample of the soil was allowed to become wetter on a suction plate operating at approximately *pF* 0·8. The suction was then increased in stages to *pF* 7 (oven-dryness), to give the drying curve A. The soil was then wetted progressively to *pF* 0·5, to give the wetting curve B. The difference between the two curves A and B is not solely attributable to hysteresis ; the difference between the moisture contents at *pF* 0·8 indicates the extent of re-packing of the particles induced by the drying process. Curve C shows the next drying cycle, *pF* 0·5 to oven-dryness. The subsequent wetting curve followed the first wetting curve, B, almost exactly. Curves B and C therefore form a closed hysteresis loop, suggesting that once the intense consolidation pressure associated with oven-drying has been applied to the soil, any subsequent increase of moisture content, unaccompanied by mechanical disturbance, causes the particles to move apart without any material modification to their arrangement. This concept is compatible with the plate-like form attributed to the clay particles. The hysteresis between the limited curves B and C, which is relatively smaller than that between the corresponding curves for incompressible soils, cannot in this case be due to differences in the suctions at which the pores fill or lose their moisture, since over the greater part of the moisture range the heavy clay soil would be saturated. It is more likely associated with the resistance to deformation that the soil structure offers during the wetting and shrinking process.

Fig. 5

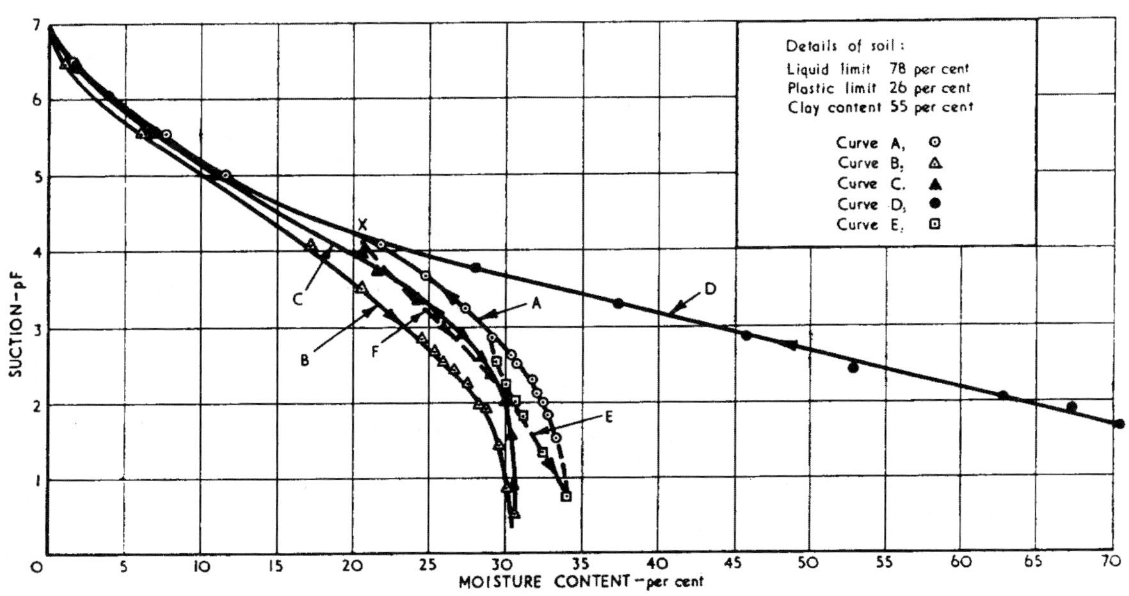

Relationship between suction and moisture content for disturbed and undisturbed samples of a heavy clay soil

The small hysteresis loop on the initial drying curve A was obtained (using another sample of the same soil) when the suction was decreased to *pF* 0·8 by stages, after a suction of *pF* 2·9 had been reached on the drying curve. The fact that this loop is closed indicates that no re-orientation of the particles resulted from the increase of the suction to *pF* 2·9. This raises the question of what suction is required to modify the particle packing. The most likely possibility is that no material modification takes place until the suction in the soil water exceeds the maximum suction experienced during the geological history of the soil. This point could be investigated further by subjecting the soil to suction cycles of increasing amplitude, and establishing at what maximum suction the loop just failed to close. In practice protracted tests of this type are difficult to conduct with accuracy, owing to errors introduced by loss of weight during repeated handling of the sample. Some confirmation of this pre-consolidation theory is provided by the drying curve D obtained on the same soil. For this test the soil was initially slurried to a high moisture content, the object being completely to destroy the structure of the clay. The suction applied to the soil was then progressively increased and the corresponding decrease of moisture content observed. The coincidence of the two curves A and D for suctions exceeding *pF* 4·4, indicates that at this suction the structure created in the slurried soil, by the drying process, is similar to that of undisturbed natural soil. The point at which the two curves meet would appear to indicate the maximum suction the soil moisture had reached during its history as a result of overburden and water-table conditions.

Partially compressible soils. Between the incompressible and fully compressible soil types there is a range of soils which are normally not saturated, but which exhibit some shrinkage when the moisture content is reduced. Such soils can be referred to as partially compressible. When the suction is increased in an initially saturated, partially compressible soil, the structure remains saturated over a limited range of suctions but the moisture content falls owing to shrinkage. Air enters the soil at a suction depending mainly on the clay content of the soil, but some shrinkage continues to occur as the suction is increased. The general characteristics of the complete relationship are therefore intermediate between those of a fully compressible, saturated soil, and an incompressible soil. Fig. 6 shows the relationship for a sandy soil with a low clay content (grading inset on the figure). This relationship applies to one condition of dry density. Owing to the low clay content any re-orientation of the clay fraction during drying is insufficient to prevent closure of the loop on subsequent re-wetting.

SOIL SUCTION AND THE DISTRIBUTION OF MOISTURE IN SOIL

If the suction at any point in a soil initially in a state of moisture equilibrium is changed, water will re-distribute itself until a new equilibrium condition is established. Even in a homogeneous soil, an equilibrium moisture condition may not mean a uniform moisture content. The previous discussion of the suction characteristics of soils has shown that if one part of the soil reaches equilibrium as a result of an increase in moisture content, it may have a lower equilibrium moisture content than an adjacent part of the soil which has reached equilibrium from the drying condition. A small moisture gradient will also be present owing to the mass of the water itself.

In the field the moisture in exposed soil is never in a state of true equilibrium. Rain water absorbed into the soil by the suction forces is subsequently partially or wholly removed, as a result of vapour transfer between the soil and the atmosphere, or by transpiration from vegetation. Both of these processes create suction gradients within the soil itself. If, however, the soil is covered by an impervious surfacing which can be regarded as having an infinite area, an equilibrium moisture distribution will be reached. This distribution will be determined in general by the suction properties of the soil and the position of the water-table. Further, if the moisture distribution at the time when the soil is covered is known, the extent to which the moisture content of the soil will change after covering can be estimated. Methods which are being developed at the Road Research Laboratory for calculating the equilibrium moisture distribution in soils under an infinite impermeable surfacing are discussed in detail in the next section of the article. The extent to which the conclusions can be applied directly to roads and airfields is considered later.

Fig. 6

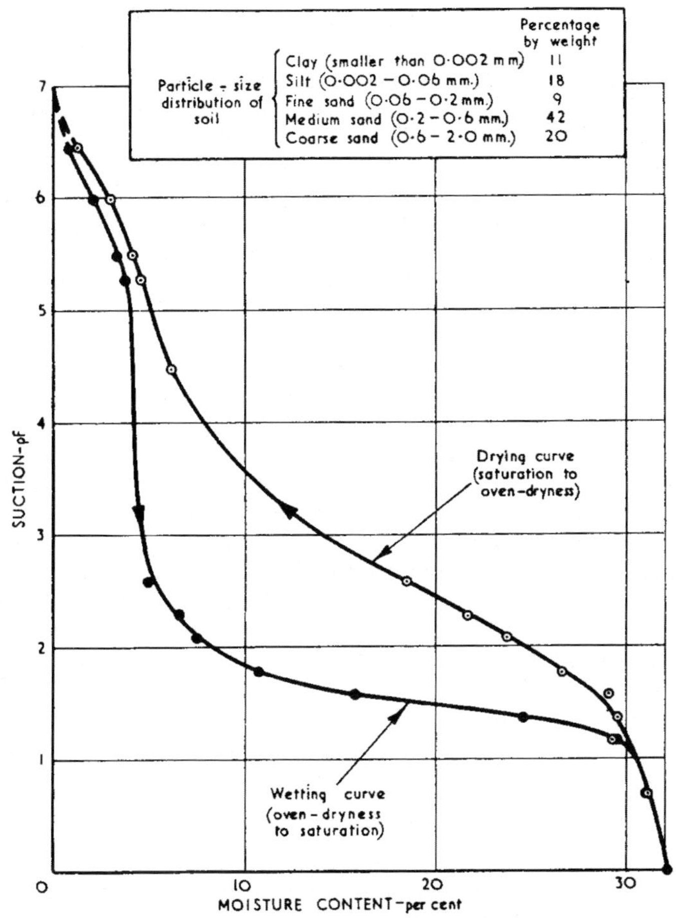

Relationship between suction and moisture content for a sandy soil having a small clay content

THE ESTIMATION OF THE EQUILIBRIUM MOISTURE DISTRIBUTION WITH DEPTH UNDER AN INFINITE IMPERVIOUS PAVEMENT

For materials of rigid structure, or for incompressible soils, the problem is simplified by the fact that no part of the overburden pressure is carried by the pore water. Below the water-table the material is saturated and, assuming uniform density conditions, the moisture content is constant with depth.

At a height x centimetres above the water-table the equilibrium suction is x centimetres of water, and the equilibrium pF of the material is $\log_{10}x$. Hence the moisture content at any height above the water-table can be read directly from the appropriate suction/moisture-content relationship. Suction curves of the type shown in Figs 2 and 4, which represent the limiting conditions for the material fluctuating in moisture content between saturation and oven-dryness, could only be used, however, to give limiting moisture distributions between which the equilibrium distribution beneath an impervious surface would lie. Curves A and

B of Fig. 7 show limiting moisture distributions calculated in this manner for the soil referred to in Fig. 4, and for a water-table at a depth of 4 feet.

The actual equilibrium moisture distribution reached under an impervious surface would depend, in practice, on the moisture conditions prevailing at the time when the soil was covered. During winter, under the climatic conditions of Great Britain, the presence of gravitational water is likely to increase the moisture content in the upper layers of unsaturated

Fig. 7

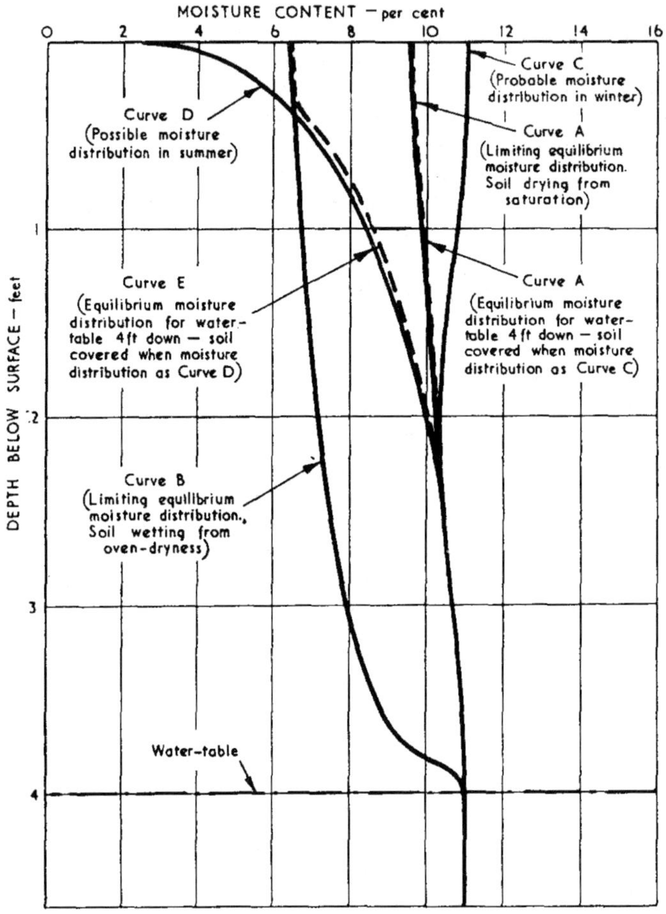

Moisture distribution curves for incompressible soil

soils to a value equivalent to a condition close to saturation. If, therefore, the surface is covered during winter, the suction/moisture-content relationship followed by the soil will be represented closely by the drying curve from saturation. If the soil referred to in Fig. 7 were covered when the moisture distribution was as curve C, the equilibrium distribution reached would be given by curve A. During summer the extent to which the upper layers of uncovered soil dry out depends on the air temperature and humidity conditions ; a state approximating to oven-dryness could never be reached except perhaps in the extreme surface. If the soil is surfaced under dry summer conditions the soil will become wetter, but the moisture contents reached in the upper soil layers will be less than those for the same soil surfaced in winter. Referring again to Fig. 7, suppose the moisture distribution in the uncovered soil in summer were given by curve D. On covering the surface, each horizon of the soil would become wetter along a different wetting curve of the suction/moisture-content relationship, depending on the extent to which the soil had dried along the drying curve. From a family of wetting curves, each originating from a different point on the drying curve of suction/moisture-content relationship (saturation to oven-dryness), and a knowledge of the actual distribution of moisture with depth at the time of surfacing, the equilibrium moisture distribution under the surfacing can be obtained. Since a family of wetting curves can be drawn with reasonable accuracy if one experimentally determined curve is available, the complexity of the problem of estimating the equilibrium distribution is reduced. Curve E on Fig. 7 shows the equilibrium moisture distribution calculated in this manner for the soil wetting from the moisture condition represented by curve D. The conclusion is, therefore, that if an incompressible soil is covered in winter, it will become drier until the moisture distribution corresponds to the upper limiting value corresponding to the prevailing water-table condition. Under summer conditions, any part of the soil which has a moisture content lower than that given by the lower limiting moisture distribution will become wetter, to reach a value given by that distribution. The moisture content of any part of the soil which has not dried to the lower limiting moisture distribution will become wetter only to a limited extent.

The problem of estimating the equilibrium moisture distribution in saturated clays under an impervious surfacing of infinite extent, is complicated because the over-burden in such soils is carried by the soil water and not by the solids. The moisture content in any zone depends therefore both on the suction properties of the soil and on the overburden above the zone. If the moisture content at any point in the soil is known, the suction associated with this moisture content can be obtained directly from the suction/moisture-content relationship. The actual pressure in the soil water is the algebraic sum of this suction and the pressure imposed by the overburden. As in the incompressible soil, the resultant pressure (positive or negative) is, under equilibrium conditions, determined by the position of the water-table. The moisture under an impervious surface will therefore adjust itself until the algebraic sum of the suction at every point and the corresponding overburden pressure is equal to the equilibrium pore-water pressure, as determined by the position of the water-table. This condition can be expressed by the equation :—

$$P - s = u \qquad (1)$$

where P denotes the pressure due to overburden ;

s ,, the suction (expressed as a positive quantity) corresponding to the moisture content at the point under consideration (as deduced from the suction/moisture-content relationship) ; and

u ,, the pore-water pressure determined by the position of the water-table.

This equation, which corresponds to the consolidation equation $\sigma_e = \sigma_n - u$, enables a new interpretation to be made of the significance of the effective pressure σ_e, in relation to the normal pressure σ_n. At the water-table, the suction of the soil is exactly neutralized by the over-burden pressure ; above the water-table there is a residual negative pore-water pressure, and below the water-table there is a residual positive pore-water pressure.

The manner in which equation (1) is used to obtain the equilibrium moisture distribution is discussed in detail in another article (Croney, Lewis, and Coleman, 1950). Since the over-burden pressure P is a function of the moisture content of the soil, a successive approximation method must be used to obtain the distribution. The moisture content is first considered to be constant with depth, at an average value chosen from a preliminary inspection of the suction/moisture-content relationship. Using this value of moisture content, the bulk density of the soil is calculated and used to give a first approximation for P at each depth in the soil. Using equation (1) the suction at each depth is calculated and hence the moisture content of the soil. This procedure gives a first approximation to the moisture distribution which is sufficiently accurate for most purposes. If required a second approximation can be made, using the moisture distribution obtained, to calculate more accurately the overburden pressure acting at each level and hence a more accurate picture of the moisture distribution itself. Since any superimposed load increases the value of P by a fixed amount at every level in the soil, equation (1) can also be used to determine the effect which surface loading has on the equilibrium moisture distribution in saturated clay soils.

As for incompressible soils, the final moisture distribution in a saturated clay must depend to some extent on the moisture conditions at the time when the surface is covered, since the previous moisture history of the soil determines which suction/moisture-content curve should be used in arriving at the distribution. During the geological period when the overburden was in the process of formation, it is probable that the suction/moisture-content relationship for the heavy clay, referred to in Fig. 5, followed a drying curve such as curve D to a point X, corresponding to the minimum moisture content and maximum suction attained by the soil. During the removal of the overburden by subsequent geological processes the clay would take up moisture, following a wetting curve such as curve F shown chain-dotted on Fig. 5.

Neglecting the effect of surface moisture changes, the condition of the soil in the ground would therefore be represented by the curve F, and a theoretical moisture distribution based on curve F can be deduced for any position of the water-table, using the method already discussed. Curve a of Fig. 8 shows this distribution for the water-table at a depth of 4 feet. Assuming that the water-table was raised to the surface the distribution would move to the right to give curve b, which would thus represent the wettest condition which could be present in the soil. In practice this condition would never be realized during the short period of one winter, and curve d represents a more probable winter moisture distribution. Once the soil had wetted it would not return to the theoretical moisture distribution on sealing the surface, since the soil would dry along a suction curve to the right of the curve along which it wetted, i.e. it would dry along a curve of the type A on Fig. 5, the actual curve depending on the point reached on the wetting curve F, before drying commenced. By drawing a family of such drying curves it is possible to estimate the extent of the drying out which would occur for any given initial moisture distribution. Curves c and e in Fig. 8 show equilibrium distributions obtained in this manner for the initial distributions b and d. (No attempt has been made to draw a family of drying curves on Fig. 5.)

During summer the upper few feet of the soil dry out owing to evaporation and transpiration from vegetation, to give a moisture distribution of the type shown by curve f of Fig. 8. During the drying process the suction condition of the soil would be represented by a family of drying curves of the type already discussed, the actual curve followed by the soil at a given depth depending on the extent to which the soil at that depth had previously wetted along curve F of Fig. 5. On sealing the surface and thereby preventing evaporation, the soil would re-wet following a family of wetting curves, of which curve E of Fig. 5 is an example. Since these wetting curves would all re-join the main wetting curve F, it follows that the soil would, on sealing, become wetter again to reach the theoretical distribution curve a of Fig. 8. If, however, during the drying process any part of the soil reached a moisture content lower than that represented by the point X on Fig. 5, some modification of the particle packing would have occurred in that part of the soil, which would prevent complete re-wetting to the curve A.

The conclusion is that if a compressible soil is covered in winter with an impermeable surface of infinite extent, when the moisture content of the surface soil is above the theoretical equilibrium value, the moisture content will decrease only to a limited extent, the final equilibrium being above the theoretical value. If, on the other hand, the soil is covered in summer when the moisture content is low, wetting to the theoretical equilibrium will occur.

Increasing the overburden on a saturated clay has an effect on the equilibrium moisture distribution similar to lowering the water-table. If the surface load is increased by n lb. per square foot, without any change of water-table level, the moisture content of the soil must decrease so that the suction at every point is increased by n lb. per square foot. Hence

Fig. 8

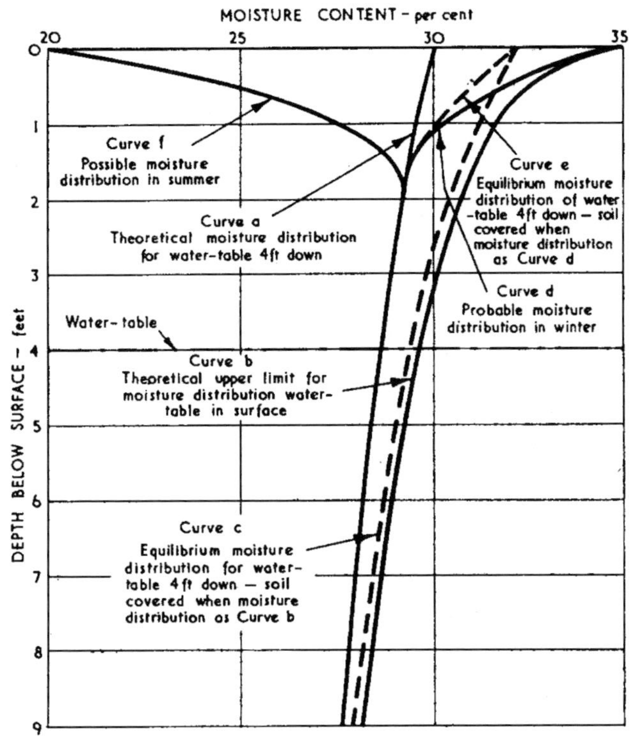

Moisture distribution curves for saturated compressible soil

the addition of the surface surcharge has the same effect on the moisture distribution as lowering the water-table a distance of $n/62{\cdot}5$ feet.

The change of volume which accompanies any change in the moisture distribution in a saturated clay, can be deduced from the initial and final moisture distribution curves. The change of volume per unit volume (dV/V) which accompanies any change of moisture content can be readily determined. This quantity can be integrated over any finite depth of soil to give the total volume change occurring within that depth.

A method of calculating the equilibrium moisture distribution in unsaturated partially compressible soils, i.e. the case intermediate between the incompressible soil and the saturated clay, is at present being developed at the Road Research Laboratory. In this case the overburden is carried partly by the pore water and partly by the soil particles, the former fraction decreasing as the soil becomes drier. From the progress of the work to date, it appears that the fraction of the overburden carried by the soil water can be deduced from the shrinkage characteristics of the soil. For soils of this type, therefore, the volume/moisture-content relationship is required in addition to the suction/moisture-content relationship. Possible methods of studying these relationships simultaneously are under review.

If the fraction α of the overburden carried by the pore water can be deduced in this manner, equation (1) can be expressed in a more general form :—

$$\alpha P - s = u \quad . \; . \; . \; . \; . \; . \; . \; . \; . \; . \; . \; . \quad (2)$$

This equation will then be applicable to all soils, since, for the incompressible soil $\alpha = 0$ and equation (2) becomes $-s = u$, the relationship which in effect was used in the previous discussion of the distribution of moisture in incompressible soils. In a saturated clay $\alpha = 1$, and equation (2) reverts to equation (1) used in the analysis of the case for saturated clays.

**

MOISTURE DISTRIBUTION IN NON-UNIFORM SOILS

Soil deposits are seldom, if ever, uniform either with depth or area. Whilst it is possible from suction tests carried out on the surface soil to estimate the equilibrium moisture content in the surface, the change of moisture content with depth is likely to be influenced by variations in the soil type. In soils which are not uniform with depth, the equilibrium moisture distribution can be estimated from a number of suction tests carried out on samples taken from different depths.

The possibility of relating the suction properties of soils to soil type and density is being investigated at the Road Research Laboratory. If a reliable basis for a correlation of this kind can be found it should be possible to estimate the equilibrium moisture content beneath an impermeable surface and the equilibrium moisture distribution with depth, from the results of tests simpler and more quickly carried out than the suction/moisture-content determination.

For cohesive soils, a preliminary investigation has been made into the possibility of basing equilibrium moisture contents in undisturbed soils solely on the results of the index tests. Since the suction properties of soils are affected by the geological and chemical history of the soil as well as by mechanical disturbance, a unique relationship between these properties and the liquid and plastic limits would not be expected. However, for many clay soils there does appear to be a correlation between the amount of water which the undisturbed soil will hold at a given suction, and the liquid and plastic limits of the soil. The suction properties of a number of clay soils obtained from southern England have been determined, and the amount of water held in the soil (in the wetting condition) has been expressed, in terms of the index test results, by a family of curves each of which corresponds to a suction between *pF* zero and *pF* 3. Fig. 9 (a) shows a typical curve for *pF* 2. The scatter of the points may in part be attributable to the experimental techniques used for determining the suction relationships for the soils. These techniques have since been considerably improved. The scatter is too great to be accounted for wholly in this manner. The geological factors already mentioned are probably partly responsible for some of the discrepancies. If the equilibrium

Figs 9

(a) (b)

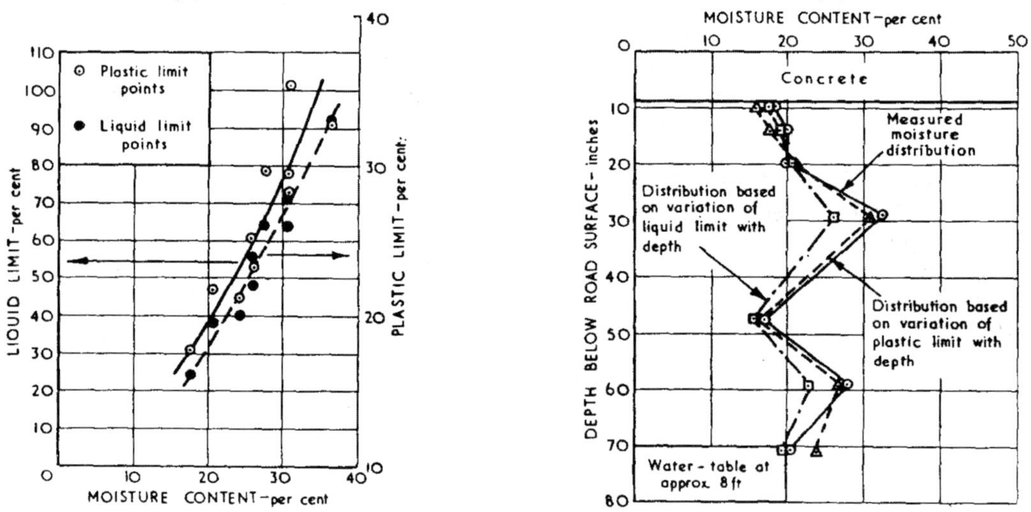

Relationship between moisture content of clay soils at *pF*2 and liquid and plastic limits (soils wetting)

Measured change in moisture content with depth, beneath concrete road, compared with distributions estimated from the change in liquid and plastic limits with depth

suction at different depths in the soil is calculated using the methods already discussed, and the liquid or plastic limits of the soil at those depths are known, a family of curves of the type shown in Fig. 9 (a) can be used to determine the equilibrium moisture distribution. Fig. 9 (b) shows the actual measured distribution of moisture beneath an impervious surface, and the distributions calculated in the above manner from both the liquid and the plastic limit profiles. The distribution based on the plastic limit agrees more closely with the experimental results than does that deduced from liquid limits.

A great deal more research would, of course, be required to ascertain to which soils curves of the type shown in Fig. 9 (a) are applicable before the curves could be used with confidence in practice.

APPLICATION OF THE EQUILIBRIUM MOISTURE DISTRIBUTION CONCEPT TO ROADS AND AIRFIELDS

The methods outlined in this article can be used to estimate the moisture content beneath road or airfield pavements if :—

(a) The surfacing is impervious to water in the liquid and vapour phases.

(b) The moisture transfer between the soil and the atmosphere, which occurs at the boundary of the covered area, has a negligible effect on the moisture distribution beneath the pavement (i.e. if the pavement can be regarded effectively as having an infinite area).

A concrete or bituminous surfacing will normally fulfil the first of these conditions. Whether or not the second is fulfilled will depend primarily on the prevailing climatic conditions. At the boundary of the road or airfield, if the uncovered soil at a given depth is drier or wetter than that at the same depth under the pavement, suction and vapour pressure gradients will cause water to move into or out of the subgrade, preventing the attainment of a true equilibrium distribution. In Great Britain, where seasonal fluctuations of moisture content affect only the top few feet of soil and where water-table conditions are generally high, the influence of boundary effects on the moisture distribution beneath pavements is small. During prolonged periods of drought, however, moisture conditions near the edges of roads may be affected sufficiently to cause settlement and cracking of roads constructed on clay subgrades.

Under desert conditions, where the water-table may be at a very considerable depth and where the soil is naturally very dry, an equilibrium moisture distribution of the type just discussed would never be realized under a road, the width of which might represent only a negligible fraction of the depth of the water-table. Under such conditions, temperature gradients and the associated movements of water in the vapour phase are likely to play a more important part in determining the moisture content beneath the pavement than the hydrostatic suction gradients just considered.

It is difficult to say, without experimental evidence, to what extent the moisture distribution beneath structures can be estimated using these methods under other climatic conditions. To obtain this evidence the Road Research Laboratory is encouraging, in the British Colonies and elsewhere, field studies into the moisture distribution under covered and uncovered areas of soil and the seasonal changes which occur in these distributions.

EFFECT OF VEGETATION ON THE DISTRIBUTION OF MOISTURE IN SOIL

The suction exerted by the roots of vegetation can have a considerable effect on the moisture distribution under pavements. Under normal conditions the water removed by vegetation is replaced by rainfall. If, however, the supply of rainfall is removed, e.g. as a result of prolonged drought or due to paving of the soil round the vegetation, the roots are capable of extracting moisture until the suction of the soil rises to a value exceeding pF 4. This may mean a considerable reduction of moisture content, accompanied by shrinkage if the soil is compressible. During drought the hair roots through which the water is extracted develop rapidly under structures on account of the comparative availability of the water. Rapid changes of soil moisture content result, and, in the case of a road, the riding qualities may be affected. When heavy clays are dried by trees to a moisture content lower than that previously reached in the soil, the soil will not re-assume its initial moisture condition after removal of the trees, and some permanent settlement results. The wetting process after the removal of vegetation is necessarily slow, since the complicated root structure by which the moisture is removed is not available to assist the wetting process.

EXPERIMENTS TO INVESTIGATE THE DISTRIBUTION OF MOISTURE IN SOILS

Three approaches are being used at the Road Research Laboratory to verify the moisture distribution theories outlined in this article :—

(1) During the investigation of several road foundation failures on clay subgrades, the moisture distribution with depth has been studied by boring and sampling. Suction tests have been carried out on undisturbed samples taken from beneath the road surface. The moisture content beneath the pavement has generally been found to agree closely with the value calculated for the water-table and surcharge conditions. Where the soil has been reasonably uniform with depth, there has been satisfactory agreement between the actual and estimated moisture distributions with depth.

(2) A programme of research is in hand in which the seasonal variations in moisture content are being measured, both beneath paved areas and in adjacent uncovered soil. In addition to moisture content it is proposed to record changes in pore-water pressure and soil temperature at different depths. Experiments with plaster of Paris moisture gauges (Croney, Coleman, and Currer, 1951) carried out in connexion with this investigation, have shown that such gauges are suitable for measuring suctions between *pF* 2·5 and *pF* 4, but they do not provide a reliable method of measuring changes of soil moisture content.

(3) Experiments with clay blocks have been in progress for some time at the Laboratory. The blocks, approximately 12 inches square and 18 inches deep, were excavated from a depth of about 20 feet. They are being allowed to become wetter in a bed of sand with a water-table 2 feet below the upper surface. Changes of pore-water pressure are being observed by tensionmeters at different points in the blocks, and, in the case of one block, swelling is being measured by dial gauges. The suction properties of each block have been studied independently.

After approximately 15 months the blocks are approaching an equilibrium condition. Preliminary measurements of the moisture distribution with depth indicate that the clay has become wetter by an average of 4 to 5 per cent of moisture content, and that the equilibrium moisture distributions are likely to be in reasonably close agreement with those predicted from the theory relating to saturated clay soils.

ACKNOWLEDGEMENTS

The work described in this article was carried out at the Road Research Laboratory of the Department of Scientific and Industrial Research as part of the programme of the Road Research Board. The article is published by permission of the Director of Road Research.

The Author's thanks are due to Mr D. J. Maclean, head of the Soils Section of the Road Research Laboratory, and to Mr J. D. Coleman who has been closely associated with the work.

REFERENCES

Croney, D., and Coleman, J. D., 1948. Soil thermodynamics applied to the movement of moisture in road foundations. *Proc. 7th Int. Congr. Appl. Mech.* 3 : 163–177.

Croney, D., Lewis, W. A., and Coleman, J. D., 1950. Calculation of the moisture distribution beneath structures. *Civ. Engg.* (*L.*) 45 (524) : 103–106.

Croney, D., Coleman, J. D., and Currer, E. W. H., 1951. The electrical resistance method of measuring soil moisture. *Brit. J. Appl. Phys.*, 2 (4) : 85–91.

Schofield, R. K., 1935. The *pF* of the water in soil. *Trans. 3rd Int. Congr. Soil. Sci.* 2 : 37–48.

Bishop, A. (1955). *Géotechnique* **5**, No. 1, 7–17

First Technical Session : General Theory of Stability of Slopes

***Tuesday morning*, 21 *September*, 1954**

Session 1/1

THE USE OF THE SLIP CIRCLE IN THE STABILITY ANALYSIS OF SLOPES

by

ALAN W. BISHOP, M.A., Ph.D., A.M.I.C.E.

INTRODUCTION

Errors may be introduced into the estimate of stability not only by the use of approximate methods of stability analysis, but also by the use of sampling and testing methods which do not reproduce sufficiently accurately the soil conditions and state of stress in the natural ground or compacted fill under consideration. Unless equal attention is paid to each factor, an elaborate mathematical treatment may lead to a fictitious impression of accuracy.

However, in a number of cases the uniformity of the soil conditions or the importance of the problem will justify a more accurate analysis, particularly if this is coupled with field measurements of pore pressure, which is the factor most difficult to assess from laboratory data alone. Two classes of problem of particular note in this respect are :

(i) The design of water-retaining structures, such as earth dams and embankments, where failure could have catastrophic results, but where an over-conservative design may be very costly.

(ii) The examination of the long-term stability of cuts and natural slopes where large scale earth movements may involve engineering works and buildings

THE USE OF LIMIT DESIGN METHODS

It has been shown elsewhere by a relaxation analysis of a typical earth dam (Bishop, 1952) that, even assuming idealized elastic properties for the soil, local overstress will occur when the factor of safety (by a slip circle method) lies below a value of about 1·8. As the majority of stability problems occur in slopes and dams having lower factors of safety than this, a state of plastic equilibrium must be considered to exist throughout at least part of the slope.

Under these conditions a quantitative estimate of the factor of safety can be obtained by examining the conditions of equilibrium when incipient failure is postulated, and comparing the strength necessary to maintain limiting equilibrium with the available strength of the soil. The factor of safety (F) is thus defined as the ratio of the available shear strength of the soil to that required to maintain equilibrium. The shear strength mobilized is, therefore, equal to s, where :

$$s = \frac{1}{F}\left\{ c' + (\sigma_n - u)\tan\phi' \right\} \quad . \quad . \quad . \quad . \quad . \quad . \quad . \quad . \quad (1)$$

where c' denotes cohesion, } in terms of effective stress.*
ϕ' denotes angle of shearing resistance, }
σ_n denotes total normal stress,
u denotes pore pressure.

* Measured either in undrained or consolidated-undrained tests with pore-pressure measurement, or in drained tests carried out sufficiently slowly to ensure zero excess pore pressure.

Failure along a continuous rupture surface is usually assumed, but, as the shape and position of this surface is influenced by the distribution of pore pressure and the variation of the shear parameters within the slope, a generalized analytical solution is not possible and a numerical solution is required in each individual case. The rigorous determination of the shape of the most critical surface presents some difficulty (see, for example, Coenen, **1948**), and in practice a simplified shape, usually a circular arc, is adopted, and the problem is assumed to be one of plane strain.

MECHANICS OF THE CIRCULAR ARC ANALYSIS

In order to examine the equilibrium of the mass of soil above the slip surface it follows from equation (1) that it is necessary to know the value of the normal stress at each point on this surface, as well as the magnitude of the pore pressure. It is possible to estimate the value of the normal stress by following the method developed by Fellenius (**1927**, **1936**), in which the conditions for the statical equilibrium of the slice of soil lying vertically above each element of the sliding surface are fully satisfied.

A complete graphical analysis of this type is most laborious, and this may result in the examination of an insufficient number of trial surfaces to locate the most critical one. Several simplified procedures have been developed using the friction circle method (Taylor, **1937**, **1948**; Fröhlich, **1951**), in which an assumption is made about the distribution of normal stress, but their use is limited to cases in which ϕ' is constant over the whole of the failure surface. More generally the " slices " method is used, with a simplifying assumption about the effect of the forces between the slices.

The significance of this assumption may be examined by considering the equilibrium of the mass of soil (of unit thickness) bounded by the circular arc ABCD, of radius R and centre at O (Fig. 1(a)). In the case where no external forces act on the surface of the slope,

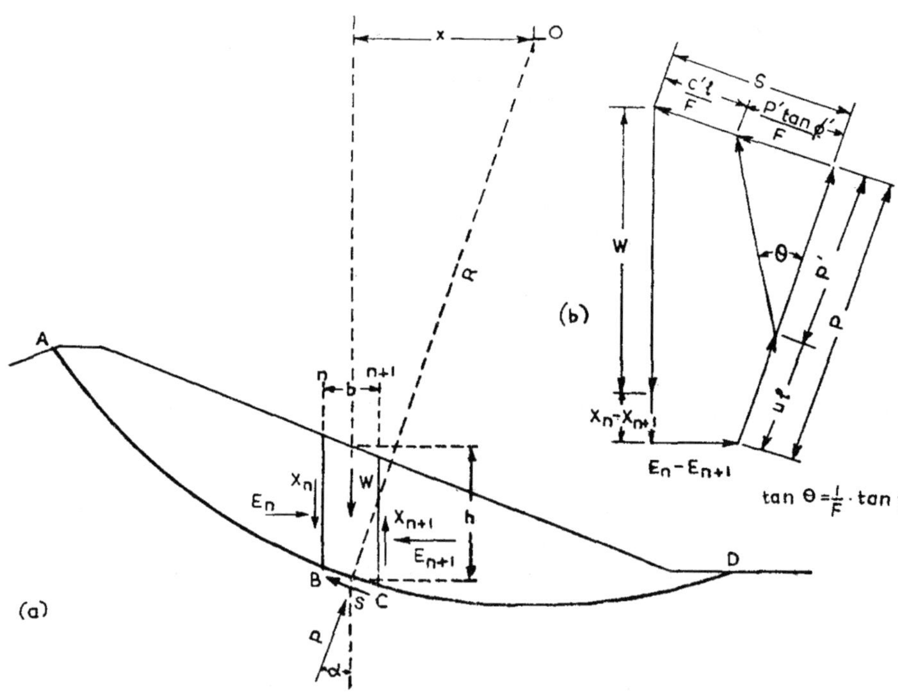

Fig. 1. Forces in the slices method

equilibrium must exist between the weight of the soil above ABCD and the resultant of the total forces acting on ABCD.

Let E_n, E_{n+1} denote the resultants of the total horizontal forces on the sections n and $n + 1$ respectively,

and	X_n, X_{n+1}	,,	the vertical shear forces,
	W	,,	the total weight of the slice of soil,
	P	,,	the total normal force acting on its base,
	S	,,	the shear force acting on its base,
	h	,,	the height of the element,
	b	,,	the breadth of the element,
	l	,,	the length BC,
	α	,,	the angle between BC and the horizontal,
	x	,,	the horizontal distance of the slice from the centre of rotation.

The total normal stress is σ_n, where

$$\sigma_n = \frac{P}{l} \quad . \quad . \quad . \quad . \quad . \quad . \quad . \quad . \quad . \quad . \quad . \quad (2)$$

Hence, from equation (1), the magnitude of the shear strength mobilized to satisfy the conditions of limiting equilibrium is s where :

$$s = \frac{1}{F}\left\{ c' + \left(\frac{P}{l} - u\right) \tan \phi'\right\} \quad . \quad . \quad . \quad . \quad . \quad . \quad . \quad (3)$$

The shear force S acting on the base of the slice is equal to sl, and thus, equating the moment about O of the weight of soil within ABCD with the moment of the external forces acting on the sliding surface, we obtain :

$$\Sigma Wx = \Sigma SR = \Sigma slR \quad . \quad . \quad . \quad . \quad . \quad . \quad . \quad . \quad (4)$$

It follows, therefore, from equation (3) that :

$$F = \frac{R}{\Sigma Wx} \cdot \Sigma\left[c'l + (P - ul) \tan \phi'\right] \quad . \quad . \quad . \quad . \quad . \quad . \quad (5)$$

From the equilibrium of the soil in the slice above BC, we obtain P, by resolving in a direction normal to the slip surface :

$$P = (W + X_n - X_{n+1}) \cos \alpha - (E_n - E_{n+1}) \sin \alpha \quad . \quad . \quad . \quad . \quad (6)$$

The expression for F thus becomes :

$$F = \frac{R}{\Sigma Wx} \cdot \Sigma\Big[c'l + \tan \phi' . (W \cos \alpha - ul) + \tan \phi' . \left\{(X_n - X_{n+1}) \cos \alpha - (E_n - E_{n+1}) \sin \alpha\right\}\Big] \quad . \quad . \quad . \quad . \quad . \quad (7)$$

Since there are no external forces on the face of the slope, it follows that :

$$\Sigma(X_n - X_{n+1}) = 0 \quad . \quad . \quad . \quad . \quad . \quad . \quad . \quad . \quad . \quad (8a)$$

$$\Sigma(E_n - E_{n+1}) = 0 \quad . \quad . \quad . \quad . \quad . \quad . \quad . \quad . \quad . \quad (8b)$$

However, except in the case where ϕ' is constant along the slip surface and α is also constant (*i.e.*, a plane slip surface), the terms in equation (7) containing X_n and E_n do not disappear. A simplified form of analysis, suggested by Krey (1926) and Terzaghi (1929) and also presented by May (1936) as a graphical method, implies that the sum of these terms

$$\Sigma \tan \phi' . \left\{(X_n - X_{n+1}) \cos \alpha - (E_n - E_{n+1}) \sin \alpha\right\}$$

may be neglected without serious loss in accuracy. This is the method at present used, for example, by the U.S. Bureau of Reclamation (Daehn and Hilf, 1951).

Putting $x = R \sin \alpha$, the simplified form may be written :

$$F = \frac{1}{\Sigma W \sin \alpha} \cdot \Sigma \Big[c'l + \tan \phi' . (W \cos \alpha - ul) \Big] \quad . \; . \; . \; . \; . \quad (9)$$

In earth dam design the construction pore pressures are often expressed as a function of the total weight of the column of soil above the point considered, i.e.

$$u = \bar{B} \left(\frac{W}{b} \right) \quad . \; . \; . \; . \; . \; . \; . \; . \; . \; . \quad (10)$$

where $\bar{B}$ is a soil parameter based either on field data or laboratory tests.*

In this case, putting $l = b \sec \alpha$, the expression for factor of safety can be further simplified to :

$$F = \frac{1}{\Sigma W \sin \alpha} \cdot \Sigma \Big[c'l + \tan \phi' . W (\cos \alpha - \bar{B} \sec \alpha) \Big] \quad . \; . \; . \; . \quad (11)$$

This expression permits the rapid and direct computation of the value of F which is necessary if sufficient trial circles are to be used to locate the most critical surface. However, as can be seen from the examples quoted later, the values of F are, in general, found to be conservative, and may lead to uneconomical design. This is especially marked where conditions permit deep slip circles round which the variation in α is large.

To derive a method of analysis which largely avoids this error it is convenient to return to equation (5). If we denote the effective normal force $(P - ul)$ by P' (see Fig. 1(*b*)), and resolve the forces on the slice vertically, then we obtain, on re-arranging :

$$P' = \frac{W + X_n - X_{n+1} - l \left(u \cos \alpha + \frac{c'}{F} \sin \alpha \right)}{\cos \alpha + \frac{\tan \phi' . \sin \alpha}{F}} \quad . \; . \; . \; . \; . \; . \quad (12)$$

Substituting in equation (5) and putting $l = b \sec \alpha$ and $x = R \sin \alpha$, an expression for the factor of safety is obtained :

$$F = \frac{1}{\Sigma W \sin \alpha} \cdot \Sigma \left[\left\{ c'b + \tan \phi' . (W(1 - \bar{B}) + (X_n - X_{n+1})) \right\} . \frac{\sec \alpha}{1 + \frac{\tan \phi' . \tan \alpha}{F}} \right] \quad . \quad (13)$$

The values of $(X_n - X_{n+1})$ used in this expression are found by successive approximation, and must satisfy the conditions given in equations (8). In addition, the positions of the lines of thrust between the slices should be reasonable, and no unbalanced moment should be implied in any slice. The factor of safety against sliding on the vertical sections should also be satisfactory, though since the slip surface assumed is only an approximation to the actual one, overstress may be implied in the adjacent soil.

The condition $\Sigma(X_n - X_{n+1}) = 0$ in equation (8*a*) can be satisfied directly by selecting appropriate values of X_n, etc. The corresponding sum $\Sigma(E_n - E_{n+1})$ can be readily computed in terms of the expression used in equation (13). Resolving the forces on a slice tangentially, we obtain the expression :

$$(W + X_n - X_{n+1}) \sin \alpha + (E_n - E_{n+1}) \cos \alpha = S$$

or

$$(E_n - E_{n+1}) = S \sec \alpha - (W + X_n - X_{n+1}) \tan \alpha \quad . \; . \; . \quad (14)$$

* In practice the value of B will vary along the slip surface, though for preliminary design purposes it is convenient to use a constant average value throughout the impervious zone.

Now, if equation (13) is written :

$$F = \frac{1}{\Sigma W \sin \alpha} . \Sigma[m] \qquad (15)$$

then

$$S = \frac{m}{F} \qquad (16)$$

and hence

$$\Sigma(E_n - E_{n+1}) = \Sigma\left[\frac{m}{F}\sec\alpha - (W + X_n - X_{n+1})\tan\alpha\right] \qquad (17)$$

The X values must, therefore, also satisfy the condition that :

$$\Sigma\left[\frac{m}{F}\sec\alpha - (W + X_n - X_{n+1})\tan\alpha\right] = 0 \qquad (18)$$

In practice, an initial value of F is obtained by solving equation (13) on the assumption that $(X_n - X_{n+1}) = 0$ throughout. By suitably tabulating the variables a solution can be obtained after using two or three trial values of F. The use of $(X_n - X_{n+1}) = 0$ satisfies equation (8*a*), but not equation (18). Values of $(X_n - X_{n+1})$ are then introduced in order to satisfy equation (18) also. These values can then be finally adjusted, either graphically or analytically, until the equilibrium conditions are fully satisfied for each slice.

From the practical point of view it is of interest to note that, although there are a number of different distributions of $(X_n - X_{n+1})$ which satisfy equation (18), the corresponding variations in the value of F are found to be insignificant (less than 1% in a typical case).

It should also be noted that, since the error in the simplified method given in equations (9) and (11) varies with the central angle of the arc, it will lead to a different location for the most critical circle than that given by the more rigorous method. It will, therefore, generally be necessary to examine a number of trial circles by this latter method.

PARTIALLY SUBMERGED SLOPES

The total disturbing moment is now that of the soil above ABCD, less the moment about O of the water pressure acting on DLM (Fig. 2 (*a*)). If we imagine a section of water bounded by a free surface at MN and outlined by NDLM, and similarly take moments about O, the normal forces on the arc ND all pass through O and the moment of the water pressure on DLM is therefore equal to the moment of the mass of water NDLM about O.

Since the weight of a mass of saturated soil less the weight of water occupying the same volume is equal to its submerged weight, the resultant disturbing moment due to the mass of soil above ABCD and to the water pressure on DLM is given by using the bulk density of the soil above the level of the external free water surface and the submerged density below.

It should be noted that the boundary MN implies nothing about the magnitude of the pore pressures inside the slope and is only used to obtain the statically equivalent disturbing moment.

The disturbing moment is therefore $\Sigma(W_1 + W_2)x$ . . . . (19)

where W_1 denotes full weight of the soil in the slice above MN,

W_2 ,, submerged weight of soil in the part of the slice below MN.

It is also convenient to obtain the expression for the effective normal force P' in terms of W_1 and W_2. It can be seen from Fig. 2 (*b*) and (*c*) that, resolving vertically as before, it follows that :

$$P' = \frac{W_1 + W_2 + X_n - X_{n+1} - l\left(u_s \cos\alpha + \frac{c'}{F}\sin\alpha\right)}{\cos\alpha + \frac{\tan\phi' . \sin\alpha}{F}} \qquad (20)$$

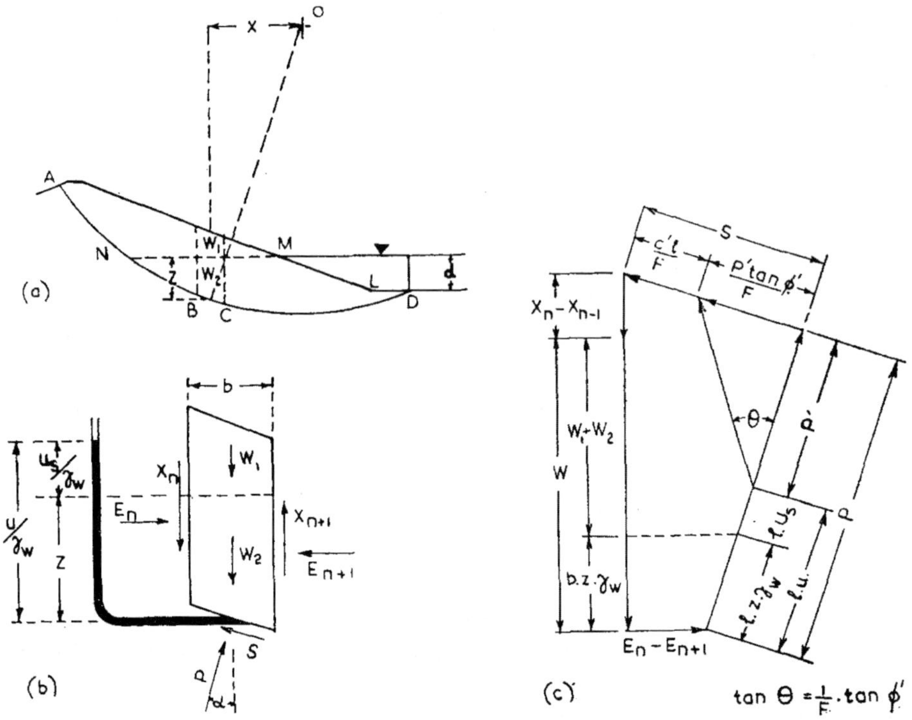

Fig. 2. Partially Submerged Slopes

where u_s is the pore pressure expressed as an excess over the hydrostatic pressure corresponding to the water level outside the slope; that is

$$u_s = u - \gamma_\omega \, . \, z \qquad (21)$$

where γ_ω denotes the density of water,
and z ,, the depth of slice below MN. (If no part of the slice is submerged, $u_s = u$.)

It follows, therefore, that for partial submergence the expression for factor of safety * becomes :

$$F = \frac{1}{\Sigma(W_1 + W_2)\sin\alpha} \, . \, \Sigma\left[\left\{c'b + \tan\phi' . (W_1 + W_2 - bu_s + \overline{X_n - X_{n+1}})\right\}\frac{\sec\alpha}{1 + \dfrac{\tan\phi' . \tan\alpha}{F}}\right] \qquad (22)$$

The forces between the slices now have to satisfy the conditions :

$$\Sigma(X_n - X_{n+1}) = 0 \qquad (23a)$$

$$\Sigma(E_n - E_{n+1}) = -\frac{1}{2} \cdot \gamma_w d^2 \qquad (23b)$$

* A similar treatment of the simplified method leads to the expression

$$F = \frac{1}{\Sigma(W_1 + W_2)\sin\alpha}\Sigma\left[c'l + \tan\phi' . (\overline{W_1 + W_2}\cos\alpha - u_s b \sec\alpha)\right] \qquad (22a)$$

This treatment of the pore pressure leads to consistent results throughout the full range of submergence (Bishop, 1952).

where d is the depth of water at the toe of the slip, which causes a horizontal water thrust on the vertical section with which the slip terminates.

To obtain an expression for $\Sigma(E_n - E_{n+1})$, we resolve tangentially as before :

$$(W_1 + W_2 + bz\gamma_\omega + X_n - X_{n+1}) \sin\alpha + (E_n - E_{n+1}) \cos\alpha = S$$

or
$$(E_n - E_{n+1}) = S \sec\alpha - (W_1 + W_2 + X_n - X_{n+1}) \tan\alpha - \gamma_w zb \tan\alpha \quad . \quad . \quad (24)$$

Writing equation (22) as :

$$F = \frac{1}{\Sigma(W_1 + W_2) \sin\alpha} . \Sigma[m] \quad . \quad . \quad . \quad . \quad . \quad . \quad . \quad . \quad (25)$$

then
$$S = \frac{m}{F} \quad . \quad . \quad . \quad . \quad . \quad . \quad . \quad . \quad . \quad . \quad . \quad . \quad . \quad . \quad (26)$$

Hence

$$\Sigma(E_n - E_{n+1}) = \Sigma\left[\frac{m}{F} \sec\alpha - (W_1 + W_2 + X_n - X_{n+1}) \tan\alpha\right] - \Sigma\left[\gamma_\omega zb \tan\alpha\right] \quad . \quad (27)$$

Now
$$\Sigma\left[\gamma_\omega zb \tan\alpha\right] = \frac{1}{2}\gamma_\omega . d^2 \quad . \quad . \quad . \quad . \quad . \quad . \quad . \quad (28)$$

Thus equation (23b) is satisfied when

$$\Sigma\left[\frac{m}{F} \sec\alpha - (W_1 + W_2 + X_n - X_{n+1}) \tan\alpha\right] = 0 \quad . \quad . \quad . \quad . \quad (29)$$

It can be seen, therefore, that for partially submerged slopes an analogous method is obtained by using submerged densities for those parts of the slices which lie below the level of the external free water surface, and by expressing the pore pressures there as an excess above the hydrostatic pressure corresponding to this water level.

PRACTICAL APPLICATION

Only a limited number of stability analyses have so far been carried out in which both the simplified and the more rigorous methods have been used. From these it appears, however, that the major part of the gain in accuracy can be obtained by proceeding with the solution of equations (13) or (22) only as far as the starting value of F given by taking $(X_n - X_{n+1}) = 0$. This avoids the more time-consuming stages of the solution, and provides a convenient routine method in which the time required for the analysis of each trial circle is from one to two hours or about twice that required using the simplified methods.

This may be illustrated by the two following cases.

(i) *Pore pressures set up in a boulder clay fill during construction.*—The cross-section of the dam and the soil properties used in the analysis are given in Fig 3 (a). From laboratory tests and other data it is considered that an average pore pressure equal to 40% of the weight of the overlying soil might be expected, *i.e.* $\bar{B} = 0{\cdot}4$.

For this condition the simplified method indicates a factor of safety of 1·38. The value given by equation (13), with $(X_n - X_{n+1}) = 0$, is 1·53.

The two methods do not lead to the same critical circle. The circle having a factor of safety of 1·38 by the simplified method gives a value of 1·59 using equation (13) with $(X_n - X_{n+1}) = 0$. The introduction of trial values of $(X_n - X_{n+1})$ makes only a small further change in the factor of safety, a value of 1·60 or 1·61 being obtained depending on the assumed distribution of X_n.

The form of tabulation used for the routine solution with $(X_n - X_{n+1}) = 0$ is given in Table 1.

It is useful from the design point of view to know the influence of possible variations in construction pore pressure on the factor of safety, and for this purpose the factor of safety may be plotted directly against average pore pressure ratio as in Fig. 3 (b).

Material	Density lb/cu. ft	c' lb/sq. ft	ϕ'
Clay fill	145	350	37·5°

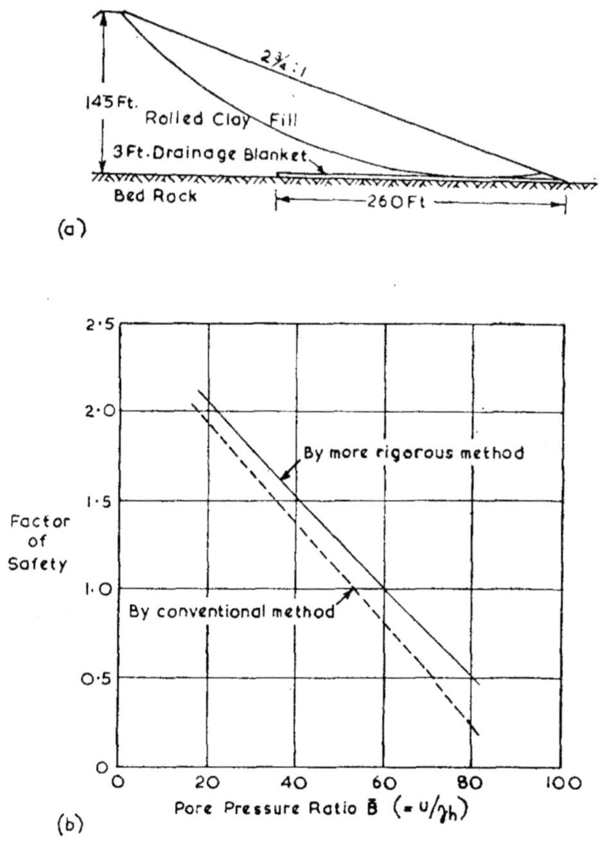

Fig. 3. End of construction case: relationship between factor of safety and pore pressure

(ii) *Pore pressures set up in a moraine fill during a partial rapid drawdown.*—The cross-section of the dam and the soil properties used in the analysis are given in Fig. 4. The basis of the method by which the draw-down pore pressures are calculated has been described by the author elsewhere (Bishop, 1952), and their values are indicated by a line representing stand-pipe levels above different points on the slip surface.

Using the average values of the soil properties (Case I), the simplified method (equation (22*a*)) gives a factor of safety of 1·50, and the more rigorous solution (equation (22), with $(X_n - X_{n+1}) = 0$) gives a value of 1·84.* As before, the two methods lead to different critical circles. The inclusion of trial values of $(X_n - X_{n+1})$ raises the value from 1·84 to 1·92, the value by the simplified method for this particular circle being 1·53. About 80% of the gain in accuracy in this case is thus achieved without introducing the X_n terms.

The form of tabulation for the routine solution in the case of partial submergence is given in Table 2.

CONCLUSION

Errors may be introduced into an estimate of stability by both the sampling and testing procedure, and in many cases an approximate method of stability analysis may be considered adequate. Where, however, considerable care is exercised at each stage, and, in particular, where field measurements of pore pressure are being used, the simplified " slices " method of analysis is not sufficiently accurate.

This is illustrated by the two examples given above. The error is likely to be of particular importance where deep slip circles are involved. This is illustrated by Fig. 5 in which the collected results of the two stability analyses are plotted in terms of the value of the central angle of the arc. It will be seen that the value of the factor of safety given by the simplified method, expressed as a percentage of that given by the more rigorous method [using $(X_n - X_{n+1}) = 0$], drops rapidly as the central angle of the arc increases. This effect is particularly marked for the higher values of excess pore pressure.

* For Case II the corresponding values are 1·14 and 1·48.

Material	Density—lb/cu. ft		Case I		Case II	
	Submerged	Above W.L.	c' lb/sq. ft	ϕ'	c' lb/sq. ft	ϕ'
Moraine	72	135	450	37°	0	40°
Rock	72	118	0	45°	0	40°

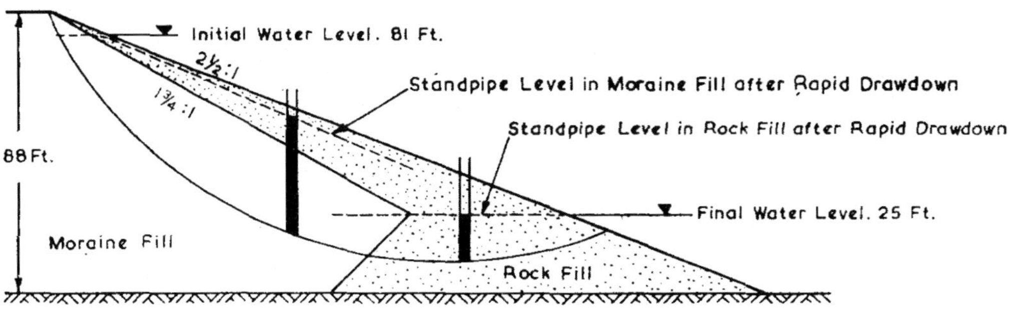

Fig. 4. Drawdown analysis of an upstream slope

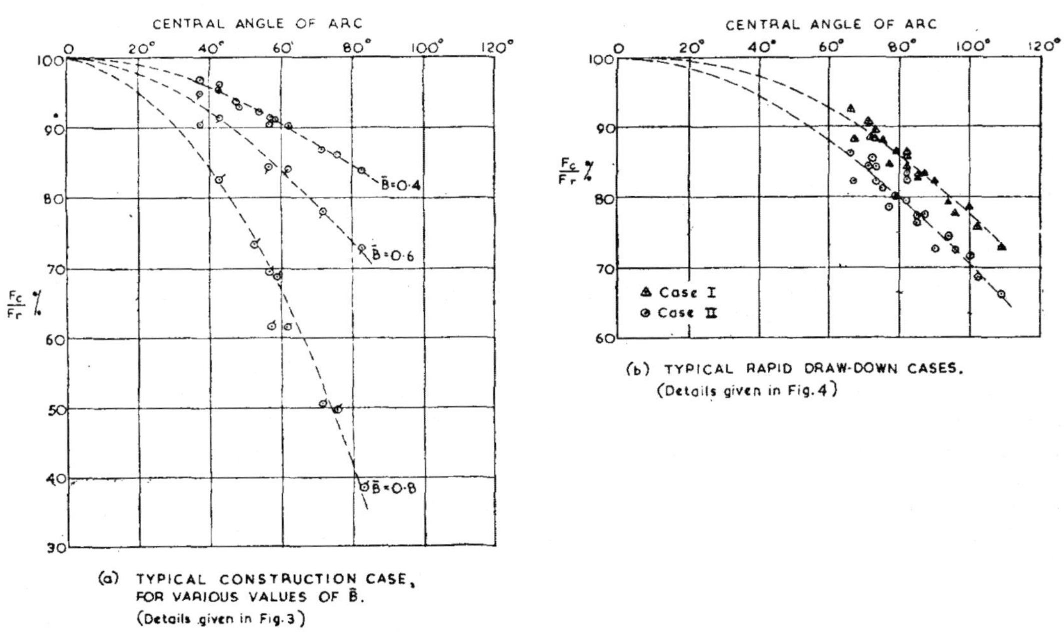

Fig. 5. Influence of central angle of arc on accuracy of conventional method

F_c is factor of safety by conventional method.

F_r is factor of safety by more rigorous method on same sliding surface.

Circle No. 1 **Table 1**

Slice No.	b ft	h ft	W lb. $\times 10^3$	α	$\sin\alpha$	$W\sin\alpha$ (1)	$c'b$ (2)	$W(1-\bar{B})\tan\phi'$ (3)	$(2)+(3)$ (4)	$\sec\alpha$	$\tan\alpha$	$\dfrac{\sec\alpha}{1+\dfrac{\tan\phi'.\tan\alpha}{F}}$ (5) $F=1{\cdot}4$	(5) $F=1{\cdot}6$	$(4)\times(5)$ (6) $F=1{\cdot}4$	(6) $F=1{\cdot}6$
1	50	23	167	42·2°	·672	112	17	77	94	1·350	·907	·902	·940	85	88
2	50	54	381	34·6°	·568	217	17	175	192	1·215	·690	·883	·912	169	175
⋮	⋮	⋮	⋮	⋮	⋮	⋮	⋮	⋮	⋮	⋮	⋮	⋮	⋮	⋮	⋮
8	40	11	64	−9·2°	−·160	−10	14	29	43	1·013	−·162	1·110	1·097	47	47
						777								1240	1262

$$F=\frac{1240}{777}=1{\cdot}60 \qquad F=\frac{1262}{777}=1{\cdot}62$$

Circle No. **Table 2**

Slice No.	b ft.	h_r (rock fill)	h_m (moraine)	h_s (submerged)	W_r } W_1	W_m } W_1	W_2	(W_1+W_2) lb. $\times 10^3$	α	$\sin\alpha$	$(W_1+W_2)\sin\alpha$ (1)	h_1 (excess standpipe ht.)	$u_s=\gamma_w\,.\,h_1$	$b\,.\,u_s$	$(W_1+W_2-b\,.\,u_s)$	$(W_1+W_2-b\,.\,u_s)\tan\phi'$ (2)	$c'\,.\,b$ (3)	$(2)+(3)$ (4)	$\sec\alpha$	$\tan\alpha$	$\dfrac{\sec\alpha}{1+\dfrac{\tan\phi'.\tan\alpha}{F}}$ (5)	$(4)\times(5)$ (6)
1 2 etc.																						

$$F=\frac{\Sigma(6)}{\Sigma(1)}$$

In these cases the use of the modified analysis outlined above is to be recommended, which, if carried only as far as the $(X_n - X_{n+1}) = 0$ stage, can readily be used for routine work.

Little field evidence is yet available for checking the overall accuracy of the stability calculation except in those cases where the " $\phi = 0$ " analysis has been applicable, and this emphasizes the need for failure records complete with pore-pressure data for both natural slopes and trial embankments.

ACKNOWLEDGEMENTS

The second example is included with the kind approval of the North of Scotland Hydro-Electric Board and their consulting engineers, Sir Alexander Gibb & Partners. The Author is also indebted to Mr D. W. Lamb for carrying out a number of the other calculations.

REFERENCES

Bishop, A. W., 1952. The Stability of Earth Dams. *Univ. of London. Ph.D. Thesis.**

Coenen, P. A., 1948. Fundamental Equations in the Theory of Limit Equilibrium. *Proc. 2nd Int. Conf. Soil. Mech.* 7 : 15.

Daehn, W. W., and Hilf, J. W., 1951. Implications of Pore Pressure in Design and Construction of Rolled Earth Dams. *Trans. 4th Cong. Large Dams.* 1 : 259.

Fellenius, W., 1927. Erdstatische Berechnungen mit Reibung und Kohaesion. Ernst, Berlin.

Fellenius, W., 1936. Calculation of the Stability of Earth Dams. *Trans. 2nd Cong. Large Dams.* 4 : 445.

Fröhlich, O. K., 1951. On the Danger of Sliding of the Upstream Embankment of an Earth Dam. *Trans. 4th Cong. Large Dams.* 1 : 329.

Krey, H., 1926. Erddruck, Erdwiderstand und Tragfaehigkeit des Baugrundes. *Ernst, Berlin.*

May, D. R., and Brahtz, J. H. A., 1936. Proposed Methods of Calculating the Stability of Earth Dams. *Trans. 2nd Cong. Large Dams.* 4 : 539.

Taylor, D. W., 1937. Stability of Earth Slopes. *J. Boston Soc. Civ. Eng.* 24 : 197.

Taylor, D. W., 1948. Fundamentals of Soil Mechanics. *John Wiley, New York.*

Terzaghi, K., 1929. The Mechanics of Shear Failures on Clay Slopes and Creep of Retaining Walls. *Pub. Rds.*, 10 : 177.

* A discussion of the pore pressures set up in earth dams has since been published in *Géotechnique*, 4 : 4 : 148.

Bjerrum, L. & Eide, O. (1956). *Géotechnique* **6**, No. 1, 32–47

STABILITY OF STRUTTED EXCAVATIONS IN CLAY

by

L. Bjerrum and O. Eide

SYNOPSIS

The present Paper describes a procedure for estimating the danger of a base failure in a strutted excavation in clay. Terzaghi and Tschebotarioff have previously given formulae for such calculations, but these are limited to shallow excavations.

On the basis of the familiar bearing capacity calculations for foundations on clay a new approach to the problem is developed, in which the critical depth of an excavation is given by :

$$D_c = N_c \cdot \frac{s}{\gamma}$$

where s and γ are the undrained shear strength and density of the clay, respectively. N_c is a dimensionless coefficient depending on the form of the excavation, i.e., the width/length ratio and the depth/width ratio. The values of N_c are identical to those given by Skempton (1951) for bearing capacity formulae.

The reliability of the formula is indicated by comparisons with fourteen excavations. In seven of these excavations complete failure occurred, the theoretical safety factor having an average of 0·96. In four excavations a partial failure was obtained, with an average theoretical safety factor of 1·11. The last three excavations were successfully performed, but site observation indicated that, with further depth, danger of failure would be serious. These excavations had an average theoretical safety factor of 1·08.

These investigations therefore indicate that the danger of base failure in excavations in clay may be estimated from the above theoretical approach with sufficient accuracy for most practical purposes.

L'article suivant donne une description de la méthode théorique pour la calculation du danger d'un soulèvement du fond d'une fouille cadragée dans l'argile. Terzaghi et Tschebotarioff ont déjà donné des formules pour un calcul pareil, mais ces formules ne sont valables que pour des fouilles de petite profondeur

Basée sur les formules de la capacité portante des fondations dans l'argile, une nouvelle solution du problème de soulèvement du fond a été développée. La profondeur critique d'une fouille est donnée par :

$$D_c = N_c \cdot \frac{s}{\gamma}$$

où s et γ sont respectivement la résistance au cisaillement et le poids d'unité de l'argile, N_c étant un coefficient sans dimensions et dependant de la forme de la fouille, c'est-à-dire de la proportion largeur-longueur et de la proportion profondeur-largeur. Les valeurs de N_c, identiques aux valeurs données par Skempton (1951) pour la capacité portante, sont montrées graphiquement dans la Figure 2.

La validité de ces formules a été examinée par comparaison de quatorze fouilles. Dans sept de ces fouilles, un soulèvement total du fond a eu lieu, les coefficients théoriques de sécurité donnant comme valeur moyenne de 0,96. Quatre fouilles, où on constatait un soulèvement partiel du fond, ont fourni une valeur moyenne des coefficients théoriques de sécurité de 1,11. Les trois dernières fouilles ont été accomplies sans évènements spéciaux, mais des observations sur place indiquaient que, avec une profondeur agrandissante, on aurait dû craindre un soulèvement du fond, la valeur moyenne des coefficients théoriques de sécurité étant de 1,08.

Les résultats obtenus ci-dessus montrent, ainsi, que la méthode décrite nous permet d'estimer avec une précision suffisante le danger d'un soulèvement du fond d'une fouille dans l'argile, et que la méthode suffit pour la plupart des usages pratiques.

INTRODUCTION

While excavating in soft Norwegian clays base failure sometimes occurs, accompanied by a consequent sinking of the surrounding ground. Such failures have occurred in excavations for basements, in trenches for water and disposal pipes, and in a number of shafts for piers for the foundations of buildings.

For shallow excavations of great width compared to the depth the danger of base failure may be estimated by a conventional stability analysis. For deep excavations and shafts, however, experience shows that a conventional stability analysis will lead to unreliable estimates.

The problem of calculating the stability of strutted excavations in clay has become of increasing importance in Norway during the last few years. The areas available for new buildings are limited and, consequently, there is a big demand for basement floors in new structures. Again, the increasing use of floating foundations for buildings on soft clay has posed the problem of estimating the danger of failure in the excavations for the basements. Finally, the stability problem has become of particular importance in Oslo, in connexion with the design of a subway through the town centre.

The Norwegian Geotechnical Institute commenced work on this problem shortly after its establishment in 1952. This work has consisted mainly of collecting case records of base failures, together with a theoretical treatment of the problem. The results of this work are presented in the present Paper, and a comparison is made between field experience and theoretical analysis.

METHOD OF CALCULATION

The problem to be treated is confined to the stability of excavations which are sufficiently strutted to prevent horizontal displacements of the walls. Furthermore, only temporary excavations are considered; that is, no attention is directed to changes in shear strength of the clay which, in permanent excavations, may occur with time.

During removal of the clay shear stresses are set up underneath an excavation, since the weight of the surrounding soil masses will tend to press the clay towards, and up into, the excavation. If the shear strength of the clay is sufficient to withstand the imposed shear stresses, then the excavation is stable. If, however, the shear strength of the clay is smaller than the applied shear stresses, the bottom of the excavations will be forced upwards and the soil masses on one or several sides of the excavation will sink vertically. The problem is thus one of stability, the solution of which requires a consideration of the direction of movement and the extent of the failure zone.

In the literature, this stability problem was discussed by Terzaghi (1943), later by Terzaghi and Peck (1948), and by Tschebotarioff (1951).

Terzaghi, as shown in Fig. 1, considers a clay mass alongside the excavation as exerting a certain pressure on the clay on a plane level with the bottom of the excavation. The bearing

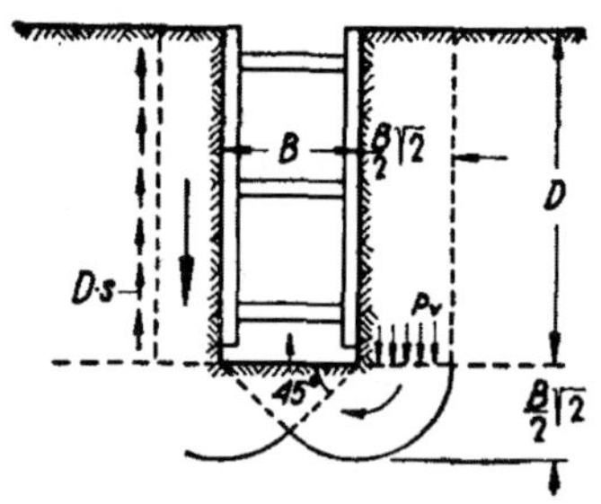

Fig. 1. Calculation of the stability of excavation according to Terzaghi (1943)

capacity q_f is calculated by the usual methods, and for an excavation of great length compared to the width the following equation is applicable:

$$q_f = 5{\cdot}7s$$

where s is the undrained shear strength of the clay (assumed to be saturated).

When calculating the pressure applied by the surrounding mass to a plane level with the bottom of the excavation, Terzaghi considers a section of clay of width $\frac{B}{2}\sqrt{2}$, this width being determined by the position of the failure zone. The weight of the clay block is reduced by the shear strength $D \cdot s$ on the plane of section. Thus the vertical stress is:

$$p_v = \gamma D - \sqrt{2}\frac{D \cdot s}{B}$$

where γ is the density of the clay.

When the excavation reaches the depth where failure occurs, then:

$$p_v = q_f$$

Thus the critical depth D_c of the excavation is given by:

$$D_c = \frac{5{\cdot}7s}{\gamma - \frac{s}{B}\sqrt{2}} \quad \ldots\ldots\ldots\ldots \quad (1)$$

Tschebotarioff's considerations are in principle similar to those of Terzaghi, assuming only a different shape of the failure zone. He also considers the ratio of the length L to the breadth B of the excavation, and his expression for the critical depth is:

$$D_c = \frac{5{\cdot}1s}{\gamma - \left(1 + \frac{2B}{L}\right)\frac{s}{B}} \quad \ldots\ldots\ldots\ldots \quad (2)$$

On the basis of observations from excavations in Oslo where bottom heave has occurred, it can be stated that Terzaghi's equation (1) gives a reliable estimate of the stability of an excavation, provided that (*a*) the width of the excavation is large compared with the depth, and (*b*) the clay is comparatively homogeneous with no stiff upper layer (weathering crust).

In cases of deep excavations, where the ratio depth/width is great, or in cases where the surface clay has a stiff drying crust, equation (1) gives unreliable results.

In other words, failure occurs at considerably smaller depths than indicated by equation (1); and corresponding divergencies between theory and field observations are found with Tschebotarioff's equation (2).

These divergencies can be explained by shortcomings of the theoretical assumptions regarding the shape of the failure zone. Both equations (1) and (2) assume a circular sliding surface underneath the bottom of the excavation, and that this surface extends vertically up to the ground surface. The formulae thus assume that the shear strength of the clay is fully mobilized right up to the surface. For very deep excavations this assumption will not hold true. Underneath a deep excavation, such as a shaft for a pier, failure will occur without fully mobilizing the shear strength of all the upper clay layers.

In order to obtain a reliable estimate of the stability of deep excavations it is thus necessary to consider the danger of local failure below the bottom of the excavation. An exact treatment of this problem is not possible at present since the necessary plasticity theory is not available. The calculation of the bearing capacity of foundations, however, presents a corresponding problem, for which a satisfactory approximate solution has recently been found. The solution of this problem may be applied to the calculation of the stability of excavations.

The bearing capacity of a footing on clay is expressed by the equation:

$$q_f = \gamma D + N_c \cdot s \quad \ldots\ldots\ldots\ldots \quad (3)$$

where q_f = the bearing capacity of the footing,
γ = the density of the clay,
s = undrained shear strength of the clay,
D = depth of footing,
N_c = coefficient depending on the shape and depth of the footing.

For foundations of great width compared to the depth, N_c may be calculated by assuming that the sliding surface is extended up to the ground level. For shallow footings N_c will thus increase approximately linearly with depth. If the depth of the foundation is increased N_c will, however, not increase correspondingly. When the depth exceeds about three times the breadth N_c will remain constant. This is because, when a deep footing is loaded to failure, a local failure zone will be produced underneath the footing and the surrounding soil masses will be pressed aside without mobilizing the full shear strength of clay in the upper layers. The value of N_c will thus remain constant for depths greater than about three times the width of the footing.

A complete discussion of the problem of calculating the bearing capacity of deep foundations was given by Skempton (1951). In this Paper he mentions the investigations carried out by Meyerhof and Wilson, but his main treatment is based on a calculation method developed by Bishop, Hill, and Mott for metals and later extended by Gibson to include clays. According to this treatment, failure is caused under a deep circular footing by the formation of a spherical plastic zone underneath the footing whilst the surrounding clay remains in an elastic state.

For this case N_c, Gibson derives the equation :

$$N_c = \frac{4}{3}\left[\log\frac{E}{s} + 1\right] + 1$$

where E is the elastic modulus of the clay for undrained deformation. E may be obtained from the inclination of the stress/strain curve of the clay, determined by unconfined compression tests. For ordinary clays, the ratio E/s varies between 50 and 200. This corresponds to a variation of N_c from 7·6 to 9·4. In the sensitive Norwegian clays, E/s will vary between 100 and 200 and N_c will consequently vary between 8·5 and 9·4.

On the basis of all existing theoretical determinations and a number of model tests, Skempton finally sums up the present knowledge of this problem by giving a diagram showing N_c as a function of the depth and the shape of the foundation. This diagram is shown in Fig. 2.

A comparison between calculated and observed bearing capacities has confirmed the reliability of this diagram, although loading tests seem to indicate that the values for N_c may be slightly conservative for sensitive Norwegian clays (Bjerrum, 1955).

Professor Skempton pointed out to the Authors in 1952, that it was probable that the above calculation method for the bearing capacity of foundations may, without any adaptations, be used to calculate the stability of excavations ; in fact, the two problems are analogous. In the first case we have to determine the additional load which, at a certain depth, will cause failure in the clay. In the second case, the question is to what limit the clay, by an excavation, may be unloaded before failure occurs. As the shear stresses in the two cases simply have opposite signs, one can directly derive from equation (3) the following expression for the critical depth of excavation:

$$D_c = N_c\frac{s}{\gamma} \quad . \quad . \quad . \quad . \quad . \quad . \quad . \quad . \quad . \quad . \quad . \quad (4)$$

More generally the safety factor against base failure may be expressed by the equation :

$$F = N_c\frac{s}{\gamma D + q} \quad . \quad . \quad . \quad . \quad . \quad . \quad . \quad . \quad . \quad . \quad (5)$$

where F = the safety factor,
D = the depth of the excavation,
γ = the density of the clay,
s = the undrained shear strength of the clay in a zone underneath and immediately around the bottom of the excavation,
p = the surface surcharge,
N_c = a coefficient depending upon the dimensions of the excavation.

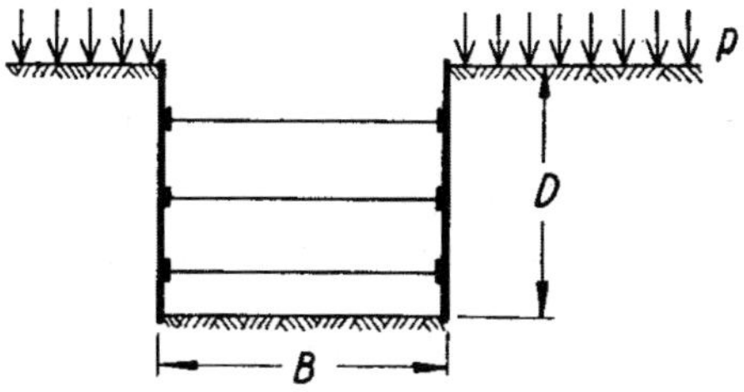

$$F = N_c \frac{s}{\gamma D + p}$$

D = depth of excavation
B = width of excavation
L = length of excavation
p = surcharge
s = undrained shear strength of clay
γ = density of clay
N_c = coefficient
F = safety factor

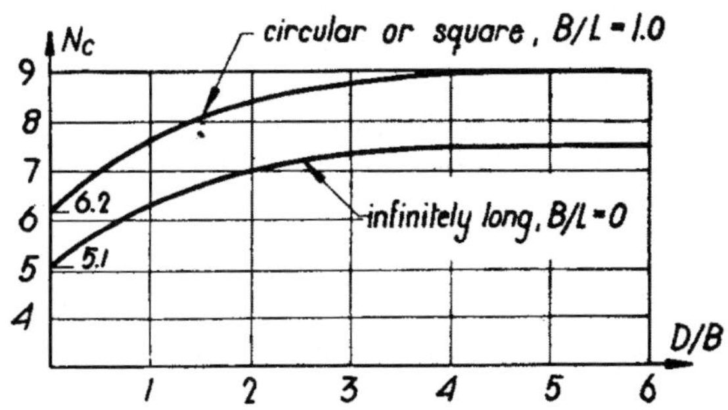

$$N_{c\ rectangular} = (0.84 + 0.16\ B/L)\ N_{c\ square}$$

Fig. 2. The critical depth of excavations calculated on the basis of the bearing capacity theory. Values of N_c from Skempton (1951).

As a working hypothesis the above-mentioned analogy is assumed to be so complete that the N_c-values for the bearing capacity of footings on clay may be directly used in equations (4) and (5). But the accuracy of this theory must be investigated by field observations. Indeed, one cannot ignore the fact that the N_c-values for an excavation may differ from those in Fig. 2, the latter being valid for the calculation of the bearing capacity of a rigid foundation with uniform deformations, whereas the deformations at the bottom of an excavation may take place freely and therefore will be greater in the middle of the excavation than along the edges.

The reliability of the above theory is investigated by field observation from fourteen deep excavations, given in the following examples.

EXAMPLES OF COMPLETE FAILURES IN EXCAVATIONS

Pumping station, Fornebu, Oslo

A typical example of base failure occurred in October 1954 during the excavations for a sewage tank in Oxenövegen, Fornebu, Oslo.

The dimensions of the excavation were 5 × 5 m and it was designed to be excavated to a depth of 3·20 m below ground level. The excavations were made within heavy wooden sheet piling, with tongue and groove connexions. The sheet piling itself was securely strutted. As the work proceeded the excavated soil was removed from the vicinity of the excavation so that no surface surcharge loads were acting.

When the excavation reached a depth of 3·0 m (Fig. 3) a total failure occurred. The bottom rose about 0·5 m and there was a simultaneous sinking of the surrounding ground of

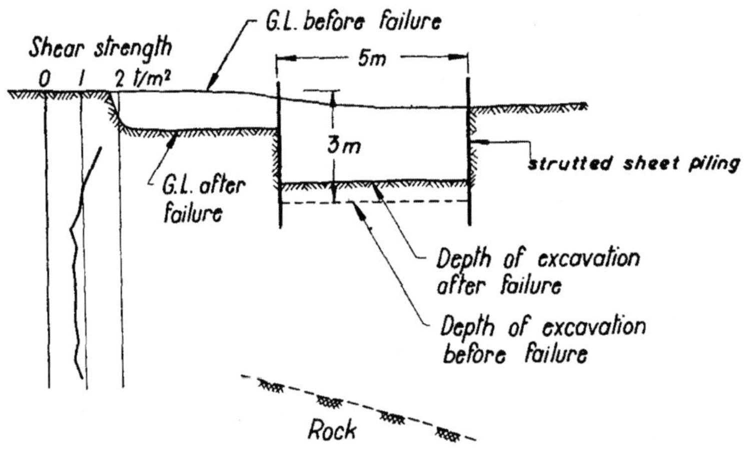

Fig. 3. Pumping Station, Fornebu, Oslo

about 1 m on that side where the ground was at its highest. Furthermore, cracks appeared in the ground around the excavation. Apart from small displacements the sheet piling remained intact during the heave.

Shortly after the failure field investigations were carried out by the Norwegian Geotechnical Institute. Four vane borings were made around the excavation, and one in the middle. Furthermore, in one boring undisturbed samples were taken for detailed laboratory investigations.

Underneath a fill of thickness 0·4–0·7 m there was a relatively stiff and dry clay crust down to about 1·4 m below ground level. Vane borings and samples showed that below this crust there was a layer of very soft and very quick clay which went down to rock, which was found at a depth of 8–9 m below ground level. The shear strength of the quick clay was nearly constant throughout the depth and equalled 0·75 ton/sq. m. A vane boring performed in the excavation showed that the clay was completely disturbed to a depth of 2–2·5 m below the bottom of the excavation.

A stability calculation indicated an N_c-value of 7·0 corresponding to a safety factor of 1·0. The theoretical value corresponding to a depth/width ratio of the excavation equal to 0·60 is 7·2. When the failure occurred the theoretical safety factor was thus 1·03.

Store house, Drammen

A failure occurred during the excavation for a store house in Drammen in August 1954, just before the required depth of excavation of 2·5 m had been reached. Further work proved to be fruitless and the basement floor was abandoned since, with further excavation, clay masses were forced up in the bottom of the excavation.

The work was performed by a mechanical excavator, leaving the sides of the excavation vertical and not strutted. In spite of the bottom heave the unstrutted sides remained intact and vertical. Furthermore, there was no appreciable sinking of the ground immediately around the excavation, but at a distance of 4–5 m from the work marked cracks and settlements were observed. The continued excavation after the base failure had occurred led to considerable sinking of a nearby road fill about 1·5–2·0 m high.

According to earlier field investigation of this area the ground may be described as having a firm clay crust down to a depth of 2 m. Underneath this crust was a layer of very soft and very sensitive clay. Rock was found to a depth of 6–7 m. The shear strength of the clay was, to a depth of 3–4 m, 1·0–1·6 ton/sq. m. The average value of the shear strength was estimated to be 1·2 ton/sq. m. The weight of the excavator together with other loads amounted to a surcharge of approximately 1·5 ton/sq. m. As the density of the clay was 1·9 ton/cu. m, the coefficient N_c will be 5·1 for a critical depth of excavation of 2·4 m. The depth/width ratio was approximately 0·5, leading to a theoretical N_c-value of 5·9. The calculated safety factor was thus 1·16. Such a high figure may possibly be explained by the load from the nearby road filling, but it is difficult to consider this factor in the calculation.

Pier shaft, Nordenskioldsgatan, Göteborg

In 1930 a block of flats was built in Nordenskioldsgatan in Göteborg. The building was founded on steel piles driven down to rock. The structure was built on a slope with a comparatively steep gradient, and that may be the reason for the slow creep which has been observed in the slope. Shortly after the building was completed horizontal movements were observed, causing cracks and other damage in the building.

In 1954 the damage had become so comprehensive that the building had to be pulled down. A new building on the same site was designed, but this time the foundations were to be piers down to rock. Circular pier shafts were excavated to rock at a depth of 25 m. Sheet piling, strutted by concrete rings, was successively driven down as excavating proceeded. At the surface the shaft had a diameter of 1·30 m, this diameter decreasing gradually to 0·90 m at the rock.

The foundation work was, on the whole, successfully performed in the above manner. Only one pier shaft had to be abandoned because of base failure. In this shaft the excavation had almost reached the rock at a depth of 25 m. At this point the rock sloped steeply. Just before the failure occurred the distance down to the rock was approximately 0·3 m on one side of the shaft and 1·3 m on the other. The workman who was in the excavation relates that the bottom rose so quickly that he only just managed to climb up before the clay masses reached him. The bottom rose about 5 m, and repeated attempts to complete the excavation failed since, when further clay was removed, new clay masses continued to flow into the shaft.

Field investigations carried out in connexion with the planning of the building showed the existence of clay layers from ground level down to rock. The top layer was a post-glacial clay containing some organic matter, whereas the bottom layer was a marine clay of high sensitivity (S_t = 20–50).

The water content of the clay decreased from about 80% in the upper layer to about 55% at a depth of 20 m. The average density of the clay was 1·54 ton/cu. m. The undrained shear strength increased linearly with depth. Underneath a weathering crust 3–4 m thick

(a)

(b)

Fig. 5 *(a)* and *(b)*. Test shaft (N), Ensjöveien, Oslo. Photographs taken after failure

the shear strength was 1·5 ton/sq. m, increasing to approximately 3·5 tons/sq. m at a depth of 20–25 m.

As the depth of excavation was 25 m when failure occurred, the total overburden at the bottom of the excavation was 38·5 tons/sq. m. The shear strength of the clay being 3·5 tons/sq. m, N_c will in this case be 11. According to the theory for the D/B ratio of 28, an N_c-value of 9·0 is indicated. The high actual value of N_c may be partly due to the close proximity of the rock surface to the bottom of the excavation.

Sewage tank, Drammen

During the construction of a nursing-home in Drammen in August 1952, a failure occurred in the excavation for a sewage tank. The dimensions of the excavation were 5·5 × 8·0 m, with a designed depth of 4·5 m. Down to a depth of 3·5 m, a stiff clay was encountered, but as soon as the underlying soft, quick clay was reached the bottom of the excavation started to rise, and continued during the subsequent attempt to complete the excavation. The excavation was not braced, and in spite of the bottom failure the walls of the upper stiff clay remained intact and vertical. Several attempts to complete the excavation within sheet piles driven well below the final depth proved to be unsuccessful, and only after the removal of the upper soil layers surrounding the excavation was the intended depth reached.

Soil investigations carried out after the failure showed that below an upper drying crust there was a soft, quick clay of shear strength approximately 1 ton/sq. m, with a density of 1·8 tons/cu. m. Based on the collected data an N_c-value is calculated for the conditions immediately before the first failure, which took place at a depth of 3·5 m. A surcharge of 1 ton/sq. m on the surface around the excavation was taken into account. The result of this calculation is that N_c should be 7·2 to obtain the safety factor 1·0. The theoretical value is 6·7, but it is emphasized that this excellent agreement between observation and theory may be somewhat fortuitous since there are some uncertainties in the estimate of the shear strength and the surcharge.

Test shaft (N), *Ensjöveien, Oslo*

In connexion with the design of a new subway in Oslo, three trial shafts were excavated in 1953 for the purpose of determining the reliability of the theoretical analysis of the stability of excavations.

For all three shafts thorough soil investigations were made including sampling, vane tests, and pore pressure measurements. During the excavation large samples were cut out at different depths for a detailed laboratory investigation. At different stages of the excavation vane tests were made from the bottom in order to investigate possible changes in shear strength of the clay.

The excavation was started by digging a 1·0–1·5 m-deep circular hole with a diameter just exceeding 1·5 m. A number of 1·5-m ϕ rings, made by welding 30-mm ϕ steel bars, were placed in the bottom of the hole and 2-in. timber planks were driven down as sheet piles outside the rings. The sheet piles were driven successively as the clay was excavated, splicing each plank to the preceding one as they were driven into the ground. The rings were lowered in steps each time leaving the top ring behind. In this way the rings were distributed along the shaft at a distance of 0·5 m from each other, thus forming a regular support for the sheet piles (Fig. 4).

The northern shaft (N) was excavated through 1·5 m of fill and then through 2 m of stiff drying crust. Below the drying crust a soft—and in parts a very soft—quick clay was encountered. The quick clay continued down to a depth of 10–11 m, and from this depth, continuing down to rock at a depth of 15 m, was a sandy clay.

The shaft (N) was terminated at a depth of 7 m, at which point failure was expected. The

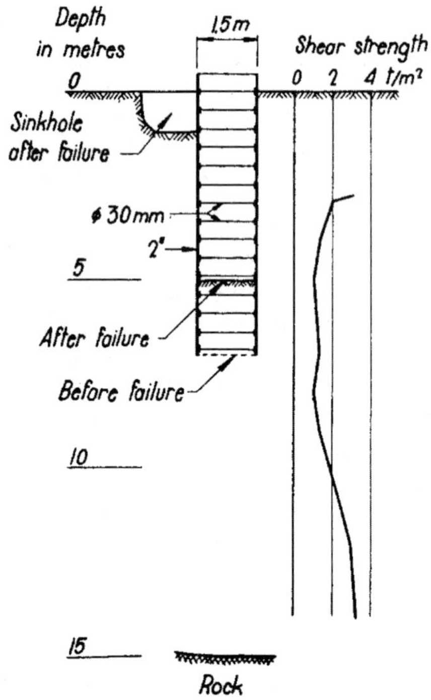

Fig. 4. Test shaft (N), Ensjöveien, Oslo. Cross-section after failure

failure did not occur, however, until two weeks after the excavation had been completed. The bottom rose about 2 m and, as shown in Fig. 5, a sinkhole appeared in the ground close to the sheet piling. Two vane tests carried out after the failure showed that the clay was remoulded to 1·0–1·5 m below the original shaft bottom.

The clay below the bottom of the shaft had a shear strength of 1·2 ton/sq. m, and the average density of the soil was 1·85 ton/cu. m. A calculation based upon these values shows that N_c should be 10·8 in order to obtain the safety factor 1·0. This is 20% in excess of the theoretical value.

Excavation, Grey Wedels Plass, Oslo

In 1913 a total failure occurred in an excavation in Oslo. The excavation was 5·8 × 8·1 m in area and was designed to reach a depth of 5 m, but a base failure took place at a depth at 4·5 m. Even after the driving of heavy sheet piling the bottom continued to rise as the clay was removed. The work was finally completed by excavating small sections which were covered with concrete and then loaded.

Investigations carried out in 1954 by the Norwegian State Railways showed the soil to consist of a soft clay. The shear strength of the clay layer from 4·5 m—at which depth the failure took place—down to 7·5 m was determined, by vane tests, to be 1·4 ton/sq. m.

A calculation of the stability of the excavation immediately before failure led to a safety factor of 1·08. A surcharge loading of approximately 1 ton/sq. m due to building materials was included in the calculation.

" Kronibus " shaft, Tyholt

During the last war the German occupation authorities started the construction of a railway tunnel in clay near Trondheim (Hartmark, 1955). A German contractor, who was supposed to be a specialist, made an attempt to sink a 20-m-deep shaft midway on the tunnel length. At this site the soil consists of a very sensitive clay. When the shaft had reached a depth of 19·7 m a failure occurred and the shaft was partly filled with liquid clay. The work was thereupon abandoned and the tunnel was not completed until after the war, when it was built by the shield-driving method.

The dimensions of the shaft were 2·7 × 4·4 m. The shear strength of the clay at the bottom of the shaft was approximately 3·5 tons/sq. m, as determined by vane tests carried out by the Norwegian State Railways. The density of the clay was 1·8 ton/cu. m.

A calculation of the stability indicates a theoretical safety factor of 0·84. The N_c-value which gives a safety factor of 1·0 is 10·1, compared with the theoretical value 8·5.*

EXAMPLES OF PARTIAL FAILURES IN EXCAVATIONS

Pumping station, Jernbanetorget, Oslo

In 1943 a deep excavation was made for a pumping station in Oslo. During the work a movement in the surrounding ground was experienced, and Mr S. Skaven-Haug, civil engineer of the Norwegian State Railways, was consulted. His report, and additional information required for the calculations, was placed at the disposal of the Norwegian Geotechnical Institute.

The dimensions of the excavation were 12·2 × 8·5 m with an intended depth of about 9 m below ground level. The top 2·5 m of the excavation was strutted, and from this depth slender and unsatisfactory sheet piles about 10 m long were driven.

When the excavation reached a depth of 6·3 m below ground level cracks appeared around the hole at a distance of about 8–10 m, i.e., at a distance relatively large compared with the depth. Mr Skaven-Haug explained the cracks as being caused by movements in the underlying clay masses, following slight base failure in the excavation. A complete failure did not occur.

The existing struts were strengthened, and the work was completed by the excavation of small sections which were then loaded by casting the base of the footing.

Site investigations were made in advance by a private firm, and vane tests to a depth of 19 m were carried out by the Norwegian Geotechnical Institute in 1952. Under 2 m of fill, soft to medium-firm clay was encountered to great depth. The shear strength of the clay as determined by vane tests was about 2–3 tons/sq. m, the lowest values being experienced at a depth of about 12 m below ground level. The shear strength near the bottom of the excavation, when the cracks occurred, was about 2·2 tons/sq. m. The average density of the clay was 1·9 ton/cu. m. For a failure at a depth of 6·3 m an N_c-value of 5·5 may be expected.

A calculation according to the above theory indicates an N_c-value of 6·9. This means that deformations occurred at a depth at which the calculated safety factor is 1·25. It is probable that the excavation could have been deepened without total failure occurring. The critical depth according to the theory is about 8 m.

Store house and office block, A/S Freia, Oslo

During the summer and autumn of 1954, excavations were dug for a new store house in Oslo. The building was supported partly by steel piles to rock, and partly it rested directly

* The theoretical N_c-value 8·5 is Skempton's value. The present Authors suggest that the effect of the width/length ratio on the N_c-value may be questioned as far as deep shafts are concerned. For shafts with depth/width as well as depth/length ratios greater than 4 (see Fig. 2) it may be that the theoretical N_c-value should be 9 instead of 8·5 as used in this case.

on the rock, which was encountered at depths varying from 5–20 m below ground level. The depth to rock, of interest in the present Paper, was about 9 m at a side wall, where the rock sloped somewhat towards the centre of the building. The depth of excavation was about 5 m, as shown in Fig. 6.

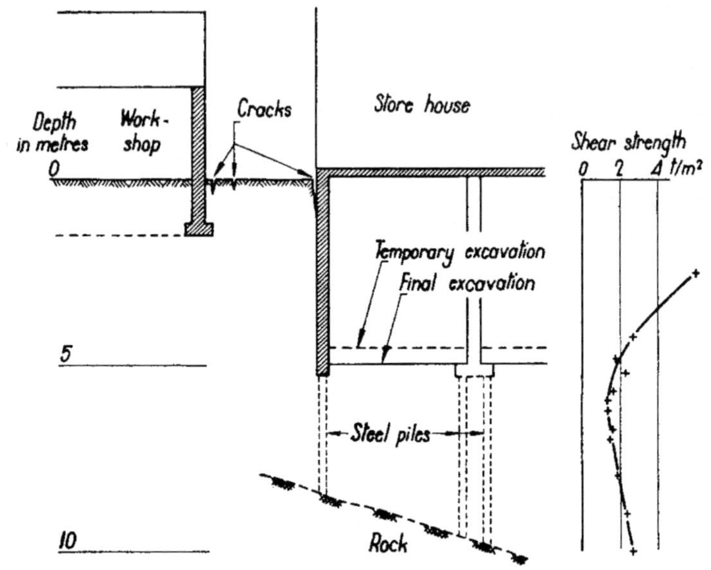

Fig. 6. Store house, A/S Freia, Oslo. Cross-section through excavation

The work was carried out with a side slope of 1 : 1 to a depth of 0·5 m above the required depth. At this stage a small slide occurred in the side slope, but this had no effect on the underlying clay. The excavation was widened to vertical walls in small sections, which were strutted. After the sides were strutted and the piles driven, the foundations for the outside walls and columns were cast. Excavation of the remaining 0·5 for the basement, and the casting of the basement floor, was carried out after completion of the walls, columns, and ground floor of the building.

The placing of the foundation and walls up to ground-floor level, and also a little higher, was completed by May 1955, and then work on the basement commenced. Before the casting of the basement floor the remaining 0·5 m was excavated and at that time a settlement of the ground outside the wall was noticed.

The gable wall of a single-storey brick workshop without a basement was situated about 3 m from the side wall. The gable wall had settled slightly during the first part of the excavation, but the settlements increased noticeably with the excavations for the basement, to an extent of about 10 cm. Cracks were noticed on the ground, and the settlement near the side wall was less than that near the gable wall. Those settlements, and the cracks, indicated that the safety factor against base failure in the basement was very small, and the basement slab was cast as soon as possible.

The Norwegian Geotechnical Institute carried out two vane borings and took samples before the excavation commenced. The top 3 m consisted of a hard weathering crust and, under that, the shear strength of the clay decreased with depth to a minimum value of 1·4 ton/sq. m at a depth of 6 m, and then increased to 2·5 ton/sq. m at a depth of 10 m.

At a depth of 5 m, with a density of clay of 1·9 ton/cu. m, together with a surcharge load of 0·5 ton/sq. m due to the workshop, the pressure was 10 tons/sq. m. The average shear strength of the clay was 1·6 ton/sq. m, which gave an N_c-value of 6·25 at that depth.

The excavation was very long compared to the width, and since the depth to rock was limited the depth/breadth ratio was approximately equal to 1·0 which gave a theoretical N_c-value of 6·4 with a safety factor of only 1·02.

Shaft, Shellhaven, England

In connexion with the construction of a pumping station at Shellhaven, in the Thames Estuary clays, a shaft of 5·5 m dia. was excavated and, at a depth of 5·2 m, a slight failure occurred. Messrs John Mowlem kindly supplied the relevant facts and gave permission for their publication in this Paper. The excavation was supported by cast-iron rings. A total failure was not experienced and it was thought that the work could possibly have proceeded a further 0·5 m. But a movement of the surrounding ground had been noticed and, for safety, the work was stopped for a while, and then completed under air pressure.

Site investigations have been described by Skempton and Ward (1952) and Skempton and Henkel (1953) in connexion with adjacent works at Shellhaven. Under a hard brown clay a soft-to-fairly-soft sensitive clay is found, with occasional layers of peat and silt. The shear strength of the clay in the vicinity of the bottom of the shaft at the depth of 5·2 m is about 1·6 ton/sq. m.

The pressure due to the clay mass, together with a surcharge load of 1 ton/sq. m, gives a pressure of about 9 tons/sq. m at the base of the excavation. The average shear strength of the clay being 1·6 ton/sq. m, N_c-values of 5·6 and 6·1 are obtained, corresponding to depths of 5·2 and 5·7 m.

According to theory, an N_c-value of 7·5 and a critical depth of 7·6 m are indicated. A comparison between the theoretical values and the field experience indicates that a partial failure occurred when the safety factor was 1·34. Part of this divergence between theory and practice may be explained by the decrease in shear strength due to remoulding at the bottom of the shaft.

Test shaft, Sörligaten, Oslo

In a similar manner to the test shaft at Ensjöveien a test shaft was excavated near the tunnel at Sörligaten, Oslo. Here also, extensive site investigations, including vane borings and samplings, were performed, and several blocks of the clay were obtained for laboratory investigations. The shaft is shown in Fig. 7.

The ground conditions may be described as follows. Under a 1·5-m-thick top layer of fill, silt is encountered to a depth of 3 m, and then a fairly sensitive clay is found to a depth of 10 m, the shear strength of which decreases from 4 tons/sq. m to 1·5–2 tons/sq. m. At a depth of 9–11 m the clay contains some sand and gravel, and below this—to rock at 17 m—is a quick clay of shear strength about 2 tons/sq. m.

Excavation proceeded to a depth of 13 m without failure. Tests performed during the work showed a decrease in the shear strength of the clay underneath the bottom of the shaft when a depth of 12 m was reached. That may be considered as the critical depth, as the decrease in the shearing strength was due to large shear deformations. The decrease in shear strength could be seen very clearly from a vane boring carried out, after the excavation, about 0·5 m away. That local failure did not, however, lead to a complete failure.

At a depth of 12 m the shear strength of the clay is 2·1 tons/sq. m and, with an average density of 1·85 ton/cu. m, an N_c-value of 10·6 is obtained. The depth/diameter ratio for the shaft is 8·0 and the theoretical N_c-value is 9, giving a safety factor of 0·85.

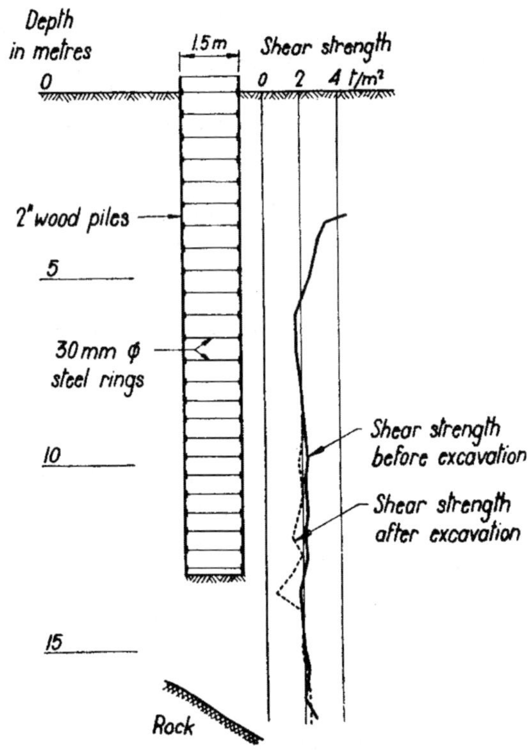

Fig. 7. Test shaft, Sörligaten, Oslo

In spite of the partial failure at a depth of 12 m it was possible to excavate to a depth of 13 m where the safety factor is 0·75.

It is possible that total failure was retarded by the relatively hard top layer preventing the development of deformations.

EXAMPLES OF SUCCESSFUL EXCAVATIONS

Basement excavation, Storgaten 5, Oslo

In the summer of 1952 a 6–7-m-deep excavation for two basement floors was made in the central part of Oslo. The excavation was carried out in sections within sheet piles, and as soon as a section was excavated concrete was placed and loaded with sand.

Careful soil investigations were made by the Norwegian Geotechnical Institute. The shear strength of the soft and partly quick clay was 1·75 ton/sq. m and the density was 1·9 ton/cu. m. The depth to rock was 13–15 m.

The above data are valid for the place where the most critical section was excavated. The dimensions of that section were 8 × 4 m and the depth was 6·7 m. Corresponding to the depth/width ratio of 1·7, the theoretical N_c-value is 8·0. The safety factor is thus only 1·10.

During the excavation careful observations were made, including an attempt to measure the rise of the bottom of the excavation. Those measurements proved to be very difficult, and were only partly successful. From the results it was, however, concluded that the bottom of the critical section rose about 2 cm. The excavation was situated close to neighbouring buildings and some settlements and cracks of secondary importance were observed.

Subway in Chicago

One of the examples used by Tschebotarioff (1951) in his treatment of the problem was a deep excavation for the Chicago subway. A detailed description of the work and of the site conditions is given by Peck (1943), but it must be remarked that Peck does not direct attention to the question of stability. At one particular point in the subway the excavation was 16·0 m wide and 11·3 m deep. The excavation was strutted, but a maximum horizontal deformation of 6 cm occurred.

During excavation a maximum heave of 10 cm was observed together with a sinking of the adjacent ground of about 3–15 cm, the largest values being nearest the excavation. Although a complete failure did not occur it is clear that the safety factor was low.

The clay underneath the excavation had, according to Tschebotarioff, a shear-strength of 3·5 tons/sq. m. A calculation of the stability of the excavation shows a theoretical N_c-value of 6·1 with a safety factor of 1·0.

Test shafts (S), *Ensjöveien, Oslo*

As described earlier, two test shafts were excavated at Ensjöveien, Oslo, in the summer of 1953. Whilst a total failure was obtained in the northern shaft, the southern shaft was excavated to a depth of 11 m, which was not sufficient to produce failure. The dimensions and method of excavation of the test shafts have already been described.

The shear strength of the clay at the bottom of the shaft was 2·7 tons/sq. m, the average density of the clay was 1·93 ton/cu. m, the depth/diameter ratio was 7·3, and a theoretical N_c-value of 9·0 was obtained. The theoretical safety factor was 1·14. The test shaft was kept open for 12 months without any sign of failure.

CONCLUSIONS

The data of the investigated excavations are summarized in Table 1, which also gives the factors of safety against base failure calculated by the proposed method. From the data presented in Table 1 the following conclusions may be drawn :

(1) The theoretical safety factor for the seven cases in which *complete failure* occurred varies between 0·82 and 1·16. The average value is 0·96. The field observations therefore confirm the reliability of the theoretical analysis, and the accuracy with which the observed total failures could have been predicted by this method is considered to be satisfactory.

(2) Also the four excavations in which *partial failures* were observed support the theoretical approach, since the calculated safety factors vary between 0·85 and 1·25 with an average of 1·11.

(3) The three *successful excavations* are selected cases in which the safety factors against base failure are judged to be small. The theoretical safety factors of these cases vary between 1·0 and 1·14 with an average value of 1·08. If compared with the field observations made during the excavations, the calculated safety factors are believed to give a correct evaluation of the stability. It is very likely that a further excavation of the two cases which show the lowest safety factors would have led to failure.

(4) Included among the excavations with a total or partial base failure are four *shafts of great depth* compared to the width. It is immediately seen from Table 1 that the safety factors of these four shafts are remarkably low compared with the other cases. The safety factors of the four shafts vary between 0·82 and 0·85 with an average of 0·84. These results may be explained by a too low theoretical N_c-value for deep excavations, and it is thus possible that Skempton's N_c-values, which are confirmed by field observations of foundation failures or loading tests

Table 1.
Data concerning instability of excavations

Site	Dimensions $B \times L$: m	Depth D : m	Surcharge p : tons/sq. m	Density γ : tons/cu. m	Shear strength S : tons/sq. m	Sensitivity	B/L	D/B	N_c theoretical	Safety factor F	Type of failure	Average safety factor
1. Pumping station, Fornebu, Oslo	5·0 × 5·0	3·0	0·0	1·75	0·75	50	1·0	0·60	7·2	1·03	Total failure	
2. Storehouse, Drammen	4·8 × ∞	2·4	1·5	1·90	1·2	5–10	0·0	0·50	5·9	1·16	do.	
3. Pier shaft, Göteborg	ϕ 0·9	25·0	0·0	1·54	3·5	20–50	1·0	28·0	9·0	0·82	do.	
4. Sewage tank, Drammen	5·5 × 8·0	3·5	1·0	1·80	1·0	20	0·69	0·64	6·7	0·93	do.	0·96
5. Test shaft (N), Ensjöveien, Oslo	ϕ 1·5	7·0	0·0	1·85	1·2	140	1·0	4·7	9	0·84	do.	
6. Excavation, Grev Vedels pl., Oslo	5·8 × 8·1	4·5	1·0	1·80	1·4	5–10	0·72	0·78	7·0	1·08	do.	
7. " Kronibus shaft," Tyholt, Trondheim	2·7 × 4·4	19·7	0·0	1·80	3·5	40	0·61	7·3	8·5	0·84	do.	
8. Pumping station, Jernbanetorget, Oslo	12·2 × 8·5	6·3	0·0	1·90	2·2	6	0·7	0·74	6·9	1·25	Partial failure	
9. Storehouse, Freia, Oslo	5·0 × ∞	5·0	0·0	1·90	1·6	12	0·0	1·00	6·4	1·02	do.	1·11
10. Shaft, Shellhaven	ϕ 5·5	5·2	1·0	1·54	1·6	6	1·0	0·95	7·5	1·34	do.	
11. Test shaft, Sörligaten Oslo	ϕ 1·5	12·0	0·0	1·85	2·1	60	1·0	8·00	9·0	0·85	do.	
12. Basement excavation, Storgaten Oslo	8·0 × 4·0	6·7	0·0	1·90	1·75	10–20	0·5	1·70	8·0	1·10	No failure	
13. Subway, Chicago	16 × ∞	11·3	0·0	1·90	3·5	—	0·0	0·70	6·1	1·00	do.	1·08
14. Test shaft (S), Ensjöveien, Oslo	ϕ 1·5	11·0	0·0	1·93	2·7	200	1·0	7·30	9·0	1·14	do.	

on English clays, are somewhat smaller than the values valid for the sensitive Norwegian clays with a high value of E/s. A better agreement would result if N_c was assumed to be 10 for deep shaft excavations instead of 9.

The observations made during excavation of the test shaft in Sörligaten indicate, however, another explanation of the low safety factors of the deep shafts. This test shaft proved that under certain conditions it is possible to excavate a shaft to a greater depth than that calculated theoretically without obtaining a total failure. If a shaft is dug through a stiff clay down to soft clay, the upper stiff-clay layer will prevent, at least for a time, a local overstressing of the soft clay below the bottom of the shaft from extending to a total failure. This effect of an upper stiff-clay layer was observed in the test shaft in Sörligaten which was excavated down to almost 2 m below the calculated critical depth. Moreover, this effect may explain why the total failure in the test shaft (N) at Ensjöveien occurred a certain time after the excavation was completed.

(5) The scattering of the calculated safety factors for some of the excavations seems to indicate that the danger of failure will increase if the excavation is not sufficiently braced to prevent excess horizontal deformations. If the upper clay layers are stiff, and the depth of the excavation is small, the danger of a base failure seems to be less affected by the bracing.

(6) Experience cannot yet indicate exactly how much the danger of a base failure is reduced if sheet piles are driven below the bottom of an excavation. The stabilizing effect of sheet piling seems, however, to be small for the cases where the shear strength of the clay does not increase with depth.

ACKNOWLEDGEMENTS

Four of the described fourteen case records were collected by Mr Skaven-Haug, chief engineer of the Norwegian State Railways. The Authors are indebted to Mr Skaven-Haug for permission to use and publish his field observations. The Authors are, moreover, indebted to the Planning Office for Suburban Lines and Subways, Oslo, whose support made it possible to excavate the three test shafts. The Authors also wish to express their particular gratitude to Professor A. W. Skempton and to Dr A. W. Bishop for valuable discussions on the problem, and to the Directors of John Mowlem & Company for permission to use and publish the Shellhaven case records. Finally, the Authors acknowledge the assistance of their colleagues, especially the work done by Mr I. Johannessen, during the laboratory and field investigations.

REFERENCES

BJERRUM, L., 1955. Stability of Natural Slopes in Quick Clay. *Géotechnique*, 5 : 101.

CUEVAS, I. A., 1936. Foundation Conditions in Mexico City. *Proc. 1st Int. Conf. Soil Mech.* (Cambridge), 3 : 233.

HARTMARK, H., 1955. Geotekniske Observasjoner fra utförelsen av Tyholdt Jordtunnel. *Teknisk Ukeblad*, 27 : 565.

JANBU, N., 1952. Stabilitet av skjaeringer i åpen sjakt. *Norges Geotekniske Institutt. Int. rapport F* 13, *Oslo.*

PECK, R. B., 1943. Earth pressure measurements in open cuts, Chicago (Ill.) subway. *Trans. Amer. Soc. Civ. Engrs*, 108 : 1008.

SKEMPTON, A. W., 1951. The Bearing Capacity of Clays. *Bldg Res. Cong., London*, 1951. Papers presented in Div. 1, pp. 180–189.

SKEMPTON, A. W., *and* WARD, W. H., 1952. Investigations Concerning a Deep Cofferdam in the Thames Estuary Clay at Shellhaven. *Géotechnique*, 3 : 119.

SKEMPTON, A. W., *and* HENKEL, D. J., 1953. The Post-glacial Clays of the Thames Estuary at Tilbury and Shellhaven. *Proc. 3rd Int. Conf. Soil Mech.* (Zurich), 1 : 302.

TERZAGHI, K., 1943. Theoretical soil mechanics. *Wiley, New York.*

TERZAGHI, K., *and* PECK, R. B., 1948. Soil mechanics in engineering practice. *Wiley, New York.*

TSCHEBOTARIOFF, G. P., 1951. Soil mechanics, foundations, and earth structures. *McGraw-Hill, New York.*

Roscoe, K. H., Schofield, A. N. & Wroth, C. P. (1958). *Géotechnique* **8**, No. 1, 22–53

ON THE YIELDING OF SOILS

by

K. H. Roscoe, M.A., A. N. Schofield, M.A., and C. P. Wroth, M.A.

SYNOPSIS

Hvorslev's equation for the shear strength of clay is shown to define a surface in a space of three variables σ', e and τ. The progressive yielding of a sample defines a loading path in this space, and the paths taken by samples in differing tests can be correlated if a boundary energy correction is applied. The final portions of all paths then lie in a unique surface, and the paths end at a unique critical voids ratio line. At the critical voids ratio state unlimited deformation can take place while σ', e and τ remain constant.

The two concepts of the existence of such a surface and such a critical voids ratio line are verified by an analysis of results of triaxial tests on a clay. Similar tests on silt and sand also confirm the concepts, although some inaccuracy is inherent in these tests. Finally results of tests on granular media in a simple shear apparatus are presented which confirm closely the applicability of both concepts to the yielding of granular media.

On montre que l'équation de Hvorslev sur la force de cisaillement de l'argile détermine une surface dans un espace à trois variables, σ', e et τ. Le tassement progressif d'un échantillon détermine un parcours de charge dans cet espace, et on peut établir une corrélation entre les parcours pris par les échantillons dans des essais différents si l'on applique une correction de l'énergie limite. Les parties finales de tous les parcours se trouvent alors dans une surface unique, et les parcours se terminent à une ligne critique unique des indices de vide. A l'état critique des indices de vide il peut se produire une déformation illimitée tandis que σ', e et τ restent constants.

Les deux conceptions de l'existence d'une telle surface et d'une telle ligne critique des indices de vide sont vérifiées par une analyse des résultats d'essais triaxiaux d'une argile. Des essais similaires sur du limon et du sable confirment aussi les conceptions, quoique quelque inexactitude soit inhérente à ces essais. Finalement, on présente des résultats d'essais effectués sur des milieux granulaires dans un appareil de cisaillement simple qui confirment exactement l'applicabilité des deux conceptions au tassement des milieux granulaires.

INTRODUCTION

The classic work of Hvorslev (1937)[1] on the shear resistance of remoulded saturated cohesive soils at failure contains a clear statement of the fundamentals upon which the present knowledge of the subject is based. He showed that the peak shear stress at failure* τ_f of such a soil is a function of the effective normal stress σ'_f *on*, and of the voids ratio e_f *in, the plane of failure at the moment of failure* and this function is independent of the stress history of the sample. Subsequent publications, such as Gibson (1953), have all tended to confirm the findings of Hvorslev.

However, from Hvorslev's work it is only possible to predict the strength at *failure* if the values of the normal stress and voids ratio are known at *failure*. In many problems what is known is the *initial* voids ratio *prior* to the application of any shear stress. To develop Hvorslev's work to cover such problems it is necessary to have a test apparatus which imposes uniform strain throughout the whole of any sample of the soil for the entire duration of the test. Some soil samples need to be subjected to large strains before developing their peak shear resistance τ_f, and in tests such as those of the triaxial, shear box, or torsion ring apparatus, there may be cases in which the volume of the zone of failure within the sample is small and of unknown magnitude. Hence Roscoe (1953) designed an apparatus to impose uniform simple shear strain to soil samples. With the uniformity of strain imposed by this apparatus it is possible to determine the voids ratio with some reliability at all times during a test by measuring the average voids ratio of the whole sample. A feature of this apparatus is that it

1 The references are given on p. 53.

* Throughout this Paper a sample subjected to increasing strain will be said to be in a state of failure at that moment when the total applied shear stress reaches a maximum or peak value. Such values are given the suffix f.

measures directly the quantities τ_f, σ'_f, and e_f used in Hvorslev's failure criterion. Being strain-controlled it is also suitable for determining the conditions of the soil when subjected to deformations beyond those required for failure.

This simple shear apparatus has been used to investigate the yielding of cohesionless media such as sands, glass beads, and steel balls. It has also been used to investigate concepts regarding the critical voids ratio of these media. In addition to this work it was found from a study of previously published triaxial tests on several clays that they yield in a very similar manner to the cohesionless media but, in fact, the analysis is even simpler. Furthermore, the concepts of critical voids ratios apply to both types of soil.

One object of the Paper is to show the similarities between the behaviour of cohesionless and cohesive soils when subjected to shear stresses. Part I deals with the simpler conditions of cohesive soils and Part II deals with cohesionless media.

Part I: Remoulded saturated clays

THE HVORSLEV SURFACE AND DOMAIN FOR CLAY

1(*a*). *The Hvorslev surface*

The Hvorslev criterion of failure stated verbally in the introduction above may be expressed, for a given soil, by the equation* :

$$\tau_f = \mu_0 \sigma'_f + \nu \exp.(- B e_f) \quad . \quad . \quad . \quad . \quad . \quad . \quad . \quad . \quad (1)$$

where μ_0, ν, and B are constants of the particular soil.

This equation (1) is satisfied by all possible states of stress and voids ratio when the soil is at failure in a drained test. It is a unique function connecting the three independent variables σ'_f, e_f, and τ_f. If equation (1) is plotted three dimensionally with right-handed co-ordinate axes (σ', e, τ) it will define a unique surface in (σ', e, τ) space, which will be called the *Hvorslev surface*. Fig. 1 shows two general isometric views of this surface for Kleinbelt Ton which have been plotted from the results quoted by Hvorslev from his tests on that clay in a standard shear box. For convenience of presentation in this Paper, the $\tau = 0$ plane is considered horizontal with the positive direction of the τ-axis vertically upwards. A, B, C, F, E, D represent points on the Hvorslev surface and all points below or on this surface will be within the *Hvorslev domain*. A point above the surface represents a condition which the soil cannot experience.

The Hvorslev domain is bounded, in directions other than the vertical, by additional surfaces. It is assumed that both $\sigma' \geqslant 0$ and $\tau \geqslant 0$ and hence the planes $\sigma' = 0$ and $\tau = 0$ form two further boundaries to the domain, as shown in Fig. 1. By definition $e \geqslant 0$, but in practice for any given soil a boundary will be formed by the surface $e = e_{min}$ where e_{min} is the minimum voids ratio that the soil can experience. Before discussing the final boundary of the domain, it is necessary to consider the normal or virgin consolidation curve of the soil as shown by $A_1B_1C_1$ in Fig. 2. This curve represents the limit of the loosest packing or greatest voids ratio for any given value of σ', so that values of (σ', e) lying further from the origin than this curve are not permissible when $\tau = 0$. It is assumed here that the same limitation applies to (σ', e) when $\tau > 0$, so that a curved vertical wall through the normal consolidation curve becomes a boundary for the domain as shown in Fig. 1. Hvorslev's experimental work did, in fact, show a more stringent limitation to the possible values of (σ', e) at which failure can occur. However, the last assumption is convenient for the description of the surface given in the next section, and it can only be modified in the light of later sections.

The Hvorslev surface and the Hvorslev domain are based only upon results of fully-drained tests on saturated clays in shear boxes. Many refinements will be introduced as other types of test are considered, but it is important at this stage to consider the geometry of the Hvorslev surface in some detail.

* The exponential function is denoted throughout by exp. (x).

1(*b*). *Shape of the Hvorslev surface*

Any point on the Hvorslev surface denoted by a letter in Fig. 1 has projections on the $\tau = 0$, $e = 0$ and $\sigma' = 0$ planes which will be denoted by the same letter as the point but with the suffix 1, 2, and 3, respectively.*

The Hvorslev surface can best be obtained by taking any point P_1 in the (σ', e) plane and plotting P above it where PP_1 is the value of τ_f obtained from equation (1). Figs 3 and 4 show projections of the Hvorslev surface on the $e = 0$ and $\sigma' = 0$ planes, respectively.

The points A_1, B_1, C_1 on the normal consolidation curve in Fig. 2 satisfy the equation :

$$e - e_0 = -\frac{1}{B}\log_e\left(\frac{\sigma'}{\sigma'_0}\right) \quad . \quad . \quad . \quad . \quad . \quad . \quad . \quad (2)$$

where e_0, σ'_0 are constant and B is the same as in equation (1). The corresponding points A, B, C on the Hvorslev surface must satisfy both (1) and (2) when $\sigma' = \sigma'_f$. Combining these equations :

$$\tau_f = \mu_0\sigma'_f + \xi\sigma'_f \quad . \quad . \quad . \quad . \quad . \quad . \quad . \quad . \quad (3)$$

where $\xi = \dfrac{\nu}{\sigma'_0}\exp.\,(-Be_0)$ which is a constant for any one soil. Equation (3) is that of a plane intersecting the $e = 0$ plane in the straight line $OA_2B_2C_2$ which passes through the origin and is of slope $\psi = \tan^{-1}(\mu_0 + \xi)$. This line is the projection on the $e = 0$ plane of the curved edge line ABC, which contains the points on the Hvorslev surface corresponding to the normal consolidation curve.

Points on the straight line F_1C_1 in Fig. 2 are of constant voids ratio, $e = e_c$ (say), and the corresponding points on FC in Fig. 1 must from equation (1) therefore satisfy the equation :

$$\tau_f = \mu_0\sigma'_f + \zeta \quad . \quad . \quad . \quad . \quad . \quad . \quad . \quad . \quad (4)$$

where $\zeta = \nu\exp.\,(-Be_c)$ which is constant since e_c is constant. Since equation (4) is the equation of FC in the plane $e = e_c$ it follows that FC is a straight line. Hence the projection F_2C_2 on the plane $e = 0$ is also a straight line and both FC and F_2C_2 are at an angle $\phi_0 = \tan^{-1}\mu_0$ to the $\tau = 0$ plane.

Similarly EB and DA are straight lines contained in the Hvorslev surface which therefore is a ruled surface of parallel straight lines inclined at ϕ_0 to the $\tau = 0$ plane.

1(*c*). *The edge of the Hvorslev surface*

The assumption of the curved edge line ABC, vertically above the normal consolidation curve, was made in order to simplify the presentation of the Hvorslev surface. Hvorslev was, in fact, unable to obtain test results corresponding with the edge ABC. He observed, in agreement with unpublished work of Casagrande and Albert, that when a normally consolidated sample was sheared, it decreased in volume and expelled water from its voids. The sample reached what is, in effect, an over-consolidated state during the process of shear. The failure values of such samples give an edge to the Hvorslev surface with a projection on the $\tau = 0$ plane lying inside the normal consolidation curve. To trace the boundary which must replace the assumed vertical wall, the concept of progressive yielding before and after shear failure must now be developed.

LOADING PATHS AND YIELD SURFACES

2(*a*). *Failure, progressive yielding and test paths*

The shear tests carried out by Hvorslev were stress controlled, which prevented any investigation of the conditions of samples after the peak shear stress had been attained. With the advent of strain-controlled shear tests, the failure of a soil no longer becomes catastrophic with unlimited deformation, since the shear stress can pass through a peak value without any sudden collapse of the sample. In order to plot the progressive yielding of such samples the parameters σ', e, and τ are often plotted separately against strain, or deformation.

* D, E, F lie in the $\sigma' = 0$ plane and coincide with their projections D_3, E_3, F_3.

However, at any stage in the test these parameters (σ', e, τ) define a point in the Hvorslev domain. As the test progresses this point (σ', e, τ) defined by the condition of any one element* will trace out a path in the domain. The direction of this test path is governed by the conditions of drainage allowed during the test, which restricts the degrees of freedom from three to two. In general, tests are carried out either with full drainage or with no drainage, and the differences between these extremes must be considered in detail.

2(*b*). *Fully-drained tests in shear-boxes*

In tests with *full drainage*, such as those carried out by Hvorslev, the effective normal stress σ' remains constant throughout. The test path is therefore constrained to lie in this $\sigma' =$ constant plane. Such a plane is shown by $BPP_2P_1B_1$ in Fig. 1 and in elevation in Fig. 5. As the test progresses the volume of the soil element will be changing, and this will require work to be done *on* or *by* the external stresses acting on the boundaries of the element. The shear stress, τ, as measured at any instant, can be split into components τ_e and τ_i. The component τ_e is that part of the shear stress which would be just sufficient to provide the work used by the soil externally in changing its volume against the confining pressure; and τ_i is the remainder which provides the work absorbed internally by the soil (usually assumed to be purely frictional).

When the element is expanding τ_e is positive, and when contracting negative. With this convention:

$$\tau = \tau_e + \tau_i \qquad (5)$$

For a shear-box:

$$\tau_e = \sigma' \frac{dy}{dx} \qquad (6)$$

where y = sample thickness and x = relative displacement of shear-box halves. If $\dfrac{dy}{dx}$ is observed for all points on the test path it is possible to calculate τ_e from equation (6), so that from equation (5), τ_i can be deduced. τ_i will represent the shear stress corrected for the work done in changing the boundaries of the sample; and the necessary correction will be called the *boundary energy commission.*

2(*c*). *Undrained tests in shear-boxes*

In tests where no drainage is allowed the test path is constrained to lie in an e = constant plane, while the effective normal stress σ' changes according to the excess pore-water pressure u set up in the soil. Such a plane is shown by BEB_1 in Fig. 1 and by the parallel plane through P drawn in elevation in Fig. 6. Since the volume of the small soil element remains constant $\tau_e = 0$ for the whole test path. Equation (5) reduces to $\tau = \tau_i$, and the entire work provided by the shear stress is absorbed internally.

2(*d*). *Correlation of drained and undrained tests*

Since τ_e is identically zero throughout undrained tests, and is dependent on the rate of dilatation in drained tests, it is necessary to compare the values not of $\tau = (\tau_e + \tau_i)$ but of τ_i obtained in the two types of test. The test paths traced out by successive values of (σ', e, τ_i) will be called *loading paths.*

An implicit assumption in the definition of τ_i in sections 2(*b*) and 2(*c*) is that the work absorbed *internally* by the sample is not affected by any volume change. It is possible that in a drained test when a sample changes its volume extra work has to be done, not only externally in altering the boundaries but also internally in changing the voids ratio. In an

* In shear tests an attempt is made to create uniform conditions throughout a sample. The analysis can be extended to regions in which conditions vary, provided that an element can be defined, which is of a size small enough, that throughout any test the conditions can be considered uniform *within* this element, yet large enough compared with the soil particles that its voids ratio can be defined sensibly.

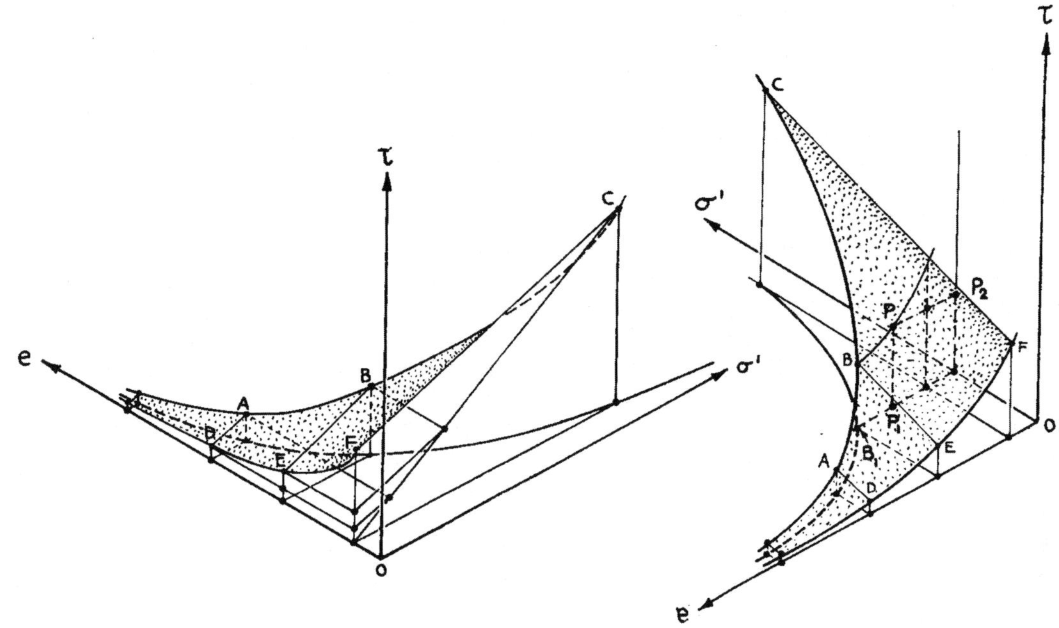

Fig. 1. Two isometric views of the Hvorslev surface for Kleinbelt Ton

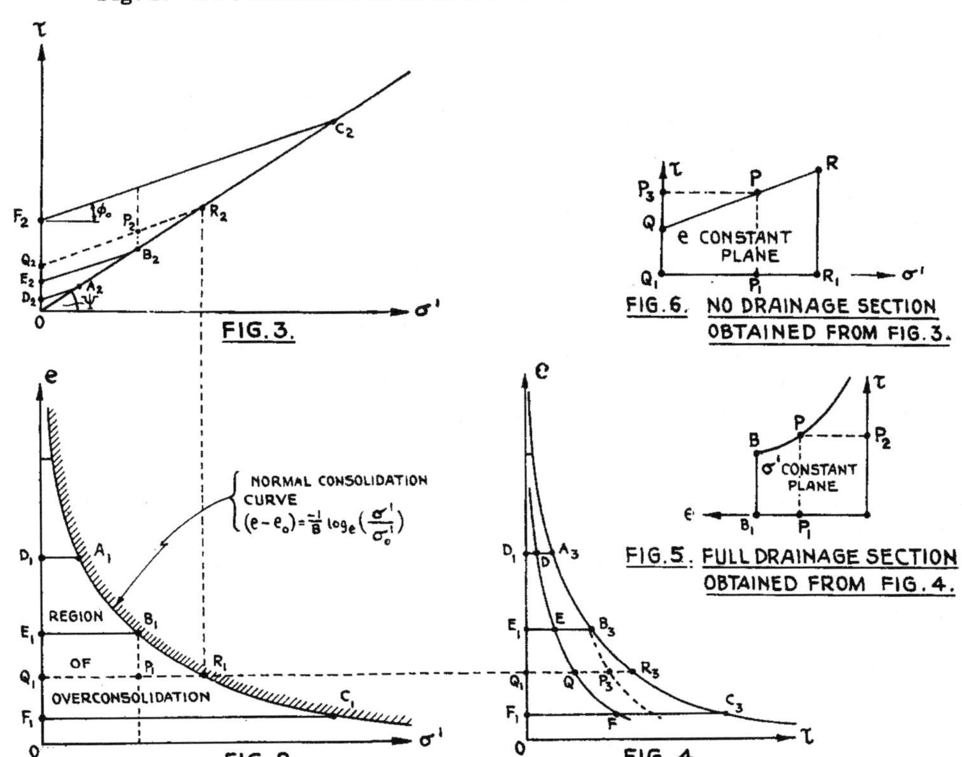

Figs 2–6. Geometry of Hvorslev surface for Kleinbelt Ton

undrained test a sample is distorted in shear, so that the relative positions of the particles are shifted without the average distance between the centres of particles changing. In a *drained* test the relative positions of the particles are again shifted, and this certainly involves absorption of work internally ; but in addition the average distance between the centres of particles changes and it is possible that additional internal work is absorbed if there is a marked inelastic hysteresis in the compressibility of the grain structure. If this is the case, then values of τ_i obtained from the two tests will not be comparable even after applying the above boundary energy correction to the results from the drained tests. But if, as proves to be the case, the two surfaces obtained from the two sets of test results coincide, then within the limits of experimental accuracy the work internally absorbed will be independent of the rate of dilatation and, in addition, τ_i will be purely a function of σ' and e.

Experimental evidence (discussed in detail in later sections) suggests that in each type of test the loading paths have a surface as their envelope in (σ', e, τ_i) space, and as yield continues, the loading paths continue to *lie on that surface.* The surfaces so defined will consequently be called the *undrained yield surface* and the *drained yield surface.* The common surface, when established, will then be the *yield surface,* and the space it contains the *yield domain.*

It is important to note at this stage the extent to which the concepts of the last paragraph have developed beyond those of section 1(*a*). Each test now gives a succession of results which form a loading path, in place of a single result which forms a failure point. The family of possible loading paths lies within the yield domain or on its surface. In contrast the Hvorslev surface is only defined by a set of failure points, each of which is given by the peak shear stress τ_f from a drained test, with no boundary energy correction. The Hvorslev surface is quite distinct from both the drained and undrained yield surfaces, and cannot be obtained experimentally with the same accuracy.

3(*a*). *Extension of method to results from other apparatuses*

If shear tests on saturated remoulded clays are carried out in any apparatus other than the shear-box, the results can be plotted in an exactly similar and consistent fashion to give three-dimensional surfaces reached by the various loading paths. These surfaces will differ according to the choice of apparatus and to the choice of variable parameters by which the condition of a sample is defined.

3(*b*). *Choice of variables for analysis of standard triaxial test results*

Throughout all tests in the standard triaxial apparatus it will be assumed that σ_1 is the stress on the end caps of the sample and $\sigma_2 = \sigma_3$ is the cell pressure. Since for the Hvorslev domain τ must be > 0 it will be assumed for the triaxial domains that $\sigma'_1 \geqslant \sigma'_3$, thereby excluding consideration of conventional extension tests.

The stress parameters chosen for plotting the triaxial yield surfaces will be the *mean principal effective stress* $p = \frac{1}{3}(\sigma'_1 + 2\sigma'_3)$, and the *deviator* stress $(\sigma_1 - \sigma_3) = (\sigma'_1 - \sigma'_3)$ where (σ'_1, σ'_2, σ'_3) are the *effective* stresses, defined as usual. The parameter p is the first invariant of the stress tensor while $(\sigma'_1 - \sigma'_3)$ is derived from the second invariant of the stress deviation tensor. These variable parameters both satisfy the condition that they are independent of the choice of reference axes in a triaxial system of stress.* The choice of the third variable is essentially unchanged being the voids ratio, the porosity or the water content, whichever is most convenient.

Three surfaces can be obtained from the analysis of standard triaxial test results : (*a*) the " peak " surface from values of the variables obtained at failure in drained tests (analogous to the Hvorslev surface), (*b*) the drained yield surface, and (*c*) the undrained yield surface.

* See, for example, Prager and Hodge (1951) " Theory of perfectly plastic solids ", Chapter 1, Wiley. The theory of plasticity is concerned with plastic *incompressible* media, but the discussion of invariants of stress tensors is still relevant to dilatant media.

To obtain the drained-yield surface a boundary energy correction must be applied to $(\sigma'_1 - \sigma'_3)$.

As in equation (5) :
$$(\sigma'_1 - \sigma'_3) = (\sigma'_1 - \sigma'_3)_e + (\sigma'_1 - \sigma'_3)_i \quad \quad (7)$$

into which is substituted :
$$(\sigma'_1 - \sigma'_3)_e = \sigma_3 \frac{dv}{d\epsilon_1} \quad \quad (8)$$

where v = volumetric strain and ϵ_1 = axial strain, because the boundary work is done by or against the cell pressure σ_3 (Bishop 1954). In the case of tests where the sample is made to fail by increasing the pore-water pressure, the boundary energy correction has to allow for the work supplied by this increase of pressure, becoming :

$$(\sigma'_1 - \sigma'_3)_e = (\sigma_3 - u) \frac{dv}{d\epsilon_1} = \sigma'_3 \frac{dv}{d\epsilon_1} \quad \quad (9)$$

This reduces to equation (8) for fully drained tests when $u = 0$. To emphasize the importance of this correction, and for ease of distinction between the uncorrected and corrected values, the *corrected* deviator stress $(\sigma'_1 - \sigma'_3)_i$ will be denoted by the symbol q.

CRITICAL VOIDS RATIO LINE

3(c). *The concept of critical voids ratio*

The continuous yielding of a sample can be represented by a loading path which rises to the yield surface and then remains on the surface. The problem arises as to whether the path ends at any specific point. If such a point exists, then it will be appropriate to say that a sample is in a *critical voids ratio state* when the loading path reaches that point.

However, the words " critical voids ratio " have generally been applied to a particular state of a sand. This state has not been precisely defined, although there are two alternative sorts of definition ; one is concerned with changes of volume in drained tests (Casagrande, 1938), and the other with changes of effective stress, and consequently of strength, in undrained tests (Taylor, 1948). These definitions can be made more explicit if the concept of loading paths in (p, e, q) space is used. It will be shown later that the critical voids ratio concept is valid for clays as well as for sands.

In a *drained* test the critical voids ratio state can be defined as that ultimate state of a sample at which any arbitrary further increment of shear distortion will not result in any change of voids ratio. In any series of drained tests the set of critical voids ratio points so defined can be expected to lie in or near a line on the drained yield surface.

In an *undrained* test the sample remains at a constant voids ratio, but the effective stress p will alter to bring the sample into an ultimate state such that the particular voids ratio, at which it is compelled to remain during shear, becomes a critical voids ratio. In any series of undrained tests the set of critical voids ratio points so defined can be expected to lie in or near a line (not necessarily the same as above) on the undrained yield surface.

If the results of drained and undrained tests show that there is, in fact, one unique line to which all loading paths in (p, e, q) space converge, then this will be called the *critical voids ratio line* (C.V.R. line, for short). The drained and undrained yield surfaces will then coincide along the C.V.R. line. These two surfaces may be identical and, if so, the common surface would then form the yield surface.

In testing the two hypotheses of the existence of a yield surface in (p, e, q) space, and of the existence of a C.V.R. line on that surface, it becomes evident that the first hypothesis is more easily tested than the second. Loading paths reach the yield surface after only a slight distortion of the sample, and the magnitudes of the parameters p, e, q as determined from mean conditions in the sample can be expected to be accurate. Loading paths only reach the C.V.R. line after a severe distortion of the sample and, unless this distortion is a large controlled shear distortion, mean conditions in the sample cannot be expected to define accurately a point in (p, e, q) space. This may mean that the loading paths converge on some bounded region without closely defining a C.V.R. line.

Many sets of test results which have been examined tend to confirm the hypotheses stated within the limitations outlined above, but there are few sets of sufficiently comprehensive test results available for detailed study. One such set, which was kindly made available during the preparation of the Paper, has proved extremely comprehensive and most suitable, and for the sake of simplicity the following analysis will be confined to these tests. It is the set of tests on samples of Weald Clay carried out at Imperial College, London, by Gilbert (1954) and summarized by Henkel (1956).

ANALYSIS ON UNDRAINED TESTS ON CLAY

4(*a*). *Introduction*

The theoretical work in the following sections will be applied to the analysis of triaxial test results on saturated remoulded samples of Weald Clay. The properties of this clay are, according to Henkel :

LL	PL	PI	% clay fraction	Activity
43	18	25	40	0·6

The triaxial test results are given in terms of water content w% which is used here, for convenience, in place of the voids ratio e : the other variables p and q can readily be calculated from quoted values of σ_1, σ_3, u, etc. At any stage in a test a small increment of the stress parameters will be denoted by $\delta\sigma_1$, $\delta\sigma_3$, δu, etc., and the total change since the start of the test by $\Delta\sigma_1$, $\Delta\sigma_3$, Δu, etc.

To facilitate the discussion of loading paths in (p, w, q) space, the $q = 0$ plane will be considered to be horizontal with the q axis rising vertically from it. The last *observed* point on a loading path will be called the *end point* and will be denoted by Q. There may be tests which have not been carried sufficiently far beyond the *failure* state for the sample to have reached the *ultimate* state. In such cases the point on the C.V.R. line to which, by hypothesis, the loading path is expected to converge will be denoted by X. On the other hand in many tests Q will coincide with X and throughout the Paper values corresponding to X will be given the suffix u to denote the *ultimate* state.

4(*b*). *Theoretical relationship between overconsolidation ratio, consolidation pressure, and equivalent pressure*

Before considering the Weald Clay in detail, a simple theoretical relationship which will be of use later is now derived. Fig. 7 represents the consolidation lines in a semi-logarithmic plot for a clay sample which has been normally consolidated from A to C and then allowed to swell back to D. If the slope of each line is assumed to be constant and of magnitude $-\lambda$ for ABC and $-\kappa$ for CD then a simple theoretical relationship can be established between the over-consolidation ratio $n = \dfrac{p_2}{p_3}$, the final consolidation pressure p_3 and the equivalent pressure p_1. This pressure p_1 is that of the point B on the normal consolidation line having the same water content as D.

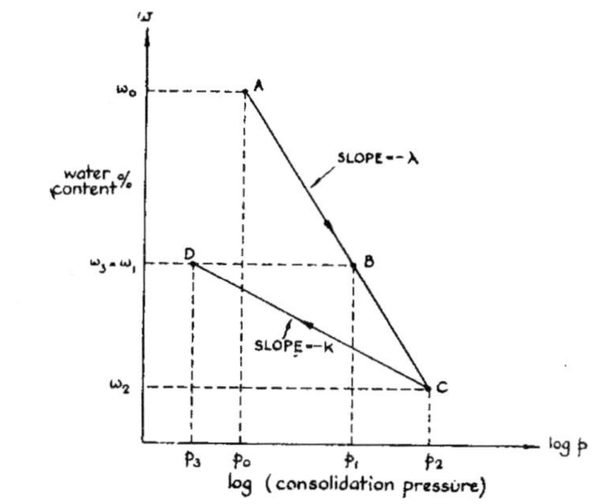

Fig. 7. Simplified relation between water content and consolidation pressure for a clay

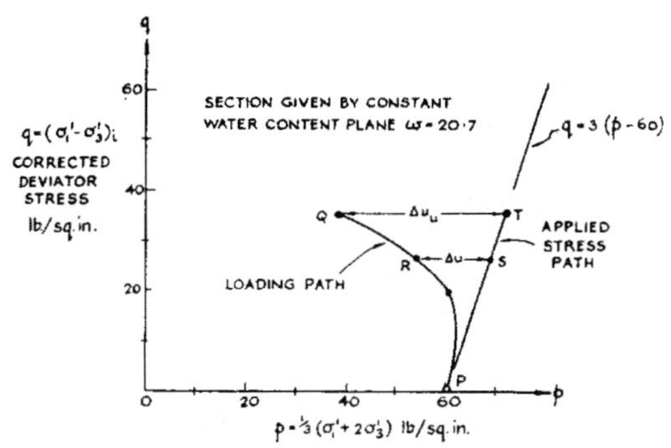

Fig. 8. Loading path for undrained triaxial test on normally consolidated Weald Clay

The normal consolidation line is :

$$w - w_0 = -\lambda \log_e (p/p_0) \quad \ldots \ldots \quad (10)$$

The swelling or overconsolidation line is :

$$w - w_2 = -\kappa \log_e (p/p_2) \quad \ldots \ldots \quad (11)$$

Point B given by (p_1, w_1) satisfies equation (10) :

$$\therefore w_1 - w_0 = -\lambda \log_e (p_1/p_0) \quad \ldots \ldots \quad (12)$$

Point C given by (p_2, w_2) satisfies equation (10) :

$$\therefore w_2 - w_0 = -\lambda \log_e (p_2/p_0) \quad \ldots \ldots \quad (13)$$

Point D given by (p_3, w_1) satisfies equation (11) :

$$\therefore w_1 - w_2 = -\kappa \log_e (p_3/p_2) \quad \ldots \ldots \quad (14)$$

Equations (12), (13), and (14) give :

$$0 = -\lambda \log_e (p_1/p_0) + \lambda \log_e (p_2/p_0) + \kappa \log_e (p_3/p_2) \quad \ldots \quad (15)$$

But by definition $n = p_2/p_3$, and equation (15) becomes :

$$\lambda \log_e (np_3/p_1) = \kappa \log_e n \quad \ldots \ldots \quad (16)$$

and finally :

$$p_3/p_1 = n^{\left(\frac{\kappa}{\lambda}-1\right)} n = n^{-\Lambda} \quad \ldots \ldots \quad (17)$$

where $-\Lambda = \left(\frac{\kappa}{\lambda} - 1\right)$.

Equation (17) is valid for any clay satisfying the assumption that the consolidation lines are straight. For the Weald Clay approximate values from Fig. 1 of Henkel's Paper give $\frac{\kappa}{\lambda} = 0{\cdot}35$; so that the ratio of the final consolidation pressure to the equivalent pressure $= p_3/p_1 = n^{-0{\cdot}65}$.

4(*c*). *Undrained tests on normally consolidated samples*

The loading path of an undrained test will lie in a section of the (p, w, q) domain made by a $w =$ constant plane. From Gilbert's results, when a sample of Weald Clay is normally consolidated under an all round pressure of 60 lb/sq. in., the equilibrium moisture content is $w = 20{\cdot}7$. This condition of the sample is represented by point P in Fig. 8, and all points in this diagram lie in the plane $w = 20{\cdot}7$.

Any system of applying external stresses to the boundaries of the sample can be represented by a line in the plane of Fig. 8 which will start at P, provided (*i*) the initial pore-water pressure $u_0 = 0$,* and (*ii*) the applied stresses are plotted on the same axes as the effective stresses p and q. Such a line will be called the *applied stress path.*

If the sample, in a condition represented by P, is subjected to a conventional undrained test ($\sigma_3 =$ constant, σ_1 increasing) the applied stress path is a straight line PST making an angle $\tan^{-1} 3$ with the p axis. (This is because $\delta\sigma_3 = \Delta\sigma_3 = 0$ and every increment of applied stress in the p direction of $\delta\left\{\frac{1}{3}(\sigma_1 + 2\sigma_3)\right\} = \frac{1}{3}\delta\sigma_1$ is accompanied by an increment in the q direction which is $\delta(\sigma_1 - \sigma_3) = \delta\sigma_1$.) The loading path PRQ for this sample will also start at P. The point R on the loading path has the same ordinate as the corresponding point S on the applied stress path since $(\sigma_1 - \sigma_3) \equiv (\sigma'_1 - \sigma'_3)$, but the abscissa of R is $\frac{1}{3}(\sigma'_1 + 2\sigma'_3) = \frac{1}{3}(\sigma_1 + 2\sigma_3) - \Delta u$, where Δu is the positive excess pore-water pressure corresponding to the applied stress system S. Hence RS $= \Delta u$ in magnitude, and if RS is measured algebraically from R to S in the positive direction of the p axis it will represent positive values of Δu, and in the reverse direction negative values.

Loading paths for conventional tests on six normally consolidated samples are shown in Fig. 9. Each path lies in a different water-content plane, and they are seen in elevation (or projection on the $e = 0$ plane) giving a family of similar curves.† The end points Q_1, $Q_2 \ldots Q_6$ lie on a line in space which has a straight line projection $Q_{12}, Q_{22} \ldots Q_{62}$ in Fig. 9 passing through the origin. Fig. 10 is a plan (or projection on the $q = 0$ plane) showing in addition the normal consolidation curve given by the starting points $P_1, P_2 \ldots P_6$. Fig. 11 is the other elevation (or projection on the $p = 0$ plane). Hence the end points $Q_1, Q_2 \ldots Q_6$ lie on a curved line in space exactly analogous to the curved edge line ABC on the Hvorslev surface except that it is not vertically above the normal consolidation curve.

4(*d*). *Undrained tests on overconsolidated samples*

If the loading paths for a series of tests for one given overconsolidation ratio n are plotted then a family of similar curves is obtained. It is formed in the same way as the family shown in Fig. 9, where $n = 1$, but the shape of each family radically alters with the value of n. In tests on overconsolidated samples the positive change in pore-water pressure Δu is less than in tests on normally consolidated samples. For heavily overconsolidated samples Δu may become negative. The loading path will cross the applied stress path at the point for which $\Delta u = 0$, and will lie on the other side for negative values of Δu. Except for the very heavily overconsolidated series, the end points of paths of each family lie very close to the same curved line in space as that described by $Q_1, Q_2 \ldots Q_6$. Figs 12, 13, and 14 show the end points of all the thirty-one undrained tests with $n \leqslant 8$ which are quoted by Gilbert (1954).

Since in Fig. 13 the abscissae of both curves for any given w bear a constant ratio to each other, and because in Fig. 12 the relation of p to q is linear, it follows that if the locus of Q is

* For the case when $u_0 \neq 0$, see section 10(*b*).

† Points on the yield surface are denoted by a single suffix, e.g., $Q_1, P_3 \ldots$ Projections of these points on the $q = 0$, $e = 0$, and $p = 0$ reference planes in Figs 9, 10, and 11 are denoted by a second suffix 1, 2, and 3, respectively, e.g. $Q_{11}, P_{32} \ldots$

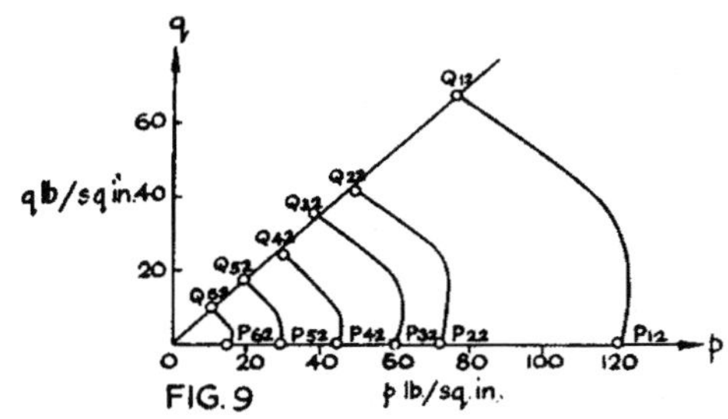

FIG. 9

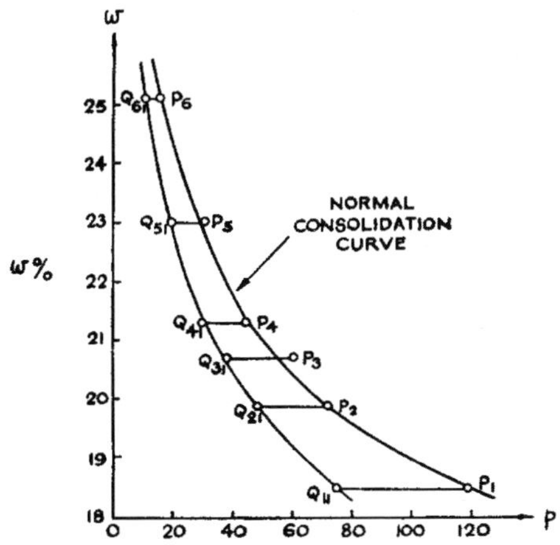

FIG. 10

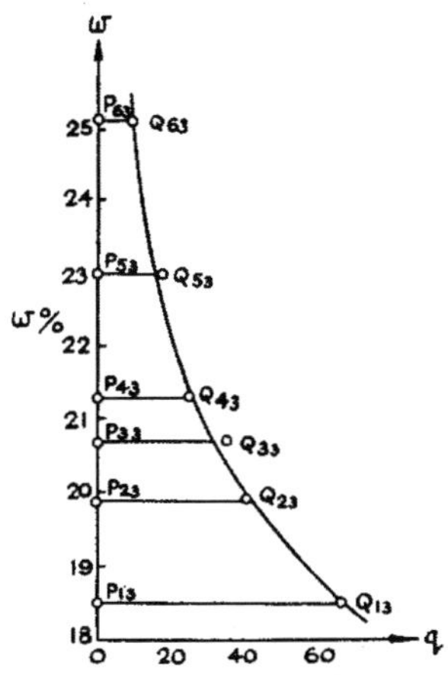

FIG. 11

Figs 9–11. Family of loading paths for undrained triaxial tests on normally consolidated Weald Clay

plotted with logarithmic scales for p and q and arithmetic for w, then it will be a straight line in this space. Hence the actual dimensions of the members of any one family vary logarithmically with their water contents. Fig. 15 shows the appropriate member of each family for the plane given by $w = 20{\cdot}7$: the representative loading path for each value of n has been obtained by scaling up or down test loading paths from other water content planes, by means of this logarithmic relationship.

The paths for tests with *all* values of $n > 3$ become tangential to a straight line, shown by VUQ in Fig. 15, and PRQUV form the boundary for all loading paths in this one plane. As w varies, giving sections of the same shape but of different size, the boundary PRQUV will form the ***undrained yield surface.*** The part of the surface formed by the set of lines parallel to VUQ is a ruled surface and is similar in shape to the Hvorslev surface. The other part of the surface is formed by the loading paths of normally consolidated samples, and has a curved profile (PRQ). This replaces the vertical wall through the normal consolidation curve which had to be assumed for the Hvorslev surface (section 1(a)) for want of more detailed information.

The end points plotted in Figs 12, 13, and 14 closely define a line on the undrained yield surface and this line is therefore a line of critical voids ratios. An isometric view of the undrained yield surface for Weald Clay is shown in Fig. 16. The section PRQUV and the line PST in this view correspond to $w = 20{\cdot}7$ and are the same as those shown in Fig. 15 in which the same letters are used. The C.V.R. line in Fig. 16 corresponding to all the undrained test results is shown by X′QX″.

The loading paths for heavily overconsolidated samples ($n > 8$) appear to rise to the line

VQ as shown in Fig. 15, but do not appear to reach Q in this w = constant plane and instead fall below the line VQ. The Authors suggest that ultimately the moisture content is not uniform throughout the sample, and that only a portion of the sample achieves a state of continuous deformation at constant shear stress. If so, the moisture content of this portion will be greater than the measured mean value, and that of the rest of the sample will be smaller. The *true* loading paths for this portion and for the rest of the sample would then diverge on opposite sides of the w = constant plane, and the Authors believe that the former path would reach the C.V.R. line at some other point such as X_1, shown in Figs 15 and 16.

However, the evidence available to the Authors is felt to be inconclusive, because :

(*i*) Much higher values of strain are required to reach the ultimate state for high values of n than for low values of n. This means that the error in the estimate of the stresses σ_1 and σ_3 (and consequently p and q) due to the assumption that the triaxial sample remains cylindrical, becomes worse.

(*ii*) The effective stresses involved for high values of n are small because of the relatively low value of the maximum pressure that a conventional triaxial cell will withstand. The experimental errors due to corrections for membrane and filter drain strengths and plunger friction become thereby accentuated.

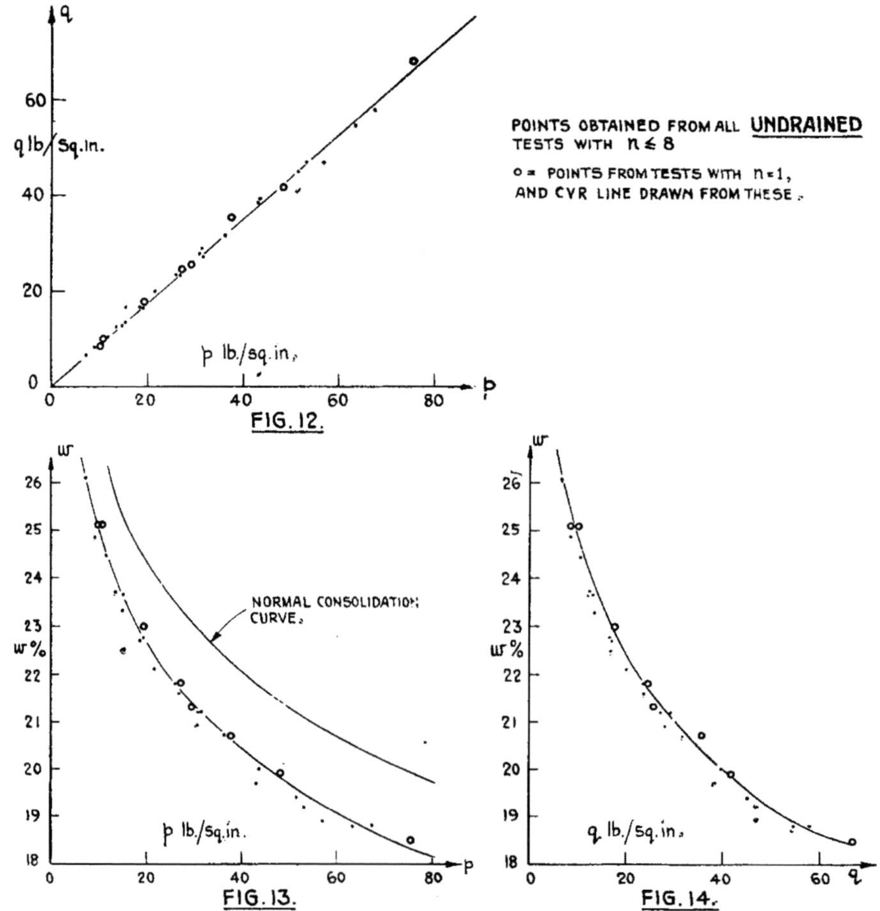

Figs 12–14. Critical voids ratio line from undrained triaxial tests on Weald Clay

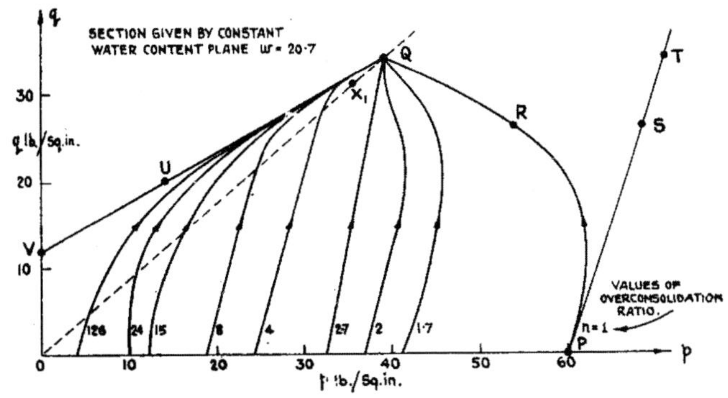

Fig. 15. Section of yield domain for undrained tests on Weald Clay showing loading paths for various values of *n*.

4(*e*). *Theoretical relationship between the overconsolidation ratio n and the pore-pressure coefficient A*

In the case of fully saturated clays the pore-pressure coefficient, A, defined by Skempton (1954) is given by the equation :

$$\Delta u = \Delta\sigma_3 + A\ (\Delta\sigma_1 - \Delta\sigma_3) \qquad . \quad . \quad . \quad . \quad . \quad . \quad . \qquad (18)$$

For conventional undrained tests where σ_3 is held constant $\Delta\sigma_3 = 0$ and equation (18) can be rewritten as :

$$A = \frac{\Delta u - \Delta\sigma_3}{\Delta\sigma_1 - \Delta\sigma_3} = \frac{\Delta u}{\Delta\ (\sigma_1 - \sigma_3)} = \frac{\Delta u}{\Delta\ (\sigma'_1 - \sigma'_3)} = \frac{\Delta u}{\Delta q} \qquad . \quad . \quad . \qquad (19)$$

Fig. 17 shows a typical section of the yield domain by a w = constant plane. Consider a sample with overconsolidation ratio n, having reached equilibrium at J under a final consolidation pressure p_3. For this sample JS_nT_n is the applied stress path and JR_nQ the loading path.

When the sample has reached the state, represented by R_n, then from section 4(*c*), $+\Delta u =$ R_nS_n, $q = JM$ and $A = \dfrac{\Delta u}{\Delta q} = \dfrac{R_nS_n}{JM}$. When the sample has reached the C.V.R. line at Q:

$$A_u = \frac{\Delta u_u}{\Delta q_u} = \frac{QT_n}{JN} = \frac{JL}{KQ} = \frac{p_3 - p_4}{q_u} \qquad . \quad . \quad . \quad . \quad . \quad . \qquad (20)$$

Eliminating p_3 from equations (17) and (20) :

$$A_u = \frac{p_1 n^{-\Lambda} - p_4}{q_u} \qquad . \quad . \quad . \quad . \quad . \quad . \quad . \quad . \quad . \quad . \qquad (21)$$

Now $\dfrac{p_1}{q_u} = a$ and $\dfrac{p_4}{q_u} = b$ are geometrical properties of the section, and therefore since all sections w = constant are similar, a and b will be dimensionless constants for the particular clay. Equation (21) becomes :

$$(A_u + b) = an^{-\Lambda} \qquad . \quad . \quad . \quad . \quad . \quad . \quad . \quad . \quad . \qquad (22)$$

so that if log $(A_u + b)$ is plotted against log n a straight line of slope $-\Lambda$ should be obtained.

This *predicted* relationship is shown for the Weald Clay by the broken straight line in Fig. 18. The values used for the soil constants are $\Lambda = +\ 0{\cdot}65$ (from section 4(*b*)) $a = 1{\cdot}77$ and $b = 0{\cdot}824$ (both obtained from Fig. 15). For low values of n the failure conditions of a sample should be the same as the ultimate conditions and the predicted values of $(A_u + b)$ should agree with the experimental values of $(A_f + b)$ which may be determined from Fig. 7 of Henkel's (1956) Paper. These values of $(A_f + b)$ are plotted as the full line in Fig. 18, and while agreement is good for low values of n the divergence, as expected, increases with n. An additional reason to those given in section 4(d) for this divergence, is that the swelling line in Fig. 7 is not straight for high values of n as assumed in section 4(b).

In an exactly similar manner relationships can be predicted between n and functions of the values of the variables at *the ultimate state* in undrained tests, comparable with the experimental relationships quoted by Henkel *for failure*.

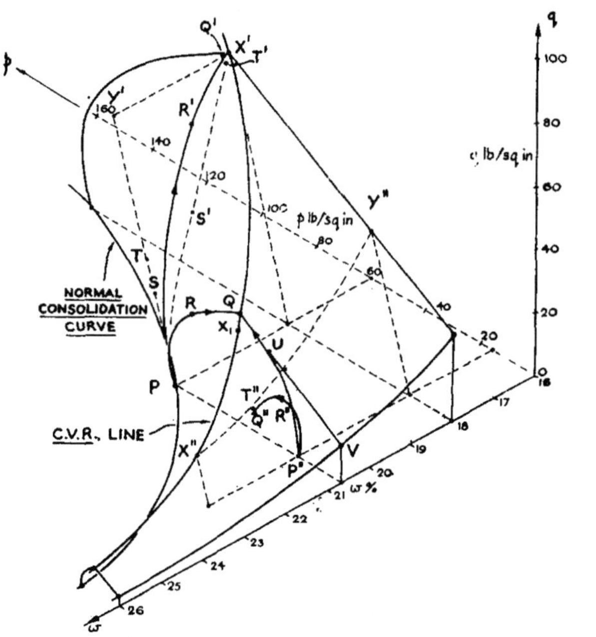

Fig. 16. Isometric view of yield surface for Weald Clay

These are :

$$\frac{(\sigma_1 - \sigma_3)_u}{\sigma_{3c}} = \frac{n^{\Lambda}}{a} \quad \text{(Henkel, Fig. 5)} \qquad (23)$$

$$1 - \frac{\Delta u_u}{\sigma_{3c}} = \frac{bn^{\Lambda}}{a} \quad \text{(Henkel, Fig. 5)} \qquad (24)$$

$$\frac{(\sigma_1 - \sigma_3)_u}{\sigma'_{3u}} = \frac{1}{b} \quad \text{(Henkel, Fig. 14)} \qquad (25)$$

The last relationship, equation (25), predicts that $\frac{(\sigma_1 - \sigma_3)_u}{\sigma'_{3u}}$ is independent of the overconsolidation ratio with a constant value of $\frac{1}{b} = 1{\cdot}21$. This is a direct result of the fact that the C.V.R. line has a straight line projection on the $e = 0$ plane in Fig. 12.

ANALYSIS OF DRAINED TESTS ON CLAY

5(a). *Drained tests on normally consolidated samples*

Consider a sample which has been normally consolidated under an all-round pressure of 60 lb/sq. in. having reached equilibrium at P in Fig. 16 and in Fig. 20. In a conventional drained test σ'_3 is held constant so that $\delta\sigma'_3 = 0$ and if *uncorrected* values of the deviator stress were plotted $\delta p = \frac{1}{3}(\delta\sigma'_1 + 2\delta\sigma'_3) = \frac{1}{3}\delta\sigma'_1$ and the increment in the q direction would be $(\delta\sigma'_1 - \delta\sigma'_3) = \delta\sigma'_1 = 3\delta p$. So that the uncorrected stress path, i.e. the applied stress path, PS'T' in Fig. 16 would be confined to a plane given by :

$$q = 3(p - 60) \qquad (26)$$

This plane given by equation (26) will be called an *applied stress plane* and is represented by the plane containing X'Y'P in Fig. 16 making an angle $\tan^{-1} 3$ with the $q = 0$ plane. The

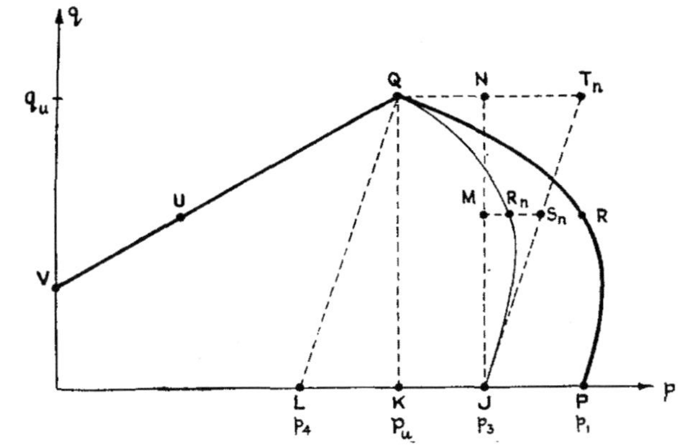

Fig. 17. Typical section of yield surface by constant water-content plane

projection of PS'T' in the $w =$ constant plane containing P is the straight line PSTY' in Fig. 16. This is shown in elevation in Fig. 19 by $P_2S'_2T'_2$.

The loading path PR'Q' in Fig. 16 is obtained from the applied stress path by correcting the deviator stress at each point by the boundary energy correction, represented by R'S', which is given by equations (7) and (8). In this test the sample is contracting, so that $(\sigma'_1 - \sigma'_3)_e$ is negative, and thus q is greater than the applied deviator stress. The particular test quoted here from Gilbert's work, was still showing a very small decrease in volume when the test was stopped. At this stage the paths had only reached Q' and T', and had not arrived at X' where they would be expected to coincide.

Fig. 20 is a plan view with the vertical projections of the two paths necessarily coincident. It will be observed that for this particular test the initial water content at P differs from that of the corresponding point W on the mean normal consolidation curve by a small fraction of 1%. This discrepancy PW is due to experimental scatter, for it is difficult to measure initial water contents accurately. Fig. 21 is the other projection on the $p = 0$ plane.

In order to be able to determine how close the loading path for this *drained* test PR'Q' lies to the *undrained* yield surface the broken lines $WZ_1X'_1$ and $W_3Z_3X'_3$ are drawn in Figs 20 and 21. These broken lines are obtained as follows. The loading path for the *drained* test PR'Q' in Fig. 16 ($P_2R'_2Q'_2$ in Fig. 19) is first projected parallel to the w axis on to the *undrained* yield surface, thereby making a line lying on this surface. This line is then projected on to the $q = 0$ and $p = 0$ planes to give $WZ_1X'_1$ and $W_3Z_3X'_3$ in Figs 20 and 21, respectively.

If the water content discrepancy represented by PW, R'_1Z_1 and $Q'_1X'_1$ is ignored as experimental error the curves PR'Q' and WZX' in space are seen to be coincident; so that the *drained* loading path PR'Q' climbs up the *undrained* yield surface. Moreover, the point X' lies very close to the C.V.R. line defined by the *undrained* tests. Within such limits of accuracy, the same is true for all other *drained* tests on normally consolidated samples.

5(*b*). *Drained tests on overconsolidated samples*

Drained tests on overconsolidated samples have to be investigated individually, because it is not possible to consider the series given by any one value of the overconsolidation ratio collectively as a family, as proved to be so convenient in section 4(*d*). Before considering any test in detail some general features which the Authors would expect in the behaviour of samples will be discussed.

(*a*) *Samples which decrease in volume.*—In all cases when the sample contracts the tests are well conditioned in the following respects :

(*i*) The sample increases in strength throughout the test and the peak value of the uncorrected deviator stress, defining failure, occurs when the sample has reached its ultimate state.

(*ii*) The whole sample will tend to deform uniformly, since if a localized region of greater shear deformation develops, this region will decrease in volume, increase in strength, and " work harden ".

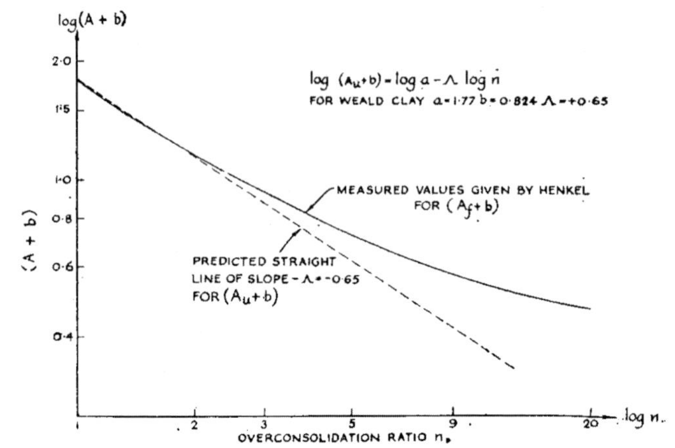

Fig. 18. Relation between A and overconsolidation ratio n for Weald Clay

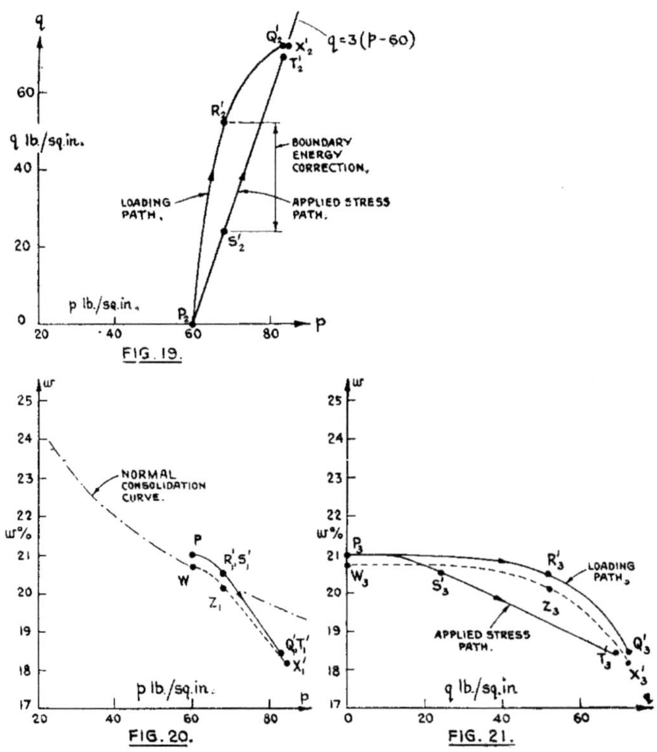

Figs 19–21. Paths for a drained test on normally consolidated Weald Clay

(*iii*) Due to the uniformity of deformation the average values of the water content and boundary energy correction obtained from the test results remain representative of the whole sample and give reliable information.

(*b*) *Samples which increase in volume.*—In contrast to (*a*) the tests in this case are ill-conditioned :

(*i*) The peak value of the uncorrected stress occurs well before the ultimate state is reached.

(*ii*) If a localized region of greater shear deformation develops, this region will increase in volume, decrease in strength, and become weaker than the rest of the sample. Work is used externally in changing the volume. The smaller the region of deformation, the smaller the boundary energy correction, and the smaller the amount of extra work that has to be supplied. Hence the sample will be encouraged to confine its deformation to some localized region, if this is possible, and restrict the extra work necessary to a minimum.

(*iii*) As a result of any non-uniformity of deformation, the values of the water content and boundary energy correction obtained from measurements made on the whole sample, are not representative of the *actual* region which is failing, and, in fact, both must necessarily be under-estimates.

Consequently the Authors would expect in general that a sample which decreases in volume would give reliable results, and a sample which increases in volume would give unreliable results, and they have therefore treated results of tests on heavily overconsolidated samples with caution.

However, the results of *drained* tests with an overconsolidation ratio of $n \leqslant 8$ show good

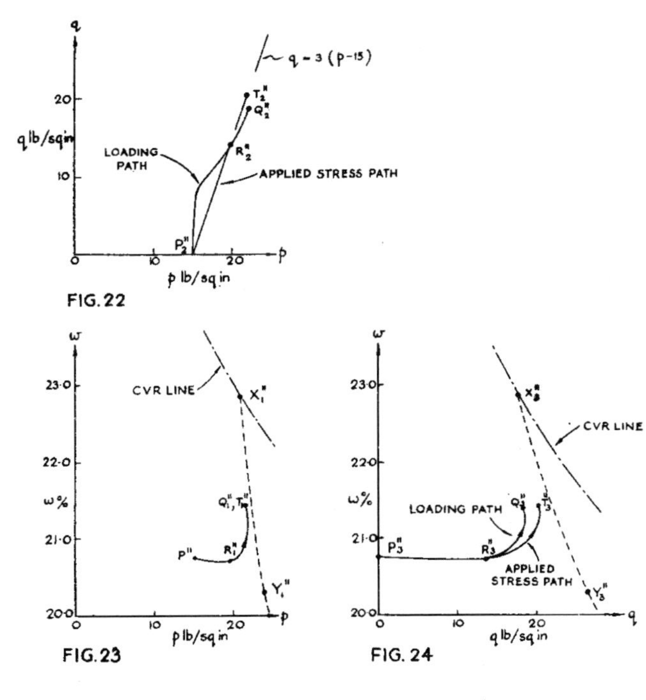

Figs 22–24. Paths for drained test on heavily overconsolidated Weald Clay

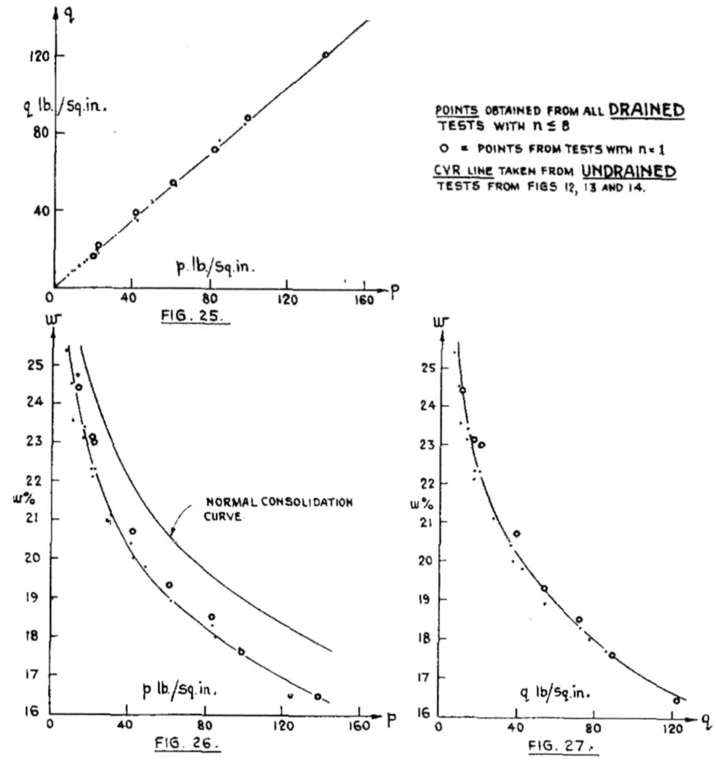

Figs 25–27. Drained-triaxial-test results on Weald Clay with C.V.R. line from undrained tests superimposed

agreement with those obtained from *undrained* tests. The results of one drained test, which is typical of the others, will now be described in some detail. In Figs 16 and 23, the over-consolidated sample, for which $n = 8$, is initially in equilibrium at P″ under a final consolidation pressure of 15 lb/sq. in. The applied stress path P″R″T″ lies, throughout the test, in the applied stress plane X″Y″P″ which makes an angle of $\tan^{-1} 3$ with the $q = 0$ plane. The loading path P″R″Q″, however, does not lie in the applied stress plane. Both paths are shown projected on the reference planes in Figs 22, 23, and 24. In the initial phases of the test the sample shows a negative volume change, or contraction, which makes q greater than the applied deviator stress, and puts the loading path above the applied stress path. At the moment that $\frac{dv}{d\epsilon_1} \equiv 0$, there is no correction and both paths coincide at R″. Thereafter the sample expands causing the loading path to be beneath the applied stress path. When the test was stopped shortly after failure, the sample was still expanding and the paths had reached the points given by Q″ and T″.

The intersection of the applied stress plane X″Y″P″, given by $q = 3\,(p - 15)$, with the undrained yield surface is shown in Fig. 16 by the broken line X″Y″, where X″ is on the undrained C.V.R. line. Had the test been continued the loading path could be expected to become tangential to X″Y″ and end at X″, when the sample would have reached an ultimate state.

In view of the limitations stated above, the correlation between the drained and undrained yield surfaces is reasonable. More surprising is the extent to which the end points of drained tests confirm the same C.V.R. line that applied to the undrained tests. This is shown in

Figs 25, 26, and 27 where the end points Q of all the twenty-five *drained* tests with $n \leqslant 8$ quoted by Gilbert have been plotted. The projections of the C.V.R. line shown in these diagrams are *not* drawn through the mean of the points, each of which represents the end of a *drained* test, but have been taken directly from the *undrained* test results plotted in Figs 12, 13, and 14.

5(*c*). *The line of critical overconsolidation ratio*

Fig. 17 shows a typical section of the yield surface by a constant w-plane. The point L was chosen so that QL should be at a slope of three (parallel to JT_n). Consider a sample that has reached equilibrium at L under a final consolidation pressure p_4. If an undrained test is carried out on the sample, LQ will be the applied stress path. Since Q is on the C.V.R. line the loading path will also end at Q, and so for this sample $\Delta u_u = 0 = A_u$.

If a drained test is carried out on the sample, LQ becomes the intersection of the w-plane with the applied stress plane. This latter plane cuts the C.V.R. line in the point Q which is also the common end point of the loading and applied stress paths. Hence the loading paths for drained and undrained tests should both end at Q. As L and Q are in the same w-plane, the sample will have no net change of water content and $(\Delta w)_u = 0$.

For differing values of w, P traces out the natural consolidation curve on the $q = 0$ plane. Because of the similarity of all sections, L and K (the vertical projection of Q) will also trace out curves on the $q = 0$ plane exactly similar to the natural consolidation curve. These are straight lines both parallel to the normal consolidation line, on the semi-logarithmic plot of Fig. 28. The curve traced out by L will be called the *critical overconsolidation ratio line*, or critical n-line for short. It must be emphasized that this critical n-line only applies to conventional tests (σ_3 constant, σ_1 increasing) ; there will be different critical n-lines for other triaxial tests, all of which will be similar in shape. All samples on the " wet " side of this line (i.e. further from the origin) will contract in drained tests, or show a positive pore-pressure change in undrained tests ; whereas all those on the " dry " side will dilate or develop a negative pore-pressure change. The possibility that such a curve existed for a clay, was suggested by Hvorslev (1937), and confirmed by Haefeli (1951).

With the same assumption that the swelling lines in Fig. 28 are straight, as was made in Fig. 7 and section 4(*b*), relationships can be predicted between the ultimate values of the variables for the drained tests in addition to those for the undrained tests.

In Fig. 28* consider a sample that has consolidated down the line ABC and has swollen back to E with overconsolidation ratio n where it has reached equilibrium under a final consolidation pressure p_e which is equal to that of B, namely p_b. Since B and C lie on the natural consolidation line :

$$w_c - w_b = -\lambda \log_e (p_c/p_b) \qquad (27)$$

Since C and E lie on a swelling line :

$$w_e - w_c = -\kappa \log_e (p_b/p_c) \qquad (28)$$

Equations (27) and (28) give :

$$w_e - w_b = (\kappa - \lambda) \log_e (p_c/p_b) \qquad (29)$$

But $p_c = np_e = np_b$ and equation (29) becomes :

$$w_b - w_e = (\lambda - \kappa) \log_e n \qquad (30)$$

L is the point where BE intersects the critical n-line. A drained or undrained loading path starting from L reaches the C.V.R. line at the point Q vertically above K. (See also Fig. 17.) But L has the same applied stress plane as B, E, or any point on BEL. Therefore a drained loading path starting from any point on BEL should also end at Q.

* For this piece of analysis the suffixes are those of the points to which the parameters refer.

Fig. 28. C.V.R. line and critical n line

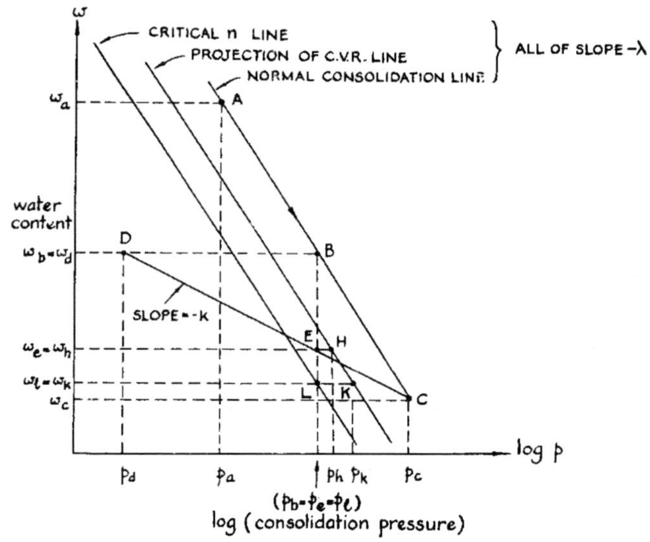

Consequently a sample with a drained loading path starting from E and reaching Q shows an ultimate change in water content of :

$$w_k - w_e = w_l - w_e = (w_l - w_b) + (w_b - w_e)$$

Now $(w_l - w_b)$ is constant for the soil, say, $-c$; so that:

$$(\Delta w)_u = (w_k - w_e) = -c + (\lambda - \kappa)\log_e n \quad . \quad . \quad . \quad . \quad . \quad (31)$$

Equation (31) is a predicted relationship between the ultimate change in water content in a drained test and the overconsolidation ratio. For the particular case when $n = 1$, the failure state coincides with the final state so that $(\Delta w)_f = (\Delta w)_u = -c$ giving the surprising result that drained tests on normally consolidated samples starting from *any* water content show the same decrease in water content at *failure.* This phenomenon was observed by Hvorslev (1937). In addition this will be true for other families given by any fixed value of n, provided that the failure and ultimate states are synonymous. Experimental evidence of this has been published by Haefeli (1951).

6. *Tests on other remoulded saturated clays*

The results of shear box tests by Hvorslev (1937) on Wiener Tegel give a Hvorslev surface of similar shape to that found for the Kleinbelt Ton. They also confirm the concepts of a C.V.R. line (see especially his Figs 40 and 44) and a critical n line for the Wiener Tegel.

Unpublished triaxial tests on Supreme Kaolin carried out at the Building Research Station (privately communicated), supplemented by further tests at Cambridge, confirm that the findings for the Weald Clay also apply to Kaolin. A similar series of tests described by Plant (1956) and summarized by Henkel (1956) suggest that the same is true for London Clay.

THE YIELD SURFACE AND C.V.R. LINE FOR CLAY

7(*a*). *Mathematical equations for the C.V.R. line and the yield surface*

Making the same assumptions as in previous mathematical sections, formulae are derived below which are applicable to any clay which has a yield surface similar in shape to that of the Weald Clay.

7(*b*). *Equation of the C.V.R. line*

From Fig. 28, the equation of the projection of the C.V.R. line on the $q = 0$ plane, provided $p_h = 1$, is :

$$p = \exp.\left(\frac{w_h - w}{\lambda}\right) \quad . \quad . \quad . \quad . \quad . \quad . \quad . \quad . \quad (32)$$

From Figs 12 or 25 the projection of the C.V.R. line on the $w = 0$ plane is :

$$q = Mp \quad . \quad . \quad . \quad . \quad . \quad . \quad . \quad . \quad . \quad . \quad (33)$$

M (= capital μ) is used here to avoid confusion with other μ's or ϕ's in current use.

The C.V.R. line is the line of intersection of the surfaces given by equations (32) and (33) and hence is given by :

$$\exp.\left(\frac{w_h - w}{\lambda}\right) = p = \frac{q}{M} \quad . \quad . \quad . \quad . \quad . \quad . \quad . \quad . \quad (34)$$

where w_h, λ and M are fundamental constants for a given soil.

7(*c*). *The part of the yield surface generated by the line VUQ in Fig.* 17

By analogy to the Hvorslev surface, expressed in equation (1), the corresponding portion of the yield surface generated by the line VUQ in Fig. 17 is given by an equation of the form :

$$q = \mu'_0 p + \nu' \exp.\left(-\frac{w}{\lambda}\right) \quad . \quad . \quad . \quad . \quad . \quad . \quad . \quad . \quad (35)$$

But this surface passes through the C.V.R. line and hence, by eliminating w from equations (32) and (35), any point on this line is given by :

$$q = \mu'_0 p + \nu' p \exp.\left(-\frac{w_h}{\lambda}\right)$$

This must be the same as equation (33), so that

$$\nu' = (M - \mu'_0) \exp.\left(\frac{w_h}{\lambda}\right)$$

Substituting in equation (35) gives :

$$q = \mu'_0 p + (M - \mu'_0) \exp.\left(\frac{w_h - w}{\lambda}\right) \quad . \quad . \quad . \quad . \quad . \quad (36)$$

which is the equation of the part of the surface formed by the line VUQ. μ'_0 is a further fundamental constant for a given soil.

7(*d*). *The curved part of the yield surface generated by the line PRQ in Fig.* 17

For the Weald Clay and the Supreme Kaolin the curve shown by PRQ in Fig. 17 can be represented approximately by a parabola. This is equivalent to assuming that the rate of change of Skempton's pore-pressure coefficient A with respect to q is a constant throughout a test on a normally consolidated sample. The initial portion of the loading path PRQ is difficult to obtain experimentally and it has been plotted on the basis of somewhat scanty information, all obtained for an axial strain $\epsilon_1 < 2\%$. For the above two clays the parabola may be assumed to have its vertex at P and be symmetrically placed about the p axis, and these two simplifications are adopted in the analysis below. Any error introduced by the second assumption can easily be corrected by introducing a further term giving the slope of PRQ at P.

Fig. 17 is a section of the yield surface formed by the plane w = constant = w_1 say. K is a point on the projection of the C.V.R. line on the $q = 0$ plane. This point is also shown in Fig. 28, and taking $p_h = 1$:

$$w_1 - w_h = -\lambda \log_e p_u \quad . \quad . \quad . \quad . \quad . \quad . \quad . \quad . \quad (37)$$

where p_u corresponds to the points K and Q.

But P lies on the normal consolidation curve, and without loss of generality choosing $p_a = 1$ also, this satisfies :

$$w_1 - w_a = -\lambda \log_e p_1 \quad . \quad . \quad . \quad . \quad . \quad . \quad . \quad . \quad (38)$$

Equations (37) and (38) give :

$$\log_e \frac{p_1}{p_u} = \left(\frac{w_a - w_h}{\lambda}\right) = d, \text{ say,} \quad . \quad . \quad . \quad . \quad . \quad . \quad . \quad (39)$$

where d is a constant for the soil.

But from Fig. 17, the parabola PRQ is of the form :

$$p - p_1 = -Fq^2 \quad . \quad . \quad . \quad . \quad . \quad . \quad . \quad . \quad . \quad (40)$$

where F is a constant for the *particular section* $w = w_1$. But Q, with co-ordinates (p_u, q_u) related by the equation $q_u = Mp_u$, lies on the parabola ; hence by simple algebra :

$$p_1 F = \frac{(\text{exp. } d - 1) \text{ exp. } d}{M^2} = G, \text{ say,} \quad . \quad . \quad . \quad . \quad . \quad (41)$$

Since d and M are constants for all sections of the yield domain, G must also be a constant for all sections and is dependent upon the fundamental soil constants (w_a, w_h, λ, M).

Equation (40) therefore becomes :

$$p_1 (p - p_1) + Gq^2 = 0 \quad . \quad . \quad . \quad . \quad . \quad . \quad . \quad . \quad (42)$$

Eliminating p_1 from equations (38) and (42), and dropping the suffix 1 from w, the equation for the curved portion PRQ of the yield surface becomes :

$$\left.\begin{aligned} Gq^2 + p \text{ exp.} \left(\frac{w_a - w}{\lambda}\right) - \text{exp. } 2 \left(\frac{w_a - w}{\lambda}\right) = 0 \\ \text{where} \quad G = \frac{1}{M^2}\left\{ \text{exp.} \left(\frac{w_a - w_h}{\lambda}\right) - 1 \right\}.\text{exp.} \left(\frac{w_a - w_h}{\lambda}\right) \end{aligned}\right\} \quad . \quad . \quad . \quad (43)$$

7(*e*). *Fundamental soil constants*

It is apparent from equations (36) and (43) that, subject to the above assumptions, both parts of the yield surface are given by five independent fundamental soil constants w_a, λ, w_h, M and μ'_0. From equation (34), three of these constants are required to define the C.V.R. line. In the above work a sixth constant has also been introduced ; this is κ, or alternatively Λ.

8. *Summary and discussion on the yielding of clays*

It has been shown that a remoulded saturated clay, when tested under triaxial compression conditions of either full drainage or no drainage, has a characteristic yield surface which contains a C.V.R. line in (p, w, q) space, to which loading paths converge at high strains. The yield surface has been confirmed experimentally for Weald Clay and Supreme Kaolin. The concept of a C.V.R. line has been verified experimentally for Weald Clay, Supreme Kaolin. Wiener Tegel, Kleinbelt Ton, London Clay and Haefeli's clays. It is suggested that this surface and line are characteristic of a clay under *all* conditions of drainage.

When the yield surface and C.V.R. line have been determined experimentally, or alternatively the five fundamental parameters have been obtained from the surface or from a selected series of simple tests, it is possible to predict (geometrically in the former case and mathematically in the latter) the ultimate state of any soil sample when stresses are imposed upon it provided (*i*) the initial state of the sample is known and (*ii*) the drainage conditions are known. These two conditions together with the yield surface determine the loading path for the test. The loading path after reaching the yield surface traverses this surface until the C.V.R. line is reached. The soil then continues to deform at constant stress and constant voids ratio.

The drained yield surface has been obtained by applying a boundary energy correction and the form of this correction has been specified in detail. It is possible that additional corrections should be applied, but since the drained and undrained yield surfaces for Weald Clay coincide, within the limits of experimental observation, the error entailed by using the present form of correction does not appear to be important for this series of tests. The same is true for the test results on Supreme Kaolin.

It must be emphasized that any number of uncorrected yield surfaces can be obtained for a soil depending upon the test apparatus employed, the parameters used for plotting results and the conditions of full, partial, or no drainage imposed upon the soil. The customary parameters, such as the large number of ϕ's, used in current soil mechanics literature to define the shear properties of a soil can probably be represented by lines ruled on these surfaces or within the domain. The only feature that is likely to be common to all these surfaces (for any one test apparatus) is the C.V.R. line, since the boundary energy correction is zero for all points on this line. For the same reason tests which permit no drainage whatever, give the yield surface directly and by carrying out a few such tests on relatively highly overconsolidated samples ($n > 4$ for Weald Clay) the fundamental soil constant μ'_0 can be obtained. The angle $\tan^{-1}\left(\frac{\mu'_0}{2}\right)$ is probably nearer to being a true angle of internal friction for the soil than that defined by Hvorslev as ϕ_0.

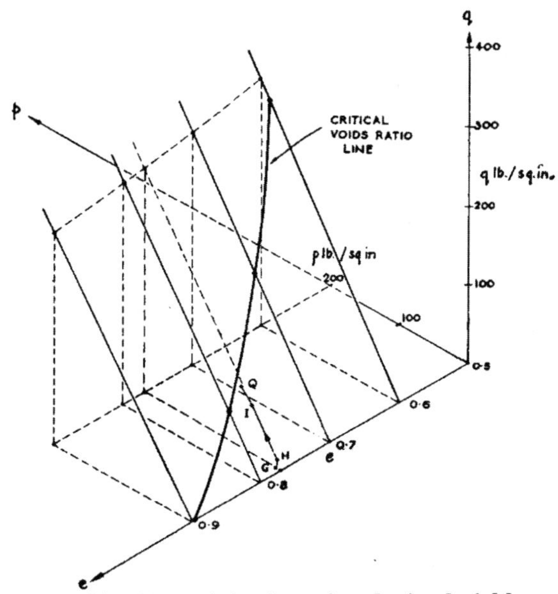

Fig. 29. Isometric view of undrained yield surface for silt

The concept of a critical n-line has also been shown to apply to Weald Clay, Supreme Kaolin, Wiener Tegel, Kleinbelt Ton, and Haefeli's clays. However, this concept must be used with caution for two reasons. First, because the position of the critical n-line depends upon the nature of the proposed test as well as on the soil properties ; secondly, because results obtained by Hvorslev (1937) show in the upper part of his Fig. 41 that, for Kleinbelt Ton, Δw, as defined in his Fig. 40, is not quite constant for all values of σ'. Hence the critical n-line for this clay may not be parallel to the normal consolidation curve. This may mean that the analysis of sections 5(c) and 7 above would require slight modification before being applied directly to Kleinbelt Ton.

Finally, from the work of Hilf and Gibbs (1957), it would appear possible that the concepts outlined above for remoulded saturated clays also apply to *unsaturated* clays.

Part II : Cohesionless granular media

TRIAXIAL TESTS ON COHESIONLESS MEDIA

9. *Triaxial tests on granular cohesionless soils*

The difficulty in obtaining reliable information at the end of a triaxial test on a clay when the sample has reached its critical voids ratio state has already been noted. In the case of silts, sands, and other granular media the difficulty is magnified in the following respects :

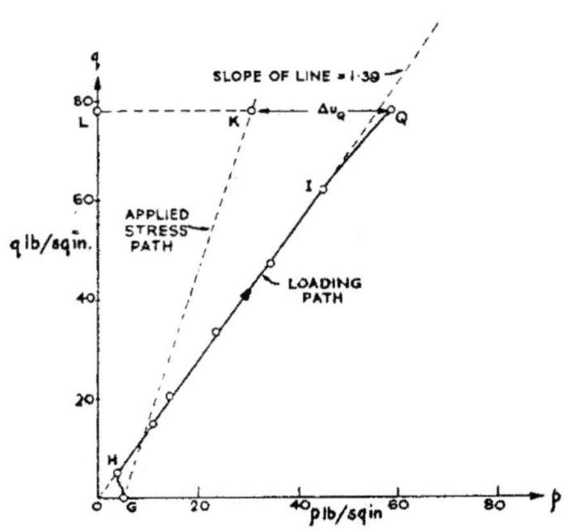

Fig. 30. **Loading path for undrained test on silt**

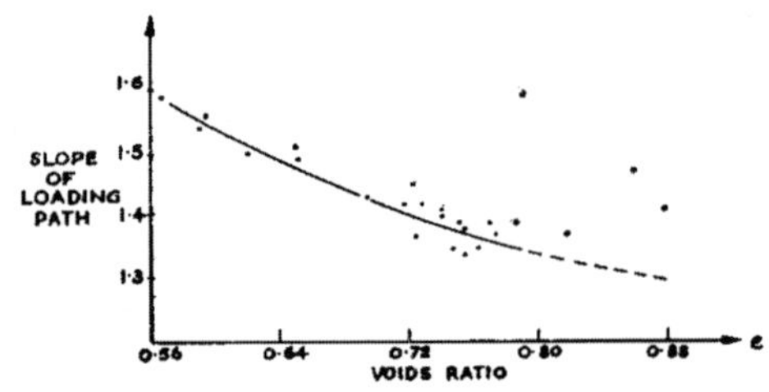

Fig. 31. **Slope of undrained yield surface for silt**

(*i*) *Undrained tests.*—Generally in an undrained compression test on a cohesionless granular sample the pore-water pressure u shows a small positive increment at the start of the test, which rapidly becomes negative and steadily decreases (in the mathematical sense) until failure occurs. If the value of u itself drops below atmospheric pressure gas or vapour may be liberated from the sample. Even if the pore-water has been perfectly de-aired, vapour is probably formed when the absolute value of u corresponds to the saturation pressure of water at the temperature of the sample. Under these conditions the sample will no longer remain at constant volume, and although the measured values of the stresses may be correct, the supposed values of e or w will not be so, and the loading path will leave the undrained section of the yield domain given by the w = constant plane which contained the original undrained part of the test. It will no longer be possible to trace this path from the data that are usually observed. If, however, the volume change is measured by means, say, of the change in volume of the cell water which it directly causes, then the loading path can be determined after gas or vapour has been formed.

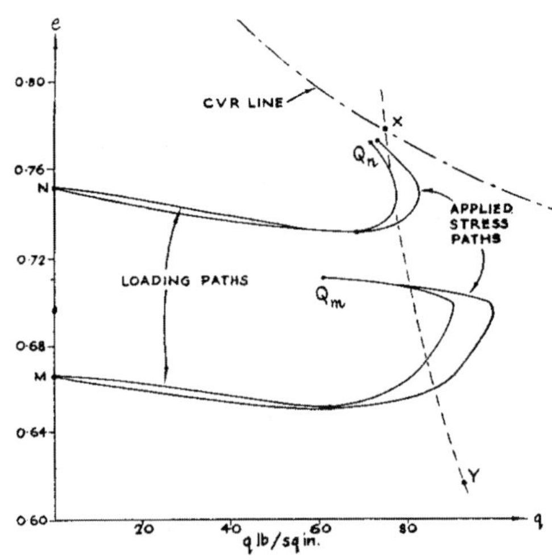

Fig. 32. **Paths for drained tests on silt**

By raising the cell pressure σ_3, and consequently u, after the sample has reached its initial state of equilibrium prior to the test, the possible range of values of Δu can be considerably increased. Even so, the test apparatus imposes a limit upon the maximum values of σ_3 and it may be in certain tests that the change in u is so great that the difficulties created by the formation of gas or vapour cannot be avoided without strengthening the apparatus.

(*ii*) *Drained tests.*—As will be seen later, large volume changes may be needed for the whole of a cohesionless granular sample to reach its critical voids ratio, and in such cases the inaccuracies stated in section 5(*b*) sub-section (*b*) are greatly increased. So much so that it is impossible to establish a ***drained*** yield surface or complete C.V.R. line from triaxial tests.

10(*a*). *Triaxial tests on Braehead Silt*

A brief analysis is now made of the results of tests on samples of saturated silt by Penman (1953). These published results have been supplemented by others privately communicated by Penman. The silt came from the site of the Braehead Power Station and had a liquid limit of 24% and contained no particles finer than 3 microns equivalent diameter.

The parameters used are p, q, and the voids ratio e. Fig. 12 of Penman's Paper shows no unique normal consolidation curve for the soil, so that it is not possible to divide the tests into families according to an initial overconsolidation ratio.

10(*b*). *Undrained tests on Braehead Silt*

The undrained tests carried out by Penman cover a wide range of voids ratio. Each test reached equilibrium under an all round pressure $\sigma_{3c} = 5$ lb/sq. in. Prior to the test, the cell pressure was raised to obviate the difficulty mentioned in section 9.

The results of one particular test are plotted as the loading path GHIQ, shown in the general isometric view of Fig. 29 and in projection on the $e = 0$ plane in Fig. 30. The major part HI lies on a straight line through the e axis. The divergence of the final part IQ is probably caused by the inaccuracy of readings at high strains. If the initial rise in cell pressure (u_0) is subtracted from the results throughout the test, then the applied stress path can be considered to be GK, given by $q = 3\,(p - 5)$, and to start from the same point as the loading path. Under these conditions KQ represents $-\Delta u_Q$.

All other loading paths are similar except that the slope and length of the straight portion HI vary with the voids ratio of the sample. The relationship between slope and voids ratio is shown on a large scale in Fig. 31 where the slope is seen to increase with the density of the sample. The undrained yield surface is formed by these straight lines of varying slope. These all pass through the e axis and so the " cohesion intercept " of the surface is everywhere zero.

For all the undrained tests the applied stress paths lie in the one plane $q = 3\,(p - 5)$ and KQ $= -\Delta u_Q$, so that the end point Q can be defined uniquely by the parameters $(-\Delta u_Q, q_Q)$ in place of its co-ordinates (p_Q, q_Q). For the series of tests, Penman plotted these two parameters separately against e in his Figs 14 and 18, with marked experimental scatter. The resulting curves both give the ***undrained*** C.V.R. line for the silt which is also shown in Fig. 29. These diagrams provided the first experimental evidence confirming that the C.V.R. concept developed from Hvorslev's work on clays might also be applied to cohesionless media.

It is suggested that this scatter is caused partly by high negative values of Δu, and partly by the difficulty of measuring the voids ratio of a sample with consistent accuracy. Although the surface is well defined on the " dry " side of the undrained C.V.R. line, there is no information about the surface on the " wet " side.

10(*c*). *Drained tests on Braehead Silt*

It is difficult to set up very loose samples of cohesionless soils for a triaxial test and the majority of drained tests on silts and sands will start with the sample in a relatively dense

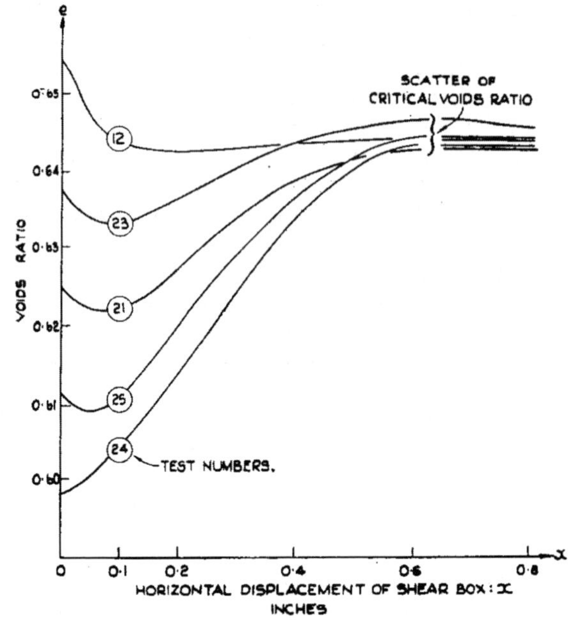

Fig. 33. Results of simple shear tests on 1-mm steel balls with normal stress—20 lb/sq. in.

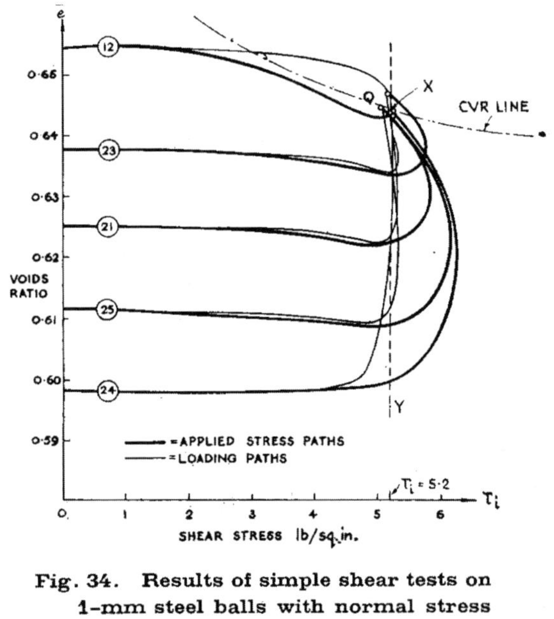

Fig. 34. Results of simple shear tests on 1-mm steel balls with normal stress —20 lb/sq. in.

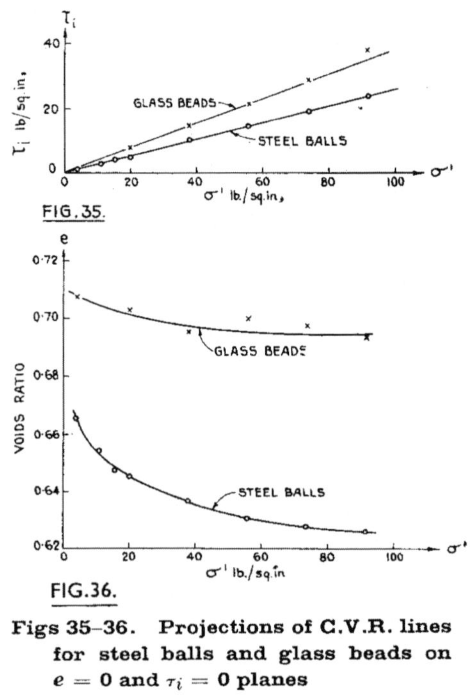

Figs 35–36. Projections of C.V.R. lines for steel balls and glass beads on $e = 0$ and $\tau_i = 0$ planes

state so that loading paths approach the undrained C.V.R. line from the " dry " side, and the sample must expand for them to do so. The extent to which this introduces errors into the measured values of e and the boundary energy correction (obtained from equations (7) and (8) and not by the method suggested in Penman's Paper) is clearly shown in Fig. 32.

This diagram shows the projection, on the $p = 0$ plane, of the paths of two different drained tests, with both samples consolidated and tested under a cell pressure of 30 lb/sq. in. One has started in a dense state at M and the other in a loose state at N. The broken line XY is the intersection of the applied stress plane $q = 3(p - 30)$ (common to both samples) with the undrained yield surface and X is the point where this line cuts the undrained C.V.R. line.

If the C.V.R. concept holds for silts the loading and applied stress paths for both samples should all become tangential to XY and end at X. There is a clear discrepancy here, which suggests that the errors involved increase with the initial density of the sample. Other drained tests confirm this tendency and emphasize the lack of correlation between the drained tests and the undrained yield surface.

In fact, were it possible to obtain the true values of (p, e, q) for the localized region of the sample that is *actually* deforming, the paths might well end at X. The discrepancy demonstrated here does not invalidate the hypothesis of the C.V.R. concept, but rather shows the limitations of the triaxial apparatus in this particular case, that is of a drained test on a dense (or heavily overconsolidated) sample.

11. *Triaxial tests on sands and other granular media*

Triaxial tests on sands, glass beads, steel balls, and other granular materials carried out at Cambridge and elsewhere give results of an identical form to those for the silt. Undrained tests give a well-defined undrained yield surface with large experimental scatter for a C.V.R. line. This scatter appears to increase with the irregularity of shape and variety of sizes of the particles in the material being tested. The drained tests correlate with the undrained yield surface no better than in the case of the silt.

Eldin (1951) made an extensive series of tests on Brasted Sand and his results do show a part of the undrained C.V.R. line well defined for very loose samples (in which the value of $-\Delta u_u$ is small). The beginning of this line is plotted in elevation in Eldin's thesis (Fig. 6), in a similar manner to that shown in Fig. 18 of Penman's Paper. For samples in denser states his results have a similar scatter to that of the silt.

The results obtained by Fraser (1957), for a large variety of tests on the same sand, may be shown to confirm the existence of the undrained yield surface. His results indicate that there is probably a drained yield surface, which coincides with the undrained surface. As he was interested in conditions at the start of a test, few were carried sufficiently far to supply information about a C.V.R. line.

The general pattern of results for the silt, sand, and other granular media indicate that accurate definition of drained yield surfaces can only be made in some other apparatus such as the simple shear apparatus. The results of such tests are discussed in the next section.

SIMPLE SHEAR TESTS ON COHESIONLESS MEDIA

12(*a*). *Tests in the simple shear apparatus*

The simple shear apparatus designed by Roscoe (1953) has been developed and used for shear tests on various granular media. The full details of this development, the methods of measurement, and of the test procedure cannot be included here, and reference must be made to Wroth (1958). The granular media, referred to in the following sections, are steel balls of 1 mm dia. ($\pm$ 0·015 mm), glass beads of diameters between 0·75 and 0·80 mm, a graded sample of standard rounded Leighton Buzzard sand ($<$ 18 and $>$ 25 B.S. sieve), and a graded sample of a local, angular, unweathered sand from Hauxton ($<$ 14 and $>$ 25 B.S. sieve).

Prior to test these materials had been oven dried and subsequently exposed to the atmosphere as little as possible.

The tests now to be described are all of the fully drained type since these allow a direct check to be made of the C.V.R. hypothesis for the particular case in which the triaxial test cannot give reliable information.

In early tests on sands and on glass beads it was not found possible to achieve a *loaded* sample which, prior to the start of the test, was sufficiently loose to contract on shearing and approach the C.V.R. line from the loose or " wet " side. This restricted the value of the tests, and as this limitation did not apply to steel balls the results of a series of drained tests on the latter medium will be described in some detail. The use of the steel balls had a further advantage in that there was no danger of individual particles breaking when subjected to shear stress under the higher normal stresses of the order of 100 lb/sq. in.

The parameters used for plotting are the effective normal stress σ', the voids ratio e and the shear stress τ_i, corrected for boundary energy according to equations (5) and (6).

12(*b*). *Specimen set of tests on steel balls under one normal stress*

Samples were placed in the apparatus with a range of initial densities but always in a

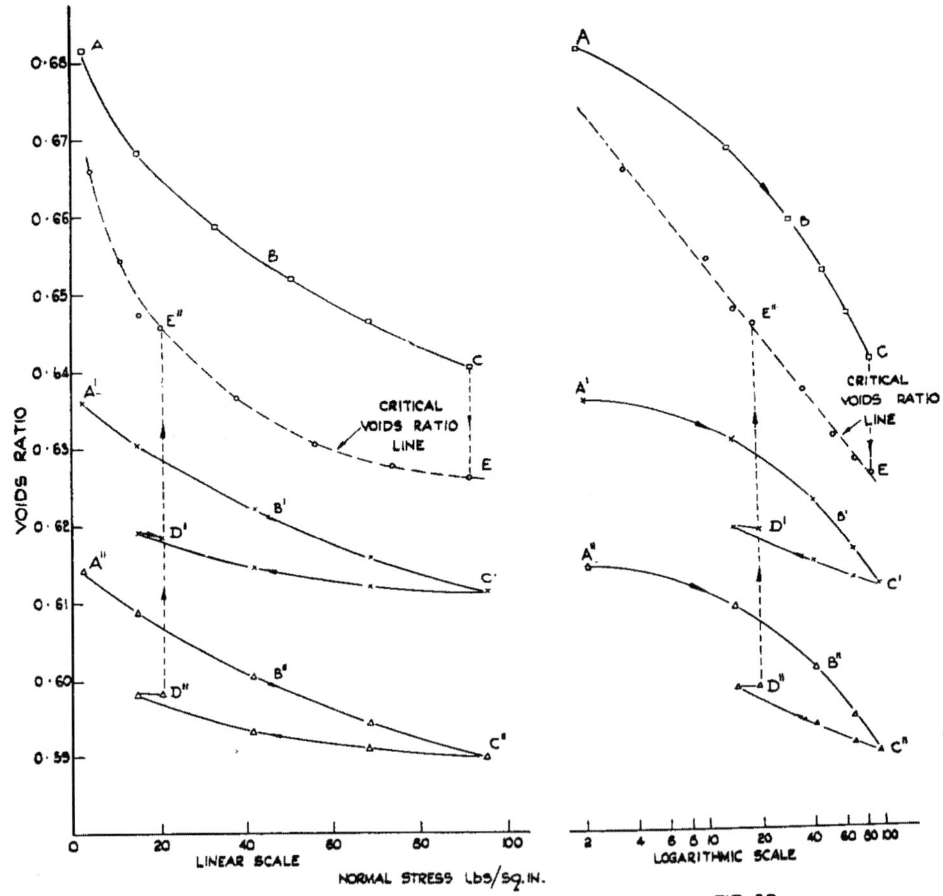

FIG. 37 FIG. 38

Figs 37–38. Representative consolidation curves and projection of C.V.R. line from tests on 1-mm steel balls

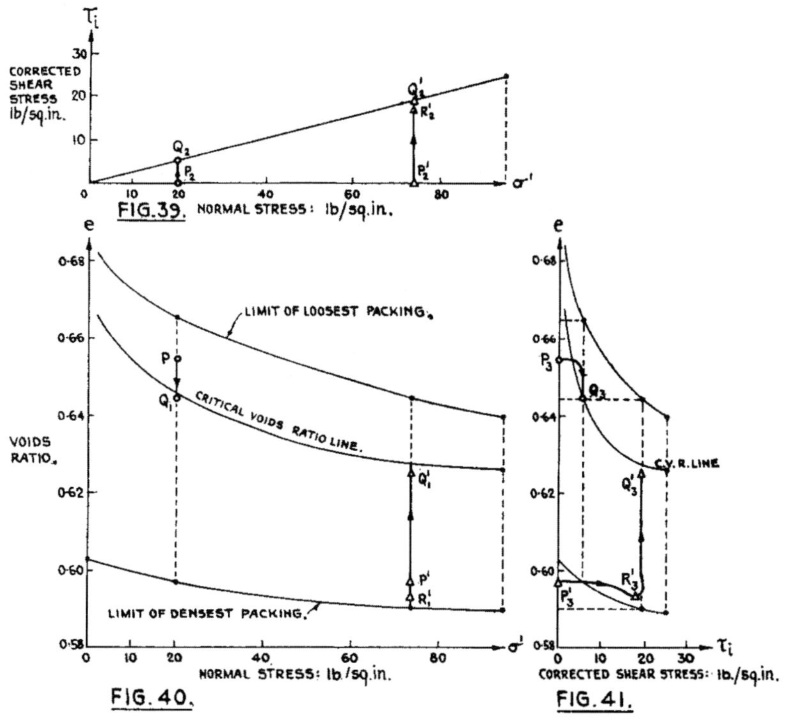

Figs 39–41. Drained yield surface for 1-mm steel balls, from results of simple shear tests

state of random packing. The loosest state was achieved by means of a special chute. Denser states were obtained by tamping since vibratory methods caused regular packing in localized regions of the sample. The steel balls showed a great facility for becoming regularly packed, and under these conditions gave artificially low values of the shear strength.

Fig. 33 shows the voids ratio plotted against the relative horizontal displacement of the top and bottom surfaces of the sample, for five tests with different initial voids ratio but all with the same value of $\sigma' = 20$ lb/sq. in. The tests reach a final state within a very narrow band of voids ratio giving, for this fixed σ', a mean C.V.R. of 0·646 with a total scatter of less than 0·005.

The loading and applied stress paths for all these samples remain throughout each test in the plane $\sigma' = 20$ lb/sq. in. as shown in Fig. 34. Within small limits of experimental error all paths are seen to end at one unique point Q in this plane. The loading paths have a common envelope XY given by the straight line $\tau_i = 5{\cdot}2$ lb/sq. in., which is the section of the $\sigma' = 20$ lb/sq. in. plane with the drained yield surface. These curves can be compared with those obtained in drained triaxial tests for silt in Fig. 32.

12(*c*). *Series of tests on steel balls with different normal stresses*

Series of tests with a range of initial voids ratios, under other fixed normal stresses gave similar results to those shown in Figs 33 and 34. For each value of σ' the co-ordinates (e_u, τ_{iu}) of Q differ, but the envelope XY given by the loading paths is always a straight line parallel to the e axis. Fig. 35 shows τ_{i_u} plotted against σ' giving a straight line through the e axis, which is an elevation of the C.V.R. line and, in fact, of the complete drained yield

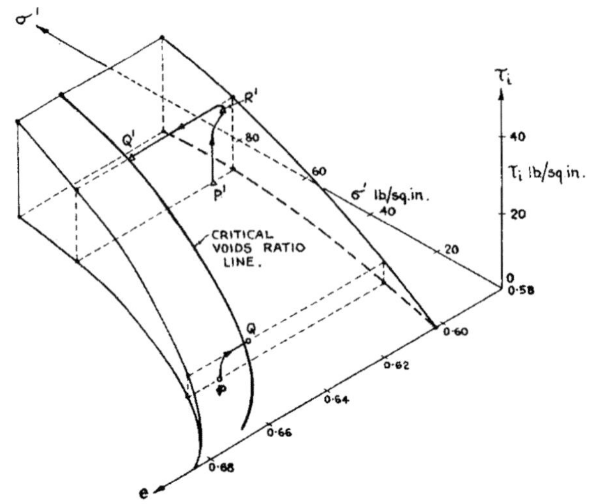

Fig. 42. Drained yield surface for 1-mm steel balls

surface. Fig. 36 shows e_u plotted against σ' giving a very exactly defined curve, which is a plan of the C.V.R. line.

Fig. 37 shows the plan of the C.V.R. line in relation to three representative " consolidation " curves. The top curve ABC is the maximum voids ratio or loosest packing obtainable. A sample starting a shear test at C on the loose side of the C.V.R. line contracts throughout the test and reaches the C.V.R. line at a point vertically above E. The minimum voids ratio is achieved by starting a sample at A″, consolidating it to C″, and relieving it back to an " overconsolidated " state at D″. A sample starting a shear test from D″ expands throughout the test and reaches the C.V.R. line at a point vertically above E″.

Fig. 38 shows the same curves with σ' plotted on a logarithmic scale. It shows the plan of the C.V.R. line to be straight. The consolidation curves have a resemblance to the cyclic overconsolidation curves of a clay, although they appear not to join a unique normal consolidation curve.

12(*d*). *Drained yield surface and C.V.R. line for steel balls*

The drained yield surface and C.V.R. line have been very closely defined for the steel balls from thirty-nine tests under eight different values of the normal stress σ'. These results, including those already discussed, have been used to obtain Figs 39, 40, and 41 which show the projections on the reference planes of the drained yield surface and the corresponding C.V.R. line. These are also plotted isometrically in Fig. 42, where two specimen loading paths are shown. The first starts from a loose state at P, rises almost vertically to the yield surface and then traverses the surface, as the sample contracts, to a point Q on the C.V.R. line. The second starts from a dense state at P′, attains a minimum voids ratio at R′ and then traverses the surface with the sample expanding until it also reaches the C.V.R. line at Q′. Figs 39 and 42 show that the yield surface is a plane containing the e-axis and inclined at a constant angle to the $\tau_i = 0$ plane.

13. *Tests on other granular media*

Of the other granular media mentioned in section 12(*a*), the glass beads gave similar results to the steel balls, defining a drained yield surface and C.V.R. line, but with larger scatter. This scatter can be seen from the projection of the surface and C.V.R. line on the $e = 0$ plane shown in Fig. 35 and the plan of the C.V.R. line shown in Fig. 36.

Both the sands show a drained yield surface containing a C.V.R. line but the scatter of these results was greater than for the glass beads particularly for the higher values of σ'. This is attributed to the breaking of the particles under these high stresses, especially in the case of the Hauxton Sand. After each such test it was found that about 1% of the sample passed through the 25 B.S. sieve.

14. *Summary and discussion of tests on cohesionless granular media*

By means of the simple shear apparatus, it has been shown that a well-defined drained yield surface containing a C.V.R. line exists for steel balls over a range of σ' from 0 to 100 lb/sq. in. The same has been found to be true for glass beads. The C.V.R. line lies within a relatively small range of voids ratio, being approximately $0{\cdot}67 > e_u > 0{\cdot}63$ for the steel balls and $0{\cdot}71 > e_u > 0{\cdot}69$ for the glass beads. For these media, as σ' increases, the C.V.R. line becomes approximately parallel to an $e =$ constant plane. Sands, when tested in this apparatus, behave in a similar manner but all results have a greater scatter, due in part to fracture of the particles. Both drained yield surfaces make constant angles with the $e = 0$ plane and their slopes give the fundamental soil constant μ'_0 for the respective media. It must be emphasized that each drained yield surface has been obtained from tests on *dry* media and it is possible that a different surface will be found for drained tests on *saturated* media

Triaxial test results have also confirmed that the concepts of a yield surface and C.V.R. line are applicable to silts, sands, and glass beads. The experimental evidence is not so conclusive as that obtained in the simple shear apparatus, and this is shown to be due to the limitations of the triaxial test procedure.

The main, and perhaps most surprising, conclusion to be drawn from the work outlined above is the remarkable similarity between the behaviour of the clays and the cohesionless granular media, and it might be relevant to refer to a comment by Terzaghi (1956) on the mechanical concept of the behaviour of clays.

ACKNOWLEDGEMENTS

The Authors wish to thank Dr L. F. Cooling and his colleagues of the Building Research Station, Watford, especially Mr A. D. M. Penman, for generously placing at their disposal the unpublished results on Supreme Kaolin and on Braehead Silt; the latter lending support to previously published work. These unpublished results are quoted by permission of the Director of Building Research. The Authors are indebted to Professor A. W. Skempton of Imperial College, London, for so kindly granting permission to use the excellent experimental data obtained in his laboratory, particularly those published in the thesis of Dr G. D. Gilbert. The Authors also wish to acknowledge that the following diagrams have been derived from experimental data obtained by other workers listed below:

Figs 1 to 6 — Dr M. J. Hvorslev (1937),
Figs 8 to 16, and 19 to 27 — Dr G. D. Gilbert (1954),
Figs 29 to 32 — Mr A. D. M. Penman (1953).

These data were originally published in a different form from that used in the present Paper. Figs 31 and 32 contain additional points which have been derived from unpublished experimental work by Penman.

The Authors would also like to express their appreciation of the support for their work in soil mechanics given by Professor J. F. Baker, Head of the Department of Engineering, Cambridge University.

REFERENCES

BISHOP, A. W., 1954. *Correspondence* on " Shear characteristics of a saturated silt measured in triaxial compression ", by A. D. M. Penman, *Géotechnique*, 4 : 43–44.

CASAGRANDE, A., 1938. " Compaction tests and critical density investigations of cohesionless materials for Franklin Falls dam." *U.S. Engineer Corps. Boston District* (from Taylor, D. W., 1948).

ELDIN, A. K. GAMAL, 1951. " Some fundamental factors controlling the shear properties of sand." *Ph.D. Thesis. London.*

FRASER, A. M., 1957. " The influence of stress ratio on compressibility and pore pressure coefficients in compacted soils." *Ph.D. Thesis. London.*

GIBSON, R. E., 1953. " Experimental determination of the true cohesion and true angle of internal friction in clays." *Proc. 3rd Int. Conf. Soil Mech.*, 1 : 126–130.

GILBERT, G. D., 1954. " Shear strength properties of Weald Clay." *Ph.D. Thesis. London.*

HAEFELI, R., 1951. " Investigations and measurement of the shear strengths of saturated cohesive soils." *Géotechnique*, 2 : 186–208.

HENKEL, D. J., 1956. " The effect of overconsolidation on the behaviour of clays during shear." *Géotechnique*, 6 : 139–150.

HILF, J. W., *and* GIBBS, H. J., 1957. " Triaxial shear tests holding effective lateral stress constant." *Proc. 4th Int. Conf. Soil Mech.* 1 : 156–159.

HVORSLEV, M. J., 1937. " Über die Festigkeitseigenschaften gestörter bindiger Böden " (" On the physical properties of disturbed cohesive soils ") Ingeniorvidenskabelige Skrifter. A. No. 45. 159 pp.

PENMAN, A. D. M., 1953. " Shear characteristics of a saturated silt, measured in triaxial compression." *Géotechnique*, 3 : 312–328.

PLANT, J. R., 1956. " Shear strength properties of London Clay." *M.Sc. Thesis. London.*

ROSCOE, K. H., 1953. " An apparatus for the application of simple shear to soil samples." *Proc. 3rd Int. Conf. Soil Mech.*, 1 : 186–191.

SKEMPTON, A. W., 1954, " The pore-pressure coefficients *A* and *B*." *Géotechnique* 4, : 143–147.

TAYLOR, D. W., 1948. " Fundamentals of soil mechanics." *Wiley, New York*, pp. 354–9.

TERZAGHI, K., 1956. *Correspondence* on the " Physico-chemical analysis of the compressibility of pure clays ", by G. H. Bolt. *Géotechnique*, 6 : 191–192.

WROTH, C. P., 1958. " The behaviour of soils and other granular media when subjected to shear." Forthcoming *Ph.D. Thesis. Cambridge.*

Roscoe, K. H., Schofield, A. N. & Wroth, C. P. (1958). *Géotechnique* **8**, No. 3, 134–136

CORRESPONDENCE

The Secretary,
The Institution of Civil Engineers.

DEAR SIR,

In their Paper "On the yielding of soils" (*Géotechnique*, 8 : 1 : 22), Messrs Roscoe, Schofield, and Wroth have drawn extensively on the test results on Weald Clay obtained at the Imperial College of Science and Technology in order to illustrate their hypothesis that the end points of all compression tests on a particular clay lie on a unique "critical void ratio" line. The Writer has been concerned with the testing of saturated clays for a number of years and in his opinion the conclusions drawn from the test results by the Authors are in some instances misleading and present an over-simplified picture of the actual behaviour of clay.

It is very probable that the end points of drained tests, which for heavily overconsolidated samples are reached only after very large shear strains, lie on a "critical void ratio" line. This concept has been discussed by Casagrande (1936) and provides the basis of the $c' = 0$ approach to the long-term stability of slopes in heavily overconsolidated fissured clays (Henkel and Skempton, 1955).

However, when undrained tests are considered, the experimental evidence obtained at the Imperial College conflicts with the hypothesis that a common "critical void ratio" line is reached irrespective of the consolidation history of the sample. In order to illustrate this point the strengths measured in various types of compression tests on Weald Clay have been plotted against water content in Fig. 1. It can readily be seen that a unique line is obtained for both drained tests and undrained tests on normally consolidated specimens (U.S. War Department, 1947) and that another unique curve can be drawn through the results of the drained and the undrained tests on samples having a maximum consolidation pressure of 120 lb/sq. in. The C.V.R. line proposed for this clay by Roscoe, Schofield, and Wroth is also shown.

In arriving at their C.V.R. line the Authors rejected all data obtained from undrained tests on samples having overconsolidation ratios greater than 8 on the grounds that the measured

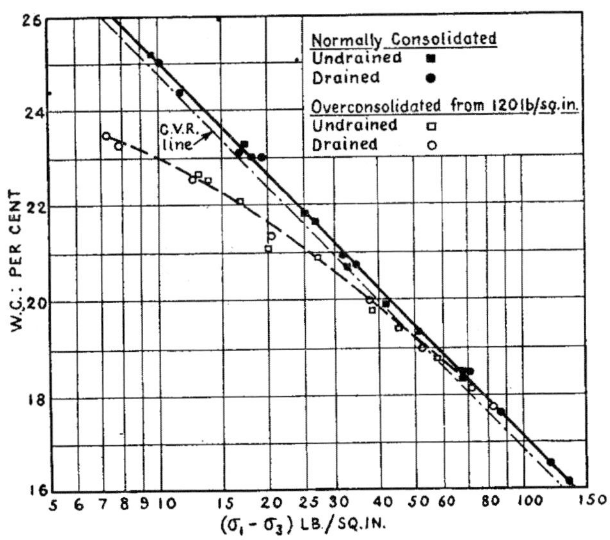

Fig. 1. Water content-strength relationship for Weald Clay showing C.V.R. line.

values of the water content did not represent the water content of that portion of the sample in which continuous deformation under constant shear stress occurred. This possibility was appreciated at an early stage in the testing programme and a series of detailed tests was run in order to investigate the point. It was found that in no test did failure occur in a narrow zone but that uniform bulging accompanied the axial compression. The water contents measured at various points in the specimen after failure showed scatters of up to 0·3%, but there was no evidence of appreciably higher water contents in the central zone of the sample where the shear deformation was largest. Undrained tests on specimens having a maximum consolidation pressure of 120 lb/sq. in. and an overconsolidation ratio of 24 gave a compression strength of about 13 lb/sq. in. with a water content of 22·6%. If the strength measurement is considered to be correct, the water content in the failure zone of this sample, if the result is to lie on the Authors' proposed C.V.R. line, would have to be 23·8% and the error in the measurement of water content would thus be 1·2%. Errors of this magnitude are considered to be too large to be possible and the Writer does not think that the result of this undrained test could ever approach the proposed C.V.R. line.

The Authors also suggest that the results of the tests on the heavily overconsolidated samples are inconclusive on the grounds that at the large strains required to reach the ultimate state the measurement of stress in the triaxial apparatus is inaccurate and the corrections to be applied for the rubber membrane, the filter paper drains and plunger friction are accentuated owing to the low effective stresses. If the experimental evidence of essentially uniform water content in the samples is used it can be seen from Fig. 1 that at a water content of 22·6% the strength required for the result of the test to lie on the proposed C.V.R. line is about 19 lb/sq. in. compared with a measured value of 13 lb/sq. in. Here again it is believed that errors of this magnitude are too large to be possible.

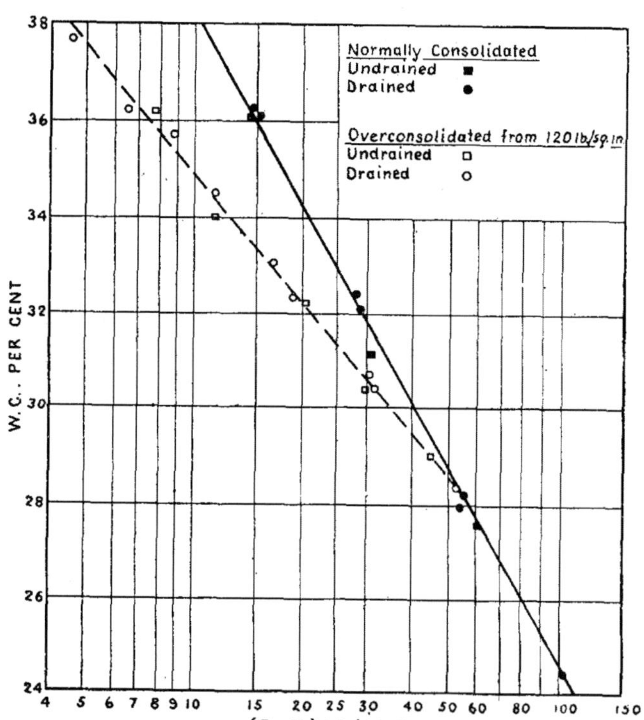

Fig. 2. Water content-strength relationship for London Clay.

In view of this wide differences between the experimental evidence and the implications of the Authors' hypothesis of the existence of a C.V.R. line, it is considered that their hypothesis takes far too simple a view of the behaviour of clay during shear. Their C.V.R. line is in fact an average line for the results of tests on samples having an overconsolidation ratio less than 8, and further evidence of the differences between the results of their theory and the experimental data is apparent in their Fig. 18 (page 37) where measured A values are compared with their theoretical predictions.

The more important and fundamental link between drained tests and undrained tests is that unique water content–strength relationships are obtained for normally consolidated samples and also for samples with the same maximum consolidation pressure. These relationships are shown for Weald Clay in Fig. 1 and for London Clay in Fig. 2. It can be shown that these particular relationships form part of more general relationships between water content, strength, and the effective stresses similar to those first postulated by Rendulic (1937). The consequences of these general relationships for normally and overconsolidated clays have been studied in detail by the Writer (Henkel, 1958) and it has been possible to correlate volume changes and strengths in drained tests with pore-pressure changes and strengths in undrained tests. Papers dealing with these general relationships are being prepared.

Yours faithfully,
DAVID J. HENKEL, B.Sc., A.M.I.C.E.

Talbot Laboratory,
Urbana, Illinois.
7 July, 1958.

REFERENCES

CASAGRANDE, A., 1936. " Characteristics of cohesionless soils affecting the stability of slopes and earth fills ". *J. Boston Soc. civ. Engrs*, 23 : 13–32.

HENKEL, D. J., 1958. " The correlation between deformation, pore-water pressure and strength characteristics of saturated clays ". *Ph.D. Thesis, University of London.*

HENKEL, D. J., *and* SKEMPTON, A. W., 1955. " A landslide at Jackfield, Shropshire, in a heavily overconsolidated clay ". *Géotechnique*, 5 : 131–137.

RENDULIC, L., 1937. " Ein grundgesetz der tonmechanik und sein experimentaller beweis " (" The fundamental law of clay mechanics and the experimental proof "). *Bauingenieur*, 18 : 459–467.

U.S. WAR DEPARTMENT, 1947. " Triaxial shear research and pressure distribution studies on soils. *U.S. Waterw. Expt. Sta.*, 332 pp.

Roscoe, K. H., Schofield, A. N. & Wroth, C. P. (1958). *Géotechnique* **8**, No. 4, 183–187

CORRESPONDENCE

The Secretary,
The Institution of Civil Engineers.

DEAR SIR,

ON THE YIELDING OF SOILS

The Paper by Messrs Roscoe, Schofield and Wroth (*Géotechnique*, 8:1:22) raises several points of interest. Since the subject is so important to all engineers, the Writer would like to put some findings on record.

The concept of a critical voids ratio for clays is supported by a wide range of triaxial tests carried out at the Imperial College of Science and Technology, London, on Weald Clay and London Clay. These tests were carried out on fully saturated, remoulded, samples giving a wide range of overconsolidation ratios, and failed both in compression and extension. A description of these clays, together with the testing programmes and test results have been published (Gilbert, 1954; Plant, 1955; Henkel, 1956; Parry, 1956). These test results are used with the kind permission of Professor A. W. Skempton. A number of the test results given by Gilbert have already been used by Roscoe, Schofield and Wroth to demonstrate the critical voids ratio for Weald Clay.

The majority of the tests were not continued to the ultimate condition but, as shown here, a clear indication of the C.V.R. line can be obtained from a study of the way in which the volume is changing at failure.

The behaviour of clay at failure

The behaviour of a clay soil at failure depends on the consolidation history before testing and the shear stress history at failure. A normally consolidated sample, tested in either compression or extension in the triaxial cell, will show a zero rate of volume change or a continuing decrease in volume at failure under drained conditions. Under undrained conditions the pore pressure will show either a zero rate of change or a continuing increase at failure. A heavily overconsolidated sample will show a continuing increase in volume under drained conditions and a continuing decrease in pore pressure under undrained conditions. A lightly overconsolidated sample may behave in any one of the above ways.

In Fig. 1 the water content at failure (w_f) has been plotted against the mean effective stress at failure (p_f) for fully saturated Weald Clay and London Clay, respectively. For Weald Clay the points are shown for compression and extension tests carried out under both drained and undrained conditions. For London Clay the results of compression tests only are shown as extension tests have not been carried out on this clay.*

In Fig. 1(a) the plotted points are for samples showing either zero rate of volume change or continuing decrease in volume at failure for drained tests, or zero rate of pore-pressure change or a continuing increase in pore pressure at failure for undrained tests. In Fig. 1(b) the plotted points are for samples showing a continuing increase in volume at failure or a continuing decrease in pore pressure. It can be seen that a line may be drawn which represents a lower bound to the plotted points in Fig. 1(a) and an upper bound for the points in Fig. 1(b).

The further importance of this dividing line is shown in Fig. 2. In Fig. 2(a) the rate of unit volume change with change in maximum shear strain has been plotted against the ratio of the mean effective stress given by the dividing line at the water content considered to the average principal effective stress acting within the clay structure $\left(\frac{pcr}{pf}\right)$. The plotted

* See Appendix for modification to London Clay test results.

points represent the conditions at failure for drained tests on Weald Clay and London Clay. As these rates of volume change are of comparatively small magnitude they are very sensitive to experimental variations, in particular the rates of axial strain up to and at failure and the distortion of the samples at failure. This results in considerable scattering of the plotted

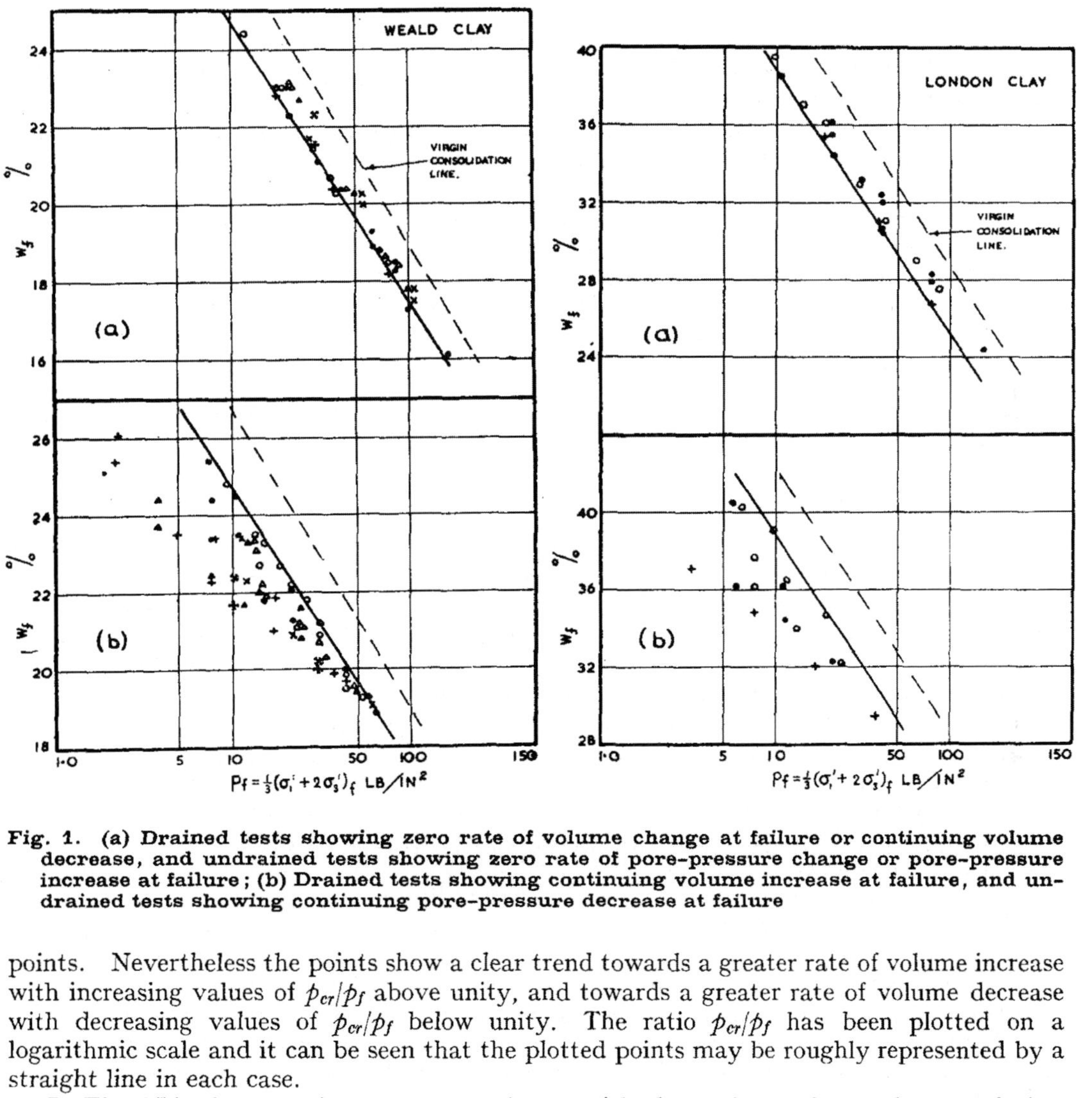

Fig. 1. (a) Drained tests showing zero rate of volume change at failure or continuing volume decrease, and undrained tests showing zero rate of pore-pressure change or pore-pressure increase at failure; (b) Drained tests showing continuing volume increase at failure, and undrained tests showing continuing pore-pressure decrease at failure

points. Nevertheless the points show a clear trend towards a greater rate of volume increase with increasing values of p_{cr}/p_f above unity, and towards a greater rate of volume decrease with decreasing values of p_{cr}/p_f below unity. The ratio p_{cr}/p_f has been plotted on a logarithmic scale and it can be seen that the plotted points may be roughly represented by a straight line in each case.

In Fig. 2(b), the rate of pore-pressure change with change in maximum shear strain has been plotted against the ratio p_{cr}/p_f. The plotted points represent failure conditions for undrained tests, and the rate of pore-pressure change has been divided by the mean effective stress to give a dimensionless quantity. Again the comparatively small rates of pore-pressure change at failure lead to considerable scatter of the plotted points, which is emphasized by the fact that undrained tests tend to lie closer to the dividing line in Fig. 1 than drained tests, so that only a limited spread of p_{cr}/p_f values is obtained. However, there is an obvious trend towards a greater rate of pore-pressure decrease with increasing values of p_{cr}/p_f above unity, and towards a greater rate of pore-pressure increase with decreasing values of p_{cr}/p_f below unity. Adopting a logarithmic scale for p_{cr}/p_f enables the points to be roughly represented by a straight line for each clay.

Predicted behaviour after failure

The experimental evidence given in Figs 1 and 2, suggests that the volume change at failure under drained test conditions and the pore-pressure change at failure under undrained test conditions are controlled, both in rate and sense, by the relation between the mean effective stress and the water content for any clay. The volume change or pore-pressure change at failure is such that the samples tend to move towards the dividing line, but do

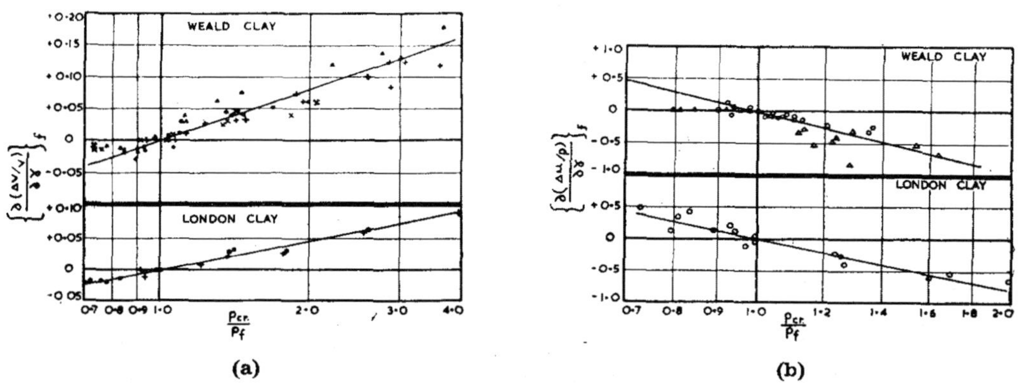

Fig. 2. (a) Drained tests: rate of volume change at failure; (b) Undrained tests: rate of pore-pressure change at failure

not pass beyond it. The further evidence that the rate of volume change or pore-pressure change at failure is a function of p_{cr}/p_f suggests that under continuing shear strain after failure all samples, regardless of the type of test, will eventually lie on this line, when the rate of volume change or pore-pressure change will be zero. The samples have then reached a condition of volume equilibrium and this dividing line is the C.V.R. line.

Yours faithfully,

R. H. G. PARRY, Ph.D., Research Officer

C.S.I.R.O. (Soil Mechanics),
Victoria, Australia.
12 June, 1958.

APPENDIX

Modification of London Clay test results

The drained and undrained compression tests on London Clay, with increasing axial stress and constant radial stress, were carried out by Plant during 1953–54. The drained compression tests with constant axial stress and decreasing radial stress were carried out by the Writer during 1956, on clay taken from the same bulk sample as that used by Plant. However, the clay used by the Writer showed slightly different consolidation characteristics, the water content for a sample normally consolidated at 30 lb/sq. in. being 36·5% whereas Plant obtained 35·8%. At 120 lb/sq. in. the water contents were 29·0% compared with 27·6%. Repeats of a test carried out by Plant showed that the deformation characteristics at a particular consolidation pressure were sensibly the same in each case. For this reason, the water contents for the test carried out by the Writer were adjusted by an amount equal to the difference between the water contents for the virgin consolidation curves to enable these tests to be fitted into the pattern of Plant's results.

SYMBOLS USED FOR PLOTTING TEST RESULTS

Drained tests

●—Radial stress constant, axial stress increased.
+—Radial stress decreased, axial stress constant.
+—Radial stress decreased and axial stress increased, so that the mean effective stress remains constant.
▲—Radial stress constant, axial stress decreased.
×—Radial stress increased, axial stress constant.
×—Radial stress increased and axial stress decreased so that the mean effective stress remains constant.

Undrained tests

o—Radial total stress constant, axial total stress increased, and radial total stress decreased, axial total stress constant.
△—Radial total stress constant, axial total stress decreased, and radial total stress increased, axial total stress constant.

REFERENCES

GILBERT, G. D., 1954. "The basic shear strength properties of Weald Clay." Ph.D. Thesis, London.
HENKEL, D. J., 1956. "The effect of overconsolidation on the behaviour of clays during shear." *Géotechnique*, 6 : 139.
PARRY, R. H. G., 1956. "Strength and deformation of clay." Ph.D. Thesis, London.
PLANT, J. R., 1956. "Shear strength properties of London Clay." M.Sc. Thesis, London.
ROSCOE, K. H., SCHOFIELD, A. N., *and* WROTH, C. P., 1958. "On the yielding of soils." *Géotechnique*, 8 : 22.

The Secretary,
The Institution of Civil Engineers.

DEAR SIR,

Although in Berlin engineers seldom deal with problems of cohesive soils, they follow with great interest your excellent publications on this subject. The Paper "On the yielding of soils" (*Géotechnique*, 8 : 1 : 22) by Messrs Roscoe, Schofield and Wroth gave a very thoughtful view of the problem of yielding, trying to get a comprehensive concept for both cohesive and cohesionless soils. Implicit, however, the Authors presuppose the validity of the generalized Mohr-Coulomb yield condition :

$$\{(\sigma_1 - \sigma_2)^2 - \sin^2\rho\ (\sigma_1+\sigma_2)^2\}\{(\sigma_1-\sigma_3)^2 - \sin^2\rho\ (\sigma_1+\sigma_3)^2\}\{(\sigma_2 - \sigma_3)^2 - \sin_2\rho\ (\sigma_2 + \sigma_3)^2\} = 0$$

The physical significance of this condition is that, owing to this, the hypothesis yielding of a material characterized by it is only probable in one of the principal stress planes. Although in a triaxial test following to $\sigma_2 = \sigma_3$ all vertical planes are principally of equal importance, this means that yielding happens only in one special and unpredictable vertical plane of fracture. *Vice versa*, one may conclude by saying : if in a triaxial test plane yielding happens, then the Mohr-Coulomb condition is obviously valid.

In this connexion the Writer would like to comment on Mr Henkel's letter (*Géotechnique*, 8 : 3 : 134) pointing out that the conclusions of Roscoe, Schofield and Wroth would not represent the behaviour of soil samples with a large overconsolidation ratio. In such cases there was no distinct failure zone to be observed, and the water content was uniform throughout the sample. The failure occurred in the form of a uniform bulging. It is this experience, however, that made the Writer think of the probability that in such cases the Mohr-Coulomb condition cannot be assumed to describe the initial state of yielding. One should rather examine, whether it might not lead to better agreement with experience when using a completely symmetrical condition, such as the extended von Mises condition :

$$(\sigma_1 - \sigma_2)^2 + (\sigma_1 - \sigma_3)^2 + (\sigma_2 - \sigma_3)^2 = \text{const}\,.\,(\sigma_1 + \sigma_2 + \sigma_3)^2$$

To the Writer it would appear that the conclusions of Roscoe, Schofield and Wroth are limited to cases in which one is allowed to assume the validity of the Mohr-Coulomb hypothesis. Contrary to those, cases of uniform bulging seem to be better analysed by means of the adequate von Mises condition, which maintains that a material reacts equally to an increase or a decrease of the first principal stress in relation to the combined spherical state of stress.

Yours faithfully,

Technische Universität Berlin,
Grundbauinstitut.
6 October, 1958.

HANS-U. SMOLTCZYK, Dr.-Ing.

Roscoe, K. H., Schofield, A. N. & Wroth, C. P. (1959). *Géotechnique* **9**, No. 2, 71–83

CORRESPONDENCE

The Secretary,
The Institution of Civil Engineers.

DEAR SIR,

In the correspondence pages of *Géotechnique* (8 : 3 : 134) Henkel has challenged the conclusions drawn by Messrs Roscoe, Schofield, and Wroth in their Paper "On the yielding of soils" (*Géotechnique,* 8 : 1 : 22). The Writers have recently completed a series of tests on a dry clay powder, the results of which may have some bearing on this question.

Samples of the powder were first consolidated under a selected cell pressure and then axially loaded until the point of maximum consolidation, which was close to the point of failure, was reached. The axial strain required was approximately 15% and the voids ratio attained corresponded to a cell pressure nearly four times the original cell pressure. From this point on, both drained and undrained tests were carried out at cell pressures chosen to give over consolidation ratios ranging from 1 to about 60. In the undrained tests, the sample

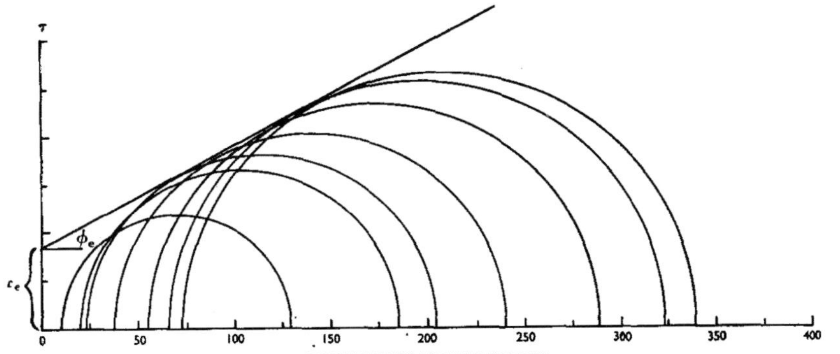

(*a*) HEAVILY OVERCONSOLIDATED. p_c = 100 lb/sq. in; initial σ_3 = 5 lb/sq. in.

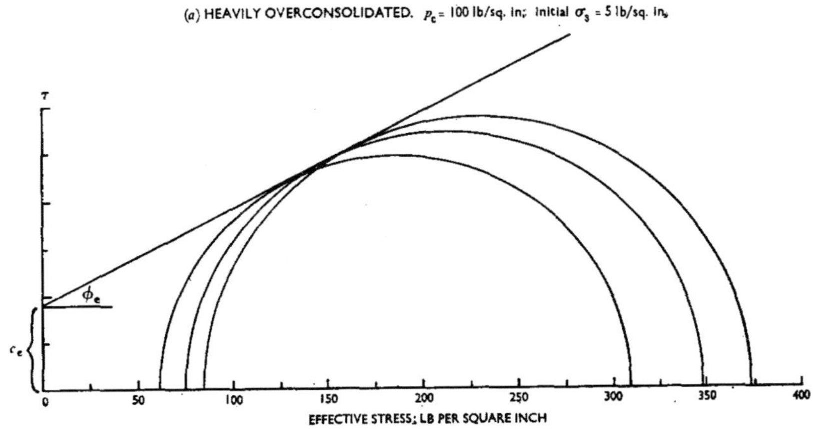

(*b*) LIGHTLY OVERCONSOLIDATED. p_c = 75 lb/sq. in; initial σ_3 = 55 lb/sq. in.

Fig. 1 (a) and (b). Results of constant volume tests on a dry clay powder at a voids ratio $e = 0{\cdot}82$

was maintained at constant volume by raising or lowering the cell pressure so that the level in a 1-mm bore capillary connected to the base of the sample was maintained constant.

The above technique of first axially loading all samples to incipient failure was done in an attempt to eliminate many of the undesirable features associated with the comparison of drained and undrained tests carried out in the normal way; in particular the state of anisotropy and the cross-sectional area.

The results of a single undrained test on a heavily overconsolidated sample are shown in Fig. 1 (*a*), while Fig. 1 (*b*) shows those from a single undrained test on a lightly overconsolidated sample. In both cases, the additional axial strain by the end of the test was about 15%. It will be seen that the second test is to all intents and purposes a continuation of the first, which was terminated when distortion of the sample was becoming excessive. Differential expansion and contraction of samples can be ruled out on account of the linear envelope to the Mohr circles.

Of further interest is the fact that a normally consolidated sample in a drained test with a cell pressure of 74 lb/sq. in. fails at the same voids ratio as the above tests with a Mohr strength circle (not shown) which is tangential to the same envelope. Furthermore, it was noted that normally consolidated samples were in the process of expanding a small but definite amount at failure and that to maintain constant volume beyond this point it was necessary to increase the cell pressure. Thus, at least in this particular set of tests, undrained strengths are obtained which are greater than drained strengths at the same voids ratio.

There are several other interesting and novel features arising out of this research and the work is being prepared for publication.

Yours faithfully,
P. L. Newland
B. H. Alley

Soil Mechanics Laboratory, D.S.I.R.,
Auckland, New Zealand.
28 November, 1958.

The Secretary,
The Institution of Civil Engineers.

Dear Sir,

We, the Authors "On the yielding of soils" (*Géotechnique*, 8:1:22), welcome the opportunity to reply in detail to correspondence from Mr D. J. Henkel and Drs R. H. G. Parry and Hans-U Smoltczyk which discusses certain aspects of our Paper. Each Writer will be answered separately in order of publication, despite the fact that Dr Parry's contribution is dated prior to Mr Henkel's and answers some of his points.

(A) D. J. Henkel (*Géotechnique*, 8:3:134)

1. *Introduction*

Mr Henkel, while confining his comments on the Paper strictly to the section on Weald Clay and London Clay, generalizes in such a way as to imply that our deductions are "misleading and present an oversimplified picture of the actual behaviour of clay". In order to support our belief that this is not so, we propose to analyse in detail the systematic errors in the triaxial test. We have previously stated in brief our reasons for these beliefs (§§ 4(*d*) and 5(*b*) of the Paper), and we hope to show here that it is the triaxial test itself that can be misleading and not the "simple picture".

The principal difference of opinion between Mr Henkel and ourselves appears to be in the interpretation of undrained triaxial test results on Weald Clay and London Clay. In Figs 1 and 2 of his letter Mr Henkel glosses over the point that, on the one hand we are estimating the state of stress and voids ratio that is *ultimately* reached by the yielding soil, and with due allowance for inevitable errors we believe that our C.V.R. line represents that state; while on the other hand Mr Henkel is concerned with the state of stress and voids ratio (averages throughout a triaxial sample) at which the observed deviator stress (after correction for bulging) passes through a maximum value and, to use his term, "failure" occurs. He accepts our point that in *drained* tests overconsolidated samples will pass through this "failure" state when on the "dry" side of the C.V.R. line and would continue to the C.V.R. state in nature although the samples do not always appear to do so in triaxial tests (see Fig. 32 of the Paper for triaxial tests and compare this with Fig. 34 for simple shear tests). But, if we understand his contention correctly, he is not prepared to concede that there should be systematic errors in *undrained* triaxial tests that make his line for "failure" of normally consolidated samples lie to the "wet" side of the C.V.R. line, and his line for "failure" of overconsolidated samples curve away from the "dry" side of the C.V.R. line as shown in his Figs 1 and 2. To explain this one must consider in detail the errors in the determination of strain (including voids ratio) and stress, in a triaxial sample. The discussion will be confined to undrained tests, for which we believe Mr Henkel's "failure" should theoretically coincide with our "ultimate" state. Much of what is discussed can also be applied to drained tests.

2. *Errors of strain and stress in undrained triaxial tests*

(*a*) *Errors of strain.*

(i) *Area corrections.*—The estimation of the cross-sectional area of the sample is subject to errors that increase progressively throughout the test as the axial strain is increased. The conventional assumption, as described by Henkel and Bishop (1957)[1] is that the sample is and remains a right circular cylinder. But, in fact, the end planes of the sample do not expand and an alternative assumption might be that the sample bulges with a side of parabolic section. The relative magnitudes of the corrections determined from these two assumptions for various values of the axial strain ϵ_1 are shown in Table 1 below. The symbol a_0 denotes the original central horizontal cross-sectional area and a_1 the value at an axial strain ϵ_1.

The Table shows clearly the unreliability of either of these corrections for strains in excess of 10%.

Table 1.

Axial strain ϵ_1%	Area correction $\frac{a_1 - a_0}{a_0} \times 100\%$	
	Conventional	Parabolic
10	11·1	16·8
20	25·0	38·1
30	42·9	65·9

(ii) *Voids ratio.*—In undrained tests on normally consolidated samples of remoulded clays and on loose cohesionless media we have always found that the sample bulges uniformly with increase of ϵ_1 and continues this uniform bulging well beyond the "failure" point. Such samples are "work hardening" and hence the formation of planes or thin zones of shear is not to be expected. For our undrained tests on overconsolidated samples and dense cohesionless media we have observed that the samples appear to bulge uniformly up to "failure" but at

[1] See references on p. 82.

"failure" a distinct plane or planes of shear became visible. This phenomenon was described by Rendulic (1937) and excellent photographs can be seen in Fig. 6a of Schaerer, Schaad and Haefeli (1948), Fig. 13 of Haefeli and Schaerer (1946), Fig. 14.4, p. 334 of Taylor (1948), or Figs 13 and 14 of Newland and Allely (1956)—to quote only a few of the many available references.

The samples in which such planes or thin zones of shear appear are all "work softening", and the extent of these zones will be a minimum consistent with the distortion imposed by the boundaries. The voids ratio within these zones may differ considerably from that of the remainder of the sample and admirable evidence for the existence of such zones in the field is provided by Henkel (1956(*b*) in Fig. 24 of his contribution). Further evidence of migration of water within a sample during an undrained triaxial test may be found in the Massachusetts Institute of Technology Report (1951); the difference between the moisture content at the centre and at the ends of the samples was of the order of 1·5% in 30%.

(*b*) *Errors of stress.*

(i) *Theoretical.*—Let the error in estimating any stress variable x, due solely to errors in estimating the area of cross-section of the sample (e.g. § 2(*a*) (i) above) be denoted by err $[x]$. In an undrained test the recorded values of the cell pressure σ_3 and the excess pore-water pressure u are not affected by this error and hence err $[\sigma_3]$ = err $[u] = 0$, so that err $[\sigma'_3]$ = err $[\sigma_3]$ − err $[u] = 0$. In particular err $[p] = \frac{1}{3}$ err $[\sigma'_1 + 2\sigma'_3] = \frac{1}{3}$ err $[\sigma'_1] = \frac{1}{3}$ err $[\sigma'_1 - \sigma'_3] = \frac{1}{3}$ err $[q]$.

Therefore the error in q at any instant is three times the error in p. This would mean in practice that once a sample had reached a point in its loading path in an undrained test where the excess pore-water pressure remained constant (i.e. had reached the C.V.R. state) then on further straining, the loading path (which is confined to a (p, q) plane), would be expected to deviate progressively from the C.V.R. state along a line of slope $\tan^{-1} 3$. A series of undrained triaxial tests will now be described to substantiate this theory which was due to Wroth (1958).

(ii) *Experimental evidence.*—"Normally consolidated kaolin". Fig. 1 shows, to a large scale, the end portions of two loading paths for undrained tests on normally consolidated samples of kaolin, which had been strained well beyond their "failure" points. A number near a point on either loading path gives the value of ϵ_1 at that point. It can be seen that the peak (and in this case also ultimate) deviator stresses q_f, corresponding to points Q_1 and Q_2 on the C.V.R. line are reached at strains of 12·2 and 13·5% respectively. For further strains the loading paths proceed down lines of slope $\tan^{-1} 3$ until the tests were stopped at the end points Q'_1 and Q'_2. Since $Q_1Q'_1$ and $Q_2Q'_2$ are parallel to the applied stress paths (*see* Fig. 8, p. 30, of "On the yielding of soils") there will be no change in excess pore-water pressure for strains beyond those required to reach Q_1 and Q_2 respectively.

Fig. 2 shows how the observed value of q diminishes beyond "failure" due to error resulting from the use of the conventional area correction.

We suggest that the error *starts at the beginning of a test* and has already caused a small but significant deviation of the observed loading path from the expected loading path by the time that "failure" is reached. Hence it is likely that the C.V.R. line as observed from tests on normally consolidated samples will be below the true C.V.R. line.

"Dense steel balls". Fig. 3(a) shows the loading path for an undrained test on a sample of densely packed 1 mm dia. steel balls with silicone fluid in the pores. This test was one of a series carried out at Cambridge by Mr H. Poorooshasb. Initial consolidation was permitted under a cell pressure of 10 lb/sq. in. but after the drain cock had been closed the cell pressure was raised to 100 lb/sq. in. The loading path starting at A reaches the undrained yield surface at B after a strain ϵ_1 of 2%. The path then continues along the straight line BC which passes through the origin O. For strains beyond 6% the path begins to fall below the straight line

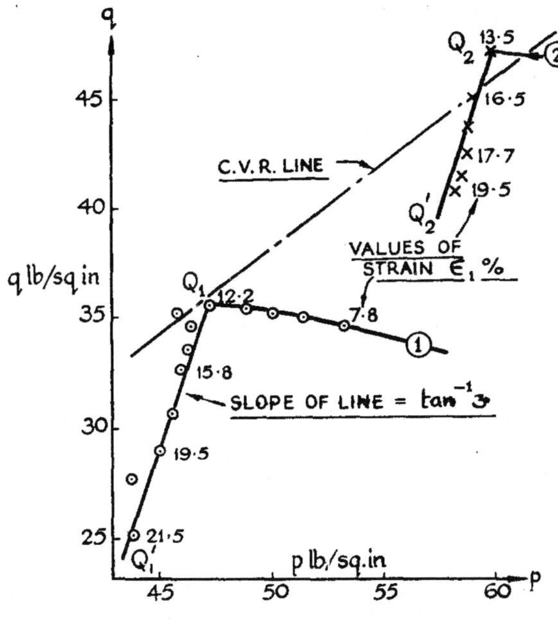

Fig. 1. Portions of loading paths at high strain for undrained tests on kaolin

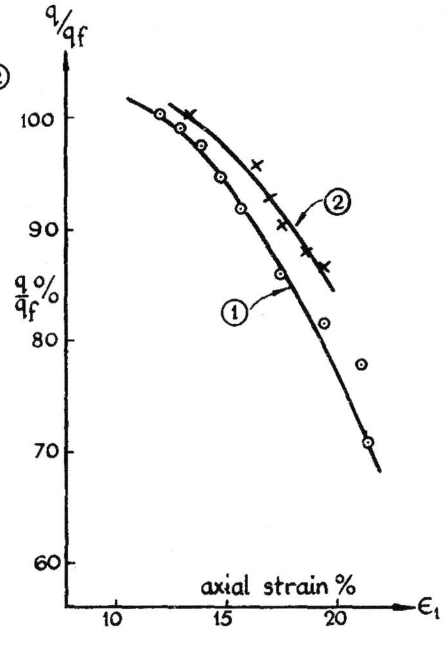

Fig. 2. Relationship between shear stress and strain occurring after failure

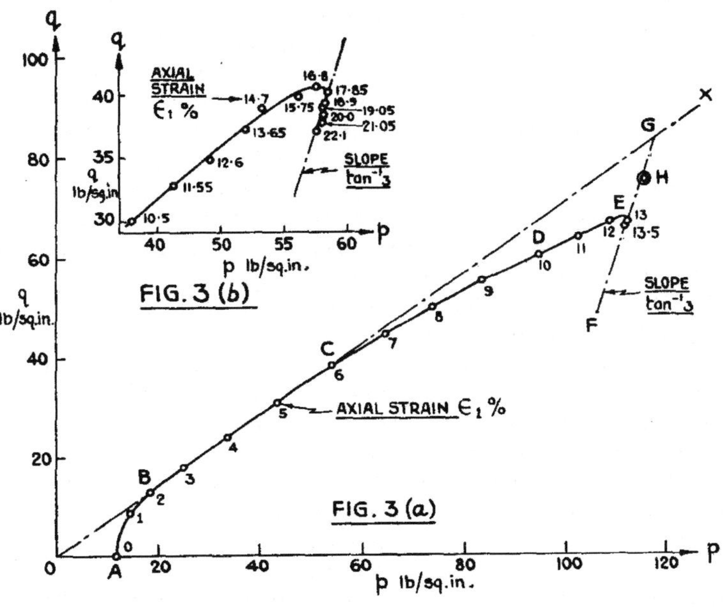

Fig. 3. (a) Loading path for undrained triaxial test on dense (e = 58·9%) "shiny" steel balls; (b) End of loading path for undrained triaxial test on dense ($e \simeq$ 61%) "rusty" steel balls

OCX until at D (for which $\epsilon_1 \simeq 10\%$) the sample, which had shown an initially slight but apparently uniform bulge, was beginning to develop a zone of shear of pronounced inclination. The peak stresses were recorded at E for which $\epsilon_1 = 13\%$. Further strain caused the loading path to move down the line EF, of slope $\tan^{-1} 3$. Since the error in locating B is small we would suggest that the point G where the line OB intersects FE would give a better estimate of the stresses at the C.V.R. state than would point E. The point H represents the estimated C.V.R. state for the voids ratio corresponding to this undrained test but determined by extrapolation from the mean of sixteen undrained tests on loose samples of steel balls which all "failed" with uniform bulging at an average strain ϵ_1 of about 3·5%. In Fig. 3(b) the end points are plotted of the loading path of a similar test on steel balls to that shown in Fig. 3(a) but continued to even greater strains. There can be little doubt that beyond the peak point the loading path turns down a straight line of slope $\tan^{-1} 3$.

It is important to note in Fig. 3(a) that the loading path of the "heavily overconsolidated" sample strikes the observed yield surface and proceeds along it for quite a distance before stress correction errors and *internal* changes of voids ratio cause it to appear to deviate from this surface. The same phenomenon has been observed for heavily overconsolidated samples of silt, *see* Fig. 30, p. 45 of "On the yielding of soils".

"Overconsolidated kaolin". Fig. 4 shows the complete loading paths for undrained triaxial tests on kaolin samples with overconsolidation ratios of $n = 1{\cdot}7$, $3{\cdot}3$, and 10. They are plotted from observations taken during the preliminary part of an extensive programme at the Building Research Station, Watford, which have been privately communicated and are quoted here by kind permission of the Director of Building Research. The projection of the C.V.R. line represented by OX in Fig. 4 is drawn through the mean peak points of twenty

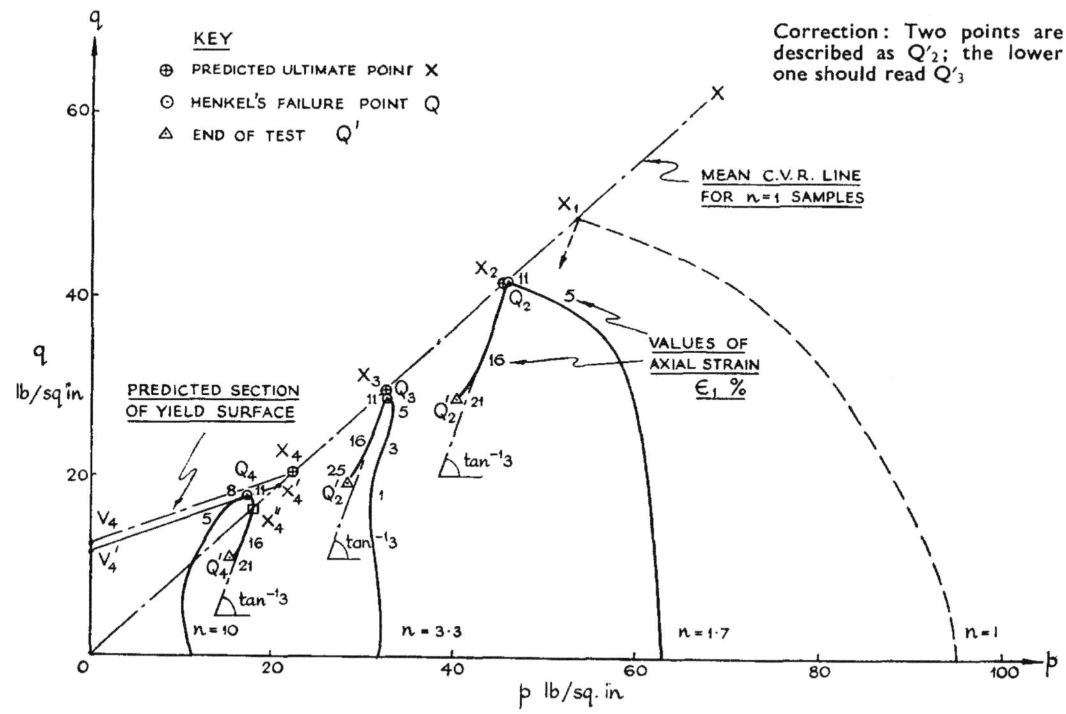

Fig. 4. Loading paths for undrained triaxial tests on overconsolidated samples of kaolin

undrained tests on normally consolidated samples, of which only one has been plotted to indicate the shape of these $n = 1$ curves. The value of ϵ_1 at which these normally consolidated samples reached their "failure" points varied between 5% and 13% with an average of 11%; this may be compared with the strain figures on the curves for the overconsolidated samples. From these results of tests on normally consolidated samples it is possible to predict the point at which an undrained sample, of known initial voids ratio, might be expected to reach the C.V.R. state provided it is assumed that there are no errors in the triaxial test. Such points are represented by X in Fig. 4 and correspond to our predicted "ultimate" states. The letter Q is used to represent the "failure" state as actually *observed* in a test and the letter Q′ represents the final *observed* point in a test. The discrepancy between the points Q and Q′ and the point X represents *an error of measurement in the triaxial test* and does not indicate any serious deviation of the behaviour of the samples from the "simple picture".

It will be noted that for lightly overconsolidated samples $n \leqslant 3{\cdot}3$ the predicted points X_2 and X_3 agree well with the observed points Q_2 and Q_3, respectively, and also that the strains to reach "failure" are the same as for the normally consolidated samples, i.e. 11%. For the more heavily overconsolidated sample the same strain is required to reach Q_4 but this point is somewhat distant from X_4. The predicted section of the yield surface for this last sample is X_4V_4 but apparently the sample never quite reaches this surface. At a strain of 8% it reaches a line $X'_4V'_4$ which is parallel to X_4V_4 and then proceeds along this line to Q_4 before falling away on a line of slope $\tan^{-1} 3$ to Q'_4. The line $X'_4V'_4$ corresponds to a predicted voids ratio of 1·498 while X_4V_4 corresponds to 1·478 which was the measured value for the actual test. Hence if the possibility is admitted that the voids ratio in the failure zone differs from the average for the whole sample by 0·020 in 1·478 (or 1·35%) or alternatively that there had been an error in the determination of q of 0·75 lb/sq. in. in 18 lb/sq. in., then the sample would have reached the predicted yield surface and travelled along it while the strain increased from 8% to 11%. Since the strain for "failure" at Q_4 is the same as for the normally consolidated samples we suggest that the lack of coincidence between $X'_4V'_4$ and X_4V_4 is due to error in the assessment of voids ratio, as the errors in assessing the stress parameters of X_4 will be of the same order as of Q_4. In all our tests on heavily overconsolidated samples we have found that the line X′V′ lies below and not above XV.

When the loading path has reached Q_4 we suggest that it should proceed to X'_4 but in fact it appears to move down to Q'_4. Since we believe that a shear "plane" had already formed at Q_4 the additional errors with further strain beyond this point will be predominantly errors in stress. This is because the area correction will continue to be applied despite the fact that there is no longer any continuation of the uniform bulging. The more the test is prolonged the further the sample *appears* to move from the C.V.R. state.

3. *Undrained triaxial tests on overconsolidated Weald Clay*

Fig. 5 shows examples of typical loading paths for undrained triaxial tests on three samples of remoulded Weald Clay with overconsolidation ratios of $n = 8$, 15, and 24·5; the "failure" points of these tests have been used by Mr Henkel in Figs 1 and 2 of his letter. The line OX is the projection of the mean C.V.R. line for all undrained tests for which $n = 1$. These *normally* consolidated samples reached the C.V.R. state at axial strains ϵ_1 of about 10% but on increasing ϵ_1 to even 20% only two of the tests appeared to move down a line of slope $\tan^{-1} 3$ to any appreciable extent. Fig. 6 shows the end of the loading path of one of these normally consolidated samples plotted to a very large scale.

Each of the three loading paths for the overconsolidated samples in Fig. 5 reached a line parallel to the predicted sections of the yield surface at a strain of about 10% and proceeded along this line until the strain reached about 15%. The $n = 8$ sample reached "failure" at 15% strain (see Q_1) and on straining to $\epsilon_1 = 20\%$ the loading path curved downwards a little with a slight decrease in deviator stress to Q'_1. The $n = 15$ sample behaved in a similar

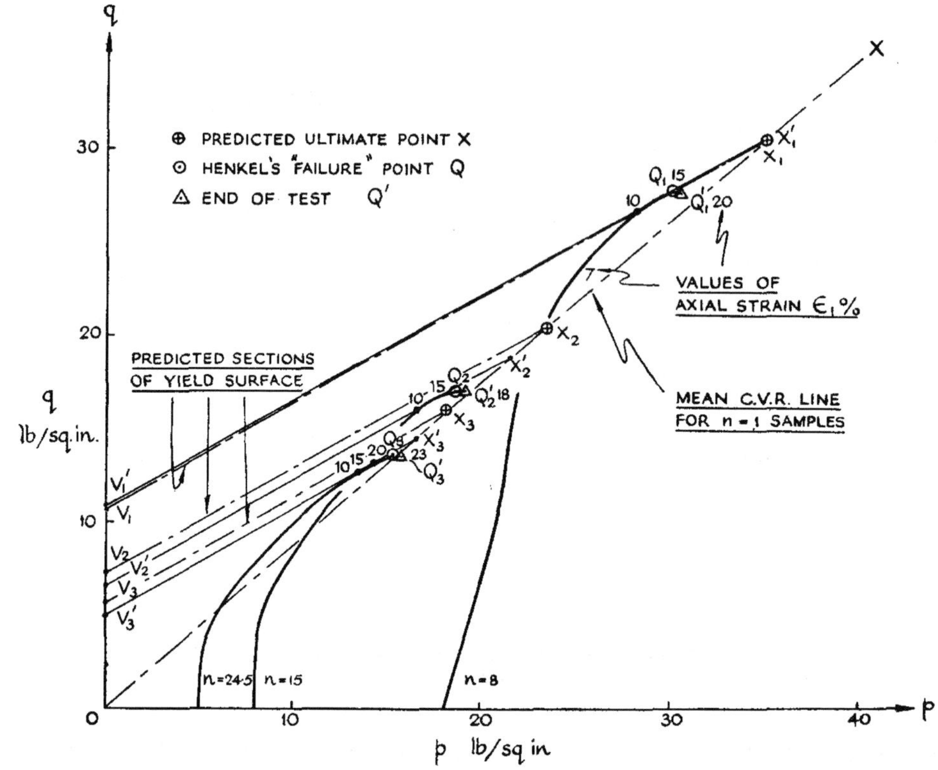

Fig. 5. Loading paths for undrained triaxial tests on overconsolidated samples of Weald Clay

way but the $n = 24$ sample did not reach "failure" until $\epsilon_1 = 20\%$ and the deviator stress remained constant while ϵ_1 was increased to 23% when the test was stopped. These three loading paths are merely examples of an extensive series of tests and it is interesting to note that in *none* of these tests on samples for which $n \geqslant 4$ did the loading path finally turn down a line of slope $\tan^{-1} 3$: this despite strains of as much as 32·5% for a sample for which $n = 126$.

In all these tests the final observed value of q was the same as the "failure" value, or varied only slightly from it as shown in Fig. 5; but the pore-water pressure was still decreasing at "failure". Typical stress-strain ($q - \epsilon_1$) curves for undrained tests on remoulded Weald Clay have been published, Henkel (1956(*a*)). His Fig. 4 shows the stress-strain curve for an overconsolidated sample ($n = 24$) and it is noticeable that the pore-water pressure was still decreasing when the test was stopped at a strain of 27%. At this same strain of 27% the value of q (after applying an area correction, presumably of the conventional type) *appears* already to have passed a peak value. A glance at Table 1 will show the potential reliablity of q at this strain ϵ_1 of 27%. We suggest that an undrained sample has truly failed *when and only when* the pore-water pressure has reached a constant value, i.e. were it possible to perform a perfect triaxial test free of any error, the failure condition would coincide with the ultimate. Strains in excess of 30% are required to achieve this on heavily overconsolidated samples of Weald Clay. At such strains both the apparent magnitude of the peak deviator stress and the choice of the value of the axial strain at which this peak occurs can be varied at will, between wide limits, depending upon the type of area correction adopted. It is important to notice at this phase that the strains required to achieve even the "failure" state increase progressively with increase of overconsolidation ratio n. Since the errors of measurement of stress also

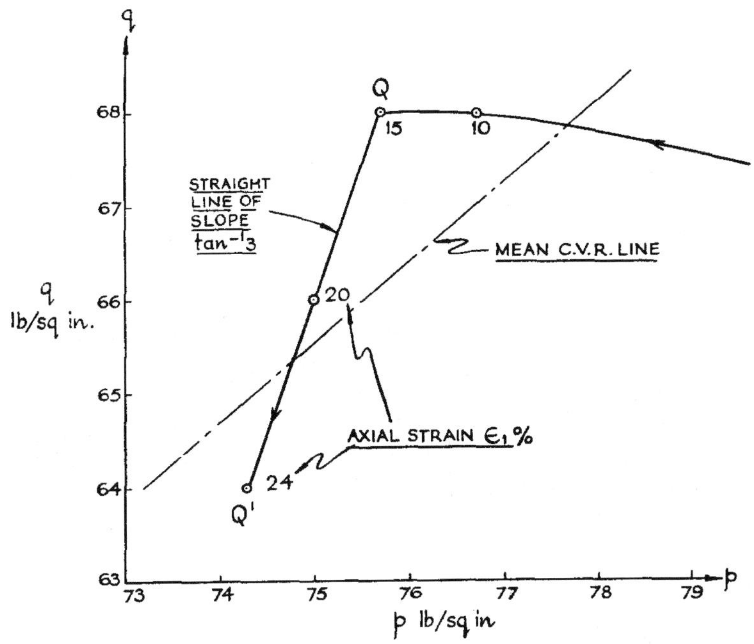

Fig. 6. End of loading path for undrained triaxial test on normally consolidated sample of Weald Clay

increase progressively with increase of strain we regarded the "failure" and end points of the loading paths for samples for which $n > 8$ with suspicion and hence did not show such points on our curves. Though we did not trust the later stages of these loading paths we found that the earlier stages showed that the samples were behaving in the same way as all other soils or similar granular media of which we had any detailed knowledge.

4. *Conclusions on the undrained triaxial tests on Weald Clay*

In addition to the points already specified above we would suggest that:—

(i) Weald Clay requires very large strains to achieve "failure" when in the overconsolidated state.*

(ii) The tests on Weald Clay were not carried to sufficient strains to achieve "proper" failure and this effect became more marked the greater the degree of overconsolidation. This explains why Mr Henkel's "failure" line for tests with a given maximum consolidation pressure diverges from the C.V.R. line. It may also be the reason why a shear plane was never seen to develop in any of the tests.

(iii) The order of magnitude of the error in assessing moisture content in the shear zone may be assessed from Fig. 5 by the difference between the moisture contents corresponding to a point X and the point X′. For the three samples for which $n \geqslant 8$ these errors in moisture content are shown in Table 2.

* Further evidence is provided by Bishop (1953), who plotted stress-strain curves for undrained triaxial tests on undisturbed and remoulded samples of Weald Clay. His Fig. 2 shows such a curve for an undisturbed sample in which the "apparent" deviator stress is still increasing and the observed excess pore-water pressure is still falling rapidly despite an axial strain ϵ_1 of 50%.

Table 2.

Overconsolidation ratio n	8·0	15·0	24·5
Initial moisture content %	20·9	22·1	22·9
Suggested final error in moisture content %	0·0	0·3	0·4

The value of 0·4% for the error in moisture content for the $n = 24{\cdot}5$ sample may be compared with the figure of 1·2% which Mr Henkel states in his letter is the error we postulate.

(iv) The order of the error in the final observed deviator stress is represented by the difference of ordinate in Fig. 5 of a point Q′ and the corresponding point X′. For the three examples these errors are shown in Table 3.

Table 3.

Overconsolidation ratio n	8·0	15·0	24·5
Axial strain at end of test ϵ_1%	20·0	17·5	23·0
Final observed deviator stress q lb/sq. in.	27·7	17·0	13·6
Suggested error in $q = \Delta q$ lb/sq. in.	2·8	1·7	1·0
$\frac{\Delta q}{q} \times 100\%$	10·0	10·0	7·5

The value of the error of 1 lb/sq. in. (in 13·6 lb/sq. in.) for the $n = 24{\cdot}5$ sample should be compared with the 6 lb/sq. in. that Mr Henkel states is the error we postulate. It is important to realize that we have located the X points in Fig. 5 from the (e, p) projection rather than the (e, q) projection of the C.V.R. line. We did this because the error in p is only one-third the error in q. Hence it would be fairer to say that we postulate a combined error of the order of 0·4% in moisture content and of 2·4 lb/sq. in. in the deviator stress for this particular sample.

(v) It is unlikely that the difference between the moisture content within a thin zone or plane of shear and within the remainder of the sample can be assessed with any accuracy by cutting up the sample and then determining the moisture content of each portion.

(vi) It is the relatively large error in q_f for high values of n that is the major cause of the deviation of Mr Henkel's $(A_f + b)$ curve from our $(A_u + b)$ line in Fig. 18, p. 37, of our Paper.

5. *Drained tests and the C.V.R. line concept*

The experimental evidence that the loading paths for samples of Weald Clay, if strained sufficiently in a drained triaxial test, reach a common C.V.R. line for all values of n is, in fact, less good than for the undrained tests. For overconsolidated samples the peak deviator stress "failure" point in a drained test occurs at much smaller strains than are required to reach the ultimate C.V.R. state. The apparent inability of triaxial samples to reach this C.V.R. state has already been indicated in our Paper in Figs 22–24 for Weald Clay and in Fig. 32 for silt, and is confirmed by all the triaxial tests that we have done on soils and similar media.

Mr Henkel suggests that the C.V.R. line concept was "discussed by Casagrande (1936) and provides the basis of the $c' = 0$ approach to the long-term stability of slopes in heavily overconsolidated fissured clays, Henkel and Skempton (1955)." We have read Professor Casagrande's Paper again with care and confirmed our impression that the Paper states that there was only *one* critical porosity (n_0 in his Fig. 2) which remained constant for all pressures. It is relevant to quote the section on critical voids ratio in the U.S. War Department Report (1947) on Triaxial Shear Research. After indicating that the critical voids ratio had not been defined precisely, and then summarizing the views of Professor A. Casagrande and Professor D. W. Taylor, the final paragraph opens with the following statement:—

"The only definite conclusion that can be derived from the critical void ratio studies in this programme is that positive results have not been obtained. At the beginning of this programme it was believed that, for a given sand under given field conditions, a critical void ratio that would be a definite measure of the degree of liquefaction could be established by tests. A superficial examination of the test results indicates that this belief is not confirmed by the test critical void ratios that vary widely with the method of test. On closer examination these results do not prove that the original conception of the action of masses of saturated sand is unsound. It is probable that the non uniform distortion of the specimens and the closely allied and heretofore unsuspected test inaccuracies disclosed by analysis of the strength results have large effects on observed test volume changes. . . ."

Concerning the long-term stability of slopes, we cannot find any mention of the C.V.R. concept for clays in the Paper by Henkel and Skempton (1955) but we are indeed aware that this concept can be used to rationalize the otherwise empirically deduced $c' = 0$ hypothesis.

6. *Conclusion*

It has been our wish throughout to find some simplicity in the yielding of soils. We believe that some of the present complexity is due not only to the inevitable scatter in the results of any series of tests on a soil but also to the consistent errors that are inherent in the test technique. In the triaxial test these systematic errors are likely to be large when the axial strains exceed 10%. An excellent example of their occurrence is to be found in the U.S. War Department Report (1947) in which the results determined for triaxial samples 1·4 in. dia. differed consistently from those obtained using 2·8-in.-dia. samples by about 0·05 in a mean voids ratio at "failure" of about 0·75 (see p. 165 of the Report).

We realized the value and also the limitations of the triaxial test at an early date and it was not until we obtained the results from the simple shear apparatus that we felt confident that soils when yielding followed a simple general pattern. It is this very pattern that can be seen in the results of nearly every research project on soil strength published during the past 25 years. It is as foolish to assert that there are "wide differences" between the experimental evidence and this pattern, as it is to call this general pattern merely "our hypothesis".

(B) R. H. G. Parry (*Géotechnique,* 8:4:183)

Dr Parry's letter lucidly presents new evidence in support of the "simple picture". His method of plotting results in his Figs 1(a) and 1(b) can leave little doubt of the validity of the C.V.R. line concept. It is particularly to be noted that all points on the "dry" side of this line apply to samples which are showing continuing volume increase at "failure" in drained tests or continuing pore pressure decrease at "failure" in undrained tests, i.e. the loading paths are going *towards* the C.V.R. line. Failure must be interpreted here as the peak deviator stress as determined after the conventional area correction.

The lines in Fig. 2 of Dr Parry's letter afford a measure of the lack of completeness of the tests on Weald Clay and London Clay from the point of view of reaching the C.V.R. state. High values of $\frac{p_{cr}}{p_f}$ correspond to high values of n and they clearly show that such samples were not strained sufficiently to reach the C.V.R. line. We would like to suggest that it would have been slightly more logical to use $\frac{w_{cr}}{w_f}$ as abscissa in Fig. 2(a) instead of $\log \frac{p_{cr}}{p_f}$.

Dr Parry's conclusion that "the further evidence that the rate of volume change or pore pressure change at failure is a function of $\frac{p_{cr}}{p_f}$ suggests that under continuing shear strain after failure *all* samples, *regardless of the type of test*, will eventually lie on this (the C.V.R.) line, when the rate of volume change or pore pressure change will be zero" made very welcome reading and we fully endorse this view.

(C) HANS-U. SMOLTCZYK (*Géotechnique*, 8:4:186)

Dr Smoltczyk correctly comments that we have made an implicit assumption about the yield criterion in choosing our stress variables. It would have been appropriate for us to have worked in principal stress space if we had been studying the small discrepancy between an isotropic criterion such as that of Von Mises, and a shear stress criterion of the Tresca/Mohr/Coulomb type. It is not correct, however, that we have assumed the existence of the coefficient of internal friction ϕ (or ρ) which is implicit in the Mohr/Coulomb criterion. Our isotropic parameters (p, q) are derived without it. Other workers, such as Casagrande and Wilson (1953), do introduce this coefficient of internal friction but their parameters still have a similar character to ours and the general form of the curves of Fig. 9 of our Paper has already been shown in the excellent publications of these and other American workers. We chose to work in terms of the simplest possible isotropic parameters p and q since it was part of our intention to investigate the worth of soil constants such as ϕ.

We were not primarily concerned with the selection of *stress* variables. We wished to extend the excellent work of Hvorslev by plotting test results in such a way that both the changes in *voids ratio* and their importance could no longer be ignored. Since stresses alone do not suffice to define the behaviour of soils, it would seem more logical to include a strain variable as an axis. We have taken the voids ratio, or moisture content, or porosity, as this strain variable.

It is true that the choice of stress variables is important, but as yet it seems that worthwhile conclusions cannot be drawn, since the differences of the various yield criteria are small compared with the errors introduced by the present inability to measure voids ratios with precision.

Yours faithfully,

K. H. ROSCOE
A. N. SCHOFIELD
C. P. WROTH

University of Cambridge,
Department of Engineering.
25 March, 1959.

REFERENCES

BISHOP, A. W., 1953. "Pore pressure changes during shear in two undisturbed clays". *Proc. 3rd Int. Conf. Soil Mech.*, 1:94–99.

BISHOP, A. W., *and* HENKEL, D. J., 1957. "The measurement of soil properties in the triaxial test". *Arnold, Lond.*, pp. 115 and 116.

CASAGRANDE, A., 1936. "Characteristics of cohesionless soils affecting the stability of slopes and earth fills". *J. Boston Soc. civ. Engrs*, 23:13–32.

CASAGRANDE, A., *and* WILSON, S. D., 1953. "Prestress induced in consolidated-quick triaxial tests". *Proc. 3rd Int. Conf. Soil Mech.*, 1:106–110.

CUNDALL, P. A. & STRACK, O. D. L. (1979). *Géotechnique* **29**, No. 1, 47–65

A discrete numerical model for granular assemblies

P. A. CUNDALL* and O. D. L. STRACK†

The distinct element method is a numerical model capable of describing the mechanical behaviour of assemblies of discs and spheres. The method is based on the use of an explicit numerical scheme in which the interaction of the particles is monitored contact by contact and the motion of the particles modelled particle by particle. The main features of the distinct element method are described. The method is validated by comparing force vector plots obtained from the computer program BALL with the corresponding plots obtained from a photoelastic analysis. The photoelastic analysis used for the comparison is the one applied to an assembly of discs by De Josselin de Jong and Verruijt (1969). The force vector diagrams obtained numerically closely resemble those obtained photoelastically. It is concluded from this comparison that the distinct element method and the program BALL are valid tools for research into the behaviour of granular assemblies.

La méthode des éléments distincts est un modèle numérique capable de décrire le comportement mécanique de l'assemblage de disques et de sphères. La méthode est basée sur l'utilisation d'un système numérique explicite dans lequel l'interaction des particules est contrôlée contact par contact et le mouvement des particules simulé particule par particule. Les caracteristiques principales de la méthode des eléments distints sont décrites. La méthode est validée en comparant les tracés de vecteur de force obtenus par le programme sur ordicateur BALL avec les tracés correspondants obtanus à l'aide d'une analyse photo-élastique. L'analyse photo-élastique utilisée pour la comparaison est celle appliquée sur un assemblage de disques par De Josselin de Jong et Verruijt (1969). Les diagrammes de vecteur de force obtenus numériquement sont très voisins de ceux obtenus photo-élastiquement. Cette comparaison permet de conclure que la methode des éléments distincts et le programme BALL sont des instruments valables pour la recherche du comportement des assemblages granulaires.

INTRODUCTION

A granular medium is composed of distinct particles which displace independently from one another and interact only at contact points. The discrete character of the medium results in a complex behaviour under conditions of loading and unloading and to date no satisfactory constitutive relationships have been established. Extensive programmes of laboratory tests are necessary to generate new constitutive relations or to validate existing ones.

Interpretation of tests on real granular media, such as sand, is difficult because the stresses inside the sample cannot be measured and must be estimated from the boundary conditions. New experimental techniques using x-ray photography make measurements of strains possible but no such techniques exist for the determination of stresses within a sand sample.

The uncertainties regarding interior stresses in samples of sand have led to the development of models of granular media that are geometrically simpler than sand and for which the stresses and displacements can either be calculated or measured. These models consist of assemblies of discs or spheres and may be analytical, physical or numerical.

An analytical model for cubic arrays of spheres of uniform size was proposed by Deresiewicz (1958). The resulting expressions predict a non-linear and hysteretic stress–strain behaviour. Ultimate failure is also accommodated in the formulation. The analytic approach, however, is restricted in that the array is cubic, the spheres are uniform in size and the loading sequence is limited.

Discussion on this Paper closes 1 June, 1979. For further details see inside back cover.

* Principal Engineer, Advanced Technology Group, Dames & Moore, London.
† Assistant Professor, Department of Civil and Mineral Engineering, University of Minnesota, USA.

A testing technique for assemblies of discs that makes possible the direct determination of contact forces between the particles was proposed by Dantu (1957) and Wakabayashi (1957). These investigators suggested the use of optically sensitive material for the discs. Analysis of the force distribution in such a test was described by De Josselin de Jong and Verruijt (1969). Although testing of assemblies of photoelastic discs is quite general and allows for an accurate determination of contact forces and of displacements and rotations of the individual discs, the analysis is time-consuming. Drescher and De Josselin de Jong (1971) performed a series of such tests in order to verify De Josselin de Jong's double-sliding free-rotating model. The information extracted from these tests was sufficient to confirm the main features of the continuum model.

Possibly the most powerful way of modelling assemblies of discs and spheres is by numerical techniques. Numerical modelling is more flexible in application than analytical modelling and has the advantage over physical modelling that any data are accessible at any stage of the test. The flexibility of numerical modelling extends to loading configurations, particle sizes, size distributions and physical properties of the particles.

Serrano and Rodriguez-Ortiz (1973) and Rodriguez-Ortiz (1974) developed a numerical model for assemblies of discs and spheres. Contact forces and displacements are calculated for equilibrium conditions assuming that increments of contact forces are determined by incremental displacements of the particle centres. Herzian-type contact compliances are used for normal forces, the effects of tangential forces are considered according to the theories of Mindlin and Deresiewicz (1953) and Nayak (1972), and shape changes are negligible. A major drawback of the method used to solve the equations is that only a relatively small number of particles can be processed by the present computers. In addition, the calculation is time-consuming because the matrix representing the contact stiffnesses must be reformulated whenever a contact is made or broken.

The distinct element method is a numerical model that is capable of handling particles of any shape and was developed by Cundall (1971 and 1974) for the analysis of rock mechanics problems. In the distinct element method, the interaction of the particles is viewed as a transient problem with states of equilibrium developing whenever the internal forces balance. The numerical scheme is explicit and the consequent efficiency in the use of memory space is such that problems involving about 1500 particles can be accommodated in the core memory of a mini computer with a memory of 64 K of 32-bit words.

It is the main objective of the Paper to show that the distinct element method models the behaviour of assemblies of discs realistically. This is done by the independent numerical reproduction of a test on an assembly of discs reported by De Josselin de Jong and Verruijt (1969). A comparison between the numerical results and the results of the photoelastic disc tests indicates that the numerical model may be used to replace tests on photoelastic discs. The numerical model is proposed as a new tool for fundamental research into the behaviour of two-dimensional assemblies of discs.

THE DISTINCT ELEMENT METHOD

The principles on which the distinct element method is based and the nature of the fictitious material which may be modelled by this method are discussed below. The reader is referred to Cundall (1978) for a more detailed description of the method and of the computer program BALL developed for modelling two-dimensional assemblies of discs.

In the distinct element method, the equilibrium contact forces and displacements of a stressed assembly of discs are found through a series of calculations tracing the movements of the individual particles. These movements are the result of the propagation through the medium of disturbances originating at the boundaries: a dynamic process. The speed of propagation is a function of the physical properties of the discrete medium.

In describing the above dynamic behaviour numerically, time steps are taken over which velocities and accelerations are assumed to be constant. The distinct element method is based upon the idea that the time step chosen may be so small that during a single time step disturbances cannot propagate from any disc further than its immediate neighbours. Then, at all times the resultant forces on any disc are determined exclusively by its interaction with the discs with which it is in contact. It is this key feature of the distinct element method which makes it possible to follow the non-linear interaction of a large number of discs without excessive memory requirements or the need for an iterative procedure. It should be noted that all analyses presented in this Paper are for dry assemblies, and deal with total contact forces. A technique that has been devised for accommodating pore pressures will be described in a later publication.

The calculation cycle

The calculations performed in the distinct element method alternate between the application of Newton's second law to the discs and a force–displacement law at the contacts. Newton's second law gives the motion of a particle resulting from the forces acting on it. The force–displacement law is used to find contact forces from displacements.

The deformations of the individual particles are small in comparison with the deformation of a granular assembly as a whole. The latter deformation is due primarily to the movements of the particles as rigid bodies. Therefore, precise modelling of particle deformation is not necessary to obtain a good approximation of the mechanical behaviour. In the current version of BALL, the particles are allowed to overlap one another at contact points. This overlapping behaviour takes the place of the deformation of the individual particles. The magnitude of the overlap is related directly to the contact force in the way explained below. It should be noted, however, that these overlaps are small in relation to the particle sizes.

To illustrate how forces and displacements are determined during a calculation cycle, the case represented in Fig. 1(a) will be considered. Two weightless discs, labelled as disc x and disc y, are squashed between a pair of rigid walls. The walls move toward each other at a constant velocity v. Initially, at time $t = t_0$, the walls and discs are touching and no contact forces exist. A time Δt later, the walls have moved inward over distances $v\Delta t$. In accordance with the assumption that the disturbances cannot travel beyond a single disc during one time step, both discs are assumed to maintain their initial positions during the time interval from $t = t_0$ to $t = t_0+\Delta t$. Overlaps therefore exist at time $t_1 = t_0+\Delta t$ at contacts A and C (see Fig. 1(b)) and are of magnitude $\Delta n = v\,\Delta t$.

Points $A_{(D)}$ and $A_{(w)}$ (see Fig. 1(b)) are points of the disc and the wall, respectively, lying on the line drawn perpendicular to the wall and through the centre of the disc. The contact A is defined as the point halfway between $A_{(D)}$ and $A_{(w)}$. The relative displacement $(\Delta n_{(A)})_{t1}$ at the contact (the overlap) is defined as the displacement of point $A_{(w)}$ relative to that of point $A_{(D)}$ occurring over one time increment. The subscript $t1$ in $(\Delta n_{(A)})_{t1}$ refers to time.

The relative displacements occurring at contacts A and C at time $t_1 = t_0+\Delta t$ (see Fig. 1(b)) are used in a force–displacement law for the calculation of contact forces. In the present version of the computer program BALL, an incremental force–displacement law of the following form is used

$$\Delta F_n = k_n(\Delta n)_{t1} = k_n v\,\Delta t \quad . \quad . \quad . \quad . \quad . \quad . \quad . \quad . \quad (1)$$

where k_n is the normal stiffness and ΔF_n represents the increment in normal force.

Defining the postive 1 direction as pointing from disc x to disc y (see Fig. 1(b)) the sums of

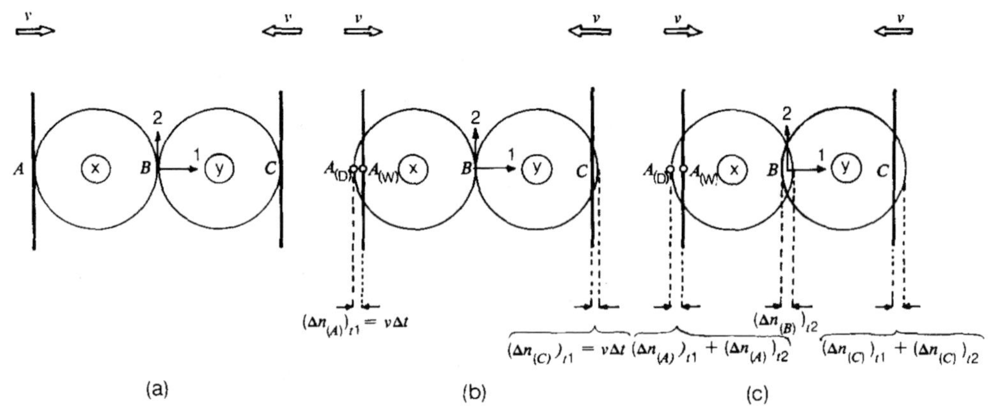

Fig. 1. Two discs compressed between rigid walls (the overlaps are exaggerated); (a) $t = t_0$: (b) $t = t_1 = t_0 + \Delta t$: (c) $t = t_2 = t_0 + 2\Delta t$

force $F_{(x)1}$ and $F_{(y)1}$ for discs x and y at time $t_1 = t_0 + \Delta t$ become, taking $F_{(x)1}$ and $F_{(y)1}$ to be positive in the positive 1 direction,

$$F_{(x)1} = k_n(\Delta n)_{t1}, \quad F_{(y)1} = -k_n(\Delta n)_{t1} \quad \ldots \ldots \quad (2)$$

These forces are used to find new accelerations using Newton's second law

$$\ddot{x}_1 = F_{(x)1}/m_{(x)}, \quad \ddot{y}_1 = F_{(y)1}/m_{(y)} \quad \ldots \ldots \quad (3)$$

where $\ddot{x}_1$ and $\ddot{y}_1$ stand for the respective accelerations of discs x and y in the 1 direction and where the subscripts (x) and (y) in the masses $m_{(x)}$ and $m_{(y)}$ refer to discs x and y. The accelerations found from equation (3) are assumed to be constant over the time interval from $t_1 = t_0 + \Delta t$ to $t_2 = t_0 + 2\Delta t$ and may be integrated to yield velocities

$$[\dot{x}_1]_{t2} = [F_{(x)1}/m_{(x)}]\Delta t, \quad [\dot{y}_1]_{t2} = [F_{(y)1}/m_{(y)}]\Delta t \quad \ldots \quad (4)$$

The relative displacement increments at contacts A, B and C at time $t_2 = t_0 + 2\Delta t$ are found from

$$(\Delta n_{(A)})_{t2} = (v - [F_{(x)1}/m_{(x)}]\Delta t)\Delta t \quad \ldots \ldots \quad (5)$$

$$(\Delta n_{(B)})_{t2} = ([F_{(x)1}/m_{(x)}]\Delta t - [F_{(y)1}/m_{(y)}]\Delta t)\Delta t \quad \ldots \quad (6)$$

$$(\Delta n_{(C)})_{t2} = ([F_{(y)1}/m_{(y)}]\Delta t - [-v])\Delta t \quad \ldots \ldots \quad (7)$$

where $\Delta n_{(A)}$, $\Delta n_{(B)}$ and $\Delta n_{(C)}$ are taken as positive for compression.

This cycle may be repeated again and again: forces corresponding to the displacements are found using the force displacement law, equation (2), and the force sums for the two discs are substituted in Newton's second law, equation (3), to obtain displacements.

The above example is an illustration of the cycling through a force–displacement law and the law of motion. In the general case of an assembly of many discs, the force–displacement law is applied at each contact of any disc and the vectorial sum of these contact forces is determined to yield the resultant force acting on that disc. When this has been accomplished for every disc, new accelerations are calculated from Newton's second law. The equations used in a calculation cycle for a general assembly are presented in the following section.

The force–displacement law

The force–displacement law will be presented for the case of two discs in contact, discs x and

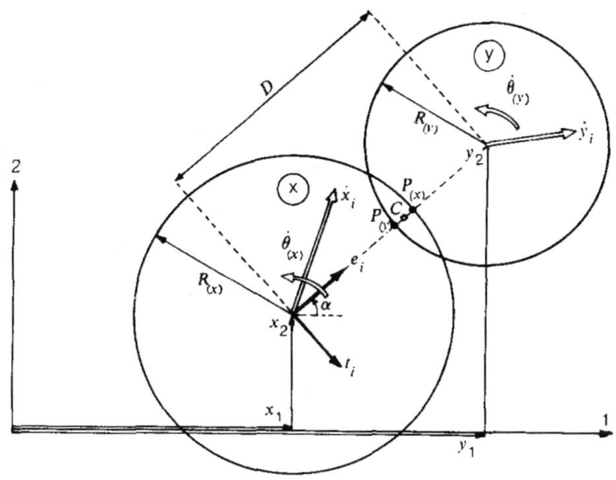

Fig. 2. The force–displacement law

y in Fig. 2. The co-ordinates of the disc centres are represented as $x_i = (x_1, x_2)$ and $y_i = (y_1, y_2)$ where the indices 1 and 2 refer to the co-ordinates of a Cartesian co-ordinate system as indicated in Fig. 2. The components of the velocity vectors of discs x and y are $\dot{x}_i = (\dot{x}_1, \dot{x}_2)$, $\dot{y}_i = (\dot{y}_1, \dot{y}_2)$ and the angular velocities are $\dot{\theta}_{(x)}$ and $\dot{\theta}_{(y)}$, taken positive in counterclockwise direction, where the dots stand for differentiation with respect to time. Discs x and y have radii $R_{(x)}$ and $R_{(y)}$ and masses $m_{(x)}$ and $m_{(y)}$. Points $P_{(x)}$ and $P_{(y)}$ are defined as the points of intersection of the line connecting the disc centres with the boundaries of discs x and y, respectively.

Two discs are taken to be in contact only if the distance D between their centres is less than the sum of their radii, i.e. if

$$D < R_{(x)} + R_{(y)} \quad . \quad . \quad . \quad . \quad . \quad . \quad . \quad . \quad . \quad . \quad . \qquad (8)$$

If this condition is met, the relative displacement at the contact C (see Fig. 2) is determined by integration of the relative velocity. The relative velocity at the contact is defined as the velocity of point $P_{(x)}$ with respect to $P_{(y)}$. The unit vector $e_i = (\cos\alpha, \sin\alpha)$ (see Fig. 2) is introduced as pointing from the centre of disc x to the centre of disc y, i.e.

$$e_i = \frac{y_i - x_i}{D} = (\cos\alpha, \sin\alpha) \quad . \quad . \quad . \quad . \quad . \quad . \quad . \quad . \quad . \qquad (9)$$

and the unit vector t_i is obtained by a clockwise rotation of e_i through 90°, i.e.

$$t_i = (e_2, -e_1) \quad . \quad . \quad . \quad . \quad . \quad . \quad . \quad . \quad . \quad . \quad . \qquad (10)$$

The relative velocity of point $P_{(x)}$ with recpect to $P_{(y)}$ now may be expressed as $\dot{X}_i$ with

$$\dot{X}_i = (\dot{x}_i - \dot{y}_i) - (\dot{\theta}_{(x)} R_{(x)} + \dot{\theta}_{(y)} R_{(y)}) t_i \quad . \quad . \quad . \quad . \quad . \quad . \qquad (11)$$

The normal ($\dot{n}$) and tangential ($\dot{s}$) components of the relative velocities are the projections of $\dot{X}_i$ onto e_i and t_i, respectively,

$$\dot{n} = \dot{X}_i e_i = (\dot{x}_i - \dot{y}_i) e_i - (\dot{\theta}_{(x)} R_{(x)} + \dot{\theta}_{(y)} R_{(y)}) t_i e_i = (\dot{x}_i - \dot{y}_i) e_i \quad . \quad . \quad . \qquad (12)$$

and

$$\dot{s} = \dot{X}_i t_i = (\dot{x}_i - \dot{y}_i) t_i - (\dot{\theta}_{(x)} R_{(x)} + \dot{\theta}_{(y)} R_{(y)}) t_i t_i = (\dot{x}_i - \dot{y}_i) t_i - (\dot{\theta}_{(x)} R_{(x)} + \dot{\theta}_{(y)} R_{(y)}) \quad . \qquad (13)$$

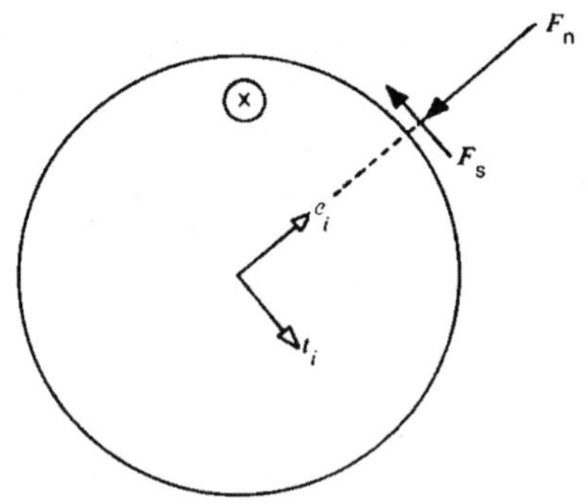

Fig. 3. Sign convention for F_n and F_s

where the Einstein summation convention is adopted for the index i; i.e. the summation is performed over the index that occurs twice in each expression (indices in parentheses are not summed). Integration of the relative velocity component with respect to time gives the components Δn and Δs of the relative displacement increment

$$\Delta n = (\dot{n})\Delta t = \{(\dot{x}_i - \dot{y}_i)e_i\}\Delta t \quad . \; . \; . \; . \; . \; . \; . \; . \; . \qquad (14)$$

and

$$\Delta s = (\dot{s})\Delta t = \{(\dot{x}_i - \dot{y}_i)t_i - (\dot{\theta}_{(x)}R_{(x)} + \dot{\theta}_{(y)}R_{(y)})\Delta t \quad . \; . \; . \; . \; . \qquad (15)$$

These relative displacement increments are used with the force–displacement law to calculate increments of the normal and shear forces, ΔF_n and ΔF_s

$$\Delta F_n = k_n \, \Delta n = k_n\{(\dot{x}_i - \dot{y}_i)e_i\}\Delta t \quad . \; . \; . \; . \; . \; . \; . \; . \qquad (16)$$

and

$$\Delta F_s = k_s \, \Delta s = k_s\{(\dot{x}_i - \dot{y}_i)t_i - (\dot{\theta}_{(x)}R_{(x)} + \dot{\theta}_{(y)}R_{(y)})\}\Delta t \quad . \; . \; . \; . \qquad (17)$$

where k_n and k_s represent the normal and shear stiffnesses, respectively.

Finally, at each time step the force increments ΔF_n and ΔF_s are added into the sum of all force increments, F_n and F_s, determined for previous time steps:

$$(F_n)_N = (F_n)_{N-1} + \Delta F_n; \; (F_s)_N = (F_s)_{N-1} + \Delta F_s \quad . \; . \; . \; . \; . \qquad (18)$$

where the indices N and $N-1$ refer to times t_N and t_{N-1} such that $t_N - t_{N-1} = \Delta t$. The sign convention for the normal and shear forces acting on disc x is as indicated in Fig. 3: F_n and F_s are taken as positive in the directions opposite to e_i and t_i.

A Coulomb-type friction law is incorporated as follows. The magnitude of the shear force F_s found from equation (18) is checked against the maximum possible value $(F_s)_{max}$ defined as

$$(F_s)_{max} = F_n \tan \phi_\mu + c \quad . \; . \; . \; . \; . \; . \; . \; . \; . \qquad (19)$$

where ϕ_μ is the smaller of the interparticle friction angles of the two discs in contact and c the smaller of their cohesions.[1] If the absolute value of $(F_s)_N$ as found from equation (18) is larger than $(F_s)_{max}$, $(F_s)_N$ is set equal to $(F_s)_{max}$, preserving the sign obtained from equation (18).

[1] This scheme was used for programming convenience when running problems with only two different friction angles: one for particle-to-particle contact, the other for particle-to-wall contact. For problems involving many dissimilar particles the program could be modified quite easily to associate each value of ϕ_μ and c with a particular type of contact rather than with a particular type of particle.

Once the normal and shear forces have been determined for each contact of a disc, for example disc x, they are resolved into components in the 1 and 2 directions. The sum of these contact force components gives the resultant forces $\Sigma F_{(x)1}$ and $\Sigma F_{(x)2}$. The resultant moment acting on disc x, $\Sigma M_{(x)}$, is taken positive if acting in the counterclockwise direction and is found from $\Sigma M_{(x)} = \Sigma F_s R_{(x)}$ where summation is taken over all contacts of disc x.

The resultant moments and forces acting on disc x are used with Newton's second law to determine the new accelerations $\ddot{x}_i$ and $\ddot{\theta}_{(x)}$.

Motion

The velocities $\dot{x}_i$ and $\dot{\theta}_{(x)}$ used in the force–displacement law in equations (16) and (17) are obtained as follows. The current resultant force and the moment at time t_N are assumed to act on disc x during the interval Δt from $t_{N-\frac{1}{2}}$ to $t_{N+\frac{1}{2}}$. Newton's second law applied to disc x is

$$m_{(x)}\ddot{x}_i = \Sigma F_{(x)i} \quad . \; . \; . \; . \; . \; . \; . \; . \; . \quad (20)$$

$$I_{(x)}\ddot{\theta}_{(x)} = \Sigma M_{(x)} \quad . \; . \; . \; . \; . \; . \; . \; . \; . \quad (21)$$

where $I_{(x)}$ represents the moment of inertia of disc x. Taking $\ddot{x}_i$ and $\ddot{\theta}_{(x)}$ constant over the time step Δt, equations (20) and (21) lead to the following expressions for the velocities

$$(\dot{x}_i)_{N+\frac{1}{2}} = (\dot{x}_i)_{N-\frac{1}{2}} + [\Sigma F_{(x)i}/m_{(x)}]_N \,\Delta t \quad . \; . \; . \; . \; . \; . \; . \quad (22)$$

$$(\dot{\theta}_{(x)})_{N+\frac{1}{2}} = (\dot{\theta}_{(x)})_{N-\frac{1}{2}} + [\Sigma M_{(x)}/I_{(x)}]_N \,\Delta t \quad . \; . \; . \; . \; . \; . \quad (23)$$

These equations are applied to each disc in turn. These new values for the velocities now may be used in the force–displacement law and the cycle repeated for a new time increment.

The new values for velocities are used also to update the positions and rotations of the discs by a further numerical integration

$$(x_i)_{N+1} = (x_i)_N + (\dot{x}_i)_{N+\frac{1}{2}} \,\Delta t \quad . \; . \; . \; . \; . \; . \; . \; . \quad (22a)$$

$$(\theta_{(x)})_{N+1} = (\theta_{(x)})_N + (\dot{\theta}_{(x)})_{N+\frac{1}{2}} \,\Delta t \quad . \; . \; . \; . \; . \; . \; . \quad (23a)$$

The finite difference equations (22), (22a), (23), (23a) represent a time–centred system.

Body forces, such as gravitational forces, may be incorporated, if so desired. In that case a term $m_{(x)}g_i$ is added to the force sum $\Sigma F_{(x)i}$ in equation (22), where $g_i = (g_1, g_2)$ represents the two components of the acceleration vector due to the body force.

Damping

Friction damping occurs during sliding when the absolute value of the shear force at any contact is $(F_s)_{max}$.

The program BALL further contains the option to include two forms of viscous damping to be referred to as contact damping and global damping. Contact damping operates on the relative velocities at the contacts and may be envisioned as resulting from dashpots acting in the normal and shear directions at the contacts. The viscous damping in the shear direction is not applied when sliding occurs. In this case friction damping alone operates.

The coefficients of viscous contact damping in the normal and shear directions are represented by c_n and c_s. If contact damping is taken into account, the damping forces must be included in the force–sums in equations (22) and (23), which become

$$(\dot{x}_i)_{N+\frac{1}{2}} = (\dot{x}_i)_{N-\frac{1}{2}} + \{\Sigma\,[F_{(x)i} + D_{(x)i}]/m_{(x)}\}_N \,\Delta t \quad . \; . \; . \; . \; . \quad (24)$$

and

$$(\dot{\theta}_{(x)})_{N+\frac{1}{2}} = (\dot{\theta}_{(x)})_{N-\frac{1}{2}} + \{\Sigma M_{(x)}/I_{(x)}\}_N \,\Delta t \quad . \; . \; . \; . \; . \; . \quad (25)$$

where $\Sigma\, D_{(x)i}$ represents the sum of the components of the contact damping forces and where $\Sigma\, M_{(x)}$ now includes the contribution of the contact damping forces to the moment sum. The global components $D_{(x)i}$ are found from the normal (D_n) and shear (D_s) components of the damping force at the contacts, which are obtained from the following equations (see equations (12) and (13))

$$(D_n)_N = c_n \dot{n} = c_n[\dot{x}_i - \dot{y}_i]_{N-\frac{1}{2}}\, e_i \quad \ldots \quad (26)$$

and

$$(D_s)_N = c_s \dot{s} = c_s[(\dot{x}_i - \dot{y}_i)_{N-\frac{1}{2}} t_i - (\dot{\theta}_{(x)} R_{(x)} + \dot{\theta}_{(y)} R_{(y)})_{N-\frac{1}{2}}] \quad \ldots \quad (27)$$

There is an error of half a time-step in the calculation of D_n and D_s, but its effects have been found to be negligible.

In the current version of BALL, the contact damping coefficients c_n and c_s are taken to be proportional to the stiffnesses k_n and k_s with proportionality constant β,

$$c_n = \beta k_n, \quad c_s = \beta k_s \quad \ldots \quad (28)$$

Global damping operates on the absolute velocities of the discs and is introduced in the calculations of motion. Global damping as incorporated in BALL may be envisioned as the effect of dashpots connecting each particle to ground. These dashpots operate both on the velocity vector components and on the rotational velocity.

If global damping is included in addition to contact damping, the equations of motion, (20) and (21), become (cf. equations (24) and (25)),

$$m_{(x)} \ddot{x}_i = \Sigma\, [F_{(x)i} + D_{(x)i}] - C\dot{x}_i \quad \ldots \quad (29)$$

$$I_{(x)} \ddot{\theta}_{(x)} = \Sigma\, M_{(x)} - C^* \dot{\theta}_{(x)} \quad \ldots \quad (30)$$

where C and C^* are the coefficients of global damping operating, respectively, on $\dot{x}_i$ and $\dot{\theta}_{(x)}$. A central difference scheme is used to integrate equations (29) and (30), whereby velocities are evaluated halfway through the time step, i.e.

$$(\dot{x}_i)_N = \tfrac{1}{2}[(\dot{x}_i)_{N-\frac{1}{2}} + (\dot{x}_i)_{N+\frac{1}{2}}], \quad (\dot{\theta}_{(x)})_N = \tfrac{1}{2}[(\dot{\theta}_{(x)})_{N-\frac{1}{2}} + (\dot{\theta}_{(x)})_{N+\frac{1}{2}}] \quad \ldots \quad (31)$$

Writing $\ddot{x}_i$ as $[(\dot{x}_i)_{N+\frac{1}{2}} - (\dot{x}_i)_{N-\frac{1}{2}}]/\Delta t$ and $\ddot{\theta}_{(x)}$ as $[(\dot{\theta}_{(x)})_{N+\frac{1}{2}} - (\dot{\theta}_{(x)})_{N-\frac{1}{2}}]/\Delta t$ and using equation (31), equations (29) and (30) may be solved for $(\dot{x}_i)_{N+\frac{1}{2}}$ and $(\dot{\theta}_{(x)})_{N+\frac{1}{2}}$.

$$(\dot{x}_i)_{N+\frac{1}{2}} = \left\{(\dot{x}_i)_{N-\frac{1}{2}}\left[1 - \frac{C}{m_{(x)}}\frac{\Delta t}{2}\right] + \Sigma\,(F_{(x)i} + D_{(x)i})_N \frac{\Delta t}{m_{(x)}}\right\} \Big/ \left\{1 + \frac{C}{m_{(x)}}\frac{\Delta t}{2}\right\} \quad . \quad (32)$$

and

$$(\dot{\theta}_x)_{N+\frac{1}{2}} = \left\{(\dot{\theta}_{(x)})_{N-\frac{1}{2}}\left[1 - \frac{C^*}{I_{(x)}}\frac{\Delta t}{2}\right] + (\Sigma\, M_{(x)})_N \frac{\Delta t}{I_{(x)}}\right\} \Big/ \left\{1 + \frac{C^*}{I_{(x)}}\frac{\Delta t}{2}\right\} \quad . \quad . \quad (33)$$

In BALL, the coefficients C and C^* of global damping are taken to be proportional to mass and moment of inertia, respectively

$$C = \alpha m_{(x)}, \quad C^* = \alpha I_{(x)} \quad \ldots \quad (34)$$

Using equation (34), equations (32) and (33) may be simplified to

$$(\dot{x}_i)_{N+\frac{1}{2}} = \left\{(\dot{x}_i)_{N-\frac{1}{2}}\left[1 - \alpha\frac{\Delta t}{2}\right] + \Sigma\,(F_{(x)i} + D_{(x)i})_N \frac{\Delta t}{m_{(x)}}\right\} \Big/ \left\{1 + \alpha\frac{\Delta t}{2}\right\} \quad . \quad . \quad (35)$$

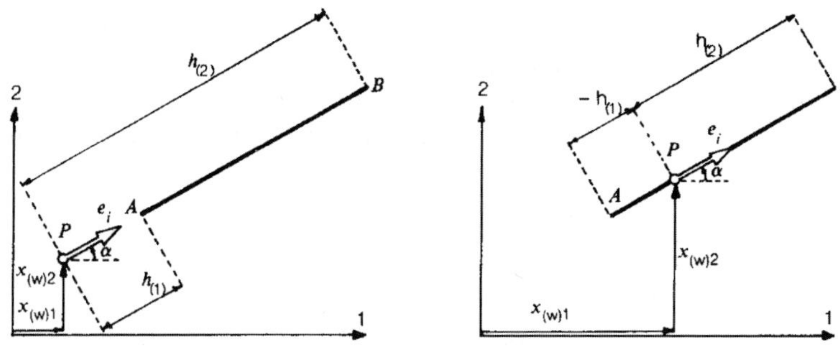

Fig. 4. Possible wall definitions

and

$$(\dot{\theta}_{(x)})_{N+\frac{1}{2}} = \left\{(\dot{\theta}_{(x)})_{N-\frac{1}{2}}\left[1-\alpha\frac{\Delta t}{2}\right]+(\Sigma\, M_{(x)})_N\frac{\Delta t}{I_{(x)}}\right\}\bigg/\left\{1+\alpha\frac{\Delta t}{2}\right\} \quad . \quad . \quad . \quad (36)$$

These equations for force–displacement and motion fully describe the BALL model of a granular material. Energy is dissipated in the model only through friction, contact and global damping. The use of damping other than by friction is necessary in order that the assemblies reach a state of equilibrium for all conditions. If neither contact nor global damping were included, the assemblies would not be able to reach exact equilibrium, but in practice the departure from equilibrium may be made as small as desired by reducing the applied loading rate. It may be noted that the linear force–displacement law discussed above is not inherent in the method. A non-linear law, such as that of Herzian contact, could equally well be employed.

The numerical scheme will be stable only if the time step Δt is taken as a fraction less than one of the critical time step, a result of the explicit nature of the model. An estimated value for the critical time step is determined in the program (see Cundall, 1978) and the user specifies the fraction of this time step that is to be used for the calculations. The critical time step is estimated on the basis of the single degree-of-freedom system of a mass m connected to ground by a spring of stiffness k, for which the critical time step equals $2\sqrt{(m/k)}$.

To illustrate how the program calculates and what effect variations in the physical properties specified by the user have on the output of the program, the results of some simple tests on nine discs are presented after a brief discussion of the input parameters.

Input parameters

The input parameters that must be specified to run numerical tests with BALL may be divided into two groups. One group consists of geometrical data and the other of physical properties data.

The geometrical data describe the positions and orientations of the straight rigid boundaries (the walls) and the positions and radii of the discs, with respect to a global co-ordinate system (1, 2). A typical wall is defined in terms of a fixed point P with co-ordinates $x_{(w)i} = (x_{(w)1}, x_{(w)2})$ (see Fig. 4) and a unit vector $e_i = (\cos\alpha, \sin\alpha)$ such that the wall is a portion of the line passing through P and of the direction of e_i. Scalar multipliers of e_i, $h_{(1)}$ and $h_{(2)}$ define the wall endpoints A and B with co-ordinates $x_{(A)i} = x_{(w)i}+h_{(1)}\, e_i$ and $x_{(B)i} = x_{(w)i}+h_{(2)}\, e_i$, respectively.

The boundary conditions are of the strain-controlled type. The motion of the wall is specified rather than the forces applied to it. Movement of the wall is defined in terms of the velocity of

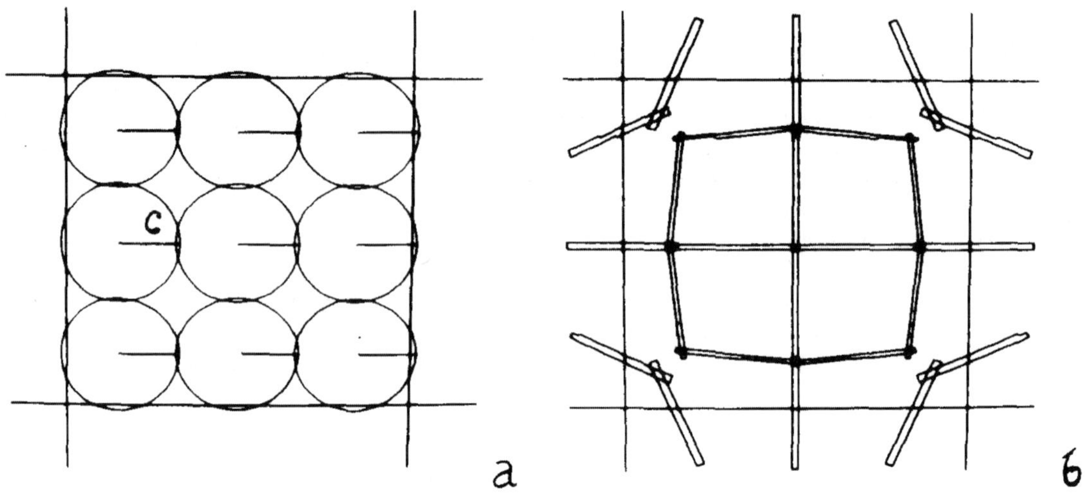

Fig. 5. Nine disc test

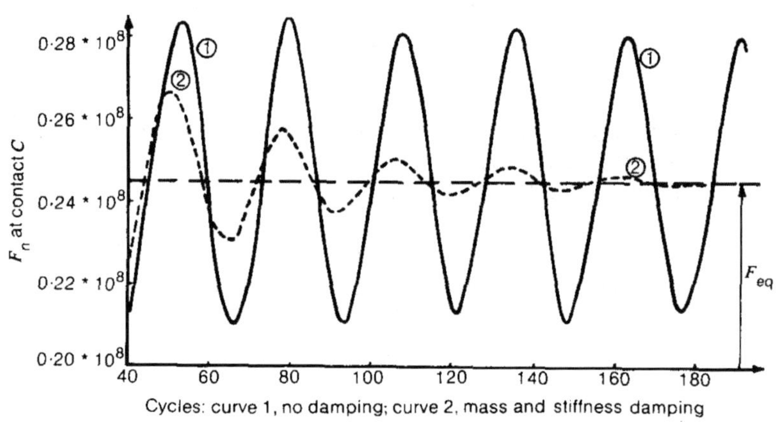

Fig. 6. Damping effects

point P, $\dot{x}_{(w)i} = (\dot{x}_{(w)1}, \dot{x}_{(w)2})$ and the angular velocity of the wall about P, $\dot{\alpha}$. Defining a disc requires only that the disc centre and radius be specified.

For the purpose of assigning physical properties to the discs and walls, they are divided into groups such that all entries of a group have the same properties. Assigning a single type number to each disc or wall is then sufficient to associate with it the appropriate set of properties. The physical properties treated in this way include disc radius, density, cohesion, interparticle friction coefficient tan ϕ_μ, shear stiffness and normal stiffness (see footnote 1). There are other parameters that apply to the field as a whole, defining the contact damping, global damping and the fraction of the critical time step.

Units

The numerical values for density, stiffness and damping coefficients have been selected such that the overlaps are small in relation to the particle sizes, that the numerical process is stable and that the numerical values calculated in the program are in a range that can be handled with accuracy. No attempts have been made to relate the units used in BALL to physical units in the examples presented here; however, if the physical quantities are known, they may be used in BALL. To interpret the output of BALL, contact forces are displayed graphically. This is done by plotting rectangles through the contacts and with the orientation of the contact forces (see Fig. 5(b)). The lengths of the rectangles are equal to the sum of the radii of the discs in contact and the widths are proportional to the magnitudes of the contact forces. Displays produced in this way resemble the force diagrams obtained from tests on photoelastic discs (see De Josselin de Jong and Verruijt, 1969).

The values of two parameters are chosen with a regard for the properties of physical discs: the interparticle friction angle, ϕ_μ, and the ratio of shear and normal stiffnesses, k_s/k_n. According to analytical results obtained by Mindlin (1949), the ratio k_s/k_n for elastic bodies in contact with elliptical contact areas is in a range from $\frac{2}{3}$ to 1.[2] On the basis of this result, the values of k_s/k_n have been restricted to this range in the tests.

NINE DISC TEST

A test on a regular loosely packed assembly of nine discs is performed as an example. The radii of the discs are 50 units and the density of the disc material is 1000 units. The assembly is enclosed in a square formed by four rigid walls, see Fig. 5(a). The purpose of the test is to illustrate how contact forces develop with time as a function of input parameters and loading. The display of contact forces obtained by moving the walls towards each other uniformly is reproduced in Fig. 5(b).

The normal force at contact C of Fig. 5(a) is selected for a closer examination. The development of this force for various tests is discussed below, firstly for volume decrease without distortion and secondly, for distortion.

Undistorted compression

The boundary conditions for the tests on undistorted compression are that the walls move inward at a constant velocity until 40 calculation cycles have been completed. The wall movement is stopped at this point, but the test is continued, to observe subsequent behaviour of the contact force.

Damping effects

The effect of damping is shown in Fig. 6 where the contact force at C is plotted against the calculation cycles. The plot starts at 40 cycles, the cycle count corresponding to the termination of the wall movement. Curve 1 in Fig. 6 is associated with a case in which neither global nor contact damping operates. The force oscillates about the equilibrium value F_{eq} with an amplitude of approximately $0{\cdot}15F_{eq}$. Curve 2 in Fig. 6 applies to a case in which both contact and global damping are used. Similar curves are obtained if contact and global damping are used separately. It appears from the figures that the dampling does not affect the equilibrium value of the force. It does, however, reduce the number of calculation cycles needed to reach equilibrium. An interesting feature of the test illustrated in Fig. 5 is that shear forces appear along the walls as a result of relative movement between discs and walls, even though the walls move towards each other uniformly.

Speed of loading

The effect of the speed of loading is illustrated in Fig. 7 for two different stiffnesses. Figure 7(a) corresponds to the cases that the normal, k_n, and shear, k_s, stiffnesses are both $1{\cdot}35*10^9$.

---

[2] This range applies if the shear force is in the direction of the minimum diameter of the contact area.

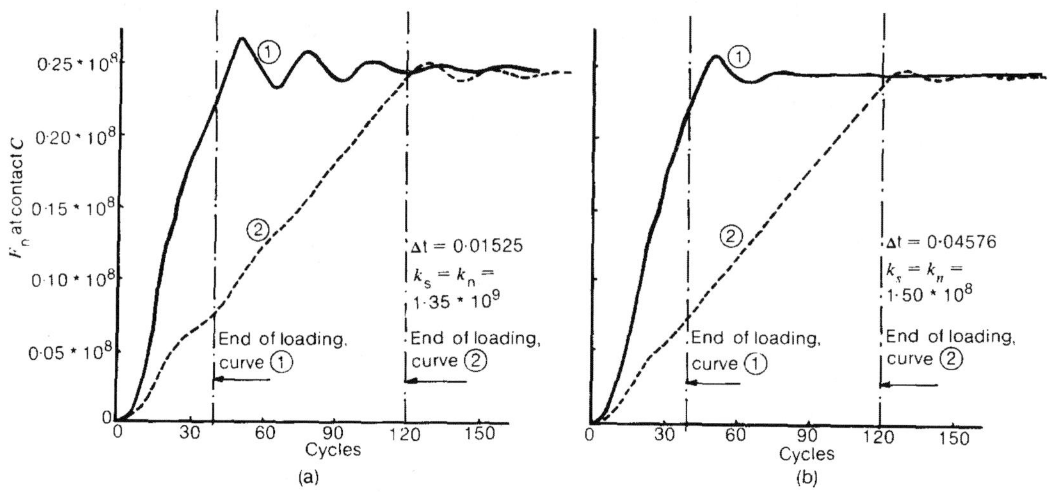

Fig. 7. (a) and (b) Speed of loading

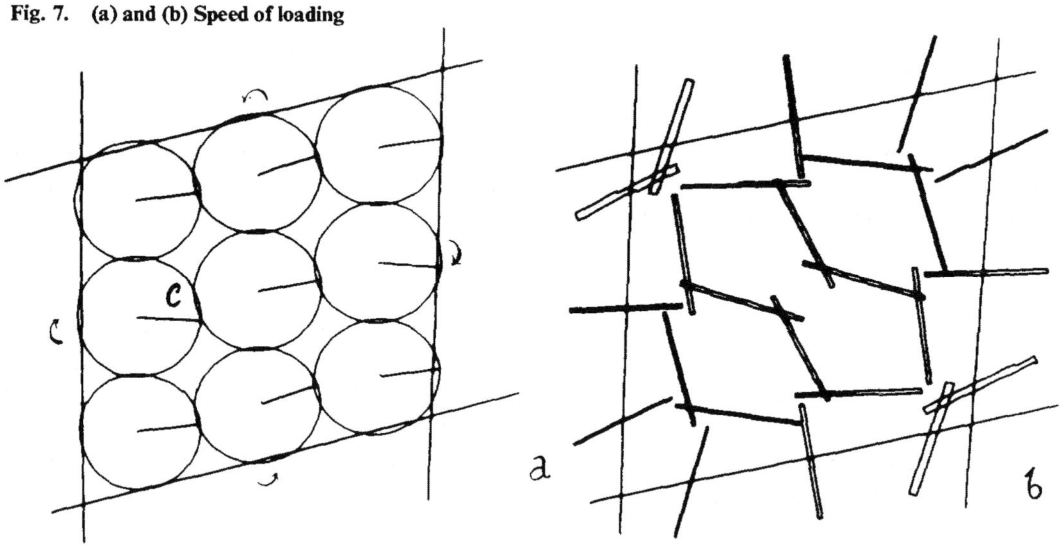

Fig. 8. Constant volume strain

Curves 1 and 2 apply to wall velocities of 0·12 and 0·04, respectively. The deformations for the two tests have been chosen such that about the same equilibrium force results; the number of calculation cycles at the end of loading for the wall velocity of 0·04 is three times that for the wall velocity of 0·12. It is seen that the amplitude of the oscillations occurring at the end of loading is much larger for curve 1 (high speed) than for curve 2 (low speed), whereas the frequencies are about the same. Thus, the amplitude of the oscillations occurring at the end of loading increases with the speed of loading. The same effect is observed in the graphs of Fig. 7(b), corresponding to $k_n = k_s = 1{\cdot}50*10^8$. Curve 1 applies to a loading speed of 0·36 and curve 2 to a loading speed of 0·12. The loading has been continued up to a compression that yields the same equilibrium force as for the low-stiffness case of Fig. 7(a).

The oscillations are smaller for a low speed than for a high speed of loading because other accelerations induced in the particles decrease with decreasing speed of loading. If the inertial

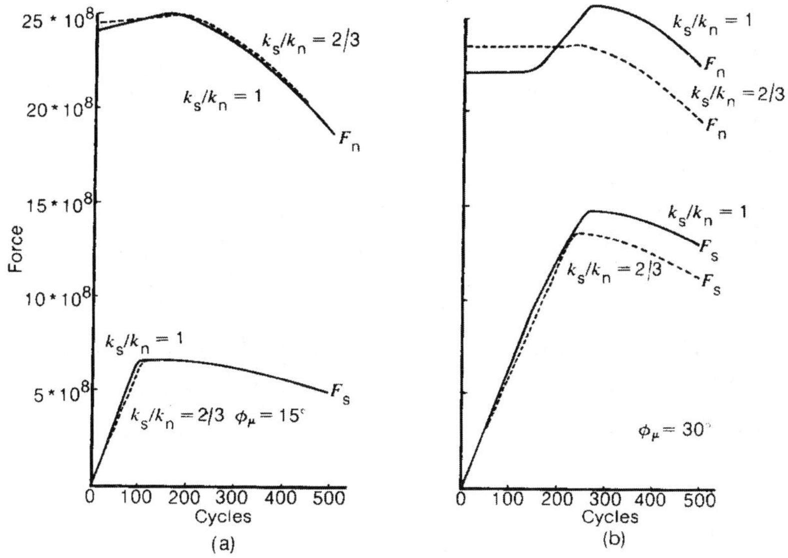

Fig. 9. The effect of k_s/k_n and ϕ_μ on the contact force at c

forces are kept low compared to the contact forces, the system will remain close to equilibrium at all times. For a more detailed discussion of the effects of the various forms of damping and of the speed of loading on the behaviour of the assembly see Cundall (1978).

Distortion

To illustrate the effects of the ratio k_s/k_n and the friction angle ϕ_μ on the contact force at point *C*, the compressed assembly is subjected to distortion, at constant volume. This is done by specifying a constant angular velocity about the wall centres, as indicated in Fig. 8(a). Both global and contact damping are applied and the speed of loading is kept low to ensure that the assembly remains close to a state of equilibrium at all times. The contact forces generated in the assembly are reproduced in Fig. 8(b).

Shear to normal stiffness ratio and friction angle

The effects of the ratio k_s/k_n and the friction angle ϕ_μ on the normal (F_n) and shear (F_s) forces at contact *C* are illustrated in Figs 9(a) and (b), corresponding to 15° and 30° for ϕ_μ, respectively. The normal and shear forces are plotted against the calculation cycles and the value for k_n for all curves is $1{\cdot}35*10^9$, i.e. the ratio k_s/k_n is changed by varying k_s only. The curves for $k_s/k_n = \frac{2}{3}$ and $k_s/k_n = 1$ nearly coincide for the case that $\phi_\mu = 15°$ (see Fig. 9(a)). The explanation that the influence of the value of k_s/k_n upon F_n and F_s for small friction angles is minor may be that slip sets in early and controls the shear force for a major portion of the test; as soon as slip sets in, the shear force is determined solely by the normal force. For the case of Fig. 9(b) however, where $\phi_\mu = 30°$, the shear stiffness controls the behaviour over a longer portion of the test and the curves for $k_s/k_n = \frac{2}{3}$ and for $k_s/k_n = 1$ are seen to deviate significantly. For the case of the high shear stiffness, $k_s/k_n = 1$, the value for F_n resulting from undistorted compression is less than for the case that $k_s/k_n = \frac{2}{3}$ (see Fig. 9(b)). Even for undistorted compression, shear stresses develop at the boundaries as a result of movement of discs along the walls. These

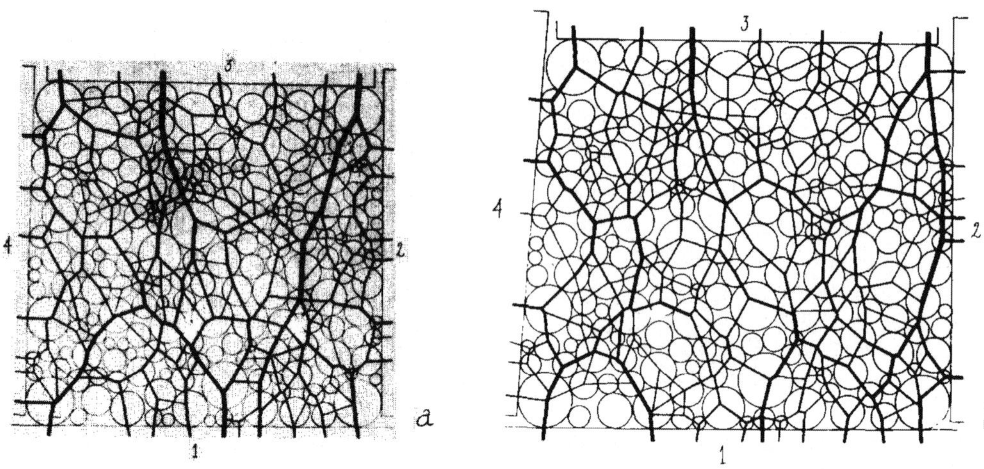

Fig. 10. Force vector plots obtained by De Josselin de Jong and Verruijt (1969); (a) Stage A, ratio $F_H/F_v = 0{\cdot}39$: (b) Stage B, ratio $F_H/F_v = 0{\cdot}33$

shear stresses carry part of the load and are higher, the higher the ratio of k_s to k_n. This explains the difference in initial values for F_n. However, as shearing proceeds, the normal force for the case that $k_s/k_n = 1$ increases beyond the one for $k_s/k_n = \frac{2}{3}$ to reach a maximum of about $25{\cdot}6{*}10^8$ after which slip sets in and the force falls back.

It follows from the above results that the parameters k_s/k_n and ϕ_μ have a considerable influence upon the behaviour of the assembly. Therefore, care has been taken in estimating friction angles for the comparison against results of tests on photoelastic discs. In addition, the ratio k_s/k_n has been varied between $\frac{2}{3}$ and 1 and tests with different friction angles have been performed.

In addition to the experiments discussed above as illustrative examples, experiments were performed to investigate the effect of changes in the values of k_s and k_n, keeping the ratio k_s/k_n constant, upon the behaviour of the assemblies. It appears that assemblies of very stiff particles behave quite differently from assemblies of soft particles. This may be understood physically: a material of particles that deform easily (i.e. particles with a small value for k_n) is expected to respond in a different way to loading than a material of stiff particles (e.g. sand).

It should be noted that the force–displacement law currently used in BALL is a rather simple one. Investigation into the effect of changes made in these laws upon the behaviour of the assembly will indicate whether it will be necessary to refine the model. However, the results of the verification tests discussed below, qualitative as they are, indicate that the model is quite realistic in its present form.

VERIFICATION TEST

An attempt is made to duplicate numerically the test on photoelastic discs described by De Josselin de Jong and Verruijt (1969). The assembly of discs was placed between two vertical Plexiglas plates and was loaded by means of two horizontal and two vertical rigid beams which formed the side boundaries. The horizontal beams were subjected to vertical forces and the vertical beams to horizontal forces. All beams were free to rotate. De Josselin de Jong and Verruijt present figures with disc locations and contact forces for two stages of the test, referred to in this Paper as stages A and B. These figures are reproduced as Figs 10(a) and (b). At stage A the ratio F_H/F_v of the forces applied in the horizontal (F_H) and vertical (F_v) directions was

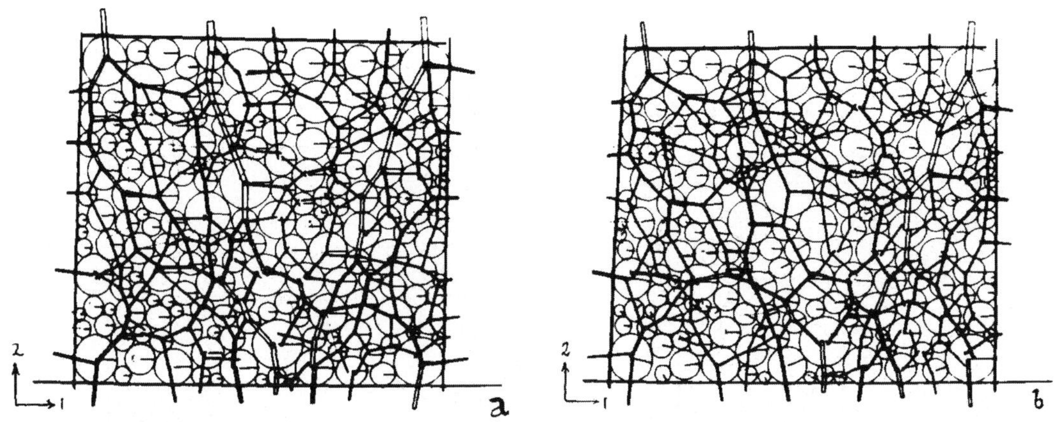

Fig. 11. **Numerical reproduction of Fig. 10, with data from Table 1; (a) Stage A, ratio $F_H/F_v = 0{\cdot}43$; (b) Stage B, ratio $F_H/F_v = 0{\cdot}40$**

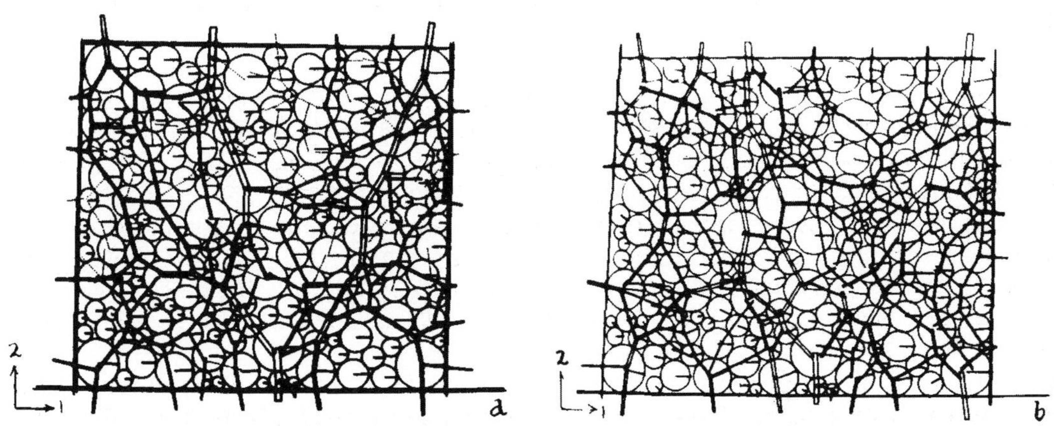

Fig. 12. **Numerical reproduction of Fig. 10, with data from Table 2; (a) Stage A, ratio $F_H/F_v = 0{\cdot}41$: (b) Stage B, ratio $F_H/F_v = 0{\cdot}40$**

0·39. The transition from stage A to stage B was made by increasing the vertical force by 4·348 % while decreasing the horizontal force by the same amount, yielding a ratio F_H/F_v of 0·33.

The purpose of the verification is to compare force vector plots produced by BALL with those presented by De Josselin de Jong and Verruijt. The verification is primarily qualitative, and, in interpreting the results, heavy emphasis is placed upon visual comparison.

The discs' centre locations and radii were obtained from Fig. 10(a) by use of a digitizer. The centre points of the discs, marked by cross-hairs in the original test, were located with reference to the grid in the background of the photograph. It may be noted that use of the above procedure results in an approximate geometry of the assembly as errors are caused by occasional invisibility of the cross-hairs and by distortions in the figure.

The density used for the discs is taken as 2000 units and the sizes of the discs are as follows: there are 6 discs of a radius of 40, 7 of 35, 16 of 30, 33 of 25, 33 of 20, 36 of 18, 28 of 15 and 38 of 10 units.

The disc-to-disc and disc-to-wall friction angles are determined by measuring the largest obliquity of the contact forces. This is done under the assumption that the largest obliquity

Table 1. Data for Fig. 11. The wall numbers are as labelled in Fig. 10

Stage	A				B			
Wall No.	1	2	3	4	1	2	3	4
F_1 (→)	$+0{\cdot}2836*10^8$	$-0{\cdot}3040*10^9$	$+0{\cdot}3790*10^8$	$+0{\cdot}2386*10^9$	$-0{\cdot}6240*10^8$	$-0{\cdot}8445*10^9$	$+0{\cdot}2848*10^8$	$+0{\cdot}8770*10^9$
F_2 (↑)	$+0{\cdot}6337*10^9$	$-0{\cdot}1843*10^7$	$-0{\cdot}6260*10^9$	$-0{\cdot}8954*10^7$	$+0{\cdot}2111*10^{10}$	$+0{\cdot}5217*10^8$	$-0{\cdot}2169*10^{10}$	$+0{\cdot}1092*10^6$
$(e/\frac{1}{2}L)*100\%$[1]	7·95%	1·52%	5·25%	0·84%	14·5%	4·88%	16·5%	4·19%
$(\Delta F/F)*100\%$[2]	$(0{\cdot}1872*10^7/0{\cdot}2009*10^9)*100\% = 0{\cdot}9\%$				$(0{\cdot}5872*10^6/0{\cdot}7365*10^9)*100\% = 0{\cdot}1\%$			

Key: $k_n = 1{\cdot}5*10^9$, $k_s = 1{\cdot}0*10^9$, $(\tan\phi_\mu)_{dd} = 0{\cdot}45$, $(\tan\phi_\mu)_{dw} = 0{\cdot}17$.

[1] $e/(\frac{1}{2}L)$ represents the eccentricity of the force acting normal to the wall divided by half the wall length.

[2] $\Delta F/F$ represents the ratio of the largest force-sum on any disc to the largest contact force.

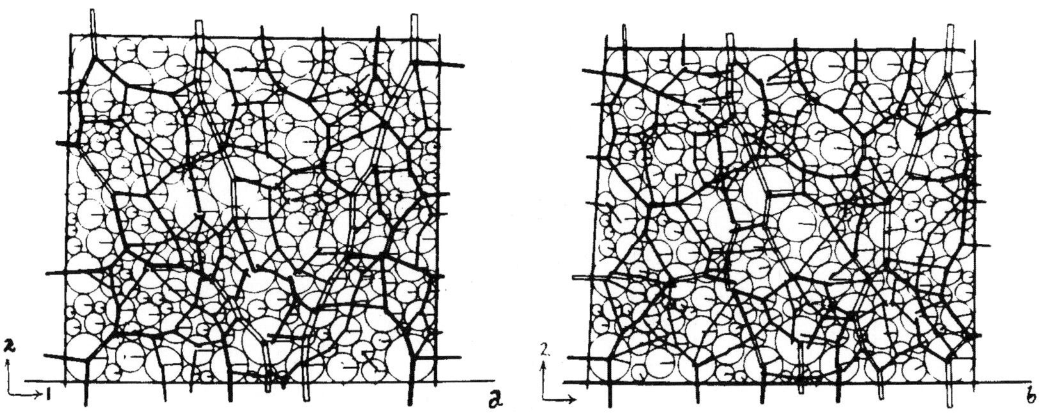

Fig. 13. Numerical reproduction of Fig. 10, with data from Table 3; (a) Stage A, ratio $F_H/F_v = 0{\cdot}46$; (b) Stage B, ratio $F_H/F_v = 0{\cdot}47$

observed equals the maximum possible obliquity. The resulting value for all disc-to-disc contacts is $(\tan \phi_\mu)_{dd} = 0{\cdot}45$ and the value for all disc-to-wall contacts is $(\tan \phi_\mu)_{dw} = 0{\cdot}17$.

A difficulty in reproducing the photoelastic disc test is that the geometry of the assembly and the wall locations as they existed before loading are unknown. Therefore, the geometry of Fig. 10(a) is used as the initial configuration. The four walls are translated inward in such a way that the ratio of vertical and horizontal resultant wall forces approximates to the original value. The transition from stage A to stage B is made by moving the four walls to the positions indicated in Fig. 10(b). It may be noted that the above procedure differs from the one used for the photoelastic analysis in two respects. Firstly, the initial conditions deviate somewhat from those of the original test and secondly, the numerical test is strain-controlled whereas the original is stress-controlled.

The vector plot and disc locations obtained numerically for stages A and B are reproduced in Figs 11(a) and (b) for the case that $(\tan \phi_\mu)_{dd} = 0{\cdot}45$, $(\tan \phi_\mu)_{dw} = 0{\cdot}17$ and $k_s/k_n = \frac{2}{3}$. Keeping in mind the approximations made, the resemblance is good. The boundary forces obtained numerically are listed in Table 1. The ratios of horizontal over vertical forces deviate somewhat from the original ones as a result of the discrepancy in boundary and initial conditions. The boundary forces are given in terms of components F_1 and F_2, taken positive in the directions of the 1 and 2 axes indicated in the figure.

To investigate the effect of the ratio k_s/k_n upon the force vector plots, the test is repeated for the case that $k_s = k_n$. The force vector plots and disc locations are reproduced in Figs 12(a) and (b).

Finally, the results are presented in Figs 13(a) and (b) of a test with friction angles that are slightly less than for the above test: $(\tan \phi_\mu)_{dd} = 0{\cdot}40$, $(\tan \phi_\mu)_{dw} = 0{\cdot}08$. The shear and normal stiffnesses are both taken as $1{\cdot}5*10^9$. The boundary forces and moments are listed in Table 3.

CONCLUSION

The above verification is primarily qualitative and therefore subjective. Nonetheless, the correspondence between numerically and photoelastically obtained force vector plots is sufficiently good to conclude that the distinct element method is a valid tool for fundamental research into the behaviour of granular assemblies.

Table 2. Data for Fig. 12

Stage	A				B			
Wall No.	1	2	3	4	1	2	3	4
F_1 (→)	$-0{\cdot}2093*10^7$	$-0{\cdot}2981*10^8$	$-0{\cdot}1469*10^7$	$+0{\cdot}2962*10^8$	$-0{\cdot}5063*10^8$	$-0{\cdot}5536*10^9$	$-0{\cdot}2592*10^7$	$+0{\cdot}6039*10^9$
F_2 (↑)	$+0{\cdot}7372*10^8$	$-0{\cdot}1542*10^6$	$-0{\cdot}6987*10^8$	$-0{\cdot}1493*10^7$	$+0{\cdot}1437*10^{10}$	$+0{\cdot}2766*10^8$	$-0{\cdot}1446*10^{10}$	$+0{\cdot}1809*10^8$
$(e/\frac{1}{2}L)*100\%$	5·90%	6·60%	1·44%	0·06%	4·63%	2·72%	9·03%	1·03%
$(\Delta F/F)*100\%$	$(0{\cdot}7462*10^6/0{\cdot}2835*10^8)*100\% = 2{\cdot}6\%$				$(0{\cdot}4494*10^6/0{\cdot}4136*10^9)*100\% = 0{\cdot}1\%$			

Key: $k_n = k_s = 1{\cdot}5*10^9$, $(\tan\phi_\mu)_{dd} = 0{\cdot}45$, $(\tan\phi_\mu)_{dw} = 0{\cdot}17$.

Table 3. Data for Fig. 13

Stage	A				B			
Wall No.	1	2	3	4	1	2	3	4
F_1 (→)	$+0{\cdot}1774*10^7$	$-0{\cdot}1468*10^9$	$+0{\cdot}8015*10^7$	$+0{\cdot}1254*10^9$	$-0{\cdot}3588*10^8$	$-0{\cdot}8002*10^9$	$+0{\cdot}8067*10^7$	$+0{\cdot}8182*10^9$
F_2 (↑)	$+0{\cdot}2963*10^9$	$-0{\cdot}4272*10^6$	$-0{\cdot}2972*10^9$	$-0{\cdot}4042*10^7$	$+0{\cdot}1725*10^{10}$	$+0{\cdot}1003*10^8$	$-0{\cdot}1694*10^{10}$	$-0{\cdot}4016*10^8$
$(e/\frac{1}{2}L)*100\%$	12·79%	5·22%	10·92%	0·66%	13·26%	5·29%	13·23%	3·54%
$(\Delta F/F)*100\%$	$(0{\cdot}1617*10^7/0{\cdot}9375*10^8)*100\% = 1{\cdot}7\%$				$(0{\cdot}2902*10^7/0{\cdot}5037*10^9)*100\% = 0{\cdot}6\%$			

Key: $k_n = k_s = 1{\cdot}5*10^9$, $(\tan\phi_\mu)_{dd} = 0{\cdot}40$, $(\tan\phi_\mu)_{dw} = 0{\cdot}08$.

ACKNOWLEDGEMENTS

The Authors appreciate the valuable assistance of Mr John Christenson who skilfully performed the numerical experiments and assisted in preparing this Paper, and thank Mike Voegele for producing the photographs.

The research work reported in this Paper and the development of the program BALL has been supported by a grant from the Program for Structural, Materials and Geotechnical Engineering of the National Science Foundation. Also, much support has been given by Dames & Moore in the form of Research and Development funding, administered by the Advanced Technology Group.

The Interdata minicomputer dedicated to this project and used for the numerical experiments has been purchased on an equipment grant from the above Program of the National Science Foundation. Peripheral computer interactive graphics equipment has been obtained on grants from the Office of the Vice-President for Academic Affairs, the Institute of Technology and the Graduate School of the University of Minnesota.

REFERENCES

Cundall, P. A. (1971). A computer model for simulating progressive, large-scale movements in blocky rock systems. *Proc. Symp. Int. Soc. Rock Mech., Nancy* **2,** No. 8.

Cundall, P. A. (1974). *A computer model for rock-mass behavior using interactive graphics for the input and output of geometrical data.* Report MRD–2–74 prepared at the University of Minnesota under Contract number DACW 45–74–C–006, for the Missouri River Division, US Army Corps of Engineers, available from National Technical Information Service—report number AD/A–001 602.

Cundall, P. A. (1978). *BALL—A program to model granular media using the distinct element method.* Technical Note, Advanced Technology Group, Dames & Moore, London.

Dantu, P. (1957). Contribution á l'étude mécanique et géomtrique des milieux pulvérulents. *Proc. 4th Int. Conf. Soil Mech. Fdn Engng, London* **1,** pp. 144.

De Josselin de Jong, G. & Verruijt, A. (1969). Etude photo-élastique d'un empilement de disques. *Cah. Grpe fr. Etud. Rhéol.* **2,** 73–86.

Deresiewicz, H. (1958). Mechanics of granular material. *Advd. appl. Mech.* **5,** 233–306.

Drescher, A. & De Josselin de Jong, G. (1972). Photoelastic verification of a mechanical model for the flow of a granular material. *J. Mech. Phys. Solids* **20,** 337–351.

Mindlin, R. D. (1949). Compliance of elastic bodies in contact. *J. appl. Mech.* **71,** No. 3, 259–268.

Mindlin, R. D. & Deresiewicz, H. (1953). Elastic spheres in contact under varying oblique forces. *J. Appl. Mech.* **20,** 327–344.

Nayak, P. R. (1972). Surface roughness effects in rolling contact. *Trans. Am. Soc. mech. Engrs* Series E, *J. Appl. Mech.* **39,** No. 2, 456–460.

Rodriguez-Ortiz, J. M. (1974). Estudio del comportamiento de medios granulares heterogéneos mediante modelos discontinuos analógicos y matemáticos. PhD thesis, Universidad Politécnica de Madrid.

Serrano, A. A. & Rodriguez-Ortiz, J. M. (1973). A contribution to the mechanics of heterogeneous granular media. *Proc. Symp. Plasticity and Soil Mech., Cambridge.*

Strack, O. D. L. & Cundall, P. A. (1978). *The distinct element method as a tool for research in granular media.* Report to National Science Foundation from Department of Civil and Mineral Engineering, University of Minnesota, USA.

Wakabayashi, T. (1957). Photoelastic method for determination of stress in powdered mass. *Proc. 7th Jap. Nat. Congr. Appl. Mech.* 153–192.

Cundall, P. A. & Strack, O. D. L. (1980). *Géotechnique* **30**, No. 3, 331–336

A discrete numerical model for granular assemblies

CUNDALL, P. A. & STRACK, O. D. L. (1979). *Géotechnique* **29**, No. 1, 47–65.

B. C. Burman, Coffey & Partners Pty Ltd

The Authors have produced an interesting paper on the application of numerical modelling of discontinua to the investigation of constitutive relationships for granular materials. The Writer (Burman, 1972) was involved in similar work between 1969 and 1971, the results of which are belatedly published (Trollope & Burman, 1980).

The senior Author and the Writer have independently developed similar but different approaches to the numerical analysis of discontinua with a view to mathematically modelling the behaviour of jointed masses in problems of rock mechanics. While Cundall chose to

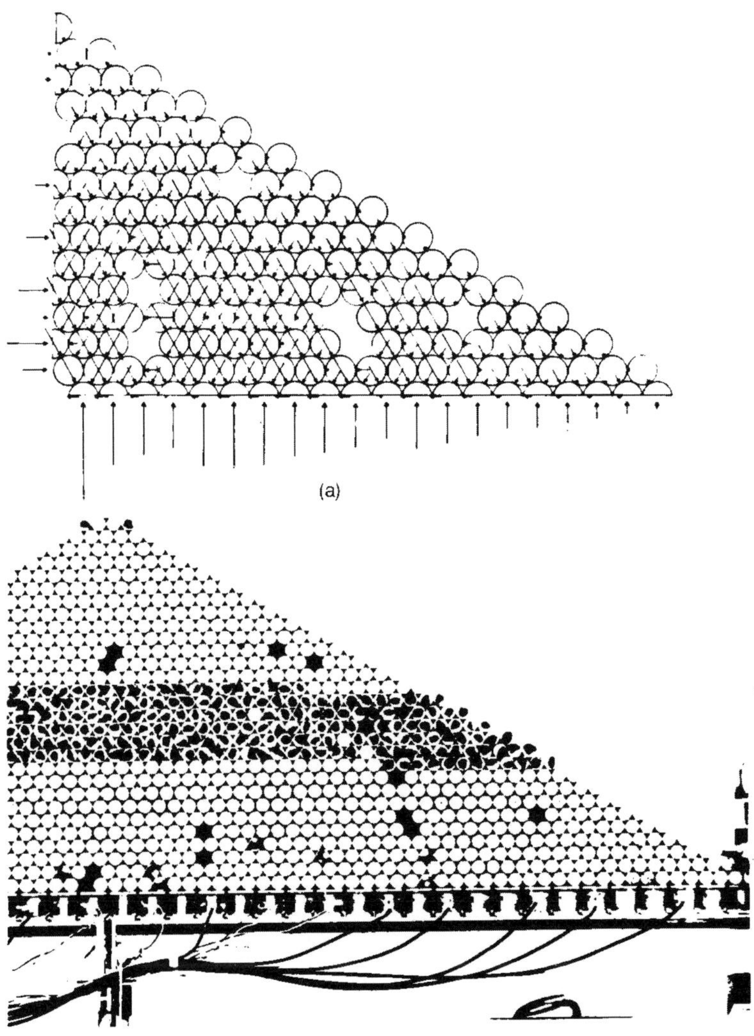

Fig. 1. (a) Numerical model with load distribution for rigid base condition; (b) numerical approximation and physical model for 30° wedge of sorted cylindrical blocks

employ a dynamic finite difference-based technique and the Writer developed a static finite element-based technique, the central common feature of these developments has been the assumption of negligibly small deformability of the individual units of the discontinuum. This assumption results in a considerable reduction in the scale of the numerical problems (by at least a factor of 3) and allows the essential mechanics of discontinua to be more readily appreciated. While there are undoubtedly advantages associated with either technique, both parties seek to solve basically similar problems. The Writer is therefore inclined to the simple view that, in spite of differences in detail, similar computational overheads are involved in solving the similar problems. The different approaches are as much an expression of individual backgrounds and training as they are of relative computing efficiencies. Any further discussion of the particular merits of the technique is incidental to the main topic of the behaviour of idealized granular masses. It is preferable to concentrate discussion on the aspects of model verification and use of numerical modelling in studying constitutive relations for granular masses.

The Authors' approach to the important question of model verification is convenient but unconvincing. As they note, the results of their verification test are 'primarily qualitative and therefore subjective'. Verification is difficult for discontinuum modelling and the Writer knows of no simple or complete answer. There is certainly a most useful area for further research in this topic. The Writer's approach was to test an instrumented physical model consisting of steel cylinders formed into a half wedge configuration with a deformable base (Fig. 1). However, due to the significant effects of dimensional tolerance in a regular granular array, the Writer's results were also in part qualitative. Indeed, it may be that verification will remain circumstantial until a major effort perhaps along the lines of photoelastic modelling pioneered by Chappell (1972) is undertaken specifically to verify discontinuum numerical models.

Although the Paper deals essentially with the mathematical derivation of the model and peripherally with verification, it suggests that the technique has application in investigating the mechanics of granular assemblies.

The Writer has carried out a study of the behaviour of idealized granular masses through an extensive series of numerical experiments involving a regular hexagonal packing of uniform cylinders arranged into a wedge configuration (Fig. 3 of Trollope & Burman, 1980). The wedge was subject to simulated gravity loading with a variety of subsequent basal deformations so that other than homogenous conditions could be evaluated. These tests lead the Writer to the conclusion that numerical experimentation with regular uniform assemblies of particles is a logical starting point in studying the mechanics of real granular materials. This belief is based on recognition of the necessity for understanding the respective roles of the various deformational mechanisms available to discontinuous systems before attempting to investigate the more general but more complex question of irregular, non-uniform assemblies. In addition to the conceptual advantage associated with regular assemblies, there is also a real practical advantage in being able to determine continuous stress and strain fields throughout the assemblage from the results of numerical experiments. This advantage derives from the ease with which repeatable units suitable for determining averaged stress and strain values can be defined. Figure 4

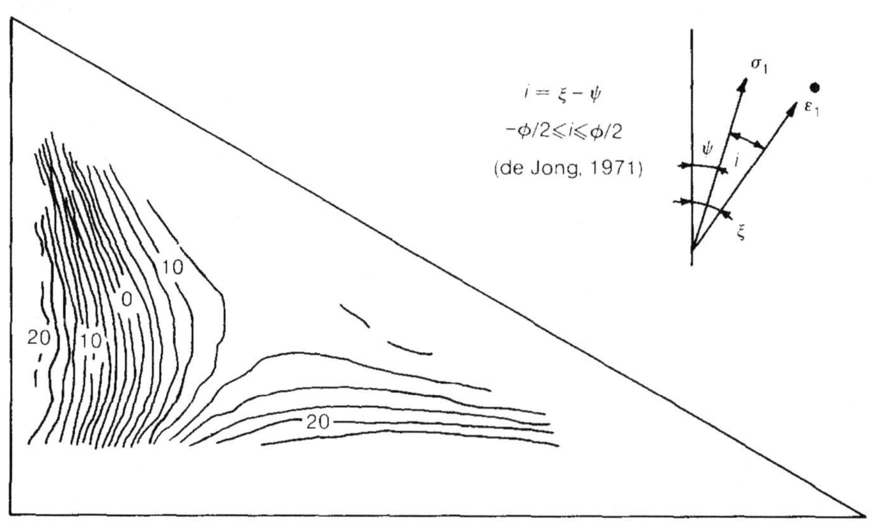

Fig. 2. Contour plot of actual deviation angle, *i*, between stress and strain rate axes for convex base deflexion of 30° numerical wedge with shearless contacts

of Trollope & Burman (1980) shows the repeatable units which were used to determine normal and shear stresses for the regular hexagonal array. Stress fields thus determined were shown to satisfy equilibrium conditions while strain fields similarly determined were shown to satisfy continuity requirements. The definition of consistent stress and strain fields for irregular assemblies would be much less easily established.

An advantage which arises from the ability to define continuous stress and strain fields through an idealized granular mass is that equivalent continuum based deformational parameters can be derived throughout the mass. The Writer, for instance, was able to investigate the developing relationship between stress, strain and rate vectors as the idealized mass was progressively deformed. Details are presented by Burman (1971) but it is worthwhile noting a result predicted analytically by de Josslin de Jong (1971) in respect of the double-sliding, free-rotating model. Results from the Writer's numerical experiments (Fig. 2) show an independence of the stress and strain rate vectors with increasing distortion of the mass that corresponds to a non-coaxial plasticity model in which the angle of divergence is a function of distortion through the mass. The Writer would be interested in investigation of these analytic and numeric predictions for idealized granular arrays by physical experiments with real granular materials. This proposition is an example of the potential benefits available to geomechanics from the development and application of numerical models of idealized discontinua.

The Writer appreciates the Authors' efforts to understand the behaviour of granular material through the application of what are often thought to be over-complicated numerical techniques. This field seems to offer considerable potential.

REFERENCES

Burman, B. C. (1971). *A numerical approach to the mechanics of discontinua*. PhD thesis, James Cook University of North Queensland.

Chappell, B. A. (1972). *The mechanics of blocky material*. PhD thesis, Australian National University, Canberra.

de Josselin de Jong, G. (1971). The double sliding, free rotating model for granular assemblies. *Géotechnique* **21**, No. 2, 155–163.

Trollope, D. H. & Burman, B. C. (1980). Physical and numerical experiments with granular wedges. *Géotechnique* **30**, No. 2, 137–157.

Professor M. E. Harr and Professor R. D. Holtz, School of Civil Engineering, Purdue University, USA

The distinct element method and its use to describe the mechanical behaviour of an assembly of discs are indeed intriguing developments. The Authors are to be congratulated for the innovative use of interactive computer graphics. Can the method realistically describe the behaviour of real collections of particles?

Contrary to popular opinion, in a transverse section through a soil mass, even for 'homogeneous soils', few particles will be in contact with each other in that plane. To illustrate this, consider the idealized packing of perfect spheres (only 4 shown) in Fig. 1 (which is similar to Fig. 5 in the Paper). The trace of the plane AA′ will exhibit the same circular array as shown in the figure. However, all other planes, such as BB′, CC′ and DD′, passing through the collection will show the particles not to be in contact. For real granular soils a similar situation can be expected. This observation can provide the basis for understanding, for example, how rupture propagates through a granular mass.

Conventional soil mechanics failure solutions assume that the rupture surface progresses within the plane of consideration. However, as noted, because there are few contacts to be broken in a given vertical plane, the propagation of rupture, should it continue, is much more likely to occur in the third dimension rather than in the plane of consideration. It would seem

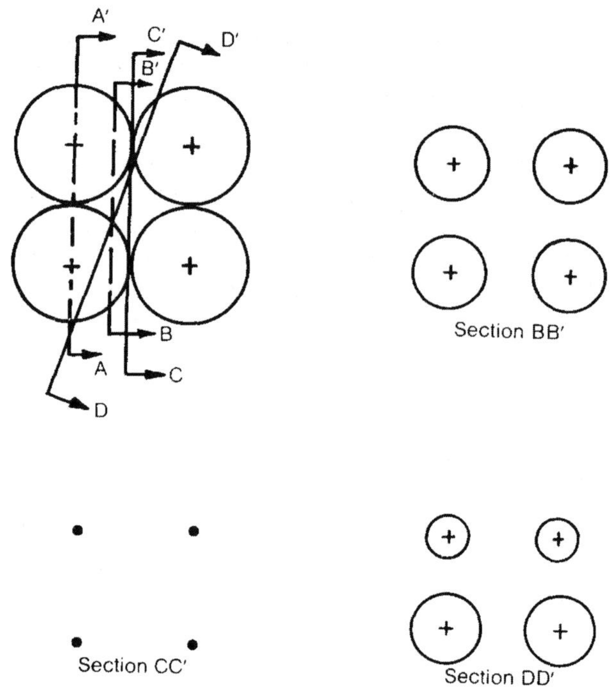

Fig. 1

that similar effects must affect how boundary energy is transmitted through particulate media.

Can the Authors indicate how the distinct element method may be extended to model the more realistic situation existing in real granular assemblies in which only a few particles are in contact in a plane?

Authors' reply

Professor Harr and Professor Holtz express the opinion that the number of particles that are in contact in a plane affects the behaviour of a granular mass, particularly the propagation of rupture surfaces. The Authors would like to emphasize that the technique of numerical testing is intended to verify opinions such as the one expressed by Harr and Holtz, rather than build them into the model. The distribution of contacts in a plane is not used in the method; contacts are formed and broken on the basis of physical laws that govern the interaction of particles.

The model presented in the Paper is a two-dimensional one and suffers from the drawback that all contacts are lines of contacts between rods rather than points of contacts between spheres. The development of a two-dimensional model should be viewed as a first step. An elementary three-dimensional version of the program BALL discussed in the Paper has been developed, and the first results of three-dimensional modelling have been published (see Cundall & Strack, 1979).

Numerical methods of the type used by Dr Burman idealize a granular assembly subjected to a prescribed loading history as a sequence of static, boundary value problems. However, it is not known that a unique set of contact forces and displacements exists for any given boundary value problem of a granular assembly. If such uniqueness does not exist, an implicit, static method will produce solutions that depend on the particular numerical algorithm used, because there is no physical way in which the algorithm can select the correct path from the several alternatives. Therefore the Authors prefer to use an explicit, time-marching method whereby the move from one state to the next is determined by physical laws of motion. The problem of non-uniqueness may then be studied by varying the physical parameters.

Dr Burman mentions that he prefers to consider regular assemblies because of the conceptual advantage and because continuous stress and strain fields may be defined readily. He further reports that his access to such fields enabled him to demonstrate the independence of stress and strain rate vectors as predicted analytically by de Josselin de Jong (1971). It may be noted that Drescher & de Josselin de Jong (1972) achieved the same result on an irregular assembly of photoelastic discs. They used average stress and strain rate tensors. The Authors utilize similar average tensors for the irregular assemblies modelled by the program BALL (see Strack & Cundall, 1978).

REFERENCES

Cundall, P. A. & Strack, O. D. L. (1979). The distinct element method as a tool for research in granular media. *Part II, Report to the National Science Foundation from Department of Civil and Mineral Engineering, University of Minnesota.*

Drescher, A. & de Josselin de Jong, G. (1972). Photoelastic verification of a mechanical model for the flow of a granular material. *J. Mech. Phys. Solids* **20**, 337-351.

Strack, O. D. L. & Cundall, P. A. (1978). The distinct element method as a tool for research in granular media. *Part I, Report to the National Science Foundation from Department of Civil and Mineral Engineering, University of Minnesota.*

SIMPSON, B., O'RIORDAN, N. J. & CROFT, D. D. (1979). *Géotechnique* **29**, No. 2, 149–175

A computer model for the analysis of ground movements in London Clay

B. SIMPSON,* N. J. O'RIORDAN* and D. D. CROFT*

A computer model for the stress–strain behaviour of London Clay has been developed for use in a non-linear finite element program. The behaviour of London Clay is divided into three ranges of strain: elastic, intermediate and plastic. In the elastic range strains are very small and the material is very stiff. The intermediate range represents the larger strain behaviour normally observed in laboratory tests; it is not truly elastic, but in many respects exhibits anisotropic elastic behaviour. As yield is approached, plastic behaviour can accompany the intermediate phase and is governed by a flow rule and a state boundary curve. In all phases the stiffnesses of the model increase with mean normal effective stress. It is shown that the model can be used to compute the results of both laboratory and in situ tests. The movements around two excavations in London Clay, including the New Palace Yard Underground Car Park, have also been computed retrospectively with considerable success.

Un modèle informatique destiné à refléter le comportement contraintes-déformations de l'Argile de Londres a été mis au point en vue de son emploi dans un programme non linéaire à éléments finis. Le comportement de l'Argile de Londres se divise en trois domaines de déformation: élastique, intermédiaire et plastique. Dans le domaine élastique, les déformations sont très faibles et le matériau est très rigide. Le domaine intermédiaire représente les déformations plus importantes que l'on observe normalement dans les essais en laboratoire; il n'est pas véritablement élastique mais présente à bien des égards un comportement élastique anisotrope. Lorsque l'on approche de la limite d'élasticité, on peut observer un comportement plastique dans la phase intermédaire, et ce comportement est régi par une loi de fluage et une courbe de limite entre états. Dans toutes les phases, la rigidité du modèle augmente en fonction de la contrainte efficace moyenne normale. L'article montre que le modèle peut servir à déterminer les résultats obtenus aussi bien en laboratoire qu'*in situ*. Le calcul rétrospectif des mouvements enregistrés autour de deux excavations effectuées dans l'Argile de Londres, dont le Parking Souterrain du New Palace Yard, a également été effectué avec beaucoup de succès.

INTRODUCTION

The work described in this Paper has been undertaken to aid the design of the proposed British Library, adjacent to St Pancras Station in London. A number of deep basements are required for this project, the largest of which will be 24 m deep and about 100 m square, extending through the London Clay and 4 m into the Woolwich and Reading Beds.

In order to design the necessary diaphragm walls and to predict the likely movements of the nearby tunnels and buildings, it is important to understand the deformation behaviour of the London Clay and of the underlying strata. Previous workers (Cole and Burland, 1972; St John, 1975; Burland and Hancock, 1977) have analysed the movements around deep excavations in London Clay on the basis of linear elasticity, and have proposed suitable values of Young's modulus for use in finite element or other computations. Although a fair degree of agreement with field measurements has been obtained, a number of problems have been noted as follows:

(a) It has been difficult to achieve good predictions of both the movements of diaphragm walls and those of the ground around the excavation from the same analysis.

(b) In order to obtain satisfactory results, the stiffnesses assumed for the London Clay have to be very much greater than those measured in laboratory tests.

Discussion on this Paper closes 1 September, 1979. For further details see inside back cover.

* Ove Arup and Partners, 13 Fitzroy Street, London W1.

NOTATION

c'	cohesion
c_u	undrained shear strength
E_u	undrained Young's modulus
E_v'	vertical Young's modulus for effective stresses
E_H'	horizontal Young's modulus for effective stresses
F	normalized shear stress ($t/s \sin \phi' + c' \cos \phi'$)
F_0	normalized shear stress at zero plastic shear strain
G_{HH}	horizontal to horizontal shear modulus
G_{HV}	vertical to horizontal shear modulus
K_0	coefficient of earth pressure at rest
M	critical state stress ratio in a triaxial test
p	mean normal effective stress in the triaxial cell = $(\sigma_1'+\sigma_2'+\sigma_3')/3$
p	a suffix indicating plastic
q	deviatoric effective stress in the triaxial cell = $(\sigma_1'-\sigma_2')$
s	mean normal effective stress: $(\sigma_1'+\sigma_2')/2 = (\sigma_2'+\sigma_y')/2$
t	$(\sigma_1'-\sigma_2')/2$
v_p	plastic volumetric strain
x, y, z	co-ordinate axes: y vertical; strains in z direction (horizontal) zero or small
γ_p	plastic shear strain = $\Sigma(\delta\varepsilon_{1p}-\delta\varepsilon_{2p})$
γ_{ptriax}	plastic shear strain in the triaxial test = $\frac{2}{3}(\varepsilon_1-\varepsilon_2)$
$\varepsilon_1, \varepsilon_2$	major and minor principal strains in the x, y plane
$\varepsilon_x, \varepsilon_y, \varepsilon_z$	direct strains in the x, y, z directions
ε_α	radius of the kinematic yield surface
ν_{HV}	Poisson's ratio for effect of vertical effective stress on horizontal strains
ν_{HH}	Poisson's ratio for effect of horizontal effective stress on horizontal strains
σ_1, σ_2'	major, minor principal effective stresses in the x, y plane
$\sigma_x', \sigma_y', \sigma_z'$	effective stress in x, y, z directions
ϕ'	drained angle of friction

(c) It is apparent that the material inside the diaphragm walls will progressively yield passively as the excavation proceeds. St John (1975) has suggested that improved predictions could be achieved if this was taken into account. This requires a non-linear model.

(d) Linear elastic theory is not generally applicable to the behaviour of clay in the drained condition. The change in stiffness that occurs with change in volume involves a high degree of material non-linearity. Long-term predictions are therefore usually restricted to simple calculations of overall consolidation settlement and heave, etc.

Although the above factors are not of overriding importance, it was thought that they must be considered since analysis for the British Library requires an extrapolation beyond the experience gained from other, but significantly smaller, excavations. A non-linear elastic–plastic model of London Clay (called Model LC) has therefore been developed to enable these effects to be quantified and their importance assessed.

NON-LINEAR EFFECTS OBSERVED IN LONDON CLAY

Laboratory and field tests

It is well known that the results of laboratory stiffness tests are difficult to correlate with stiffness back-figured from measurements on full-scale structures. Typically the ratio E_u/C_u is

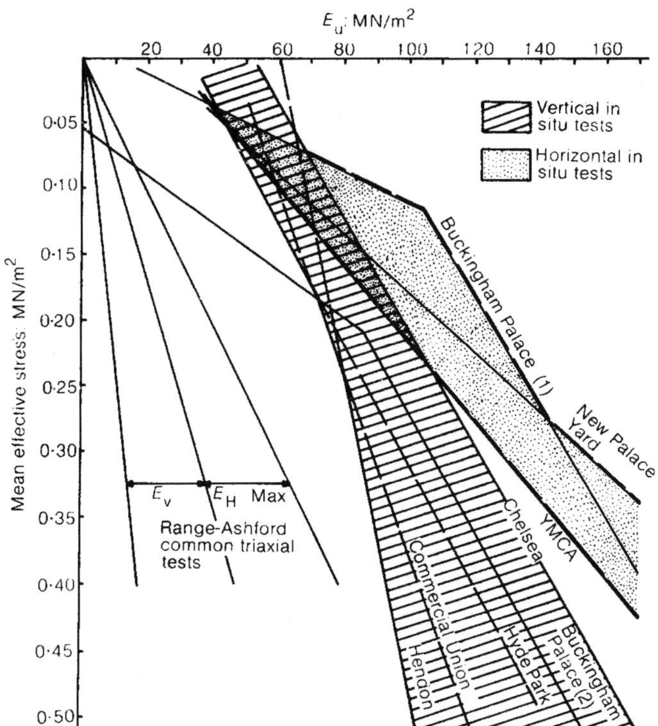

Location	Type of test	Source
Buckingham Palace Brittanic House New Palace Yard YMCA	Back analysis of ground movements	Cole and Burland (1972) St. John (1975)
Hyde Park Commercial Union	Settlement records	Hooper (1974) Green (1971)
Chelsea Hendon	Plate tests	Marsland (1971 a) Marsland (1971 b)

Fig. 1. Variation of *E* (undrained) with mean effective stress after St John (1975)

found in laboratory triaxial tests to be about 100, but in the field to be between 500 and 1000 (Butler, 1975).

Marsland (1971a, b) has claimed that stiffnesses measured in large (865 mm) diameter plate-bearing tests are fairly close to the back-figured values. In Fig. 1 the comparison between triaxial, plate test and large scale results is shown, reproduced from St John (1975).

Marsland (1971c) and Atkinson (1973) have suggested that the reasons for the much lower stiffnesses measured in the laboratory are sample disturbance (especially the opening of fissures) and bedding of the samples in the test apparatus. However, the results of very careful laboratory work show that remarkably consistent and repeatable results can be obtained. Atkinson (1973) at Imperial College prepared triaxial specimens from block samples obtained at the Barbican site. He found that for strains up to about 1 % the material behaviour appeared to be anisotropic elastic for a large variety of stress paths. For a given initial stress state, the behaviour was linear up to 1 % shear strain.

In contrast to the consistency found by Atkinson, it would be expected that the effects of sample disturbance and bedding would be fairly random. It therefore seems appropriate to investigate other reasons for the consistent low stiffnesses measured in the laboratory.

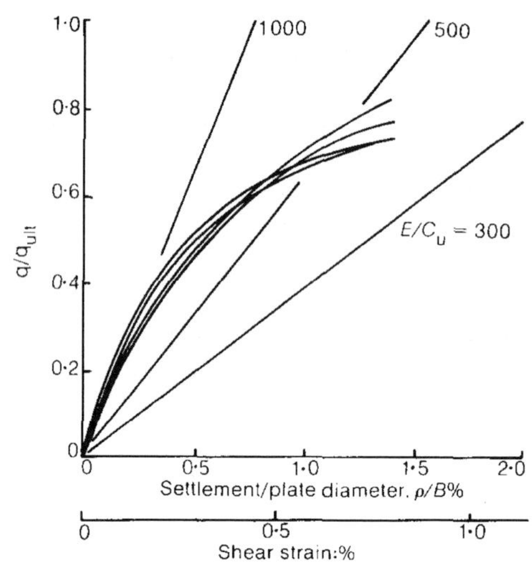

Fig. 2. Results of 865 mm diameter plate bearing tests on London Clay after Marsland (1971). Values of C_u were obtained from the plate tests and are probably 30% lower than laboratory values. Assumptions: $q_u = 10C_u$ (including allowance for depth); $E = q\pi B(1-\nu^2)/4\rho$, $\nu = 0{\cdot}4$

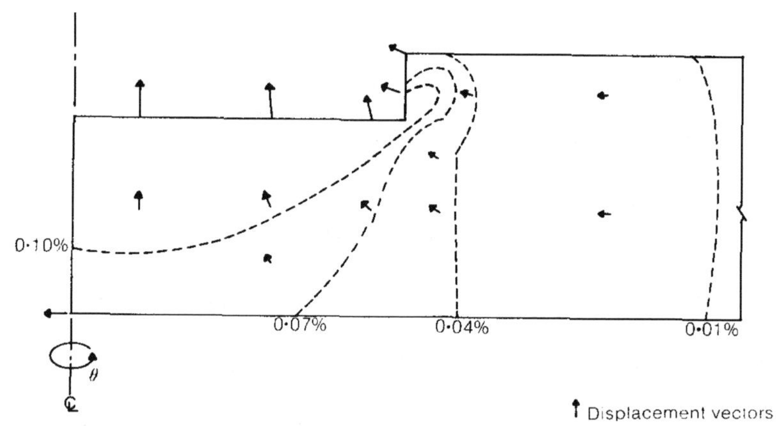

Fig. 3. Strain contours around an excavation. Shear strain contours for a maximum wall displacement 0·2% of wall height, linear elastic model

Threshold effects

Many workers, including Lewin (1970) and Som (1968), have reported threshold effects in London Clay and other soils. In some of his oedometer tests, Som held samples at constant stress for many days and then recorded exceptionally high stiffness as the next small increment of strain was applied. Lewin reported a similar high stiffness when the direction of straining was changed.

Reviewing these data, Atkinson's (1973) conclusions were that:

'These tests, and others like them, seem to indicate that where a delay at constant stress state is involved or where the direction of the stress probe differs from the previous stress path the strains produced by a small stress increment may be very much smaller than the strains produced by a larger increment or those where there was no rest period or where the direction of the stress probe did not deviate much from the previous loading path. This apparent stiffening of a soil at small stress increments may be likened to a threshold effect; that is to say that there exists a threshold of the probing stress increment below which the soil is apparently very much stiffer than for a larger probe. This threshold effect, which has been variously attributed to a rigid bond (Terzaghi, 1941), secondary consolidation (Bjerrum, 1967) or to structural development, must be regarded as a soil property rather than a testing error, but in any case the effect must be recognized if test results are to be properly interpreted.

'In the field a natural soil has remained at a constant stress state for a very long period—since the last period of erosion, deposition, or tectonic movement; on the other hand sampling and reloading in the testing machine implies a cycle of shear stress . . . and consequently any threshold effects may be reduced. This implies that for small stress increments in the field, where the threshold value is not exceeded, laboratory measured parameters may overestimate the actual strains.'

It is useful to compare the magnitudes of the strains recorded in laboratory tests, plate bearing tests and full-scale measurements.

(a) The stiffness parameters reported in laboratory tests are usually based on a range of shear strain up to about 1%. Atkinson (1973) states that his measurements at strains of less than 0·2% are not reliable and therefore quotes Young's moduli based on tangent moduli at 0·25% strain.

(b) In Fig. 2, the results of typical plate bearing tests reported by Marsland (1971a, b) are reproduced. An approximate scale for the average shear strains occurring within one diameter beneath the plate has been added to the horizontal axis. At very small strains (<0·1%), the tangent modulus is similar in magnitude to the values found from back-analysis of full-scale excavations ($E_u/C_u = 1000$). As shear strains increase beyond 0·5% however, the secant modulus approaches that measured in the laboratory.

(c) The shear strains occurring behind a typical retaining wall as an excavation is formed are shown in Fig. 3. For a movement of 0·2% of the wall height (an upper bound for retaining walls in firm soil according to Peck (1969)), the shear strain in the ground behind the wall exceeds 0·1% only very locally.

Hence, it appears likely that when London Clay is subjected to very small strain increments following a period of rest or a change of the direction of straining, its stiffness is much greater than that at the larger strains normally measured in the laboratory. This change of stiffness as straining proceeds is an important feature of Model LC.

Elastic–plastic behaviour

At large strains, beyond about 1% shear strain, London Clay exhibits the characteristics of plastic flow. The basic assumptions adopted for the elastic–plastic behaviour are listed as follows.

(a) The total strain increment is the sum of elastic and plastic strains, and changes of effective stress can be expressed as functions of the elastic strains alone (or of the intermediate strains defined later).

(b) The relative proportions of the components (volumetric–shear) of a plastic strain increment are dependent on the current stress state and not on the applied stress increment; that is, a plastic flow rule is operative.

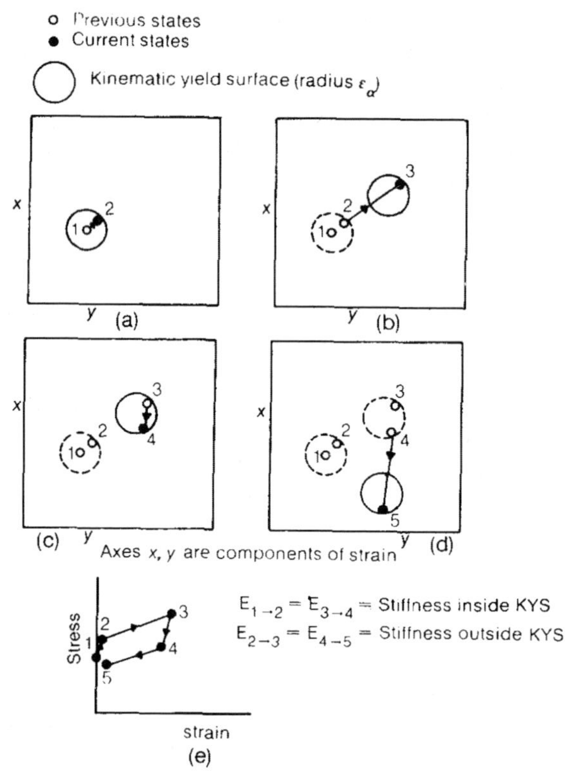

Fig. 4. The kinematic yield surface effect

(c) The overall magnitude of a plastic strain increment is dependent on both the current stress state and the stress increment; a hardening law is used.
(d) The principal axes of plastic strain increment are coincident with those of the current stress state and not the applied stress increment.
(e) The stiffnesses increase linearly with the mean effective stress in the plane of deformation.

DESCRIPTION OF MODEL

General

Model LC relates increments of effective stress to increments of strain, given the current stress state. Attention was concentrated initially on plane strain conditions, but it is considered that the model can be used for axisymmetric work and for plane stress provided that strains are relatively small in the hoop or constant stress direction respectively.

In this Paper the (x, y) plane is the plane of major deformation, with y vertical. It is assumed that strains in the z direction are not large enough to influence the plastic behaviour in the (x, y) plane, though they do influence the elastic and 'intermediate' behaviour.

Three ranges of strain are considered:

(a) Very small strains (within the strain threshold): elastic behaviour.
(b) Moderate strains (the linear range measured in the laboratory): intermediate behaviour.
(c) Large strains, approaching the limiting shear strength of the material: plastic behaviour accompanying further intermediate behaviour.

The assumption of a threshold property is the dominant feature of the model. This makes possible predictions of both laboratory and field behaviour using the same model.

Numerical parameters and governing equations for the model are listed together below and the implementation of the model in the SAFE finite element program developed by Ove Arup and Partners is described in the Appendix. This has required that the model be used to compute stress increments from strain increments. The reverse process has not been developed but would be rather simpler.

Elastic strains: kinematic yield surface

The concept of a higher stiffness at very small strains can conveniently be represented using a kinematic yield surface (KYS). This defines a small zone in stress or strain space such that the higher stiffness applies to an element of soil until its state reaches the KYS (e.g. Mroz, 1967). For Model LC, the KYS is defined in terms of strains. Straining within the KYS is purely elastic, though non-linear (Fig. 4(a)). As the element is strained further the yield surface moves (hence 'kinematic') and the effective stiffness is reduced (Fig. 4(b)). If the direction of straining is reversed the higher stiffness again applies (Fig. 4(c)) until the KYS starts to move again (Fig. 4(d)).

The most appropriate shape for the KYS is not known. A spherical shape has been assumed in strain axes defined by $(\varepsilon_x+\varepsilon_y, \varepsilon_x-\varepsilon_y, \gamma_{xy})$.

It is assumed that the in situ state for London Clay (defined in computations as zero strain) is at the centre of the initial KYS. It is possible that, due to creep effects, the centre of the KYS will move to coincide with a new strain state if that state is maintained for a long period of time. This could model the threshold effects after time lapses, reported by Som.

Data are not available to allow direct assessment of the magnitude of strain within which the higher stiffness is applicable, nor is it clear what are the correct stiffness moduli within this range. The numerical values adopted in Model LC have been chosen to give a good fit to field data. They are, however, consistent with Som's data which suggest that the KYS should correspond to a change of stress of about 10%.

The work of St John (see Fig. 1) suggests that the ratio of horizontal to vertical stiffness may be similar at very small strains to that found in laboratory tests at larger strains. All the stiffnesses used for the small strain range have been assumed to be ten times those obtained from laboratory test results (ie the stiffnesses during intermediate behaviour).

Intermediate strains

The strains during intermediate behaviour represent the range normally measured in laboratory tests and are accompanied by movement of the kinematic yield surface. Atkinson (1973) reported that London Clay behaved like an anisotropic elastic material in this range of strain, except that strains were not necessarily recoverable. Thus the behaviour has the appearance of elasticity but could be regarded as a combination of elastic and plastic effects, hence the term intermediate. The plastic (irrecoverable) strains are not however governed by a plastic flow rule as are those described in the section below. The term plastic will therefore be reserved for strains governed by a plastic flow rule, as distinct from the intermediate strains described here.

The stiffness moduli adopted for intermediate straining have been chosen to fit Atkinson's data with the assumption that all intermediate stiffnesses are proportional not simply to mean normal stress but to $(s+c' \cot \phi')$. The item $c' \cot \phi'$ has been added to s, the mean normal stress in the plane of deformation, in order to ensure that stiffness and strength both fall to zero at the same stress state ($s = -c' \cot \phi'$).

Zytinski *et al.* (1976) have shown that the assumption that shear stiffness varies linearly with mean normal stress violates basic energy criteria. However, it is found in practice that this assumption gives reasonable results.

The vertical to horizontal shear stiffness, G_{HV} cannot be found from Atkinson's work. Gibson (1974) found in laboratory tests that G_{HV}/E_u was about 0·38 in triaxial tests. Assuming the values of other elastic parameters are as listed below, this gives $G_{HV}/E_H' = 0{\cdot}35$. Alternatively, it could be assumed that $G_{HV} = G_{HH}$, and G_{HH} may be derived from the other elastic parameters. This gives $G_{HV} = G_{HH} = 0{\cdot}42E_H'$. A value of $0{\cdot}4E_H'$ has been adopted for G_{HV} in the program.

Plastic strains

The behaviour proposed for larger strains is complicated, but less important in the computation of ground deformations around stable excavations. It is assumed that at large strains (>1%) laboratory test results are relevant and curve-fitting has been based on Atkinson's results. It has been found that the curve shown in Fig. 5 can be used as a state boundary curve, ie, a curve separating attainable stress–strain states from those which cannot be attained. Movement along the state boundary determines the magnitude of plastic strains (ie, it provides the required hardening law).

The abscissa in Fig. 5 is γ_p, the sum of all incremental plastic shear strains, irrespective of direction; this is not, strictly, a state property. If a flow rule is assumed it can be shown that γ_p is uniquely related to the plastic volumetric strain, v_p. Thus γ_p is a reasonable directionless parameter which, with suitable manipulation and choice of origin, can be related to voids ratio, which is a state property.

A modified form of the Cam-clay flow rule (Schofield and Wroth, 1968) has been found to fit Atkinson's data when used in conjunction with the state boundary curve. The Cam-clay flow rule for soil behaviour in the triaxial test, derived from energy considerations, was

$$\frac{\delta v_p}{\delta \gamma_{p\,\mathrm{triax}}} = M\left(1 - \frac{q}{Mp}\right)$$

For plane strain conditions, Roscoe and Burland (1969) showed that it was reasonable to use a similar equation written in terms of plane strain parameters. This has been adopted for Model LC, except that $(s + c' \cot \phi')$ has been substituted for s, the mean normal stress in the plane of deformation. The flow rule adopted may be written as

$$\frac{\delta v_p}{\delta \gamma_p} = F_0 \sin \phi' \left(1 - \frac{F}{F_0}\right)$$

where $F = t/(s \sin \phi + c' \cos \phi')$.

To implement this plastic behaviour in the computer program, it is postulated that the large strains are a combination of plastic flow with intermediate behaviour. As plastic strain develops the state of an element of soil will progress along the state boundary curve shown in Fig. 5 unless it is unloaded, in which case it will fall below the curve. On reloading it will rejoin the curve at the point where it left it.

It is postulated that although elastic and intermediate strains may take place in any direction (using full x, y, z axes), plastic flow as described above is confined to the (x, y) plane (ie one, chosen, horizontal direction and the vertical direction). In the computer program, the stress in the second horizontal direction (z) is also limited to remain within the active and passive limits.

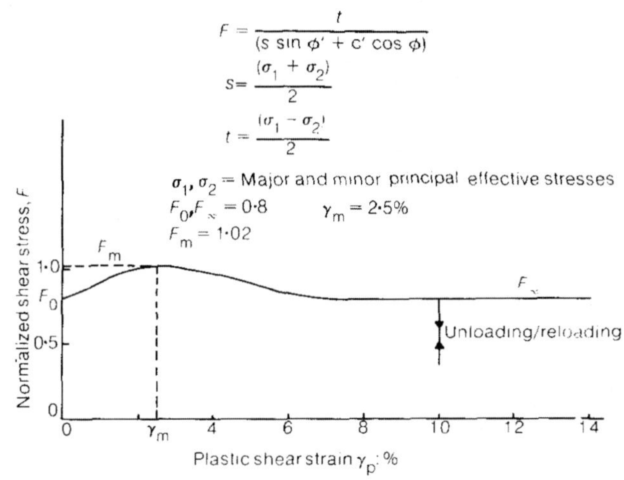

Fig. 5. The state boundary curve

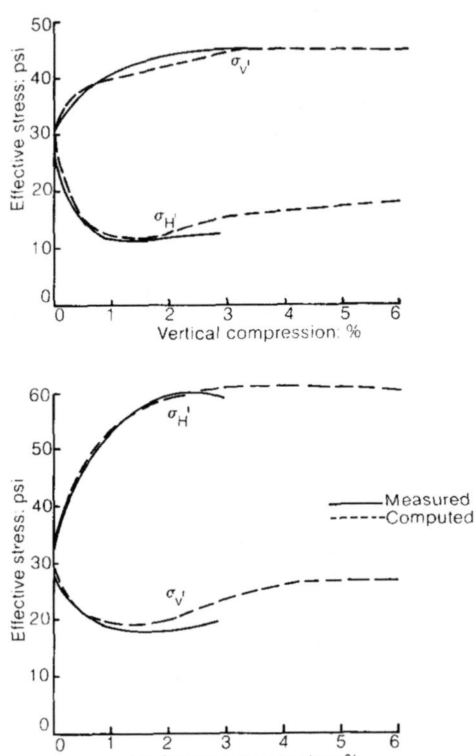

Fig. 6. Plane strain tests

Numerical values

The basic equations of the model and numerical values are presented below. Unless stated otherwise, the following numerical values were used in all computations. As outlined above, the source of almost all the parameters was Atkinson's laboratory work.

Elastic range. The kinematic yield surface is a sphere in $(\varepsilon_x+\varepsilon_y, \varepsilon_x-\varepsilon_y, \gamma_{xy})$ space. Its radius = $\varepsilon_\alpha = 0{\cdot}02\%$ (200 microstrain). The stiffness in this range is ten times the intermediate stiffness.

Intermediate range. The parameters of anisotropic elasticity are adopted.

$$E_v' = \bar{E}'_v\,(s+c'\cot\phi')$$
$$\bar{E}_v' = 50 \text{ (a dimensionless material constant)}$$
$$E_H' = 1{\cdot}67E_v'$$
$$G_{HV} = 0{\cdot}4\,E_H'$$
$$\nu_{HV} = 0{\cdot}16$$
$$\nu_{HH} = 0{\cdot}2$$

Plastic range. Plastic shear strains (γ_p) are defined by the curve in Fig. 5, and there is no plastic straining for states below the curve. In computations this curve is represented by an equation of the form

$$F = c_1+c_2(\gamma_p-c_3)\exp\,[c_4(\gamma_p-c_3)^2]$$

where c_1 to c_4 are material constants.

Plastic volumetric strains (υ_p) are defined by the flow rule

$$\delta\upsilon_p = \delta\gamma_p \sin\phi'(F_0-F)$$

where

$$F = t/(s\sin\phi'+c'\cos\phi')$$
$$\left.\begin{aligned} c' &= 20\text{ kN/m}^2 \\ \phi' &= 26^\circ \end{aligned}\right\}\text{for British Library London Clay}$$
$$\left.\begin{aligned} c' &= 35\text{ kN/m}^2 \\ \phi' &= 22^\circ \end{aligned}\right\}\text{for Atkinson's London Clay}$$

COMPARISON WITH LABORATORY AND FIELD TESTS

Undrained tests

Predictions of the model have been compared with the results of plane strain and triaxial undrained tests carried out by Atkinson (1973). In Fig. 6 the computed predictions for two undrained plane strain tests are compared with Atkinson's results, and a good correlation is achieved in both cases. The inclusion of a kinematic yield surface of radius 100 or 200 microstrain has little effect on the computed behaviour, but 500 microstrain causes some divergence from the measurements. However, it is possible that, as suggested by Atkinson, the threshold effects modelled by the KYS have been destroyed during the sampling process.

The triaxial test is much more difficult to model. Two approaches have been tried and the results are shown in Fig. 7.

Plane stress assumptions. If plane stress is assumed, the results agree fairly well with Atkinson's measurements. The failure stress is lower and occurs at smaller strain than that measured, possibly because there is less restriction in the third principal direction than occurs in the laboratory test.

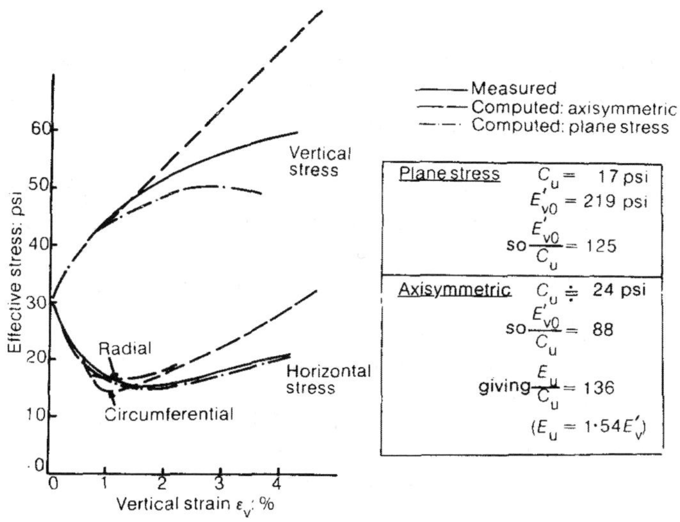

Fig. 7. Triaxial tests

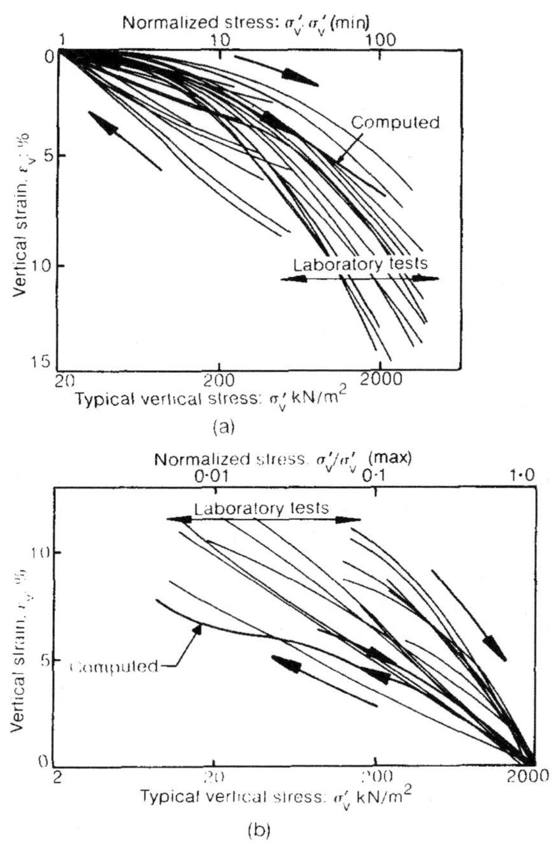

Fig. 8. Oedometer tests: (a) swelling followed by consolidation; (b) consolidation followed by swelling

Axisymmetric assumptions. A single finite element was tested in axisymmetric conditions. Up to about 2% vertical strain the computations agreed well with measurements. However, no failure was predicted and the shear stress increased slowly. This may be because in the laboratory the specimen does not fail in an axisymmetric manner. Alternatively, it could be that the model dilates rather too much during shearing. In this use of the model, the assumption of relatively small strains in the *z* direction is violated.

The ratio E_u/C_u found from the triaxial test computations is found to be 136 for 'intermediate' strains, which is close to the measured values. The corresponding ratio within the strain threshold is therefore 1360 which is larger than that back-figured from field observations (compare with Butler, 1975).

Oedometer tests

In Fig. 8 the results of a large number of oedometer tests have been plotted to a normalized scale. The results are taken from the British Library site investigation. Predictions from Model LC have been superimposed onto these plots.

In studying Fig. 8 it is necessary to compare the gradients of the curves in the various ranges of stress. On this basis it can be seen that the computed and measured results agree fairly well in the stress range of major interest, 100 to 400 kN/m^2. In general, the stiffness of the model is greater than that of the soil being tested.

Insufficient compression is predicted at high stresses. It is possible that at these stresses the London Clay is subject to some plastic volumetric strains as the normally consolidated state is approached. This is not allowed for in the model.

Insufficient swelling is predicted at low stresses. This may be due to the inclusion of $c' \cot \phi'$ in the stress function which determines the stiffness. There are data to suggest that the value of c' should be reduced at low stresses (Bishop, Webb and Lewin, 1965).

The laboratory test results show more hysteresis when the stress path is reversed than do the predictions. This could be modelled more closely by increasing the radius of the kinematic yield surface.

A compression or swelling index in the range 0·03 to 0·06 is predicted by the model, depending on the ratio of vertical to horizontal stress. These predictions are in reasonable agreement with both laboratory observations and field measurements.

Plate bearing tests

As stated above, Marsland (1971a, b) has found that the stiffnesses maesured in 865 mm dia. plate bearing tests are similar to those derived from back-analysis of full-scale excavations. It is therefore valuable to compare the predictions of the model with the results of one of Marsland's tests, and the boring and test procedure of a test at Hendon was modelled, assuming a value of K_0 of 2·0 and undrained deformation of the clay.

In Fig. 9 the measured load–displacement curve is compared with predictions made using three different values for the radius of the kinematic yield surface (ε_α). The computed load–deflection curve is close to that measured at stresses up to about 80% of failure stress. Rather better results could be achieved using a higher value of $\varepsilon_\alpha = 500\mu\varepsilon$, rather than 200µε.

As with the triaxial test computations, failure was not correctly modelled. Again this is probably due to the restriction in the program to axisymmetric displacements, but it should also be noted that the mesh used here may be too coarse to permit the proper development of failure surfaces.

BACK ANALYSES OF TWO EXCAVATIONS IN LONDON CLAY

Two excavations in London Clay have been analysed using Model LC: New Palace Yard

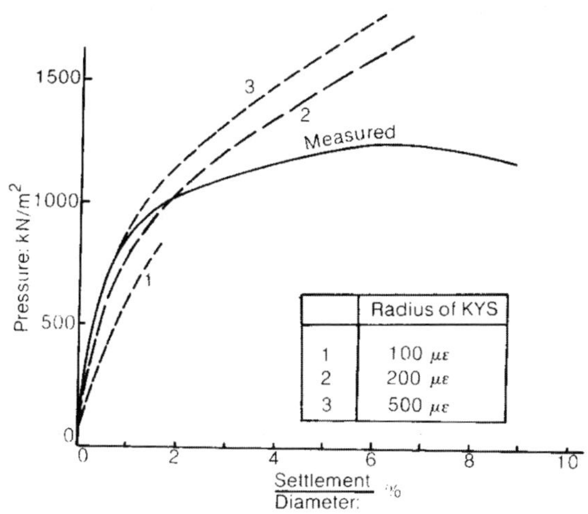

Fig. 9. Predictions for plate bearing test

Underground Car Park (which conforms approximately to axial symmetry) and Neasden Underpass (which conforms to a plane strain assumption).

The two construction methods were completely different. The diaphragm wall at New Palace Yard was supported by floor slabs cast progressively at about 3 m centres; excavation proceeded by mining beneath the slabs. At Neasden Underpass the diaphragm wall was restrained by four rows of ground anchors.

Both excavations were instrumented and closely monitored as construction proceeded. For the New Palace Yard excavation, a parametric study has been carried out to ascertain the relative importance of the Model LC parameters in the modelling of excavations. Undrained behaviour of the clay is assumed in all the following computations.

New Palace Yard Car Park

Burland and Hancock (1977) describe the site, together with the design, construction method and monitoring of the New Palace Yard excavation. Figures 10 and 11 are reproduced from their paper and show the site investigation results and a plan and section of the work. St John (1975) provides further information on the excavation process and its effect on the ground.

The magnitude and direction of the movements of the diaphragm wall and of the ground around the excavation were found to be governed by three construction processes: the construction of the piles and diaphragm walls; piling by percussion boring in the sands and gravels for a minor excavation to the south of the principal excavation; (this has a particular effect on measurements along the south wall); and the principal excavation sequence itself.

The finite element idealization of the excavation and surrounding ground is shown in Fig. 12. The variation in K_0 with depth in London Clay is taken from that suggested by Burland and Hancock (1977). In the absence of other information the Woolwich and Reading Beds were modelled as a continuation of the London Clay.

St John (1975) has compared results from computations using plane strain and axisymmetric assumptions with those of a full three-dimensional analysis, concluding that the axi-symmetric assumptions model the behaviour perpendicular to the mid-side of the square better than do plane strain assumptions. Consequently, axial symmetry has been used for most of the computations presented here.

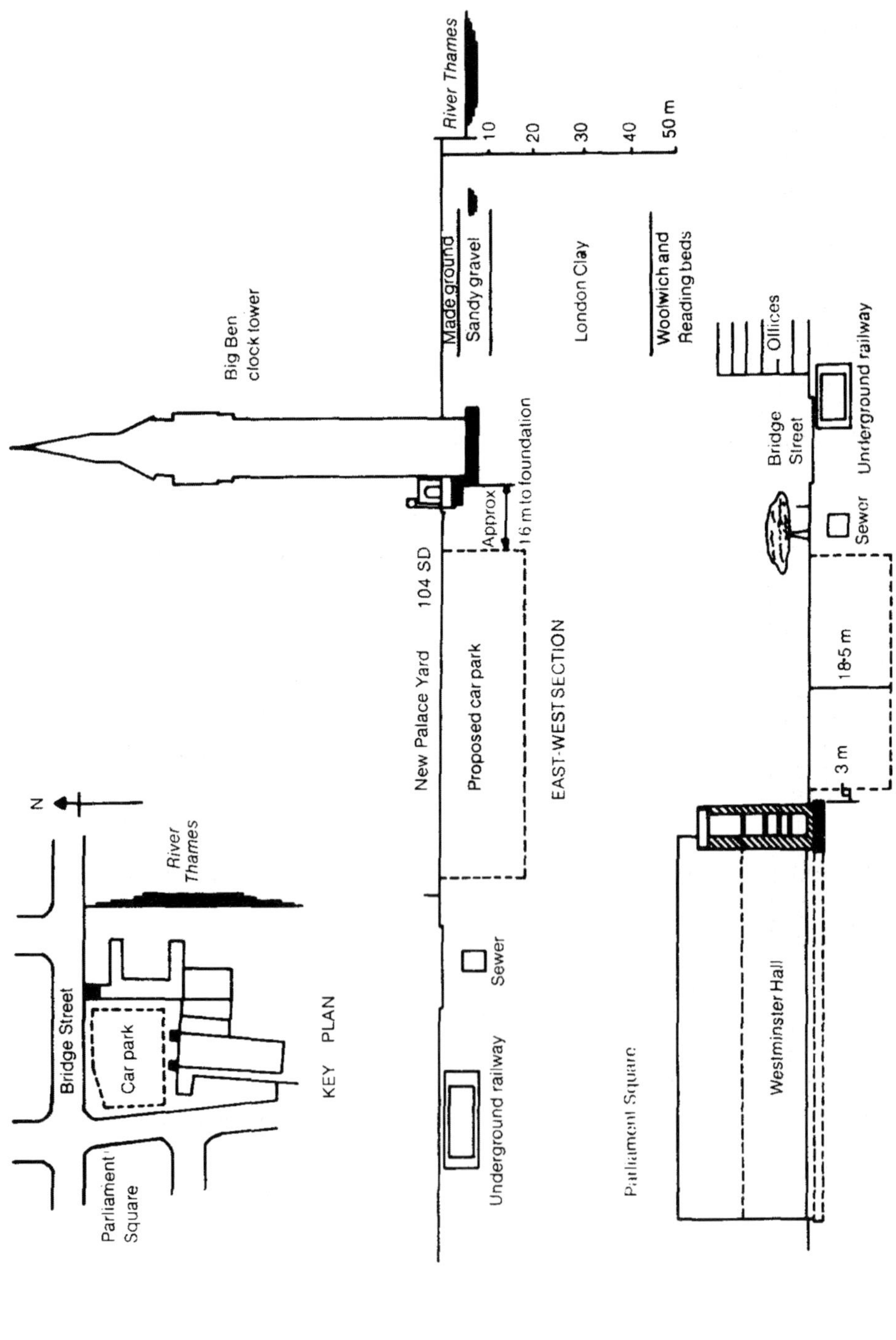

N
River Thames
Bridge Street
Car park
Parliament Square
KEY PLAN
Big Ben clock tower
River Thames
Made ground
Sandy gravel
London Clay
Woolwich and Reading beds
10
20
30
40
50 m
Approx 16 m to foundation
104 SD
New Palace Yard
Proposed car park
EAST-WEST SECTION
Sewer
Underground railway
Parliament Square
Offices
Bridge Street
Underground railway
Sewer
18·5 m
3 m
NORTH-SOUTH SECTION
Westminster Hall

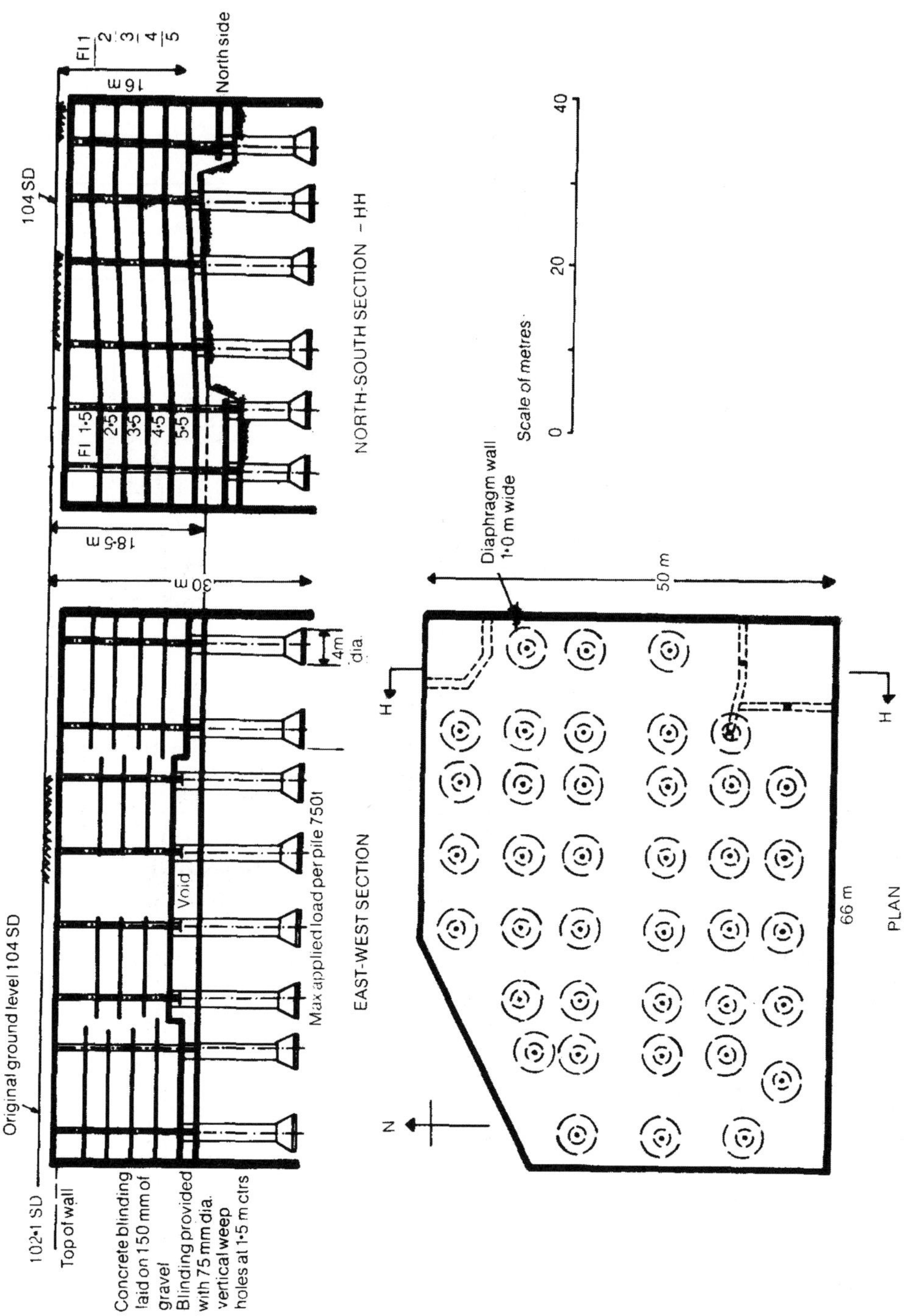

Fig. 10. (Pages 162 and 163) New Palace Yard: layout of project. Reproduced from Burland and Hancock (1977)

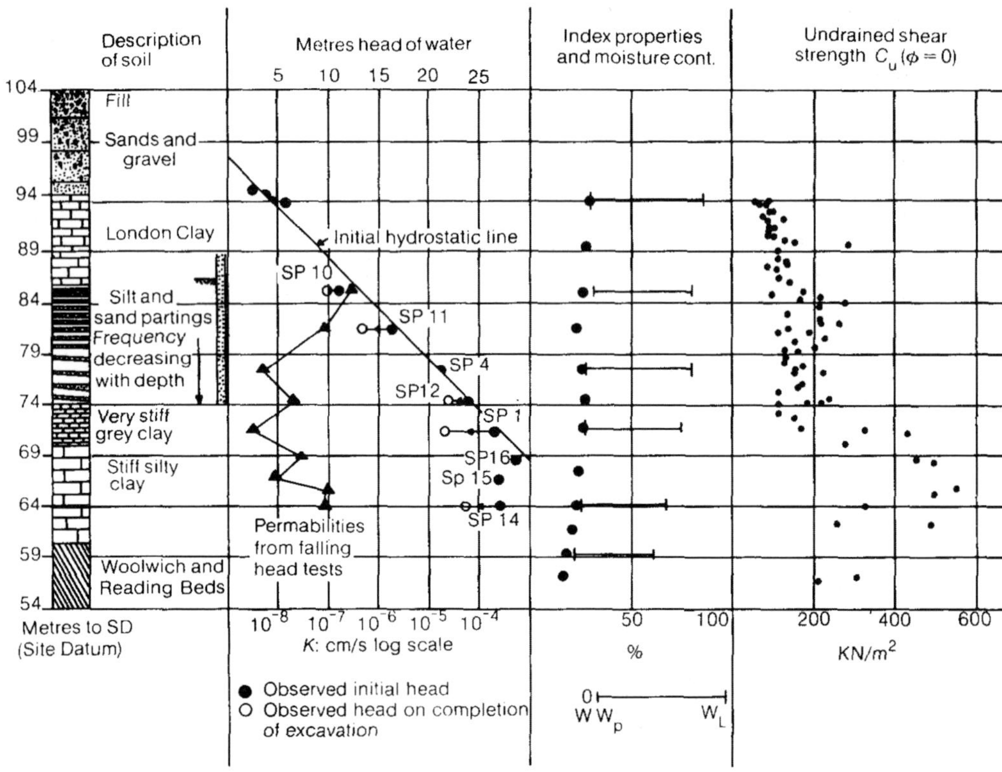

Fig. 11. New Palace Yard: site investigation. Reproduced from Burland and Hancock (1977)

To prevent the diaphragm wall from having an unrealistic ring stiffness, its stiffness in the circumferential direction was set to zero.

A study of the effect of piles within the excavation was carried out and it was concluded that the vertical stiffness of the piled zone should be modelled as 100 MN/m^2 greater than that of the clay alone.

The results of the computations are shown on Figs 13 to 15. Also shown are the movements computed by Ward and Burland (1973) before the excavation was carried out. The following points may be noted.

Vertical movements at ground level. The computed settlements are close to those measured perpendicular to the east wall and to those which occurred during excavation perpendicular to the south wall. As noted by Burland and Hancock (1977), the movements behind the south wall were complicated by the percussion boring outside the principal excavation and the total movements for this wall were somewhat larger.

The computed distribution of settlements is close to that measured. It shows a peak at about 10 m behind the wall and reduces to zero about 50 m away.

Horizontal movements at ground level. The computed and measured horizontal movements of the ground surface are compared in Fig. 13. The results lie between the total movements measured for the east and south walls, with a peak, somewhat sharper than was measured, at about the right distance, 15 to 20 m from the wall.

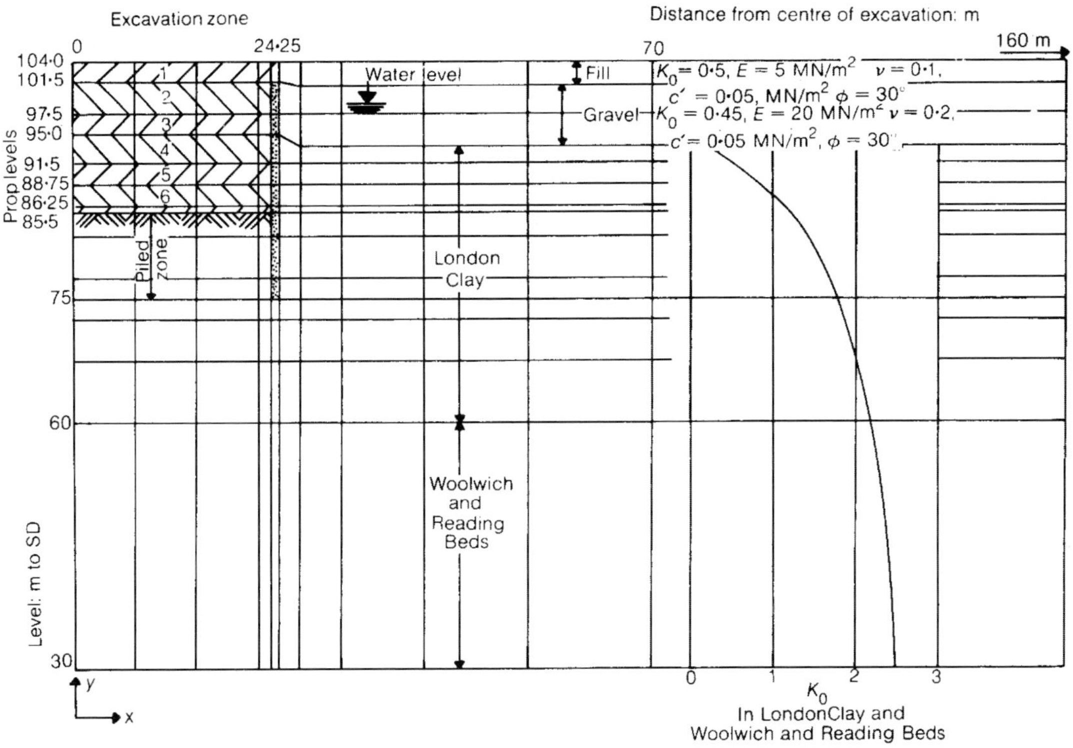

Fig. 12. New Palace Yard finite element mesh. Boundary $y = 30$ m SD fully restrained, $x = 0$ and 160 m restrained in x direction only

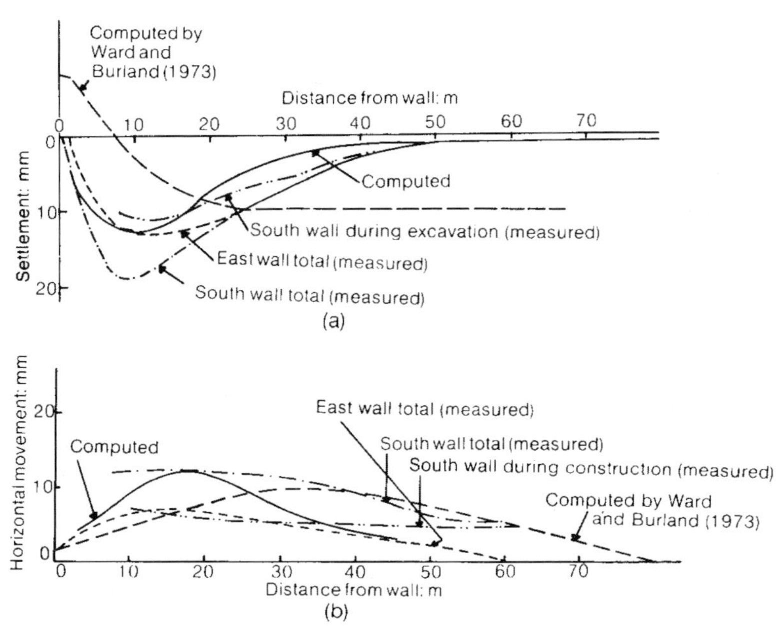

Fig. 13. New Palace Yard movement at ground level: (a) settlement; (b) horizontal movements

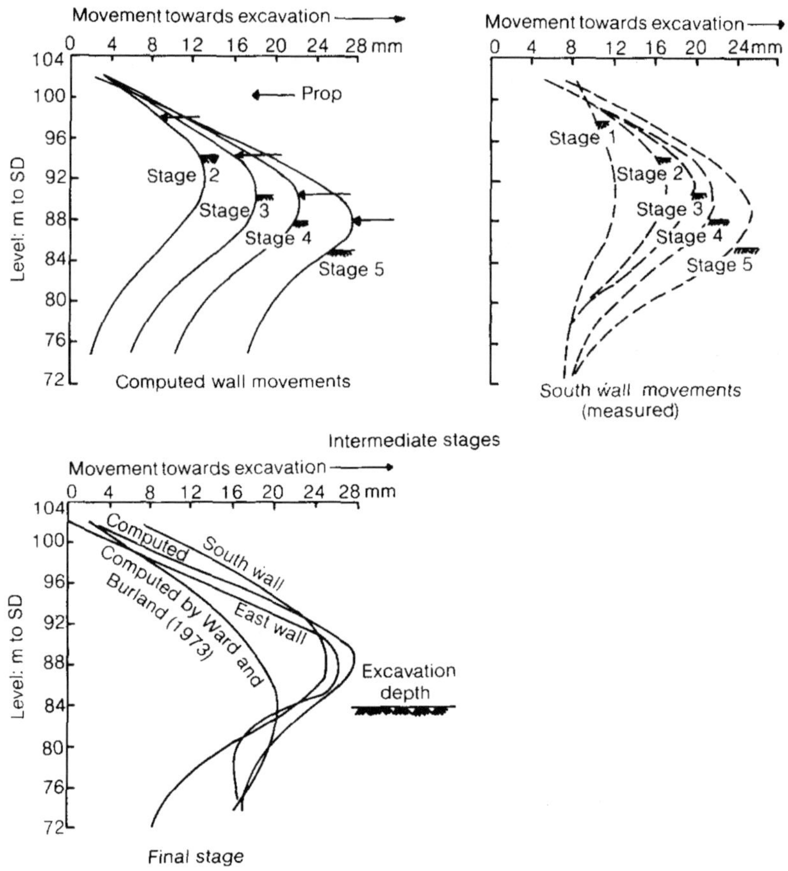

Fig. 14. New Palace Yard: movements of diaphragm walls

Diaphragm wall movements. St John (1975) records large variation between the measured wall movements at different sections around the circumference. Since the computations presented here are intended to apply to the mid-side of the walls, attention will be concentrated on inclinometers situated at the mid-sides of the east and south walls.

St John also observed that although the distortion of each wall is known with reasonable accuracy from the inclinometer profile, the absolute movements were difficult to determine because there were insufficient surface surveys to fix the position of the top of each wall during excavation.

The computed and measured movements of the wall at various stages during excavation are shown on Fig. 14. The correlation is generally good, and the following points may be noted.

(a) The computed magnitude of the peak displacement of the wall is close to the measured value and is correctly predicted to occur just above the final excavation level.

(b) The computed movement at the base of the wall agrees well with that measured at the east wall but not at the south wall.

Pore pressures. Figure 15 shows the computed and measured pore pressures at final excavation stage. The computed distribution is complicated and a detailed discussion would be somewhat speculative. However, it is considered that the computed and measured values are consistent if some allowance is made for short-term local seepage.

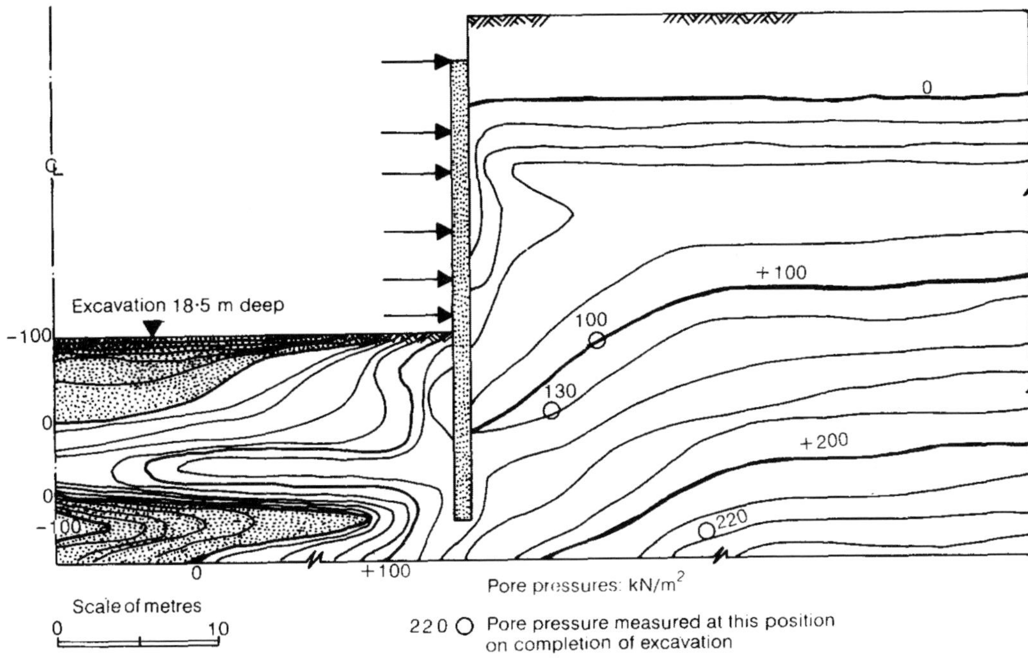

Fig. 15. New Palace Yard: computed pore pressures

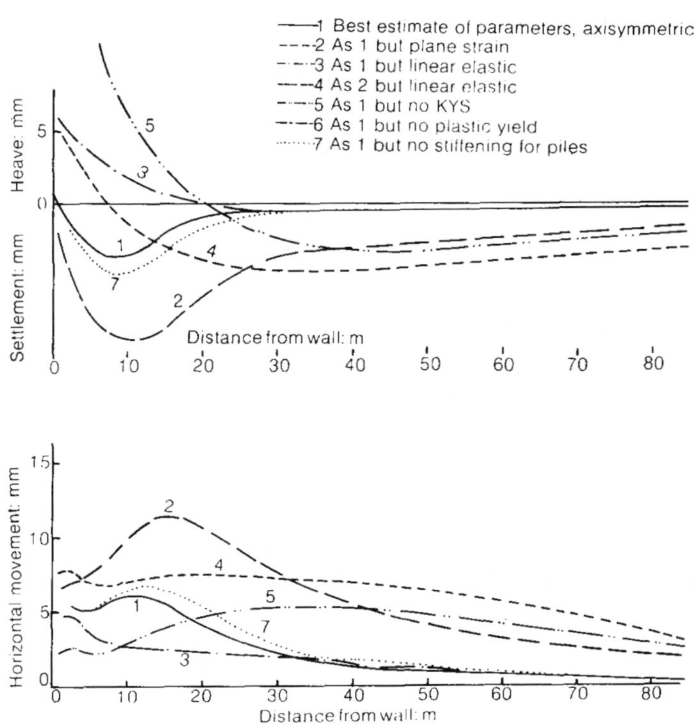

Fig. 16. Parametric study: movements at ground level at third stage of excavation

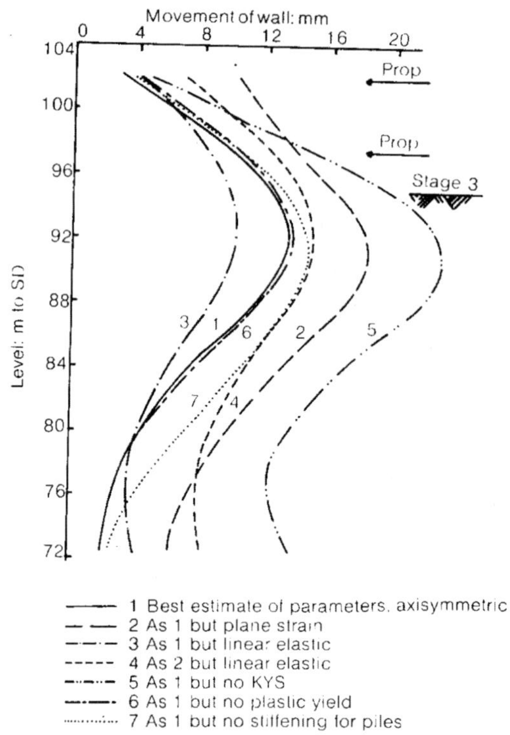

Fig. 17. Parametric study: movements of diaphragm wall at third stage of excavation

Comparison with linear elastic computations. Ward and Burland (1973) computed the expected movements around the excavation before work began on site. They assumed linear elasticity and plane strain conditions, and their predictions, which are shown on Figs 13 and 14, were in many respects close to the measured values.

The computations using Model LC give results which are at least as good, and in particular the observed settlement close to the wall is correctly predicted. This important difference between the linear models and Model LC is also evident in Fig. 16.

Parametric studies. In order to check the importance of the various characteristics of Model LC on the ground behaviour around the New Palace Yard excavation, six parametric studies were carried out. The results of these are shown in Figs 16 and 17. These runs were only taken to the third stage of excavation, equivalent to excavation below the second prop floor. The results may be compared against the primary run for which the movements at the end of excavation compared well with site measurements. The following points may be noted.

(a) The movement at the top of the diaphragm wall is largely independent of London Clay parameters (compare profiles 1, 3, 5 and 6).

(b) The pattern of ground movements around the excavation is modelled more closely if axisymmetric conditions are assumed (compare profiles 1 and 2).

(c) The threshold effect of the KYS greatly reduces the localized heave behind the excavation and results in smaller diaphragm wall movements (compare profiles 1, 3 and 5).

(d) The effects of the plastic flow in Model LC are minimal for this problem (compare profiles 1 and 6). The stiffening effect of the piles only slightly reduces the movement of the toe of the wall at this stage of the excavation (compare profiles 1 and 7).

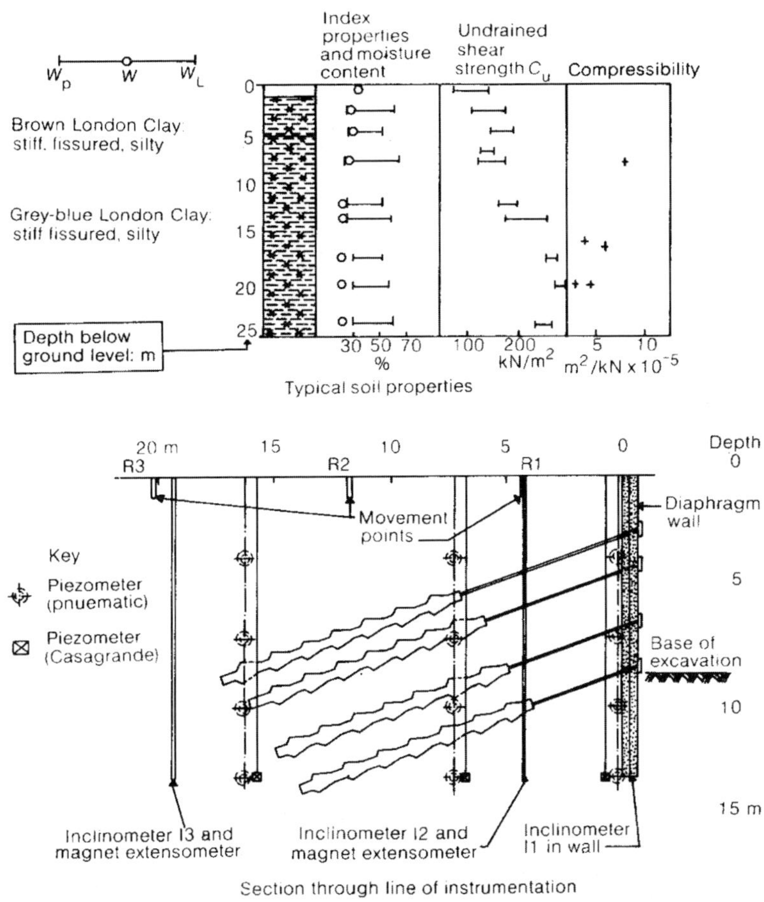

Fig. 18. Neasden Underpass: section and soils data. After Sills, Burland and Czechowski (1977)

Neasden Underpass

The site at Neasden Underpass and the design and monitoring of the diaphragm wall are described by Sills, Burland and Czechowski (1977) from which Fig. 18 is taken, summarizing the site investigation and the excavation details.

Figure 19 shows the finite element idealization incorporating the data as published by Sills *et al.* (1977), except for the inclination of the ground anchors. The diaphragm wall was restrained by 40° anchors except at the instrumented section described by Sills *et al.*, where the anchors were at 20°. It is considered that the wall behaviour was controlled by the 40° anchors rather than those inclined at 20°.

The simulated excavation sequence comprises seven steps:

(1) Initialization (calculates initial stresses)
(2) Excavation to 3 m below ground level
(3) First row of anchors installed
(4) Excavation to 5·5 m below ground level
(5) Second row of anchors installed
(6) Excavation to 8·5 m below ground level and third row of anchors installed
(7) Final row of anchors installed

The results of the computations are summarized in Figs 20 to 22. Figure 20 shows the

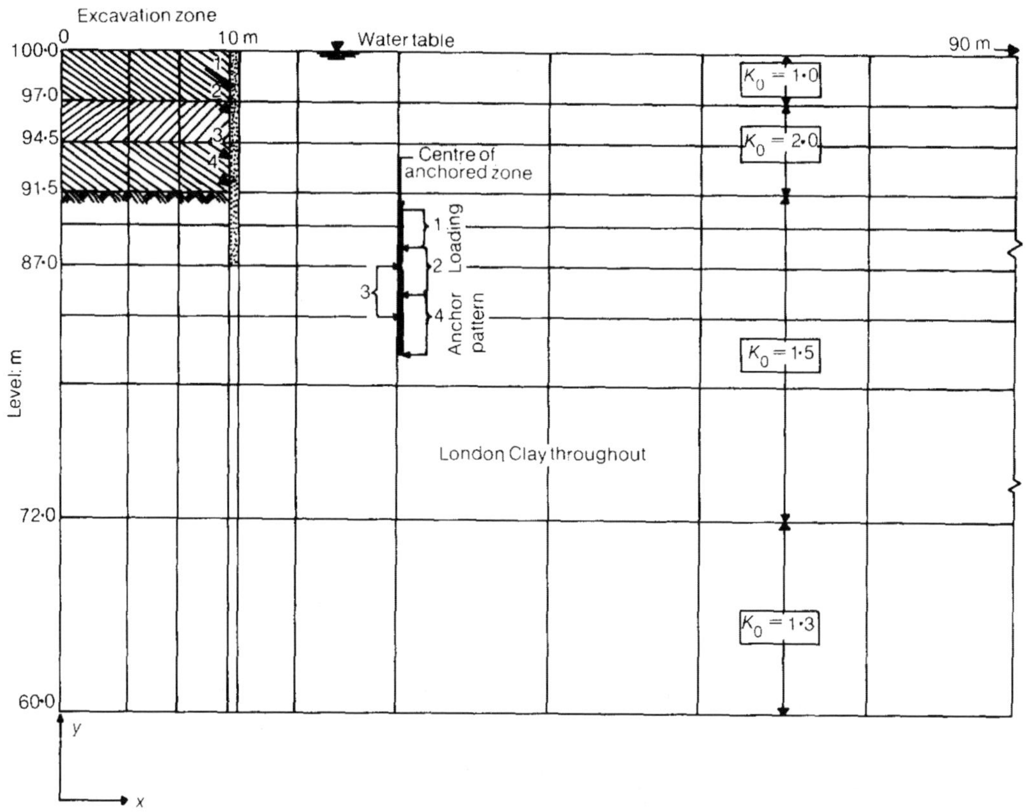

Fig. 19. Neasden Underpass: finite element mesh and modelling of anchors. Anchors modelled by inclined point loads each representing 197 kN/m run at the diaphragm wall and by pressure loads applied at 40° in the centre of the anchored zone within the ground. Boundary $y = 60$ m level fully restrained, $x = 0$ and 90 m restrained in x direction only

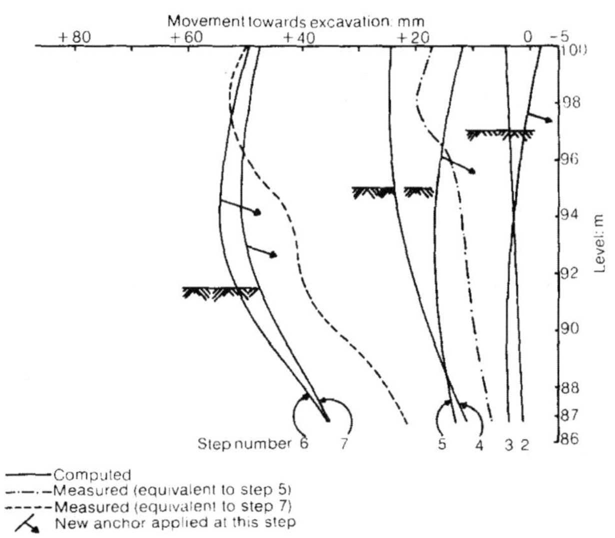

Fig. 20. Neasden Underpass: movements of diaphragm wall

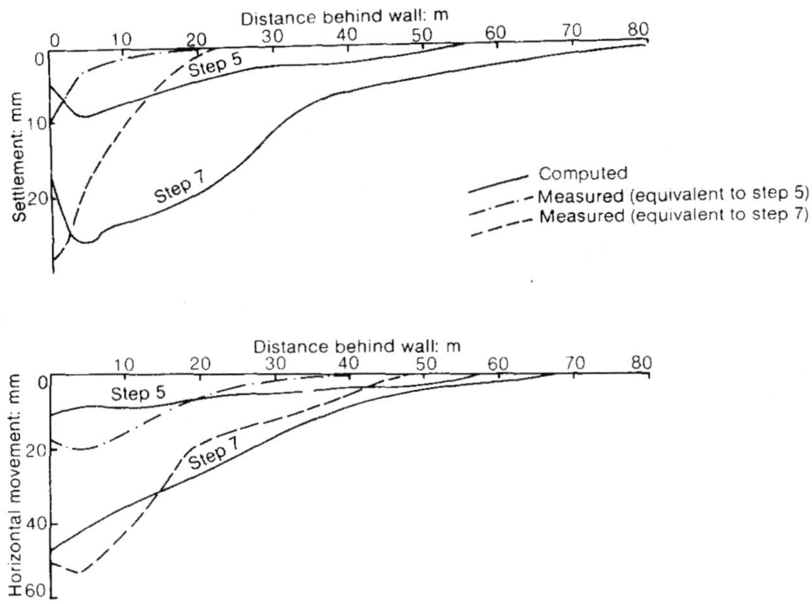

Fig. 21. Neasden Underpass: movement at ground level

predicted displacements of the diaphragm wall together with those presented by Sills *et al.* corresponding to steps 4 and 7 above. Although the displacements of the top of the wall show good correlation to the observed values, the predictions give up to 15 mm too much movement at the base of the diaphragm wall. On site, this movement may have been restrained by the stiffer soil indicated by the higher shear strength recorded below 12 m depth (Fig. 18).

Figure 21 shows the profile of predicted and observed horizontal movements behind the excavation. The predicted movements generally conform to those observed, and indicate over 2 mm lateral translation at distances at and beyond those corresponding to the surface survey datum points. Figure 21 also indicates the settlement profiles behind the wall. As the material is undrained and therefore effectively at constant volume, the overprediction of vertical movement at distances greater than 5 m from the excavation is probably associated with the excessive predicted horizontal movement of the base of the diaphragm wall as noted above.

The distribution of pore pressures around the excavation is shown in Fig. 22. The pattern is complex, as was found in the New Palace Yard back-analysis, but certain points can be noted as follows.

(a) Negative pore pressures are predicted in the centre of the excavation, together with moderate positive pore pressures close to the wall on the passive side.
(b) Immediately behind and below the wall there is a zone of comparatively high pore pressures corresponding to the local effect of the ground anchors.
(c) The anchors embedded 10 m from the wall induce moderately high excess pore pressures in a block of anchored soil between 12 m and 16 m below ground level.
(d) In general the predicted values over-estimate pore pressures when compared with standpipe readings eight months after the excavation (compare with Sills *et al.* 1977).

CONCLUSIONS

A mathematical model of the practical range of stress–strain behaviour of London Clay has been developed and used in the finite element program SAFE. The most important feature of

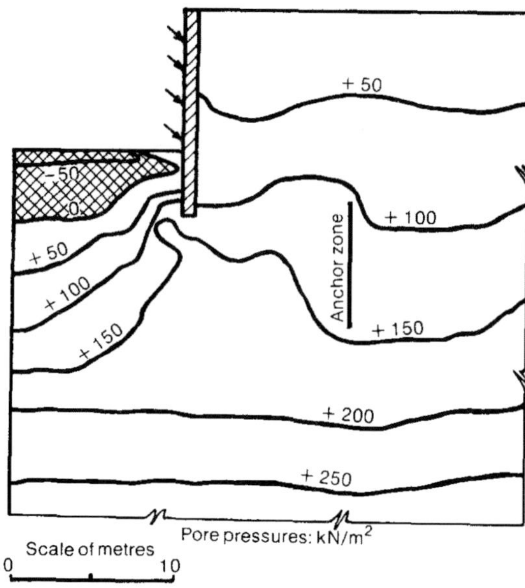

Fig. 22. Neasden Underpass: computed pore pressures

the model is the assumption of a threshold property. This means that at the very small strains often encountered in the field the material is much stiffer than at the larger strains to which laboratory samples are subjected. The model also takes account of variation of stiffness with mean normal stress and of plastic flow at large strains.

Computations made using the model correlate well with measurements of undrained and drained behaviour in the laboratory, in field tests and in full-scale excavations. In particular, the model has been used successfully to compute, in retrospect, the movements of both the diaphragm wall and the surrounding ground which occurred during construction of the New Palace Yard Underground Car Park. It was confirmed that axisymmetric assumptions were more appropriate to this type of excavation.

It is likely that for most field situations the predictions of the model will be significantly better than would be obtained using assumptions of linear elasticity. The model is considered suitable for analysis of both short- and long-term movements around excavations in London Clay and will be used for the prediction of the movements associated with the British Library excavations.

The close agreement with observations over a wide range of stress and strain conditions suggests that its use is not restricted to the analysis of excavations alone. The model may, with suitable changes in material parameters, be used for many soils other than London Clay.

ACKNOWLEDGEMENTS

The work has been carried out in connection with the design of the new British Library for the Department of the Environment. The Authors are grateful to their colleagues in Ove Arup and Partners who have aided the development of Model LC by their comments and suggestions.

APPENDIX: IMPLEMENTATION OF MODEL LC IN 'SAFE' FINITE ELEMENT PROGRAM

Introduction

The finite element program SAFE developed by Ove Arup and Partners is used for two-dimensional finite element computations under plane stress, plane strain or axisymmetric conditions.

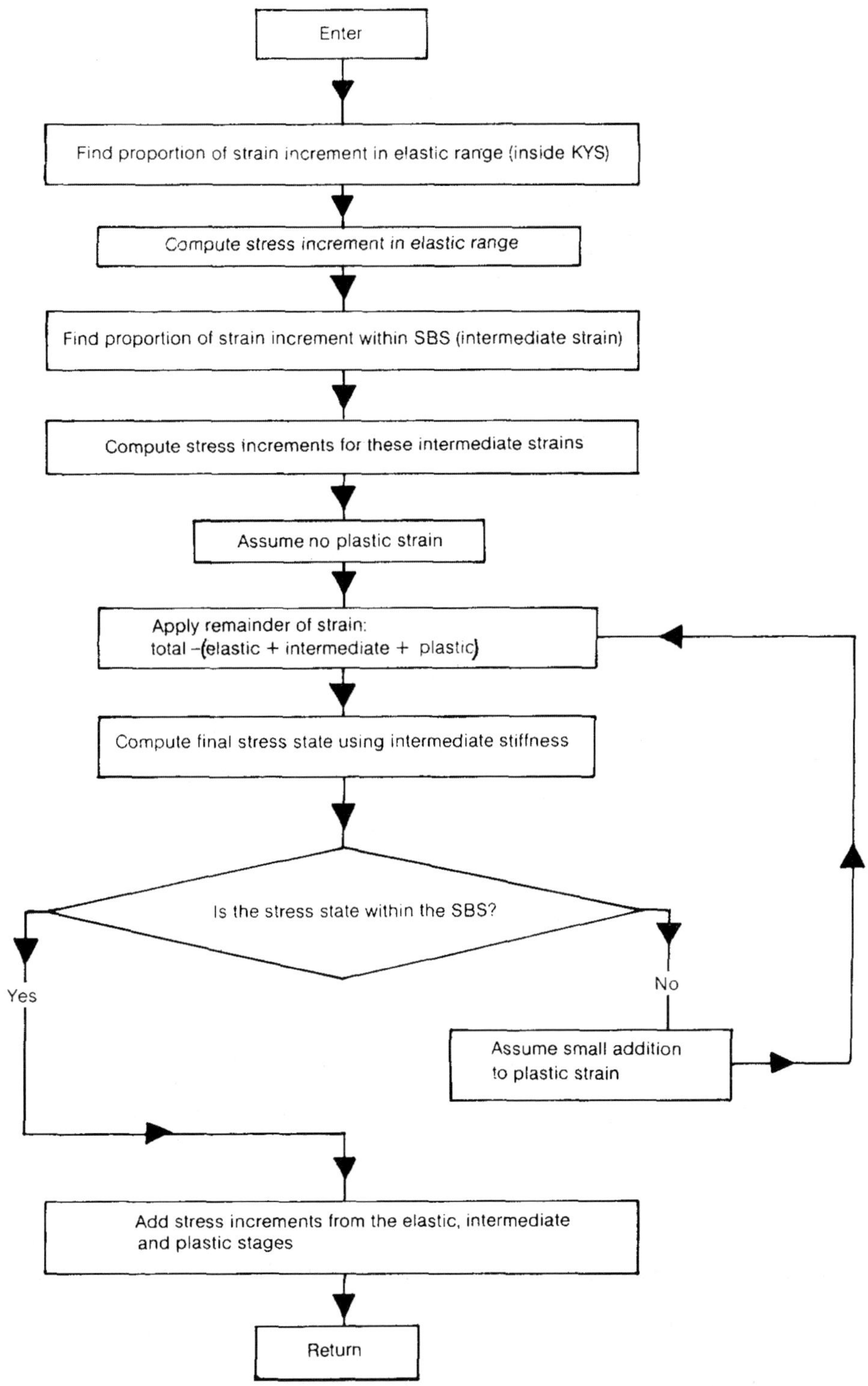

Fig. 23. Model LC: derivation of stress increment from strain increment

The program is suited to geotechnical problems and to non-linear behaviour in particular. Pore pressures and effective stresses are identified separately, and gravitational loads and initial stresses may be specified. Non-linear problems are dealt with by the initial stress technique, which may be used incrementally or otherwise, and a facility is also available for changing material types in order to simulate excavations or embankments incrementally.

The implementation of Model LC in the SAFE system is conveniently described in terms of its incremental behaviour in the elastic/intermediate stage and the plastic stage (a flow chart is shown in Fig. 23).

Elastic and intermediate ranges

Behaviour is non-linear at all stages, but within the elastic and intermediate stages, taken separately, the equations can be integrated and stress and strain increments are directly related. This makes it possible to find at what stage within an increment the strain path intersects the kinematic yield surface and the stress path intersects the state boundary curve.

Plastic range

In the plastic stage of the increment both intermediate and plastic strains occur. Their sum is known, and equal to the total strain in the increment minus the strain required to reach the state boundary curve. The estimated amount of plastic strain is gradually increased from zero, so that the estimated intermediate strain falls and the state progresses along the state boundary curve. A state is eventually found where the stress increment, computed from the intermediate stiffness and strains, is consistent with the point reached on the state boundary. If more than 20% plastic shear strain is required in one increment, the process is stopped and a reasonable, though inaccurate, stress increment is adopted. During this process it is difficult to find the correct principal directions for the plastic component of strain increment. An attempt is made to do this iteratively as the size of the plastic shear strain is incremented. The validity of this process in the rare situation where principal directions change rapidly during plastic shearing is uncertain.

The third principal stress, perpendicular to the main plane of deformation is governed by elastic and intermediate behaviour. It is also restricted to remain within the active and passive limits.

Drained and undrained behaviour

In the SAFE program, pore pressures are identified separately from effective stresses. Fully drained behaviour can be modelled by holding pore pressure constant. The undrained state, ie when there is no macroscopic flow of water relative to the soil skeleton, is more complicated; in this condition there are large changes of pore pressure and changes in volume are governed by the compressibility of the pore water.

For undrained elements, the bulk modulus of the water is included in the finite element formulation. Simpson (1973) and Naylor (1973) have reported that this can lead to unreasonable fluctuations in the computed pore pressures but a technique has been developed in the SAFE program to overcome this. The result is a set of computed pore pressures which is smoothly distributed and also conforms to overall equilibrium. Mean volumetric strain in each element is very small but significant volumetric strains may occur locally within each element, particularly in areas of high stress gradient.

Increment size

The theory developed above assumes that the increments of stress and strain are small. However, in order to reduce this restriction as far as possible, the program for Model LC has been written so that, provided the changes are monotonic, large increments will not introduce errors. The only exception to this rule is the rare event of large changes of principal directions during plastic straining, for which approximations have been adopted.

The iterative procedure used for non-linear work within the SAFE finite element program, however, converges faster if increment sizes are restricted to some extent.

REFERENCES

Atkinson, J. H. (1973). *The deformation of undisturbed London Clay*. PhD thesis, University of London.

Bishop, A. W., Webb, D. L. & Lewin, P. I. (1965). Undisturbed samples of London Clay from the Ashford Common Shaft: strength–effective stress relationships. *Géotechnique* **15**, No. 1, 1–31.

Bjerrum, L. (1967). Seventh Rankine Lecture: Engineering geology of Norwegian normally consolidated clays as related to settlements of buildings. *Géotechnique* **17**, No. 2, 81–118.

Burland, J. B. & Hancock, R. J. R. (1977). Underground car park at the House of Commons: geotechnical aspects. *Struct. Engr*, **55**, 87–100.

Butler, F. G. (1975). Heavily over-consolidated clays. *Conference on settlement of structures, Cambridge*, Review Paper, Session III. London: Pentech Press.

Cole, K. W. & Burland, J. B. (1972). Observation of retaining wall movements associated with a large excavation. *Proc. 5th Eur. Conf. Soil Mech. Fdn Engng, Madrid* **1**, 445–453.

Gibson, R. E. (1974). Fourteenth Rankine Lecture: The analytical method in soil mechanics. *Géotechnique* **24**, No. 2, 115–140.

Green, P. (1971). Some aspects of the foundation design of the Commercial Union Building. *Proc. Symp. Interaction of Structures and Foundations*, Midland Soil Mechanics and Foundation Engineering Society, Birmingham.

Hooper, J. A. (1974). Analysis of a circular raft in adhesive contact with a thick layer. *Géotechnique* **24**, No. 4, 561–580.

Lewin, P. I. (1970). *Stress deformation characteristics of a saturated soil.* MSc thesis, University of London.

Marsland, A. (1971a). Large in-situ tests to measure the properties of stiff fissured clays. *Proc. First Aust. N.Z. Conf. Geomechanics, Melbourne* **1**, 180–189.

Marsland, A. (1971b). Clays subjected to in situ plate tests. *Ground Engng* **5**, No. 7, 24–26, 29–31.

Marsland, A. (1971c). Laboratory and in situ measurements of the deformation moduli of London Clay. *Symposium on the interaction of structure and foundations.* Midland Soil Mechanics and Foundation Engineering Society, Birmingham.

May, J. (1974). Heave on a deep basement in the London Clay. *Conference on settlement of structures, Cambridge.* Pentech Press.

Mroz, Z. (1967). On the description of anisotropic hardening. *J. Mech. Phys. Solids* **15**, 163–165.

Peck, R. B. (1969). Deep excavations and tunnelling in soft ground. *Proc. Seventh Conf. Soil Mech. Fdn Engng, Mexico.* State of the Art Report, 225–290.

Roscoe, K. H. & Burland, J. B. (1968). On the generalized stress–strain behaviour of 'wet' clay. *Engineering plasticity* (eds Heyman & Leckie), 535–609. Cambridge University Press.

St John, H. J. (1975). *Field and theoretical studies of the behaviour of ground around deep excavations in London Clay.* PhD thesis, University of Cambridge.

Schofield, A. N. & Wroth, C. P. (1968). *Critical state soil mechanics.* London: McGraw-Hill.

Sills, G. C., Burland, J. B. & Czechowski, M. K. (1977). Behaviour of an anchored diaphragm wall in stiff clay. *Proc. Ninth Int. Conf. Soil Mech. Fdn Engn, Tokyo* **2**, 147–155.

Simpson, B. (1973). *Finite elements for problems in soils.* PhD thesis, University of Cambridge.

Som, N. N. (1968). *The effect of stress path on the deformation and consolidation of London Clay.* PhD thesis, University of London.

Terzaghi, K. (1941). Undisturbed clay samples and undisturbed clays. *J. Boston Soc. civ. Engrs* **28**, No. 3.

Ward, W. H. & Burland, J. B. (1973). The use of ground strain measurements in civil engineering. *Phil. Trans. R. Soc.* **A274**, 421–428.

Zytinski, M., Randolph, M. F., Nova, R. & Wroth, C. P. (1976). *On modelling the unloading–reloading behaviour of soils.* Internal Report CUED/C—Soils TR37, University of Cambridge.

Simpson, B., O'Riordan, N. J. & Croft, D. D. (1980). *Géotechnique* **30**, No. 3, 336–339

A computer model for the analysis of ground movements in London clay

SIMPSON, B., O'RIORDAN, N. J. & CROFT, D. D. (1979). *Géotechnique* **29**, No. 2, 149–176.

L. F. Costa-Filho and P. R. Vaughan, Department of Civil Engineering, Imperial College of Science and Technology

It seems that the model proposed by the Authors is able to predict satisfactorily the behaviour of excavations in London clay as well as the initial stages of loading of conventional laboratory tests and of in situ large diameter plate load tests. However, except for the laboratory triaxial and plane strain tests, the accuracy of the predictions of the model relies heavily on the assumption of a threshold of stress (or strain) changes below which the material is assumed to behave elastically with a stiffness equal to ten times that generally quoted for laboratory tests for London clay. The importance of this initial, very stiff, threshold stage is clearly shown in Figs 16 and 17 of the Paper in connection with the parametric study performed by the Authors for the New Palace Yard excavation.

Laboratory triaxial tests have been recently performed in the soil mechanics laboratory of Imperial College specially designed to investigate the behaviour of London clay in the small strain range. Axial displacements have been measured directly on the specimen on its central part, therefore free from end effects and bedding errors. Overall displacements have also been measured directly between the top cap and the pedestal. The equipment is described by Daramola (1978) and Costa Filho (1979) but at this stage it is sufficient to note that axial strains of about 0·05% can be measured with an accuracy of ±0·005% in the central part of a specimen 76 mm high.

Block samples were taken from an open pit in South Ockendon, east London. This site has

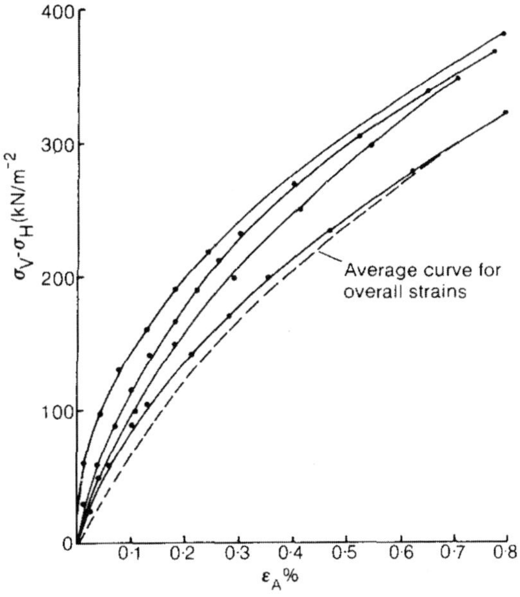

Fig. 1. Unconsolidated undrained tests

already been investigated by Apted (1977) in connection with a study of the effects of weathering on the geotechnical characteristics of London clay and a considerable amount of information was available on the properties of London clay and their variation with depth. The samples were taken from a depth of about 18 m in apparently intact blue London clay. At this depth, however, the clay is very silty and sandy seams were frequently observed in some specimens after a few moments of drying in the soil mechanics laboratory. Undrained compression tests were performed on apparently intact 38 mm diameter specimens unconsolidated (UU) and anisotropically consolidated (CAU) to assumed field effective stresses. A value of K_0 equal to 2 has been adopted as being typical of London clay for the depth from which the samples were obtained, and this value was confirmed by suction measurements in the specimens. A fast rate, about 4% per hour, was used in most tests so that a negligible equalization of pore water pressure occurred in the centre part of the specimen, where the axial displacements were being measured.

Typical results for the UU and CAU tests are shown in Figs 1 and 2 respectively using the axial strains measured directly on the specimen. Only that portion of the stress–strain curves up to axial strains of the order of 1% is shown since, as mentioned by the Authors and by Burland, Simpson & St John (1979), except locally the shear strains mobilized in most field situations seldom exceed 0·25%. Both sets of curves show significant non-linearity and, what is more important, the tests starting from the assumed field stress show an initial behaviour much stiffer than that observed in the tests starting from an isotropic state of stress. Stiffer behaviour of specimens of London clay subjected to an extension state of stress before compression shearing has also been reported by Som (1968) and Simons & Som (1970). Average stress–strain curves deduced from the overall compression of the specimens, according to conventional practice, are also shown in Figs 1 and 2. These curves substantially underestimate the initial stiffness. An initial very stiff behaviour has also been observed by the Writers in tests on reconstituted London clay consolidated anisotropically to a K_0 equal to 1·5 and sheared in undrained compression. The same series of tests on reconstituted specimens

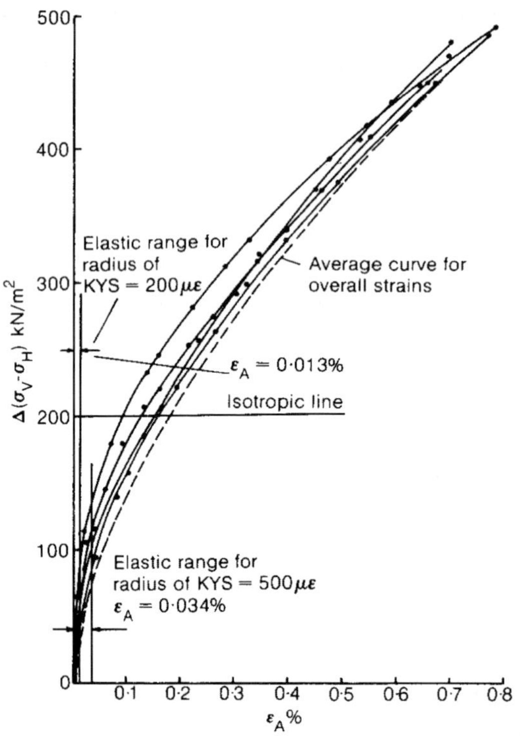

Fig. 2. Consolidated anisotropically undrained tests

indicated that the reconsolidation to field stresses, with $K_0 > 1{\cdot}0$, to which the intact specimens of London clay were subjected with the corresponding two reversions in direction of shearing, is likely to induce an increase of stiffness of the material of only about 10% if compared with a specimen that has not been subjected to any sampling cycle.

The high initial stiffness observed in the laboratory CAU tests confirms the type of model proposed by the Authors and the stiffness measured in the laboratory is comparable with that invoked by the Authors in explaining field observations. A similar but more muted behaviour is also observed in UU tests when internal strains are measured.

The marked non-linearity observed in all tests at small strains implies that different stiffnesses can be obtained for London clay for different strain levels. If meaningful comparisons are to be made between laboratory tests, in situ tests and observed field behaviour, proper allowance must be made for the strains involved as well as for the stress history of the material.

It is encouraging that the much increased stiffness at small strains, inferred by the Authors as necessary to explain field deformations, can be measured independently in the laboratory when anisotropic unconsolidated tests and accurate measuring techniques are used.

The findings of the Authors, together with the laboratory results reported here, suggest that the laboratory and field give similar results, and that previous discrepancies (e.g. Atkinson, 1973) have resulted mainly from incorrect measurement and interpretation.

REFERENCES

Apted, J. P. (1977). *Effects of weathering on some geotechnical properties of London clay.* PhD thesis, University of London.

Atkinson, J. H. (1973). Discussion on Interpretation. *Field instrumentation in geotechnical engineering*. Butterworths, London, 1974, 677–679.
Bishop, A. W., Webb, D. L. & Lewin, P. I. (1965). Undisturbed samples of London clay from the Ashford Common shaft-strength-effective stress relationships, *Géotechnique* **15**, No. 1, 1–30.
Burland, J. B., Simpson, B. & St John, H. D. (1979). Movements around excavations in London clay. *Proc. 7th Eur. Conf. Soil Mech., Brighton* **1**.
Costa-Filho, L. de M. (1979). *A laboratory investigation of the small strain behaviour of London clay*. PhD thesis, University of London (in preparation).
Daramola, O. (1978). *The influence of stress history on the deformation of a sand*. PhD thesis, University of London.
Som, N. N. (1968). *The effect of stress path on the deformation and consolidation of London clay*. PhD thesis, University of London.
Simons, N. E. & Som, N. N. (1970). *Settlement of structures of clay, with particular emphasis on London clay*. Construction Industry Research and Information Association, London, report 22.

Authors' reply

The Authors have only become aware of Costa-Filho's work since publication of the Paper. His data strongly support the concept, embodied in the Authors' numerical model, that London clay displays much greater stiffness at very small strains than at the larger strains normally measured in the laboratory.

Costa-Filho's work also supports the idea that this 'threshold' effect is more noticeable when the direction of the stress path is abruptly changed, and good, possibly fortuitous, agreement has been found to exist between the Authors' model and the results shown in the Writers' Fig. 2.

It is interesting that the Writers also show a smaller threshold effect in Fig. 1 for tests which have not had such an abrupt change of stress path. This raises the important question, touched on by Parry (1979), as to what leads to the threshold effect. Is it (a) change of stress path, (b) ageing when the clay is allowed to stand, (c) the current stress ratio, (d) some combination of all these?

This question could usefully be the subject of further research, together with a study of the inter-particle effects which cause the phenomenon to occur.

REFERENCE

Parry, R. H. G. (1979). *Proc. 7th Eur. Conf. Soil Mech., Brighton*, Discussion to Session 4.

Alonso, E. E., Gens, A. & Josa, A. (1990). *Géotechnique* **40**, No. 3, 405–430

A constitutive model for partially saturated soils

E. E. ALONSO*, A. GENS* and A. JOSA*

The Paper presents a constitutive model for describing the stress–strain behaviour of partially saturated soils. The model is formulated within the framework of hardening plasticity using two independent sets of stress variables: the excess of total stress over air pressure and the suction. The model is able to represent, in a consistent and unified manner, many of the fundamental features of the behaviour of partially saturated soils which had been treated separately by previously proposed models. On reaching saturation, the model becomes a conventional critical state model. Because experimental evidence is still limited, the model has been kept as simple as possible in order to provide a basic framework from which extensions are possible. The model is intended for partially saturated soils which are slightly or moderately expansive. After formulating the model for isotropic and triaxial stress states, typical predictions are described and compared, in a qualitative way, with characteristic trends of the behaviour of partially saturated soils. Afterwards, the results of a number of suction-controlled laboratory tests on compacted kaolin and a sandy clay are used to evaluate the ability of the model to reproduce, quantitatively, observed behaviour. The agreement between observed and computed results is considered satisfactory and confirms the possibilities of reproducing the most important features of partially saturated soil behaviour using a simple general framework.

KEYWORDS: compaction; constitutive relations; partial saturation; plasticity.

L'article présente un modèle constitutif pour décrire le comportement contrainte/déformation des sols partiellement saturés. Le modèle est formulé dans le cadre de la plasticité qui s'endurcit, en utilisant deux jeux indépendants de variables de contrainte, c'est-à-dire l'excès de la contrainte totale sur la pression de l'air et la succion. Le modèle peut représenter d'une façon logique et unifiée beaucoup des caractéristiques fondamentales du comportement des sols partiellement saturés qui avaient été traités séparément dans des modèles proposés précédemment. Lorsque la saturation est atteinte le modèle devient un modèle conventionnel d'état critique. Comme les données expérimentales restent encore limitées, le modèle d'été simplifié le plus que possible, afin de fournir un cadre de base à partir duquel des extensions sont possibles. Le modèle convient à des sols partiellement saturés qui sont expansifs à un degré faible ou modéré. Le modèle est d'abord formulé pour des états de contrainte isotropes et triaxiaux. Puis l'article décrit des prédictions typiques et les compare de façon qualitative avec des tendances caractéristiques du comportement des sols partiellement saturés. Les résultats de quelques essais régis par la succion effectués dans le laboratoire sur du kaolin compact et sur une argile sablonneuse sont utilisés afin d'évaluer la capacité du modèle de reproduire de façon quantitative le comportement observé. On considère que les résultats observés et calculés s'accordent bien et qu'ils confirment les possibilités de reproduire les caractéristiques les plus importantes du comportement des sols partiellement saturés par l'utilisation d'un cadre général très simple.

NOTATION

- a, b, c, d parameters of the state surface for specific volume
- e void ratio
- G shear modulus
- K_0 lateral stress coefficient
- k parameter describing the increase in cohesion with suction
- M slope of critical state lines
- $N(s)$ specific volume at $p = p^c$
- m, n parameters of the state surface for degree of saturation
- p excess of mean stress over air pressure (net mean stress)
- p_{at} atmospheric pressure
- p_0 preconsolidation stress
- $p_0{}^*$ preconsolidation stress for saturated conditions
- p^c reference stress
- q deviatoric stress ($= \sigma_1 - \sigma_3$)
- r parameter defining the maximum soil stiffness

Discussion on this Paper closes 4 January 1991; for further details see p. ii.

* Civil Engineering School, Technical University of Catalunya, Barcelona, Spain.

S_r degree of saturation
s suction ($= u_a - u_w$)
s_0 hardening parameter of the suction increase yield curve
u_a air pressure
u_w water pressure
v specific volume
β parameter controlling the rate of increase of soil stiffness with suction
$\varepsilon_1, \varepsilon_3$ principal strains
$\varepsilon_1^e, \varepsilon_3^e$ elastic principal strains
$\varepsilon_1^p, \varepsilon_3^p$ plastic principal strains
$\varepsilon_s, \varepsilon_s^e, \varepsilon_s^p$ total, elastic and plastic deviatoric strains
$\varepsilon_v, \varepsilon_v^e, \varepsilon_v^p$ total, elastic and plastic volumetric strains
ε_{vp}^e elastic volumetric strain induced by changes in net mean stress
ε_{vp}^p plastic volumetric strain associated with the LC yield surface
ε_{vp} $(\varepsilon_{vp}^e + \varepsilon_{vp}^p)$
ε_{vs}^e elastic volumetric strain induced by changes in suction
ε_{vs}^p plastic volumetric strain associated with the SI yield surface
ε_{vs} $(\varepsilon_{vs}^e + \varepsilon_{vs}^p)$
φ' saturated friction angle
κ elastic stiffness parameter for changes in net mean stress
κ_s elastic stiffness parameter for changes in suction
$\lambda(s)$ stiffness parameter for changes in net mean stress for virgin states of the soil
λ_s stiffness parameter for changes in suction for virgin states of the soil
$\sigma_1, \sigma_2, \sigma_3$ total principal stresses
σ_{ij} total stresses
σ_m total mean stress ($= (\sigma_1 + \sigma_2 + \sigma_3)/3$)

INTRODUCTION

The description of the stress–strain behaviour of partially saturated soils has been closely linked with efforts to isolate the relevant stress variables concerning soil behaviour. Difficulties in finding a single effective stress equation, illustrated in the works of Aitchison and Donald (1956), Bishop and Donald (1961), Bishop and Blight (1963) and Burland (1965), progressively led to the acceptance of two effective stress fields as a necessary framework to describe the observed features of partially saturated soil behaviour under paths involving the variation of total stress and pore water pressure deficiency (or suction).

Important contributions to this topic were the early works of Bishop & Donald (1961), Coleman (1962), Matyas & Radhakrishna (1968) and the null tests reported by Fredlund & Morgenstern (1977) which led to a formal proposition of any pair of stress fields among the following three stress states: $(\sigma_{ij} - u_a\delta_{ij})$; $(\sigma_{ij} - u_w\delta_{ij})$ and $(u_a - u_w)\delta_{ij}$ as a suitable stress framework to describe stress–strain—strength behaviour of partially saturated soils. In these expressions, σ_{ij} is the total stress, u_a and u_w the air and water pressure and δ_{ij} the Kronecker delta. Under conditions of partial saturation the term $u_a - u_w$ is a positive quantity and will be denoted as the suction term s.

Adopting the suction and the excess of total stress over air pressure as the relevant stress variables, several aspects of the mechanical behaviour of partially saturated soils have been modelled in the past through constitutive relationships. Some of these models which cover partial aspects of the stress–strain behaviour are

(*a*) The incremental stress–strain relations suggested by Coleman (1962) to relate changes in s and $\sigma_{ij} - u_a\delta_{ij}$ to the strain response of the soil. This is essentially an incremental elastic type of formulation and it was also proposed by Fredlund (1979) who recognized the nonlinear character of the moduli involved in the incremental relations.

(*b*) The description of volumetric behaviour by means of state surfaces in a three-dimensional $[e, (\sigma - u_a), s]$ space. These surfaces are able to describe the transitions from swelling to collapse behaviour when the total stress increases. They were formally introduced by Matyas & Radhakrishna (1968). Based on experimental results, Fredlund (1979) and Lloret & Alonso (1980, 1985) proposed mathematical expressions for these surfaces, for a class of imposed stress paths.

(*c*) An extension of the Mohr-Coulomb failure criteria to include partial saturation, through a new angle of friction relating changes in shear strength to changes in suction (Fredlund, Morgenstern & Widger, 1978). The original linear expression has been recently questioned in view of increasing experimental evidence (Escario & Sáez, 1986, 1987; Fredlund, Rahardjo & Gan, 1987).

These valuable efforts are, however, unable to provide an integrated scheme for a stress–strain characterization of partially saturated soils. In addition, some basic features of partially saturated soil behaviour such as its strong stress path dependence are poorly represented by the existing models.

In devising the model, simplicity was considered a desirable feature. Nevertheless, fundamental tendencies of observed behaviour of partially saturated soils are believed to be represented by the model. The model, which uses well established concepts of the theory of hardening plasticity and reduces to a critical state model when the soil reaches full saturation, offers room for improvement when more experimental evidence becomes available. A qualitative description of this model and its relationship with observed behaviour is given in Alonso, Gens & Hight (1987).

The model described in this Paper is intended for partially saturated soils which are slightly or moderately expansive such as partially saturated sands, silts, clayey sands, sandy clays and clays of low plasticity. It is also thought that some features of the behaviour of expansive clays are suitably represented by the proposed model as well. However, a comprehensive description of this type of soil will probably require significant modifications of the model described below.

FEATURES OF BEHAVIOUR OF PARTIALLY SATURATED SOIL

Some patterns of the behaviour of non-plastic and moderately expansive partially saturated soils have been established in the past three decades through experimental work involving mainly oedometer testing (with or without suction control) and, to a lesser extent, triaxial and direct shear testing. The main features of this behaviour are summarized in the following paragraphs. Additional details are given in Alonso, Gens & Hight (1987).

Suction contributes to stiffening the soil against external loading. This is a well proved result when compacted samples either dry or wet of optimum are loaded in compression (Rico & Del Castillo, 1976). The oedometric results of samples tested at different moisture conditions can also be interpreted as an increase in the apparent preconsolidation stress as suction increases (Dudley, 1970).

If the natural depositional processes or the compaction method induce an open structure in the soil, a reduction in suction (wetting) for a given confining stress may induce an irrecoverable volumetric compression (collapse). For a certain range of the confining stress (which varies from soil to soil) the amount of collapse increases with the intensity of the confining stress (Blight, 1965; Barden, Madedor & Sides, 1969; Dudley, 1970; Booth, 1975; Jennings & Knight, 1975; Erol & El Ruwaih, 1982; Yudhbir, 1982). However, as the confining stress is increased, the amount of collapse may reach a maximum followed by decreasing values. Results compiled by Yudhbir (1982) for different residual soils show that the confining stress leading to the maximum collapse varies widely for different initial conditions. Similar qualitative results for compacted samples of a low plasticity residual clay have been reported by Booth (1975). Finally, once wetted, the collapsed sample follows, approximately, the stress–strain path of an initially saturated sample in compression (Erol & El Ruwaih, 1982, testing a natural loess; Maswoswe, 1985, testing a sandy clay).

Any partially saturated soil included in the types mentioned (and even active clays) may either expand or collapse on wetting if the confining stress is sufficiently low (expansion) or high (collapse). Among many others, results of this type may be found in Maswoswe (1985) (testing a statically compacted sandy clay) and Justo, Delgado & Ruiz (1984) (testing compacted samples of expansive clay).

Suction controlled experiments have revealed, however, that a reversal in volumetric behaviour may also occur during the wetting process. In fact, potentially collapsible microstructures subjected to a suction reduction from a relatively high initial value have been reported to experience first an expansion and then a compressive strain (Escario & Sáez, 1973; Cox, 1978; Josa, Alonso, Lloret & Gens, 1987).

The volumetric response of partially saturated soils depends not only on the initial and final stress and suction values but on the particular path followed from the initial to the final state. It has been often reported (Matyas & Radhakrishna, 1968; Barden *et al.*, 1969; Lloret & Alonso, 1985; Josa *et al.*, 1987) that paths which involve loading steps and non-increasing suction are essentially path independent. However, significant departures from this result are obtained when the stress path involves an increase in suction and an increase in external load. The state surfaces described by some of the authors listed above are only meaningful if stress-path independence is guaranteed.

Changes in suction may induce irreversible volumetric deformations. This behaviour may be investigated by subjecting samples of soil to drying-wetting cycles. Tests of this type have been reported by Yong, Japp & How (1971) for a kaolinite and a natural low activity clay and by Josa *et al.* (1987) for a moderately plastic kaolin. In both cases plastic volumetric strains were measured as a consequence of a drying process.

For a limited range of confining stress and applied suction, shear strength tests reported by Fredlund *et al.* (1978), Gulhati and Satija (1981), Ho and Fredlund (1982) and Fredlund & Rahardjo (1985) may be interpreted in the sense that an increase in suction results in an increase in effective cohesion while maintaining constant the (saturated) friction angle φ'. Departures from this pattern have been reported by Escario and Sáez (1986) and Delage, Suraj de Silva & De Laure (1987). The increase in strength cannot continue indefinitely with applied suction. Some recent results reported by Escario and Sáez (1987) and Fredlund *et al.* (1987), which extend the suction ranges used in previously published results, indicate a non-linear increase of strength with suction and a tendency towards a maximum at some given (high) suction.

Most of this experimental evidence has been used to develop a model which could offer a unified approach to some fundamental features of the behaviour of partially saturated soils. The conceptual bases for the model and the basic formulation are given in the next sections. Examples of the response of the model to some imposed stress and strain conditions are then presented. Finally, a comparison between predicted and measured response for some suction controlled tests will be made. This comparison will also allow a more detailed presentation of some features of the behaviour of partially saturated soils which have only been outlined in the summary above.

FORMULATION OF MODEL FOR ISOTROPIC STRESS STATES

Stress states inducing yield

A suitable stress space to describe isotropic states is (p, s) where $p = \sigma_m - u_a$ (net mean stress) is the excess of mean stress σ_m over air pressure and s is the suction. The purpose of this section is to derive a yield locus in the (p, s) space using as background information the behaviour of saturated soils (which impose a limiting condition to partially saturated soils) and observed features of isotropic and oedometer testing under controlled suction.

It is not strictly necessary to derive a yield function in the way described below. It may simply be proposed, on the basis of observed yield behaviour, as a mathematical relationship that involves constitutive parameters. Stress path testing may later be used to check the overall behaviour of the model and its ability to reproduce measured stress–strain response of the soil. However, it is convenient in this case to present a formal derivation of the yield function to underline 'ab initio' the relationship of yield states with observed volumetric behaviour of partially saturated soils and to incorporate the well established features of isotropic or oedometer compression of saturated soils. In this way the developed framework will be hopefully well rooted in commonly observed experimental behaviour.

Consider an isotropic test in which a soil sample at a given suction (subsequently maintained) is subjected to p-load increments along virgin states. In full correspondence with the behaviour of saturated soils the specific volume, $v = 1 + e$, will be given by

$$v = N(s) - \lambda(s) \ln \frac{p}{p^c} \tag{1}$$

where p^c is a reference stress state for which $v = N(s)$. On unloading and reloading (at constant s) the soil is supposed to behave elastically:

$$dv = -\kappa \frac{dp}{p} \tag{2}$$

In these equations the stiffness parameter $\lambda(s)$ depends on suction in accordance with observations reported in the previous section. A dependence of $N(s)$ on s will be maintained for the sake of generality. Although some dependence of κ on s is also likely, a constant value has been adopted to ensure that the elastic part of the model is conservative. A schematic representation of Equations (1) and (2) is given in Fig. 1.

Consider in Fig. 2(a) the response to isotropic loading of two samples subjected to different suctions: $s = 0$ (saturated case) and a larger suction s. The saturated preconsolidation stress is labelled p_0^* (point 3 in Fig. 2(a)). The second sample will yield at a larger isotropic stress p_0 (point 1).

If both points, 1 and 3, belong to the same yield curve in a (p, s) stress plane (Fig. 2(b)), a relationship between the generic yield stress, p_0 and the saturated value p_0^* can now be obtained by relating the specific volumes at points 1 and 3 through a virtual path which involves an initial unloading, at constant suction, from p_0 to p_0^*, and a subsequent reduction in suction, from s to zero, at constant stress (p_0^*). The sample, initially at point 1 (p, s) follows the path 1-2-3 (Figs 2(a)

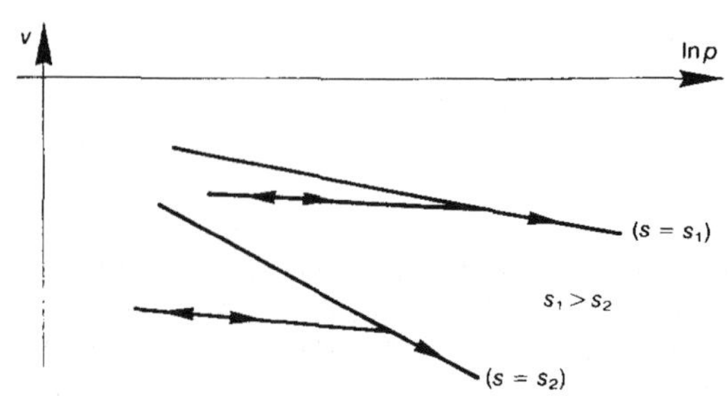

Fig. 1. Proposed variation of specific volume with net mean stress and suction along virgin and unloading–reloading paths

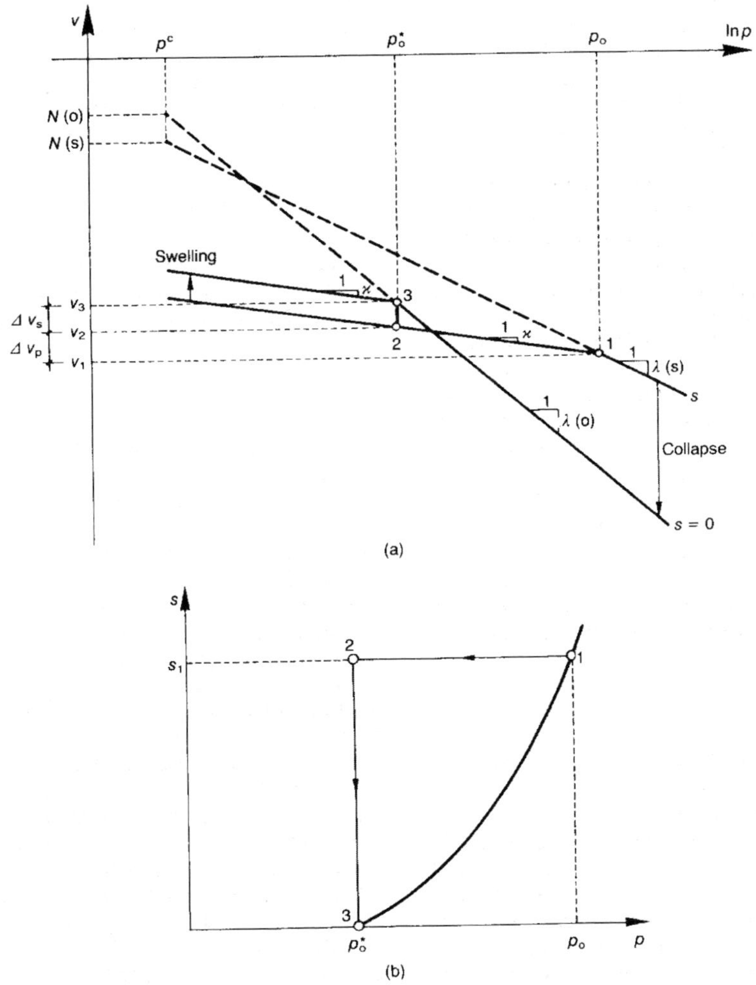

Fig. 2. Relationship between preconsolidation stresses p_0 and $p_0{}^*$: (a) compression curves for saturated and unsaturated soil; (b) stress path and yield curve in (p, s) stress plane

and 2(b)). The following identity may be established

$$v_1 + \Delta v_p + \Delta v_s = v_3 \quad (3)$$

where the different quantities are indicated in Fig. 2(a).

The suction unloading (wetting) from 2 to 3 occurs in the elastic domain. Under these circumstances a reversible swelling Δv_s takes place which is given by a logarithmic expression similar to Equation (2)

$$dv = -\kappa_s \frac{ds}{(s + p_{at})} \quad (4)$$

p_{at} has been added to s to avoid infinite values as s approaches zero.

Taking into account Equations (1), (2) and (4), Equation (3) becomes

$$N(s) - \lambda(s) \ln \frac{p_0}{p^c} + \kappa \ln \frac{p_0}{p_0{}^*} + \kappa_s \ln \frac{s + p_{at}}{p_{at}} = N(0) - \lambda(0) \ln \frac{p_0{}^*}{p^c} \quad (5)$$

which provides a relationship between p_0 and s as a function of some reference stress values ($p_0{}^*$, p^c) and some soil parameters ($N(s)$, $\lambda(s)$, κ, κ_s). A convenient choice of p^c and $N(s)$ to simplify Equation

(5) is to assume (Fig. 2(a))

$$\Delta v(p^c)|_s{}^0 = N(0) - N(s) = \kappa_s \ln \frac{s + p_{at}}{p_{at}} \qquad (6)$$

In other words, p^c is the net mean stress at which one may reach the saturated virgin state, starting at a partially saturated condition, through a wetting path which involves only (elastic) swelling.

If Equation (6) is introduced into Equation (5), the following relationship is obtained

$$\left(\frac{p_0}{p^c}\right) = \left(\frac{p_0{}^*}{p^c}\right)^{[\lambda(0) - \kappa]/[\lambda(s) - \kappa]} \qquad (7)$$

This Equation defines the set of yield p_0 values for each associated suction (it can be considered therefore as a family of yield curves in a (p, s) space). In order to isolate a single yield curve it is necessary to specify the preconsolidation net mean stress for saturated conditions ($p_0{}^*$) which may be viewed as the hardening parameter in Equation (7). This Equation plays a central role in the model developed and explains, as discussed at length later, not only the apparent increase in preconsolidation stress associated with increasing suction, but also the collapse phenomena observed in wetting paths. For this reason, the yield curves (7) will be named LC yield curves (after loading–collapse). An LC yield curve through points 1 and 3 has been sketched in Fig. 2(b). Note that when $p_0{}^* = p^c$ the LC yield curve becomes a straight line ($p_0 = p^c$). In this case, changes in s do not result in plastic deformations. Only the elastic component, as given by Equation (4), is maintained.

Existing information on the volumetric behaviour of partially saturated soils, as revealed by suction controlled tests, may be used to support some of the model assumptions. The results of a number of isotropic and oedometer tests of this type were compiled by Lloret & Alonso (1985) and reported in terms of state surfaces (the void ratio was expressed as a function of net vertical or net confining stress and suction). Surface fitting analysis was performed in order to find the most appropriate analytical expression for the analysed data. The tests considered involved different soil types (low plasticity clays, silty clay and clayey sand) and suction values ranging from zero to 100 kPa in most of the cases.

Most of the reported state surfaces have been obtained through stress paths which involve reductions in suction from an initial value and a subsequent increment in net mean (or vertical) stress p. A path of this kind is shown in Fig. 3. For this particular path, the final specific volume for the (generic) stress state (p, s) may be written:

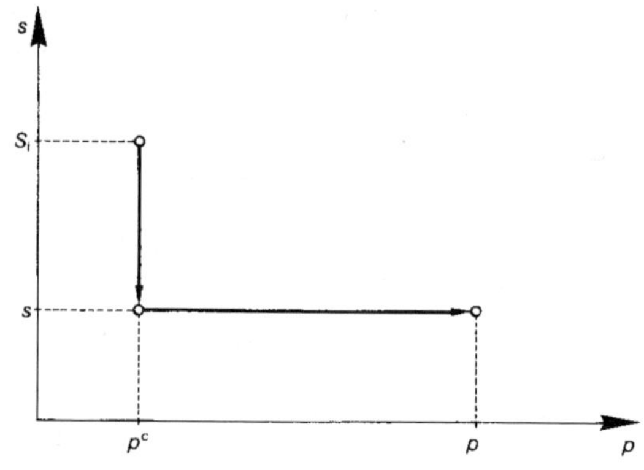

Fig. 3. Wetting and loading stress path in (p, s) stress plane

$$v(p, s) = v(p^c, s_i) + \Delta v(p^c)\Big|_{s_i}^{s} + \lambda(s) \ln \frac{p}{p^c} \qquad (8)$$

where the p-loading was supposed to take place along virgin states. This expression may be compared with the following empirical equation which was found by Lloret & Alonso (1985) to fit well experimental results

$$v = a + b \ln \frac{p}{p^c} + c \ln s + d \ln \frac{p}{p^c} \ln s$$

$$= a + c \ln s + (b + \mathrm{d} \ln s) \ln \frac{p}{p^c} \qquad (9)$$

where a, b, c and d are constants.

Identifying Equations (8) and (9) suggests that

$$v(p^c, s_i) + \Delta v(p^c)\Big|_{s_i}^{s} = a + c \ln s \qquad (10)$$

which supports the logarithmic variation of v with suction adopted in Equation (6).

In addition Equation (9) provides an expression describing the increase in soil stiffness with suction:

$$\lambda(s) = b + \mathrm{d} \ln s \qquad (11)$$

This expression may adequately predict an increase in soil stiffness with suction if $d < 0$. It predicts, however, an unlimited increase in stiffness for increasing s and this is a limitation for wide ranges of suction changes.

An asymptotic maximum stiffness, which is probably closer to real behaviour, is predicted by the alternative equation

$$\lambda(s) = \lambda(0)[(1 - r) \exp(-\beta s) + r] \qquad (12)$$

where r is a constant related to the maximum stiffness of the soil (for an infinite suction), $r = \lambda(s \to \infty)/\lambda(0)$, and β is a parameter which controls the rate of increase of soil stiffness with suction. Equation (12) has been adopted in the remainder of this Paper given its superior predicting capabilities as compared with Equation (11). Different expressions may be used for the (virgin) coefficient of compressibility when more experimental evidence becomes available. Fig. 4 illustrates the shape of the LC yield curves, when $\lambda(s)$ is given by Equation (12), for selected values of parameters $p_0{}^*$, r and β. Other parameters used in drawing Fig. 4 are $p^c = 0{\cdot}1$ MPa, $\lambda(0) = 0{\cdot}2$ and $\kappa = 0{\cdot}02$.

As reviewed in a previous section, an increase in suction may also induce irrecoverable strains.

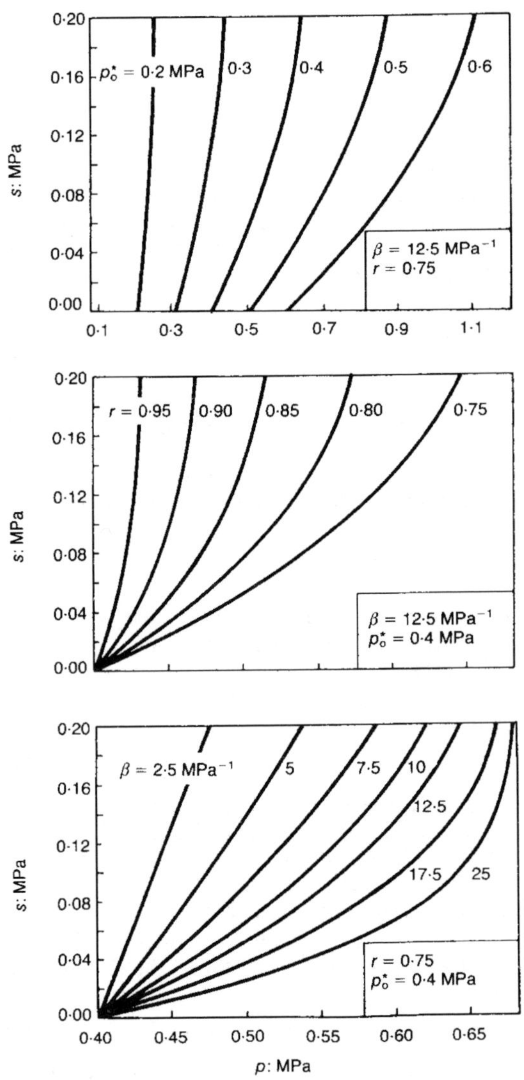

Fig. 4. Shapes of the LC yield curve for different values of the parameters $p_0{}^*$, r and β

It is proposed that whenever the soil reaches a maximum previously attained value of the suction s_0, irreversible strains will begin to develop. As a preliminary and simple choice, before more experimental evidence becomes available, the following yield condition

$$s = s_0 = \text{constant} \tag{13}$$

is adopted. s_0 has the meaning of the maximum past suction ever experienced by the soil and bounds the transition from the elastic state to the virgin range when suction is increased (Fig. 5). This yield locus is named SI (after suction increase). Both, LC and SI yield loci enclose an elastic region in the (p, s) plane (Fig. 6). Adopting a linear dependence between v and $\ln(s + p_{at})$, both in the elasto-plastic and elastic range, one may write, for virgin states

$$dv = -\lambda_s \frac{ds}{(s + p_{at})} \tag{14}$$

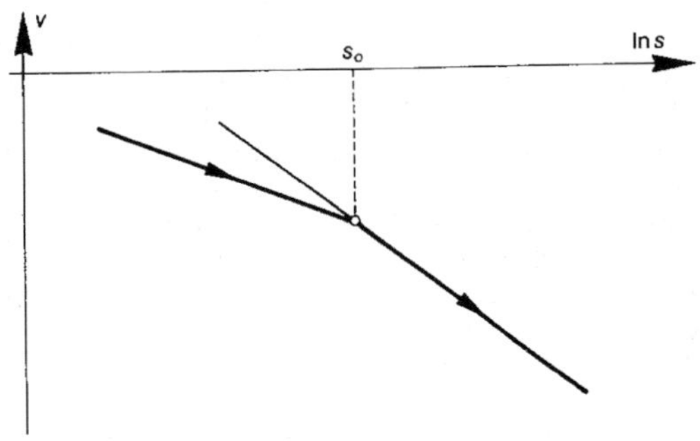

Fig. 5. Definition of yield suction s_0

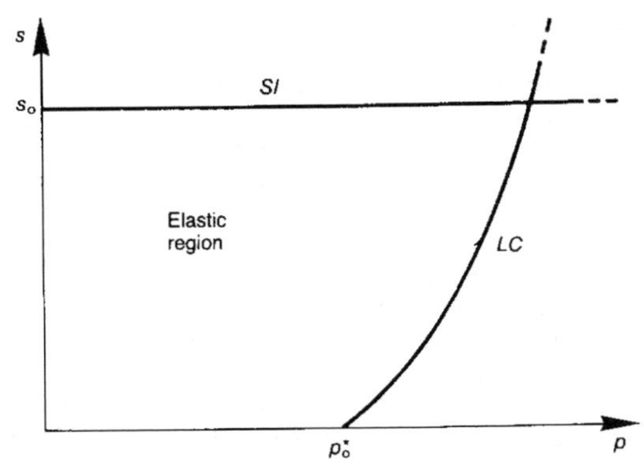

Fig. 6. Loading–collapse (LC) and suction increase (SI) yield curves

For drying and wetting reversals

$$dv = -\kappa_s \frac{ds}{(s + p_{at})} \tag{15}$$

which is, basically, Equation (4). However, some dependence of λ_s and κ_s on the net mean stress may be suspected but, for simplicity, they will be taken as constants.

Hardening laws

According to Equation (2), an increase of p in the elastic region will induce a compressive (positive value) volumetric deformation given by

$$d\varepsilon_{vp}{}^e = -\frac{dv}{v} = \frac{\kappa}{v}\frac{dp}{p} \tag{16}$$

Once the net mean stress p reaches the yield value p_0 the total volumetric deformation may be computed from Equation (1)

$$d\varepsilon_{vp} = \frac{\lambda(s)}{v}\frac{dp_0}{p_0} \tag{17}$$

and, therefore, the plastic component of volumetric strain will be given by

$$d\varepsilon_{vp}{}^p = \frac{\lambda(s) - \kappa}{v}\frac{dp_0}{p_0} \tag{18}$$

Taking into account Equation (7) for the LC yield locus, it is simple to show that the plastic volumetric deformation (18) is also given by

$$d\varepsilon_{vp}{}^p = \frac{\lambda(0) - \kappa}{v}\frac{dp_0{}^*}{p_0{}^*} \tag{19}$$

Similarly, an increase in suction within the elastic region will result in the volumetric strain (Equations (4) or (14))

$$d\varepsilon_{vs}{}^e = \frac{\kappa_s}{v}\frac{ds}{(s + p_{at})} \tag{20}$$

and, if the yield locus $s = s_0$ is reached, the following total and plastic deformations will be induced (Equation 13)

$$d\varepsilon_{vs} = \frac{\lambda_s}{v}\frac{ds_0}{(s_0 + p_{at})} \tag{21}$$

$$d\varepsilon_{vs}{}^p = \frac{\lambda_s - \kappa_s}{v}\frac{ds_0}{(s_0 + p_{at})} \tag{22}$$

Irreversible deformations control the position of the LC and SI yield surfaces through Equations (19) and (22). This type of hardening implies an independent motion of both yield curves in the (p, s) stress space. However, some experimental evidence (Josa *et al.*, 1987) suggests a definite coupling between them (see also Fig. 20 and its associated discussion in a later section). However, the different treatment given so far to the plastic deformations induced by suction or stress changes have only a formal meaning which has been useful to derive the appropriate relationships. It could be argued, as a first approximation, that both sets of plastic deformations have similar effects. In this manner, a simple way to couple both yield curves results if their position is controlled by the total plastic volumetric deformation, $d\varepsilon_v{}^p = d\varepsilon_{vs}{}^p + d\varepsilon_{vp}{}^p$. Then, from Equations (19) and (22) the proposed hardening laws are as follows

$$\frac{dp_0{}^*}{p_0{}^*} = \frac{v}{\lambda(0) - \kappa}\, d\varepsilon_v{}^p \tag{23}$$

$$\frac{ds_0}{s_0 + p_{at}} = \frac{v}{\lambda_s - \kappa_s}\, d\varepsilon_v{}^p \tag{24}$$

MODEL FOR TRIAXIAL STRESS STATES

A third stress parameter, $q = (\sigma_1 - \sigma_3)$, has to be incorporated to include the effect of shear stresses. The strain state is defined by $\varepsilon_v(= \varepsilon_1 + 2\varepsilon_3)$, volumetric strain, and $\varepsilon_s(= \frac{2}{3}(\varepsilon_1 - \varepsilon_3))$, shear strain. For consistency, the model must predict saturated behaviour when the value of suction reduces to zero. In accordance with the aim of simplicity, the saturated model adopted as a limit condition is a version of the modified Cam-clay model which gives an adequate qualitative description of saturated soil behaviour. The unsaturated formulation presented here could be coupled to more complex saturated models, achieving thus a better representation of some aspects of soil behaviour. However, given that the main aim of the model is the reproduction of the general behaviour of unsaturated soil, the selected saturated model is adequate for the purpose. Accordingly, it is proposed that the yield curve for a sample at constant suction s will be described by an ellipse which will exhibit an isotropic preconsolidation stress given by the previously defined p_0 value which lies on the loading–collapse (LC) yield curve (Fig.7).

In order to define the ellipse it is also necessary to specify the failure states. In parallel with the saturated condition, a critical state line (CSL) for non-zero suction will represent the increased strength induced by suction. As a first hypothesis, the effect of suction will be represented by an increase in cohesion, maintaining the slope M of the CSL for saturated conditions. As discussed previously this may be a suitable approach for a limited range of stress changes. Other choices are, however, possible.

If the increase in cohesion follows a linear relationship with suction, the ellipses will inter-

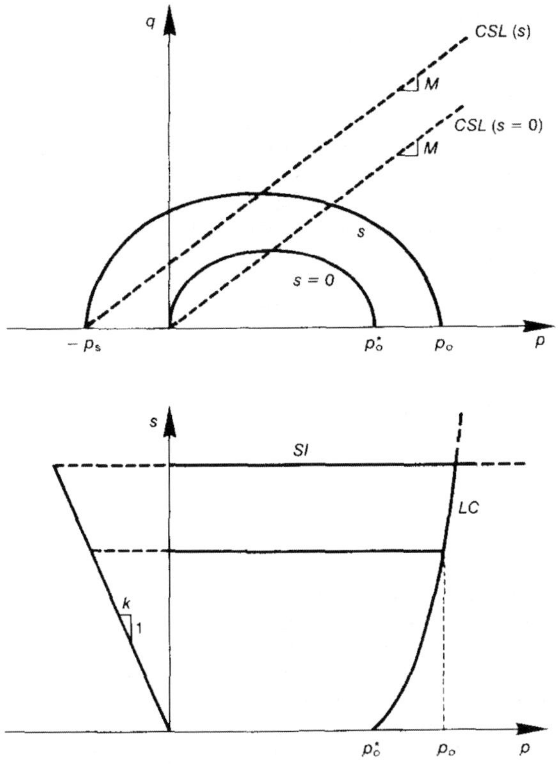

Fig. 7. Yield surfaces in (p, q, s) space

sect the p axis at a point for which

$$p = -p_s = -ks \tag{25}$$

where k is a constant. The major axis of the ellipse will span the segment $-p_s(s)$ to $p_0(s)$ and its equation will be given by

$$q^2 - M^2(p + p_s)(p_0 - p) = 0 \tag{26}$$

It is also proposed that the yield locus SI extends into the region $q > 0$ by means of a plane parallel to the q axis, so that the Equation (13) is maintained in the (p, q, s) space. A three-dimensional view of the yield surfaces in the (p, q, s) space is given in Fig. 8.

Regarding the direction of plastic strain increments, associated with the yield surface (26), a non-associated flow rule in the planes $s =$ constant is suggested. In fact, it is known (Gens & Potts, 1982a) that conventional critical-state models often overestimate K_0 values. To avoid this shortcoming, the expression for the associated flow rule is modified by introducing a parameter α (Ohmaki, 1982) resulting in the following equation

$$\frac{d\varepsilon_s{}^p}{d\varepsilon_{vp}{}^p} = \frac{2q\alpha}{M^2(2p + p_s - p_0)} \tag{27}$$

where α is chosen in such a way that the flow rule predicts zero lateral strain for stress states corresponding to Jaky's (1948) K_0 values ($K_0 = 1 - \sin\varphi' = (6 - 2M)/(6 + M)$). Given the relationship between K_0 and M the new flow rule does not introduce any additional constitutive parameter. The components of the plastic strain associated with this yield surface will be given by $(d\varepsilon_{vp}{}^p, d\varepsilon_s{}^p)$. The expressions for these strain components can be derived by the application of standard procedures of strain hardening plasticity. For the second yield surface (Equation (13)), the vector of plastic strain increment induced by suction increase will be $(d\varepsilon_{vs}{}^p, 0)$, where $d\varepsilon_{vs}{}^p$ is given by Equation (22).

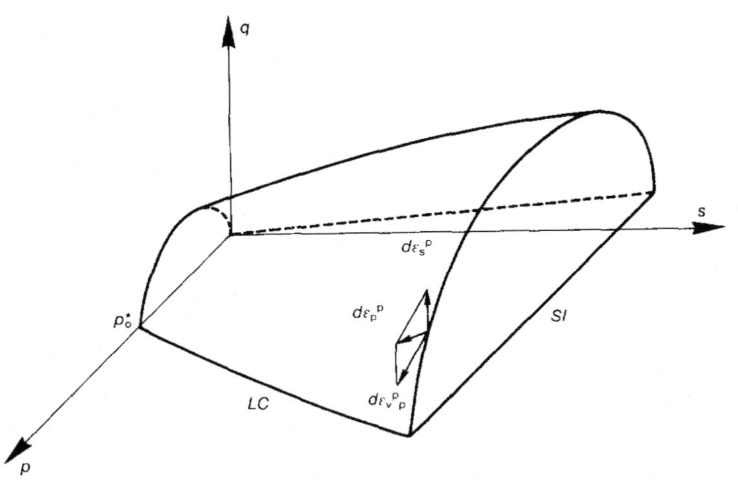

Fig. 8. Three-dimensional view of the yield surfaces in (p, q, s) stress space

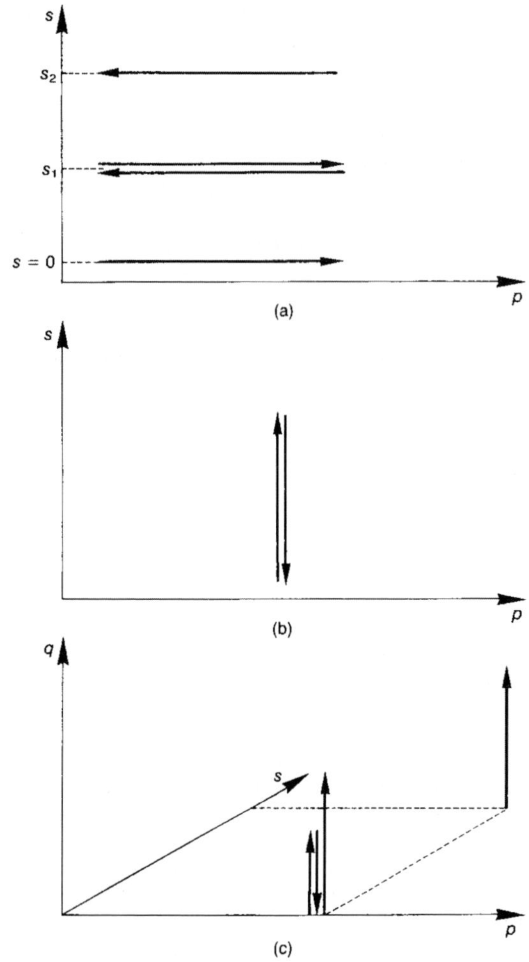

Fig. 9. Suitable stress paths to determine the model parameters

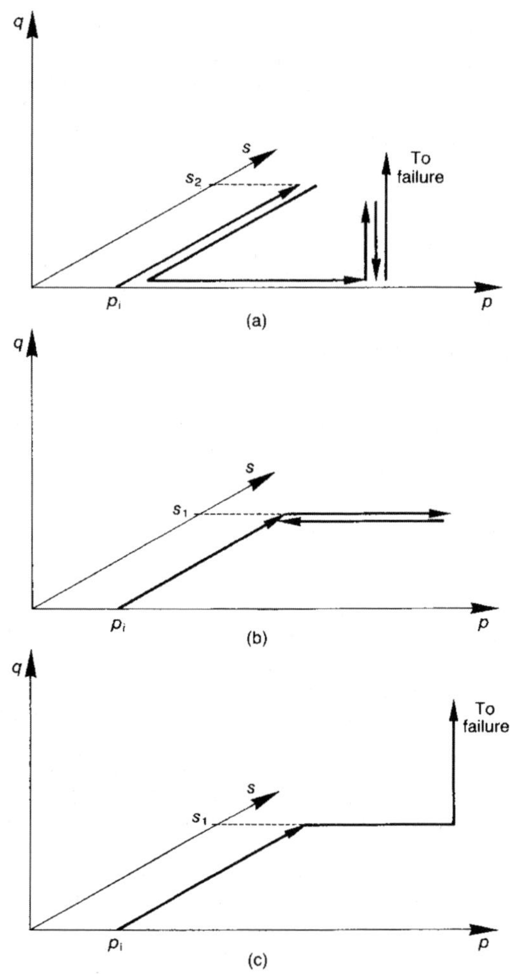

Fig. 10. Suggested tests to determine model parameters

The elastic strains, $d\varepsilon_s^e = \frac{2}{3}(d\varepsilon_1^e - d\varepsilon_3^e)$ induced by changes in q will be computed through a shear modulus G

$$d\varepsilon_s^e = (\tfrac{1}{3}G)\, dq \qquad (28)$$

An ordered summary of the mathematical expressions defining the model are presented in Appendix 1. Extension to the three-dimensional principal stress space can be easily formulated (Gens & Potts, 1988).

PARAMETERS OF THE MODEL AND THEIR DETERMINATION

The application of the model requires information on the following stress states and parameters.

(*a*) Initial state: initial stresses (p_i, q_i, s_i), initial specific volume v_0 and initial reference stress variables (strain hardening parameters) defining the initial position of the yield surfaces (p_{0i}^*, s_{0i}).

(*b*) Parameters directly associated with the LC yield curve (isotropic stress): p^c, reference stress; $\lambda(0)$, compressibility coefficient for the saturated state along virgin loading; κ, compressibility coefficient along elastic (unloading–reloading) stress paths; r, establishes the minimum value of the compressibility coefficient (virgin states) for high values of suction; β, controls the rate of increase in stiffness (virgin states) with suction.

(*c*) Parameters directly associated with changes in suction and the SI yield curve: λ_s, compressibility coefficient for increments of suction across virgin states; κ_s, compressibility coefficient for changes in suction within the elastic region.

(*d*) Parameters directly associated with changes in shear stress and the shear strength: G, shear modulus within the elastic domain; M, slope of the critical state lines; k, parameter that controls the increase in cohesion with suction.

A rational engineering approach requires the use of the simplest model compatible with the desired accuracy of results. A particular analysis of a given problem will involve some soil type and initial conditions and an estimation of expected stress paths. It may well happen that one of the proposed yield surfaces will never be reached. For instance, the behaviour of partially saturated non-expansive soils under paths which typically involve wetting and increase or decrease in mean and shear stress may be modelled by the LC yield curve and the associated yield surface in the (p, q, s) space. The SI yield plane will not play any role in this case.

An example of this type may be found in the analysis of embankment dams in which the initial 'as-compacted' suction will not likely be increased. In this case the model requires nine parameters (p^c, $\lambda(0)$, κ, r, β, κ_s, G, M, k), five more than the parameters involved in a critical state model ($\lambda(0)$, κ, G and M). The five additional parameters (p^c, r, β, κ_s, k) allow for the modelling of typical patterns of the behaviour of partially saturated soil such as increasing stiffness and strength with increasing suction, collapse phenomena and (moderate) volumetric expansion when the soil is wetted. If drying processes are expected to influence the soil, a new parameter (λ_s) and the consideration of the SI yield surface may become necessary.

The determination of the model parameters will, in general, require suction-controlled testing. Suggested stress paths for different sets of parameters are the following (they have been indicated in Fig. 9)

(*a*) tests that involve isotropic drained compression (loading and unloading) at several constant suction values; they provide data to find p^c, $p_0{}^*$, $\lambda(0)$, κ, r and β (Fig. 9a)
(*b*) tests that involve a drying–wetting cycle at a given net mean applied stress; they provide data to find s_0, λ_s and κ_s (Fig. 9b)
(*c*) drained shear strength tests at different suction values; they provide data to determine G, M and k (Fig. 9c).

These basic stress paths may be combined into a few different tests necessary to obtain all the model parameters. Three tests which could lead to a full identification of the model are represented in Fig. 10. This is, of course, a minimum experimental programme. A more reliable determination of parameters will certainly require more tests.

TYPICAL FEATURES OF PREDICTED BEHAVIOUR

Before a comparison is made with some experimental results, it is perhaps convenient to show the capabilities of the model by computing its predictions along a number of selected triaxial stress paths. In all the cases presented a single set of model parameters has been selected to describe a reference 'soil'. They are

(*a*) parameters directly associated with changes in p and with the LC yield curve:

$$\lambda(0) = 0{\cdot}2; \quad \kappa = 0{\cdot}02; \quad r = 0{\cdot}75;$$

$$\beta = 12{\cdot}5\ \text{MPa}^{-1}; \quad p^c = 0{\cdot}10\ \text{MPa}$$

(*b*) compressibility coefficients for suction changes:

$$\lambda_s = 0{\cdot}08; \quad \kappa_s = 0{\cdot}008$$

(*c*) parameters directly associated with changes in shear stress and shear strength:

$$G = 10\ \text{MPa}; \quad M = 1; \quad k = 0{\cdot}6$$

In addition, the initial void ratio was assumed to be 0·9 ($v = 1{\cdot}9$). This ideal soil is a moderately compressible material. For large values of suction the compressibility against changes in p is reduced to 75% of the saturated values. The angle of shearing resistance for saturated conditions is 25·4° and the change in cohesion due to suction increase is moderate ($k = 0{\cdot}6$). For each of the examples presented the initial position of the yield curves LC and SI is indicated.

Case 1, volumetric deformation induced by wetting at increasing confining stress

The stress paths and the initial position of the yield LC and SI curves are indicated in Fig. 11a ($p_0{}^* = 0{\cdot}20$ MPa; $s_0 = 0{\cdot}3$ MPa). Three wetting paths—AB, CD and EF (1, 2 and 3)—starting at an initial suction $s_i = 0{\cdot}2$ MPa and three different confining stresses (0·15; 0·35 and 0·6 MPa) lead to the volumetric deformations plotted in Fig. 11(b). The first wetting path (AB) takes place within the elastic region and induces a continuous expansion (segment AB in Fig. 11(b)). The second wetting path, CD, starts once a compression AC is applied to the sample. During wetting, a collapse of the sample starts at the beginning of suction reduction. The slight elastic expansion during path CD is smaller than the compression induced by the motion of the LC curve to its final position and therefore a net collapse is computed. Finally, under path EF at a larger confining

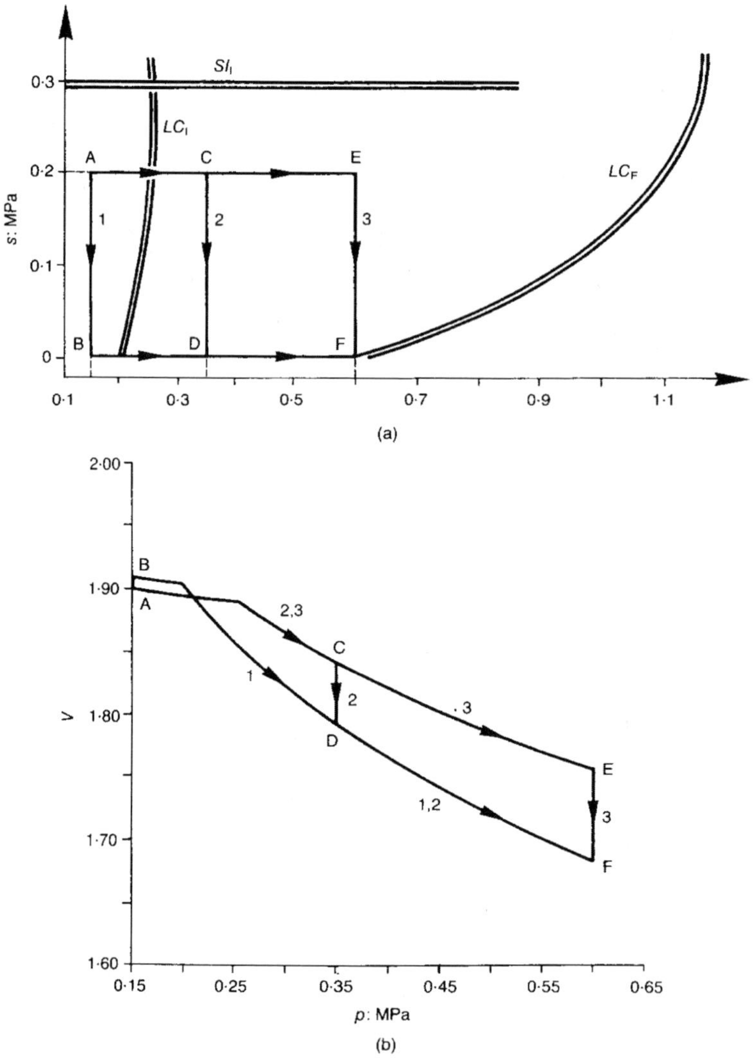

Fig. 11. Response of model to decreasing suction (wetting) at increasing net mean stress p: (a) stress paths; (b) specific volume-net mean stress relationships

stress, a larger collapse takes place (segment EF in Fig. 11(b)). Also, the compression curves, once the wetting is completed, follow a unique path for saturated conditions (path DF after wetting 2 follows the saturated compression curve BDF). Experimental results which favour this uniqueness have been reported by Jennings and Burland (1962) and Blight (1965) for collapsing sands, by Erol and El-Ruwaih (1982) testing loess and by Maswoswe (1985) in the case of a low plasticity sandy clay.

Case 2, effect of alternate application of load and suction changes

A first example is illustrated in Fig. 12(a). Beginning at the initial stress point A ($s_A = 0{\cdot}2$ MPa, $p_A = 0{\cdot}15$ MPa), three alternative loading stress paths (1, 2, 3) which always involve a non-increasing suction and a common final stress point, $s_F = 0$ and $p_F = 0{\cdot}6$ MPa, are represented. The initial position of the yield curves is given by $p_0{}^* = 0{\cdot}20$ MPa and $s_0 = 0{\cdot}3$ MPa. The different combinations of p-loading and suction reduction, inducing either expansion (A → C → E → F) or collapse (A → B → D → F), take the LC yield curve to a final common position LC_F. The final volumetric deformation is the same under the three stress paths. These types of loading and suction change sequences lead to volume changes which are stress path independent and they have been used by some authors, quoted before, to formulate the volumetric behaviour of partially satu-

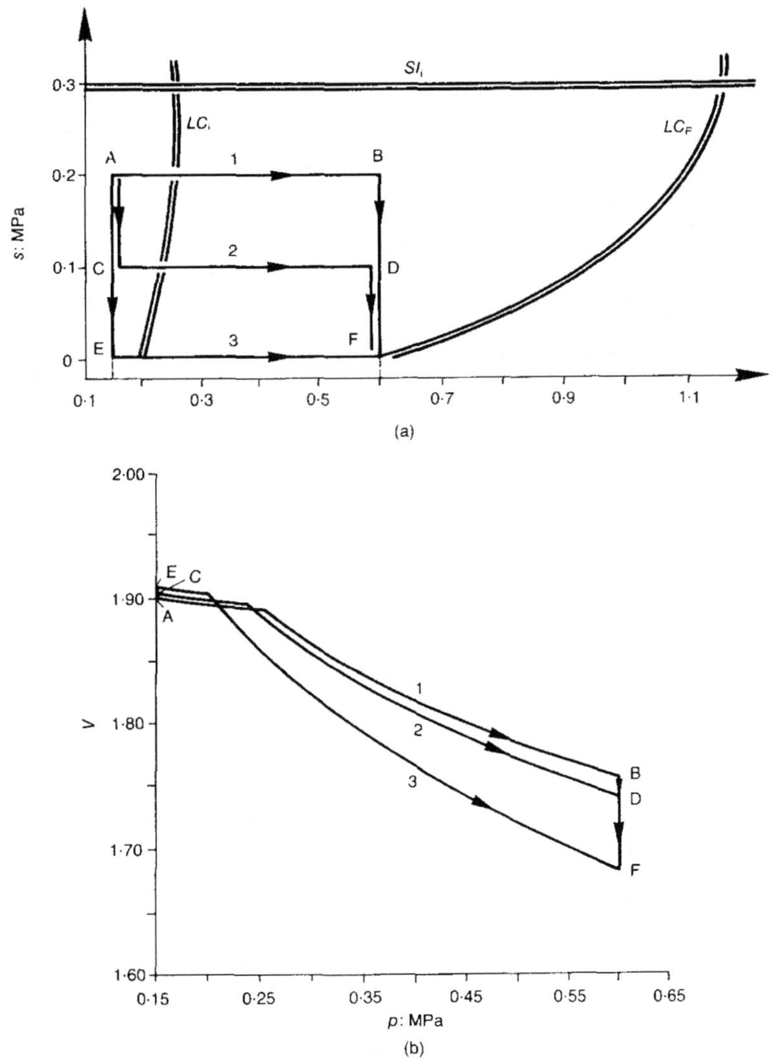

Fig. 12. Response of model to increments of net mean stress *p* and decreasing suction (wetting): (a) stress paths; (b) specific volume-net mean stress relationships

rated soils in terms of state surfaces. Experimental results that involved stress paths qualitatively similar to the paths indicated in Fig. 12(a) were also reported by Barden *et al.* (1969) (group 2 tests) for low-plasticity clay. The pattern of model predictions agrees well with measured results.

Consider, however, stress paths which involve drying steps (Fig. 13). Three samples (1, 2, 3) start now at a common stress point A ($s_A = 0$, $p_A =$ 0·15 MPa) and are loaded and dried in three different sequences towards the final point F ($s_F =$ 0·2 MPa; $p_F = 0{\cdot}6$ MPa). The same initial preconsolidation state of the sample ($s_0 = 0{\cdot}3$ MPa; $p_0{}^* = 0{\cdot}2$ MPa) as before is adopted in this case. Sample 1, loaded under saturated conditions, displaces the yield LC curve to the position LC_B and then experiences an elastic decrease in volume when suction is increased from B to F, without any further change of the position of the LC curve. This sample experiences the largest volumetric deformation of the three. Under path 3, the suction applied from A to E induces some compression of the sample and an increase in stiffness against the subsequent increase of confining stress. The stress path EF displaces the LC yield curve to the final position LC_F at a reference $p_{0F}{}^* < p_{0B}{}^* = 0{\cdot}6$ MPa. It will experience the smallest volumetric deformation as Fig. 13(b)

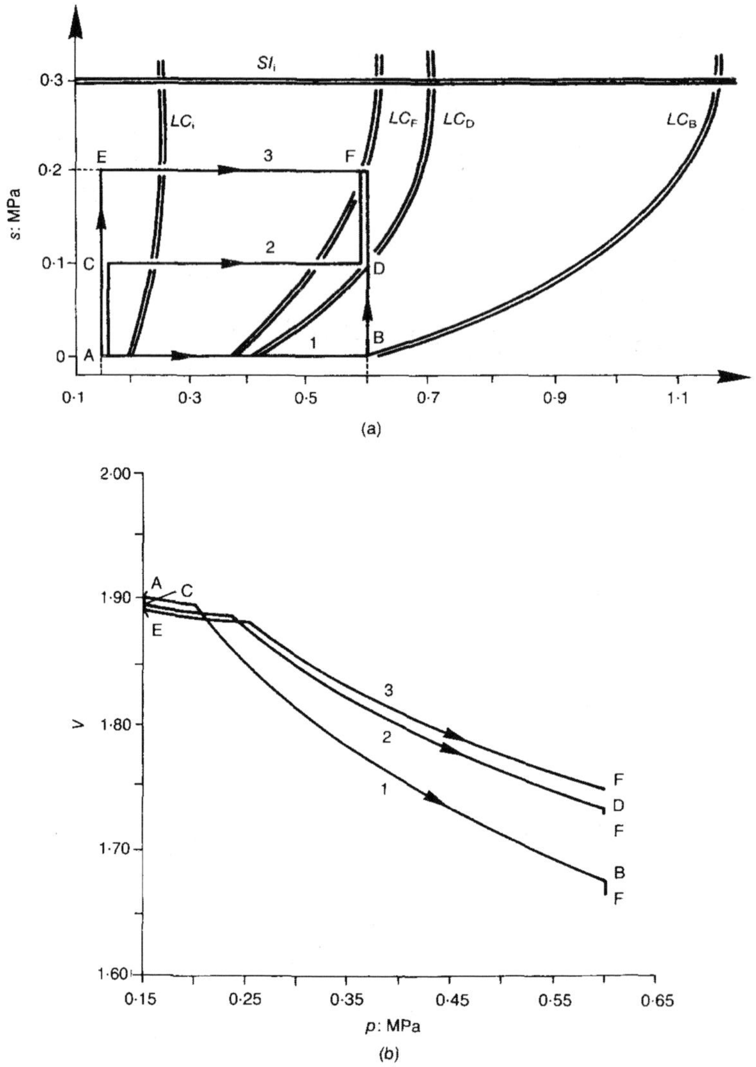

Fig. 13. Response of model to increments of net mean stress p and increasing suction (drying): (a) stress paths; (b) specific volume-net mean stress relationships

indicates. The paths which involve drying before subsequent p-loading are therefore stress path dependent. This conclusion was also reached in the experimental work reported by Barden *et al.* (1969) testing low plasticity clay, Matyas & Radhakrishna (1968) testing a low plasticity sandy clay and Josa *et al.* (1987) testing kaolin.

Case 3, coupling of the SI and LC yield curves

In the preceding two cases the SI yield curve did not play any role in the model predictions since the applied suctions were smaller than s_0. In Fig. 14, a p-loading path AB at a constant suction ($s_A = 0$MPa) is compared with a similar loading (DE) preceded in this latter case by a drying–wetting cycle (ACD) in which a maximum suction of 0·3 MPa is applied. The initial yield curves are characterized by $p_0^* = 0{\cdot}2$ MPa and $s_0 = 0{\cdot}025$ MPa. The drying–wetting cycle induces a net compaction of the sample and the irreversible path of this deformation moves LC_i to the position LC_D shown in Fig. 14(a). On subsequent p-loading, path DE, the sample experiences a larger preconsolidation stress than the sample subjected to path AB (Fig. 14(b)). Experimental indication of this type of behaviour may

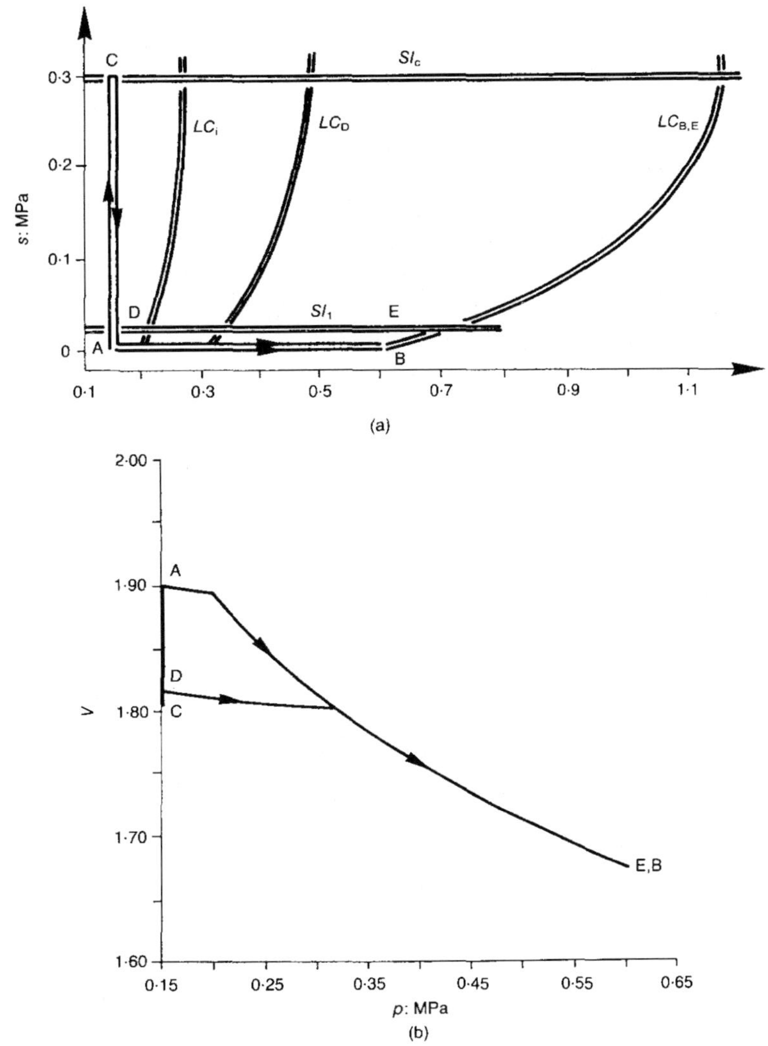

Fig. 14. Effect of initial increase of suction: (a) stress paths; (b) specific volume-net mean stress relationships

be found in the experimental work of Josa *et al.* (1987).

Case 4, shear tests for different suctions

Three shear tests maintaining a constant net mean stress $p_A = 0{\cdot}15$ MPa, for three different suctions ($s_1 = 0{\cdot}10$ MPa, $s_2 = 0{\cdot}20$ MPa, $s_3 = 0{\cdot}30$ MPa) are indicated in Fig. 15. The initial yield curves LC and SI ($p_0{}^* = 0{\cdot}15$ MPa, $s_0 = 0{\cdot}4$ MPa) are shown in Fig. 15(a). The shear stress–shear strain relationships computed in the three cases are plotted in Fig. 15(b). These curves show an increase in strength with suction. Results of this kind have been reported by Delage *et al.* (1987) testing silt in a suction controlled triaxial cell.

Case 5, failure under decreasing suction

A stress path to illustrate the possibility of failure of a stressed partially saturated soil due to suction reduction is indicated in Fig. 16. This type of failure is probably significant in slopes located in partially saturated soils (i.e. slopes in residual soils in tropical regions) subjected to wetting due to infiltration from rainfalls (Brand, 1981). In the case analysed the suction was first increased at constant net mean stress (path AB), then the net mean stress was increased at constant suction (BC), a deviatoric stress was applied (CD) and the suction was finally reduced until failure was achieved (DE). The initial positions of the yield curves LC and SI are plotted in Fig.

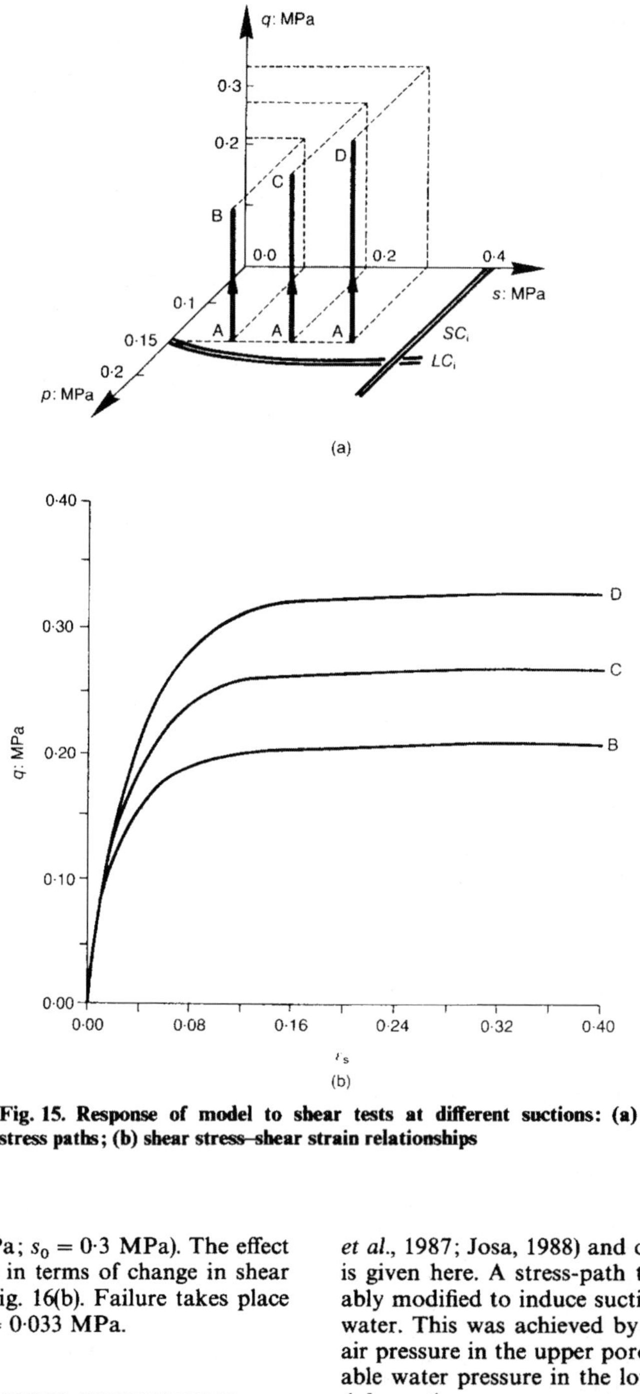

Fig. 15. Response of model to shear tests at different suctions: (a) stress paths; (b) shear stress–shear strain relationships

16(a) (p_0* = 0·15 MPa; s_0 = 0·3 MPa). The effect of final wetting (DE) in terms of change in shear strain is plotted in Fig. 16(b). Failure takes place at a final suction s_F = 0·033 MPa.

COMPARISON OF MODEL PREDICTIONS WITH EXPERIMENTAL RESULTS

Tests on partially saturated compacted kaolin reported by Josa (1988)

Details of the tests performed on partially saturated compacted kaolin are given elsewhere (Josa *et al.*, 1987; Josa, 1988) and only a brief summary is given here. A stress-path triaxial cell was suitably modified to induce suction states in the pore water. This was achieved by applying a constant air pressure in the upper porous stone and a variable water pressure in the lower stone. Load and deformation measurements were made inside the testing chamber. Pore and confining pressures as well as vertical load were applied by means of mercury pots whose position was controlled by stepper motors. A computer, linked to the system, controlled the predefined stress and suction path

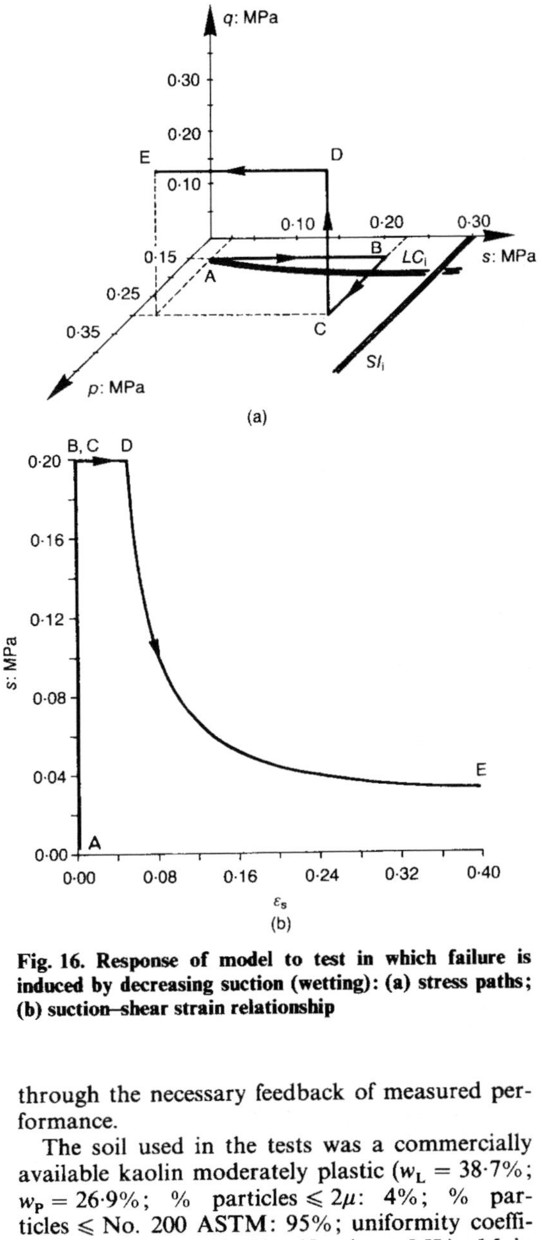

Fig. 16. Response of model to test in which failure is induced by decreasing suction (wetting): (a) stress paths; (b) suction–shear strain relationship

through the necessary feedback of measured performance.

The soil used in the tests was a commercially available kaolin moderately plastic ($w_L = 38{\cdot}7\%$; $w_P = 26{\cdot}9\%$; % particles $\leqslant 2\mu$: 4%; % particles $\leqslant$ No. 200 ASTM: 95%; uniformity coefficient: 2; Universal Classification: ML). Main mineral components were quartz and kaolinite. A moderate percentage of illite was also found in X-ray diffraction tests. Samples were compacted by kneading and then allowed to consolidate to an initial stress state given by $p = 0{\cdot}045$ MPa; $s = 0{\cdot}01$ MPa and $q = 0$. Once consolidated, they reached $e_0 = 0{\cdot}915$ and $S_{r0} = 0{\cdot}875$. From this state the subsequent application of a desired stress path was carried out.

The parameters controlling the variation of soil stiffness with suction ($\lambda(0)$, r and β; see Equation (12)) can be obtained by means of isotropic compression tests at different constant suctions. Tests of this kind are shown in Fig. 17(a). In Fig. 17(b) the variation of $\lambda(s)$, computed through Equation (17) was plotted in terms of p. Three $\lambda(s)$ values were obtained ($\lambda(0{\cdot}04) = 0{\cdot}09$; $\lambda(0{\cdot}06) = 0{\cdot}075$; $\lambda(0{\cdot}09) = 0.06$) from the final horizontal sections of the curves which relate to samples loaded beyond their preconsolidation pressure. They allow the determination of the three unknowns $\lambda(0)$, r and β through Equation (12). It was found $\lambda(0) = 0{\cdot}14$; $\beta = 16{\cdot}4$ MPa^{-1} and $r = 0{\cdot}26$. Also, an average value $\kappa = 0{\cdot}015$, was measured under unloading paths.

In order to obtain λ_s and κ_s, tests involving the application of suction cycles at a constant net mean stress p, are ideally suited, provided the

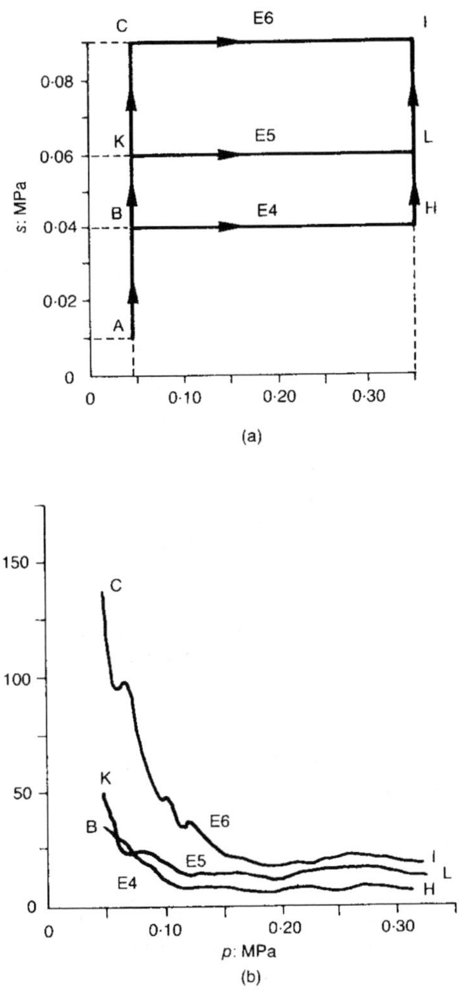

Fig. 17. Isotropic compression tests of partially saturated compacted kaolin: (a) stress paths; (b) variation of soil compressibility

suction decrease path does not intersect the LC yield locus. This may be achieved if the suction is first increased and then decreased, at a constant net mean stress. A test of this kind is represented in Fig. 18. It shows a quasi-elastic response of the soil beyond the first suction increase, as suggested by the model proposed. Several tests of this kind provided the following average parameter values: $\lambda_s = 0{\cdot}05$ and $\kappa_s = 0{\cdot}01$.

The shear modulus G was obtained in drained tests involving the application of a deviatoric stress q at different values of net mean stress and suction. The stress paths selected and the measured shear stress–strain relationships are given in Figs 22(a) and (b) respectively (these tests will be used subsequently to compare with model predictions). As a consequence of the flow rule assumed, no plastic shear strains occur at the start of the test, so the initial part of the stress–strain curves can be used to determine the elastic shear modulus G. A value of 3·3 MPa was obtained.

Several shear strength tests were performed at different values of p and s (Fig. 19). Within the

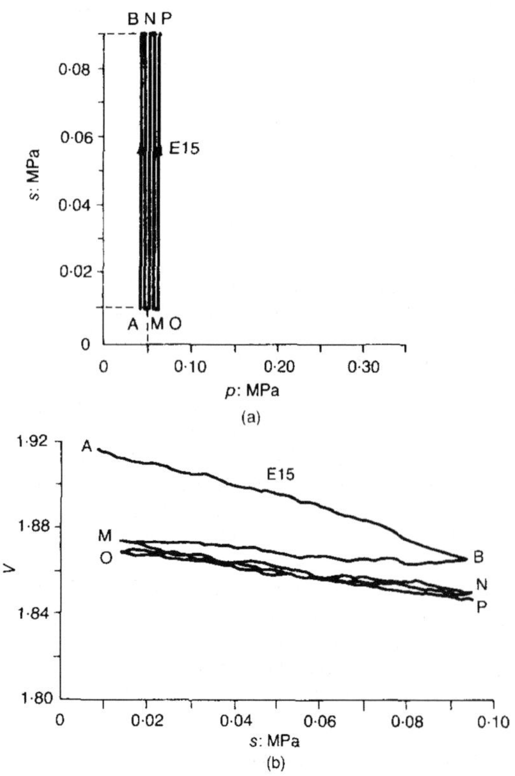

Fig. 18. Effect of cycles of applied suction on changes in specific volume of isotropically confined partially saturated compacted kaolin: (a) stress paths; (b) experimental results

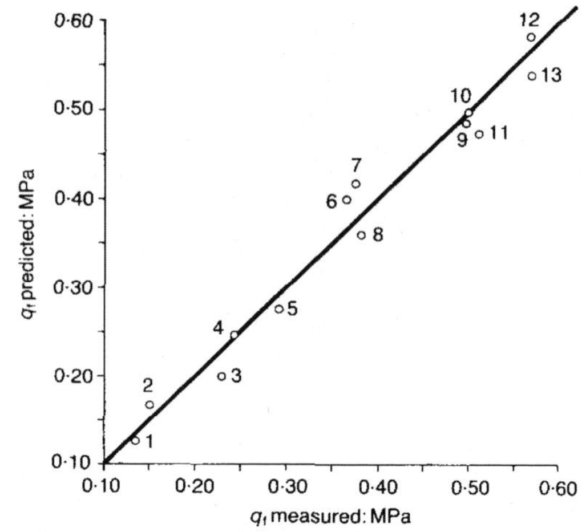

Test	p: MPa	s: MPa
1	0·142	0·0097
2	0·142	0·0142
3	0·137	0·0950
4	0·280	0·0160
5	0·340	0·0000
6	0·475	0·010
7	0·390	0·095
8	0·325	0·091
9	0·544	0·040
10	0·557	0·041
11	0·558	0·016
12	0·590	0·097
13	0·540	0·095

Fig. 19. Comparison between measured q_f values and predicted shear strength assuming critical state plane in the (p, q, s) space

range of stresses applied ($s = 0$ to 0·097 MPa and $p = 0{\cdot}137$ to 0·59 MPa) the plane

$$q = Mp + Mks = 0{\cdot}821p + 1{\cdot}022s \quad (29)$$

fits the results with a standard deviation of 6·8% (Fig. 19). This plane provides the constants $M = 0{\cdot}821$ and $k = 1{\cdot}245$.

The initial position of the yield surface is governed by the reference stress variables $p_0{}^*$ and s_0. $p_0{}^*$ may be obtained if there is information concerning the yield stress under increasing p. In our case, this information is provided by the tests reported in Fig. 17. A relatively sudden change in $1/\lambda(s)$ with p provides an estimation of p_0. Although it is difficult to pinpoint exactly the position of the yield point, the following values can perhaps be identified from Fig. 17(b): $p_0(0{\cdot}04) = 0{\cdot}065$ MPa; $p_0(0{\cdot}06) = 0{\cdot}07$ MPa; $p_0 = (0{\cdot}09) = 0{\cdot}085$ MPa. These values have been used to compute $p_0{}^*$ and p^c through Equation (7). A least squares error prediction procedure leads to $p_0{}^* = 0{\cdot}055$ MPa and $p^c = 0{\cdot}043$ MPa. In addition, tests in which suction was increased, thus maintaining a constant net mean stress p, equal to the initial value, could be interpreted in a similar way (test E1 in Fig. 20). It was found $s_0 = 0{\cdot}03$ MPa. A second test (E3) represented in Fig. 20 shows an apparent increase in s_0 when the sample has undergone a previous increase in net mean stress p (path AC). This result suggests a coupling between the yield loci LC and SI as proposed in the model.

A summary of model parameters and initial and reference conditions is given in Table 1.

The stress paths followed by two tests (E5 and E6) in the plane $q = 0$ are represented in Fig. 21. The measured volumetric deformations of the two samples (in terms of v) and the model predictions using the parameters given in Table 1 are plotted in Fig. 21. In these two tests the samples are subjected to ever increasing changes in net mean

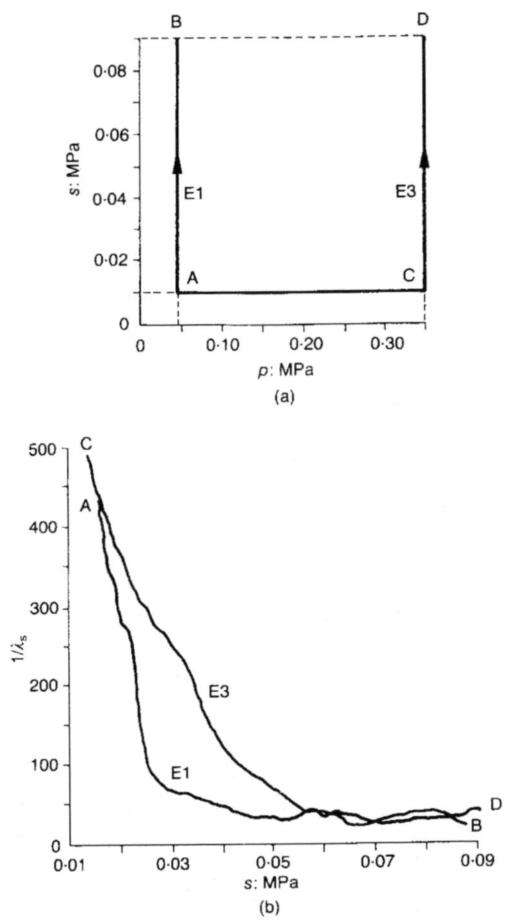

Fig. 20. Tests on partially saturated compacted kaolin involving increase in suction: (a) stress paths; (b) Variation of soil compressibility with applied suction

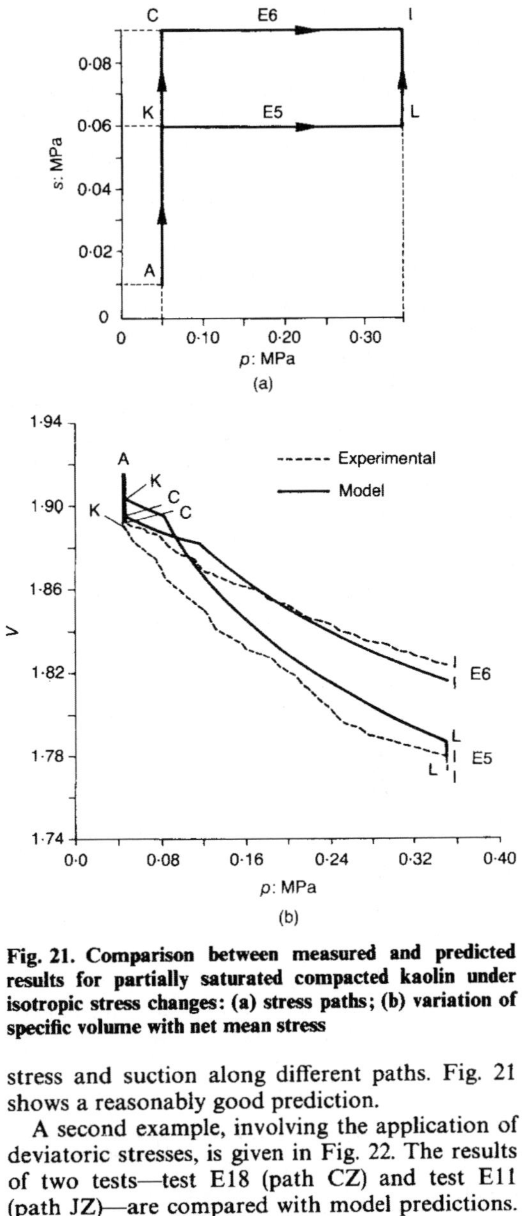

Fig. 21. Comparison between measured and predicted results for partially saturated compacted kaolin under isotropic stress changes: (a) stress paths; (b) variation of specific volume with net mean stress

stress and suction along different paths. Fig. 21 shows a reasonably good prediction.

A second example, involving the application of deviatoric stresses, is given in Fig. 22. The results of two tests—test E18 (path CZ) and test E11 (path JZ)—are compared with model predictions. Test E18 involves the application of q maintaining $dp = 0$. In test E11, $d\sigma_3 = 0$. In both tests the samples reached failure. The comparison between measured and predicted shear stress–shear strain relationships is given in Fig. 22(b). In general terms, a good correspondence is obtained.

Tests on partially saturated compacted sandy clay reported by Maswoswe (1985)

Maswoswe (1985) reports several suction controlled oedometer and triaxial tests of low plasticity sandy clay (Lower Cromer till) having the

Table 1. Model parameters, initial state and reference stress values for three soils indicated

Model parameters		CK1*	LCT*	CK2*
Compressibility under mean stress changes	$\lambda(0)$	0·14	0·066	0·065
	$\kappa(\kappa)$	0·015	0·0077	0·011
	β: MPa^{-1}	16·4	20	20
	r	0·26	0·25	0·75
	p^c: MPa	0·043	0·012	0·01
Compressibility under suction changes	λ_s	0·05	—	0·025
	κ_s	0·01	0·001	0·005
Shear modulus	G: MPa	3·3	7	—
Shear strength	M	0·82	1·2	—
	k	1·24	0·8	—

Initial state	CK1	LCT	CK2
p: MPa	0·045	0·0066	0·02
s: MPa	0·01	0·005	0
q: MPa	0	0·24	0·05
v	1·915	1·66	2·098

Reference stress parameters	CK1	LCT	CK2
p_0*: MPa	0·055	0·02	0·04
s_0: MPa	0·03	0·3	0·07

CK1: Compacted kaolin (Josa *et al.*, 1987).
LCT: Lower Cromer till (Maswoswe, 1985).
CK2: Compacted kaolin (Karube, 1986).

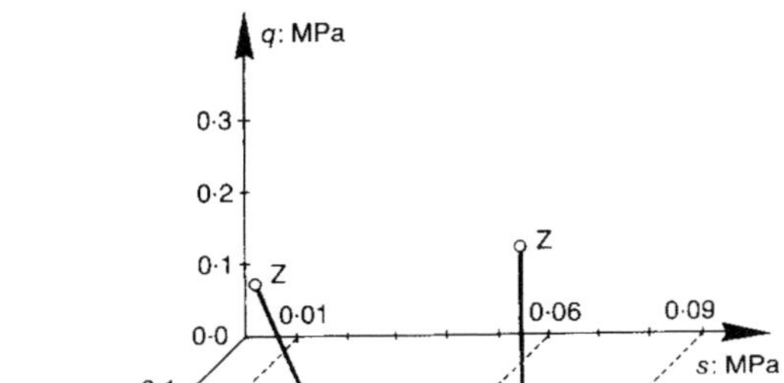

Fig. 22. Shear tests on partially saturated compacted kaolin: (a) stress paths; (b) comparison of measured and predicted shear stress–shear strain relationships

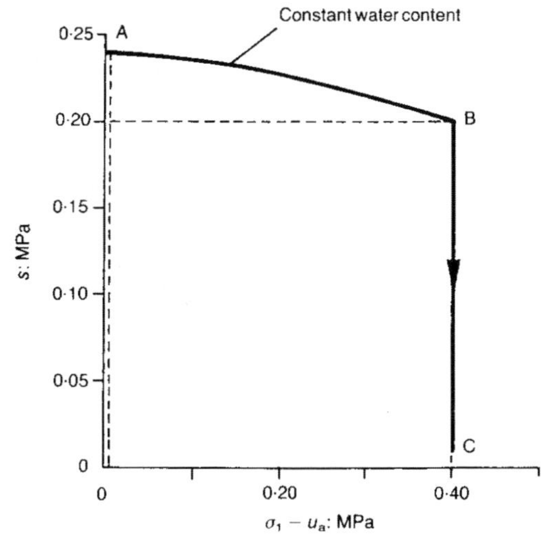

Fig. 23. Stress path followed by sample of sandy clay (Lower Cromer till) tested by Maswoswe (1985) (test SK3)

following characteristics: $w_L = 25\%$; $w_P = 12\%$; clay content: 17%; percentage of sand $>50\%$; clay activity = 0.71. Samples were statically compacted, dry of optimum.

An important aim of the research work reported by Maswoswe (1985) was to determine the stress path followed by the soil during collapse under oedometric conditions (no lateral deformation). Accordingly, the radial deformation of the samples in the triaxial tests was constantly monitored and the stress state was modified to ensure zero lateral deformation. The results of a particularly relevant test in which a sample was first subjected to an increase in vertical stress under constant water content and subsequently wetted under constant vertical stress were selected for comparison with the model developed. The stress path followed by the sample is shown in Fig. 23 in a plane $(\sigma_1 - u_a, s)$. Measured behaviour of the sample in terms of variation of void ratio with vertical stress and in terms of the stress path followed by the soil in the planes $((\sigma_1 + \sigma_3)/2 - u_w, (\sigma_1 - \sigma_3)/2)$ and $((\sigma_1 + \sigma_3)/2 - u_a, (\sigma_1 - \sigma_3)/2)$ are plotted in Fig. 24(a and b).

The tests performed by Maswoswe (1985) are not well suited to determine, in a direct way, the parameters of the proposed model. In particular, the loading test under constant water content are difficult to interpret as the saturation of the sample increases with the load and therefore 'pure' compression behaviour and collapse phenomena may coexist in the sample along the same stress path. Fortunately, some basic parameters of saturated compacted Lower Cromer till have been reported by independent studies (Gens, 1982; Gens & Potts, 1982b). The following values were taken from this work: $\lambda(0) = 0{\cdot}066$; $\kappa = 0{\cdot}0077$; $M = 1{\cdot}2$. They have been selected for the purposes of the present comparison as they appear to fit well the behaviour of the soil saturated after compaction.

Other model parameters could not be obtained directly from Maswoswe's results. However, the observed trends of behaviour have been used to obtain approximate values for the unknown parameters. For instance, the collapsing properties of the soil tested are relatively important and therefore a low value for $r (r = 0{\cdot}25)$ was adopted. On the other hand, the increase in stiffness with suction is relatively marked and this is reflected in the selected values $\beta = 20$ MPa^{-1}, and $k = 0{\cdot}8$. The test to be modelled (test SK3, Fig. 24(a)) indicated an apparent preconsolidation net mean stress $p_0 = 0{\cdot}2$ MPa for a suction close to

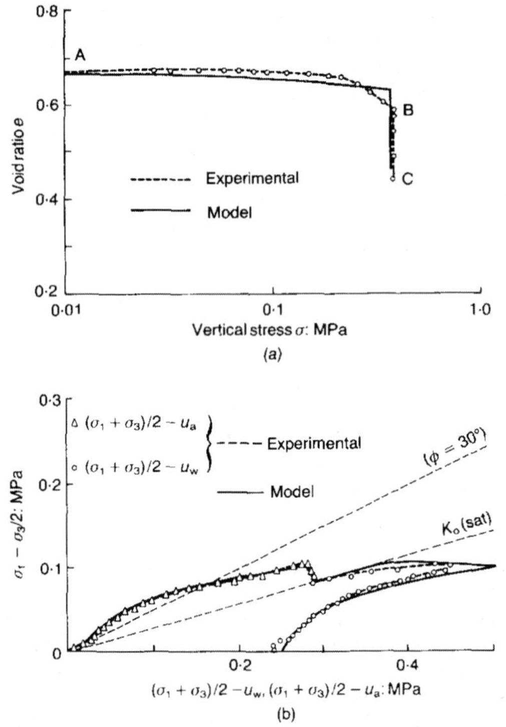

Fig. 24. Results reported by Maswoswe (1985) (Test SK3) and model predictions: (a) variation of void ratio with vertical stress; (b) stress paths in the triaxial plane

$s = 0{\cdot}22$ MPa. Oedometer tests on saturated samples prepared in the same way and with the same initial conditions showed preconsolidation stresses significantly lower, around $p_0{}^* = 0{\cdot}02$ MPa. These results, interpreted through Equation (7), provide a value $p^c = 0{\cdot}012$ MPa for the reference stress. Again due to the significant collapsing characteristics of the tested soil, it is difficult to isolate recoverable volumetric strains along the wetting paths. It is thought that, given the type of soil and the rigid open structure likely to develop in the process of static compaction dry of optimum, the value of κ_s will probably be very small. Therefore κ_s was selected equal to 0·001.

In his experimental work on saturated compacted Lower Cromer till, Gens (1982) proposed a variation of G with the maximum previous effective preconsolidation stress $p_0{}'$, $G = 93{\cdot}4 p_0{}'$. It is unclear to what extent could this expression, valid for saturated conditions, be used for a partially saturated soil. Thinking in terms of the proposed model, given an apparent preconsolidation stress p_0, the wetter the soil (i.e. the lower the suction), the smaller its specific volume. This probably indicates that the preceding expression for G in terms of $p_0{}'$ will probably overestimate the shear modulus if $p_0{}'$ is substituted by p_0. The value of G used in the present analysis of Maswoswe's results, $G = 7$ MPa, which provides a good fit for the experimental results, may be found if the expression for G is computed for a value of $p_0{}'$ intermediate between $p_0 (= 0{\cdot}2$ MPa) and $p_0{}^* (= 0{\cdot}02$ MPa). It is also believed that secondary plastic strains take place within the elastic region bounded by the LC and SI yield loci and this will tend to underestimate the shear modulus which better fits the experimental results.

A summary of the model parameters discussed here and the initial stress conditions of the test analysed is given in Table 1. Although the chosen parameters are consistent with the experimental evidence, it is recognized that other sets of parameter values might lead to similar model predictions.

The first part of the test represented in Figs 23 and 24 involves the application of vertical stress at constant water content. Suction was not controlled and its variation is a response of the sample to the boundary conditions. This type of test cannot be reproduced with the proposed model unless a relationship between water content (or, alternatively, degree of saturation S_r) and the stress state (p, q, s) is established. A simple way to model this relationship is to use the concept of state surface as described, for instance, in Lloret and Alonso (1985). In general S_r depends markedly on s whereas the influence of applied stress is small (Alonso, Gens & Hight, 1987). Accordingly, the following relationship

$$S_r = 1 - m \tanh(ns) \qquad (30)$$

where m and n are constants, has been used. Lloret and Alonso (1985) found that this expression fitted well published experimental results for soils with degrees of saturation over 0·4. Maswoswe's results for test SK3 can be approximated with the values $n = 5{\cdot}38$ MPa^{-1}, $m = 0{\cdot}64$. The loading process at constant water content could now be modelled in an iterative way by imposing that the S_r values derived from the computed deformations of the sample should match the values given by Equation (30).

Using the parameters and the initial and reference stress values indicated in Table 1 the computed model predictions, plotted in Fig. 24, show a satisfactory agreement with the measured results. The model correctly predicts that the K_0 value after collapse corresponds to K_0 for saturated specimens.

More than the success in predicting accurately a particular test in this case should be stressed the capability of the model to reproduce the basic features of a relatively complicated stress path such as the loading and collapse under K_0 conditions.

Tests on partially saturated compacted kaolin reported by Karube (1986)

The results of several tests on partially saturated compacted kaolin in a triaxial cell were reported by Karube (1986). The following basic properties of the soil are provided in this reference: $w_L = 37\%$; IP = 28%; clay fraction: 22%; maximum size of particles: 0·04 mm. Among the different results reported, two tests corresponding to the stress paths shown in Fig. 25(a) have been selected for comparison with the model developed.

These two tests are particularly interesting since they reveal most of the patterns of behaviour of partially saturated soil, discussed in previous sections of this Paper. The samples used in the tests were compacted at a water content 6% dry of optimum and they had $S_r = 0{\cdot}627$ and $v = 2{\cdot}114$. They were subsequently consolidated under the initial stress state ($p = 0{\cdot}02$ MPa; $s = 0{\cdot}05$ MPa) and they reached $S_{r0} = 0{\cdot}645$ and $v_0 = 2{\cdot}098$. Two alternative isotropic stress paths, shown in Fig. 25(a), starting and ending at the same stress point were followed by Karube (1986). The variation of specific volume is plotted in Fig. 25(b).

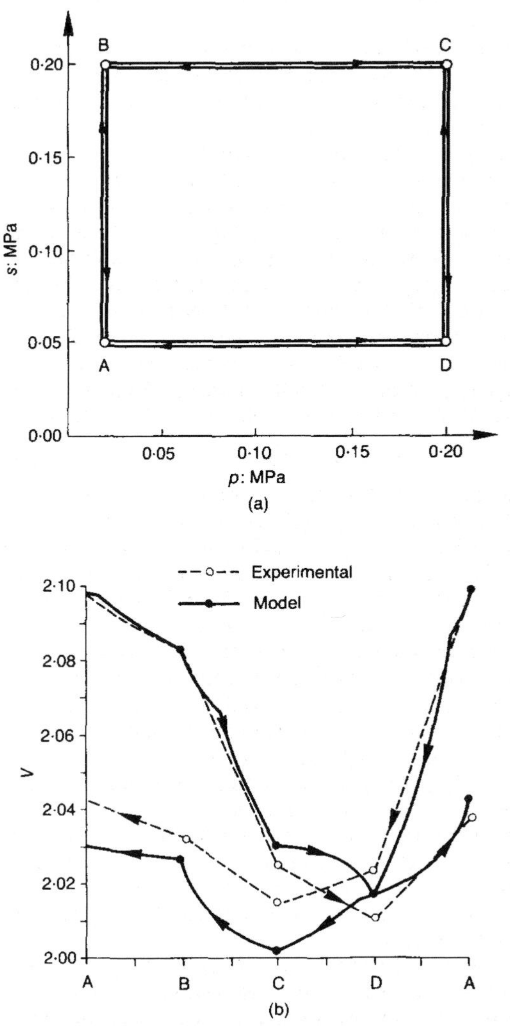

Fig. 25. Tests on compacted kaolin reported by Karube (1986): (a) stress paths; (b) comparison between experimental and computed results

In order to derive the model parameters, the results reported by Karube have been used in the following way.

The value of κ is directly related to the behaviour observed in the path steps C to B and D to A which involve unloading of the net mean stress. The behaviour in the steps B to C and A to D is controlled by the parameters $\lambda(0)$, p^c, β, r and p_0^* as well as κ. Given the relatively low water content used in compaction, the soil should have a low compressibility. The values $\lambda(0) = 0{\cdot}065$ and $p^c = 0{\cdot}01$ MPa were first selected and the remaining parameters mentioned were subsequently determined from the measured results along the stress path steps indicated. It was estimated: $\kappa = 0{\cdot}011$; $\beta = 20$ MPa^{-1}; $r = 0{\cdot}75$ and $p_0^* = 0{\cdot}04$ MPa. The value of κ_s is directly related to the wetting path under low confining stress (B to A) and it was found $\kappa_s = 0{\cdot}005$. In the drying paths A to B and D to C both λ_s and s_0, as well as κ_s, are relevant. Using the reported results it was found $\lambda_s = 0{\cdot}025$ and $s_0 = 0{\cdot}07$ MPa.

All of these parameters and the initial and reference stress states have been summarized in Table 1. Model predictions using these values are plotted in Fig. 25(b). The quantitative agreement is in this case satisfactory. Even more important is, again, the ability of the model to reproduce the main features of the measured behaviour.

CONCLUDING REMARKS

A review of the main characteristics of the behaviour of partially saturated soils has shown that existing models cover only limited aspects of the stress–strain response of these soils. Using as effective stress fields two independent stress variables—the excess of total stress over air pressure and the water pressure deficiency (or suction)—an elastoplastic hardening model has been developed and is described in the Paper. It accounts for the stiffness changes of the soil induced by suction changes, reproduces the irreversible response of the soil against stress and suction reversals, provides the conditions for collapsibility and links the amount of collapse to the stiffness changes of the soil induced by changes in suction. For isotropic stress conditions the model is characterized by two yield curves (LC; loading–collapse and SI, suction–increase) whose hardening laws are controlled by the total plastic volumetric deformation. These curves become yield surfaces when a third axis (accounting for shear stress) is introduced to model triaxial stress states. The behaviour of saturated soils is a particular case of the model developed which is reached when suction becomes zero. Different constitutive models for saturated conditions can be coupled to the unsaturated formulation presented in this Paper. For simplicity, a critical state model has been used as the reference model for saturated soil.

For applications that involve stress paths which imply monotonic wetting of the soil (which may be the case of embankment dam construction and subsequent water impounding) the model is characterized by nine constants, five more than the critical state model. These additional five constants account for phenomena typically associated with the behaviour of partially saturated soils such as decrease in stiffness and strength with suction decrease, collapse and (moderate) expansion of the soil when wetted. An additional parameter is required when the stress path implies suction increases beyond the maximum past suction value. Suitable testing procedures for obtaining the model parameters have been indicated.

As described in the Paper the model provides a simple representation of swelling, but is unable to approximate the behaviour of highly expansive soils. It is therefore intended for use with partially saturated soils of moderate to low plasticity, such as sandy clays, clayey sands and silts, and granular soils.

Extension and further elaboration of the proposed model may be envisaged to improve the description of partially saturated soil behaviour. Some aspects not specifically included in the model are the irreversible nature of expansion in swelling clays, the decrease in collapse potential when the confining stress becomes larger than a critical or threshold value and the non-linear character of the strength envelope. It is believed, however, that before such extensions are made more experience should be gained with the model in its simplest forms.

The comparison made between the results of several suction controlled tests on low plasticity kaolin and a sandy clay and model predictions is encouraging. The examples analysed demonstrate that the model is capable of making accurate predictions. More significant perhaps is the possibility of reproducing relevant features of partially saturated soil behaviour in a unified context.

ACKNOWLEDGEMENTS

The authors express their gratitude for the support received from the CAICYT (Comisión Asesora de Investigación Científica y Técnica) through research projects 2673/83 and PB86-0379.

APPENDIX 1. MATHEMATICAL FORMULATION OF CONSTITUTIVE MODEL

An ordered summary of the proposed model is presented here. The plastic strain increments can be obtained following standard procedures of the theory of hardening plasticity.

Yield surfaces

Two yield surfaces have been defined (in terms of the following stress variables: $p = (\sigma_1 + 2\sigma_3)/3 - u_a$; $q = \sigma_1 - \sigma_3$; $s = u_a - u_w$):

$$f_1(p, q, s, p_0{}^*) \equiv q^2 - M^2(p + p_s)(p_0 - p) = 0 \qquad (31)$$

where

$$p_s = ks \qquad (32)$$

$$\frac{p_0}{p^c} = \left(\frac{p_0{}^*}{p^c}\right)^{[\lambda(0) - \kappa]/[\lambda(s) - \kappa]} \qquad (33)$$

$$\lambda(s) = \lambda(0)[(1 - r)\exp(-\beta s) + r] \qquad (34)$$

$$f_2(s, s_0) \equiv s - s_0 = 0 \qquad (35)$$

p_0 and $p_0{}^*$ are preconsolidation p stresses for suction s and for saturated conditions respectively.

Flow rules

The plastic strain increment associated with yield surface f_1 is $(\mathrm{d}\varepsilon_{vp}{}^p, \mathrm{d}\varepsilon_s{}^p)$,

$$\mathrm{d}\varepsilon_{vp}{}^p = \mu_1 n_p \qquad (36)$$

$$\mathrm{d}\varepsilon_s{}^p = \mu_1 n_q \qquad (37)$$

where

$$n_p = 1 \qquad (38)$$

$$n_q = [2q\alpha/M^2(2p + p_s - p_0)] \qquad (39)$$

In Equation (39) α is a constant which can be derived by requiring that the direction of plastic strain increment for zero lateral deformation,

$$\frac{\mathrm{d}\varepsilon_s{}^p}{\mathrm{d}\varepsilon_{vp}{}^p} = (2/3)\{1/[1 - \kappa/\lambda(0)]\} \qquad (40)$$

(where for simplicity $\mathrm{d}\varepsilon_s{}^e$ has been assumed to be zero), is found for stress states satisfying K_0 conditions:

$$(q/p + p_s) = 3(1 - K_0)/(1 + 2K_0) \qquad (41)$$

Equations (6), (39–41) provide

$$\alpha = \frac{M(M - 9)(M - 3)}{9(6 - M)}\{1/[1 - \kappa/\lambda(0)]\} \qquad (42)$$

For the yield surface f_2 the associated plastic strain vector is $(d\varepsilon_{vs}{}^{p}, 0)$ where

$$d\varepsilon_{vs}{}^{p} = \mu_2 \quad (43)$$

Both μ_1 and μ_2 can be obtained through plastic consistency conditions.

Hardening laws

The evolution of the yield surfaces is controlled by the hardening parameters $p_0{}^*$ and s_0. They depend on the total plastic volumetric strain increment $d\varepsilon_v{}^{p}$ as follows

$$\frac{dp_0{}^*}{p_0{}^*} = \frac{v}{\lambda(0) - \kappa} d\varepsilon_v{}^{p} \quad (44)$$

$$\frac{ds_0}{(s_0 + p_{at})} = \frac{v}{\lambda_s - \kappa_s} d\varepsilon_v{}^{p} \quad (45)$$

Elastic strains

The volumetric and shear elastic components of strain are given by

$$d\varepsilon_v{}^{e} = \frac{\kappa}{v}\frac{dp}{p} + \frac{\kappa_s}{v}\frac{ds}{(s + p_{at})} \quad (46)$$

$$d\varepsilon_s{}^{e} = (1/3G)\, dq \quad (47)$$

REFERENCES

Aitchison, G. D. & Donald, I. B. (1956). Some preliminary studies of unsaturated soils. (b) Effective stresses in unsaturated soils. *Proc. 2nd Austr.-N. Zealand Conf. Soil Mech. Fdn Engng, Christchurch,* 192–199.

Alonso, E. E., Gens, A. & Hight, D. W. (1987). Special problem soils. General report. *Proc. 9th European Conf. Soil Mech. Fdn Engng, Dublin,* **3,** 1087–1146.

Barden, L., Madedor, A. O. & Sides, G R. (1969). Volume change characteristics of unsaturated clays. *J. Soil Mech. Fdn Engng, Am. Soc. Civ. Engrs* **95,** SM1, 33–51.

Bishop, A. W. & Blight, G. E. (1963). Some aspects of effective stress in saturated and partly saturated soils. *Géotechnique* **13,** No. 3, 177–197.

Bishop, A. W. & Donald, I. B. (1961). The experimental study of partly saturated soils in the triaxial apparatus, *Proc. 5th. Int. Conf. Soil Mech. Fdn Engng* **1,** 13–21.

Blight, G. E. (1965). A study of effective stress for volume change. In *Moisture equilibria and moisture changes in soils beneath covered areas,* pp. 259–269. Sydney: Butterworths.

Booth, A. R. (1975). The factors influencing collapse settlement in compacted soils. *Proc. 6th Regional Conf. for Africa on Soil Mech. Fdn Engng, Durban,* **1,** pp. 57–63.

Brand, D. (1981). Some thoughts on rain-induced slope failures. *Proc. 10th Int. Conf. Soil Mech. Fdn Engng, Stockholm,* **3,** pp. 373–376.

Burland, J. B. (1965). Some aspects of the mechanical behaviour of partially saturated soils. In *Moisture equilibria and moisture changes beneath covered areas,* pp. 270–278. Sydney: Butterworths.

Coleman, J. D. (1962). Stress strain relations for partly saturated soil. Correspondence to *Géotechnique* **12,** No. 4, 348–350.

Cox, D. W. (1978). Volume change of compacted clay fill. *Clay fills,* pp. 79–86, London: ICE.

Delage, P., Suraj de Silva, G. P. R. & De Laure, E. (1987). Un nouvel appareil triaxial pour les sols non-saturés. *Proc. 9th European Conf. Soil Mech. Fdn Engng, Dublin,* **1,** pp. 25–28.

Dudley, J. H. (1970). Review of collapsing soils. *J. Soil Mech. Fdn Engng, Am. Soc. Civ. Engrs* **96,** SM3, 925–947.

Erol, O. A. & El-Ruwaih, I. A. (1982). Collapse behaviour of desert loess. *Proc. 4th Cong. IAEG* **1,** 443–448.

Escario, V. & Sáez, J. (1973). Measurement of the properties of swelling and collapsing soils under controlled suction. *Proc. 3rd Int. Conf. Expansive Soils, Haifa,* 196–200.

Escario, V. & Sáez, J. (1986). The shear strength of partly saturated soils. *Géotechnique* **36,** No. 13, 453–456.

Escario, V. & Sáez, J. (1987). Shear strength of soils under high suction values. Written discussion. Session 5. *Proc. 9th European Conf. Soil Mech. Fdn Engng, Dublin,* **3,** 1157.

Fredlund, D. G. (1979). Appropriate concepts and technology for unsaturated soils. *Can. Geotech. J.* **16,** 121–139.

Fredlund, D. G. & Morgenstern, N. R. (1977). Stress state variables for unsaturated soils. *J. Geotech. Engng Div., Am. Soc. Civ. Engrs* **103,** GT5, 447–446.

Fredlund, D. G., Morgenstern, N. R. & Widger, R. S. (1978). The shear strength of unsaturated soils. *Can. Geotech. J.* **15,** No. 3, 313–321.

Fredlund, D. G. & Rahardjo, H. (1985). Theoretical context for understanding unsaturated residual soil behaviour. *Proc. 1st Int. Conf. on Geomech. in Tropical, Lateritic and Saprolitic Soils, Brasilia,* **1,** 295–306.

Fredlund, D. G., Rahardjo, H. & Gan, J. K. M. (1987). Nonlinearity of strength envelope for unsaturated soils. *Proc. 6th Int. Conf. Expansive Soils, New Delhi,* 49–54.

Gens, A. (1982). *Stress–strain and strength characteristics of a low plasticity clay.* PhD thesis, London University.

Gens, A. & Potts, D. M. (1982a). A theoretical model for describing the behaviour of soils not obeying Rendulic's principle. *Proc. 1st Int. Symp. Numer. Mod. Geomech., Zurich,* 24–32.

Gens, A. & Potts, D. M. (1982b). Application of critical state soil models to the prediction of the behaviour of a normally consolidated low plasticity clay. *Proc. 1st Int. Symp. Numer. Mod. Geomech., Zurich,* 312–323.

Gens, A. & Potts, D. M. (1988). Critical state models in computational geomechanics. *Engineering Computations* **8,** No. 3, 178–197.

Gulhati, S. K. & Satija, B. S. (1981). Shear strength of partially saturated soils, *Proc. 10th Int. Conf. Soil Mech. Fdn Engng, Stockholm,* **1,** 609–612.

Ho, D. Y. F. & Fredlund, D. G. (1982). Increase in

strength due to suction for two Hong Kong soils. *Proc. Conf. Engng and Construction in Tropical and Residual Soils, Honolulu*, 263–295.

Jaky, J. (1948). Pressure in silos. *Proc. 2nd Int. Conf. Soil Mech. and Fdn Engng* **1**, 103–107.

Jennings, J. E. & Burland, J. B. (1962). Limitations to the use of effective stresses in partly saturated soils. *Géotechnique* **12**, No. 2, 125–144.

Jennings, J. E. & Knight, K. (1975). A guide to construction on or with materials exhibiting additional settlement due to 'collapse' of grain structure. *Proc. 6th Regional Conf. for Africa on Soil Mech. Fdn Engng, Durban*, **1**, 99–105.

Josa, A. (1988). *Un modelo eslastoplástico para suelos no saturados*. Tesis Doctoral. Universitat Politècnica de Catalunya, Barcelona.

Josa, A., Alonso, E. E., Lloret, A. & Gens, A. (1987). Stress–strain behaviour of partially saturated soils. *Proc. 9th European Conf. Soil Mech. Fdn Engng, Dublin*, **2**, 561–564.

Justo, J. L., Delgado, A. & Ruiz, J. (1984). The influence of stress-path in the collapse–swelling of soils at the laboratory. *Proc. 5th Int. Conf. Expansive Soils, Adelaide*, 67–71.

Karube, D. (1986). New concept of effective stress in unsaturated soil and its proving tests. *ASTM Symp. on Advanced Triaxial Testing of Soil and Rock*, ST 977, 539–552.

Lloret, A. & Alonso, E. E. (1980). Consolidation of unsaturated soil including swelling and collapse behaviour. *Géotechnique* **30**, No. 4, 449–477.

Lloret, A. & Alonso, E. E. (1985). State surfaces for partially saturated soils. *Proc. 11th Int. Conf. Soil Mech. Fdn Engng, San Francisco*, **2**, 557–562.

Maswoswe, J. (1985). *Stress path for a compacted soil during collapse due to wetting*. PhD thesis, Imperial College, London.

Matyas, E. L. & Radhakrishna, H. S. (1968). Volume change characteristics of partially saturated soils. *Géotechnique* **18**, No. 4, 432–448.

Ohmaki, S. (1982). Stress–strain behaviour of anisotropically, normally consolidated cohesive soil. *Proc. 1st Int. Symp. Num. Mod. Geomech., Zurich*, 250–269.

Rico, A. & Del Castillo, H. (1976). *La ingeniería de suelos en las vías terrestres*. **1**, Limusa, México.

Yong, R. N., Japp, R. D. & How, G. (1971). Shear strength of partially saturated clays. *Proc. 4th Asian Reg. Conf. Soil Mech. Fdn Engng, Bangkok*, **2**, 12: 183–187.

Yudhbir, (1982). Collapsing behaviour of residual soils. *Proc. 7th Southeast Asian Geotech. Conf., Hong-Kong*, **1**, 915–930.

Alonso, E. E., Gens, A. & Josa, A. (1991). *Géotechnique* **41**, No. 2, 273–275

DISCUSSION

A constitutive model for partially saturated soils

E. E. ALONSO, A. GENS and A. JOSA (1990) *Géotechnique* **40**, No. 3, 405–430

T. M. P. de Campos and E. A. Vargas, Jr, *Catholic University of Rio de Janerio*, **and S. A. S. Aguilar**, *formerly Catholic University of Rio de Janerio, now University of Costa Rica*

Recently we had the opportunity to test the Authors' model in laboratory experiments on a collapsible, non-active undisturbed clay soil, submitted to isotropic stress changes under suction controlled paths (Aguilar, 1990; de Campos, Aguilar & Vargas, 1991). Table 2 presents the model parameters obtained in the experimental programme, which included five suction-controlled triaxial tests.

As exemplified in Figs 26 and 27, the Authors' model was able to predict most of the experimental results quite well. The poorest quantitative comparisons between model and experimental data were related to suction decrease under different levels of constant isotropic stresses p $(= \sigma - u_a)$, the experimental collapse deformations being greater than the predicted ones.

A large collapse was observed in one of the tests when the sample was saturated, from its natural stress state, by the application of a small differential water pressure between its bottom and top, while the cell pressure was kept constant. According to the model, this collapse should not occur as the sample was within the assumed initial elastic region.

In our opinion, before questioning the model assumptions, two experimental aspects should be considered

(*a*) difficulty in defining the initial stress state of the unsaturated soil, in particular the initial p_0* (conventional saturated overconsolidation pressure)
(*b*) lack of knowledge of the stress path followed by the sample in the conventional saturation process.

As air pressures are not measured, the variation of p is unknown in case (*b*). (All other wetting paths under constant p steps, followed in the experimental programme, were controlled by known decreases in suction.)

The model parameters r, β and p^c are obtained from

$$\left(\frac{p_0}{p^c}\right) = \left(\frac{p_0{}^*}{p^c}\right)^{[\lambda(0)-\kappa]/\{\lambda(0)[(1-r)\exp(-\beta s)+r]-\kappa\}} \quad (48)$$

which is a combination of equations (7) and (12).

An unlimited number of numerical values for these parameters can, in principle, satisfy equation (48). Our experience indicated that p^c is the most sensitive parameter for the definition of the LC yielding surface.

Owing to the importance of the parameter p^c, could the Authors comment further on its definition? Is there any physical meaning behind it (as, for instance, the Hvorslev p_e normalization parameter for overconsolidated saturated soils)?

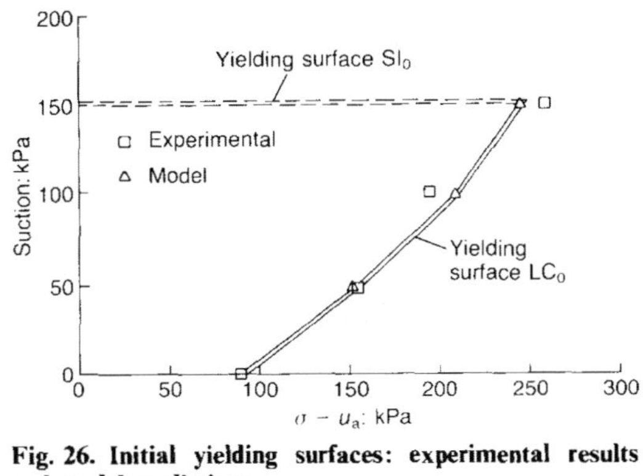

Fig. 26. Initial yielding surfaces: experimental results and model predictions

Table 2. Model parameters

$\lambda(0)$†	κ†	κ_s	λ_s	p_0*: kPa	s_0: kPa	p^c: kPa	r	β: kPa^{-1}
0·217	0·006	0·008	0·108	80	150	18·1	0·565	0·0215

† Average values from isotropic consolidation and confined tests.

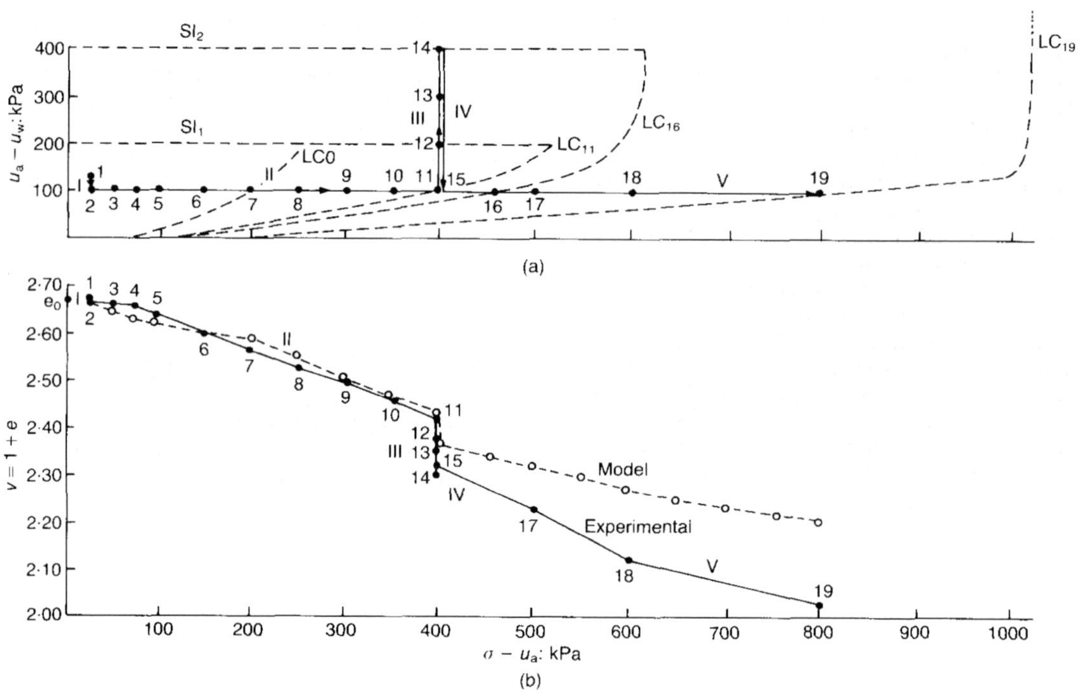

Fig. 27. (a) Stress and suction paths followed on experiments and corresponding yielding surfaces; (b) isotropic stress × volumetric deformations: experimental results and model predictions

Authors' reply

The discussion on some of the points raised is hampered by the lack of more background information on the testing programme carried out.

The Authors share the opinion expressed regarding the difficulties of defining precisely the preconsolidation (or yield) stress p_0^*. This is a general difficulty found in the interpretation of test results under both saturated and unsaturated conditions for many soil types.

The meaning of 'lack of knowledge of the stress path followed by the sample in a conventional saturation process' and 'air pressures are not measured' is not totally clear. If it is the air entrapment phenomenon, which may occur near saturation, which is being referred to, it is agreed that some unknown air pressure may develop within the air bubbles. At this point a transition to a saturated condition takes place. This is a difficult stage for modelling purposes and it is not particularly addressed by the model presented. However, for most of the range before the final saturated condition the permeability of the soil to air allows the easy escape of air from the specimen and, therefore, the air pressure is in general known and equal to the atmospheric value.

The Writers also report the collapse, on wetting, of a sample whose stress state was apparently in the assumed elastic region. As no figures or data concerning this test are given the Authors cannot comment in detail.

Regarding the apparent freedom in the selection of a combination of model parameters to reproduce a given measured result, it should be stressed that, as discussed in the Paper, the parameters have a physical meaning and they can, in principle, be unambiguously determined from specific tests. For instance, p^c of a given unsaturated specimen is a consolidation net mean stress belonging to its virgin loading curve for which only elastic expansion results when wetted. When $p_0 = p^c$ (equation (48)) all the preconsolidation stresses $p_0(s)$ become equal to p^c. The LC yield curve then becomes a straight line parallel to the s axis at $p_0 = p^c$.

As indicated in the Paper direct physical significance can also be ascribed to parameters β and r; these also control the shape of the LC yield surface.

The influence of p^c on the shape of the LC yield curves is similar, in quantitative terms, to the variations observed in Fig. 4 for changes in p_0^*.

With regard to the test result plotted in Fig. 27, the calculated value of the virgin compressibility parameter $\lambda(s)$ (for $s = 100$ kPa) is shown as 0·25 (model curve). This value cannot be found from the model parameters reported in Table 2. In fact $\lambda(s)$ (for $s = 100$ kPa) $= 0{\cdot}13$. Similarly, a value of

$\kappa = 0{\cdot}033$ is derived for the model curve in Fig. 27 whereas the value of $\kappa = 0{\cdot}006$ is reported in Table 1. In view of the shape of the experimental loading curve plotted in Fig. 27 it is surprising that the initial preconsolidation stress p_0 (yield curve LC_0) is located at $p_0 = 200$ kPa (point 7) when the test seems to indicate a much lower value (point 4 of the stress path).

The behaviour of the specimen after the suction cycle applied at $p = 400$ kPa (a sudden increment in the compressibility parameter which varies from 0·15 to 0·45) suggests that the natural soil tested may have cementation bonds between particles which have been destroyed by the plastic strains induced during the previous desiccation. As a result of adopting the modified Cam clay model as the saturated boundary of the model, the particular version presented is not intended for soils exhibiting strong structural effects.

The Authors would like to encourage further applications of the model to the interpretation and prediction of experimental results concerning unsaturated soils of various types and under different stress conditions. In this way the strong points and shortcomings of the model can be identified and appropriate modifications could be obtained.

REFERENCES

Aguilar, S. A. S. (1990). *Deformability of an unsaturated collapsible clay under suction controlled isotropic consolidation.* MSc thesis, PUC-Rio (in Portuguese).

de Campos, T. M. P., Aguilar, S. A. S. & Vargas, Jr, E. A. (1991). Testing and modeling of a porous unsaturated clay. *Proc. 9th Pan-Am. Conf. Soil Mech.*, Chile (in press).